Lecture Notes in Artificial Intelligence　　10752

Subseries of Lecture Notes in Computer Science

More information about this series at http://www.springer.com/series/1244

Ngoc Thanh Nguyen · Duong Hung Hoang
Tzung-Pei Hong · Hoang Pham
Bogdan Trawiński (Eds.)

Intelligent Information and Database Systems

10th Asian Conference, ACIIDS 2018
Dong Hoi City, Vietnam, March 19–21, 2018
Proceedings, Part II

 Springer

Editors
Ngoc Thanh Nguyen
Wrocław University of Science
 and Technology
Wrocław
Poland

Duong Hung Hoang
Quang Binh University
Dong Hoi City
Vietnam

Tzung-Pei Hong
National University of Kaohsiung
Kaohsiung
Taiwan

Hoang Pham
Rutgers University
Piscataway, NJ
USA

Bogdan Trawiński
Wrocław University of Science
 and Technology
Wrocław
Poland

ISSN 0302-9743   ISSN 1611-3349 (electronic)
Lecture Notes in Artificial Intelligence
ISBN 978-3-319-75419-2   ISBN 978-3-319-75420-8   (eBook)
https://doi.org/10.1007/978-3-319-75420-8

Library of Congress Control Number: 2018931885

LNCS Sublibrary: SL7 – Artificial Intelligence

Printed on acid-free paper

This Springer imprint is published by the registered company Springer International Publishing AG part of Springer Nature
The registered company address is: Gewerbestrasse 11, 6330 Cham, Switzerland

Preface

ACIIDS 2018 was the tenth event in a series of international scientific conferences on research and applications in the field of intelligent information and database systems. The aim of ACIIDS 2018 was to provide an international forum for scientific research in the technology and application of intelligent information and database systems. ACIIDS 2018 was co-organized by Quang Binh University (Vietnam) and Wrocław University of Science and Technology (Poland) in co-operation with the IEEE SMC Technical Committee on Computational Collective Intelligence, European Research Center for Information Systems (ERCIS), University of Newcastle (Australia), Bina Nusantara University (Indonesia), Yeungnam University (South Korea), Leiden University (The Netherlands), Universiti Teknologi Malaysia (Malaysia), Ton Duc Thang University (Vietnam), and Vietnam National University, Hanoi (Vietnam). It took place in Dong Hoi City in Vietnam during March 19–21, 2018.

The conference series ACIIDS is already well established. The first two events, ACIIDS 2009 and ACIIDS 2010, took place in Dong Hoi City and Hue City in Vietnam, respectively. The third event, ACIIDS 2011, took place in Daegu (South Korea), followed by the fourth event, ACIIDS 2012, in Kaohsiung (Taiwan). The fifth event, ACIIDS 2013, was held in Kuala Lumpur in Malaysia while the sixth event, ACIIDS 2014, was held in Bangkok in Thailand. The seventh event, ACIIDS 2015, took place in Bali (Indonesia), followed by the eight event, ACIIDS 2016, in Da Nang (Vietnam). The last event, ACIIDS 2017, was held in Kanazawa (Japan).

We received more than 400 papers from 42 countries all over the world. Each paper was peer reviewed by at least two members of the international Program Committee and international reviewer board. Only 133 papers with the highest quality were selected for an oral presentation and publication in the two volumes of the ACIIDS 2018 proceedings.

Papers included in the proceedings cover the following topics: knowledge engineering and Semantic Web; social networks and recommender systems; text processing and information retrieval; machine learning and data mining; decision support and control systems; computer vision techniques; advanced data mining techniques and applications; multiple model approach to machine learning; intelligent information systems; design thinking-based R&D; development techniques; and project-based learning; modelling, storing, and querying of graph data; computational imaging and vision; computer vision and robotics; data science and computational intelligence; data structures modelling for knowledge representation; intelligent computer vision systems and applications; intelligent and contextual systems; intelligent biomarkers of neurodegenerative processes in the brain; sensor networks and Internet of Things; intelligent applications of Internet of Thing and data analysis technologies; intelligent systems and algorithms in information sciences; intelligent systems and methods in biomedicine; intelligent systems for optimization of logistics and industrial applications; analysis of image, video, and motion data in life sciences.

The accepted and presented papers highlight new trends and challenges facing the intelligent information and database systems community. The presenters showed how new research could lead to novel and innovative applications. We hope you will find these results useful and inspiring for your future research work.

We would like to extend our heartfelt thanks to Jarosław Gowin, Deputy Prime Minister of the Republic of Poland and Minister of Science and Higher Education, for his support and honorary patronage of the conference.

We would like to express our sincere thanks to the honorary chairs, Prof. Cezary Madryas (Rector of Wrocław University of Science and Technology, Poland), Prof. Huu Duc Nguyen (Vice-Rector of National University Hanoi, Vietnam), and Dr. Tien Dung Tran (Vice-President of Quang Binh Province, Vietnam), for their support.

Our special thanks go to the program chairs, special session chairs, organizing chairs, publicity chairs, liaison chairs, and local Organizing Committee for their work for the conference. We sincerely thank all the members of the international Program Committee for their valuable efforts in the review process, which helped us to guarantee the highest quality of the selected papers for the conference. We cordially thank the organizers and chairs of special sessions who contributed to the success of the conference.

We would like to express our thanks to the keynote speakers: Thomas Bäck from University of Leiden, The Netherlands, Lipo Wang from Nanyang Technological University, Singapore, Satoshi Tojo from Japan Advanced Institute of Science and Technology, Japan, and Nguyen Huu Duc from Vietnam National University, Hanoi, Vietnam, for their world-class plenary speeches.

We cordially thank our main sponsors, Quang Binh University (Vietnam), Wrocław University of Science and Technology (Poland), IEEE SMC Technical Committee on Computational Collective Intelligence, European Research Center for Information Systems (ERCIS), University of Newcastle (Australia), Bina Nusantara University (Indonesia), Yeungnam University (South Korea), Leiden University (The Netherlands), Universiti Teknologi Malaysia (Malaysia), Ton Duc Thang University (Vietnam), and Vietnam National University, Hanoi (Vietnam). Our special thanks are due also to Springer for publishing the proceedings and sponsoring awards, and to all the other sponsors for their kind support.

We wish to thank the members of the Organizing Committee for their excellent work and the members of the local Organizing Committee for their considerable effort.

We cordially thank all the authors, for their valuable contributions, and the other participants of this conference. The conference would not have been possible without their support.

Thanks are also due to many experts who contributed to making the event a success.

March 2018

Ngoc Thanh Nguyen
Duong Hung Hoang
Tzung-Pei Hong
Hoang Pham
Bogdan Trawiński

Organization

Honorary Chairs

Cezary Madryas   Rector of Wrocław University of Science
and Technology, Poland

Huu Duc Nguyen   Vice-President of Vietnam National University, Hanoi,
Vietnam

Tien Dung Tran   Vice-President of Quang Binh Province, Vietnam

General Chairs

Ngoc Thanh Nguyen   Wrocław University of Technology, Poland

Duong Hung Hoang   Quang Binh University, Vietnam

Program Chairs

Tzung-Pei Hong   National University of Kaohsiung, Taiwan

Hoang Pham   Rutgers University, USA

Edward Szczerbicki   University of Newcastle, Australia

Bogdan Trawiński   Wrocław University of Science and Technology, Poland

Special Session Chairs

Manuel Núñez   Universidad Complutense de Madrid, Spain

Andrzej Siemiński   Wrocław University of Science and Technology, Poland

Quang Thuy Ha   Vietnam National University, Hanoi (VNU), Vietnam

Publicity Chairs

Van Dung Hoang   Quang Binh University, Vietnam

Marek Kopel   Wrocław University of Science and Technology, Poland

Marek Krótkiewicz   Wrocław University of Science and Technology, Poland

Liaison Chairs

Quang A. Dang   Vietnam Academy of Science and Technology, Vietnam

Ford Lumban Gaol   Bina Nusantara University, Indonesia

Mong-Fong Horng   National Kaohsiung University of Applied Sciences,
Taiwan

Dosam Hwang   Yeungnam University, South Korea

Ali Selamat   Universiti Teknologi Malaysia, Malaysia

Organizing Chairs

Thi Dung Vo	Quang Binh University, Vietnam
Hoai Thu Le Thi	Quang Binh University, Vietnam
Adrianna Kozierkiewicz	Wrocław University of Science and Technology, Poland

Local Organizing Committee

Maciej Huk	Wrocław University of Science and Technology, Poland
Marcin Jodłowiec	Wrocław University of Science and Technology, Poland
Rafał Kern	Wrocław University of Science and Technology, Poland
Marcin Pietranik	Wrocław University of Science and Technology, Poland
Krystian Wojtkiewicz	Wrocław University of Science and Technology, Poland
Xuan Hau Pham	Quang Binh University, Vietnam
Tuan Nha Hoang	Quang Binh University, Vietnam
Van Thanh Phan	Quang Binh University, Vietnam
Van Chung Nguyen	Quang Binh University, Vietnam
Lan Phuong Pham Thi	Quang Binh University, Vietnam
Xuan Hao Nguyen	Quang Binh University, Vietnam

Webmaster

Marek Kopel	Wrocław University of Science and Technology, Poland

Steering Committee

Ngoc Thanh Nguyen (Chair)	Wrocław University of Technology, Poland
Longbing Cao	University of Science and Technology Sydney, Australia
Suphamit Chittayasothorn	King Mongkut's Institute of Technology Ladkrabang, Thailand
Ford Lumban Gaol	Bina Nusantara University, Indonesia
Tu Bao Ho	Japan Advanced Institute of Science and Technology, Japan
Tzung-Pei Hong	National University of Kaohsiung, Taiwan
Dosam Hwang	Yeungnam University, South Korea
Lakhmi C. Jain	University of South Australia, Australia
Geun-Sik Jo	Inha University, South Korea
Hoai An Le-Thi	University of Lorraine, France
Zygmunt Mazur	Wrocław University of Science and Technology, Poland
Toyoaki Nishida	Kyoto University, Japan
Leszek Rutkowski	Częstochowa University of Technology, Poland
Ali Selamat	Universiti Teknologi Malaysia, Malaysia

Keynote Speakers

Thomas Bäck	University of Leiden, The Netherlands
Lipo Wang	Nanyang Technological University, Singapore
Satoshi Tojo	Japan Advanced Institute of Science and Technology, Japan
Nguyen Huu Duc	Vietnam National University, Hanoi, Vietnam

Special Sessions Organizers

1. *Special Session on Multiple Model Approach to Machine Learning (MMAML 2018)*

Tomasz Kajdanowicz	Wrocław University of Science and Technology, Poland
Edwin Lughofer	Johannes Kepler University Linz, Austria
Bogdan Trawinski	Wrocław University of Science and Technology, Poland

2. *Special Session on Analysis of Image, Video, and Motion Data in Life Sciences (IVMLS 2018)*

Konrad Wojciechowski	Polish-Japanese Academy of Information Technology, Poland
Marek Kulbacki	Polish-Japanese Academy of Information Technology, Poland
Jakub Segen	Polish-Japanese Academy of Information Technology, Poland
Andrzej Polanski	Silesian University of Technology, Poland

3. *Special Session on Intelligent and Contextual Systems (ICxS 2018)*

Maciej Huk	Wrocław University of Science and Technology, Poland
Keun Ho Ryu	Chungbuk National University, South Korea
Thai-Nghe Nguyen	Cantho University, Vietnam
Nguyen Hong Vu	Ton Duc Thang University, Vietnam
Goutam Chakraborty	Iwate Prefectural University, Japan

4. *Special Session on Intelligent Systems for Optimization of Logistics and Industrial Applications (ISOLIA 2018)*

Farouk Yalaoui	University of Technology of Troyes, France
Habiba Drias	University of Science and Technology (USTHB), Algeria
Taha Arbaoui	University of Technology of Troyes, France

5. *Special Session on Intelligent Applications of Internet of Thing and Data Analysis Technologies (IoT&DAT 2018)*

Shunzhi Zhu	Xiamen University of Technology, Xiamen, P.R. China
Rung Ching Chen	Chaoyang University of Technology, Taiwan
Yung-Fa Huang	Chaoyang University of Technology, Taiwan

6. *Special Session on Intelligent Systems and Methods in Biomedicine (ISaMiB 2018)*

Jan Kubicek	Technical University of Ostrava, Czech Republic
Marek Penhaker	Technical University of Ostrava, Czech Republic
Ondrej Krejcar	University of Hradec Kralove, Czech Republic
Kamil Kuca	University of Hradec Kralove, Czech Republic

7. *Special Session on Intelligent Systems and Algorithms in Information Sciences (ISAIS 2018)*

Martin Kotyrba	University of Ostrava, Czech Republic
Eva Volna	University of Ostrava, Czech Republic
Ivan Zelinka	VSB - Technical University of Ostrava, Czech Republic

8. *Special Session on Design Thinking-Based R&D, Development Techniques, and Project-Based Learning (2DT-PBL 2018)*

Shinya Kobayashi	Ehime University, Japan
Keiichi Endo	Ehime University, Japan

9. *Special Session on Modelling, Storing, and Querying of Graph Data (MSQGD 2018)*

Jaroslav Pokorny	Charles University, Prague, Czech Republic
Bela Stantic	Griffith University, Australia

10. *Special Session on Data Science and Computational Intelligence (DSCI 2018)*

Veera Boonjing	King Mongkut's Institute of Technology Ladkrabang, Thailand
Ronan Reilly	Maynooth University, Ireland

11. *Special Session on Computer Vision and Robotics (CVR 2018)*

Van-Dung Hoang	Quang Binh University, Vietnam
Chi-Mai Luong	University of Science and Technology of Hanoi, Vietnam

| My-Ha Le | Ho Chi Minh City University of Technology and Education, Vietnam |
| Kang-Hyun Jo | University of Ulsan, South Korea |

12. *Special Session on Intelligent Biomarkers of Neurodegenerative Processes in the Brain (InBinBRAIN 2018)*

| Andrzej Przybyszewski | Polish-Japanese Academy of Information Technology, Poland |

13. *Special Session on Data Structures Modelling for Knowledge Representation (DSMKR 2018)*

| Marek Krotkiewicz | Wrocław University of Science and Technology, Poland |

14. *Special Session on Computational Imaging and Vision (CIV 2018)*

Jeonghwan Gwak	Seoul National University Hospital, South Korea
Manish Khare	Dhirubhai Ambani Institute of Information and Communication Technology Gandhinagar, India
Jong-In Song	Gwangju School of Electrical Engineering and Computer Science, South Korea

15. *Special Session on Intelligent Computer Vision Systems and Applications (ICVSA 2018)*

Dariusz Frejlichowski	West Pomeranian University of Technology, Szczecin, Poland
Leszek J. Chmielewski	Warsaw University of Life Sciences, Poland
Piotr Czapiewski	West Pomeranian University of Technology, Szczecin, Poland

16. *Special Session on Advanced Data Mining Techniques and Applications (ADMTA 2018)*

Bay Vo	Ho Chi Minh City University of Technology, Vietnam
Tzung-Pei Hong	National University of Kaohsiung, Taiwan
Chun-Hao Chen	Tamkang University, Taiwan

International Program Committee

Salim Abdulazeez	College of Engineering, Trivandrum, India
Muhammad Abulaish	South Asian University, India
Waseem Ahmad	Waiariki Institute of Technology, New Zealand
Toni Anwar	Universiti Teknologi Malaysia, Malaysia
Ahmad Taher Azar	Benha University, Egypt
Amelia Badica	University of Craiova, Romania
Costin Badica	University of Craiova, Romania
Kambiz Badie	ICT Research Institute, Iran
Hassan Badir	École Nationale des Sciences Appliquées de Tanger, Morocco
Emili Balaguer-Ballester	Bournemouth University, UK
Zbigniew Banaszak	Warsaw University of Technology, Poland
Dariusz Barbucha	Gdynia Maritime University, Poland
Ramazan Bayindir	Gazi University, Turkey
Leon Bobrowski	Bialystok University of Technology, Poland
Bülent Bolat	Yildiz Technical University, Turkey
Veera Boonjing	King Mongkut's Institute of Technology Ladkrabang, Thailand
Mariusz Boryczka	University of Silesia in Katowice, Poland
Urszula Boryczka	University of Silesia in Katowice, Poland
Zouhaier Brahmia	University of Sfax, Tunisia
Stephane Bressan	National University of Singapore, Singapore
Peter Brida	University of Zilina, Slovakia
Andrej Brodnik	University of Ljubljana, Slovenia
Piotr Bródka	Wrocław University of Science and Technology, Poland
Grażyna Brzykcy	Poznan University of Technology, Poland
The Duy Bui	University of Engineering and Technology, VNU Hanoi, Vietnam
Robert Burduk	Wrocław University of Science and Technology, Poland
Aleksander Byrski	AGH University of Science and Technology, Poland
David Camacho	Universidad Autonoma de Madrid, Spain
Tru Cao	Ho Chi Minh City University of Technology, Vietnam
Frantisek Capkovic	Institute of Informatics, Slovak Academy of Sciences, Slovakia
Dariusz Ceglarek	Poznan High School of Banking, Poland
Zenon Chaczko	University of Technology, Sydney, Australia
Altangerel Chagnaa	National University of Mongolia, Mongolia
Goutam Chakraborty	Iwate Prefectural University, Japan
Kuo-Ming Chao	Coventry University, UK
Somchai Chatvichienchai	University of Nagasaki, Japan
Rung-Ching Chen	Chaoyang University of Technology, Taiwan
Shyi-Ming Chen	National Taiwan University, Taiwan
Suphamit Chittayasothorn	King Mongkut's Institute of Technology Ladkrabang, Thailand

Quang-Thuy Ha	VNU University of Engineering and Technology, Vietnam
Sung Ho Ha	Kyungpook National University, South Korea
Dawit Haile	Addis Ababa University, Ethiopia
Pei-Yi Hao	National Kaohsiung University of Applied Sciences, Taiwan
Tutut Herawan	University of Malaya, Malaysia
Marcin Hernes	Wroclaw University of Economics, Poland
Bogumila Hnatkowska	Wrocław University of Science and Technology, Poland
Huu Hanh Hoang	Hue University, Vietnam
Quang Hoang	Hue University of Sciences, Vietnam
Van-Dung Hoang	Quang Binh University, Vietnam
Tzung-Pei Hong	National University of Kaohsiung, Taiwan
Mong-Fong Horng	National Kaohsiung University of Applied Sciences, Taiwan
Eklas Hossain	Oregon University, USA
Jen-Wei Huang	National Cheng Kung University, Taiwan
Maciej Huk	Wrocław University of Science and Technology, Poland
Zbigniew Huzar	Wrocław University of Science and Technology, Poland
Dosam Hwang	Yeungnam University, South Korea
Roliana Ibrahim	Universiti Teknologi Malaysia, Malaysia
Dmitry Ignatov	National Research University Higher School of Economics, Russia
Lazaros Iliadis	Democritus University of Thrace, Greece
Hazra Imran	University of British Columbia, Canada
Agnieszka Indyka-Piasecka	Wrocław University of Science and Technology, Poland
Mirjana Ivanovic	University of Novi Sad, Serbia
Sanjay Jain	National University of Singapore, Singapore
Jaroslaw Jankowski	West Pomeranian University of Technology, Szczecin, Poland
Chuleerat Jaruskulchai	Kasetsart University, Thailand
Khalid Jebari	LCS Rabat, Morocco
Joanna Jedrzejowicz	University of Gdansk, Poland
Piotr Jedrzejowicz	Gdynia Maritime University, Poland
Janusz Jezewski	Institute of Medical Technology and Equipment ITAM, Poland
Geun Sik Jo	Inha University, South Korea
Kang-Hyun Jo	University of Ulsan, South Korea
Janusz Kacprzyk	Systems Research Institute, Polish Academy of Sciences, Poland
Tomasz Kajdanowicz	Wrocław University of Science and Technology, Poland
Nadjet Kamel	University Ferhat Abbes Setif1, Algeria
Mehmed Kantardzic	University of Louisville, USA
Mehmet Karaata	Kuwait University, Kuwait
Nikola Kasabov	Auckland University of Technology, New Zealand
Arkadiusz Kawa	Poznan University of Economics and Business, Poland

Rafal Kern	Wrocław University of Science and Technology, Poland
Manish Khare	Dhirubhai Ambani Institute of Information and Communication Technology, India
Chonggun Kim	Yeungnam University, South Korea
Marek Kisiel-Dorohinicki	AGH University of Science and Technology, Poland
Attila Kiss	Eotvos Lorand University, Hungary
Jerzy Klamka	Silesian University of Technology, Poland
Goran Klepac	Raiffeisen Bank, Croatia
Shinya Kobayashi	Ehime University, Japan
Marek Kopel	Wrocław University of Science and Technology, Poland
Jozef Korbicz	University of Zielona Gora, Poland
Jerzy Korczak	Wroclaw University of Economics, Poland
Raymondus Kosala	Bina Nusantara University, Indonesia
Leszek Koszalka	Wrocław University of Science and Technology, Poland
Leszek Kotulski	AGH University of Science and Technology, Poland
Adrianna Kozierkiewicz	Wrocław University of Science and Technology, Poland
Bartosz Krawczyk	Virginia Commonwealth University, USA
Ondrej Krejcar	University of Hradec Kralove, Czech Republic
Dalia Kriksciuniene	Vilnius University, Lithuania
Dariusz Krol	Wrocław University of Science and Technology, Poland
Marek Krótkiewicz	Wrocław University of Science and Technology, Poland
Marzena Kryszkiewicz	Warsaw University of Technology, Poland
Adam Krzyzak	Concordia University, Canada
Tetsuji Kuboyama	Gakushuin University, Japan
Elżbieta Kukla	Wrocław University of Science and Technology, Poland
Julita Kulbacka	Wroclaw Medical University, Poland
Marek Kulbacki	Polish-Japanese Academy of Information Technology, Poland
Kazuhiro Kuwabara	Ritsumeikan University, Japan
Halina Kwasnicka	Wrocław University of Science and Technology, Poland
Mark Last	Ben-Gurion University of the Negev, Israel
Annabel Latham	Manchester Metropolitan University, UK
Bac Le	University of Science, VNU-HCM, Vietnam
Hoai An Le Thi	University of Lorraine, France
Kun Chang Lee	Sungkyunkwan University, South Korea
Yue-Shi Lee	Ming Chuan University, Taiwan
Chunshien Li	National Central University, Taiwan
Horst Lichter	RWTH Aachen University, Germany
Sebastian Link	University of Auckland, New Zealand
Igor Litvinchev	Nuevo Leon State University, Mexico
Lian Liu	University of Kentucky, USA
Rey-Long Liu	Tzu Chi University, Taiwan
Edwin Lughofer	Johannes Kepler University Linz, Austria
Ngoc Quoc Ly	Ho Chi Minh City University of Science, Vietnam
Lech Madeyski	Wrocław University of Science and Technology, Poland

Bernadetta Maleszka	Wrocław University of Science and Technology, Poland
Marcin Maleszka	Wrocław University of Science and Technology, Poland
Mustafa Mat Deris	Universiti Tun Hussein Onn Malaysia, Malaysia
Takashi Matsuhisa	Karelia Research Centre, Russian Academy of Science, Russia
Tamás Matuszka	Eotvos Lorand University, Hungary
Joao Mendes-Moreira	University of Porto, Portugal
Héctor Menéndez	University College London, UK
Jacek Mercik	WSB University in Wroclaw, Poland
Radosław Michalski	Wrocław University of Science and Technology, Poland
Peter Mikulecky	University of Hradec Kralove, Czech Republic
Marek Milosz	Lublin University of Technology, Poland
Jolanta Mizera-Pietraszko	Opole University, Poland
Leo Mrsic	IN2data Ltd. Data Science Company, Croatia
Agnieszka Mykowiecka	Institute of Computer Science, Polish Academy of Sciences, Poland
Pawel Myszkowski	Wrocław University of Science and Technology, Poland
Grzegorz J. Nalepa	AGH University of Science and Technology, Poland
Mahyuddin K. M. Nasution	Universitas Sumatera Utara, Indonesia
Richi Nayak	School of Electrical Engineering and Computer Science, Australia
Fulufhelo Nelwamondo	Council for Scientific and Industrial Research, South Africa
Huu-Tuan Nguyen	Vietnam Maritime University, Vietnam
Loan T. T. Nguyen	University of Warsaw, Poland
Quang-Vu Nguyen	Korea-Vietnam Friendship Information Technology College, Vietnam
Thai-Nghe Nguyen	Cantho University, Vietnam
Agnieszka Nowak-Brzezinska	University of Silesia in Katowice, Poland
Mariusz Nowostawski	Norwegian University of Science and Technology, Norway
Alberto Núñez	Universidad Complutense de Madrid, Spain
Manuel Núñez	Universidad Complutense de Madrid, Spain
Cheol-Young Ock	University of Ulsan, South Korea
Richard Jayadi Oentaryo	Singapore Management University, Singapore
Kouzou Ohara	Aoyama Gakuin University, Japan
Shingo Otsuka	Kanagawa Institute of Technology, Japan
George Papadopoulos	University of Cyprus, Cyprus
Marcin Paprzycki	Systems Research Institute, Polish Academy of Sciences, Poland
Rafael Parpinelli	Santa Catarina State University (UDESC), Brazil
Jakub Peksinski	West Pomeranian University of Technology, Szczecin, Poland
Danilo Pelusi	University of Teramo, Italy
Hoang Pham	Rutgers University, USA

Xuan Hau Pham	Quang Binh University, Vietnam
Tao Pham Dinh	INSA Rouen, France
Maciej Piasecki	Wrocław University of Science and Technology, Poland
Bartłomiej Pierański	Poznan University of Economics and Business, Poland
Dariusz Pierzchala	Military University of Technology, Poland
Marcin Pietranik	Wrocław University of Science and Technology, Poland
Elias Pimenidis	University of the West of England, UK
Jaroslav Pokorný	Charles University, Prague, Czech Republic
Andrzej Polanski	Silesian University of Technology, Poland
Elvira Popescu	University of Craiova, Romania
Piotr Porwik	University of Silesia in Katowice, Poland
Małgorzata Przybyła-Kasperek	University of Silesia in Katowice, Poland
Andrzej Przybyszewski	Polish-Japanese Academy of Information Technology, Poland
Paulo Quaresma	Universidade de Evora, Portugal
David Ramsey	Wrocław University of Science and Technology, Poland
Mohammad Rashedur Rahman	North South University, Bangladesh
Ewa Ratajczak-Ropel	Gdynia Maritime University, Poland
Manuel Roveri	Politecnico di Milano, Italy
Przemysław Rozewski	West Pomeranian University of Technology, Szczecin, Poland
Leszek Rutkowski	Czestochowa University of Technology, Poland
Henryk Rybiński	Warsaw University of Technology, Poland
Alexander Ryjov	Lomonosov Moscow State University, Russia
Keun Ho Ryu	Chungbuk National University, South Korea
Virgilijus Sakalauskas	Vilnius University, Lithuania
Daniel Sanchez	University of Granada, Spain
Moamar Sayed-Mouchaweh	Ecole des Mines de Douai, France
Rafal Scherer	Czestochowa University of Technology, Poland
Juergen Schmidhuber	Swiss AI Lab IDSIA, Poland
Björn Schuller	University of Passau, Germany
Jakub Segen	Gest3D, USA
Ali Selamat	Universiti Teknologi Malaysia, Malaysia
S. M. N. Arosha Senanayake	Universiti Brunei Darussalam, Brunei Darussalam
Tegjyot Singh Sethi	University of Louisville, USA
Natalya Shakhovska	Lviv Polytechnic National University, Ukraine
Donghwa Shin	Yeungnam University, South Korea
Andrzej Siemiński	Wrocław University of Science and Technology, Poland
Dragan Simic	University of Novi Sad, Serbia
Bharat Singh	Universiti Teknology PETRONAS, Malaysia
Krzysztof Slot	Lodz University of Technology, Poland
Adam Slowik	Koszalin University of Technology, Poland

Vladimir Sobeslav	University of Hradec Kralove, Czech Republic
Kulwadee Somboonviwat	King Mongkut's University of Technology Thonburi, Thailand
Jong-In Song	Gwangju Institute of Science and Technology, South Korea
Zenon A. Sosnowski	Bialystok University of Technology, Poland
Bela Stantic	Griffith University, Australia
Jerzy Stefanowski	Poznan University of Technology, Poland
Serge Stinckwich	University of Caen-Lower Normandy, France
Ja-Hwung Su	Cheng Shiu University, Taiwan
Andrzej Swierniak	Silesian University of Technology, Poland
Edward Szczerbicki	University of Newcastle, Australia
Julian Szymanski	Gdansk University of Technology, Poland
Yasufumi Takama	Tokyo Metropolitan University, Japan
Maryam Tayefeh	Mahmoudi ICT Research Institute, Iran
Zbigniew Telec	Wrocław University of Science and Technology, Poland
Dilhan Thilakarathne	Vrije Universiteit Amsterdam, The Netherlands
Krzysztof Tokarz	Silesian University of Technology, Poland
Bogdan Trawinski	Wrocław University of Science and Technology, Poland
Ualsher Tukeyev	al-Farabi Kazakh National University, Kazakhstan
Aysegul Ucar	Firat University, Turkey
Olgierd Unold	Wrocław University of Science and Technology, Poland
Natalie Van Der Wal	Vrije Universiteit Amsterdam, The Netherlands
Pandian Vasant	Universiti Teknologi PETRONAS, Malaysia
Jorgen Villadsen	Technical University of Denmark, Denmark
Bay Vo	Ho Chi Minh City University of Technology, Vietnam
Gottfried Vossen	ERCIS Münster, Germany
Lipo Wang	Nanyang Technological University, Singapore
Yongkun Wang	University of Tokyo, Japan
Izabela Wierzbowska	Gdynia Maritime University, Poland
Konrad Wojciechowski	Silesian University of Technology, Poland
Michal Wozniak	Wrocław University of Science and Technology, Poland
Krzysztof Wrobel	University of Silesia in Katowice, Poland
Tsu-Yang Wu	Harbin Institute of Technology Shenzhen Graduate School, China
Marian Wysocki	Rzeszow University of Technology, Poland
Farouk Yalaoui	University of Technology of Troyes, France
Xin-She Yang	Middlesex University, UK
Lina Yao	University of New South Wales, Australia
Tulay Yildirim	Yildiz Technical University, Turkey
Slawomir Zadrozny	Systems Research Institute, Polish Academy of Sciences, Poland
Drago Zagar	University of Osijek, Croatia
Danuta Zakrzewska	Lodz University of Technology, Poland

Constantin-Bala   Lucian Blaga University of Sibiu, Romania
   Zamfirescu
Katerina Zdravkova   Ss. Cyril and Methodius University in Skopje,
                                Republic of Macedonia
Vesna Zeljkovic   Lincoln University, USA
Aleksander Zgrzywa   Wrocław University of Science and Technology, Poland
De-Chuan Zhan   Nanjing University, China
Qiang Zhang   Dalian University, China
Zhongwei Zhang   University of Southern Queensland, Australia
Dongsheng Zhou   Dalian Unviersity, China
Zhandos Zhumanov   al-Farabi Kazakh National University, Guyana
Maciej Zieba   Wrocław University of Science and Technology, Poland
Adam Ziebinski   Silesian University of Technology, Poland
Beata Zielosko   University of Silesia in Katowice, Poland
Marta Zorrilla   University of Cantabria, Spain

Program Committees of Special Sessions

Special Session on Multiple Model Approach to Machine Learning (MMAML 2018)

Emili Balaguer-Ballester   Bournemouth University, UK
Urszula Boryczka   University of Silesia, Poland
Abdelhamid Bouchachia   Bournemouth University, UK
Robert Burduk   Wrocław University of Science and Technology, Poland
Oscar Castillo   Tijuana Institute of Technology, Mexico
Rung-Ching Chen   Chaoyang University of Technology, Taiwan
Suphamit Chittayasothorn   King Mongkut's Institute of Technology Ladkrabang,
                                Thailand
Jose Alfredo F. Costa   Federal University (UFRN), Brazil
Bogustaw Cyganek   AGH University of Science and Technology, Poland
Ireneusz Czarnowski   Gdynia Maritime University, Poland
Patrick Gallinari   Pierre et Marie Curie University, France
Fernando Gomide   State University of Campinas, Brazil
Francisco Herrera   University of Granada, Spain
Tzung-Pei Hong   National University of Kaohsiung, Taiwan
Konrad Jackowski   Wrocław University of Science and Technology, Poland
Piotr Jędrzejowicz   Gdynia Maritime University, Poland
Tomasz Kajdanowicz   Wrocław University of Science and Technology, Poland
Yong Seog Kim   Utah State University, USA
Bartosz Krawczyk   Wrocław University of Science and Technology, Poland
Kun Chang Lee   Sungkyunkwan University, South Korea
Edwin Lughofer   Johannes Kepler University Linz, Austria
Hector Quintian   University of Salamanca, Spain
Andrzej Sieminski   Wrocław University of Science and Technology, Poland
Dragan Simic   University of Novi Sad, Serbia

Adam Slowik	Koszalin University of Technology, Poland
Zbigniew Telec	Wrocław University of Science and Technology, Poland
Bogdan Trawinski	Wrocław University of Science and Technology, Poland
Olgierd Unold	Wrocław University of Science and Technology, Poland
Pandian Vasant	University Technology Petronas, Malaysia
Michal Wozniak	Wrocław University of Science and Technology, Poland
Zhongwei Zhang	University of Southern Queensland, Australia
Zhi-Hua Zhou	Nanjing University, China

Special Session on Analysis of Image, Video, and Motion Data in Life Sciences (IVMLS 2018)

Artur Bąk	Polish-Japanese Academy of Information Technology, Poland
Leszek Chmielewski	Warsaw University of Life Sciences, Poland
Aldona Barbara Drabik	Polish-Japanese Academy of Information Technology, Poland
Marcin Fojcik	Sogn og Fjordane University College, Norway
Adam Gudys	Silesian University of Technology, Poland
Celina Imielinska	Vesalius Technolodgies LLC, USA
Henryk Josinski	Silesian University of Technology, Poland
Ryszard Klempous	Wrocław University of Science and Technology, Poland
Ryszard Kozera	The University of Life Sciences - SGGW, Poland
Julita Kulbacka	Wroclaw Medical University, Poland
Marek Kulbacki	Polish-Japanese Academy of Information Technology, Poland
Aleksander Nawrat	Silesian University of Technology, Poland
Jerzy Pawet Nowacki	Polish-Japanese Academy of Information Technology, Poland
Eric Petajan	LiveClips LLC, USA
Andrzej Polanski	Silesian University of Technology, Poland
Joanna Rossowska	Polish Academy of Sciences, Poland
Jakub Segen	Gest3D LLC, USA
Aleksander Sieron	Medical University of Silesia, Poland
Michat Staniszewski	Polish-Japanese Academy of Information Technology, Poland
Adam Switonski	Silesian University of Technology, Poland
Agnieszka Szczęsna	Silesian University of Technology, Poland
Kamil Wereszczynski	Polish-Japanese Academy of Information Technology, Poland
Konrad Wojciechowski	Polish-Japanese Academy of Information Technology, Poland
Stawomir Wojciechowski	Polish-Japanese Academy of Information Technology, Poland

Special Session on Intelligent and Contextual Systems (ICxS 2018)

Adriana Albu	Politehnica University Timisoara, Romania
Basabi Chakraborty	Iwate Prefectural University, Japan
Goutam Chakraborty	Iwate Prefectural University, Japan
Ha Manh Tran	Ho Chi Minh City International University, Vietnam
Hong Vu Nguyen	Ton Duc Thang University, Vietnam
Hideyuki Takahashi	RIEC, Tohoku University, Japan
Jerzy Swiątek	Wrocław University of Science and Technology, Poland
Jozef Korbicz	University of Zielona Gora, Poland
Keun Ho Ryu	Chungbuk National University, South Korea
Kilho Shin	University of Hyogo, Japan
Lina Yang	University of Macau, Macau
Maciej Huk	Wrocław University of Science and Technology, Poland
Masafumi Matsuhara	Iwate Prefectural University, Japan
Michael Spratling	University of London, UK
Nguyen Khang Pham	Can Tho University, Vietnam
Plamen Angelov	Lancaster University, UK
Qiangfu Zhao	University of Aizu, Japan
Quan Thanh Tho	Ho Chi Minh City University of Technology, Vietnam
Rashmi Dutta Baruah	Lancaster University, UK
Takako Hashimoto	Chiba University of Commerce, Japan
Tetsuji Kubojama	Gakushuin University, Japan
Tetsuo Kinoshita	RIEC, Tohoku University, Japan
Thai-Nghe Nguyen	Can Tho University, Vietnam
Yicong Zhou	University of Macau, Macau, SAR China
Yuan Yan Tang	University of Macau, Macau, SAR China
Zhenni Li	University of Aizu, Japan

Special Session on Intelligent Systems for Optimization of Logistics and Industrial Applications (ISOLIA 2018)

Zaki Sari	Abou Bakr Belkaid University of Tlemcen, Algeria
Hicham Chehade	University of Technology of Troyes, France
Yassine Ouazene	University of Technology of Troyes, France
Habiba Drias	University of Science and Technology (USTHB), Algeria
Lionel Amodeo	University of Technology of Troyes, France
Mustapha Nourelfath	University of Laval, Canada
Nathalie Sauer	University of Lorraine, France
Alexandre Dolgui	National Institute of Science and Technology Mines-Telecom, France
Olga Battaia	Institut Superieur de l'Aeronautique et d'Espace, France
Zaki Sari	Abou Bakr Belkaid University of Tlemcen, Algeria
Daoud Ait-Kadi	University of Laval, Canada

Dmitry Ivanov	Hochschule fur Wirtschaft und Recht Berlin, Germany
Lyes Benyoucef	Aix-Marseille University, France
Nidhal Rezg	University of Lorraine, France

Special Session on Intelligent Applications of Internet of Thing and Data Analysis Technologies (IoT&DAT 2018)

Goutam Chakraborty	Iwate Prefectural University, Japan
Bin Dai	University of Technology Xiamen, China
Qiangfu Zhao	University of Aizu, Japan
David C. Chou	Eastern Michigan University, USA
Chin-Feng Lee	Chaoyang University of Technology, Taiwan
Lijuan Liu	University of Technology Xiamen, China
Kien A. Hua	Central Florida University, USA
Long-Sheng Chen	Chaoyang University of Technology, Taiwan
Xin Zhu	University of Aizu, Japan
David Wei	Fordham University, USA
Qun Jin	Waseda University, Japan
Jacek M. Zurada	University of Louisville, USA
Tsung-Chih Hsiao	Huaoiao University, China
Hsien-Wen Tseng	Chaoyang University of Technology, Taiwan
Nitasha Hasteer	Amity University Uttar Pradesh, India
Chuan-Bi Lin	Chaoyang University of Technology, Taiwan
Cliff Zou	Central Florida University, USA

Special Session on Intelligent Systems and Methods in Biomedicine (ISaMiB 2018)

Jan Kubicek	Technical University of Ostrava, Czech Republic
Marek Penhaker	Technical University of Ostrava, Czech Republic
Martin Augustynek	Technical University of Ostrava, Czech Republic
Martin Cerny	Technical University of Ostrava, Czech Republic
Vladimir Kasik	Technical University of Ostrava, Czech Republic
Lukas Peter	Technical University of Ostrava, Czech Republic
Ondrej Krejcar	University of Hradec Kralove, Czech Republic
Kamil Kuca	University of Hradec Kralove, Czech Republic
Petra Maresova	University of Hradec Kralove, Czech Republic
Ali Selamat	Universiti Teknologi Malaysia, Malaysia

Special Session on Intelligent Systems and Algorithms in Information Sciences (ISAIS 2018)

Martin Kotyrba	University of Ostrava, Czech Republic
Eva Volna	University of Ostrava, Czech Republic
Ivan Zelinka	VSB-Technical University of Ostrava, Czech Republic
Hashim Habiballa	Institute for Research and Applications of Fuzzy Modeling, Czech Republic
Alexej Kolcun	Institute of Geonics, AS CR, Czech Republic
Roman Senkerik	Tomas Bata University in Zlin, Czech Republic

Zuzana Kominkova-Oplatkova	Tomas Bata University in Zlin, Czech Republic
Katerina Kostolanyova	University of Ostrava, Czech Republic
Antonin Jancarik	Charles University in Prague, Czech Republic
Igor Kostal	The University of Economics in Bratislava, Slovakia
Eva Kurekova	Slovak University of Technology in Bratislava, Slovakia
Leszek Cedro	Kielce University of Technology, Poland
Dagmar Janacova	Tomas Bata University in Zlin, Czech Republic
Martin Halaj	Slovak University of Technology in Bratislava, Slovakia
Radomil Matousek	Brno University of Technology, Czech Republic
Roman Jasek	Tomas Bata University in Zlin, Czech Republic
Petr Dostal	Brno University of Technology, Czech Republic
Jiri Pospichal	The University of Ss. Cyril and Methodius (UCM), Slovakia
Vladimir Bradac	University of Ostrava, Czech Republic

Special Session on Design Thinking-Based R&D, Development Technique, and Project-Based Learning (2DT-PBL 2018)

Tsuyoshi Arai	Okayama Prefectural University, Japan
Yoshihide Chubachi	Advanced Institute of Industrial Technology, Japan
Keiichi Endo	Ehime University, Japan
Takuya Fujihashi	Ehime University, Japan
Yoshinobu Higami	Ehime University, Japan
Tohru Kawabe	University of Tsukuba, Japan
Shinya Kobayashi	Ehime University, Japan
Hisayasu Kuroda	Ehime University, Japan
Kazuo Misue	University of Tsukuba, Japan
Katsumi Sakakibara	Okayama Prefectural University, Japan
Kiyoshi Sakamori	Advanced Institute of Industrial Technology, Japan
Hironori Takimoto	Okayama Prefectural University, Japan
Toshiyuki Uto	Ehime University, Japan
Chiemi Watanabe	Advanced Institute of Industrial Technology, Japan
Hitoshi Yamauchi	Okayama Prefectural University, Japan

Special Session on Modelling, Storing, and Querying of Graph Data (MSQGD 2018)

Michal Valenta	Czech Technical University, Czech Republic
Martin Svoboda	Charles University, Czech Republic
M. Praveen	CMI, India
Jianxin Li	University of Western Australia, Australia
Konstantinos Semertzidis	University of Ioannina, Greece
Marco Mesiti	DICO - University of Milan, Italy

Vincenzo Moscato	University of Naples, Italy
Cedric Du Mouza	CNAM, France
Jacek Mercik	WSB University in Wroclaw, Poland
Virginie Thion	University of Rennes, France

Special Session on Data Science and Computational Intelligence (DSCI 2018)

Adisak Sukul	Iowa State University, USA
Akadej Udomchaiporn	King Mongkut's Institute of Technology Ladkrabang, Thailand
Jittima Tongurai	Kobe University, Japan
Kulsawasd Jitkajornwanich	King Mongkut's Institute of Technology Ladkrabang, Thailand
Natawut Nupairoj	Chulalongkorn University, Thailand
Peeraphon Sophatsathit	Chulalongkorn University, Thailand
Peerapon Vateekul	Chulalongkorn University, Thailand
Pisit Chanvarasuth	Sirindhorn International Institute of Technology, Thailand
Ronan Reilly	Maynooth University, Ireland
Sanparith Marukatat	National Electronics and Computer Technology Center, Thailand
Sarun Intagosum	King Mongkut's Institute of Technology Ladkrabang, Thailand
Suradej Intagorn	Kasetsart University, Thailand
Turki Talal Salem Turki	King Abdulaziz University, Saudi Arabia
Veera Boonjing	King Mongkut's Institute of Technology Ladkrabang, Thailand

Special Session on Computer Vision and Robotics (CVR 2018)

Van-Dung Hoang	Quang Binh University, Vietnam
Chi- Mai Luong	University of Science and Technology of Hanoi, Vietnam
My-Ha Le	Ho Chi Minh City University of Technology and Education, Vietnam
Kang-Hyun Jo	Ulsan University, South Korea
The- Anh Pham	Hong Duc University, Vietnam
Van-Huy Pham	Ton Duc Thang University, Vietnam
Youngsoo Suh	University of Ulsan, South Korea
Danilo Caceres Hernandez	Universidad Tecnologica de Panama, Panama
Kaushik Deb	Chittagong University of Engineering and Technology, Bangladesh
Ha Nguyen Thi Thu	Electric Power University, Vietnam
Viet-Vu Vu	Thai Nguyen University of Technology, Vietnam
Lan Le Thi	Hanoi University of Science and Technology, Vietnam
Thanh Binh Nguyen	University of Science Ho Chi Minh City, Vietnam
Trung Duc Nguyen	Vietnam Maritime University, Vietnam

The Bao Pham	University of Science Ho Chi Minh City, Vietnam
Yoshinori Kuno	Saitama University, Japan
Heejun Kang	University of Ulsan, South Korea
Wahyono	Universitas Gadjah Mada, Indonesia
Do Van Nguyen	Nagaoka University of Technology, Japan
Truc Thanh Tran	Danang Department of Information and Communications, Vietnam
Long-Thanh Ngo	Le Quy Don Technical University, Vietnam

Special Session on Intelligent Biomarkers of Neurodegenerative Processes in the Brain (InBinBRAIN 2018)

Zbigniew Struzik	RIKEN Brain Science Institute, Japan
Zbigniew Ras	University of North Carolina at Charlotte, USA
Konrad Ciecierski	Warsaw University of Technology, Poland
Piotr Habela	Polish-Japanese Academy of Information Technology, Poland
Peter Novak	Brigham and Women's Hospital, USA
Wieslaw Nowinski	Cardinal Stefan Wyszynski University, Poland
Andrei Barborica	Research and Compliance and Engineering, FHC, Inc., USA
Alicja Wieczorkowska	Polish-Japanese Academy of Information Technology, Poland
Majaz Moonis	UMass Medical School, USA
Krzysztof Marasek	Polish-Japanese Academy of Information Technology, Poland
Mark Kon	Boston University, USA
Lech Polkowski	Polish-Japanese Academy of Information Technology, Poland
Andrzej Skowron	Warsaw University, Poland
Ryszard Gubrynowicz	Polish-Japanese Academy of Information Technology, Warsaw, Poland
Dominik Slezak	Warsaw University, Poland
Radoslaw Nielek	Polish-Japanese Academy of Information Technology, Warsaw, Poland

Special Session on Data Structures Modelling for Knowledge Representation (DSMKR 2018)

| Marek Krotkiewicz | Wrocław University of Science and Technology, Poland |

Special Session on Computational Imaging and Vision (CIV 2018)

Ishwar Sethi	Oakland University, USA
Moongu Jeon	Gwangju Institute of Science and Technology, South Korea
Jong-In Song	Gwangju Institute of Science and Technology, South Korea

Kiseon Kim	Gwangju Institute of Science and Technology, South Korea
Taek Lyul Song	Hangyang University, South Korea
Ba-Ngu Vo	Curtin University, Australia
Ba-Tuong Vo	Curtin University, Australia
Du Yong Kim	Curtin University, Australia
Benlian Xu	Changshu Institute of Technology, China
Peiyi Zhu	Changshu Institute of Technology, China
Mingli Lu	Changshu Institute of Technology, China
Weifeng Liu	Hangzhou Danzi University, China
Ashish Khare	University of Allahabad, India
Om Prakash	University of Allahabad, India
Moonsoo Kang	Chosun University, South Korea
Goo-Rak Kwon	Chosun University, South Korea
Sang Woong Lee	Gachon University, South Korea
Ekkarat Boonchieng	Chiang Mai University, Thailand
Jeong-Seon Park	Chonnam National University, South Korea
Unsang Park	Sogang University, South Korea
R. Z. Khan	Aligarh Muslim University, India
Sathya Narayanan	NTU, Singapore

Special Session on Intelligent Computer Vision Systems and Applications (ICVSA 2018)

Ferran Reverter Comes	University of Barcelona, Spain
Michael Cree	University of Waikato, New Zealand
Piotr Dziurzanski	University of York, UK
Marcin Iwanowski	Warsaw University of Technology, Poland
Heikki Kalviainen	Lappeenranta University of Technology, Finland
Tomasz Marciniak	UTP University of Science and Technology, Poland
Adam Nowosielski	West Pomeranian University of Technology, Szczecin, Poland
Krzysztof Okarma	West Pomeranian University of Technology, Szczecin, Poland
Arkadiusz Ortowski	Warsaw University of Life Sciences, Poland
Edward Potrolniczak	West Pomeranian University of Technology, Szczecin, Poland
Pilar Rosado Rodrigo	University of Barcelona, Spain
Khalid Saeed	AGH University of Science and Technology Cracow, Poland
Rafael Saracchini	Technological Institute of Castilla y Leon (ITCL), Spain
Samuel Silva	University of Aveiro, Portugal
Gregory Slabaugh	City University London, UK
Egon L. van den Broek	Utrecht University, The Netherlands

Special Session on Advanced Data Mining Techniques and Applications (ADMTA 2018)

Tzung-Pei Hong	National University of Kaohsiung, Taiwan
Tran Minh Quang	Ho Chi Minh City University of Technology, Vietnam
Bac Le	University of Science, VNU-HCM, Vietnam
Bay Vo	Ho Chi Minh City University of Technology, Vietnam
Chun-Hao Chen	Tamkang University, Taiwan
Chun-Wei Lin	Harbin Institute of Technology, China
Wen-Yang Lin	National University of Kaohsiung, Taiwan
Yeong-Chyi Lee	Cheng Shiu University, Taiwan
Le Hoang Son	University of Science, Vietnam
Vo Thi Ngoc Chau	Ho Chi Minh City University of Technology, Vietnam
Van Vo	Ho Chi Minh University of Industry, Vietnam
Ja-Hwung Su	Cheng Shiu University, Taiwan
Ming-Tai Wu	University of Nevada, USA
Kawuu W. Lin	National Kaohsiung University of Applied Sciences, Taiwan
Tho Le	Ho Chi Minh City University of Technology, Vietnam
Dang Nguyen	Deakin University, Australia
Hau Le	Thuyloi University, Vietnam
Thien-Hoang Van	Ho Chi Minh City University of Technology, Vietnam
Tho Quan	Ho Chi Minh City University of Technology, Vietnam
Ham Nguyen	University of People's Security Ho Chi Minh City, Vietnam
Thiet Pham	Ho Chi Minh University of Industry, Vietnam

Contents – Part II

Intelligent Systems and Methods in Biomedicine

Intelligent Biomarkers of Neurodegenerative Processes in Brain

Analysis of Image, Video and Motion Data in Life Sciences

Computational Imaging and Vision

Computer Vision and Robotics

Intelligent Computer Vision Systems and Applications

Intelligent Systems for Optimization of Logistics and Industrial Applications

Contents – Part I

Text Processing and Information Retrieval

Machine Learning and Data Mining

Decision Support and Control Systems

Computer Vision Techniques

Advanced Data Mining Techniques and Applications

Multiple Model Approach to Machine Learning

Sensor Networks and Internet of Things

Intelligent Information Systems

Data Structures Modelling for Knowledge Representation

Deep Learning Based Approach for Entity Resolution in Databases

Nihel Kooli[(✉)], Robin Allesiardo, and Erwan Pigneul

PagesJaunes - Solocal Group Rennes, Rennes, France
{nkooli,rallesiardo,epigneul}@pagesjaunes.fr

Abstract. This paper proposes a Deep Neural Networks (DNN) based approach for entity resolution in databases. This approach is mainly based on a record linkage process which aims to detect records that refer to the same entity. First, record pairs are represented by their word embedding using an N-gram embedding based method. Then, they are classified into matching or unmatching pairs using a DNN model. Three DNN architectures: Multi-Layer Perceptron, Long Short Term Memory networks and Convolutional Neural Networks are investigated and compared for this purpose. The approach is experimented on two databases. The results exceed 97% for recall and 96% for precision. The comparison with similarity measure and classical classifier based approaches shows a significant improvement in the results on the two databases.

Keywords: Entity resolution · Databases · Record linkage
Deep neural networks · Word embedding · Similarity measures

1 Introduction

A database is a repository that can merge records from several data sources with heterogeneous data formats. For example, the collection of a professional database from different telecommunication carriers or the collection of a scientific publication database from different archive websites. It may be also updated dynamically and/or managed by different users. This leads to duplicated, incomplete and erroneous data. Indeed, record attributes may be absent, may be non-normalized (abbreviations, acronyms, punctuation, etc.), may contain typographical errors or may have various entries (such as a professional which has two phone numbers).

To manage this data, such as using the database as a referential for a query-based search system, it is necessary to clean it. This is achieved by the Entity Resolution (ER) process which aims to synthesize records, to remove the redundancy and to identify the various entries of the same attribute. ER consists of detecting records that refer to the same entity, where an entity represents an existing or real thing, such as a person, a location, an organization, etc. Each entity is defined by a set of attributes. For instance, a person has the attributes: name, date of birth, social security number, etc.

© Springer International Publishing AG, part of Springer Nature 2018
N. T. Nguyen et al. (Eds.): ACIIDS 2018, LNAI 10752, pp. 3–12, 2018.
https://doi.org/10.1007/978-3-319-75420-8_1

The ER problem is also known in the databases community under the name of merge/purge or record linkage. It is often performed by the attribute comparison using similarity measures to tolerate their various representations.

Table 1. An extract of the publications database showing an example of two records referring to the same entity

Field	Record 1	Record 2
Title	Simple greedy matching for aligning large knowledge bases	SiGMa
Authors	Simon Lacoste-Julien; Konstantina Palla; Alex Davies; Gjergji Kasneci; Thore Graepel; Zoubin Ghahramani	S. Lacoste-Julien; K. Palla; A. Davies; G. Kasneci; T. Graepel; Z. Ghahramani
Affiliations	INRIA; University of Cambridge; University of Cambridge; Microsoft Research; Microsoft Research; University of Cambridge	(null)
Production date	2013-08-11	2013
Journal	(null)	(null)
Pages	pp. 572–580	572–580
Conference	The 19th ACM SIGKDD international conference on knowledge discovery and data mining	KDD 2013
Conference date	2013-08-11	2013

Table 1 shows an extract of the scientific publication database used in our experiments. It represents two records referring to a same scientific publication. These records present several attribute dissimilarities, such as the acronyms in the title and the conference name, the representation of the author first names by their initials and the lack of some attributes (represented by "(null)").

String similarity measures [1], traditionally used for the ER purpose, are insufficient to overcome problems related to acronyms and abbreviations. Indeed, the edit distances, such as Levenshtein and Jaro-Winkler, are able to overcome elementary variations on characters. The bag of words distances, such as Jarccard and Tf-idf, are able to overcome term permutations. The hybrid distances, such as Monge-Elkan and Soft-tf-idf, are able to overcome term permutations with some character variations. But, none of these measures could detect that "The 19th ACM SIGKDD International Conference on Knowledge Discovery and Data Mining" and "KDD 2013" represent the same conference name. A system that is able to detect the similarity between these non-normalized representations of attributes is then required.

Recently, deep learning models have been successfully applied to the domains of text mining, natural language processing and computer vision [2]. These models have shown their power in learning features for several tasks. In this paper, we propose an ER approach based on word embedding for records representation and Deep Neural Networks (DNN) for record linkage learning and prediction. The approach is experimented on two databases. The first database is in French language and represents contact information of professionals in our company's repository. The second one is in English language, represents meta-data of scientific publications and is extracted from public websites. Several deep neural networks are experimentally compared for this purpose. The comparison with three supervised learning classifiers (SVM, C4.5, Naives bayes) using similarity measure combination for attribute matching, shows the interest of employing a DNN model for the ER task.

The remainder of this paper is organized as follows. First, an overview of existent ER approaches is proposed in Sect. 2. Second, our DNN ER based approach is detailed in Sect. 3. Then, experiments on two real world databases are presented in Sect. 4. Finally, Sect. 5 concludes and suggests future works.

2 State of the Art

A detailed literature review of ER approaches is provided in [3]. These approaches could be categorized into deterministic or probabilistic ones.

Deterministic approaches, such as that proposed in [4], are based on a set of rules fixed by experts which determinate the matching conditions of record pairs. These rules depend generally on a set of relevant fields. Similarity measures are generally employed in the comparison of attributes in order to tolerate the typographical errors. These techniques are time-consuming because they require significant human involvement. Furthermore, the predefined rules are very dependent on the database and on the domain.

Probabilistic approaches treat the problem as a classification one. This may be classifying records into entities or classifying record pairs into matching or unmatching ones. These approaches may be unsupervised or supervised.

Unsupervised approaches are useful when there is no annotated data. Most of them are inspired by the Fellegi and Sunter model described in [5]. It proposes to estimate matching and unmatching probabilities between attributes pairs based on a statistical study of the database. These probabilities are then used to calculate a matching score. The latter is compared to a threshold to make the matching decision.

Supervised approaches use annotated data to train a classification model. Such a method is described in [6] and employs two classification levels using an SVM. The first level represents attribute comparison using the Levenshtein distance, while the second level represents record comparison based on attribute similarity learning. The work in [1] combines 7 similarity measures to compute the similarity between attributes and uses a tree based classifier (C4.5) to match record pairs. Authors in [7] use active learning based techniques to select the most informative record pairs. Users are solicited to annotate these pairs as matching or unmatching ones in order to train the classifier. The latter generates classification rules that intersect attributes and their similarity measures.

To our knowledge, there is no ER method that uses deep learning for the complete process of record linkage. Authors in [8] propose a hybrid human/machine approach. It takes as input a record and proposes to match it with its corresponding entity. First, candidate entities are selected for each record based on a single layered convolutional neural network. The input of this network is a vector representing the word embedding of its relevant words. Then, candidate entities are analyzed by human experts based on a crowd-sourcing approach. Even if the cost of the crowdsourcing is reduced by the deep learning network, this approach still requires human intervention. Furthermore, word embedding is realized using word2vec [9] which can not output a vector for a word that is not in the pre-trained model and does not take into account word syntactic variation. This approach was experimented on only 300 records.

In the domain of name disambiguation, the work in [10] proposes an approach for integrating Vietnamese author names in different publications. It uses a multilayer perceptron which takes as input record pairs and proposes to classify it into a matching or an unmatching ones. Each record pair is represented by a vector of similarity distances between attribute pairs. The problem with this method is that is dependent on the employed similarity measures. This approach was experimented on about 4300 records.

Our novel approach is completely automatic and does not require any human intervention. In addition, the attribute comparison is independent of any similarity measure choice and gets over their deficiency in particular cases. This similarity is automatically learned by the DNN.

3   Deep Learning Based Entity Resolution Approach

Our ER approach is performed by comparing record pairs in order to link those referring to the same entity. To reduce the number of comparisons, a preliminary phase called database segmentation is integrated.

In the following, we detail the database segmentation process employed in this approach. Then, we present a novel record linkage method based on DNN. Finally, we explain the generation process of the training dataset.

3.1   Database Segmentation

In large databases, record pairs comparison is expensive since it is a Cartesian product of database records. Database segmentation [11] is often used to solve this problem. It consists of grouping into blocks close records that are likely to represent the same entity. Therefore, only the records of a same block are compared two by two in the record linkage step. This grouping is performed using grouping keys. A grouping key may be a given field or a combination of given fields. The records whose attributes corresponding to these keys are similar will be grouped together. The n first characters of a company's "name" field combined with the zip code in professional database and the n first terms of the "title" field of a publication in scientific publication database are good examples of such keys.

3.2 Record Linkage

The global model of the record linkage approach is represented in Fig. 1. This model takes as input pairs of records and proposes to classify them into "matching" or "unmatching" ones. Firstly, a pre-processing step is performed. It consists of extracting a vector of key words and representing it by a numerical matrix using a word embedding step. Secondly, the deep neural network is trained. The first layers of this network consists of learning the classification features while the last one corresponds to the binary classification process. Record linkage steps will be explained in the following.

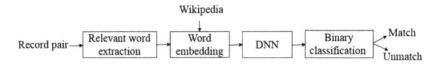

Fig. 1. Entity resolution approach model

Relevant word extraction represents a pre-processing step of the record attribute data. It consists of removing empty words, such as "the", "and", "of", etc. Indeed, these words do not contribute to the training of the DNN since they are not relevant in the process of record matching and may degrade the efficiency of the network. In addition, we propose to remove punctuation and special characters.

Word embedding is used to represent textual data in the attributes by dense real valued vectors with a notion of distance between words. One common way to achieve such embedding is to use the word2vec approach [9]. This approach is very efficient but suffer from a major drawback in our setting, as only words from the dictionary can be embedded. However, in the ER task, data is often non-normalized and many records can contain new words not previously seen in the dataset used to learn the embedding.

To overcome this issue, we instead use a N-gram based embedding method, available through the Fasttext library [12]. Fasttext is similar to word2vec but adds information about subwords (the N-grams). This additional information is available for every words, yielding to a more robust representation of new words.

In this work, we trained the Fasttext word embedding model on the wikipedia data ("wiki.fr" for the French database and "wiki.en" for the English database). The word embedding size is empirically fixed to 100.

Deep neural network is used to decide the matching of two records. Several architectures can be used to achieve this goal. We discuss some of them in this section and compare their performances later in the experiments. We generically call them DNN and consider their type as a parameter of the algorithm. Regardless of their type, every architecture takes as input a record pair $r_i r_j$ and exposes a binary classifier as their output layer.

We now formalize the data processing in input of the DNN. A record is compounded of several attributes, themselves compounded of several words. Let n be the maximum number of words by attribute; words in excess are truncated. For each attribute, the word embeddings are concatenated into a real valued vector of size $D \times n$, where D is the size of the word embedding. When the number of words is lesser than n, the end of the vector is padded with zeros. The record pair $r_i r_j$ of two entities is the matrix obtained by the concatenation of the attribute vectors of both records (one row is an attribute vector). We detail the three used DNN architectures in the following.

The Fully Connected Network is a regular Multi Layer Perceptron (MLP) [13] where each neuron of each layer is fully connected to the neurons of the previous and the next layer. For the record pairs classification, we use an MLP composed of k dense layers: the input layer, $k - 2$ hidden layers and the output layer. k is empirically fixed to 4 using cross-validation. The number of units in the input layer is $2 \times D \times n \times m$, where m is the number of database fields. The number of units in the hidden layers is empirically fixed to 100. The input and hidden layers are followed each by a Batch Normalization layer to normalize the next layer inputs. The activation layer is carried out using the rectifier function $f(x) = \max(x, 0)$ for the input and the hidden layers. Each activation layer is followed by a Dropout using a rate of 0.5. The output layer is composed of two units which corresponds to the cases of "matching" or "unmatching" record pairs. The activation layer is carried out using the sigmoid function $f(x) = sigmoid(x) = \frac{1}{1+e^{-x}}$. The gradient descent is performed by a Rooted Mean Square (RMS) back-propagation for 100 epochs with mini batches of size 100.

Long Short Term Memory Networks (LSTM) [15] were created to allow the network to maintain a memory of the previous inputs. Whereas a regular DNN only uses the current input for the prediction, the LSTM can process the input as a sequence. To our knowledge, LSTM models has never been used for the task of record linkage. In this work, we sequentially feed the LSTM with the embedding of the attributes in the record pair. The number of sequences is then equal to $2 \times m$. The used LSTM network contains 3 LSTM hidden layers with 32 hidden units. It has been trained with the Adam optimizer [16] for 70 epochs with a batch size of 200.

Convolutional Neural Networks (CNN) were popularized due to their performances on image classification [17]. The convolutional layers allow to train classifier on image with minimal pre-processing. Filters convolving on the input are being learned by the network and replace hand-engineered features. Recently, convolutional networks were successfully used on textual data [14]. In this work, we use 4 1D convolutional layers followed by a fully connected layer for the record pair classification (see Fig. 2). Each convolutional layer uses a filter of size 4 and a stride of 1. These parameters are empirically tuned. The first three convolutional layers are followed each by a max pooling layer while the last convolutional layer is followed by an average pooling layer. The CNN is regularized using a Dropout with a rate of 0.7. The activation layer is carried out using the

ReLU function for the all layers except the last one where a sigmoid is used. The gradient descent is performed by Stochastic Gradient Descent (SGD) for 100 epochs with mini-batches of size 200.

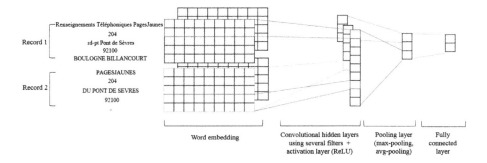

Fig. 2. Convolutional neural network model for record pair classification

3.3   Dataset Generation

The training dataset is generated based on coupling the records pairs of the same database block. Since there is much more non-similar record pairs than similar ones, we obtain a dataset with imbalanced classes distribution. To resolve this problem, we use cluster-based over sampling. This consists of clustering matching and unmatching classes independently using the k-means algorithm. Thus, record pairs samples of each cluster are duplicated such that all clusters of each class have the same number of samples and the two classes have the same number of samples. The cluster-based over sampling performed better than simple over-sampling and under-sampling methods in our experiments.

4   Experiments

4.1   Datasets

Professional database contains entities that represent professionals in our company's repository such as societies, doctors, restaurants, plumbers, etc. It is collected from telecommunication carrier databases merged with the siren public database[1]. This database is in French language. Each professional is described by the attributes: denomination, street type, street number, street, zip code, city, region, geo-localization, siret number, etc. This database is composed of 11 478 587 records and 6 897 305 entities. The dataset used for the experiments contains 17 975 367 records, decomposed as shown in Table 2.

Scientific publication database contains meta-data of scientific publications such as journals, conferences, thesis manuscripts, posters, etc. It is extracted from

---

[1] https://www.sirene.fr/.

public archive websites of scientific articles: HAL[2], ISTEX[3] and DBLP[4]. This database is in English language. The entity attributes are: title, authors, affiliations, pages, editors, conference, journal, dates, etc. The publication database is composed of 415 500 records and 286 695 entities. The datasets used for the experiments is composed of 930 800 as shown in Table 2.

Table 2. Datasets used in the experiments

Dataset	Professionals database	Publications database
Training (#records pairs)	10 785 220	558 000
Validation (#records pairs)	1 797 537	93 800
Test (#records pairs)	5 392 610	279 000
Total (#records pairs)	17 975 367	930 800

4.2   Record Linkage Results

Evaluation metrics: A matching pair is defined as a pair that refers the same entity in reality. A linked pair is defined as a record pair that is matched with our system. Recall, Precision are then defined in (1).

$$Recall = \frac{\#correctly\ linked\ pairs}{\#matching\ pairs}; \ Precision = \frac{\#correctly\ linked\ pairs}{\#linked\ pairs} \quad (1)$$

Record linkage results are reported in Table 3 for the two experimented databases. Three DNN architectures are compared for this purpose. These results show that the LSTM outperforms the two other DNN architectures on the professional database and that the CNN is the best architecture on the scientific publication database. False positives are essentially caused by different professionals having the same address. False negative are essentially caused by missing attributes in the records or the over segmentation of the database (placing similar records in different blocks discarding the possibility of comparing them).

To show the interest of employing the word embedding process, we replaced the word embedding resulting matrix by a one that contains the similarity measure values between the attributes pairs. Seven distinct similarity distances: Levenshtein, Jaro, Jaro-Winkler, Jaccard, Tf-idf, Monge-Elkan, Soft-tf-idf are used for this purpose (a study of these measures has been presented in [1]). The comparison results, reported in Table 3, show that the embedding process leads to a better performance. Indeed, learning a projection of the data to a new vector space is equivalent to learning a new similarity measure [18]. This depict the advantage of a learned similarity measure versus an arbitrary defined one.

---

[2] https://hal.archives-ouvertes.fr/.
[3] http://www.istex.fr/.
[4] http://dblp.uni-trier.de/.

Table 3. Record linkage results using DNN

Input	Model	Professionals database			Publications database		
		Recall	Precision	F-measure	Recall	Precision	F-measure
Word embedding	MLP	98.00	98.00	98.00	95.60	96.98	96.29
	LSTM	**99.05**	**98.70**	**98.87**	96.76	95.89	96.32
	CNN	98.10	97.70	97.80	**97.45**	**96.78**	**97.11**
Similarity measures	MLP	91.74	98.80	94.89	93.51	91.97	92.73
	LSTM	91.18	98.70	94.50	94.56	93.76	94.16
	CNN	85.41	97.50	90.18	94.34	94.20	94.27

Comparison: Our DNN based approach is compared with classical classifiers based approaches: SVM, C4.5 and Naive Bayes (as proposed in [1]). The used features for these classifiers are the 7 similarity measures presented above. The results are reported in Table 4. The comparison with Table 3 shows that our DNN approach significantly outperforms the classical classifiers and confirms its ability to learn the similarity between non-normalized entity attributes.

Table 4. Record linkage results using classical classifiers [1]

Similarity measures	Classifier	Professionals database			Publications database		
		Recall	Precision	F-measure	Recall	Precision	F-measure
Monge-Elkan	SVM	88.60	90.30	89.44	93.01	92.90	92.95
	C4.5	89.80	88.01	88.89	93.30	92.95	93.12
	Naive Bayes	91.25	89.45	90.34	91.00	89.96	90.48
All measures	SVM	89.57	91.76	90.65	93.40	93.00	93.20
	C4.5	91.25	92.45	91.85	94.90	94.56	94.73
	Naive Bayes	92.78	90.54	91.64	92.98	92.20	92.59
3.2cmMonge-Elkan +	SVM	89.57	91,76	90.65	93.40	93.00	93.20
Levenshtein +	C4.5	**92.00**	**92.00**	**92.00**	**94.90**	**94.56**	**94.73**
Jaro-Winkler + Tf-idf	Naive Bayes	92.78	90.54	91.64	92.98	92.20	92.59

5 Conclusion

In this paper, we proposed an ER approach based on word embedding and DNN models. The use of N-gram embedding and the automatic learn of attribute similarity by the DNN has proven to be effective in improving the record linkage process. The results on two entity databases (the first one is in the French language where the second one is in the English language) are promising and exceed 97% for recall and 96% for precision.

Our future work is to evaluate our approach on other public databases in order to investigate more attribute variability. In addition, we plan to deal with relational databases, where the entities are described by multiple tables connected by foreign keys. Another perspective is to extend the system to handle with the incremental update of databases.

References

1. Kooli, N.: Data matching for entity recognition in OCRed documents. Thesis defense, Lorraine university (2016)
2. Schmidhuber, J.: Deep learning in neural networks: an overview. Neural Netw. **61**, 85–117 (2015)
3. Christen, P.: Data Matching - Concepts and Techniques for Record Linkage, Entity Resolution, and Duplicate Detection. Data-Centric Systems and Applications Description, pp. 1–270. Springer, Heidelberg (2012). https://doi.org/10.1007/978-3-642-31164-2
4. Lee, M.L., Ling, T.W., Low, W.L.: IntelliClean: a knowledge-based intelligent data cleaner. In: Proceedings of the 6th International Conference on Knowledge Discovery and Data Mining, pp. 290–294 (2000)
5. Fellegi, I., Sunter, A.: A theory for record linkage. J. Am. Stat. Assoc. **64**, 1183–1210 (1969)
6. Bilenko, M., Mooney, R.J.: Adaptive duplicate detection using learnable string similarity measures. In: Proceedings of the Ninth ACM SIGKDD International Conference on Knowledge Discovery and Data Mining, pp. 39–48 (2003)
7. Tejada, S., Knoblock, C. A., Minton, S.: Learning domain-independent string transformation weights for high accuracy object identification. In: Proceedings of the 8th ACM SIGKDD International Conference on Knowledge Discovery and Data Mining, pp. 350–359 (2002)
8. Gottapua, R.D., Daglia, C., Ali, B.: Entity resolution using convolutional neural network. In: Procedia Computer Science, vol. 95, pp. 153–158. Elsevier (2016)
9. Mikolov, T., Chen, K., Corrado, G., Dean, J.: Efficient estimation of word representations in vector space (2013). http://arxiv.org/abs/1301.3781
10. Tran, H.N., Huynh, T., Do, T.: Author name disambiguation by using deep neural network. In: Nguyen, N.T., Attachoo, B., Trawiński, B., Somboonviwat, K. (eds.) ACIIDS 2014. LNCS (LNAI), vol. 8397, pp. 123–132. Springer, Cham (2014). https://doi.org/10.1007/978-3-319-05476-6_13
11. Bilenko, M.: Adaptive blocking: learning to scale up record linkage. In: Proceedings of the 6th IEEE International Conference on Data Mining, pp. 87–96 (2006)
12. Bojanowski, P., Grave, E., Joulin, A., Mikolov, T.: Enriching word vectors with subword. Trans. Assoc. Comput. Linguist. **5**, 135–146 (2017)
13. Rosenblatt, F.: The perceptron: a probabilistic model for information storage and organization in the brain. Psychol. Rev. **65**(6), 386–408 (1958)
14. Collobert, R.: Deep learning for efficient discriminative parsing. In: 21st International Conference on Artificial Intelligence and Statistics, pp. 224–232 (2011)
15. Hochreiter, S., Schmidhuber, J.: Long short-term memory. In: Neural computation (1997)
16. Kingma, D.P., Ba, J.: Distributed representations for biological sequence analysis. In: Data and Text Mining in Biomedical Informatics, abs/1412.6980 (2016)
17. Krizhevsky, A., Sutskever, I., Hinton, G.E.: ImageNet classification with deep convolutional neural networks. In: Advances in Neural Information Processing Systems 25 - NIPS (2012)
18. Yih, W., Meek, C.: Learning vector representations for similarity measures. Microsoft Technical Report MSR-TR-2010-139 (2010)

Generating Arbitrary Cross Table Layout in SuperSQL

Atsutomo Tabata[✉], Kento Goto, and Motomichi Toyama

Keio University, Yokohama, Kanagawa, Japan
{tabata,goto}@db.ics.keio.ac.jp, toyama@ics.keio.ac.jp

Abstract. In many Web site, various data presented in the form of a cross-table. Cross-table displays two sets of labels, typically on the top and left of the table, and attributes corresponding to the intersection of both labels in the middle. The simplest example of a cross table is a double entry table. In this work, we propose a formal model for cross-table and a way to describe cross-table from relational data. Using this model, we implement cross tables as a function of the SuperSQL system, an extension of the SQL language which generates structured documents from declarative queries on relational data. Using this method, we create elaborate and expressive cross tables that were not readily obtainable with other systems.

Keywords: SuperSQL · SQL · Database · Cross table · Pivot table

1 Introduction

In many Web site, various data is presented as the form of cross table. Cross tables display two sets of labels, typically on the top and left of the table and attributes corresponding to the intersection of both labels in the middle. Table 1 shows a cross table, the double entry table. It shows the percentage of sales in each meats at a supermarket in each year. In this work we formally define and model cross tables and use this formalization to express them in the Target Form Expression of SuperSQL.

SuperSQL is an extension of the SQL query language that generates structured documents from declarative queries ran against relational databases. While standard SQL queries only return flat results, which have to be formatted in a second step, SuperSQL directly creates structured documents such as HTML pages or PDF documents from the query. To allow this, the SQL target list is replaced by a Target Form Expression, or TFE which we describe in Sect. 3. Using the TFE, SuperSQL allows the creation of complex and long documents with minimal code and effort.

As most web content, cross tables are written using the HTML markup language, which poses several problems. Although HTML is capable of representing such tables its structure is not adapted for an easy representation of complex

© Springer International Publishing AG, part of Springer Nature 2018
N. T. Nguyen et al. (Eds.): ACIIDS 2018, LNAI 10752, pp. 13–24, 2018.
https://doi.org/10.1007/978-3-319-75420-8_2

Table 1. The example of cross table

	2015	2016	2017
Beef	45%	50%	42%
Chicken	30%	20%	38%
Pork	25%	30%	20%

cross tables. For example, the structure of the label groups and of the table contents are not separated, and many joined cells are needed. These concerns make cross table creation an effort and time consuming task in web site creation.

Similarly, cross tables could be generated using basic SuperSQL queries. However, the queries required to describe more interesting tables would rapidly become overly complex due to the many levels of nesting and separate structures involved.

To address these concerns we implement a cross_tab function in SuperSQL allowing the creation of a cross table from three TFEs that describe the header, side labels and content of the table. This allows to separate the concerns of each part of the table and to create complex tables from simple TFEs.

The rest of the paper is structured as follows: Sect. 2 presents related works, we then introduce SuperSQL in Sect. 3. We model cross tables in Sect. 4 and explain the crosstab function in Sect. 5. Finally we evaluate our system and discuss our results in Sect. 6 before concluding in Sect. 7

2   Related Works

Basic implementation of cross tables are available in current DataBase Management Systems (DBMS) but are very limited in the type of cross tables they can produce. Examples of such implementations include the *PIVOT* clause in Microsoft's SQL Server or the *crosstab* function in PostgreSQL. Both these implementations allow some freedom from the flat results of SQL queries but are basically limited to double entry tables, missing a lot of the potential of cross tables that we aim to exploit.

Similarly, Microsoft Excel allows users to create pivot tables. This tool allows for data manipulation from users who do not know the SQL query language, thus making cross tables more accessible. Still, limitations such as automatically determination of the structure of the value are imposed on the created a crosstable.

There is a technique called XSLT, in which data is specified with XPath for XML data and outputted in combination with HTML tag. By outputting the database search result by SQL as XML, it becomes possible to generate a cross table by combining with this technology. We can also generate a cross-table by doing ajax communication using javascript's library jQuery, receiving data and formatting it.

In [6], Stolte et al. introduce Polaris, a tool for data visualization that relies on cross tables. Polaris offers a visual interface for users to create cross tables featuring graphical representations of the data (such as graphs or symbols) on the table. Also, they classify the value type and the combination in the front and front side attributes.

In [7], Gray et al. propose the *CUBE* and *ROLL-UP* operators to create cross tables based on groupings and aggregated values.

In [8], Johnson et al. to aggregate in multiple dimensions they proposed the expansion of SQL by using the idea pivoting, which convert the data rows to columns.

3   SuperSQL

3.1   An Overview of SuperSQL and TFE

SuperSQL [1,5] is an extension of the SQL language and generates various structured documents such as HTML, PDF, XML. In SuperSQL the SELECT clause in an SQL query is replaced by a GENERATE clause which has the following structure: GENERATE <medium> <TFE>.

The available target medium designations are not limited to the mediums mentioned above, but also include responsive HTML, PHP, X3D, etc. The TFE is an extension of the SQL target list. Unlike the target list, which is a flat list of attributes separated with commas, the TFE uses new operators, called connectors and repeaters, to specify the structure of the generated document and operators called decorators to specify visual aspects of the document.

4   Modeling Cross Table

We define a cross table through three components: the header, or top set of labels, the side, or lateral set of table's left line labels and the set of values that populate the table. The head and side can have different structures: a *set* structure, as shown in Fig. 1, where labels refer to a single attribute; a *tree* structure, as shown in Fig. 2, where labels referring to different attributes are nesting; or a *forest* structure, as shown in Fig. 3, which concatenates multiple *set* or *tree* structures. We will refer to label sets having these structures as S-, T-, and F-sets.

The second distinction we make it along the type of data used in the table. The most direct way of creating a cross table for relational data is to directly use the raw attributes from the database, either as labels or as values. One could, for example, use the *gender* attribute of a *person* table as a label. We call this a data inclusion, or I. Another possibility is to use aggregated values from the database. In this case, we call this an aggregation, or A. Finally, attributes can be specified as enumerations, for example, male, female, we call this an enumeration, E.

Toront	Montreal

Fig. 1. The structure of SET

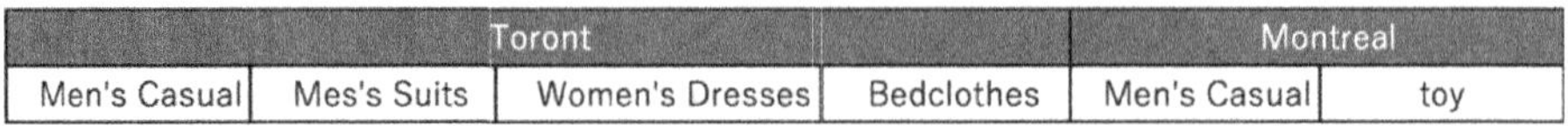

Toront				Montreal	
Men's Casual	Mes's Suits	Women's Dresses	Bedclothes	Men's Casual	toy

Fig. 2. The structure of TREE

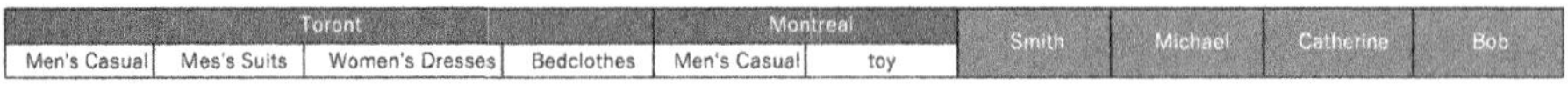

Toront				Montreal		Smith	Michael	Catherine	Bob
Men's Casual	Mes's Suits	Women's Dresses	Bedclothes	Men's Casual	toy				

Fig. 3. The structure of FOREST

5 SuperSQL Cross Table Implementation

In this section we describe the *cross_tab* function, our implementation of cross tables in SuperSQL. This function takes three TFEs as arguments and creates a cross table from data gathered from the database. We will describe the syntax of this function and the algorithms used to create the cross table from the three TFEs.

5.1 Syntax

A cross table can be seen as the sum of its three components: the header, the side, and the values. We thus define the SuperSQL *cross_tab* function as accepting three TFE arguments, that define the structure of each component. Additionally, as SuperSQL outputs visual documents, not just data structures, the *cross_tab* function can be accompanied by visual modifiers that define the looks of the cross table, such as the *side-width* modifier, that defines upper left blank or the *null-value* modifier, that defines how null values should be displayed. A cross table is thus defined in SuperSQL using a query of the form as shown below:

──────────── Structure of the *cross_tab* function ──────────

cross tab(*TFE1*, *TFE2*, *TFE3*)@{side-width=*num*, null-value='*str*'}

Figure 4 shows the correspondence between the three TFE arguments and the generated cross table.

5.2 Implementation

In this section, we describe the implementation that converts from syntax which defines Sect. 5.1 to cross table on SuperSQL. Then there was occurred a problem: the collapse of layout when there is no value corresponding to the aggregated value. We also describe this problem.

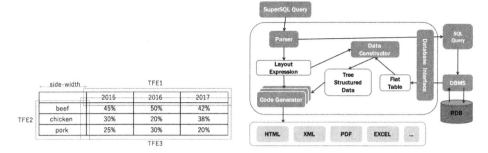

Fig. 4. The state of cross table   **Fig. 5.** SuperSQL architecture

Overall of Implementation. SuperSQl's overall architecture is shown in Fig. 5. When a SuperSQL query is received it is first parsed into an SQL query and data structure called a TFEtree representing the structure of the data. The result from the query and the TFEtree are passed to a Data Constructor that produces Tree-Structured Data which is used by the Code Generator to generate the output document.

To handle the *cross_tab* function an additional step is added between the Parser and Data Constructor, where the TFEtree representing the *cross_tab* function is converted to the TFEtree for the target table. The sorting of values and handling of null values is done in the data constructor.

TFE Conversion. We describe the process of the structure conversion from the syntax to the cross table on SuperSQL. The syntax is written by TFE. In SuperSQL the TFE is held in a tree structure. For example, a horizontal connector is represented by the tree which has the horizontal connector token as a parent and the elements to be joined in the horizontal direction as children. The cross table structure in SuperSQL is below:

Cross Table Form TFE

"" , [TFE1], ! [TFE2 , [null(TFE1) , TFE3]!]!

The Null function which is used in above TFE is to use for the attribute which we don't want to display but we want to use for grouping or sorting. The TFE1 which is applied null function does not need to display.

Process for Null Value. When the aggregate functions are used in a cross-table, depending on the cell, the corresponding value may not exist. For example like Fig. 6, we think the number of people who get grade A, B, C for each age and sex of each class at a school. In this case, there is a possibility that there is no applicable person. Then this function substitutes the character string specified by the user as its value.

		female			male		
		13	14	15	13	14	15
DBS	A	1	1	N/A	4	1	1
	B	2	3	4	N/A	1	N/A
	C	N/A	1	1	1	2	1
DM	A	1	4	1	2	2	1
	B	2	N/A	3	2	N/A	N/A
	C	N/A	1	1	1	2	1
Webapp	A	N/A	N/A	4	1	N/A	1
	B	N/A	3	N/A	1	3	1
	C	3	2	1	3	1	N/A

Fig. 6. Cross table with no corresponding value

As in the first line to the eighth line, this process saves values which are retrieved from the database as top or side attribute value. For example in case of the Fig. 6, saved values are [[female, male], [13, 14, 15]] and [[DBS, DM, Webapp], [A, B, C]].

Then in combine function, we made all combinations of saved values. But in SQL query values which are specified as same attribute name must be the same value, so we remove not same tuples. And we could make a possible list. Finally, we compare retrieved list and possible list, and if there are not applicable list, we add this list to retrieved list.

Above the process for null value is completed.

5.3   Example of Cross Table Function

In this section, we introduce an example of a cross-table function. In this example, we use the database about baseball teams and matches.

- teams (id, name, home, league)
- match (id, team1_id, team2_id, year)
- score (id, team_id, match_id, score)

The team table contains team information, the match table has team id which played against and a day, and the score table includes team id and match id and the score that is got by the team. The league is 'S league' and 'P league', and the year is from 2015 to 2017.

Algorithm 1. Process for null value

Input: Tuples which are retrieved from database and are applied aggregate process.
Output: Tuples which added lacked tuples
 1: //Get possible value of each attributes
 2: **for all** each input tuples **do**
 3:   **for all** each tuples' elements **do**
 4:     **if** The element isn't included in saved list **then**
 5:       Add to saved list
 6:     **end if**
 7:   **end for**
 8: **end for**
 9: //Make all attributes combinations from saved list.
10: combine(1, saved_list[0], saved_list[1], the size of saved_list, saved_list)
11: //In SQL query values which are specified as same attribute name must be same value, so we remove not same tuple.
12: **for all** All combined lists **do**
13:   **if** Attribute name is same but value is not same. **then**
14:     remove list
15:   **end if**
16: **end for**
17: //Compare search results and all combined lists and if there are no applicable tuples in search result, then add this list to search result.
18: **for all** All combined lists **do**
19:   **for all** Search results **do**
20:     **if** There don't exist applicable tuples **then**
21:       Add to search result
22:     **end if**
23:   **end for**
24: **end for**
25: **return** Tuples which added lacked tuples
26: //combine function
27: FUNCTION combine(num, list1, list2, list.size, list)
28: **for all** list1's elements **do**
29:   **for all** list2's elements **do**
30:     Combine two elements and add result list.
31:   **end for**
32: **end for**
33: num++
34: **if** The num is less than the size of save_list **then**
35:   combine(num, result, list[num], list.size, list)
36: **end if**
37: **return** result

Team Achievement for Each Years. We made the achievement table for each team and years, which contains total match numbers and winning numbers and losing numbers. The query is below:

─────────── Team achievement for each year ───────────

```
Generate HTML
cross_tab(
        [m1.year],,
        [t1.league,[t1.name]!]!,
        {{count[m1.id]}!{{'win:' ||w.count}, {'lose:' ||l.count}}}
)@{side-width=200}
FROM teams t1, teams t2, match m1, score s1,
score s2, win w, lose l
WHERE (
        (t1.id = m1.team1_id AND t1.id = s1.team_id
        AND m1.id = s1.match_id AND t2.id = m1.team2_id
        AND t2.id = s2.team_id AND m1.id = s2.match_id)
        OR (t1.id = m1.team2_id AND t1.id = s1.team_id
        AND m1.id = s1.match_id AND t2.id = m1.team1_id
        AND t2.id = s2.team_id AND m1.id = s2.match_id)
        )
        AND t1.name = w.name AND m1.year = w.year
        AND t1.name = l.name AND m1.year = l.year
```

			2015	2016	2017
P league	Gallops		4	4	4
			win:3 lose:1	win:2 lose:2	win:2 lose:2
	Gryphons		4	4	4
			win:2 lose:2	win:2 lose:2	win:2 lose:2
	Hounds		4	4	4
			win:2 lose:2	win:0 lose:4	win:4 lose:0
	Sharks		4	4	4
			win:1 lose:3	win:3 lose:1	win:1 lose:3
	Tigars		4	4	4
			win:2 lose:2	win:3 lose:1	win:1 lose:3
S league	Dolphins		4	4	4
			win:2 lose:2	win:1 lose:3	win:1 lose:3
	Eagles		4	4	4
			win:1 lose:3	win:2 lose:2	win:3 lose:1
	Hawks		4	4	4
			win:3 lose:1	win:2 lose:2	win:2 lose:2
	Whales		4	4	4
			win:3 lose:1	win:3 lose:1	win:1 lose:3
	Wolves		4	4	4
			win:1 lose:3	win:2 lose:2	win:3 lose:1

Fig. 7. Team achievement for each years

In this table, the top attribute is a year, side attributes are league name and team name, and the value is total numbers of a match and the number of wins and loses. The generated table is Fig. 7.

6   Evaluation

In this chapter, we compare this research with other studies in the aspect of expression and the line numbers of code. The other studies are Microsoft SQL Server, Microsoft Excel, PostgreSQL, XSLT + SQL, JavaScript + HTML5.

6.1   Expression

In this section, we compare each studies' expression based on the model; we introduced Sect. 4.

Microsoft SQL Server [3]. For the first, the top attribute of the cross table generated by Microsoft SQL Server is an enumeration. The side attribute and the value is a data inclusion. However we can use both aggregate function and values which is stored in a database in a value, we can't use an aggregate function in a side attribute. And all three attributes allow only single attribute: we can use nest form in these attributes. From above a top attribute is ES, a side attribute is IS, and a value is IAS.

PostgreSQL [2]. In PostgreSQL same as Microsoft SQL Server, a top attribute is an enumeration, and a side attribute and a value is inclusion. A value only allows aggregate function, and these three attributes allow only single attribute. Thus a top attribute is ES, a side attribute is IS, and a value is IAS.

Microsoft Excel [4]. In this way, all three attributes are inclusion, and only a value allows aggregate function. And a top attribute and a side attribute are tree form, but a value is an only single attribute. From above a top attribute and side attribute are IT, and a value is IAS.

SQL + XSLT, JavaScript + HTML5. Since we can operate a HTML Tags, by using this technique we can make a various layout cross-table. Also we can use an aggregate function in three attributes. From above these attributes are IAF.

Proposed Method. In this method, all three attributes are both enumeration and inclusion. And we can use an aggregate function in these three attributes. Then these three attributes allow forest form. From above these attributes are IAF.

In the specifications of each attribute, an inclusion is more expressive than an enumeration, and if we can use the aggregate function it is more expressive. And in the aspect of a structure, it is most expressive if it allows forest form. Thus the proposed method is more expressive than others.

6.2   Code Amount

Next, we compare these techniques from the viewpoint of code amount. We use Tables 2 and 3 as a standard for this evaluation. Table 2 represents the annual income of each employee who belongs to a company in 2016 and 2017, and Table 3 represents the average of the age of men and women members and the purchase number of each genre in a convenience store. The difference between these pictures is that an aggregate function is used in a top or a side attribute. We use Table 2 for the evaluation with Microsoft SQL Server and PostgreSQL and Microsoft Excel. Then we use Table 3 for the evaluation with SQL + XSLT and JavaScript + HTML. This distinction is based on consideration of each expressive power.

Microsoft SQL Server. From Table 4, the code amount for Table 2 by our proposed method is less than that by Microsoft SQL Server. The first reason is that we have to write all top values by using Microsoft SQL Server. Also to generate a cross table we have to specify subqueries as a data source. Because of these two reasons, the code amount increased in Microsoft SQL Server. Since in proposed method to generate a cross table we only have to attribute names and we need not make subqueries, we write less code than Microsoft SQL Server.

Table 2. The 1st standard cross table for evaluation of code amount

		2016	2017
1000	Bob	2001.4	2010.3
1001	Clark	1998.3	2000.1
1002	Daria	1805.3	1803.2
2001	Diana	2101.4	2153.7
2003	Donald	2011.3	2013.7
1005	Michael	2513.4	2688.7
3001	Minerva	1743.6	1704.5
3002	Noah	3001.4	2980.3
1006	Robin	2011.4	2109.3
1007	Smith	1995.7	1997.1

Table 3. The 2nd standard cross table for evaluation of code amount

		Snacks	Drinks	Precooked foods	Frozen foods	Commodities	Other foods
Male	26.8	508	532	307	358	419	168
Female	28.5	495	688	547	542	584	351

Table 4. The number of rows required to make Table 2

Micro SQL server	PostgreSQL	Proposed method
22	20	11

PostgreSQL. From Table 4 turns out that we can write less amount code by using the proposed method than PostgreSQL. The reasons are roughly similar to that of Microsoft SQL Server: we have to write subqueries as a data source, we have to write GROUP BY clause if you need, and we have to write all top values. About the specification, we specify side attributes and values in crosstab clause, then we define top values and a value type.

Because of the form of this statement, an error occurs when we specify types not corresponding to that of values. It is hard to recognize and specify the statement by using PostgreSQL than our proposed method.

Microsoft Excel. In Microsoft Excel, we can't refer multiple tables, so we first prepare a table that didn't be normalized. Then we select attributes that corresponding to every three attributes by using GUI. We compare in the aspect of ease of specification. First about preparations of data source, in Microsoft Excel we can't use more than two tables, so we have to prepare the table that didn't be normalized. We assumed a relational database as a data source, then it becomes a labor. Also in Microsoft Excel, we can use only GUI, so we can't change values layout. So the possible layout for value by using Microsoft Excel is more restricted than by using our proposed method. From above if we generate a cross-table from a relational database, our proposed method is easier than Microsoft Excel.

SQL + XSLT. XSLT is a technique to acquire and calculate data of XML using XPATH, and shape it with HTML tag and output it. From Table 5 when we create Table 3 by using this technique, the code amount is 90 lines. Compared with the case using the proposed method, the code amount is about 6.5 times. The cause may that we have to write two files, the one is SQL file to acquire a data from database and the other is XSLT file to shape the acquired data. As the number of files to be created increases, the amount of code naturally becomes large.

Although we can create any layouts by using this technique, as the structure of the table becomes more complicated, the amount of code that must be written increases. Also once you create it, its layout change is not easy.

JavaScript + HTML5. A cross table can be created by receiving data from a PHP file which is written to connect database using jQuery's Ajax communication which is a library of JavaScript and shaping it. We made Table 3 by this way. From Table 5 the code amount is 131 lines, which is about 9.4 times of the proposed method. The cause is that we need to write all processes by themselves, such as communication with the database and data formatting. Same as

Table 5. The number of rows required to make Table 3

SQL + XSLT	JavaScript + HTML	Proposed method
90	131	14

SQL + XSLT we can make various layouts of a cross table. But We have to fix a JavaScript file to change a layout, so by using this technique the change of a layout is harder than a proposed method.

7 Conclusion

In this paper, we implemented the cross table generation function in SuperSQL. By implementing this system it became possible to implement a cross-table easily with SuperSQL and the range of expressions of SuperSQL expanded. Since this function is implemented in SuperSQL, it is easier to specify a wider structure than other systems.

In this system, we didn't consider cell width depend on the contents, but set automatically 100px, so users have to adjust cell width if the output is not appropriate. Our future work for this system is to calculate cell width depends on data contents, and automatically make an appropriate a cross-table.

References

1. SuperSQL. http://ssql.db.ics.keio.ac.jp/
2. PostgreSQL. https://www.postgresql.org/
3. Microsoft SQL Server. https://www.microsoft.com/en-us/sql-server/
4. Microsoft Excel. https://products.office.com/en-us/excel
5. Toyama, M.: SuperSQL: an extended SQL for database publishing and presentation. ACM SIGMOD Rec. **27**, 584–586 (1998)
6. Chris, S., Diane, T., Pat, H.: Polaris: a system for query, analysis, and visualization of multidimensional relational databases. IEEE Trans. Vis. Comput. Graph. **8**, 52–65 (2002)
7. Jim, G., Surajit, C., Adam, B., Andrew, L., Don, R., Murali, V., Frank, P., Hamid, P.: Data cube: a relational aggregation operator generalizing group-by, cross-tab, and sub-total. Data Mining Knowl. Discov. **1**, 29–53 (1997)
8. Johnson S.B., Chatziantoniou D.: Extended SQL for manipulating clinical warehouse data. In: Proceedings of the AMIA Symposium, pp. 819–823 (1999)

Towards Association-Oriented Model of Lexical Base for Semantic Knowledge Base

Krystian Wojtkiewicz[1,3], Marcin Jodłowiec[1,2(✉)], and Waldemar Pokuta[2]

[1] Department of Information Systems,
Wroclaw University of Science and Technology, Wrocław, Poland
marcin.jodlowiec@gmail.com
[2] Institute of Computer Science, Opole University of Technology, Opole, Poland
[3] Institute of Control, Opole University of Technology, Opole, Poland

Abstract. Developing a knowledge-based system is a demanding task. One of the basic issues is to build a communication interface with a human. The basis for such an interface is mostly natural language. In this article, the process of building and supplementing data with one of the modules of the Semantic Knowledge Base, which is responsible for user interaction, is presented. For this purpose, the process of data acquisition from the WordNet database was carried out by remodeling it by the use of Association-Oriented Database Metamodel.

Keywords: Semantic knowledge base · WordNet
Association-oriented database · Database modeling

1 Introduction

The idea of developing an association-oriented model [9,13] of the English thesaurus stems from the work on the Semantic Knowledge Base Linguistic Module (SKM-LM) [10]. The properties of the traditional WordNet system [4,14] have proved to be insufficient and inadequate to the problem. First of all, the use of a file-based lexical database would lead to data impedances mismatch between the various modules of the system. In addition, it would also jeopardize the efficiency of thesaurus search. Due to the above, the subject of transforming WordNet into a association-oriented database, i.e. the model in which SKB was developed, was undertaken.

The main aim is to integrate the knowledge contained in WordNet with SKB in such a way that the thesaurus becomes part of SKB-LM and uses the full functionality of this module [12]. In particular, the project is expected to be replicated in other languages in the future [7]. The choice of WordNet as a data source was preceded by the thesauri building standards study that was presented in Sect. 2 as well as state of art technology implementations study presented in Sect. 3. Subsequently, a modeling process was performed that identified transitional structure needs and by the use of association-oriented design patterns [6] a database structure in AODB was proposed and presented in AML [8,16], as

© Springer International Publishing AG, part of Springer Nature 2018
N. T. Nguyen et al. (Eds.): ACIIDS 2018, LNAI 10752, pp. 25–34, 2018.
https://doi.org/10.1007/978-3-319-75420-8_3

described in Sect. 4. Section 5 presents summary, conclusions and future path of development, especially the integration process with SKB-LM.

2 Standards of Thesaurus Modeling

In the thesaurus the word is understood as a connection between form and meaning. There may be words that have the same written form but different meanings (homonyms). There are also different written identifiers that share the same or similar meaning (synonyms). Homonyms and synonyms are entries that appropriately occupy the same columns or rows (Table 1).

Table 1. Written form and meaning collocation matrix

		Written form			
		F_1	F_2	$F_{...}$	F_m
Meaning	M_1	$E_{1,1}$	$E_{1,2}$		
	M_2		$E_{2,2}$		
	$M_{...}$	...	...	...	...
	M_n				$E_{m,n}$

Another group of links are antonyms, that designate opposite meanings. Such relationships are often difficult to define, e.g. the word *poor* is an antonym of the word *rich*, but the expression *not rich* is also its antonym. However, the expressions *not rich* and *poor* do not have the same meaning. Sometimes words have similar or identical meanings, but only in a certain context. Based on information from the thesaurus, one can write a specific text, replace one word with another, without changing the meaning of the whole sentence. However, in the case of simplified information about the similarities of words (not including the context of the speech), only a person familiar with the language can do so. In the case of a person who is not fluent with a particular language, such thesaurus might be of no use. It is due to the fact, that such a person does not know whether a given synonym can be used in legal, scientific or colloquial texts; whether it is currently in use or whether it belongs to archaism or is vulgar. Some words are acceptable in conversations between children, but they would be offensive to an adult. This context is also necessary in the case of automatic generation of speech or translation of texts. If the text-translation algorithm cannot "understand" the content of a statement and describe it with certain parameters relating to the context and sense of the message, then the result of its action will require a thorough correction made by the person who understands the subject.

Thesauri are available online as well as stationary projects. Some of them are included in existing dictionaries. Some are created from scratch according to specific requirements, and are based on widely accepted standards. Following a choice of standards will be presented.

Table 2. Selected semantic relationships in NISO Standard

Relationship table	Example
Equivalency	
Synonymy	UN/United Nations
Lexical variants	Pediatrics/paediatrics
Near synonymy	Sea water/salt water
Hierarchy	
Generic or IsA	Birds/parrots
Instance or IsA	Sea/Mediterranean Sea
Whole/Part	Brain/brain stem
Associative	
Cause/effect	Accident/injury
Process/agent	Velocity measurement/speedometer
Process/counter-agent	Fire/flame retardant

NISO Standard Z39.19-2005

This standard provides guidelines and conventions on the content, display, construction, testing, maintenance and management of monolingual dictionaries. There are three main principles implemented by dictionaries: defining the scope or meaning of terms, definition of equivalence relationships for synonyms, near-synonym and homonyms. In the case of synonyms it is possible to define three types of relations e.g. equivalence, hierarchy and associativity (Table 2) [1].

ISO 25964

In this standard, every term in the glossary dictionary is represented by one preferred term in a given language and by any number of non-preferred terms. Relationships refer to the concept as a whole and not to the preferred term. On some systems, the term is identified only by the preferred term or the preferred term id, but this creates problems if the spelling of the term changes. The simplified thesaurus schema in this standard is provided in the Listing 1.1 [5].

SKOS

The Simple Knowledge Organization System is a W3C recommendation to represent thesaurus and other types of dictionaries. SKOS is part of the concept family associated with the Semantic Web, built on RDF and RDFS. It is most compatible with ISO 25964. SKOS vocabulary is based on concepts. Concepts are ideas, meanings, or events (objects or categories), and this is the basis of many knowledge management systems [13]. As such, concepts exist in the mind as abstract entities that are independent of the words that describe them. SKOS distinguishes two basic categories of semantic relations: hierarchical and associative. A hierarchical connection between two concepts implies that one of them is somehow broader than the other (narrower). Associative relationship determines that they are in relation, but one cannot establish which one is more general than another.

Listing 1.1. Part of Thesaurus Scheme in ISO 25964

```
. . .
<xsd:complexType name="PreferredTerm"></xsd:complexType>
<xsd:complexType name="ScopeNote"></xsd:complexType>
<xsd:complexType name="Definition"></xsd:complexType>
<xsd:complexType name="HistoryNote"></xsd:complexType>
<xsd:complexType name="EditorialNote"></xsd:complexType>
<xsd:complexType name="CustomNote"></xsd:complexType>
<xsd:complexType name="NodeLabel"></xsd:complexType>
<xsd:element name="lexicalValue"></xsd:element>
<xsd:element name="date" type="iso25964:date"/>
<xsd:element name="created" type="iso25964:date"/>
<xsd:element name="modified" type="iso25964:date"/>
<xsd:simpleType name="Identifier"></xsd:simpleType>
<xsd:simpleType name="date"></xsd:simpleType>
. . .
```

The properties $skos : broader$ and $skos : narrower$ define relationship of hierarchy. The triple $<A>skos : broader<B>$ means that $<B>$ is a generalization of $<A>$, while the triple $<C>skos : narrower<D>$ means that $<D>$ is a specification of $<C>$. According to the adopted convention, hierarchical relationships are defined only with the nearest hierarchy element. This means that although objects *frog* and *animal* are hierarchically linked, in the dictionary this link will be broken down into three triples:

$$<frog> skos : broader <amphibian>$$
$$<amphibian> skos : broader <vertebrate> \qquad (1)$$
$$<vertebrate> skos : broader <animal>$$

Some applications also require indirect links (e.g. to speed up search dependencies). In this case, the $skos : broaderTransitive$ and the $skos : narrowerTransitive$ properties are used to denote respectively direct or indirect generalization or specification. To specify relationships defining *relation* between terms in *SKOS* the $skos : related$ has to be used. There are several important dictionaries that implement the structure defined in the *SKOS*, e.g. EuroVoc [3], AGROVOC [2], GAMET and Library of Congress Subject Headings.

3   State of Art in Thesaurus Implementation

The previous section provided the standards in regard to thesauri design. However, the standards are often used like an approach, rather than a strict rules to follow. Therefore, the authors decided to check whether the actual implementation comply with those standards and if so, in what degree. The short summary of this research is provided in this section with some comments in regard to given solutions.

Collins Dictionary Thesaurus

Collins Dictionary thesaurus is part of the Collins English Dictionary which is available in print and online (www.collinsdictionary.com). The first edition of this dictionary was used in 1979. In its third edition, Bank of English was used to create a dictionary, using its vast 650 million running words. Currently the dictionary is already in its 12th edition published in October 2014. Since 2004 there is a discussion forum on the introduction of neologisms to the dictionary. In June 2012 the dictionary began to use Facebook as a source for early detection of language neologisms.

AGROVOC

AGROVOC is a multilingual dictionary covering all areas of interest to the Food and Agriculture Organization of the United Nations (FAO), namely food, nutrition, agriculture, fisheries, forestry and the environment. The dictionary consists of over 32,000 concepts and about 40,000 terms in 23 languages: English, French, German, Hindi, Hungarian, Italian, Japanese, Korean, Lao, Malay, Persian, Polish, Portuguese, Russian, Slovak, Spanish, Telugu, Thai, Turkish and Ukrainian. This is a joint project, created by a community of experts and coordinated by the FAO. AGROVOC was provided by FAO in the form of *RDF/SKOS-XL* and published as an element connected to 16 other dictionaries.

WordNet

WordNet [15] is a lexical database of English language. It aggregates English words into sets of synonyms called synsetes, also provides short definitions and usage examples. It specifies the relationship between these sets and the individual words. WordNet can therefore be seen as a combination of a dictionary and a thesaurus. WordNet is available to users through a web browser. But the main purpose of the system is automatic text analysis and use in artificial intelligence. Database tools and software are released under the BSD license and are available for download from the WordNet website.

4   Modeling of Association-Oriented Thesaurus

Due to ease of access and high quality of documentation, it was decided to adopt the WordNet database for the needs of the *Semantic Knowledge Base Linguistic Module* [12]. However, this solution involves the adaptation of the data structures used in the original WordNet project to a level that enables direct use in SKB. The SKB database layer is modeled in the Association-Oriented Database Metamodel, which is an approach towards data modeling based on Entity-Relationships and Object-Oriented Metamodels. The most important features of AODB are the explicit separation of competencies between the structures providing the functionality of the data storage units (collections) and the relationships definition (associations). This makes it possible to physically carry out n-ary relationships that take into account the possibility of defining multiplicity both on the side of the participant of the association and on the relationship as such [9]. It should be noted, that the move towards better definition of primitives is not unique and the similar, promising approaches can be found in other than AODB studies [17]. Considering these and other more subtle AODB properties, it was decided to carry out a two-step acquisition of the knowledge contained in WordNet into SKB Linguistic Module, what was presented on Fig. 1.

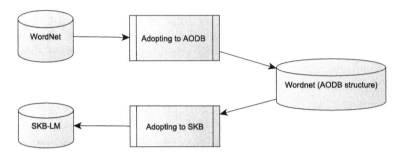

Fig. 1. Tow-step acquisition of WordNet data into SKB-LM

The first step was to model the structure in AODB, which, in the most faithful way, reproduces the structure of the original WordNet, while also enables the processing of data using object-oriented programming without the adverse effect of data impedance mismatch. The next stage involved the development of algorithms and methods that would allow the data to be imported into the SKB *Linguistic Module*.

By adapting the data contained in the WordNet database, it was assumed for the sake of proper modeling in AODB, that each term (text string) would be modeled as the object of a particular collection, and the relationships would be represented as associations with corresponding roles. Note that the basic element of relationships in WordNet is *synset*. The synset was modeled as *Synset* association, which was only a participant in the identified relationships derived from WordNet. Despite its simple construction, WordNet provides a relatively large number of relationships types that can be defined among synsets, namely speaking *Antonym, Hypernym, Instance Hypernym, Hyponym, Instance Hyponym, Member holonym, Substance holonym, Part holonym, Member meronym, Substance meronym, Part meronym, Attribute, Derivationally related form, Domain of synset, Member of this domain, Entailment, Cause, Also see, Verb Group, Similar to, Particle of verb, Pertainym/Derived* and last, but not least *synonym*. Part of abovementioned relations are in fact properties. Others are dedicated to use only with a particular part of speech.

The number of relationship types led to conclusion that building specialized structures for each of them is senseless and in general terms futile. Since presented solution aims at general use for most of national WordNet instances, one cannot be ever sure that all possible types of relations has been identified for each and every natural language. It has been decided that a general structure will be used for identification of relation type. It will be later used as input for methods that are to be developed for second stage of integration. Those methods will rely on natural language of a specific WordNet instance and will be designed for each of them individually.

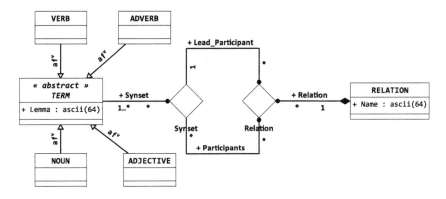

Fig. 2. Association-oriented database model of WordNet presented in AML

The transition structure for WordNet presented with AML, the dedicated AODB modeling language, is shown in Fig. 2. For the sake of simplicity, the diagrams omit specific attributes, focusing on the essence of the model presented. The main data container is an abstract collection *TERM*, which further is specialized into distinguished collections *NOUN*, *VERB*, *ADJECTIVE* and *ADVERB*. These collections correspond to the part of speech present in Word-Net. Synsets, which are the basic semantic unit of the thesaurus, are modeled as an association which, through the inherited role *Synset*, binds appropriate collections. From an intentional point of view, each synset will be represented by the corresponding Association Object, which will be bound to objects of the collection being specialization of the *TERM* collection by the means of Role Object. This is possible thanks to AODB property, which is the ability to inherit rights to participate in the role.

The relationships between the synsets are represented as a *Relation* association, with 3 distinct roles. Role *Relation* connects the *RELATION* collection, which defines the nature of this relation. The other two roles define the participants of the relationship, at the same time allowing to define the direction of the relation. The *Lead_Participant* role points to the synset for which we define the relation, while the *Participants* role is used to determine the synonyms that are in relation to the distinguished set of words.

Migration to the structure presented in Fig. 1 for English concludes the first stage of knowledge assimilation. The next step is to develop algorithms and methods of data migration from the presented structure to the structure of the SKB-LM. It is not a trivial task, since the idea that constitutes functions of *SKB* is based on the principle of unambiguous identification of the concept as a basic semantic entity. This entity in turn is identified by many terms, potentially in many languages. In the case of the knowledge contained in WordNet, it is difficult to talk about the concept as such. Therefore, at the initial stage, it will be necessary to create artificial concepts and then use the SKB built-in knowledge acquisition methods described in [11] to verify the validity of these concepts. This action is mainly aimed at preventing the explosion of the number of concepts.

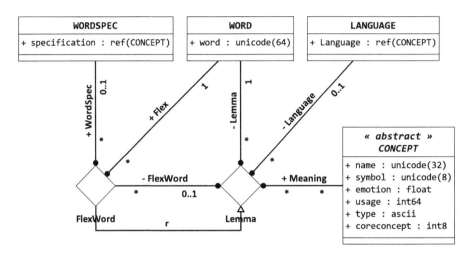

Fig. 3. Association-oriented database model of semantic knowledge base Linguistic module

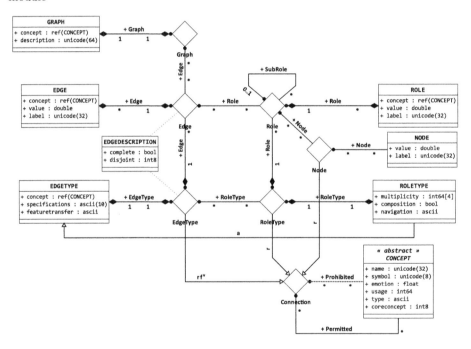

Fig. 4. Association-oriented database model of semantic knowledge base structural module

Moreover, it should be noted that the *SKB Linguistic Module* presented in Fig. 3 as AODB schemata, that is being used to describe the relationship between concepts uses the *SKB structural module* (Fig. 4). That allows for a detailed definition of relationships both in terms of their structure and semantics. For this reason, it is necessary to individually prepare the import process for each of the distinctive types of relationships between synsets. A detailed description of those algorithms that make up the essence of the second phase of WordNet migration to SKB is significantly beyond the scope of this study and will not be included in it.

5   Conclusions and Future Work

The structure presented in Sect. 4 allows for direct transfer of data from traditional WordNet file structures. It also uses the specific properties of AODB, including in particular the ability to delegate participation rigths in role through inheritance, as well as the ability to define the use of one association in another. It should be noted that as a result of the migration operation, we have obtained an association-oriented database, i.e., the base on the same level of abstraction as the target structure of the *Linguistic Module* of the *Semantic Knowledge Base*.

Association-Oriented Database Metamodel has proved to be an efficient and flexible database platform for modeling complex structures. The presented first stage of thesaurus migration process allowed for importing data from WordNet into association-oriented structures. Future work focus on developing algorithms and methods that would allow for efficient adaptation of knowledge held in transition structures into Semantic Knowledge Base and in particular into its Linguistic Module.

References

1. ANSI/NISO: guidelines for the construction, format, and management of monolingual controlled vocabularies. Technical report Z39.19-2005, Bethesda, MD, USA (2005)
2. Caracciolo, C., Stellato, A., Morshed, A., Johannsen, G., Rajbhandari, S., Jaques, Y., Keizer, J.: The AGROVOC linked dataset. Semant. Web 4(3), 341–348 (2013). http://dl.acm.org/citation.cfm?id=2786071.2786087
3. Community, E.E.: EUROVOC.: Multilingual thesaurus. EUROVOC [EN], Office for Official Publications of the European Communities (1987). https://books.google.pl/books?id=PT8mAQAAMAAJ
4. Fellbaum, C.: WordNet. Wiley, Hoboken (2012)
5. ISO - International Organization for Standardization: ISO 25964–1:2011: Information and Documentation—Thesauri and Interoperability with Other Vocabularies. Part 1: Thesauri for Information Retrieval (2011)
6. Jodłowiec, M., Krótkiewicz, M.: Semantics discovering in relational databases by pattern-based mapping to association-oriented metamodel—a biomedical case study. In: Piętka, E., Badura, P., Kawa, J., Wieclawek, W. (eds.) Information Technologies in Medicine. AISC, vol. 471, pp. 475–487. Springer, Cham (2016). https://doi.org/10.1007/978-3-319-39796-2_39
7. Krótkiewicz, M., Jodłowiec, M., Wojtkiewicz, K.: Introduction to semantic knowledge base: multilanguage support of linguistic module. In: 2016 Third European Network Intelligence Conference (ENIC), pp. 188–194, September 2016

8. Krótkiewicz, M.: A novel inheritance mechanism for modeling knowledge representation systems. Comput. Sci. Inf. Syst. (2017)
9. Krótkiewicz, M.: Association-oriented database model - n-ary associations. Int. J. Softw. Eng. Knowl. Eng. **27**(2), 281–320 (2017)
10. Krótkiewicz, M., Wojtkiewicz, K.: Conceptual ontological object knowledge base and language. In: Kurzynski, M., Puchała, E., Wozniak, M., Zołnierek, A. (eds.) Computer Recognition Systems, vol. 3, pp. 227–234. Springer, Heidelberg (2005). https://doi.org/10.1007/3-540-32390-2_25
11. Krótkiewicz, M., Wojtkiewicz, K.: An introduction to ontology based structured knowledge base system: knowledge acquisition module. In: Selamat, A., Nguyen, N.T., Haron, H. (eds.) ACIIDS 2013. LNCS (LNAI), vol. 7802, pp. 497–506. Springer, Heidelberg (2013). https://doi.org/10.1007/978-3-642-36546-1_51
12. Krótkiewicz, M., Wojtkiewicz, K.: Introduction to semantic knowledge base: Linguistic module. In: 2013 6th International Conference on Human System Interactions, HSI 2013, pp. 356–362. IEEE, Sopot, June 2013
13. Krótkiewicz, M., Wojtkiewicz, K.: Functional and structural integration without competence overstepping in structured semantic knowledge base system. J. Logic Lang. Inf. **23**(3), 331–345 (2014). http://link.springer.com/article/10.1007/s10849-014-9195-y
14. Miller, G.A.: WordNet: a lexical database for english. Commun. ACM **38**(11), 39–41 (1995). http://doi.acm.org/10.1145/219717.219748
15. Miller, G.A., Beckwith, R., Fellbaum, C., Gross, D., Miller, K.J.: Introduction to WordNet: an on-line lexical database. Int. J. Lexicography **3**(4), 235–244 (1990). http://dx.doi.org/10.1093/ijl/3.4.235
16. Wojtkiewicz, K., Jodłowiec, M., Krótkiewicz, M.: Association-oriented database metamodel: modelling language (2017)
17. Zabawa, P., Hnatkowska, B.: CDMM-F – domain languages framework. In: Świątek, J., Borzemski, L., Wilimowska, Z. (eds.) ISAT 2017. AISC, vol. 656, pp. 263–273. Springer, Cham (2018). https://doi.org/10.1007/978-3-319-67229-8_24

Integration of Relational and NoSQL Databases

Jaroslav Pokorný[(✉)]

Faculty of Mathematics and Physics, Charles University, Prague, Czech Republic
pokorny@ksi.mff.cuni.cz

Abstract. The analysis of relational and NoSQL databases leads to the conclusion that these data processing systems are to some extent complementary. In
current Big Data applications, especially where extensive analyses are needed, it
turns out that it is non-trivial to design an infrastructure involving data and
software of both types. In terms of performance, it may be beneficial to use a
polyglot persistence or multi-model approach or even to transform the SQL
database schema into NoSQL and to perform data migration between the relational and NoSQL database. The aim of the paper is to show these possibilities
and some new methods of designing such integrated database architectures.

Keywords: Relational database · NoSQL database · Big Data
Big Analytics · Database integration

1 Introduction

Recently, most large enterprises seem to be taking actually care about minimizing
application maintenance of existing production systems. This causes "bad" database
schemas to be used and "database decay" generally occurs. The authors of [14] build
the assertion on discussions with nearly twenty database administrators (DBA) at three
very large enterprises. The databases vary depending on business conditions, usually
once a quarter or more. The environment leads to the often disappearing central DBA's
roles and a more decentralized approach with more DBA groups maintaining databases
in the enterprise. NoSQL databases, a database alternative for storage and processing
so-called Big Data today, contribute to this state significantly.

The DBMS history always reflected requirements concerning new types of data to
be stored in a database way. Several database models such as Object-Oriented (OO),
Object-Relational (OR), XML, or RDF have been introduced since the relational data
model was introduced. OO and OR DBMSs responded to object-oriented approaches to
software engineering from the 1990s. However, these tools have never been competitive on the market. Reasons might be in the lack of their theoretical foundations and
the limited performance in practice. The XML databases suffer from similar problems.
Their goal is to promote the distribution of XML documents, but the use of native
XML databases is rather limited. Major vendors of relational DBMS (RDBMS) such as
Oracle, Microsoft SQL Server, and MySQL, include XML support in their products,

© Springer International Publishing AG, part of Springer Nature 2018
N. T. Nguyen et al. (Eds.): ACIIDS 2018, LNAI 10752, pp. 35–45, 2018.
https://doi.org/10.1007/978-3-319-75420-8_4

but native XML databases are not much involved in the database market. The initial enthusiasm for XML databases was based on Web application architectures and service orientations that use XML as a means to standardize data exchange format. However, this is now already possible with document-oriented NoSQL databases (see, popular JSON format), though not in such powerful languages as the XQuery in XML environment. However, the XML format has been added to the relational environment and is now the basic data type in SQL databases.

The situation in the database world today is affected by Big Data. Their V's characteristics are $Volume$, $Velocity$, and $Variety$. The author of [12] lists even 11 such V's. They fundamentally affect the storage and processing infrastructure of Big Data. Effective use of systems involving the processing of large volumes of data requires, in many application scenarios, adequate tools for storing and processing such data at a low level and analytical tools at higher levels. From the user's point of view, the most important aspect of processing large volumes of data on a computer is their analysis, as it is now called *Big Analytics*. Unfortunately, large data collections include data in different formats, such as relational tables, XML data, text data, multimedia data, or RDF triples, which may cause problems in processing data mining algorithms. Also, the growing data volume in a repository or the number of users of this repository requires a reliable solution of scaling in these dynamic environments, and more advanced means of delivering high performance than traditional database architectures offer. Moreover, traditional RDBMSs lack the dynamic data model necessary to tackle high velocity data coming in from machine-oriented systems or time series applications, as well as cases needing to manage social media data.

It is obvious that Big Analytics is also performed over a large amount of transaction data by extending the methods commonly used in Data Warehouses (DW). But DW technology has always been focused on structured data compared to the much richer variability of data types, as it is today for Big Data. Analytical processing of large data volumes therefore requires not only new database architectures but also new methods for data analysis.

To store and process Big Data today, we can choose:

- traditional DBMS (hereinafter referred to as databases, DB) - relational (SQL), OO, OR,
- traditional parallel database systems ("shared-nothing"),
- distributed file systems (e.g., HDFS),
- NoSQL databases,
- new architectures (e.g. NewSQL database).

In practice, ITC and business professionals need to determine whether NoSQL technologies are better suited than RDBMS for a particular system. The choice of technology is critical for applications that can be both transactional and analytical. They typically require different software and hardware architectures. The aim of the paper is to discuss the relation between SQL databases and NoSQL databases, modelling databases in the SQL and NoSQL polyglot world, mainly towards Big Analytics. An attention is devoted to problems of integration of such heterogeneous platforms in one architecture. In Sect. 2, we briefly describe the Big Analytics concept, i.e. the properties, processing and analysis of large volumes of data. In Sect. 3, we briefly review the NoSQL database technologies, especially their data models, architectures, and some their representatives. In Sect. 4, we show the duality between SQL databases and NoSQL databases and its reflection in various integrated database architectures. Section 5 contains conclusions and challenges for the database community.

2 Analytical Processing of Big Data

Big Analytics is used to transform information into knowledge through a combination of existing and new approaches. Related technologies include:

- data management (considering uncertainty, real-time query processing, information extraction, explicit time dimension management),
- new programming models,
- statistical methods, data mining (DM), and machine learning (ML),
- component architectures of data storage and processing systems, visualization of information.

As usual, two types of processing are distinguished:

- real-time processing (*data-in-motion*),
- batch processing of data obtained from different sources into one database (*data-at-rest*).

Batch analysis can then be:

- *small* (Small Analytics), i.e. OLAP over DW,
- *big* (Big Analytics), i.e. both DM and ML.

The problems that arise in this context are based on the fact that the requirements for Big Data are often more dynamic than the classic data processing in DWs. This concerns all 3 V's mentioned in Sect. 1. The NoSQL database is an alternative. Another issue is how to analyse Big Data coming from relational DBs.

A volume is not only a problem for data storage but also influences Big Analytics. With the increase in data complexity, its analysis is also more complex. We need to scale both the infrastructure and the standard data processing techniques for Big Data. Speed can also be a problem because the value of the analysis (and often of the data) decreases over time. If multiple data stream passes are required, data must be entered into DW where further analysis can be performed. Data can thus be stored and processed in a relatively traditional way or using cheap systems such as distributed NoSQL DB.

Big Data is often mentioned only in relation to business intelligence (BI). However, not only BI developers but, generally, data scientists are analysing large data collections. The challenge for computer professionals or data scientists is to provide people with tools that can efficiently perform complex analytics, taking into account the particular nature of processing large volumes of data. It is important to emphasize that

Big Analytics does not only include analysis and modelling phases. Often, distorted context as well as data heterogeneity and interpretation of results are taken into account. All these aspects affect scalable strategies and algorithms, so more efficient pre-processing steps (filtering and integration) and advanced parallel computing environments are needed. Data variability is now part of Big Data storage design and analytical system design. But performance is still a first order requirement.

In addition to these rather classic issues of mining large volumes of data, other interesting issues have emerged in recent years, such as recognizing named entities. The analysis of views and opinions (such as positive, negative, neutral) and their mining (sentiment analysis) are actual as topics using information retrieval methods and Web data analysis. A specific problem is the search for and characterization of discrepancies based on views and opinions. Comparison of graph patterns is commonly used in social network analysis where graphs, for example, include a billion users and hundreds of millions of links. In any case, the main problems of the current DM techniques used for Big Data come from their lack of scalability and parallelization.

3 NoSQL Databases

Large-scale data collections are often used for the storage and processing in NoSQL databases. NoSQL means "not only SQL", which makes this database category very diverse and not very clearly specified. NoSQL databases, starting in the late 1990s, provide easier scalability and performance compared to traditional RDBMS. We briefly describe their properties and classification (Sect. 3.1), followed by a discussion of their usability (Sect. 3.2). A more detailed discussion of NoSQL and, more generally, Big Data issues can be found, e.g., in [4, 8, 10, 12].

3.1 Categories of NoSQL Databases

What is the main classical approach to databases - a (logical) data model - is described in NoSQL databases rather intuitively, without any formal basis. NoSQL terminology is also very diverse, and the difference between a conceptual and database view is mostly blurred.

The most well-known NoSQL databases can be classified according to the used data model as:

- *key-value stores*, such as Redis[1],
- *column store*, e.g. CASSANDRA[2],
- *document stores*, such as MongoDB[3].

Key-value stores contain a set of pairs (key, value). The key uniquely identifies the opaque value. The choice of the key is, unlike relational DB, solved pragmatically.

---

[1] https://redis.io/.

[2] http://cassandra.apache.org/.

[3] https://www.mongodb.com/.

The goal is only quick access to data. The value can even be a list of pairs (name, value) (e.g. in Redis). Data access operations, typically get and put, only work through the key. NULL values are not required, because these databases do not use the schema. Although it is a very efficient and scalable approach to implementation, the disadvantage of a too simple data model can be substantial for such databases.

NoSQL stores can contain a set of couples (name, value) in a *column family* in a row addressed by a key. A column family in different rows can contain different columns. Then we are talking about a column-oriented NoSQL database. There is also another level of structure called, e.g., *supercolumn* in Cassandra. The supercolumn contains nested (sub)columns. Access to data using get and put is enhanced by using column names.

Most general data models belong to document-oriented NoSQL DBs. They are the same as key-value repositories, but each key is coupled with any complex data structure that resembles a semi-structured document. The JSON format is usually used to present these data structures. JSON is a typed data model that supports basic data types and objects - non-ordered sets of couples (name, value), and the value can be structured (array). JSON is similar to XML, but it is smaller, faster and easier for parsing. For example, CouchDB[4] uses the JSON format whereas MongoDB stores data in BSON (binary coded serialization of JSON documents). It is possible to query the data in a document in other ways than using a key (e.g. through indexing). Moreover, selection and projection operations on the query results can be performed.

There are also other approaches. DB-Engines Ranking server[5], e.g., considers also *search engines* as NoSQL databases, e.g., Elasticsearch[6]. They are data management systems dedicated to the search for data content. They are not typically classical document systems. They typically offer a support for complex search expressions, full text search, ranking and grouping of search results, geospatial search, and distributed search for high scalability. More generally, NoSQL databases include also graph databases [11], and others, e.g. XML and RDF ones.

The first three NoSQL categories are basically of the key-value type. They differ mainly in the possibilities of aggregating couples (key, value) and accessing these values. For our considerations, we consider only them.

3.2 Usability of NoSQL Databases

There is much debate about the role of NoSQL databases in providing information services. NoSQL camp claims that this technology is the future of databases. On the other hand, the RDBMS camp argues that the NoSQL databases have a big disadvantage of failing to provide correct data integrity. In any case, NoSQL technologies are designed with Big Data needs in mind.

NoSQL are often a part of data intensive cloud applications (mainly Web applications). Examples of such applications include Web entertainment applications,

---

[4] http://couchdb.apache.org/.

[5] https://db-engines.com/en/ranking.

[6] https://www.elastic.co/products/elasticsearch.

high-traffic Web site services, media delivery in a streamlined fashion, or typical data found in social networking applications.

NoSQL systems are more suitable for interactive data services environments. Schema enforcing and row-level locking as in relational DBs may over-complicate these applications. The absence of some ACID properties even allows significant acceleration and decentralization of NoSQL databases.

On the other hand, one of the most famous problems with NoSQL repositories is the lack of semantics caused by their underling feature – they are schema-less. The lack of metadata prevents the database system from knowing which data is stored and how it is interconnected.

NoSQL databases usually have little means of ad-hoc querying and analysis. Even a simple query requires significant programming experience, and generally used BI tools do not provide connectivity to NoSQL. NoSQL databases can also not be recommended for applications requiring enterprise level functionality (ACID properties, security, and other relational technology features). NoSQL should not be the only choice in the cloud.

Experience with the NoSQL database shows that they can be used

- even on "small" dates,
- for applications not requiring transactional semantics, such as directories, blogs, or content management systems or for analysing high-volume, real-time data (such as Web site click-streams). In the mobile data processing environment, transactions are even more technically impossible in a larger range.

Among the good properties of NoSQL databases we can find:

- massive performance in `write` operations,
- quick search in a key-value way,
- they do have no portion causing a total network failure when an error occurs,
- enable rapid prototyping and development,
- allow scalability without user intervention,
- have easy maintenance.

On the other hand, a user may find unusual and often inappropriate phenomena in NoSQL approaches:

- have different behaviour in different applications,
- no language query standards are available,
- migration from one system to another is complicated,
- join operation is missing,
- some of them are more mature than others, but each of them is trying to solve similar problems,
- checking referential integrity "over" database partition segments is missing. As the performance is crucial, an integrity control or the implementation of complex operations is limited in a distributed environment.

Table 1. Comparison of relational and NoSQL DBMSs

	Property	RDBMS	NoSQL
1	Data model	Relational	Domain-oriented
2		Data is strongly typed	Data is potentially dynamically typed
3		Data of dependent tables points to its parents (via foreign keys)	Parent's data points to children data
4		Associated entities have an identi-ty (primary key)	Environment determines identity
5		Not compositional	Compositional
6		Referential integrity based on values	Weak referential integrity based on computation or only "over" partition segments
7	Integrity	Responsibility at the DB level	Responsibility moved to the application level
8	DB schema	Expressed in SQL	Typically do not require a fixed DB schema, i.e. they have a more flexible data model
9	Detection of problems in the DB schema	At the DB level	At the application level and data access procedures
10	DB modelling management	Begins from accessible data	Patterns for data access and updates
11	Querying	SQL	Simple API, if SQL, then only its very limited version; REST, client libraries
12		Complex queries + ad hoc queries	Inappropriate for ad hoc queries and complex queries
13		Join operation	Join emulation at application level
14	Data storage	Centralized or distributed	Horizontally scaled, replications
15	Data processing	Synchronous (ACID) updates over more rows	Asynchronous (BASE) updates within single values
16		Environment coordinates changes (transactions)	Entities responsible to react to changes (eventual consistency)
17		Strong consistency and also consistency tuneable by application	Eventual consistency and also consistency tuneable in application
18		Query optimizer – responsibility by DBMS	Developer/pattern – responsibility by application

Table 1 shows a comparison of NoSQL and SQL DB in more details.

In the database world NoSQL DBs occupy a significant place. In the DB-Engines Ranking, 339 various DB-Engines were tracked in December 2017. MongoDB, Redis, and Cassandra occupied positions 5, 8, and 9, respectively, in this rating.

4   SQL and NoSQL: Towards Integrated Architectures

In the work [9], the authors argue that the NoSQL databases are rather complementary to traditional transactional DBMSs. Should not they be called "co-relational"? Maybe more natural would be to say coSQL instead of NoSQL. In Table 1, according to [9], complementary differences are given by properties 2, 3, 4, 5, 6, 16, 17, and 19. This complementarity negatively influences integration possibilities of these datastores both at the data model and data processing level.

Particularly, normalization allows single object data in a relational database to be spread over multiple relations. For example, customer data is in one table, data about the banks where his/her account are is in the second table. The interconnection is realized via foreign keys. In NoSQL database, this can be done in such a way that each bank "row" can contain data and account numbers for each customer. The basic feature of NoSQL is that they are denormalized, that is, they store copies of an object instead of the object. This, of course, leads to worse data update options.

In ICT history, different DBMSs were designed to solve different problems, considering still new and new data types. In addition to centralized RDBMSs, specialized servers, universal servers, relational DW, etc. appeared in the past. These tools were based on a fixed database schema and an associated query language (mostly SQL). OR SQL and its other extensions supported this strategy for a long time.

Concerning an integration of distributed data from different databases, two approaches based on a database schema management were at disposal:

- top-down – starting with a global schema to design schemas for data in sites,
- bottom-up – through middleware, i.e. to use schema mapping for schemas in sites into a middleware (e.g., OLE DB, JDBC) and then use a query transformation. Data is loosely integrated and managed by multiple servers.

We remind that the former concerns rather homogenous databases models, while the letter supports heterogeneous database models and consequently DBMSs.

In context of RDBMSs and NoSQL databases, it is not possible to use simply traditional approaches to data integration. The reason is the complementarity of these database types. Moreover, the problem of analysts is that the lack of data schemas (semantics) prevents them from understanding their structure and thus generating serious analyses. Now, the tendency is to create multilevel modelling approaches involving both relational and NoSQL architectures including their integration [1]. Several approaches are under a development:

Polyglot persistence. We approach particular data stores with their original data access methods [13]. The truth is that polyglot persistence is a method for data modelling problems, not a solution to them. Developers need to customize data models for an application and often need more than one, but they should not have to adopt different DBMS to get them. "Polyglot" means "able to speak many languages", not integration. As an integration architecture, polyglot persistence is its weakest form.

Multi-model approach. Maybe, it presents a more user-friendly solution of heterogeneous database integration. Multi-model represents an intersection of multiple models in one product. For example, OrientDB[7] is a multi-model DBMS including geospatial, graph, fulltext and key-valued data models. OO concepts are used for user domain modelling in OrientDB. Similarly, ArangoDB[8] is designed as a native multi-model database, supporting key-value, document and graph models. MarkLogic[9] enables to store and search JSON and XML documents and RDF triples.

NoSQL relationally. The multi-model solution [5] considers source document and column-oriented DB integrated through a middleware into a virtual SQL database.

Multilevel modelling. Despite of the fact that database schemas are mostly not used in the NoSQL world, some variations on multilevel modelling approaches exist. In relation to solution of an alternative for data processing with relational and NoSQL data in one infrastructure, common design methods for such DBs are based on the modification of the traditional 3-level ANSI/SPARC approach [7]. The approach involves not only heterogeneous data sources but also the development of a database schema in the overall infrastructure, i.e., its variability. A strong motivation for this approach is the fact that when designing a database for Big Analytics, we must consider DM/ML patterns, clustering of some attributes, etc., to ensure adequate system performance. However, the conceptual design assumes the correctness of the current knowledge of the application domain. The following examples document activities in this area:

- *Special abstract model.* A DB design methodology for NoSQL systems based on NoAM (NoSQL Abstract Model), a novel abstract data model for NoSQL databases, is presented in [2]. The associated design methodology starts with an UML class diagram, a designer identifies so called aggregates ("chunks" of related data) and maps the aggregates into NoAM blocks. These blocks are simply transformed into constructs of a particular NoSQL data model.
- *NoSQL-on-RDBMS.* A coexistence of RDBMS and a NoSQL DB includes, e.g., storing and querying JSON data in a RDBMS (see, ARGO/SQL [3]).
- *Ontology integration.* A more advanced integrating architecture including several NoSQL databases is proposed in [6]. The databases are described by several ontologies and a generated global ontology. Global SPARQL queries are transformed into query languages of sources.

Schema and data conversion. In practice, there are other options, such as the schema conversion model, in which the schema from the SQL database is converted to the NoSQL database schema [15]. Then, even a double-sided data migration between a RDBMS and a NoSQL DB can be performed.

5   Conclusions

Key issues for building Big Data processing infrastructure are in decisions concerning NoSQL databases. They include in particular

---

[7] http://orientdb.com/orientdb/.

[8] https://www.arangodb.com/.

[9] http://www.marklogic.com/.

- choosing the right (correct) product,
- designing a suitable database architecture for a given application class.

However, the role of a person is also significant especially in Big Analytics. Currently, the DM process is driven by an analyst or data scientist. Depending on the application scenario, the person determines a portion of the data from which, e.g., useful patterns can be extracted. A better solution would, however, be to have an automated DM process in place to get approximate synthetic information about both structure and content of large amounts of data. This is still a big problem for Big Data analysts.

Current challenges for database research include:

- Modelling polyglot and multi-model databases including relational and NoSQL in one infrastructure.
- Improving the quality and scalability of DM methods. Interpreting a query - especially in the schema absence - and received answers, may be non-trivial.
- Transforming content into a structured format for later analysis, because many data today is not natively in a structured format. At the same time, with a filtering we can reduce the volume of data.
- Develop a meaningful and usable formalisms for modelling NoSQL databases and a sufficiently general user-friendly query language.

Acknowledgments. This work was supported by the Charles University project Q48.

References

1. Abelló, A.: Big data design. In: Proceedings of 11th International Workshop on Data Warehousing and OLAP, DOLAP 2015, pp. 35–38. ACM (2015)
2. Bugiotti, F., Cabibbo, L., Atzeni, P., Torlone, R.: Database design for NoSQL systems. In: Yu, E., Dobbie, G., Jarke, M., Purao, S. (eds.) ER 2014. LNCS, vol. 8824, pp. 223–231. Springer, Cham (2014). https://doi.org/10.1007/978-3-319-12206-9_18
3. Chasseur, C., Li, Y., Patel, J.M.: Enabling JSON document stores in relational systems. In: 16th International Workshop on the Web and Databases (WebDB 2013), pp. 1–6 (2013)
4. Corbellini, A., Mateos, C., Zunino, A., Godoy, D., Schiaffino, S.: Persisting big-data: the NoSQL landscape. Inf. Syst. **63**, 1–23 (2017)
5. Curé, O., Hecht, R., Le Duc, C., Lamolle, M.: Data integration over NoSQL stores using access path based mappings. In: Hameurlain, A., Liddle, S.W., Schewe, K.-D., Zhou, X. (eds.) DEXA 2011. LNCS, vol. 6860, pp. 481–495. Springer, Heidelberg (2011). https://doi.org/10.1007/978-3-642-23088-2_36
6. Curé, O., Lamole, M., Duc, C.L.: Ontology Based Data Integration over Document and Column Family Oriented NOSQL, CoRR, arXiv:1307.2603 (2013)
7. Herrero, V., Abelló, A., Romero, O.: NOSQL design for analytical workloads: variability matters. In: Comyn-Wattiau, I., Tanaka, K., Song, I.-Y., Yamamoto, S., Saeki, M. (eds.) ER 2016. LNCS, vol. 9974, pp. 50–64. Springer, Cham (2016). https://doi.org/10.1007/978-3-319-46397-1_4
8. Marz, N., Warren, J.: Big Data: Principles and Best Practices of Scalable Realtime Data Systems, 1st edn. Manning Publications, New York (2015)

9. Meijer, E., Bierman, G.M.: A co-relational model of data for large shared data banks. Commun. ACM **54**(4), 49–58 (2011)
10. Pokorný, J.: NoSQL databases: a step to databases scalability in Web environment. Int. J. Web Inf. Syst. **9**(1), 69–82 (2013)
11. Pokorný, J.: Graph databases: their power and limitations. In: Saeed, K., Homenda, W. (eds.) CISIM 2015. LNCS, vol. 9339, pp. 58–69. Springer, Cham (2015). https://doi.org/10.1007/978-3-319-24369-6_5
12. Pokorný, J.: Big data storage and management: challenges and opportunities. In: Proceedings of 12th IFIP WG 5.11 International Symposium on Environmental Software Systems, IFIP AICT 507. Springer, Heidelberg (2018, to appear)
13. Sadalage, P.J., Fowler, M.: NoSQL Distilled: A Brief Guide to the Emerging World of Polyglot Persistence. Pearson Education Inc., London (2013)
14. Stonebraker, M., Deng, D., Brodie, M.L.: Database decay and how to avoid it. In: Proceedings of 2016 IEEE International Conference on Big Data, pp. 7–16. IEEE Explore (2016)
15. Zhao, G., Lin, Q., Li, L., Li, Z.: Schema conversion model of SQL database to NoSQL. In: Proceedings of the 9th International Conference on P2P, Parallel, Grid, Cloud and Internet Computing, pp. 355–362. IEEE (2014)

Complex Relationships Modeling in Association-Oriented Database Metamodel

Marcin Jodłowiec[1,2](✉)

[1] Department of Information Systems,
Wroclaw University of Science and Technology, Wroclaw, Poland
marcin.jodlowiec@pwr.edu.pl
[2] Institute of Computer Science, Opole University of Technology, Opole, Poland

Abstract. This paper elaborates upon the problem of modeling complex relationships in database models. The relationship type is complex when its definition holds IS-A or HAS-A meta-relationships between distinct relationship types. The framework of the considerations is Association-Oriented Database Metamodel, a novel solution dedicated for modeling and implementation of database layer for compound, e.g. knowledge-based systems. The contribution of the paper is a set of design constructions that enable the model designer to define polymorphic and structured relationships in Association-Oriented Database Metamodel.

Keywords: Database modeling
Association-Oriented Database Metamodel · First-class relationships
Association generalization · Association aggregation
Database design patterns

1 Introduction

Many approaches to information systems modeling introduce an *inheritance* mechanism or a *generalization-specialization* (*gen-spec*) relationship as well as the *aggregation* relationship. It is affecting metamodels of various types: system domain modeling, database layer, knowledge representation structures. *Gen-spec* relationship is a grammar element in modeling languages. Its semantics allows to create taxonomic structures, i.e. subcategory is of the same type as supercategory. In a formal way, the *gen-spec* relationship can be defined by following properties:

$$T_{sub} \prec T_{super} \wedge T_{super} :: e \vDash T_{sub} :: e \qquad (1)$$

$$T_{sub} \prec T_{super} \wedge T_{super}.m \vDash T_{sub}.m \qquad (2)$$

where: T_{super} – supercategory type, T_{sub} – subcategory type, $X :: x$ – instance x is of type X, $X.m$ – type X has (aggregates) member m, $\prec$ – *gen-spec* relationship.

Polymorphism (Eq. 1) is a fundamental consequence of the *gen-spec* relationship (*IS-A*). Other feature is transferring or copying the components (attributes, participation in relationships, etc.) – subcategories can contain components defined in supercategory (Eq. 2). Another relationship in modeling is aggregation (*HAS-A* relationship). It can be seen in between categories, in instances where one is a component of the another (part-whole), i.e. when type T_1 is a *whole* and T_2 is a *part*, then T_1 aggregates T_2 ($T_1.T_2$).

In vast majority of approaches to data modeling, only the categories representing data or type are subject to inheritance and aggregation. In this work, we deal with modeling of *IS-A* and *HAS-A* relationships between relationships (we will refer them as *meta-relationships*). Defining abstract relationships and representing their particular implementations consecutively can significantly increase semantic capacity of a model, prevent duplication of unnecessary elements of a model, and also prevent creation of artificial constructs of ambiguous semantics from the metamodel perspective [13].

The following article tackles the data modeling problem by using the concepts of relationship inheritance and relationship aggregation. The fundamental framework of the paper is the novel solution in the database modeling area – Association-Oriented Database (AODB) Metamodel [8]. The aim of this work is to show, that the metamodel has high expressiveness, when it enables to define semantically unambiguous inheritance and aggregation for relationships. The contribution of the author constitutes the technique of structurally complex relationship modeling by the use of the pattern Association Complex in AODB Metamodel.

2   Modeling Problem

Definition 1. *Relationship type is complex if it has at least one feature of the following:*

1. *It is specialization of other relationship type(s),*
2. *Its structure comprise other relationship type(s).*

To show the approach, let us consider the relationship as first-class element and analyze its layers of complexity, as we want the relationship to be more than solely a description of link between two elements. Three layers of relationships have been shown in the Fig. 1. We consider: *Metamodel layer*, which defines Relationship metatypes (relationship type between relationship types). Whilst building the model in the *Model layer* (expressing it in a metamodel), we might define relationship types, which form the structure of actual relationships (*Data layer*).

The example of the fraction of reality containing the relationship specializations shown in the Example 1.

Example 1. Let a *Lesson* be a ternary relationship, which links the following: (1) a teacher conducting the classes, (2) a group of students, (3) involved subject. Let a *Lab* be a quaternary relationship, which *is a Lesson*, but defines additional *Equipment*, which defines set of used tools.

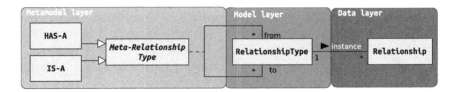

Fig. 1. An UML diagram depicting described relationship modeling layers

Natural approach to modeling Example 1 in many data metamodels, i.a. RDF, E-R and also UML would be to model *Lesson* and *Laboratory* as categories responsible for data representation (*Class* in case of UML) and to define *generalization/specialization* relationship (Fig. 2). However, it has to be noted that in case of such model, one can observe a loss of semantics: elements being actually relationship types are modeled as artificial data elements. Furthermore, participation in relationship results with creation of artificial relationship types. The aspect of data modeling concerning semantic loss in mapping mental constructions to data metamodels have been described by Krótkiewicz in [8] in the analysis of database metamodels and their possibility to express n-ary relationships.

The metastructure of UML metamodel contains *Generalization* class, which represents specification of *gen-spec* relationship between *Classifiers*. It comprise two properties: *specific* and *general* relating to corresponding classifiers. Thus, it is grammatically correct to model generalization between associations without semantics loss (Fig. 3). However, it has to be noted that it is not popular design construction and UML 2.5 formal specification [14] does not mention any example or even reference regarding association generalization.

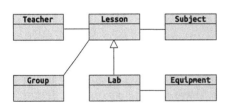

Fig. 2. UML class diagram, where relationships from Example 1 are modeled as classes.

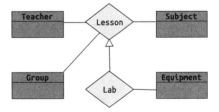

Fig. 3. UML class diagram, where relationships from Example 1 are modeled as n-ary associations

The *HAS-A* meta-relationship is based on the combination of relationships inside the others, as shown in Example 2.

Example 2. Let *Class* be a relationship, which assigns a set of *Features* to a set of *Instances*. Then, let *Static Property* be a relationship, which assigns set of *Features* to a specific *Class*.

In the example above, it is easy to notice that both *Class* and *Static Property* are conceptualised as relationships, wherein the participant of the latter is the first one. In order to model Example 2 in UML, *Class* cannot be conceptualized as association, but rather both as association and a class, thus one ought to use association class (Fig. 4). It stems from the design of UML; in order to participate in relationships, the classifier must comprise attributes defined as properties. Therefore, it stops being conceptually a relationship and becomes a category, which represents both data and relationships.

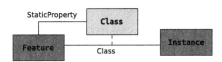

Fig. 4. An UML class diagram depicting HAS-A relationships between associations

UML is conceptual metamodel. Until programming languages will not explicitly support the association/relationship concept, the semantics loss will always occur whilst generating or writing code conforming to the model. Therefore, although the object-oriented approach distinguishes relationship as category (*first-class element*), it cannot be efficiently used to model relationship-oriented data structures. Author proposes association-oriented modeling to solve this problem. This approach assumes, that relationship (association) is in the center of interest of the model designer.

3   Aspects of Association-Oriented Database Metamodel

The AODB Metamodel, including all its components, i.e. AML (Association-Oriented Modeling Language), AFN (Association-Oriented Formal Notation), ADL (Association-Oriented Data Language), AQL (Association-Oriented Query Language) and implemented tools in regard has been precisely described in [8]. Its metastructure has been explicitly formulated in UML metamodel and presented in [7]. The description of AODB presented here is merely a brief description that should help the reader to understand foundations of the AODB metastructure.

The database (*Db*) consist of list of collections (*Coll*) and list of associations (*Assoc*). Association in the intensional sense owns a set of roles (*Role*). Moreover, it has the reference to a describing collection (*Coll*). Furthermore, both association (*Assoc*) and collection (*Coll*) are generalized to abstract basenodes (*BaseNode*) and have name, information about abstractness, navigability and inheritance. Role (*Role*) has name, information about virtuality, navigability, directionality, multiplicities, compositionality (lifetime dependency between relationship and relationship participant), uniqueness, describing collection (*Coll*) and type of participants, i.e. elements that can be role's (*Role*) destination.

Role (*Role*) is anchored in association (*Assoc*), which is its owner. Collection (*Coll*) comprise a set of attributes (*Attr*). Attribute (*Attr*) has name, information about virtuality, inheritability, type and cardinality of elements and default value. Object (*Obj*) is built of values. Furthermore, it contains unique invariant identifier. Association object (*AssocObj*) contains a set of role objects (*RoleObj*), reference to describing object (*Obj*) and unique invariant identifier. Role object (*RoleObj*) contains set of references to database elements ((*Obj*) or (*AssocObj*)) and reference to describing object (*Obj*).

3.1   Relationships Model of AODB

For the purposes of better understanding the association complex model, the fraction of AODB metastructure has been provided (Fig. 5). The diagram shows how specific metamodel categories involved in building relationships are interrelated. The metamodel implements considered meta-relationships as follows. *IS-A* is implemented in the way that resembles UML Generalization class. It is settled as two binary associations between generalization category *Inheritance* and abstract category (*BaseNode*), which classifies both associations and collections. The *HAS-A* meta-relationship can be modeled due to the fact, that (*Assoc*) is specialization of abstract category (*BaseNode*) (*Assoc* < *BaseNode*$^{\varnothing}$), which is also of property type of the (*Role*).

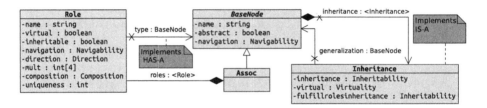

Fig. 5. An UML class diagram depicting relationship model in AODB

4   Model Pattern: Association Complex

The specification was made in accordance with the modified method for describing the design patterns proposed by [5]. The discussed pattern has been named *Association Complex* (AC), suggesting the pattern is used to define complex associations.

Applicability. Association Complex is only applicable in AODB Metamodel and cannot be concerned beyond it as such, although some of its foundations and concepts can be successfully mapped to other metamodels. Association complex is mainly applied to complex systems, e.g. knowledge-based systems, where the data semantics is crucial. The pattern has been verified practically and applied in i.a. Extended Semantic Network Module of Semantic Knowledge Base (see [10,11]).

Participants. The pattern comprise the following participants:

- ◇*RootRelationship* is abstract root of a *gen-spec* relationship hierarchy.
- ◇*ConcreteRelationship₁*, ... , *ConcreteRelationshipₙ* are the concrete specialization associations of ◇*RootRelationship*.
- ◇*SubRelationship* is relationship that is aggregated inside ◇*RootRelationship* and its children.
- Relationship participants in form of Bicompositive Association-Collection Tandem (BACT) structures, e.g.: ◇*Entity* ● ――― *+Entity* ――● □ *Entity* and ◇*OtherEntity* ● ――― *+OtherEntity* ――● □*OtherEntity*.

Implementation and diagram. The actual generalized AODB model has been shown in the AML diagram in the Fig. 6 (for the language syntax see [6,11]).

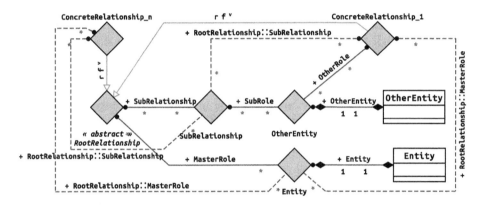

Fig. 6. An AML diagram depicting unconstrained AODB model for association complex

Usage. The pattern requires the model designer to consider the following features and constraints:

1. *Association orientation.* Each model category should be primarily conceptualized as relationships with other categories, and then which data they hold.
2. *Abstractness.* Abstract associations must have its concrete specializations in order to have instances. They serve as the base for definition concrete relationships.
3. *Dependencies between associations.* The designer has to consider the meta-relationship type:*IS-A* (association *gen-spec*), or *HAS-A* (association cascade).
4. There are two considered types of *meta-relationships* and their features and constraints:
   (a) *Association gen-spec* (implemented via ◇*ConcreteRelationsip*₍₁,...,ₙ₎ $\xrightarrow{rf^v}$ ◇ *RootAssociation*)

     – *Possibility of role inheritance* (links between Association and its destination). If *gen-spec* relationship has the property of role inheritance, each specialization obtains each inheritable roles of generalization association.

     – *Possibility of inheriting rights to fulfill roles.* If *gen-spec* relationships defines this property, specialization obtains the right to participate in each roles of generalization association.

     – *Obligation of type inheritance.* It is connected with the substitutability within database queries and emerges that each object of generalization association is an object of specialization association.

(b) *Association   cascade*   (implemented   via $\Diamond RootRelationship[*]$ $\overset{SubRelationship}{\bullet\rule{2cm}{0.4pt}}[*] \Diamond SubRelationship$):

     – *Ownership of role* determines, which one association is source of the relationship, and which one its destination. In other words, association containing role is owner in the sense of *HAS-A* relationship.

     – *Multiplicites* are implemented by four-element vector `Role::mult : int[4]`. It restricts the upper and lower limit of the owning and owned relationships.

     – *Compositionality* links the relationships with lifetime dependency. It can be one of the following: *CompositionInOwner* (when superassociation is deleted, the subassociation must also be deleted), *CompositionInDest* (when subassociation is deleted, the superassociation must also be deleted), *CompositionInBoth* (both rules are applied).

     – *Navigability* depending on the designer decision, roles can be of two storage models: *Binavigable* or *Uninavigable*. Unidirectionality provides the model with space-efficient storage, but disables information retrieval from database in the direction owned $\rightarrow$ owning element.

5. *Inheritability.* One can specify member (role) as not inheritable and avoid of its inheritance on next taxonomy levels.

6. *Other taxonomies.* One can define more generalizations of an association. Cycles are forbidden.

7. *Virtuality.* One can specify member (role) or Inheritance mode as virtual to avoid duplication of inherited members in the diamond structure.

Example. Let us consider the fraction of reality from Example 1. The AODB model (Fig. 7 is constructed as follows. The $\Diamond Lesson$ stands for *RootRelationship*), and is constructed as ternary association as shown in Eq. 3. The $\Diamond Lab$ is specialised association that stands for *ConcreteRelationship* participant.

$$\Diamond Lesson \left(\begin{array}{l} \overset{Teacher}{\bullet\rule{1.5cm}{0.4pt}} \Box Teacher \\ \overset{Group}{\bullet\rule{1.5cm}{0.4pt}} \Diamond Group \\ \overset{Subject}{\bullet\rule{1.5cm}{0.4pt}} \Box Subject \end{array} \right) \tag{3}$$

$$\Diamond Lab \langle \overset{Equipment}{\bullet\rule{1.5cm}{0.4pt}} \Box Equipment \rangle \tag{4}$$

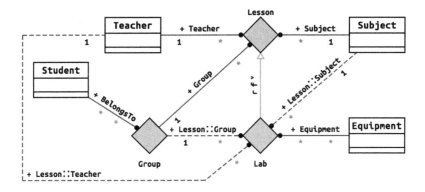

Fig. 7. AML Diagram depicting the fraction of reality from Example 1

$$\Diamond Lab \xrightarrow{rf^v} \Diamond Lesson \vDash \Diamond Lab \left(\frac{\overbrace{\bullet\rule{0pt}{0pt}\hspace{2cm}}^{Lesson::Teacher}}{} \Box Teacher \atop \frac{\overbrace{\bullet\hspace{2cm}}^{Lesson::Group}}{} \Diamond Group \atop \frac{\overbrace{\bullet\hspace{2cm}}^{Lesson::Subject}}{} \Box Subject \right) \tag{5}$$

Notice, that this model also contains the relationship aggregation semantics. The group of students ($\Diamond Group$) has been modeled as the relationship between many students ($\Box Student$). Thus, one of the participants of association $\Diamond Lesson$ is $\Diamond Group$.

The model from Example 2 could be formalised in AODB metamodel analogously (Eq. 6).

$$\Diamond Class \left(\frac{\overset{Feature}{\bullet\rule{0pt}{0pt}\hspace{1.3cm}}}{} \Box Feature \atop \frac{\overset{Instance}{\bullet\hspace{1.3cm}}}{} \Box Instance \right) \Diamond StaticProperty \left(\frac{\overset{Feature}{\bullet\rule{0pt}{0pt}\hspace{1.3cm}}}{} \Box Feature \atop \frac{\overset{Class}{\bullet\hspace{1.3cm}}}{} \Diamond Class \right) \tag{6}$$

5 Related Work

5.1 Overview of Data Models

Data metamodels treat the issues of *IS-A* and *HAS-A* relationships in various way. Metamodel categories, which implement these relationships have been collectively presented in Table 1. Shared feature of OM ODMG, EER, RDF is fact, that their construction allows to define aggregation and generalization relationships for categories that represent data structures. For UML things are different: the way the UML metastructure handles meta-relationships (relationships between relationships) has been described in Sect. 2. CDMM-F is valuable meta-metamodeling framework, that could benefit in creating the semantically rich metamodels in regard to relationships [17].

Table 1. Mechanisms implementing *aggregation* and *generalization* in entity-like categories in known data metamodels

Metamodel	Aggregating category	Generalizing category
UML [14]	Property `aggregation` of class `Property`	Generalization (between Classifiers)
OM ODMG [4]	Attribute of object type	Generalization/extends
EER [16]	Aggregation	Generalization
RDF [3]	`rdf:subPropertyOf`	`rdf:subClassOf`

There are some object-oriented approaches to domain modeling that treat relationships as *first-class elements* [1,2,15], but nevertheless they do not allow to model *meta-relationships* considered in this paper. Few of them consider polymorphic character of such relationships (e.g. RelJ [2] and other similar [12]). The significant difference between RelJ-like methods and association-oriented modeling is that the RelJ-like relationships extending is focused on their polymorphism instead of inheriting (or not) roles and extending structure on the lower level of taxonomy.

5.2   Association-Oriented Design Patterns

This article is a continuation of the publication series, which is devoted to methodology for association-oriented database modeling. In paper [6] we proposed implementations of some basic conceptual design patterns, like: *list, dictionary, n-ary relationship* and showed their implementations in AODB Metamodel. The paper [9] introduced association-oriented method for modeling recursive relationships. In the paper [7] we have compared i.a. the features of (*Assoc*) category from AODB Metamodel and UML's association in terms of *Ownership, Navigability* and *Arity* and proposed the mapping procedure $association_{UML} \mapsto Assoc_{AODB}$. The results have shown that (*Assoc*) is higher-level category, because mapping process was predominatingly reduced to creating only a new (*Role*). In the current paper we proposed the design pattern and modeling technique for complex relationships which is dedicated for AODB Metamodel.

6   Conclusions

Association inheritance combined with association aggregation enables the model designer to represent complex relationships semantics and is unique modeling technique for AODB Metamodel. Implementation of the concept of association complex in other data metamodels causes semantics loss. The necessity of model transformation by mapping association complexes to other categories (e.g. classes, relations, entities) would be unavoidable. The proposed solution is semantically rich and very expressive as the association inheritance mechanism and relationship linking mechanism avoid data definition redundancy as well as unambiguity of semantics. Moreover, the solution results with implementable models, which do not require additional transformation.

References

1. Albano, A., Ghelli, G., Orsini, R.: A relationship mechanism for a strongly typed object-oriented database programming language. In: Proceedings of the 17th International Conference on Very Large Data Bases, VLDB 1991, pp. 565–575. Morgan Kaufmann Publishers Inc., San Francisco (1991)
2. Bierman, G., Wren, A.: First-class relationships in an object-oriented language. In: Black, A.P. (ed.) ECOOP 2005. LNCS, vol. 3586, pp. 262–286. Springer, Heidelberg (2005). https://doi.org/10.1007/11531142_12
3. Brickley, D., Guha, R.: RDF Schema 1.1 - W3C Recommendation (2008)
4. Cattell, R.G., Barry, D.K., Berler, M., Eastman, J., Jordan, D., Russell, C., Schadow, O., Stanienda, T., Velez, F.: The Object Data Standard: ODMG 3.0. Morgan Kaufmann, Burlington (2000). 280 p
5. Gamma, E., Helm, R., Johnson, R., Vlissides, J.: Design patterns: abstraction and reuse of object-oriented design. In: Nierstrasz, O.M. (ed.) ECOOP 1993. LNCS, vol. 707, pp. 406–431. Springer, Heidelberg (1993). https://doi.org/10.1007/3-540-47910-4_21
6. Jodłowiec, M., Krótkiewicz, M.: Semantics discovering in relational databases by pattern-based mapping to association-oriented metamodel—a biomedical case study. In: Piętka, E., Badura, P., Kawa, J., Wieclawek, W. (eds.) Information Technologies in Medicine. AISC, vol. 471, pp. 475–487. Springer, Cham (2016). https://doi.org/10.1007/978-3-319-39796-2_39
7. Jodłowiec, M., Krótkiewicz, M.: Towards the mapping of UML class diagrams to association-oriented database metamodel schemata. In: Proceedings of 8th International Conference on Information, Intelligence and Applications (IISA2017). IEEE (2017, in press)
8. Krótkiewicz, M.: Association-oriented database model – n-ary associations. Int. J. Softw. Eng. Knowl. Eng. **27**(2), 281–320 (2017)
9. Krótkiewicz, M., Jodłowiec, M.: Modeling autoreferential relationships in association-oriented database metamodel. In: Świątek, J., Borzemski, L., Wilimowska, Z. (eds.) ISAT 2017 - Part II. AISC, vol. 656, pp. 49–62. Springer, Cham (2018). https://doi.org/10.1007/978-3-319-67229-8_5
10. Krótkiewicz, M., Jodłowiec, M., Wojtkiewicz, K.: Semantic networks modeling with operand-operator structures in association-oriented metamodel. In: Nguyen, N.T., Papadopoulos, G.A., Jędrzejowicz, P., Trawiński, B., Vossen, G. (eds.) ICCCI 2017. LNCS (LNAI), vol. 10448, pp. 24–33. Springer, Cham (2017). https://doi.org/10.1007/978-3-319-67074-4_3
11. Krótkiewicz, M., Wojtkiewicz, K., Jodłowiec, M., Pokuta, W.: Semantic knowledge base: quantifiers and multiplicity in extended semantic networks module. In: Ngonga Ngomo, A.-C., Křemen, P. (eds.) KESW 2016. CCIS, vol. 649, pp. 173–187. Springer, Cham (2016). https://doi.org/10.1007/978-3-319-45880-9_14
12. Nelson, S., Pearce, D.J., Noble, J.: First class relationships for OO languages. In: 18th ECOOP Doctoral Symposium and Ph.D. Student Workshop, p. 33 (2008)
13. Noble, J.: Basic relationship patterns. Pattern Lang. Prog. Des. **4**, 73–94 (1997)

14. Object Management Group: OMG Unified Modeling Language (OMG UML) Version 2.5. Technical report (2013). http://www.omg.org/spec/UML/2.5/
15. Shah, A.V., Hamel, J.H., Borsari, R.A., Rumbaugh, J.E.: DSM: an object-relationship modeling language. SIGPLAN Not. **24**(10), 191–202 (1989)
16. Teorey, T.J., Yang, D., Fry, J.P., Zhang, Z., Wang, S., Teorey, T.J., Yang, D., Fry, J.P.: A logical design methodology for relational databases using the extended entity-relationship model. ACM Comput. Surv. **18**(2), 197–222 (1986)
17. Zabawa, P., Hnatkowska, B.: CDMM-F – domain languages framework. In: Świątek, J., Borzemski, L., Wilimowska, Z. (eds.) ISAT 2017. AISC, vol. 656, pp. 263–273. Springer, Cham (2018). https://doi.org/10.1007/978-3-319-67229-8_24

AODB and CDMM Modeling –
Comparative Case-Study

Marek Krótkiewicz[1](✉) and Piotr Zabawa[2]

[1] Wrocław University of Science and Technology,
Wyb. Wyspiańskiego 27, 50-370 Wrocław, Poland
`marek.krotkiewicz@pwr.edu.pl`
[2] Cracow University of Technology, Warszawska 24, 31-155 Kraków, Poland
`piotr.zabawa@pk.edu.pl`

Abstract. The paper is focused on comparison of two different approaches to graph modeling. A case study, which was chosen as a common background for the comparison, is an example of modeling problems characteristic for bioinformatics. First the OMG Semantics of Business Vocabulary and Rules (SBVR) standard was used to define the domain. The Association-Oriented Database (AODB) metamodel based approach was applied to show how the appropriate domain-specific model can be created in a graph modeling language dedicated to data modeling. In contrast, the Context-Driven Meta-Modeling (CDMM) approach, illustrates how to construct a domain-specific modeling language to model the case-study as a domain. The AODB is used as a General-Purpose Modeling Language (GPML) while the CDMM is applied as a Domain-Specific Modeling Language (DSML). Both approaches constitute alternatives for MOF based languages known from OMG standards.

Keywords: Metamodel · Data model
Domain-specific modeling languages · Association-Oriented Database
AODB

1 Introduction

The paper is dedicated to a comparative study of two different approaches to creating domain-specific models. Thus, the subject of this publication is located on the border of vertical markets (domain-specific solutions) and the horizontal market (modeling techniques). Both are related to software engineering problems. Moreover, creating models of software systems is a kind of modeling knowledge. However, in contrast to artificial intelligence techniques and competitive representations known as ontologies, knowledge about software systems is usually expressed in software engineering domain in the form of modeling languages. Modeling languages are defined as graphs, which are named metamodels. Metamodels are very close to data models as they are represented in the form of interrelated data classes. In order to create a model of a software system the

© Springer International Publishing AG, part of Springer Nature 2018
N. T. Nguyen et al. (Eds.): ACIIDS 2018, LNAI 10752, pp. 57–68, 2018.
https://doi.org/10.1007/978-3-319-75420-8_6

metamodel must be instantiated. There are very well known Object Management Group (OMG) standards on the market especially designed to support modeling efforts of IT enterprises. The most suitable ones are the Unified Modeling Language (UML) [10] based on Meta-Object Facility (MOF) [11] and Model-Driven Architecture (MDA) [12] built on top of them. These standardization results are impressive and very useful, but they are also known from their limits – ambiguities in the UML specification and very long lasting standardization process, which is not able to incorporate rising IT technologies.

As the result of the identified standards disadvantages some concepts of customizing modeling languages were born. Such concepts are based on the observation that programming technologies may be modeled easier if their notions are assumed to constitute a domain. That is why domain-specific modeling languages tend to be more popular lately in software engineering domain [1,3,8]. However, all of them inherit a kind of limit from OMG standards – they are defined on the basis of MOF (or its equivalent).

In the paper two alternative approaches, which are not based on MOF, are presented and compared on a particular domain-specific case study. One is Association-Oriented Database (AODB) metamodel [7], which is a general-purpose modeling language focused on creating data models. Another one is Context-Driven Meta-Modeling (CDMM) [18] - a general approach to constructing graphs of data classes at run-time. The CDMM may be used to define general purpose modeling languages (GPML) as well as domain-specific modeling languages (DSML). In this paper the second option is explored for the CDMM.

The next subsection presents the state-of-the-art in the domain of the architectures of the both GPML and DSML illustrating them from both the OMG and CDMM perspectives. This characteristics constitutes an important element of the description of the contribution of the paper.

1.1   Metamodeling and DSL

Metamodeling and creating Domain-Specific Modeling Languages are interrelated disciplines. Their architectures are shown in Fig. 1 in the MDA context.

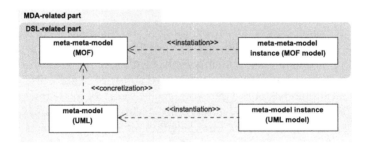

Fig. 1. MDA-related modeling.

There are two meta-layers shown in Fig. 1. The UML modeling language and its instance – a UML-compliant model are presented at the bottom of the diagram. The top-level of the diagram consists of a MOF modeling language together with its instance - the MOF-compliant model. Each UML element can be mapped to a MOF element. This mapping is denoted by «concretization» stereotype in the Fig. 1. DSLs are defined in terms of MOF while DSMLs are created as MOF instances. This fact is illustrated in Fig. 1 by dark grey rectangle labeled with "DSL-related part" placed on the light grey polygon labeled with "MDA-related part".

According to Sect. 1 there are modeling languages which are not defined in terms of MOF but there is still analogy to the architectural concept characteristic for MDA shown in Fig. 1 leading to the more general architectural concept illustrated in Fig. 2.

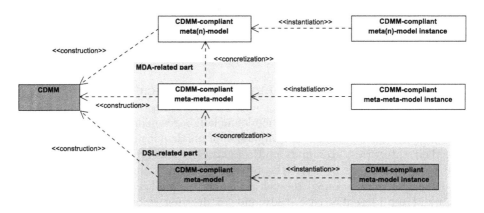

Fig. 2. CDMM-related metamodeling.

As it is shown in Fig. 2, each meta-layer can be defined in terms of CDMM in the same way any graph built of classes at run-time can be defined. This fact is represented in Fig. 2 by the dependencies stereotyped by «construction». However, this time, in the case of involving CDMM in this architecture any number of interrelated modeling languages may be defined, like for example the CDMM-compliant meta(n)-model at the top of the Fig. 2. They can be also combined with each other. As a special case the UML-like language can be defined and placed at the bottom of the diagram (as a CDMM-compliant metamodel) as well as the MOF-like language can be defined and placed one level above the UML-like language (as CDMM-compliant meta-metamodel). These languages can be also mapped – this mapping representation is stored out of them. That is why the light grey polygon labeled by "MDA-related part" was placed in the background of both modeling languages mentioned above.

The architecture introduced in this section can be implemented in the form of CDMM-F [14]. In such the case the graphs are built at run-time of precompiled elements according to the modeling language specification contained in the

CDMM-F application context file [13]. The way of defining a CDMM-compliant general-purpose modeling language was described in the paper [16] while a very simple example of a CDMM-compliant DSL was contained in the article [17].

The paper is focused on features of AODB and CDMM investigated and illustrated on a sample case-study. However, both approaches are significantly different. Their differences result from other positions they have in the general architecture shown in Figs. 1 and 2. More precisely, the AODB can be located in Fig. 1 in the place of the UML metamodel as the AODB is not defined in terms of CDMM yet. And the AODB model presented in the paper is located in the place of the UML model from the same figure. In contrast, the CDMM is already located at the left of the diagram from Fig. 2. The domain-specific model presented in this paper is in turn located in the place of the CDMM-compliant metamodel while the DSL takes the form of the CDMM-compliant metamodel instance. Thus, the CDMM is presented in the paper from the architectural perspective of the "DSL-related part" from Fig. 2.

The next section reminds fundamentals of both the AODB modeling language and the CDMM concept.

2 AODB and CDMM Languages Specification

The most important elements of the modeling languages being compared are presented in this section. The direct comparison of particular metamodels' assumptions, their categories, grammars and semantics is a very broad problem and would be characterized by strict theoretical approach. Consequently, the modeling languages are presented to such an extent that allows for correct understanding of the presented case-study and for drawing right conclusions about key features of AODB and CDMM.

2.1 AODB

AODB Metamodel [7] is a complementary and internally coherent database metamodel, which allows to model:

- on the conceptual level, which is, at the same time, the physical level of data bases based on relationships and data assuming their grammatical and semantic separation,
- in the way allowing to determine such roles in relationships that have several important mechanisms lightening logic of the system,
- taking into account inheritance both data containers and relationships.

  AODB Metamodel properties:

- atomic categories with unique responsibilities,
- high semantic power and expressiveness,

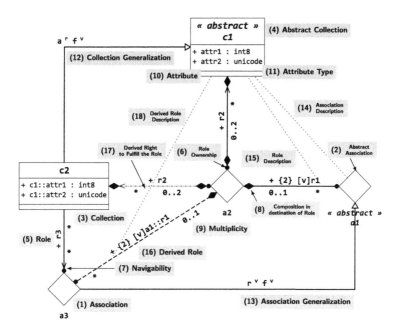

Fig. 3. Sample database schema diagram in AODB.

moreover, there are also two important models' properties:

- complete semantic unequivocality,
- direct feasibility.

Very complex data models, such as knowledge-based systems [6] are in particular interest of AODB. The metastructure of AODB Metamodel can be found in [5]. The following description addresses the most important aspects of grammar and semantics of Association-Oriented Modeling Language (AML). It is a graphical language used to design database schemata in the AODB. The graphical representation has been provided in Fig. 3.

(1) *Association* corresponds to a semantic category (*Assoc*).
(2) *Abstract Association* means that association cannot create its instances.
(3) *Collection* corresponds to a semantic category (*Coll*).
(4) *Abstract Collection* means that one collection cannot create its instances.
(5) *Role* corresponds to a semantic category (*Coll*). The graphical form in AML depends on navigability, directionality and composability.
(6) *Role Ownership*, (7) *Navigability*, (8) *Composition*, (9) *Multiplicity* is a part of the graphical form of a role (*Role*).
(10) *Attribute* is a part of the graphical form o a collection (*Coll*). Attribute (*Attr*) has a name, scope of visibility, quantity, type and default value.
(11) *Attribute Type* is a part of the graphical form of an attribute (*Attr*).
(12) *Collection Generalization* is a relationship which may link two collections.

(13) *Association Generalization* is a relationship which may link two associations. This relationship is described by an inheritance mode for roles.

(14) *Association Description* is a relationship which may link an association (*Assoc*) and a collection (*Coll*).

(15) *Role Description* is a relationship which may link a role and a collection.

(16) *Derived Role* is a relationship which may be represented in the diagram in a form similar to a role (*Role*).

(17) *Derived right to fulfill the Role* is a relationship which may be represented in the diagram analogically to the *Derived Role* case.

(18) *Derived Role Description* is a relationship which may be represented in the diagram, having an identical form as in the *Role Description*.

2.2 CDMM

The CDMM is a paradigm as it defines how the graphs of classes are built at run-time and how to place classes in the graph nodes (*Node classes*) and graph edges (*Arc classes*). That is why the first paper dedicated to CDMM introduced it as the Context-Driven Meta-Modeling Paradigm (CDMM-P) [18]. The formal definition of the CDMM-P was introduced in the paper [17] and it illustrates the key notions via the CDMM-P meta-metamodel, which is presented in Fig. 4.

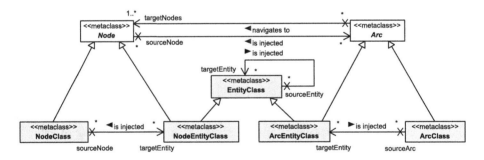

Fig. 4. Formal specification of the CDMM [17].

According to the Fig. 4 a CDMM-compliant graph (a metamodel or a domain-specific model or a data model) consists of classes located in the graph nodes (*Node classes*) and classes located in the graph edges (*Arc classes*). The main responsibility of a *Node class* is to be located in the graph node (*NodeClass classes*) while the main responsibility of *Arc classes* is to contain the representation and operations handling their targetNodes (*ArcClass classes*). *Node classes* are typically data classes (*NodeEntityClass classes*). *Arc classes* may additionally contain data attributes (*ArcEntityClass classes*). Moreover, other data classes (*EntityClass classes* or *Entity classes* as their abstraction) may be also associated to the *Node* and *Arc classes* at run-time. The *EntityClass classes* may be organized in aggregation hierarchies.

The CDMM metamodel specification may consist, in the simple case, of the metamodel elements (*Node classes* source code and *Arc classes* source code) specification and of the metamodel graph (the application context) specification. The key role of the application context for CDMM is described in the paper [13]. Both specifications may also have other forms presented in [14].

There are also other classes that can be integrated to the core concept of the CDMM-P to enrich the metamodel. These classes may constitute aggregation and/or inheritance hierarchies. The CDMM-P paradigm is not supported directly by any IT technology. That is why the paradigm has its realization in the form of a framework named CDMM-F published in [14] where technical aspects and more paradigm details are described.

The framework is extended by additional tools. One is dedicated to defining CDMM-compliant modeling languages via CDMM metamodel diagrams [15,16]. This notation is used further in the case study.

3   Case-Study

The case study is based on the Genom example originally described in the book [2] and remodelled in the paper [4]. First, the domain specification is presented in the form of OMG Semantics of Business Vocabulary and Rules (SBVR) [9] standard and then both AODB and CDMM approaches are used to model the domain according to its SBVR specification.

3.1   OMG SBVR Domain Specification

The OMG SBVR standard is used in the paper as the modeling language neutral way of defining the structure of the sample domain. It helps to avoid introducing limits resulting from the application of a modeling language for the case study specification purposes. The domain specification expressed in the form of SBVR consists of **Terms**, **Fact types** and **Rules**. The exact meaning of their contents presented below is defined in the SBVR standard.

Terms:

project
patient

sample
sequence
country

assortment
genomic_region
body_organ

Fact Types:

assortment **belongs to** project
assortment **concerns** patient
assortment **contains** sample

patient **comes from** country
sample **concerns** body_organ
sample **consists of** sequence
sample **is from** country
sequence **concerns** genomic_region

Rules:

An assortment **belongs to** exactly one project

A project may **collect** assortments

An assortment **concerns** exactly one patient

A patient may **be a source of** assortments

A patient **comes from** exactly one country

A country may **be country of origin of** patients

A sample **concerns** at most one body_organ

A sample may **consist of** sequences

A sample **is contained by** exactly one assortment

An assortment may **contain** samples

A sample **is from** exactly one country

A country may **be origin of** samples

A sequence **is an element of** exactly one sample

A sequence **concerns** at most one genomic_region

A genomic_region may **concern** sequences

3.2   AODB Model

The domain model presented in Fig. 5 illustrates a realization of the domain specified through the above SBVR expressions. The model is composed of the elements able to associate – associations (*Assoc*) and to store data – collections (*Coll*). Associations are the owners of the roles (*Role*) while the roles constitute sets of the references to the elements being associated. Collections store data in the form of the objects (*Obj*), which in turn are tuples of values. Value types were specified by the appropriate attributes (*Attr*) of the particular collections. Previous structure of this model expressed in AODB was described in [4].

Beyond the elementary categories, like: associations and roles as well as collections and attributes, the bi-compositional association-collection tandem design pattern (BACT) [5] composed of one association and one collection was used. The design pattern application examples are structures: Patient—PATIENT, Assortment—ASSORTMENT and Sequence—SEQUENCE. This pattern provides the possibility of treating such an entity as a solution, which constitutes some semantically coherent and complete unit. This unit is able to be joint to other entities and to store data.

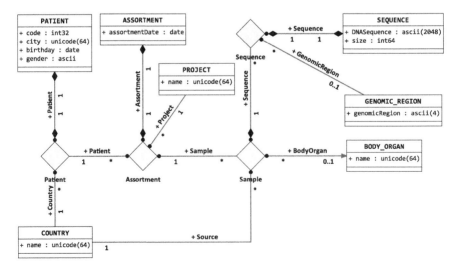

Fig. 5. Sample AODB domain-specific model.

3.3   CDMM Model

The CDMM metamodel diagram presented in Fig. 6 was created to represent the domain model specified in Sect. 3.1.

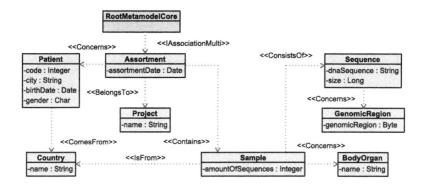

Fig. 6. Sample CDMM domain-specific model. (Color figure online)

There are *Node classes* (green rectangles) representing domain entities and interfaces of *Arc classes* (blue dotted directed lines) representing domain relationships. The convention for relationships direction was assumed in the model – the relationships are directed according to the SBVR **Fact types** reading direction. There is also the `RootMetamodelCore` model root class (red rectangle) connected to the domain classes by technical interfaces (`IAssociationMulti`).

4   Comparison of AODB and CDMM

Both analysed approaches are different, because they have different purposes. The AODB is a semantically rich modeling language with complex syntax and is dedicated to modeling data. The CDMM is an approach to constructing graphs with as simplified syntax as possible and is not dedicated to a particular field of applications.

Another, very evident difference, is connected to the modeling aspect of AODB and CDMM usage. The AODB has rich notation and diagramming capabilities. In contrast, the CDMM has a built-in notation, which is very simple and, in addition, is dedicated to defining modeling languages – not their instances that is – not models. A CDMM-compliant modeling language is expected to introduce its own, not based on MOF, user-defined notation.

Two domain specific models were created, one in AODB and one in CDMM. The SBVR domain specification presented in Sect. 3.1 was introduced to become common source of information for both models. Thus, the mapping from SBVR to AODB and from SBVR to CDMM was performed. The mapping results are presented in Figs. 5 and 6 as the models while the mapping is contained in the Table 1.

One characteristic feature of metamodels is that they are, in most cases, directed graphs. In contrast, models tend to have bi-directed graphs nature. This

Table 1. Mapping from SBVR specification to AODB and CDMM model diagrams

AODB element	SBVR element	CDMM element
Being represented by: association, collection or BACT [5] design pattern	**Term**	*Node class*
Role	**Fact type**	CDMM metamodel diagram edge
Role constraints	**Rule**	*Arc class*
		CDMM metamodel diagram edge stereotype

phenomenon is clear when looking at Fig. 5 (an AODB metamodel instance) and Fig. 6 (a CDMM metamodel like domain model). This difference can be easily observed when comparing the AODB model from Fig. 5 (a bi-directed graph) to the AODB metamodel (a uni-directed graph). The SBVR domain specification was introduced intentionally in the way giving enough freedom when defining the relationships direction – just to underline this difference. The root cause of this phenomenon is that metamodels are constructed in the manner making their traversal easy, effective and almost unambiguous. However, these criterions usually do not make sense for domain models.

A new, not published, feature of CDMM is that it is possible to introduce inheritance hierarchies of interfaces between the domain-specific interface the name of which is presented as a stereotype on the CDMM metamodel diagram and the interfaces resulting from avoiding code multiplications in the *Arc classes* implementation. These hierarchies are depicted in Fig. 7 and they constitute a mapping from domain to technical aspects of CDMM.

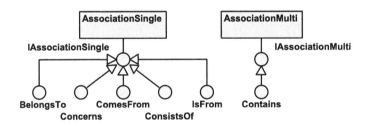

Fig. 7. Intermediate inheritance hierarchies.

There were assumed two different design criterions for AODB and CDMM:

– AODB: atomic and unique responsibility of each metamodel element,
– CDMM: well-defined responsibility of each metamodel element and maximal openness of each metamodel element for associating data both at compile-time (*NodeEntityClass classes* and *ArcEntityClass classes*) and at run-time (*EntityClass classes*).

This difference is clear when analysing possible transformations from SBVR specification of the sample domain to both AODB and CDMM domain models.

5   Conclusions

The comparative case-study presented in the paper is the first step of research leading to more detailed analysis of AODB and CDMM. Only two approaches were chosen in the paper in order to identify the nature of differences, to recognize both their advantages and limits and finally to achieve synergy as the result of the analysis and forthcoming research. It helped to identify several important differences between modeling languages being compared. Mapping between metamodel elements available in each of them has been initiated and the first result was presented in the Table 1. Application of modeling language agnostic SBVR was confirmed to be a good approach for the comparative studies. That is why the SBVR notions were included in Table 1. One attractive research goal is to define formal transformations from SBVR to AODB and CDMM.

It is worth noticing, that in the case of the AODB the association-oriented modeling covers modeling of a problem domain while the CDMM is dedicated to defining modeling languages. In the consequence, their direct comparison may lead to some misunderstandings. Nevertheless, this kind of the analysis is possible. AODB and CDMM are not competitive but complementary solutions. They are addressed to different abstraction levels and thus they complement each other. Particularly, AODB has a data storage layer both in the form of in-memory storage and in the form of persistent storage. It perfectly supplements the CDMM in this area. On the other hand the AODB is a challenging modeling language to be implemented in the CDMM.

The research also confirmed that CDMM may be used for modeling domains. One result of the paper is the concept of separating domain-specific interfaces from technical interfaces. This is a general technique useful when applying general-purpose modeling languages to modeling domains. The initial results are prospective enough to continue research in this direction.

References

1. Bettini, L.: Implementing Domain-Specific Languages with Xtext and Xtend. Packt Publishing, Birmingham (2013)
2. Ferreira, J.E., Takai, O.K.: Understanding database design. In: Gruber, A., Durham, A.M., Huynh, C., del Portillo, H.A. (eds.) Bioinformatics in Tropical Disease Research, chap. A02, pp. 69–94. National Center for Biotechnology Information (US), Bethesda (2008)
3. Fowler, M.: Domain-Specific Languages. The Addison-Wesley Signature Series. Addison-Wesley, Boston (2011)
4. Jodlowiec, M., Krótkiewicz, M.: Semantics discovering in relational databases by pattern-based mapping to association-oriented metamodel - a biomedical case study. In: Proceedings of the Information Technologies in Medicine - 5th International Conference, ITIB 2016, Kamień Śląski, Poland, 20–22 June 2016, vol. 1, pp. 475–487 (2016)

5. Jodłowiec, M., Krótkiewicz, M.: Towards the mapping of UML class diagrams to association-oriented database metamodel schemata. In: Proceedings of 8th International Conference on Information, Intelligence and Applications (IISA2017). IEEE (2017, accepted for publication)
6. Krótkiewicz, M., Wojtkiewicz, K.: Functional and structural integration without competence overstepping in structured semantic knowledge base system. J. Logic Lang. Inf. **23**(3), 331–345 (2014)
7. Krótkiewicz, M.: Association-oriented database model - n-ary associations. Int. J. Softw. Eng. Knowl. Eng. **27**(02), 281–320 (2017)
8. Microsoft: Microsoft, Getting Started with Domain-Specific Languages (2015). https://msdn.microsoft.com/en-us/library/ee943825.aspx
9. OMG: Object Management Group, Semantics of Business Vocabulary And Rules 1.4 (2014). http://www.omg.org/spec/SBVR/1.4/
10. OMG: Object Management Group, Unified Modeling Language (UML) superstructure version 2.5 (2015). http://www.omg.org/spec/UML/2.5/
11. OMG: Meta Object Facility (MOF) core specification version 2.0 (2016). http://www.omg.org/spec/MOF/2.5.1/
12. OMG: Object Management Group, Model-Driven Architecture (2017). http://www.omg.org/mda/
13. Zabawa, P.: Context-Driven Meta-Modeling Framework (CDMM-F) - context role. Tech. Trans. **112**(1-NP(19)), 105–114 (2015)
14. Zabawa, P.: Context-Driven Meta-Modeling Framework (CDMM-F) - Internal Structure (2018, accepted for publication)
15. Zabawa, P., Fitrzyk, G.: Eclipse modeling plugin for Context-Driven Meta-Modeling (CDMM)-meta-modeler. Tech. Trans. **112**(1-NP(19)), 115–125 (2015)
16. Zabawa, P., Fitrzyk, G., Nowak, K.: Context-Driven Meta-Modeler (CDMM)-meta-modeler application case-study. Inf. Syst. Manag. **5**(1), 144–158 (2016)
17. Zabawa, P., Hnatkowska, B.: CDMM-F – domain languages framework. In: Świątek, J., Borzemski, L., Wilimowska, Z. (eds.) ISAT 2017. AISC, vol. 656, pp. 263–273. Springer, Cham (2018). https://doi.org/10.1007/978-3-319-67229-8_24
18. Zabawa, P., Stanuszek, M.: Characteristics of the Context-Driven Meta-Modeling Paradigm (CDMM-P). Tech. Trans. **111**(3-NP), 123–134 (2014)

Modelling, Storing, and Querying of Graph Data

Experimental Clarification of Some Issues in Subgraph Isomorphism Algorithms

Xuguang Ren[(✉)], Junhu Wang, Nigel Franciscus, and Bela Stantic

Institute for Integrated and Intelligent Systems, Brisbane, QLD, Australia
{x.ren,j.wang,n.franciscus,b.stantic}@griffith.edu.au

Abstract. Graph data is ubiquitous in many domains such as social network, bioinformatics, biochemical and image analysis. Finding subgraph isomorphism is a fundamental task in most graph databases and applications. Despite its NP-completeness, many algorithms have been proposed to tackle this problem in practical scenarios. Recently proposed algorithms consistently claimed themselves faster than previous ones, while the fairness of their evaluation is questionable due to query-set selections and algorithm implementations. Although there are some existing works comparing the performance of state-of-the-art subgraph isomorphism algorithms under the same query-sets and implementation settings, we observed there are still some important issues left unclear. For example, it remains unclear how those algorithms behave when dealing with unlabelled graphs. It is debatable that the number of embeddings of a larger query is smaller than that of a smaller query, which further challenges the remark that the time cost should decrease for a good algorithm when increasing the size of the queries. In this paper, we conducted a comprehensive evaluation of three of most recent subgraph algorithms. Through the analysis of the experiment results, we clarify those issues.

Keywords: Subgraph isomorphism · Pattern matching

1 Introduction

The importance of graph data has long been recognized by the industry as well as the research community. Pattern matching is a fundamental processing requirement for graph data applications. Subgraph isomorphism is the most basic problem of graph pattern matching. That is, in a given data graph, retrieve all subgraphs which are isomorphic to the query graph. Subgraph isomorphism is a well known NP-complete problem whose definition can be found in any textbook of graph theory. Despite NP-completeness, many algorithms have been proposed in order to solve it in practical scenarios, such as Ullmann [12], VF2 [3], QuickSI [10], GraphQL [7], SPath [14], STW [11], TurboIso [5], BoostIso [9], VF3 [1]. Most subgraph isomorphism algorithms are based on a backtracking method which

© Springer International Publishing AG, part of Springer Nature 2018
N. T. Nguyen et al. (Eds.): ACIIDS 2018, LNAI 10752, pp. 71–80, 2018.
https://doi.org/10.1007/978-3-319-75420-8_7

computes the solutions by incrementally enumerating and verifying candidates for all vertices in a query graph [8].

Recent proposed algorithms consistently claimed themselves faster than previous ones, while the fairness of their evaluation is questionable due to query-set selections and algorithm implementations. Although there are some existing works [8] comparing the performance of state-of-the-art subgraph isomorphism algorithms under the same query-sets and implementation settings [8], we observed the following issues are left unclear.

(1) Most the performance evaluations only report the average time cost for a set of queries. While we observed that it is common that a small number of queries in a query set take much longer time than the total of the rest. The reason for this phenomenon resides in the NP-complete property of subgraph isomorphism, which may lead the time cost of two different queries to be exponentially different. Therefore it is possible that one algorithm performs better than another algorithm for most queries while performs worse for some extreme queries, and these extreme queries dominate the total processing time. In this case, we cannot simply claim one algorithm is better than another one by only considering the average performance.

(2) There are no evaluation of these algorithms conducted on unlabelled-vertex graphs. Although labelled graphs are quite common in the real world, the unlabelled graphs or graphs with very few labels are also very important to consider. For example, one may want to do pattern matching on a social network where every node is only labelled as *person*. It is intuitive that we can process unlabelled graphs by treating them as graphs with one label and using existing subgraph algorithms, while the filtering techniques based on labels of some algorithms will be unhelpful under this setting. It is unclear how those algorithms perform when not considering vertex labels.

(3) It is assumed that in [8] the number of embeddings of a smaller query is normally more than that of a larger query and thus the performance of a *good* subgraph isomorphism algorithm would decrease as we increase the query size. However consider a path query q_1 with 4 vertices and a path query q_2 with 5 vertices, if the data graph is a clique with 6 vertices, the embeddings of q_2 is much more than that of q_1 due to an extra Cartesian enumeration. The time cost of q_2 is also more than q_1. Therefore, the number of embeddings and the time cost highly depend on the structures of the query graph and data graph. It deserves experimental study to clarify this issue.

With the above issues in mind, in this paper we conducted a comprehensive evaluation for three of most recent subgraph algorithms. Through the analysis of the experiment results, we clarify those issues. To be specific, we make the following contributions:

(1) We implemented three recent state-of-the-art subgraph isomorphism algorithms under the unlabelled-vertex settings.
(2) We evaluated the performance of these algorithms with three real datasets without considering no vertex label. For each dataset, we tested three types

of queries, clique, path and subgraph. Each type contains ten queries where each query consists of 10 vertices while with increasingly larger number of edges for different queries.

(3) We clarified the aforementioned issues by analysing our experiment results. We dedicatedly present the results by reporting the results for each particular query in which way we could clearly observe the performance of each algorithm and avoid the misleading from some extreme queries.

Paper Organization. Some preliminary is given in Sect. 2. We present the related work in Sect. 3. In Sect. 4, we report our experiment results and analysis. In Sect. 5, we conclude the paper.

2   Preliminaries

In this section, we first review some fundamental concepts and assumptions used in this paper. Then we briefly present the backtracking framework that are widely followed by the subgraph isomorphism algorithms.

Data Graph and Query Graph. A *data graph* is an undirected, unlabeled graph denoted as $G = (V, E)$, where (1) V is the set of vertices; (2) E is a set of undirected edges.

A *query graph* is an undirected, unlabeled graph denoted as $G_q = (V_q, E_q)$, where V_q, E_q have the same meaning as V, E of data graph G. In most cases, the query graph is much smaller than the data graph. We assume the query graph and data graph are both connected, and will use *data vertices* (resp. *query vertices*) to refer to the vertices in the data graph (resp. query graph).

Subgraph Isomorphism. Given a query graph $G_q = (V_q, E_q)$ and a data graph $G = (V, E)$, a *subgraph isomorphism* is an injective function $f: V_q \to V$ such that for each edge $(u_i, u_j) \in E_q$, there exists an edge $(f(u_i), f(u_j)) \in E$. f is also called an *embedding*. Note that f can be represented as a set of vertex pairs (u, v) in which $u \in V_q$ is mapped to $v \in V$ (We also say v *is matched to* u).

The Generic Framework. Most subgraph isomorphism algorithms are based on a backtracking strategy which incrementally finds partial solutions by adding join-able candidate vertices. A recent survey [8] presents a generic framework for subgraph isomorphism search. We subtly modified the framework in order to reflect the techniques utilized in most recent subgraph algorithms. The framework is given as follows:

In Algorithm 1, the inputs are a query graph and a data graph, the outputs are all the embeddings. Each embedding is represented by a list f which comprises pairs of a query vertex and a corresponding data vertex (Line 2). Most recent subgraph isomorphism algorithms delicately select the starting query vertex (Line 2) instead of using a random one in contract to Ullmann [12]. The framework initializes the candidate list for the starting query vertex in Line 3. In Line 4–6, the framework first matches the starting query vertex to a candidate in f and then launches a recursive *subgraphSearch* based on the updated f.

Algorithm 1. GENERICFRAMEWORK

Input: Data graph G and query graph G_q
Output: All embeddings of G_q in G

1 $f \leftarrow \emptyset$
2 $u \leftarrow startingQueryVertex$
3 $C(u) \leftarrow candidates(G_q, G)$
4 **for** *each* $v \in C(u)$ **do**
5 $\quad$ map u to v in f
6 $\quad$ subgraphSearch(G_q, G, f)
7 subgraphSearch(G_q, G, f)

$\quad$ **Subroutine** *subgraphSearch(G_q, G, f)*
1 $\quad$ **if** $|f| = |V_q|$ **then**
2 $\quad\quad$ **report** f
3 $\quad$ **else**
4 $\quad\quad$ $u \leftarrow nextQueryVertex()$
5 $\quad\quad$ $C(u) \leftarrow candidates(G_q, G)$
6 $\quad\quad$ **for** *each* $v \in C(u)$ *and* v *is not matched* **do**
7 $\quad\quad\quad$ **if** *isJoinable(f, v, G, G_q)* **then**
8 $\quad\quad\quad\quad$ *updateState(f, u, v, G, G_q)*
9 $\quad\quad\quad\quad$ *subgraphSearch(G_q, G, f)*
10 $\quad\quad\quad\quad$ *restoreState(f, u, v, G, G_q)*

In the subroutine *subgraphSearch*, if the all the query vertices are mapped, the framework reports a found embedding. Otherwise, it first delicately selects the next query vertex u to match (Line 4) and get a list of candidate for u. For each candidate v of u, if v could pass the filtering and pruning technique in *isJoinable*, it will be added to f. *updateState* adds the newly matched pair (u, v) into f while *restoreState* restores the partial embedding state by removing (u, v) from f.

3   Related Work

In this section, we present some related work for subgraph isomorphism. Subgraph isomorphism has been investigated for many years. Existing algorithms can be divided into two classes:

(1) Given a graph database consisting of many small data graphs, retrieve all the data graphs containing a given query graph. This category includes algorithms such as GraphGrep [4], gIndex [13], FG-Index [2], C-Tree [6], Tree+△ [15] and SwiftIndex [10]. In order to speed up the search, most algorithms use a filter-and-refinement strategy based on effective indexing techniques.

(2) Given a query graph, find all embeddings in a single large graph. Existing algorithms falling into this class include Ullmann [12], VF2 [3], QuickSI [10], GraphQL [7], SPath [14], STW [11], TurboIso [5] and VF3 [1,9]. Most of

them follow a backtracking framework. The techniques used to accelerate the matching process are *matching order optimization, efficient pruning rules, pattern-at-a-time strategies* and *data compression*, as briefly surveyed below.

Matching Order Optimization. The Ullmann algorithm [12] does not define the matching order of the query vertices. VF2 [3] starts with a random vertex and selects the next vertex which is connected with the already matched query vertices. VF3 [1] is an upgraded version of VF2, which optimized both the first query selection and the candidate enumeration for each query vertex. By utilizing global statistics of vertex label frequencies, QuickSI [10] proposes a matching order which accesses query vertices having infrequent vertex labels as early as possible. In contrast to QuickSI's global matching order selection, TurboIso [5] divides the candidates into separate candidate regions and computes the matching order locally and separately for each candidate region. Both STW [11] and TurboIso [5] give higher priority to query vertices with higher degree and infrequent labels.

Efficient Pruning Rules. The Ullmann algorithm [12] only prunes out the candidate vertices having a smaller degree than the query vertex. While VF2 [3] proposes a set of feasibility rules to prune out unpromising candidates, namely, 1-look-ahead and 2-look-ahead rules. SPath [14] uses a *neighbourhood signature* to index the neighbourhood information of each data vertex, and then prunes out false candidates whose candidate signature does not contain that of the corresponding query vertex. GraphQL [7] uses a pseudo subgraph isomorphism test.

Pattern-At-A-Time Strategies. Instead of the traditional vertex-at-a-time fashion, SPath [14] proposes an approach which matches a graph pattern at a time. The graph pattern used in SPath is path. TurboIso [5] rewrites the query graph into a *NEC* tree, which matches the query vertices having the same neighbourhood structure at the same time.

Data Compression. A recent work BoostIso [9] has been proposed aiming at further speeding up the subgraph isomorphism search by utilizing data vertex relationship. Vertex equivalent relationship is used to compress the data graph and vertex containment relationship is used to optimize the loading order of the candidates for each query.

4   Experiments

In this section, we report our experiment results with analysis and clarify the aforementioned issues.

4.1   Experiment Settings

Subgraph Isomorphism Algorithms. We conducted extensive experiments to evaluate the performance of three most recent subgraph isomorphism algorithms: VF3, TurboIso, TurboIsoBoosted. TurboIso is claimed as the fastest algorithm in [5]. VF3 is an upgraded version of VF2 which is proposed later than TurboIso. While there is no comparison between VF3 and TurboIso conducted. TurboIsoBoosted is the revised version of TurboIso by [9].

Environment. All of the algorithms were implemented in C++ with VC++ 2015 as our compiler. All the experiments were carried out under 64-bit Windows 7 on a machine with an Intel 3 GHz CPU and 8 GB memory.

Dataset. We used three real datasets in our experiments: Human, Yeast and Hprd. All the three graphs are protein interaction networks. Hprd contains 37081 edges, 9460 vertices with an average degree 7.8. Yeast contains 12519 edges, 3112 vertices with an average degree 8.1. Human is a dense graph of human protein interactions, which contains 86282 edges, 4674 vertices with an average degree 36.9. We removed the label information for each data vertex from the graphs. We generated the adapted graphs following the procedure of [9]. Our data graphs are saved using the adjacency list format in plain texts, where each list is ordered following the increasing order of vertex ids. The data graph is loaded into the memory before the doing the subgraph isomorphism search.

Query Set. We generated three type of queries: cliques, paths and subgraphs as similar to [8]. In contrast, we separate the queries by various sizes. For each of clique and path query set, we have 7 queries with vertex number from 3 to 10 respectively. For the sugraph query set, we generated 10 vertices with an increasing number of edges for each query. To be specific, we first add edges to each subgraph to make it a path and then add an increasing number of random edges to them, leading to the degree of the subgraph queries to follow an increasing trend.

4.2   Results and Analysis

In contrast to previous evaluation, we report the time cost of subgraph isomorphism search for every single query. The purpose is to avoid the effects of some extreme queries that may dominate the average time cost. Since we removed the label information from the data graph, the time cost to process the subgraph isomorphism search is much longer than that of labelled data graph.

Human. We present the result for the Human dataset in Figs. 1, 2 and 3. For the clique queries, as we can see, the time cost experienced a fluctuating trend for VF3. While it showed an overall decreasing trend for TurboIso and TurboIsoBoosted. Intuitively, the degree of each query vertex is larger for a larger clique. This will dramatically reduce the number of candidates that can be matched to the query vertices of a large clique. Therefore it is easy to explain the decreasing trend of TurboIso and TurboIsoBoosted. As for VF3, in lack of the optimized order and filtering techniques proposed by TurboIso, the cost due to many deeper recursions battled the savings from the degree filtering. Thus it is not hard to understand the fluctuating trend of VF3 over the clique queries.

In terms of the path queries, all three algorithms experienced a significant increase with increasing the number of vertices in the path queries. This is easy to understand. The degrees of query vertices of large query and smaller query are almost the same. Thus the degree filter contributes the same here for both large query and smaller query. When processing the path queries, we could just follow the vertex and its adj-list without the need to verify any edges. As more vertices means more recursions which further leads to a longer processing time.

Regarding the subgraph queries, as expected, both the time cost of TurboIso and TurboIsoBoosted dropped when we increase the subgraph degrees. However, it has not shown a big difference for VF3, except for a small increase when the subgraph almost reaches a clique. The reason for this phenomenon is similar to the reason explained for clique queries. While here for VF3, the increased cost because of deep recursions totally overpowered the reduced savings from filtering due to a larger degree of query vertices.

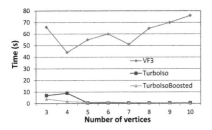

Fig. 1. Human clique

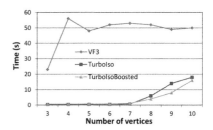

Fig. 2. Human path

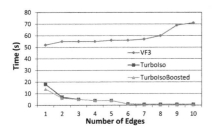

Fig. 3. Human subgraph

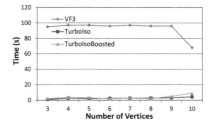

Fig. 4. Yeast clique

Yeast. We present the result for Yeast dataset in Figs. 4, 5 and 6. We could easily notice that VF3 behaves significantly different for different queries. For clique queries, VF3 dramatically dropped when the number of vertices is 10. While for the subgraph queries, the time cost of VF3 first dropped a lot and then significantly increased and after that it dropped again. In terms of TurboIso and TurboIsoBoosted, the trends are much clearer than VF3. Their time cost increased for clique queries and path queries when we increased the number of vertices while it dropped for the subgraph queries when we increased the subgraph degree. Therefore, for yeast, TurboIso and TurboIsoBoosted are much more stable than VF3.

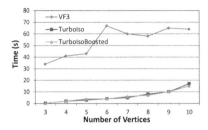
Fig. 5. Yeast path

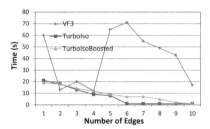
Fig. 6. Yeast subgraph

Hprd. We present the result for Hprd dataset in Figs. 7, 8 and 9. Similar to that of Yeast, VF3 behaves dramatically different for each individual query and the trend consists of multiple increases and decreases when we modify the number of vertices and the degree of subgraph queries. For TurboIso and TurboIso-Boosted, the time cost has shown a increasing trend for path queries while shown a decreasing trend for subgraph queries with the increasing degree.

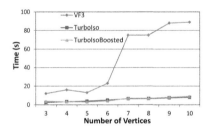
Fig. 7. Hprd clique

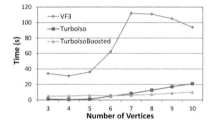
Fig. 8. Hprd path

Remarks. Based on the above results, we may draw the following conclusions:

(1) Different subgraph isomorphism queries may take significantly different time due to its property of NP-completeness. Reporting an average time elapse for a set of queries sometimes only reflect the worst performance of the algorithms on some particular queries if those queries dominate the time cost.

(2) The query structure and the degree affects the query processing time a lot. While it is not the only factor. The structure of the data graph is also an important affecting factor. Thus we may not claim that the time cost of a good algorithm should decrease when we increase the complexity of the query structure or the degree of the query graphs.

(3) Although sacrificed some memory space, the dynamic indexing structure proposed by TurboIso improved the subgraph isomorphism search even when we targeting at a specific query scale. The improvement of TurboIsoBoosted over TurboIso is heavily affected by the structure of the data graph.

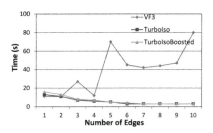

Fig. 9. Hprd subgraph

5 Conclusion

In this paper, we evaluated the performance of three most recent state-of-the-art subgraph isomorphism algorithms. To deeply understand their behaviour when we modify the structure of the queries, we used unlabelled data and query graphs. We tested three types of queries: clique, path and subgraphs. Based on our experiment results, we concluded that the query structure significantly affects the query processing time of subgraph isomorphism. And it is hard to pretend the whether time cost will decrease or increase when we modify the query structure. We believe the clarification of issues from our research will benefit the future evaluation process of subgraph isomorphism algorithms.

References

1. Carletti, V., Foggia, P., Saggese, A., Vento, M.: Introducing VF3: a new algorithm for subgraph isomorphism. In: Foggia, P., Liu, C.-L., Vento, M. (eds.) GbRPR 2017. LNCS, vol. 10310, pp. 128–139. Springer, Cham (2017). https://doi.org/10.1007/978-3-319-58961-9_12
2. Cheng, J., Ke, Y., Ng, W., Lu, A.: Fg-index: towards verification-free query processing on graph databases, pp. 857–872 (2007)
3. Cordella, L.P., Foggia, P., Sansone, C., Vento, M.: A (sub) graph isomorphism algorithm for matching large graphs. IEEE Trans. Pattern Anal. Mach. Intell. **26**(10), 1367–1372 (2004)
4. Giugno, R., Shasha, D.: GraphGrep: a fast and universal method for querying graphs, vol. 2, pp. 112–115 (2002)
5. Han, W.S., Lee, J., Lee, J.H.: TurboISO: towards ultrafast and robust subgraph isomorphism search in large graph databases. In: SIGMOD, pp. 337–348 (2013)
6. He, H., Singh, A.K.: Closure-tree: an index structure for graph queries, p. 38 (2006)

7. He, H., Singh, A.K.: Query language and access methods for graph databases. In: Aggarwal, C., Wang, H. (eds.) Managing and Mining Graph Data. ADBS, vol. 40, pp. 125–160. Springer, Boston (2010). https://doi.org/10.1007/978-1-4419-6045-0_4
8. Lee, J., Han, W.S., Kasperovics, R., Lee, J.H.: An in-depth comparison of subgraph isomorphism algorithms in graph databases. VLDB **6**, 133–144 (2012)
9. Ren, X., Wang, J.: Exploiting vertex relationships in speeding up subgraph isomorphism over large graphs. Proc. VLDB Endow. **8**(5), 617–628 (2015)
10. Shang, H., Zhang, Y., Lin, X., Yu, J.X.: Taming verification hardness: an efficient algorithm for testing subgraph isomorphism. PVLDB **1**(1), 364–375 (2008)
11. Sun, Z., Wang, H., Wang, H., Shao, B., Li, J.: Efficient subgraph matching on billion node graphs. PVLDB **5**, 788–799 (2012)
12. Ullmann, J.R.: An algorithm for subgraph isomorphism. JACM **23**(1), 31–42 (1976)
13. Yan, X., Yu, P.S., Han, J.: Graph indexing: a frequent structure-based approach, pp. 335–346 (2004)
14. Zhao, P., Han, J.: On graph query optimization in large networks. PVLDB **3**, 340–351 (2010)
15. Zhao, P., Yu, J.X., Yu, P.S.: Graph indexing: tree + delta >= graph, pp. 938–949 (2007)

Beyond Word-Cloud: A Graph Model Derived from Beliefs

Nigel Franciscus$^{(\boxtimes)}$, Xuguang Ren, and Bela Stantic

Institute for Integrated and Intelligent Systems, Nathan, QLD, Australia
{n.franciscus,x.ren,b.stantic}@griffith.edu.au

Abstract. Huge volume of text generated today poses big challenges to text mining applications. A tremendous amount of fragmented short texts are growing at a scale that makes it impossible for human to visually extract. Techniques proposed in text mining such as topic modeling dramatically improve the understanding of those unstructured and noisy texts. Naive while widely used models are word-cloud and word-bags. While we observed that these data models never considered the semantic relationship between the words, which make the results relatively hard to understand. One bright option is to organize those words semantically and generate an output of human understandable sentences. In this paper, we step our first foot into this direction by proposing a new data model: Belief Graph. We also proposed a schema to build belief graph model from short texts.

Keywords: Text modeling · Text representation · Graph of words

1 Introduction

In the era of big data, a tremendous amount of fragmented short texts is growing at a scale that makes them impossible for the human to visually extract. Many data mining applications have been developed to help understand and to summarize these collection of texts, for example, topic modelling [2] and text summarization [11]. At the same time, the huge volume of text generated today still poses big challenges to the text mining applications.

A naive while widely used model is word-cloud or unigram which naturally presents hot topics, where the topic is represented by frequent keywords. By only providing some keywords, unigram is far too limited. Bi-gram extends the unigram by capturing the relationship of two words based on the co-occurrence. Later, n-gram complements the words identification into a phrase with part-of-speech tagging. However, the n-gram is often still in bag-of-words form, which does not consider the word order in the sentence. One option is to organize those keywords according to the grammatical order and generate an output of human understandable sentences. This is a challenging task since it combines part-of-speech processing and data mining for summarization. In this paper, we step

© Springer International Publishing AG, part of Springer Nature 2018
N. T. Nguyen et al. (Eds.): ACIIDS 2018, LNAI 10752, pp. 81–90, 2018.
https://doi.org/10.1007/978-3-319-75420-8_8

our foot into this direction by proposing a new data model: *Belief Graph*. The formal definition of belief graph is presented in Sect. 3.

In order to build the belief graph, we are shifting into the text mining through graph linguistic approach, which has been the focus of multiple lines of research (see *e.g.* [11,13,14,16,17]). At the same time, there has been growing interest in adopting dependency parses tree [4] for a range of Natural Language Processing tasks, from machine translation [5,9] to question answering [1], here we refer dependency parses tree as the dependency tree. Dependency tree parses the grammatical relation to capture dependencies between words such as predicate-argument. The dependency tree is available as a popular Stanford NLP toolkit implemented in java [10].

In this paper, we present a scheme to build the belief graph by utilizing this graph dependency architecture. By using the dependency tree, we were able to store the intermediate result of text processing in a graph database which further enables extensive graph queries such as clustering and summarization. The basic idea of our architecture is to simplify the process by utilizing state of the art language processing to enable graph computation over a large amount of text.

Contribution. In this paper, we present a belief graph model based on which we are able to translate the short and sparse text into a dependency tree and store it in a graph property schema. To be specific,

(1) We proposed the reversed dependency to convert raw short text into a graph model by preserving grammatical properties and then reuse these properties to generate the synopsis of text.
(2) We designed a practical schema and storage structure to build belief graph by using NoSQL databases, MongoDB and Neo4j.
(3) We conducted extensive experiments to demonstrate the system performance and practical usage of text mining using graph queries.

Organization. The rest of the paper is organised as follows: in Sect. 2, we describe the text graph model; in Sect. 3, we present the details of building belief graph; in Sect. 4, we provide the experiment results; in Sect. 5 we give some related works and finally in Sect. 6 we conclude the paper and indicate the future work.

2   Related Work

We present some related work which is in-line with the purpose of contribution.

(1) *Natural Language Dependency Tree.* The rising of NLP processing with the dependency tree has captured the interest of research communities with the availability of NLP toolkit [10]. Two recent prominent works utilizing the natural language to interpret [9] and answering [5] query from relational

database. Another work also uses dependency tree for crowdsourcing platform [1]. Our work is similar to these concepts, however, we are targeting directly into text mining without involving SQL queries and focusing on graph approach.

(2) *Topic Modeling.* Latent Dirichlet Allocation (LDA) [2] is a well-known topic modeling technique which can be used to discover the latent information from the document. Several variances of LDA have been proposed to alleviate short and sparse text processing. Our work differs in the way we preserve the word order in contrast to using bag-of-words assumption which does not consider word order. Our work also uses graph instead of sampling to obtain keywords.

(3) *Graph-Based Model.* Graph model offers a simple, effective and interactive representation based on the relation between each vertex. Each vertex can be treated as an entity of sentences, words or characters with the edges as the label or property of the relationship between two vertices. The fundamental of the graph is based on the idea of KeyGraph [12] which converts text into a term graph based on co-occurrence relations between terms. The main strength of graph representation lies in the connection between the entity, enable user to query the relationship between keywords. Further, graph-based models explicitly consider word co-occurrence as one of the main parameters to determine the score of a keyword [13,16]. Contrast to previous work, we focus specifically on the dependency tree to capture the grammatical relationship between keywords.

3   Knowledge Graph vs Belief Graph

In this section, we present the definitions and characteristics of two classes of graph model. Before we describe the belief graph, we outline its predecessor knowledge graph and contrast the differences between two of them.

Knowledge graph has been the focus of research since the introduction of Google Knowledge Graph in 2012[1]. In practice knowledge graph has been known as a complementary for many applications. For example, assisting question answering [3], providing auxiliary information for information retrieval [8] and inferring new knowledge from its own collection [15]. However, it does not have a valid definition and therefore prone to multi-interpretations. Here, we compose the definition and characteristics of knowledge graph.

Definition 1. *A knowledge graph G is a directed labeled graph (V, E, L), where V is a set of nodes, and $E \subseteq V \times V$ is a set of edges. Each node $v \in V$ represents an entity with label $L(v)$ and each edge $e \in E$ represents a relationship $L(e)$ between two entities.*

---

[1] https://googleblog.blogspot.co.at/2012/05/introducing-knowledge-graph-things-no t.html.

Characteristics:

– *Atomity.* Each statement has a single interpretation which refers to a subject-object relationship over a specific domain. For example, Shakespeare died on 23 April 1616 in Warwickshire, England. It has exact and precise meaning.
– *Real Entity.* The content is generated from a human-curated knowledge base which contains a well-known real-world fact. It covers the broad domain of human-life aspect. For example, Wikipedia and Freebase.
– *Rich Label.* Since knowledge graph carries a lot of information, nodes and edges content may have name, type, and attribute represented as the label.
$$Shakespeare_{(person)} \xrightarrow[\text{23 April 1616}]{\text{died}} Warwickshire_{(county)}, England_{(country)}$$

On the other hand, although belief graph follows a similar schema with knowledge graph, it has significant differences in its content and source. It does not represent the grounded facts such as real-world entities and it is more focusing on the word-to-word relationship. Belief graph can be used to summarize dirty and unpredicted text which further can be extended into topic modeling, text summarization, and text visualization. We present the definition and characteristics of belief graph as follow:

Definition 2. *A belief graph G is a directed labeled graph (V, E, L, P), where V is a set of nodes, and $E \subseteq V \times V$ is a set of edges. Each node $v \in V$ represents an entity with label $L(v)$ and has properties $P(v) = id, t$ where id is the identifier and t is the grammatical property type. Each edge $e \in E$ represents a relationship label $L(e)$ between two entities and has a property $P(e) = f$ where f is the co-occurrence frequency. A belief graph is formed from multiple words-to-words relations derived from sentences.*

Characteristics:

– *Flexible.* It can be constructed from any sentence regardless the length or size of the document. It also takes any linguistic form such as active and passive voice by utilizing the language parser.
– *Dirty Text*, generated from a collection of unclassified text like opinion-based where the ground-truth identity is unreliable. Thus, it is not known whether it is a fact or fake. For example, social media and product review data.
– *Word Order.* Since belief graph is using dependency parser, it automatically preserves the word order by using grammatical information. Thus it is relatively easier to identify the subject-object relationship. For example, $LeoDicaprio_{(nsubj)} \rightarrow Environmentalist_{(noun)}$.

4   Building Belief Graph

In this section, we give an overview of our belief graph system as shown in Fig. 1. It can be divided into several inter-related components: (i) *Initial Preprocessing,*

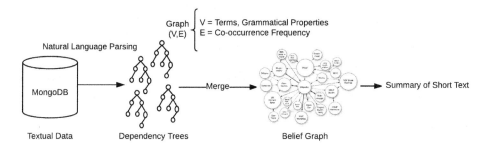

Fig. 1. System overview of the graph dependency analytic engine

where we collect the raw textual data and store them into MongoDB. (ii) *Dependency Tree (DT)*, where we map a collection of text to the dependency parser and translate them into a grammatical structured tree. (iii) *DT Merging*, where we merge each DT into a belief graph. As a case study, we demonstrate our belief graph system by using Twitter and Amazon product review data sources, and NoSQL database platforms.

4.1   Initial Preprocessing

An efficient storage architecture is essential to manage raw data and handling preprocessing. Our initial preprocessing is primarily built based on the precomputing architecture [6], where we classify unstructured data into manageable periodic content which can be drilled down or rolled up to support time-slicing query. However, in this case, we use the seed terms as the parameter to classify the collection. Hence we index the collection level according to filtered key terms. Unlike normal text preprocessing, we do not necessarily have to tokenize or stem the words. Firstly, due to the process of Part-Of-Speech (POS) tagging, the system needs to capture the original sentence including punctuations, and secondly, when we capture the grammatical relation the system requires the connection between words which cannot be stem into their root. However, the basic preprocessing such as ASCII character formatting is required to remove uninterpreted emoticons or symbols.

4.2   Dependency Tree

The dependency tree is a linguistic tree generated from the typed dependency parses of English sentences from phrase structure parses. In order to capture inherent relations occurring in corpus texts, we use noun phrase (NP) relations which are included in the set of grammatical relations used [4]. In here, typed dependencies and phrase structures have a different way of representing the structure of sentences, while a phrase structure parse produces nesting of multi-word constituents, a dependency parse produces dependencies between individual words. A typed dependency parses further labels the dependencies with grammatical relations, such as subject or indirect object. The tree structure can be found in the original paper [4].

We use dependency tree as the words segmentation process which is a process to identify the meaning of a word beyond the co-occurrence with other words. During the parsing stage, the system seeks to understand the meaning of different word relationships according to the sentence flow. We define the dependency tree as the key concept of natural language from Stanford NLP parser. In particular, we state dependency tree as:

Definition 3. *A dependency tree $T = (V, E, L)$ is a node-labeled tree where labels consist of (1) Part of Speech (POS) tagging: the syntactic role of the word, and (2) Relationship (REL): the grammatical relationship between words according to their associates in the dependency tree.*

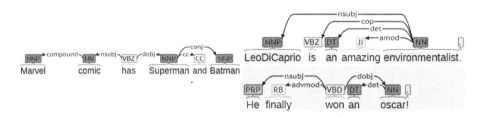

Fig. 2. Dependency parses structure

Each vertex ($v \in V$) represents word (term) while each edge ($e \in E$) represents a grammatical structure in a sentence. Each vertex has a POS-tagging type t property. This property distinguishes each word according to its classification. For example, in Fig. 2 on the sentence "Marvel comic has Superman and Batman", the word *Marvel, Superman, Batman* belong to PROPER NOUN (NNP) family while *comic* belong to NOUN (NN) family.

Another example, given two sentences "LeoDiCaprio is an amazing environmentalist. He finally won an oscar" in Fig. 2, the parser will indicate subject-object when predicate (VBZ and VBD) exist. In this way, we can virtually answer that *LeoDiCaprio* is an *Environmentalist* through the NN indicator while he also *Won* an *Oscar* via predicate-object relationship. Once we obtain the property value for each vertex, we build the belief graph as an extension of the dependency tree. Belief graph can be seen as the network of dependency tree collection. We discuss how to merge the dependency tree further in the next section.

4.3   DT Merging

Each sentence will form a dependency tree after the mapping process. Each node in the tree represents a word/term with its POS-tag type. We merge a collection

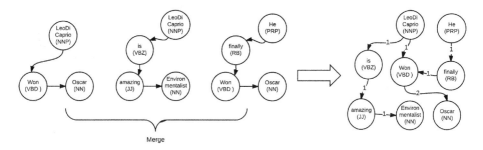

Fig. 3. Merging dependency trees

of dependency trees into a graph where the edges will record the co-occurrence between two nodes. In this process, we ignore the node(term) frequency. Note that a word may have a different POS-tag type resulting in multiple nodes with the same word. We keep the duplication of words as part of identification. For example, based on the previous illustration "LeoDiCaprio is an amazing environmentalist. He finally won an oscar" in addition to "LeoDicaprio won an Oscar", we merge each tree into the belief graph by including the co-occurrence pair of nodes (Fig. 3). In this way, we capture the significance of terms through their frequency.

5   Experiments

We used two datasets, Twitter and Amazon Product Review [7]. Both datasets are loaded into MongoDB, each tweet or review is indicated as one document. We vary the number of documents from 100 documents to 25,000 documents for each setting. Note that Twitter only allows 140 character maximum for each tweet while the product review may range from short sentences to a long paragraph. We also address the graph analytic task after the dependency graph is stored inside Neo4j. Our experiments were conducted on MongoDB version 3.4.2 and Neo4j version 3.2.5.

5.1   Results and Analysis

The results of constructing documents into Dependency Tree are given in Fig. 4. We report the time to label each document before the merging process for different document size. As we can see the time cost of converting documents grows linear with the size. We record approximately between 40–100 ms to convert each document depending on the number of words per document. Since Twitter has maximum 140 characters allowance, the processing time appears to be stable. On the other hand, Amazon review varies a lot in the number of words and tend to be double than Twitter which indicates a significant discrepancy.

Figure 5 presents the results of loading DTs into the graph database. The time is measured from merging DTs to Neo4j writing. We use transaction schema

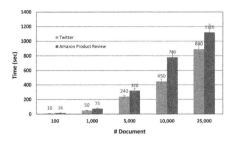
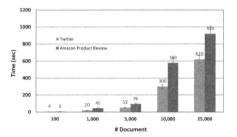

Fig. 4. Time to convert text into DT

Fig. 5. DT loading time into Neo4j

which writes per sequential word-relationship-word as we explained in Sect. 4.3. As we noted, the writing time increases gradually as the size is doubling. This is due to the fact that the database has to match existing word-relationship-word to increment the frequency property in the relationship, and one word may have different type property which leads to the increased matching time.

The overall computation time takes seconds to minutes to process due to the rich POS-tagging graph properties. Stanford dependency parser credited for more than half of the overall processing time. Note that at this point we did not consider any predefined indexing technique prior loading into the database. However, we consider this process as an offline task which can be computed as a background task. The graph analytical querying within Neo4j is somewhat real-time regardless the size of the document.

5.2   Qualitative Evaluation

Once we translate DTs into Dependency Graph, we perform the summarizing task which can be achieved by exploiting several parameters:

1. **Frequency Pattern.** We set a certain threshold to get the most frequent words relation pattern. This method is based on the relationship that co-occurs between words since the edges preserve the frequency properties. Within a specific threshold, we can filter the graph to generate a summary of text. Note that the co-occurrence frequency is generated based on the word-to-word relationship in contrast to the bag-of-word model.
2. **Core Nominals.** The other way is by filtering the core nominals property to get all the main object. Core nominals usually belong to "nsubj", "obj", or "iobj". Once we get this we can further use the frequency threshold filtering as the combination to get stronger summarization. In our observation, this combination benefits the product review dataset as the content will be toward the aspect-specific (products).

Both methods have the trade-off between each other. The limitation of the first method is that it may eliminate the main subject or object (e.g. nsubj, obj) which will not provide strong interpretations. While the second may produce excessive terms. We balance the result by taking each subset of both methods.

Table 1. List of text summarization detected by LDA, KeyGraph and BeliefGraph

Approaches	Categories	Summary
KeyGraph	Event	#theiAwards, project, #innovation, #tech, up
KeyGraph	TV-show	#MKR, kitchen, show, show, ready, win
KeyGraph	Place	Paradise, Surfers, Coast, beach, Gold, Dreamworld
LDA	TV-show	Nominate, #MKR, thought, 2016, ready, audition
LDA	Place	Paradise, Surfers, QLD, beach, bay, today
LDA	Place	Coast, Gold, summer, info, day, Dreamworld, sunshine
BeliefGraph	Place	Today, we, in, Surfers, Paradise, Beach, Gold, Coast
BeliefGraph	Event	Nominate, your, project, #theiAwards, #innovation
BeliefGraph	TV-show	Get, ready, audition, #MKR, 2016, coming

In Table 1 we measure the top three topics/summaries from KeyGraph, LDA, and Belief Graph. For LDA and KeyGraph, we aggregate the short texts into a single large document before each run. Since the dataset does not have the human-annotation, we do the qualitative measurement. Based on Table 1 we can see that belief graph has more interpretable and more cohesive structure due to preserving the word order. Due to space constraints, we only list keywords that are closely related. Belief graph does not use stopwords to retain the complete structure of the sentence.

6 Conclusion

We presented the belief graph model to mine noisy and unpredicted text. Within the model, we propose the implementation of dependency tree from Stanford dependency parser to build a dependency graph. Based on our model we are able to detect important keywords and summarize text-based their word-to-word relationship. Through the performance and practical study in our experiments and using real-world dataset, we showed the effectiveness of belief graph, especially in mining short text. Some future works including the improvement in the POS-tagging process specifically for social media content. The current tagging system is not yet effective in predicting the grammar for short text. Indexing the DT merging prior loading into the database can improve the processing time. At this point, the system is only suitable for the offline task due to the long overall processing time.

References

1. Amsterdamer, Y., Kukliansky, A., Milo, T.: A natural language interface for querying general and individual knowledge. Proc. VLDB Endow. **8**(12), 1430–1441 (2015)
2. Blei, D.M., Ng, A.Y., Jordan, M.I.: Latent dirichlet allocation. J. Mach. Learn. Res. **3**(Jan), 993–1022 (2003)

3. Chen, D., Fisch, A., Weston, J., Bordes, A.: Reading Wikipedia to answer open-domain questions. In: Proceedings of the 55th Annual Meeting of the Association for Computational Linguistics, ACL 2017, Vancouver, Canada, 30 July–4 August, vol. 1, pp. 1870–1879, Long Papers (2017)

4. De Marneffe, M.C., MacCartney, B., Manning, C.D., et al.: Generating typed dependency parses from phrase structure parses. In: Proceedings of LREC, Genoa, vol. 6, pp. 449–454 (2006)

5. Deutch, D., Frost, N., Gilad, A.: Provenance for natural language queries. Proc. VLDB Endow. **10**(5), 577–588 (2017)

6. Franciscus, N., Ren, X., Stantic, B.: Answering temporal analytic queries over big data based on precomputing architecture. In: Nguyen, N.T., Tojo, S., Nguyen, L.M., Trawiński, B. (eds.) ACIIDS 2017. LNCS (LNAI), vol. 10191, pp. 281–290. Springer, Cham (2017). https://doi.org/10.1007/978-3-319-54472-4_27

7. He, R., McAuley, J.: Ups and downs: modeling the visual evolution of fashion trends with one-class collaborative filtering. In: Proceedings of the 25th International Conference on World Wide Web, pp. 507–517. International World Wide Web Conferences Steering Committee (2016)

8. Li, C., Wang, H., Zhang, Z., Sun, A., Ma, Z.: Topic modeling for short texts with auxiliary word embeddings. In: 39th International ACM SIGIR Conference on Research and Development in Information Retrieval, pp. 165–174. ACM (2016)

9. Li, F., Jagadish, H.: Constructing an interactive natural language interface for relational databases. Proc. VLDB Endow. **8**(1), 73–84 (2014)

10. Manning, C.D., Surdeanu, M., Bauer, J., Finkel, J.R., Bethard, S., McClosky, D.: The Stanford CoreNLP natural language processing toolkit. In: ACL (System Demonstrations), pp. 55–60 (2014)

11. Mihalcea, R., Tarau, P.: TextRank: bringing order into text. In: EMNLP, vol. 4, pp. 404–411 (2004)

12. Ohsawa, Y., Benson, N.E., Yachida, M.: KeyGraph: automatic indexing by co-occurrence graph based on building construction metaphor. In: Research and Technology Advances in Digital Libraries, pp. 12–18. IEEE (1998)

13. Sayyadi, H., Raschid, L.: A graph analytical approach for topic detection. ACM Trans. Internet Technol. (TOIT) **13**(2), 4 (2013)

14. Scaiella, U., Ferragina, P., Marino, A., Ciaramita, M.: Topical clustering of search results. In: Proceedings of the Fifth ACM International Conference on Web Search and Data Mining, pp. 223–232. ACM (2012)

15. Song, Q., Wu, Y., Dong, X.L.: Mining summaries for knowledge graph search. In: 2016 IEEE 16th International Conference on Data Mining (ICDM), pp. 1215–1220. IEEE (2016)

16. Zhang, C., Wang, H., Xu, F., Hu, X.: IdeaGraph Plus: a topic-based algorithm for perceiving unnoticed events. In: 2013 IEEE 13th International Conference on Data Mining Workshops (ICDMW), pp. 735–741. IEEE (2013)

17. Zuo, Y., Zhao, J., Xu, K.: Word network topic model: a simple but general solution for short and imbalanced texts. Knowl. Inf. Syst. **48**(2), 379–398 (2016)

Meta-Modeling
Decomposition of Responsibilities

Piotr Zabawa[✉]

Cracow University of Technology, Warszawska 24, 31-155 Krakow, Poland
piotr.zabawa@pk.edu.pl

Abstract. Contemporary known and applied approaches to defining graph modeling languages are standardized. One their characteristic feature, which is fixed and static structure defined at compile time, contains generalizations and thus is difficult to change. The paper presents the results of the process of decomposing responsibilities which may be identified in meta-models. Such decomposition may be done if a meta-model is defined from the compile-time independent meta-model node and meta-model arc classes joint into meta-model graph at run-time. The Context-Driven Meta-Modeling Framework (CDMM-F) was designed to support defining such meta-model. The process of the responsibilities migration from one place they were originally concentrated to the right place is shown in the paper as well and this migration results in mapping them to the right elements of the CDMM-F.

Keywords: Modeling languages · Meta-model
Decomposition of responsibilities

1 Introduction

Meta-modeling is a process of creating rules, constraints and syntactical as well as semantical elements of the method of reality abstraction. Graph-based modeling is one possible approach to meta-modeling. Meta-modeling in the form of graph models and meta-models is a widely accepted approach in software engineering. Meta-models are used in this domain to define general-purpose modeling languages, like the Unified Modeling Language (UML) or to define Domain-Specific modeling Languages (DSLs). Their roles and motivation for leading the research are characterized below.

The role of the general-purpose modeling languages (GPML) is to support software development teams with the language focused on the specification of the software systems under development (horizontal market). The models can be also used for generating software system artifacts (UML) as well as they can be executable (BPMN2). This way of software development processes automation is very important from the economical reasons and is a common technique for improving competitiveness of the IT enterprises.

© Springer International Publishing AG, part of Springer Nature 2018
N. T. Nguyen et al. (Eds.): ACIIDS 2018, LNAI 10752, pp. 91–101, 2018.
https://doi.org/10.1007/978-3-319-75420-8_9

In contrast, the domain-specific modeling languages (DSML) are extensively used in many application domains (vertical markets) to define easy to use small textual or graphical languages. They are useful for solving simple domain-specific problems and are popular e.g. in enterprise systems [5].

The paper is related to a new approach to defining graph modeling languages. The approach is named Context-Driven Meta-Modeling (CDMM) [14]. The modeling languages defined in CDMM approach are named CDMM-compliant meta-models and they can be general-purpose graph modeling languages or domain-specific ones. Thus, the subject of the paper is related to the OMG concepts as well as to graph DSLs, both mentioned above. And the paper is focused on identification of modeling language responsibilities and constructing a query language for the transformed meta-model.

In contrast to the previous papers dedicated to CDMM-related problems this paper introduces and names different kinds of meta-modeling responsibilities not known from the existing approaches. However, none of the previous papers explained the specific nature of the run-time meta-modeling nor introduced meta-modeling responsibilities characteristic for this approach to meta-modeling. The responsibilities superposition characteristic for the run-time meta-modeling was never published before.

2   State of the Art

There are several well known Object Management Group's (OMG) standards, like Meta-Object Facility (MOF), Unified Modeling Language (UML), Business Process Model and Notation (BPMN2), both built on top of the MOF as well as other modeling standards. Some of them constitute Model-Driven Architecture (MDA) standard. The MDA is dedicated to automating software development processes in the model-driven approach. All MDA standards are general purpose standards and they are re-defined from time to time to be shared among modeling and software development communities to support model-driven software development processes.

The modeling languages defined by the OMG are created at compile-time and thus they are named compile-time meta-models or compile-time modeling languages further. In contrast, the modeling languages created at run-time are named run-time meta-models or run-time modeling languages. The paper is dedicated to run-time meta-models, nevertheless it refers also to compile-time modeling languages.

The paper contains a discussion of responsibilities that can be identified in contemporary modeling languages. However, the discussion is applicable for run-time modeling languages only. It may be, however perceived as a strong motivation to defining modeling languages this way due to many advantages.

There are known some frameworks for defining meta-models, like for example [7]. The CDMM approach presented in the article is also supported by the appropriate CDMM-F framework [13]. The paper refers both the CDMM-F and the meta-models to determine which meta-model responsibility should be mapped to the framework or to the meta-model. As the result of the analysis described in Sect. 4.1 the paper the Context-Driven Meta-Modeling Framework (CDMM-F) [11] was designed and implemented. The design was based on the concept of decomposition of responsibilities and their correct mapping to framework and meta-model elements. In consequence, the CDMM-F constitutes the feasibility case-study for the decomposition of responsibilities discussed in the paper. The CDMM-F is introduced first in the article as it forms the base for identifying responsibilities, their decomposition and mapping mentioned above. The special role of application context mentioned further in the paper is presented in [10].

3   Context-Driven Meta-Modeling Fundamentals

The Context-Driven Meta-Modeling (CDMM) approach to defining graph modeling languages (meta-models) is different from the OMG concept. The key assumption in CDMM is that meta-models are treated as data models [8]. In the consequence the generalization relationships are excluded from CDMM-compliant modeling languages. It contrasts to other approaches where generalization constitutes an important modeling element, as it was described for AODB in [4,6]. Moreover, in the CDMM a meta-model designer may introduce classes not only for meta-model graph nodes but also for graph edges. All these classes are independent one of the other at compile-time. Such a meta-model graph can be semantically enriched by introducing entity classes both to meta-model graph nodes and graph edges. The entity classes may form aggregation hierarchies. In the consequence of such assumptions, the whole such a graph structure can be defined at run-time. The formal representation of the CDMM approach is illustrated in the Fig. 1 in the form of the CDMM meta-meta-model class diagram [13].

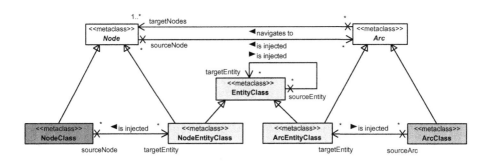

Fig. 1. CDMM meta-meta-model - from [13].

The CDMM approach to defining modeling languages offers significant ease of change introduction to modeling languages as they are defined at run-time.

Meta-models specified this way may easily become graph domain languages, that is - graph DSLs. On the other hand, the CDMM-compliant modeling languages may be created in the form of OMG modeling or meta-modeling languages customization.

The CDMM concept presented in the Fig. 1 is implemented in the form of the CDMM-F, the framework which supports defining CDMM-compliant modeling languages. The framework is implemented in the combination of Java-related technologies, namely Java, Spring, AspectJ, Guava and Apache commons.

The universality of CDMM approach manifests in the fact that the modeling language can be defined in any form not necessarily addressed to class-object paradigm. In the consequence the application of CDMM approach makes accommodating new technologies and new paradigms easier than in the case of MDA approach.

The next sections present some observations made during several stages of the evolutionary approach applied to the CDMM-F design and implementation process and resulting in the CDMM meta-meta-model presented in the Fig. 1. The special focus of this kind of research was put just on the identification and decomposition of meta-model and CDMM-F responsibilities. Then, the identified responsibilities were mapped to the right element of the CDMM-F or the meta-model.

4 Migration of Responsibilities in the CDMM-F

The analysis presented in this section leads from the identification of existing but not evident meta-model responsibilities to their mapping to the right place in the run-time meta-model or the framework. This mapping is required for the correct CDMM-F design.

The following stages were applied to identify the right place of implementation for particular responsibilities:

– STAGE ONE - the meta-model responsibilities were identified as it is described in Sect. 4.1;
– STAGE TWO - the maximal number of responsibilities was associated to the Arc classes for the meta-model while removing them from meta-model Node classes;
– STAGE THREE - the graph query language was designed in such a way that each meta-model element instance (CDMM-compliant model element) may be obtained from Arc class instances;
– STAGE FOUR - the responsibilities focused in the query language elements were moved to different elements of CDMM-compliant meta-models or to the elements of the CDMM-F.

The stages are characterized in the succeeding subsections.

4.1   STAGE ONE: Responsibilities in CDMM-Compliant Meta-Models

In this section the analysis of the CDMM meta-meta-model presented in the Fig. 1 is performed in order to identify the meta-modeling responsibilities being the subject of migration. Thus some responsibilities are inferred from the mentioned class diagram. The CDMM meta-meta-model is conceptual. In the consequence, the CDMM-compliant meta-models are not instances of the meta-meta-model. They should be rather perceived as meta-meta-model concretizations or realizations. The CDMM meta-meta-model as a conceptual one is not created in the CDMM-F in any form. The advantage of it is being just a model of roles the meta-model classes play regarding the CDMM meta-model graph.

The responsibilities inferred in this section are further mapped to the different elements of the meta-model and CDMM-F. This mapping drives the CDMM-F design decisions as well as meta-model design decisions.

The characteristic feature of MDA standards is the fixed structure (hierarchy) of their meta-models. Moreover, the meta-models of the MDA sub-standards are monolithic - their responsibilities are decomposed to packages and to generalizations from abstract classes or implementation of marker interfaces, sometimes abstract classes are used as markers. Also, the relationships between MDA sub-standard meta-model nodes are represented in the form of references (or inheritance or implementation relationships) and not in the form of classes [1–3,9]. In consequence, the set of available relationships is limited. In this section the results of the analysis of possible responsibilities that can be found in meta-models (also in the MDA meta-models) are presented from the perspective of CDMM approach. These results are inferred from the CDMM-F implementation efforts.

As it results from the CDMM meta-meta-model presented in the Fig. 1, each CDMM-compliant meta-model class can be a `Node` class or an `Arc` class or an `Entity` class. According to the CDMM assumptions these classes are not related at compile-time and the meta-model is created from these classes at run-time. The `Node`, `Arc` and `Entity` classes are defined by a meta-model designer for the purpose of a particular modeling language. They are also highly reusable both in the source code level (Java source code files) and in the byte code level (compiled class files or jar files). They can be easily exchanged between different meta-model projects and between end user applications and meta-model projects. The character of `Node`, `Arc` and `Entity` meta-model classes is determined by their role. For the `Node` classes it is enough just to exist and be empty. `Entity` classes should store data in the form of the attributes. `Arc` classes must be defined in the form resulting from the Fig. 1. More specifically, each `Arc` class must define its `targetNodes` element. This element plays role of the container for information about type of the `targetNodes` and for the object being instances of the `targetNodes` class. The `targetNodes` classes are determined when a CDMM-compliant meta-model is constructed from the meta-model definition (application context) while the `targetNodes` objects are put into the `targetNodes` container while instantiating the meta-model (model creating). Both `Node` and `Arc` classes may also contain data in attributes. In such a case they share their role with the role of `Entity` classes. Each `Arc` class is associated to a `Node` class at run time on the stage of meta-model creation. The concept of injections is applied here. The same injection mechanism is also applied for run-time `Entity` classes associating to `Arc` or `Node` classes as well as to `Entity` classes. As a special case an `Entity` class may be shared between several `Node` and several `Arc` classes by injecting the `Entity` class to each of the mentioned `sourceNode` and `sourceArc` classes. It is shown in the Fig. 1 that the cardinalities of relationships between `Node` and `Arc` classes are many-to-many. It means, that each `Node` class may have many `Arc` class injected and each `Arc` class may be injected to many `Node` classes. However, each `Arc` class must have at least one `targetNodes` end (otherwise it does not relate anything). So, the responsibility of `Arc` classes is to have a `targetNodes` end. The reflexive (that is self) association in `EntityClass` form the Fig. 1 means that the `Entity` classes may form association (aggregation) hierarchies.

All responsibilities mentioned in this section are connected to meta-model elements (user defined classes), to the meta-model itself (its representation in memory) or to the definition of the meta-model (the file which defines an application context for the CDMM-F).

4.2   STAGES TWO and THREE: CDMM-F Query Language

In this section the results of STAGE TWO and THREE are discussed together as the goal of the analysis presented in the paper is just the STAGE FOUR, which in turn is illustrated in Sect. 4.3.

The decision about constructing meta-model query language around `Arc` classes was driven by the fact that in the available modeling languages the number of different types of relationships (`Arc` classes in CDMM) is significantly smaller than the number of different types of nodes (`Node` classes in CDMM). Moreover, the `Node` classes are unique in a meta-model while the `Arc` classes may appear many times in a modeling different (say, between two pairs of `Node` classes in the case of bnary relationships) of `Node` classes and inside a set of `Node` classes (say, between two `Node` classes);

The graph query language was designed in such a form that being in a particular meta-model `Node` all `Arc` classes injected to this `Node` class the `Node` class object is casted to the `Arc` class and the right `Arc` class `targetNodes` attribute is found by the query.

The construction of a sample query `Arc` class that fulfils the concept from the last paragraph is presented below. First the CDMM-F elements referenced further are presented in the Figs. 2 and 3.

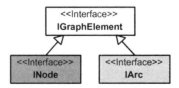

Fig. 2. CDMM-F top hierarchies for the roles in the graph.

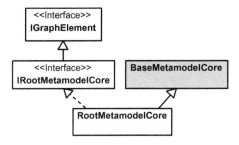

Fig. 3. CDMM-F top hierarchies for the roles in the meta-model core elements.

The CDMM-F query language was designed according to the following assumptions:

- the subject of each query is a `Node` class; this `Node` class plays the role of a `sourceNode` class of a relationship injection as was depicted in the Fig. 1
- the query contains the `sourceNode` object, the `Arc` class, the `targetNode` class as its arguments
- the query returns the list of `targetNode` objects.

The meta-model query language design rules presented above fulfill the requirement of making it possible to identify all relationship instances and return the list of all objects on the `targetNode` end of each relationship instance. It is sufficient to traverse the whole meta-model graph instance, that is the graph which is composed of meta-model objects (CDMM-compliant model) interrelated according to the injections (CDMM-compliant meta-model).

Below, the design of classes implementing this language is analysed.

The most general representation of the type of a class of the most complex relationship applied so far has the following form in the query language mentioned above:

Multimap<List<Pair<IGraphElement,Class<?>>>, BaseMetamodelCore>

This form of type reflects the superposition of the responsibilities identified in Sect. 4.1. The superposition is expressed in the Expr. 1. The appropriate elements of the type and responsibilities the elements of this type reflects are displayed in the same colors.

$$inter\text{-}object\text{-}relations \circ arity \circ \textbf{\textit{binary-relation-id}} \circ targetNode\text{-}objects \tag{1}$$

where:

- *inter-object-relations* stands for the inter-object relation instances for a particular `Arc` class injected to `Node` classes
- *arity* stands for the number of `Node` classes interrelated by an `Arc` class (N-ary nature of a relationship), thus the number of injections of an `Arc` class to the `Node` classes represented by the `sourceNode` multiplicity in the Fig. 1

- *binary-relation-id* stands for the key used to uniquely identify a binary relationship
- *targetNode-objects* stands for the number of objects located at the `targetNode` association roleName presented in the Fig. 1.

The consequences of a special approach to representing N-ary relationships, which was applied in CDMM meta-model graph query language is toched on further in the paper and will be published in separate paper.

It is worth noting that the superposition from the Expr. 1 is similar to the Decorator design pattern applied to generic types in place of the object types. Thus, the superposition of responsibilities may be seen as the Decorator design pattern moved from class-object paradigm to generic paradigm. Objects are decorated by relations, relations are decorated by N-arity, N-arity is decorated by the objects cardinality. Moreover some elements of this superposition are defined at meta-modeling level while some of them are defined at modeling level.

Another observation is that the order of superposed element must remain unchanged while responsibilities migration or while simplification of the relationship.

The Expr. 1 constitutes the result of the STAGES 2 and 3 discussed at the beginning of the current section.

4.3   STAGE FOUR: Responsibilities Mapping

The results of the STAGE 4 are presented in the Tables 1 and 2. The same coloring convention as the one used in Sect. 4.2 was applied in both tables to show the result of mapping all colored responsibilities to meta-model and CDMM-F locations.

All identified meta-model responsibilities were mapped to the right place in run-time CDMM-compliant meta-model elements according to the Table 1. They were also implemented according to the Table 2 in the CDMM-F framework both as the meta-model graph definition specified in the form of the application context and in the form of the CDMM-F software itself. The framework was then extensively tested both by manually defined meta-models oriented to testing the correctness of each responsibility mapping and in the form of semi-automatic simulation-like approach presented in [12]. In the last approach to testing the manually defined meta-models constituted the test kernels surrounded by automatically generated test context - the supergraphs for test kernels. The tests confirmed the correctness of the responsibilities decomposition.

As it was shown in the Tables 1 and 2, some responsibilities are mapped to more than one locations, like for example "being a subject of entities injections" which is distributed between `Node` class, `Arc` class and `Entity` class. Also, each location is mapped to many responsibilities. It means, that the mapping is many-to-many.

It is worthy of notice that the N-ary relationship responsibility is associated to the application context in the Table 2. In contrast to compile-time meta-models, arity is not the responsibility of the `Node` class. There are many consequences

Table 1. Mapping from the CDMM responsibilities to the meta-model elements.

Responsibility	Location
Being a meta-model node or arc	Element class
Being a meta-model graph node	Node class
Being a subject of arcs injections	
Being a subject of entities injections	
Being a meta-model graph arc	Arc class
Being able to be injected to the sourceNodes	
Being a container for targetNodes both for Node classes and Node objects (relationship end)	
Being directed from the sourceNode to targetNodes	
Being a subject of entities injections	
Being possibly bi-directed	
Storing and exposing meta-model element attributes	Entity class
Belonging to an aggregation hierarchy of entities	
Being able to be injected to the nodes, arcs or entities	
Being a subject of entities injections	

Table 2. Mapping from the CDMM responsibilities to the CDMM-F framework elements.

Responsibility	Location
Defining a scope for Elements	Application context
Defining a unique name for Elements	
Defining an arc as being a binary or N-ary relationship (arity)	
Defining an arc as a reflexive relationship	
Defining the whole meta-model graph composed of node, arc, entity classes and their injections	
Having APIs for meta-model graph traversal	CDMM-F
Having API for meta-model graph elements instantiation (creating model in memory as a graph of objects)	
Having a meta-model graph query language	
Having scope management factories	
Having configuration for a meta-model project	
Being able to instantiate the meta-model classes (creating meta-model in memory as a graph of meta-model element classes)	

of moving arity responsibility from a `Node` class to the application context. The detailed discussion of the N-ary relationships in run-time meta-models is out of scope of the paper.

5   Conclusions

One consequence of introducing the CDMM concept of defining modeling languages at run-time is the ability to identify different responsibilities, which are not clear as long as the compile-time metamodel definitions are taken into account. It was shown in the paper that many responsibilities can be uncovered, identified, taken away from meta-model graph hierarchy and, finally placed in the right places when moving from compile-time meta-models to the run-time ones. The new responsibilities were also named and mapped to the meta-model elements according to the Table 1 and meta-modeling framework elements according to the Table 2.

The right paradigm for implementing meta-modeling query language according to the concept assumed in the paper is a generic paradigm. All meta-model Elements, that is `Node` classes, `Arc` classes and `Entity` classes may be implemented in class-object paradigm as classes unrelated at compile-time. Meta-model graphs are created in CDMM-F with the aid of aspect-oriented paradigm. The combination of mentioned object-oriented paradigms is sufficient and well suited to make all CDMM concepts including the decomposition of responsibilities feasible.

References

1. Bildhauer, D.: On the relationship between subsetting, redefinition and association specialization. In: Proceedings of the 9th Baltic Conference on Databases and Information Systems, Riga (2010)
2. Diaz, I., Llorens, J., Genova, G., Fuentes, J.: Generating domain representations using a relationship model. Inf. Syst. **30**, 1–19 (2005)
3. Génova, G., del Castillo, C.R., Llorens, J.: Mapping UML associations into Java code. J. Object Technol. **2**(5), 135–162 (2003)
4. Krótkiewicz, M.: A Novel inheritance mechanism for modeling knowledge representation systems. Computer Science and Information Systems (2017)
5. Krótkiewicz, M., Jodłowiec, M., Wojtkiewicz, K., Szwedziak, K.: Unified process management for service and manufacture system—material resources. In: Burduk, R., Jackowski, K., Kurzyński, M., Woźniak, M., Żołnierek, A. (eds.) Proceedings of the 9th International Conference on Computer Recognition Systems CORES 2015. AISC, vol. 403, pp. 681–690. Springer, Cham (2016). https://doi.org/10.1007/978-3-319-26227-7_64
6. Krótkiewicz, M.: Association-oriented database model - n-ary associations. Int. J. Softw. Eng. Knowl. Eng. **27**(02), 281–320 (2017)
7. Malhotra, R.: Meta-modeling framework: a new approach to manage meta-model base and modeling knowledge. Knowl.-Based Syst. **21**, 6–37 (2008)
8. Merson, P.: Data model as an architectural view. Technical note CMU/SEI-2009-TN-024, Software Engineering Institute, Carnegie Mellon University (2009)

9. Tan, H.B.K., Yang, Y., Bian, L.: Improving the use of multiplicity in UML association. J. Object Technol. **5**(6), 127–132 (2006)
10. Zabawa, P.: Context-driven meta-modeling framework (CDMM-F) - context role. Tech. Trans. **112**(1–NP), 105–114 (2015)
11. Zabawa, P.: Context-Driven Meta-Modeling Framework (CDMM-F) - Internal Structure (2017, accepted for publication)
12. Zabawa, P.: Simulation of the CDMM-P paradigm-driven meta-modeling process. Tech. Trans. **4**, 143–154 (2017)
13. Zabawa, P., Hnatkowska, B.: CDMM-F – domain languages framework. In: Świątek, J., Borzemski, L., Wilimowska, Z. (eds.) ISAT 2017. AISC, vol. 656, pp. 263–273. Springer, Cham (2018). https://doi.org/10.1007/978-3-319-67229-8_24
14. Zabawa, P., Stanuszek, M.: Characteristics of the context-driven meta-modeling paradigm (CDMM-P). Tech. Trans. **111**(3), 123–134 (2014)

Stochastic Pretopology as a Tool for Topological Analysis of Complex Systems

Quang Vu Bui[1,2(✉)], Soufian Ben Amor[3], and Marc Bui[1,4]

[1] CHArt Laboratory EA 4004, EPHE, PSL Research University, Paris, France
`quang-vu.bui@etu.ephe.fr`
[2] Hue University of Sciences, Hue, Vietnam
[3] LI-PARAD Laboratory, University of Versailles-Saint- Quentin-en-Yvelines,
Versailles, France
[4] University Paris 8, Paris, France

Abstract. We are proposing in this paper a more general network modeling framework for complex system representation by introducing *Stochastic Pretopology*, a result of the combination of Pretopology theory and Random Sets. After giving the definition and some examples for building stochastic pretopology in many situations, we show how this approach generalizes graph, random graph, multi-relational networks and we present an application by giving *Pretopology Cascade Model* as a general model for information diffusion process that can take place in more complex networks such as multi-relational networks or stochastic graphs.

Keywords: Random set · Pretopology · Random graph
Complex network · Social network · Complex system
Multi-relational network · Information diffusion

1 Introduction

Complex system is a system composed of many interacting parts, such that the collective behavior of its parts together is more than the "sum" of their individual behaviors [10]. The topology of complex systems (who interact with whom) is often specified in terms of networks that are usually modeled by graphs, composed by vertices or nodes and edges or links. Graph theory has been widely used the conceptual framework of network models, such as random graphs, small world networks, scale-free networks [5,11].

However, having more complicated non-regular topologies, complex systems need a more general framework for their representation [10]. To overcome this issue, we propose using Stochastic Pretopology built from the mixing between Pretopology theory and Random Sets theory. Pretopology [2] is a mathematical tool for modeling the concept of proximity which allows us to follow structural transformation processes as they evolve while random sets theory [9,12] provides the good ways for handling what happens in a stochastic framework at the sets' level point ot views.

© Springer International Publishing AG, part of Springer Nature 2018
N. T. Nguyen et al. (Eds.): ACIIDS 2018, LNAI 10752, pp. 102–111, 2018.
https://doi.org/10.1007/978-3-319-75420-8_10

In this paper, after recalling basics of pretopology and the definition of graphs in the *Berge sense* [3], we first show pretopology as an extension of graph theory which leads us to the definition of pretopology networks as a general framework for network representation. Connected to random sets theory, we then give the definition of Stochastic Pretopology and propose different ways for building such a pretopology that is useful for modeling the topological structure of complex systems in different spaces such as metric space, valued or binary relation spaces. These models can be convenient to handle phenomena in which collective behavior of a group of elements can be different from the summation of element behaviors composing the group. After presenting *Independent Cascade model* [6] and *Independent Threshold model* [7] under stochastic pretopology language, we will propose pretopological information diffusion model as a general diffusion model that can take place in more complex networks such as multi-relational networks or stochastic graphs. Stochastic graphs presented in this paper are defined by extending the definition of graph in the *Berge sense* [3] $G = (V, \Gamma)$. In this approach, by considering Γ function as a finite random set defined from a degree distribution, we give a general graph-based network model in which *Erdős-Rényi model* and *scale-free networks* are special cases.

The rest of this paper is organized as follows: Sect. 2 briefly recalls basic concepts of pretopology theory prepared for building stochastic pretopology in Sect. 3; we then conclude by presenting an application of stochastic pretopology in information diffusion.

2 Pretopology as a Group Modeling in Complex Networks

For modeling the dynamic processes on complex networks, topology theory is not suitable since the idempotent property of its closure function makes it impossible for changing unless changing the topological structure. So, we propose in this section a new insight on networks modeling with pretopology theory. Pretopology [2] is considered as an extension of topology obtained by the relaxing of its axiomatic. The pretopology is a tool for modeling the concept of proximity that allows monitoring step by step the evolution of a set. It establishes a powerful tools for the structure analysis, classification, and multi-criteria clustering [4]. It also applies for group modeling in social networks. Based on set theory, by considering a group of elements as a set, pretopology formalism allows us to consider a group as a whole independent entity.

2.1 Pseudo-Closure Function

Definition 1. *We call pseudo-closure defined on a set V, any function $a(.)$ from $\mathcal{P}(V)$ into $\mathcal{P}(V)$ such as:*

(P1): $a(\emptyset) = \emptyset$;
(P2): $A \subset a(A) \quad \forall A, A \subset V$

(V, a) is then called pretopological space.

Pseudo-closure allows, for each of its applications, to add elements to a set departure according to defined characteristics. The starting set gets bigger but never reduces. There are different ways to build a pseudo-closure function.

(a) V is equipped with a metric. When space V is equipped with a metric d, we can build a *pseudo-closure* function **a(.)** on V with a closed ball of center x and radius r $(B(x, r) = \{y \in V | d(x, y) \leq r\})$:

$$\forall A \in \mathcal{P}(V), \quad a(A) = \{y \in V | B(x, r) \cap A \neq \emptyset\} \tag{1}$$

The *pseudo-closure* $a(A)$ is a set of all elements $y \in V$ such that y is within a distance of at most radius r from at least one element of A.

(b) The elements of V are linked by a valued relation. In order to model certain problems such as model in weighted graph, we often need the space V are bound by a valued relation. For instance, we can define an real value ν on relations as a function from $V \times V \to \mathbb{R}$ as: $(x, y) \to \nu(x, y)$. The *pseudo-closure* **a(.)** can build such as:

$$\forall A \in \mathcal{P}(V), \quad a(A) = \{y \in V - A | \sum_{x \in A} \nu(x, y) \geq s\} \cup A; s \in \mathbb{R} \tag{2}$$

(c) The elements of V are linked by n reflexive binary relations. Suppose we have a family $(R_i)_{i=1,\ldots,n}$ of binary reflexive relations on a finite set V. For each relation R_i, we can define pretopological structure by considering the following subset: $\forall i = 1, 2, \ldots, n, \forall x \in V, V_i(x)$ defined by:

$$V_i(x) = \{y \in V | x\, R_i\, y\}$$

We can define the *pseudo-closure* **a(.)** by:

$$\forall A \in \mathcal{P}(V), \quad a(A) = \{x \in V | \forall i = 1, 2, \ldots, n, V_i(x) \cap A \neq \emptyset\} \tag{3}$$

(d) The elements of V are equipped with a neighborhood function. Let us consider a multivalued function $\Gamma : V \to \mathcal{P}(V)$ as a neighborhood function. $\Gamma(x)$ is a set of neighborhoods of element x. We define a *pseudo-closure* **a(.)** as follows:

$$\forall A \in \mathcal{P}(V), \quad a(A) = A \cup (\bigcup_{x \in A} \Gamma(x)) \tag{4}$$

2.2 Pretopology as an Extension of Graph Theory

(a) Graphs in the Berge sense. By using the knowledge from multivalued function, Claude Berge [3] defined a graph such as:

Definition 2. *A graph, which is denoted by $G = (V, \Gamma)$, is a pair consisting of a set V of vertices or nodes and a multivalued function Γ mapping V into $\mathcal{P}(V)$.*

The pair (x, y), with $y \in \Gamma(x)$ is called an arc or edge of the graph. We therefore can also denote a graph by a pair $G = (V, E)$, which V is a set of nodes and E is a set of edges. Conversely, if we denote a graph as $G = (V, E)$, we can define the Γ function as: $\Gamma(x) = \{y \in V | (x, y) \in E\}$. $\Gamma(x)$ is a set of neighbors of node x.

(b) Pretopology as an extension of Graph theory. In this part, we show reflexive graph (V, Γ) which is a special case of pretopology. More specifically, as it is known, a finite reflexive graph (V, Γ) complies the property: $\forall A \subset V, a(A) = \cup_{x \in A} a(\{x\})$ where pseudo-closure function defined as $a(A) = \cup_{x \in A} \Gamma(x)$. For this reason, graph may be represented by a $\mathcal{V}_D$-type pretopological space. Conversely, we can build a pretopology space (V, a) presented a graph such as: $a(A) = \{x \in V | \Gamma(x) \cap A \neq \emptyset\}$ where $\Gamma(x) = \{y \in V | x\,R\,y\}$ built from a binary relation R on V. Therefore, a graph (V, Γ) is a pretopological space (V, a) in which the pseudo-closure function built from a binary relation or built from a neighborhood function in Eq. (4).

By using a graph, a network is represented with only one binary relation. In the real world, however, a network is a structure made of nodes that are tied by one or more specific types of binary or value relations. As we show in the previous, by using pretopology theory, we can generalize the definition of complex network such as:

Definition 3 *(Pretopology network). A pretopology network, which is denoted by $G^{(Pretopo)} = (V, a)$, is a pair consisting of a set V of vertices and a pseudo-closure function $a(.)$ mapping $\mathcal{P}(V)$ into $\mathcal{P}(V)$.*

3   Stochastic Pretopology (SP)

Complex systems usually involve structural phenomena, under stochastic or uncontrolled factors. In order to follow these phenomena step by step, we need concepts which allow modelling dynamics of their structure and take into account the factors' effects. As we showed in the previous section, we propose to use pretopology for modelling the dynamics of phenomena; the non idempotents of its pseudo-closure function makes it suitable for such a modelling. Then, we introduce stochastic aspects to handle the effects of factors influencing the phenomena. For that, we propose using a theory of random sets by considering that, given a subset A of the space, its pseudo-closure $a(A)$ is considered as a random set. So, we have to consider the pseudo-closure not only as a set transform but also as a random correspondence.

Stochastic pretopology was first basically introduced in Chap. 4 of [2] by using a special case of random set (the simple random set) to give three ways to define stochastic pretopology. We have also given some applications of stochastic pretopology such as: modeling pollution phenomena [8] or studying complex networks via a stochastic pseudo-closure function defined from a family of random relations [1]. Since we will deal with complex networks in which set of nodes V is a finite set, we propose in this paper another approach for building stochastic pretopology by using finite random set theory [12].

From now on, V denotes a finite set. $(\Omega, \mathcal{A}, \mathbb{P})$ will be a *probability space*, where: Ω is a set, representing the *sample space* of the experiment; $\mathcal{A}$ is a σ-algebra on Ω, representing *events* and $\mathbb{P} : \Omega \to [0, 1]$ is a *probability measure*.

3.1   Finite Random Set

Definition 4. *A finite random set (FRS) with values in $\mathcal{P}(V)$ is a map $X :$ $\Omega \to \mathcal{P}(V)$ such as*

$$X^{-1}(\{A\}) = \{\omega \in \Omega : X(\omega) = A\} \in \mathcal{A} \text{ for any } A \in \mathcal{P}(V) \tag{5}$$

The condition (5) is often called *measurability condition.* So, in other words, a FRS is a measurable map from the given probability space $(\Omega, \mathcal{A}, P)$ to $\mathcal{P}(V)$, equipped with a σ-algebra on $\mathcal{P}(V)$. We often choose σ-algebra on $\mathcal{P}(V)$ is the discrete σ-algebra $\mathcal{E} = \mathcal{P}(\mathcal{P}(V))$. Clearly, a *finite random set X* is a *random element* when we refer to the *measurable space* $(\mathcal{P}(V), \mathcal{E})$. This is because $X^{-1}(\mathcal{E}) \subseteq \mathcal{A}$ since $\forall \mathbb{A} \in \mathcal{E}; X^{-1}(\mathbb{A}) = \cup_{A \in \mathbb{A}} X^{-1}(A)$.

3.2   Definition of Stochastic Pretopology

Definition 5. *We define **stochastic pseudo-closure** defined on $\Omega \times V$, any function $a(.,.)$ from $\Omega \times \mathcal{P}(V)$ into $\mathcal{P}(V)$ such as:*

(P1): $a(\omega, \emptyset) = \emptyset \quad \forall \omega \in \Omega$;
(P2): $A \subset a(\omega, A) \quad \forall \omega \in \Omega, \forall A, A \subset V$;
(P3): $a(\omega, A)$ is a finite random set $\forall A, A \subset V$

$(\Omega \times V, a(.,.))$ *is then called Stochastic Pretopological space.*

By connecting the finite random set theory [9, 12], we can build stochastic pseudo-closure function with different ways.

3.3   SP Defined from Random Variables in Metric Space:

By considering a *random ball $B(x, \xi)$* with ξ is a non-negative random variable, we can build a *stochastic pseudo-closure* $\mathbf{a(.)}$ in metric space such as:

$$\forall A \in \mathcal{P}(V), \quad a(A) = \{x \in V | B(x, \xi) \cap A \neq \emptyset\} \tag{6}$$

3.4   SP Defined from Random Variables in Valued Space:

We present two ways to build stochastic pseudo-closure by extending the definition of pseudo-closure function on valued space presented in Eq. (2). Firstly, by considering threshold s is a *random variable η*, we can define a *stochastic pseudo-closure* $\mathbf{a(.)}$ such as:

$$\forall A \in \mathcal{P}(V), \quad a(A) = \{y \in V - A | \sum_{x \in A} v(x, y) \geq \eta\} \cup A \tag{7}$$

where threshold η is *random variable.*

Secondly, by considering the weight function $v(x, y)$ between two elements x, y as a *random variable*, we can define a *stochastic pseudo-closure* $\mathbf{a(.)}$ such as:

$$\forall A \in \mathcal{P}(V), \quad a(A) = \{y \in V - A | \sum_{x \in A} v_\Omega(x, y) \geq s\} \cup A \tag{8}$$

where $v_\Omega(x, y)$ is a *random variable.*

3.5   SP Defined from a Random Relation Built from a Family of Binary Relations

Suppose we have a family $(R_i)_{i=1,\ldots,m}$ of *binary reflexive relations* on a finite set V. We call $L = \{R_1, R_2, \ldots, R_m\}$ is a set of relations. Let us define a *random relation* $R : \Omega \to L$ as a random variable:

$$P(R(\omega) = R_i) = p_i; \quad p_i \geq 0; \sum_{i=1}^{m} p_i = 1.$$

For each $x \in V$, we can build a random set of neighbors of x with random relation R:

$$\Gamma_{R(\omega)}(x) = \{y \in V \,|\, x\, R(\omega)\, y\}$$

We can define a *stochastic pseudo-closure* **a(.,.)** such as:

$$\forall A \in \mathcal{P}(V), \quad a(\omega, A) = \{x \in V \,|\, \Gamma_{R(\omega)}(x) \cap A \neq \emptyset\} \tag{9}$$

3.6   SP Defined from a Family of Random Relations

We can extend the previous work by considering many *random relations*. Suppose we have a family $(R_i)_{i=1,\ldots,n}$ of random binary reflexive relations on a set V. For each $x \in V$, we can build a random set of neighbors of x with random relation $R_i, i = 1, 2, \ldots, n$:

$$\Gamma_{R_i(\omega)}(x) = \{y \in V \,|\, x\, R_i(\omega)\, y\}$$

We can define a *stochastic pseudo-closure* $a(.,.)$ such as:

$$\forall A \in \mathcal{P}(V), \quad a(\omega, A) = \{x \in V \,|\, \forall i = 1, 2, \ldots, n,\, \Gamma_{R_i(\omega)}(x) \cap A \neq \emptyset\} \tag{10}$$

3.7   SP Defined from a Random Neighborhood Function

Let us consider a random neighborhood function as a random set $\Gamma : \Omega \times V \to \mathcal{P}(V)$. $\Gamma(\omega, x)$ is a random set of neighborhoods of element x. We define a *stochastic pseudo-closure* **a(.,.)** as follows:

$$\forall A \in \mathcal{P}(V), \quad a(\omega, A) = A \cup \left(\bigcup_{x \in A} \Gamma(\omega, x) \right) \tag{11}$$

We have shown in this section how we construct stochastic pseudo-closure functions for various contexts. That is to say how proximity with randomness can be delivered to model complex neighborhoods formation in complex networks. In the two next sections, we will show how stochastic pretopology can be applied for modeling dynamic processes on complex networks by representing classical information diffusion models under stochastic pretopology language and then proposing Pretopology Cascade Model as a general information diffusion model in which complex random neighborhoods set can be captured by using stochastic pseudo-closure functions.

4   Stochastic Pretopology as a General Information Diffusion Model on Single Relational Networks

Information diffusion has been widely studied in networks, aiming to model the spread of information among objects when they are connected with each other. In a single relational network, many diffusion models have been proposed such as *tipping model, threshold models, cascade models,* ... [5,11]. We assume a network $G = (V, \Gamma, W)$, where $W : V \times V \to \mathbb{R}$ is weight function. $W(x, y)$ is the weight of edge between two nodes x, y in threshold model or the probability of node y infected from node x in cascade model.

The diffusion process occurs in discrete time steps t. If a node adopts a new behaviour or idea, it becomes active, otherwise it is inactive. An inactive node has the ability to become active. The set of active nodes at time t is considered as A_t. We present in the following two scenarios in which stochastic pretopology as extensions of both *Independent Cascade* (IC) model [6] and *Independent Threshold* (IT) model [7].

4.1   Stochastic Pretopology as an Extension of IC Model

Independent Cascade model. Under the IC model, at each time step t where A_{t-1}^{new} is the set of newly activated nodes at time $t-1$, each $x \in A_{t-1}^{new}$ infects the inactive neighbors $y \in \Gamma(x)$ with a probability $W(x, y)$.

Representing IC model under stochastic pretopology language: We can represent IC model by giving a definition of stochastic pretopology based on two definitions in subsection 3.4,3.7: we firstly define a random set of actived nodes from each node $x \in A_{t-1}^{new}$ and then use a random neighbor function to define the random active nodes in the time t. Specifically, A_t^{new} is defined via two steps:

i. For each $x \in A_{t-1}^{new}$, set of actived nodes from x, $\Gamma^{(active)}(x)$, defined as:

$$\Gamma^{(active)}(x) = \{y \in \Gamma(x) | W(x, y) \geq \eta\}; \quad \eta \sim U(0, 1) \tag{12}$$

ii. The set of newly active nodes, A_t^{new}, defined as:

$$A_t^{new} = a(A_{t-1}^{new}) - A_{t-1}; A_t = A_{t-1} \cup a(A_{t-1}^{new}) \tag{13}$$

where:

$$a(A_{t-1}^{new}) = A_{t-1}^{new} \bigcup \left(\bigcup_{x \in A_{t-1}^{new}} \Gamma^{(active)}(x) \right) \tag{14}$$

4.2   Stochastic Pretopology as an Extension of IT Model

Independent Threshold model: Under the IT model, each node y selects a randomly threshold $\theta_y \sim U(0, 1)$. Then, at each time step t where A_{t-1} is the set of nodes activated at time $t-1$ or earlier, each inactive node y becomes active if $\sum_{x \in \Gamma^{-1}(y) \cap A_{t-1}} W(x, y) \geq \theta_y$ where $\Gamma^{-1}(y) = \{x \in V | y \in \Gamma(x)\}$.

Representing IT model under stochastic pretopology language: We can represent IC model by giving a definition of stochastic pretopology such as:

$$A_t = a(A_{t-1}) = \{y \in V - A_{t-1} | \sum_{x \in \Gamma^{-1}(y) \cap A_{t-1}} W(x,y) \geq \eta\} \cup A_{t-1}; \quad \eta \sim U(0,1)$$

5 Pretopology Cascade Models for Modeling Information Diffusion on Complex Networks

Most of information diffusion models are defined via node's neighbors. In general, at each time step t, the diffusion process can be described in two steps:

Step 1: define set of neighbors $N(A_{t-1})$ of set of active nodes A_{t-1}.
Step 2: each element $x \in N(A_{t-1}) - A_{t-1}$ will be influenced by all elements in A_{t-1} to be active or not active node by following a diffusion rule.

We consider the way to define set of neighbors $N(A_{t-1})$ in step 1. In classical diffusion model with complex network represented by a graph $G = (V, \Gamma)$, $N(A_{t-1})$ is often defined such as: $N(A_{t-1}) = \cup_{x \in A_{t-1}} \Gamma(x)$. By using the concepts of stochastic pretopology theory introduced in the Sect. 3, the information diffusion process can be generalized by defining a set of neighbors $N(A_{t-1})$ as a *stochastic pseudo-closure* function $N(A_{t-1}) = a_\Omega(A_{t-1})$. We therefore propose the *Pretopological Cascade Model* presented in the following as a general information diffusion model which can be captured more complex random neighborhoods set in diffusion processes.

Definition 6. *Pretopological Cascade model:*
Under the Pretopological Cascade model, at each time step t, the diffusion process takes place in two steps:

Step 1: define set of neighbors $N(A_{t-1})$ of A_{t-1} as a stochastic pseudo-closure function $N(A_{t-1}) = a_\Omega(A_{t-1})$.
Step 2: each element $x \in N(A_{t-1}) - A_{t-1}$ will be influenced by A_{t-1} to be active or not active node by following a "diffusion rule".

For defining $N(A_{t-1})$ in step 1, we can apply different ways to define stochastic pseudo-closure function presented in Sect. 3. "Diffusion rule" in step 2 can be chosen by various ways such as:

- Probability based rule: element x infects the inactive elements $y \in N(A_{t-1})$ with a probability $P_{x,y}$.
- Threshold rule: inactive elements $y \in N(A_{t-1})$ will be actived if sum of all influence of all incoming elements of y greater than a threshold θ_y.

We present in the following two examples of the pretopological cascade model: the first takes place in a stochastic graph by defining random neighbors sets based on nodes' degree distribution and the second takes place in multi-relational networks where random neighbors set is built from a family of relations.

5.1   Pretopological Cascade Model on Stochastic Graphs

Definition 7 *(Stochastic Graph). A stochastic graph, which is denoted by $G^{\Omega} = (V, \Gamma_{\Omega})$ is a pair consisting of a set V of vertices and a finite random set Γ_{Ω} mapping $\Omega \times V$ into $\mathcal{P}(V)$.*

The random neighbor function Γ_{Ω} in the definition 7 can be defined in a general way from finite random set theory [12]. Since the *nodes' degree distribution* is necessary for studying network structure, we propose here the way to defining the random neighbor function Γ_{Ω} via two steps:

1. Defining *probability law* of the cardinality of Γ_{Ω} (in fact, Γ_{Ω} is a degree distribution of network).

$$Prob(|\Gamma_{\Omega}| = k) = p_k \quad for \quad k = 1, 2, \ldots, \infty \tag{15}$$

2. Assigning *probability law* on V^k for $k = 1, 2, \ldots, \infty$

$$Prob(\Gamma_{\Omega}^{(1)} = x^{(1)}, \ldots, \Gamma_{\Omega}^{(k)} = x^{(k)} || \Gamma_{\Omega}| = k) \quad for \quad x^{(1)}, \ldots, x^{(k)} \in V \tag{16}$$

We can see some classical network models are specical cases of this kind of stochastic graph. For example, we have *Erdős-Rényi* model if $|\Gamma_{\Omega}| \sim U(0, 1)$ and *scale-free networks* model when $|\Gamma_{\Omega}|$ follows a *power-law distribution*. We also have other network models by using other probability distributions such as Poisson distribution, Geometry distribution, Binomial distribution, etc.

Pretopological cascade model on Stochastic Graph. Under the *Pretopological Cascade model* on stochastic graph, at each time step t, each $x \in A_{t-1}$ generates a random number of neighbors η following a degree distribution given by the Eq. (15) and then generates random neighbors set $\Gamma_{\Omega}(x)$ following a point distribution given by the Eq. (16); after that x infects the inactive neighbors $y \in \Gamma_{\Omega}(x)$ with a probability $P_{x,y}$.

5.2   Pretopological Cascade Model on Multi-relational Networks

Multi-relational network. A *multi-relational network* can be represented as a multi-graph, which allows multiple edges between node-pairs. A *multi-relational network*, which is denoted by $G^{(multi)} = (V, (\Gamma_1, \Gamma_2, \ldots, \Gamma_m))$, is a pair consisting of a set V of vertices and a set of multivalued functions $(\Gamma_i)_{i=1,2,\ldots,m}$ mapping V into 2^V. Γ_i is a neighbor function following the relation R_i.

Defining Random neighbors set on Multi-relational network. Let us define a random index η takes values on $\{1, 2, \ldots, m\}$ such as a random variable:

$$P(\eta = i) = p_i; i = 1, 2, \ldots, m; \quad p_i \geq 0; \sum_{i=1}^{m} p_i = 1 \tag{17}$$

We define a random neighbor function Γ_{η} based on random index η such as: $\Gamma_{\eta} = \Gamma_i$ if $\eta = i$. For each $x \in V$, we can build a random set of neighbors of x: $\Gamma_{\eta}(x) = \Gamma_i(x)$ if $\eta = i, i = 1, 2, \ldots, m$.

Pretopological cascade model on Multi-relational network. Under the *pretopological cascade model* on multi-relational networks, at each time step t, each $x \in A_{t-1}$ generates a random index η given by the Eq. (17) then generates random neighbors set $\Gamma_\eta(x)$; after that x infects the inactive neighbors $y \in \Gamma_\eta(x)$ with a probability $P_{x,y}$.

We can extend this model by choosing randomly a set $S_\eta \subset \{1, 2, \ldots, m\}$ and then using interset or union operator to generate a random set of neighbors of x. For example, we can define $\Gamma_\eta(x) = \cup_{k_i \in S_\eta} \Gamma_{k_i}(x)$ or $\Gamma_\eta(x) = \cap_{k_i \in S_\eta} \Gamma_{k_i}(x)$.

6   Conclusion

In this paper, we proposed *Stochastic Pretopology* as a general mathematical framework for complex systems analysis. The advantage of this approach is that we can not only deal with uncontrolled factors by using random sets but also work with a set as a whole entity, not as a combination of elements. We illustrate our approach by introducing various ways to define a stochastic pseudo-closure function in many situations. Furthermore, we presented an application by proposing *Pretopology Cascade Model*, a general information diffusion model which can apply on diffirent kinds of complex networks such as stochastic graphs, multi-relational networks. A point not discussed in this paper which can be seen as a perspective is practical aspects of the proposed model. In future works can be developed a software library for implementing stochastic pretopology algorithms and applying the proposed model for real-world complex systems.

References

1. Basileu, C., Amor, S.B., Bui, M., Lamure, M.: Prétopologie stochastique et réseaux complexes. Stud. Inform. Univ. **10**(2), 73–138 (2012)
2. Belmandt, Z.: Basics of Pretopology. Hermann (2011)
3. Berge, C.: The Theory of Graphs. Courier Corporation (1962)
4. Bui, Q.V., Sayadi, K., Bui, M.: A multi-criteria document clustering method based on topic modeling and pseudoclosure function. Informatica **40**(2), 169–180 (2016)
5. Easley, D., Kleinberg, J.: Networks Crowds and Markets. Cambridge University Press, Cambridge (2010)
6. Goldenberg, J., Libai, B., Muller, E.: Talk of the network: a complex systems look at the underlying process of word-of-mouth. Mark. Lett. **12**(3), 211–223 (2001)
7. Granovetter, M.: Threshold models of collective behavior. Am. J. Sociol. **83**(6), 1420–1443 (1978)
8. Lamure, M., Bonnevay, S., Bui, M., Amor, S.B.: A stochastic and pretopological modeling aerial pollution of an urban area. Stud. Inform. Univ. **7**(3), 410–426 (2009)
9. Molchanov, I.: Theory of Random Sets. Springer, London (2005). https://doi.org/10.1007/1-84628-150-4
10. Newman, M.E.: Complex systems: a survey. Am. J. Phys. **79**, 800–810 (2011)
11. Newman, M.E.J.: The structure and function of complex networks. SIAM Rev. **45**, 167–256 (2003)
12. Nguyen, H.T.: An Introduction to Random Sets. CRC Press, Boca Raton (2006)

Data Science and Computational Intelligence

A Survey of Spatio-Temporal Database Research

Neelabh Pant[1], Mohammadhani Fouladgar[1], Ramez Elmasri[1], and Kulsawasd Jitkajornwanich[2(✉)] ⬛

[1] Department of Computer Science and Engineering, College of Engineering, The University of Texas at Arlington, Arlington, TX 76019, USA
{neelabh.pant, mohammadhani.fouladgar}@mavs.uta.edu, elmasri@cse.uta.edu
[2] Department of Computer Science, Faculty of Science, King Mongkut's Institute of Technology Ladkrabang, Bangkok 10520, Thailand
kulsawasd.ji@kmitl.ac.th

Abstract. The main purpose of spatio-temporal database systems is combining the spatial and temporal features of data. Almost all spatio-temporal applications—such as mobile communication systems, traffic control systems, and GIS with moving objects—have a common basis, which is the requirement to handle both space and time characteristics of the data. Similar to other data types, spatio-temporal data are required to be accurately modeled, structured, and queried efficiently. In this paper, we survey data models, related operations, data structures and access methods for spatial, temporal, and spatio-temporal data types. These access methods basically are enhanced variations of the well-known R-tree.

Keywords: Spatio-temporal database · GIS · Access method · Survey
Data model

1 Introduction

One of the most utilized applications in today's time is the global positioning system (or GPS). Through such a device, moving objects such as cell phones or vehicles can be located with reference to other fixed spatial objects like roads, gas stations, cities or continents. It also helps the user to find different routes to reach a destination dynamically or to look for traffic on a specific route to his/her workplace. All such information about the real world data objects, for example, moving vehicles or stationary objects like routes, cities, etc. are stored and retrieved from a database. The real world data objects have spatial and temporal attributes, which are nearly impossible to store in a traditional database, like RDBMS because the objects are complex and inefficient as they have spatio-temporal behaviors and are multi-dimensional in nature.

Spatio-temporal databases are used in location-based services, including geographic information systems (GIS), environmental modeling and impact assessment, resource management, decision support, real-time navigational systems and transportation scheduling. Spatio-temporal data combines spatial concepts and temporal

© Springer International Publishing AG, part of Springer Nature 2018
N. T. Nguyen et al. (Eds.): ACIIDS 2018, LNAI 10752, pp. 115–126, 2018.
https://doi.org/10.1007/978-3-319-75420-8_11

concepts together, so we first give an overview of spatial databases followed by temporal databases, then give a survey of spatio-temporal database research.

The paper is organized as follows. Section 2 gives an overview of both spatial databases and temporal databases. A survey of spatio-temporal database research is presented in Sect. 3, which were divided into three subsections: (i) spatio-temporal data modeling, query languages and operators; (ii) multivariable and multidimensional data; and (iii) spatio-temporal indexing. Finally, we conclude our survey paper in Sect. 4.

2   Overview of Spatial Databases and Temporal Databases

In this section, we review spatial database and temporal database principles as they are strongly correlated to spatio-temporal databases. There is a vast amount of research in the field of spatial and temporal databases; mostly have to do with spatial/temporal data models, query languages, indexing structures as well as ontology concepts.

2.1   Spatial Database Overview

A spatial database management system (SDBMS) is a collection of spatially referenced data that acts as a model of reality. In its most basic form, the spatial database system is used to store, compute and retrieve spatial objects such as points, lines and polygons. In today's applications, we often require and manage 2D geographic/geometric data, and in many applications from different domains, we need a storage and retrieval facility for 3D data, such as the human brain or the arrangement of molecular proteins in a human body [1].

Spatial data models. Spatial data is a term used to describe data that pertain to the spatial locations occupied by objects. The spatial objects not only have spatial aspect; they also have non-spatial attributes that describe the objects, e.g., names of all the rivers, coordinates of a particular city, etc. The geometry of an object is represented as points, lines, polygons, and other geometric shapes. Two types of spatial models are used: object model and field model. *Object model* abstracts the spatial and other information in distinct, identifiable entities, called objects. These objects have specific geometries represented by coordinates. Each object also has a set of non-spatial attributes such as name, address, etc., which are stored non-spatially in the database. In the *field model*, data that is spread over a region defined by its continuity is represented. It is continuous in nature and has function values that map each *grid* point to specific values. It sees the world as a continuous surface over which features are varied and represented through functions [2]. We give an example of how spatial data can be represented differently using these two models as well as their respective operators that are supported w.r.t. OGC (Open Geospatial Consortium) in Fig. 1.

Spatial query languages. Spatial query language is a database language which is developed to query spatial features using extensions to the traditional SQL. The spatial extension to SQL enhances the traditional relational query language and operations, with the spatial relationships which is known as a *spatial query language*. As mentioned in [3, 4], we do not develop an exclusive spatial system, but integrate spatial query attributes/operations into SQL. Different spatial query operations can be classi-fied into following groups: (1) *point query* (e.g., find all rectangle containing given point), (2) *range query* (e.g., find all points within a query rectangle), (3) *nearest neighbor* (e.g., find all lines which intersects a query rectangle), (4) *distance scan* (e.g., enumerate points in increasing distance from a query point), (5) *intersection query* (e.g., find all the rectangles and polygons intersecting a query rectangle), and (6) *containment query* (e.g., find all the polygons within a query rectangle) [5].

Spatial indexing. The efficient storage and access are among the highest priorities which are done by using sophisticated data structures. Traditional indexing methods, however, are not efficient for such purpose; many spatial index structures were pro-posed and we will discuss some of them here (this includes R-tree, one of the earliest data structures for spatial indexing, capable of handling and managing multi-dimensional objects and has many variations).

R-Tree. R-trees are an extension of B-trees, which are height-balanced and store spatial objects in such a way that spatial queries can be executed rapidly. Spatial data sets are often too large to fit in the computer primary memory and the secondary storage access time is of several orders of magnitude slower than the main memory. This is a per-formance bottleneck because data has to be shipped back and forth from the primary memory to the secondary storage. Thus the goal of good physical database design is to keep this amount of data transfer to an absolute minimum [6].

R-Tree Variations. R-tree is the most widely used index to store, manage and retrieve spatial data objects (Other multi-dimensional indexes also exist, including quadtree, K-D-B tree, to name a few). For some spatial data objects, the MBRs are prone to overlap, and in such cases the algorithm has to trace many paths in order to find the desired result. This issue consumes a lot of time. To overcome this problem, a variation of R-tree called *R+ tree* was proposed [7]. In R+ tree, MBRs of internal nodes do not overlap and the access time is noticeably less. Another modified structure, called an *R-link tree*, links the index to each other, so that range queries are much faster and more efficient as compared to R-tree or R+ tree [8]. *Hilbert R-tree* [9] maps the data in the MBRs by taking the center point, "2D-c", of each bounding rectangle, which are then sorted and aligned according to their Hilbert value. The choice of which is the best index structure depends on the shapes, sizes, and distributions of the spatial data objects, as well as the types of queries that will be requested [10].

Spatial ontology. Ontology defines formalizations of interested concepts and associ-ated relationships/operations in reality. Spatial ontology allows spatial information—especially from the Web—to be modeled, reasoned, queried, and exchanged among machines. Most spatial ontology researches involve: (i) spatial database integration and (ii) spatial ontology development. In spatial database integration, multiple spatial databases from different configurations can communicate through the use of spatial ontology [36]. In spatial ontology development, spatial ontologies were primarily created by either (1) considering a limited set of spatial databases used in the domain applications [37] or (2) generalizing the concepts and related properties of geometry objects into the ontology [38–40].

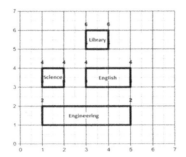

OBJECT VIEWPOINT OF THE DEPARTMENTS		
AREAID	**DEPARTMENT**	**OBJECT/ BOUNDARY**
UN1	ENGINEERING	[(1,1),(5,1),(5,2),(2,2)]
UN2	ENGLISH	[(3,3),(2,3),(5,4),(4,4)]
UN3	SCIENCE	[(1,3),(2,3),(2,4),(1,4)]
UN4	LIBRARY	[(3,5),(4,5),(4,6),(3,6)]

FIELD VIEWPOINT OF THE DEPARTMENTS	
Range	**Domain**
f(x,y) = "Engineering"	$1 = x = 5$; $1 = y = 2$
f(x,y) = "Science"	$1 = x = 2$; $1 = y = 4$
f(x,y) = "English,"	$3 = x = 5$; $3 = y = 4$
f(x,y) = "Library"	$3 = x = 4$; $5 = y = 6$

VECTOR/OBJECT DATA MODEL OPERATIONS	
Operator Type	**Sample Operations**
Set-oriented	Equals, disjoint, intersection
Topological	Boundary, interior, meets, overlap
Metric	Distance, angle, length, area
Direction	North, east, above, between
Network	Successors, ancestors, shortest-path
Dynamic	Translate, rotate, scale, merge

RASTER/FIELD DATA MODEL OPERATIONS	
Operator Type	**Sample Operations**
Local	Point-wise sums, differences, maximums, means
Focal	Slop, aspect, weighted average of neighborhood
Zonal	Sum or mean or maximum of field values in each zone

a) Vector model   b) Raster model

Fig. 1. Example of how spatial data can be represented and operated using spatial data models

2.2   Temporal Database Overview

When we talk about applications of database systems, such as accounting or banking, we can see that time issues and temporal data have significant roles in these applications. Temporal data is not limited to these applications, but almost in all applications, the time is not negligible. Therefore, temporal databases, which record time-referenced data, manifest themselves. To develop a reliable time-based application, a temporal database management system is required. Many temporal database researches have been conducted in the last two decades [12], which we will discuss next.

Temporal data models. Time is an order sequence of points in a specific granularity. These ordered points have an *origin*. The points that are before the origin are negative and the ones after the origin are positive. The granularity of points, called *chronon*, is the meaningful shortest time duration in a specific application, and is determined by the application requirements. In other words, according to the concept of time in an application, the chronon can be different. For instance, chronon for a bank transaction can be second or even millisecond, chronon for a daily or monthly sales records is day or month, and chronon for a censes in a country can be few years. It is worth mentioning that any updates to the database are done on time points and any change in time intervals is meaningless [13].

Temporal query languages. Many temporal query languages are proposed, however, only some eminent ones are discussed here. *TSQL2* (temporal structured query language) is a querying language to manipulate time varying data stored in relational database introduced in 1994. It provides a platform to express temporal datatypes and other *valid-time* information in form of SQL [20]. *TQUEL* (temporal query language) [21] is a derivative of *Quel*, a query language for Ingres RDBMS, designed to be a minimal extension in order to query temporal databases. It is considered to be the precedent to TSQL2 proposed in 1987. *T-SPARQL* is a temporal extension of the SPARQL query language for RDF graphs. It aims at embedding features from TSQL2. T-SPARQL is more suited for temporal RDF database model employing triple time stamping [22].

Temporal indexing. Since the advent of temporal data in early 90s, different algorithms and data structures have been introduced for temporal data indexing. Here, we review some of the proposed techniques. For general temporal indexes, tree-structure indexes over time intervals and versions are created based on different clustering methods according to times and key values. They also use partial replication for efficiency. Furthermore, using different storage media for historical data and current data is another approach [14] in that a better parallelization in operations can be achieved. *Time index* [15] is the first proposed indexing methods based on B+ tree over different versions, where each leaf page contains all active versions at the changes. Becker [16] proposes *multi-version B-tree*, which provides one index for both key and time with optimal I/O behavior. On the other hand, Ramaswamy [17] uses windows on time intervals as well as B+-tree in creating temporal indexes. *Aggregation indexing* is another important access method for temporal databases, by which a query like "Calculate an average GPA for each department for each academic year at UT Arlington" can be significantly accelerated [18]. A data structure, called *aggregation tree*, is built for each of the aggregation functions. In [19], *AVL trees* are used for each of start and end point. The algorithm first traverses the start index and inserts the tuples that are activated into the end point tree. It then removes the tuples, which were expired, and returns the aggregate as a result. It is also possible to create index for temporal join—a two-level index combining a B+-tree index over join attribute with a B+-tree index over the time dimension [30].

Temporal ontology. Temporal ontology depicts time-related properties of the interested objects in the application. The temporal ontology and spatial ontology are highly related as they are often analyzed together. In particular, spatial operations can be thought of as an extension of temporal operations in a 2-dimensional space. A formalization of temporal ontology is proposed in [41], which comprehensively defines temporal concepts and operations based on temporal logic developed by Allen [42]. The specified temporal ontology formalization allows temporal characteristics of the data set to be represented, reasoned about, and queried; in allowing so, a technique known as *lightweight* is used along with the ontology management software (e.g., Protégé) and reasoning module (e.g., Jess) [43].

3   Spatio-Temporal Database Research Survey

The section presented earlier in this paper gave us the basics of spatial and temporal databases and now we combine both of them to present spatio-temporal databases (see Fig. 2). This part of the paper will survey spatio-temporal databases. The following sections are arranged as follows: Sect. 3.1 discusses spatio-temporal modeling, query languages, and operators; multi-variable and multi-dimensional data are discussed in Sect. 3.2; and Sect. 3.3 explains spatio-temporal indexing.

3.1   Spatio-Temporal Modeling, Query Languages, and Operators

In spatio-temporal databases, we consider *time* as one of the dimensions along with the spatial dimensions. Spatial objects, namely points; lines; polygons and geometry collection, are the major focus in a spatio-temporal query where a user wants to measure the change in the existence, position and shape of the object [11] (see Fig. 2 for sample spatio-temporal queries). Spatio-temporal databases are used in location-based services, esp. GIS, where spatial and temporal properties of the real world data are stored and utilized.

Objects continue to change with the passage of time. Since objects change their geometries and positions with time, so the major objective of spatio-temporal databases is to store and retrieve the data in the same way as it was at a particular time interval. For such purposes, using traditional databases or relational databases would not suite the needs as they are efficient in storing one-dimensional data with discrete values. The challenge is to manage data with the data value, spatial-location and with the time of occurrence. We need to design an efficient DBMS which will consider the different cases mentioned above [23].

Spatial and temporal concepts need to be thoroughly investigated to come up with efficient techniques to manage spatial-temporal data. There are two different ways in which an object changes over time: some objects change slowly overtime, and mainly the changes are with respect to their shapes and other attributes; other rapidly changing objects, known as *moving objects*, typically do not change their shape but their position over time.

The principle of spatio-temporal databases is to capture the changes in the *geo-referenced objects*. We know that the changes are done in the non-spatial and spatial objects, but to identify what makes the changes is of more importance. Adding an identity with the objects helps us to recognize the changed and the original object. Such is the method to maintain the historical objects, which helps us to analyze the data better [25].

The spatio-temporal model extends the spatial data models with time. With the set of *object classes,* a time interval is needed to be associated, which can be in a schema of a traditional database where it would also have some objects, say tables with some attributes. This is common to all data models and is already given in the relational models [24]. Since objects can be created and destroyed at some point of time, a validity interval needs to be associated with each object. For this purpose, we can rely on time dependent attributes.

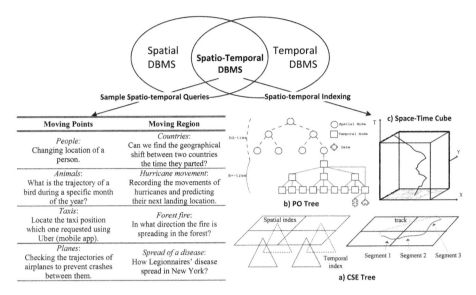

Fig. 2. Relationships between spatial, temporal, and spatio-temporal DBMSs and sample spatio-temporal queries (left) and indexing structures (right).

In Basic SQL we need to have two columns for starting point and end point of an event, and two columns for the coordinates of the objects. Query over spatio-temporal data represented by basic types in SQL can be possible, but in this way the queries are so complicated to make. However, if we use a temporal data type interval and spatial data type point, the queries can be so much easier to write. Spatio-temporal databases support both spatial and temporal data types. As mentioned earlier, temporal types include *point events* or *facts*, which are associated in a *single time point*, and *time period*, which are represented by *time intervals*. On the other hand, spatial datatypes support geometry objects including points, lines, and polygons. Spatio-temporal query language comprises temporal operators and spatial operators, as well as spatio-temporal operators.

3.2 Multi-variable and Multi-dimensional Data

We know that spatio-temporal data consists of spatial and temporal properties of objects. For example, the location of a student in the UT-Arlington according to his/her cell phone records. However, an object can have some other properties as well, determining the object's status. The multivariable data are used to express the state of an object. For example, a specific person in a specific place has some other properties other than time and location, such as age, sex, etc. All of these properties as well as location and time make a multivariable spatio-temporal data.

Different kinds of queries can be addressed to multivariable spatio-temporal data according to where (spatial data), when (temporal data), and how (multivariable data) an object is. To be able to respond to these queries, we need to study multivariable spatio-temporal, and relationship among spatial values, temporal values and multi-variable values. There are many different multivariable spatio-temporal data these days, many of which are related to diverse important issues such as global climate changes, economic development, and infectious diseases in which detecting and analyzing changes of the multivariable data are utilized [26].

Multivariable spatio-temporal can be modeled and viewed on a multidimensional data model by a *data cube*. Basically, dimensions are the attributes of an object including spatial and temporal attributes. For example for a chain store, we can keep records of things like monthly sales of items, and branches and locations where the items were sold. A multidimensional data model focuses on a specific issue such as sales in the chain store [27]. This issue has some facts, which are numerical measures. These facts are acquired from summarization and aggregation on raw data, which help us analyze the data.

3.3   Spatio-Temporal Indexing

With the advancement of technology comes the challenge to manage the huge amount of geospatial data that is generated every second. The objective is to handle such a large amount of geospatial-temporal data efficiently. Weng [28] introduced a new method of dynamic indexing of moving object trajectories. The index they proposed is made up of two major components *cube cell* and *trajectory*. Cube cell is a 3-D space, which is predefined by time interval and has (x, y) spatial coordinates. The trajectory, on the other hand, is considered and stored as the line object, which has time as an attribute. To store trajectory, we can use software like Oracle Spatial. The position data, forming moving trajectory, is stored in each cube cell and records a given object location over a specified time interval and trajectory is defined as the basic visiting unit.

Two components of geospatial object are: *static data* which consists of object description—such as Vehicle ID, Vehicle type, capacity etc.—and *dynamic data* which contains data of the real time position, which changes over time. The space-time cube (see Fig. 2(c)) is used as spatio-temporal model. There are several options to build an indexing structure and according to [29] for the moving objects; one can attempt the following three methods:

1. Use the existing model, but may be inefficient, e.g., 3-D R-Tree.
2. Invent a new access method, but that may take a very long time and may also be complex to integrate with the database management system.
3. Transform the data from higher to lower dimensionality.

Pfoser [29] made the changes to the data rather than building an indexing structure. The data from the 3-D space transformed into lower dimensional space which resulted into reduced overall size. The reduction in dimensionality allows us to use existing DBMS to store and index trajectories.

Since GPS devices record spatial and temporal data hence, R-tree and its variants would not be the optimal structure to index because of the continuously growing accumulative GPS tracks. Wang [30] designed a 2-D index structure which is known as *compressed start end tree* (CSE Tree). They divide the space that contains the entire spatial region into disjoint cells (see Fig. 2(a)). Every GPS trajectory is assigned a unique ID which is partitioned into segments by spatial grids. Each segment is then inserted into the temporal index of corresponding spatial grid. Whenever a query is encountered the indexing algorithm retrieves the spatial grids that overlap the query's spatial predicate and a candidate cell list is produced. After looking for the spatial part the temporal index of the candidate cell is searched to match the tracks that intersect the temporal predicate. Finally, the *trackIDs* are returned after being merged from different cells to remove the duplicates.

Indexing structure for spatio-temporal data can be done in several ways as defined by Pfoser [29], one of which is to integrate two different data structures which could fulfill the needs of spatio-temporal data. We can make use of individual different layers of data structures combined together which are efficient to manage spatial/multi-dimensional and temporal/single-dimensional data separately. One such data structure is *Po-Tree* which uses two different data structures to manage spatio-temporal data.

Po-Tree (see Fig. 2(b)) was initially designed by Noël [31] to study a Mexican volcano Popocatepetl where there were different stations/sensors spatially referenced at fixed positions around the volcano at a distance of 1.5 km, which were used to send measurements data to a central database.

Query done to the database is to fetch the data from a specific sensor (spatial location) at a specific time (temporal instance). The real time databases have time constraints [32]. Thus, for the database, it's very important for them to commit a transaction before a deadline. In the case of sensors, the deadline is the arrival of a new measurement. There are quite a few indexing structures for spatio-temporal data, of which Po-Tree is but one example.

1. 3D R-Tree: it takes *time* as another spatial dimension [33].
2. MVLQ: uses *multi-version tables* to track the data [33].
3. HR-Tree: it overlaps the snapshots for multi-versioning [34].
4. RT-Tree: prioritizes the dimensions [35].

The Po-Tree is based on the differentiation of the spatial and temporal data. The authors for this purpose have integrated two different kinds of data structures. For the indexing of spatial data, they use Kd-Trees, while the temporal aspect uses B+-Trees. Po-Tree structure has spatial objects linked with a different temporal sub-tree. This helps it to maintain the spatial objects and their temporal attributes. Since the recent values are always of higher interest than the older ones, the temporal data insertion is done in the right most leaf, where the newest data are found. For the fastest computation the links from the nodes of the data is added.

Queries are done by comparing the end time with the first temporal key of the last node, since the last node is directly accessible from the root of the tree. Queries for the recent data act directly on the last, but in other case, queries proceed with a B+-Tree.

4   Conclusion

In this paper we have surveyed accomplished work in the field of spatial, temporal and spatio-temporal databases. Spatio-temporal databases are used pervasively in many vast varieties of applications such as geographical information systems (GIS), computer aided design (CAD), etc. The capabilities of SDBMS and the spatial indexing techniques can also be used in many other applications where spatial data is involved for its fast access. In this paper, we mentioned the areas which are still open for research such as spatial and spatio-temporal indexing methods, query languages, ontology, and spatial data models.

References

1. Shashi, S., Sanjay, C.: Spatial Databases-A Tour. Pearson, London (2003)
2. Shekhar, S., et al.: Spatial databases-accomplishments and research needs. IEEE Trans. Knowl. Data Eng. **11**(1), 45–55 (1999)
3. Egenhofer, M.J.: Spatial SQL: a query and presentation language. IEEE Trans. Knowl. Data Eng. **6**(1), 86–95 (1994)
4. Borrmann, A., Rank, E.: Topological analysis of 3D building models using a spatial query language. Adv. Eng. Inform. **23**(4), 370–385 (2009)
5. Gandhi, V., Kang, J.M., Shekhar, S.: Technical report TR07-020. University of Minnesota (2007)
6. Guttman, A.: R-Trees: a dynamic index structure for spatial searching. In: SIGMOD 1984 (1984)
7. Sellis, T., Roussopoulos, N., Faloutsos, C.: The R+-tree: a dynamic index for multi-dimensional objects. In: VLDB 1987 (1987)
8. Ng, V., Kameda, T.: The R-link tree: a recoverable index structure for spatial data. In: Karagiannis, D. (ed.) DEXA 1994. LNCS, vol. 856, pp. 163–172. Springer, Heidelberg (1994). https://doi.org/10.1007/3-540-58435-8_181
9. Kamel, I., Faloutsos, C.: Hilbert R-tree: an improved R-tree using fractals. In: VLDB (1994)
10. Pant, N., et al.: Performance comparison of spatial indexing structures for different query types. In: Proceedings of 57th IRF International Conference (2016). ISBN 978-93-86083-35-7
11. Frank, A.U.: Chapter 2: ontology for spatio-temporal databases. In: Sellis, T.K. (ed.) Spatio-Temporal Databases. LNCS, vol. 2520, pp. 9–77. Springer, Heidelberg (2003). https://doi.org/10.1007/978-3-540-45081-8_2
12. Jensen, C.S.: Temporal database management. Ph.D. Dissertation, Aalborg University. Accessed http://people.cs.aau.dk/~csj/Thesis/
13. Dyreson, C., et al.: A consensus glossary of temporal database concepts. ACM SIGMOD Rec. **23**(1), 52–64 (1994)
14. Lomet, D., Betty, S.: Access methods for multiversion data. ACM **18**(2), 315–324 (1989). https://doi.org/10.1145/66926.66956
15. Elmasri, R., Wuu, G.T.J., Kim, Y.: The time index: an access structure for temporal data. In: VLDB 1990 (1990)
16. Becker, B., et al.: An asymptotically optimal multiversion B-tree. VLDB J. **5**(4), 264–275 (1996). https://doi.org/10.1007/s007780050028
17. Ramaswamy, S.: Efficient indexing for constraint and temporal databases. In: Afrati, F., Kolaitis, P. (eds.) ICDT 1997. LNCS, vol. 1186, pp. 419–431. Springer, Heidelberg (1997). https://doi.org/10.1007/3-540-62222-5_61
18. Kline, N., Snodgrass, R.T.: Computing temporal aggregates. In: Proceedings of the 11th International Conference on Data Engineering. IEEE (1995)
19. Böhlen, M., Gamper, J., Jensen, C.S.: Multi-dimensional aggregation for temporal data. In: Ioannidis, Y., et al. (eds.) EDBT 2006. LNCS, vol. 3896, pp. 257–275. Springer, Heidelberg (2006). https://doi.org/10.1007/11687238_18

20. Snodgrass, R.T.: TSQL2 language specification. SIGMOD Rec. **23**(1), 65–86 (1994). https://doi.org/10.1145/181550.181562
21. Snodgrass, R.T.: The temporal query language TQuel. ACM TODS **12**(2), 247–298 (1987)
22. Grandi, F.: T-SPARQL: a TSQL2-like temporal query language for RDF. In: ADBIS 2010 (2010)
23. Abraham, T., Roddick, J.F.: Survey of spatio-temporal data. Geoinformatica **3**(1), 61–99 (1999)
24. Erwig, M., et al.: Spatio-temporal data types: an approach to modeling and querying moving objects in databases. GeoInformatica **3**(3), 269–296 (1999)
25. Roshannejad, A.A., Kainz, W.: Handling identities in spatio-temporal databases. In: Proceedings of ACSM/ASPRS 1995 Annual Convention and Exposition Tech (1995)
26. Li, X., Kraak, M.J.: Explore multivariable spatio-temporal data with the time wave: case study on meteorological data. In: Yeh, A., Shi, W., Leung, Y., Zhou, C. (eds.) Advances in Spatial Data Handling and GIS. Lecture Notes in Geoinformation and Cartography. Springer, Heidelberg (2012). https://doi.org/10.1007/978-3-642-25926-5_7
27. Han, J., Kamber, M.: Data Mining: Concepts and Techniques. Morgan Kaufmann, Burlington (2001)
28. Weng, J., Wang, W., Fan, K., Huang, J.: Design and implementation of spatial-temporal data model in vehicle monitor system. In: Proceedings of 8th International Conference on GeoComputation (2005)
29. Pfoser, D., Jensen, C.S.: Trajectory indexing movement constraints. Geoinformatica **9**(2), 93–115 (2005). https://doi.org/10.1007/s10707-005-6429-9
30. Wang, L., et al.: A flexible spatio-temporal indexing scheme for large-scale GPS track retrieval. In: IEEE-Mobile Data Management, MDM 2008 (2008)
31. Noël, G., Servigne, S., Laurini, R.: Po tree-a real-time spatio-temporal data indexing structure. In: Proceedings of 11th International Symposium on Spatial Data Handling, UK (2004)
32. Kuo, T.W., Lam, K.Y.: Real-time database systems: an overview of system characteristics and issues. In: Lam, K.Y., Kuo, T.W. (eds.) Real-Time Database Systems. The International Series in Engineering and Computer Science (Real-Time Systems), vol. 593. Springer, Boston (2002). https://doi.org/10.1007/0-306-46988-X_1
33. Zhu, Q., Ging, J., Zhang, Y.: An efficient 3D R-tree spatial index method for virtual geographic environments. ISPRS J. Photogram. Remote Sens. **62**(3), 217–224 (2007). https://doi.org/10.1016/j.isprsjprs.2007.05.007
34. Nascimento, M., Silva, J.: Towards historical R-trees. In: Proceedings of the 1998 ACM Symposium on Applied Computing, Atlanta, USA, pp. 235–240 (1998)
35. Xu, X., Han, J., Lu, W.: RT-tree: an improved R-tree indexing structure for temporal spatial databases. In: Proceedings of the 4th International Symposium on Spatial Data Handling, Switzerland, Zurich (1990)
36. Bennacer, N., Aufaure, M.-A., Cullot, N., Sotnykova, A., Vangenot, C.: Representing and reasoning for spatiotemporal ontology integration. In: Meersman, R., Tari, Z., Corsaro, A. (eds.) OTM 2004. LNCS, vol. 3292, pp. 30–31. Springer, Heidelberg (2004). https://doi.org/10.1007/978-3-540-30470-8_14
37. Baglioni, M., Masserotti, M.V., Renso, C., Spinsanti, L.: Building geospatial ontologies from geographical databases. In: Fonseca, F., Rodríguez, M.A., Levashkin, S. (eds.) GeoS 2007. LNCS, vol. 4853, pp. 195–209. Springer, Heidelberg (2007). https://doi.org/10.1007/978-3-540-76876-0_13
38. Spaccapietra, S., et al.: On Spatial Ontologies. Swiss Federal Institute of Technology (2004)

39. Parent, C., Spaccapietra, S., Zimányi, E.: Conceptual Modeling for Traditional and Spatio-Temporal Applications: The MADS Approach. Springer, Heidelberg (2006). https://doi.org/10.1007/3-540-30326-X
40. Hogenboom, F., et al.: Spatial knowledge representation on the semantic web. In: ICSC 2010 (2010)
41. Hobbs, J.R., Pan, F.: An ontology of time for the semantic web. ACM Trans. Asian Lang. Inf. Process. (TALIP) **3**(1), 66–85 (2004). https://doi.org/10.1145/1017068.1017073
42. Allen, J., Kautz, H.: A model of naive temporal reasoning. Northeast Artificial Intelligence Consortium (NAIC), Review of Technical Tasks. Syracuse University, New York (1987)
43. O'Connor, M.J., Das, A.K.: A method for representing and querying temporal information in OWL. In: Fred, A., Filipe, J., Gamboa, H. (eds.) BIOSTEC 2010. CCIS, vol. 127, pp. 97–110. Springer, Heidelberg (2011). https://doi.org/10.1007/978-3-642-18472-7_8

Automatically Mapping Wikipedia Infobox Attributes to DBpedia Properties for Fast Deployment of Vietnamese DBpedia Chapter

Nhu Nguyen[1,2]([⊠]) [iD], Dung Cao[2], and Anh Nguyen[2]

[1] Haiphong University, Haiphong, Vietnam
nhunt@dhhp.edu.vn
[2] Hanoi University of Science and Technology, Hanoi, Vietnam
dungct@soict.hust.edu.vn, vietanknb@gmail.com

Abstract. DBpedia is one of the best practices for publishing and connecting structured data on the Web (Linked Data) to lead to the creation of a global data in different languages. This project extracts information from Wikipedia editions. The extraction procedure requires to manually map Wikipedia infoboxes into the DBpedia ontology. Thanks to crowdsourcing, a large number of infoboxes has been mapped in different languages. However, the number of accomplished mappings is still small and limited to most frequent infoboxes. There are many languages that have not yet mapped. In this paper, we concern about the problem of automatically mapping infobox attributes to properties into the DBpedia ontology. This task aims to identify which Wikipedia attribute should be match to any DBpedia property using instance-based approach. In this work, we perform with Vietnamese edition as our case-study. Experiments show that our method achieves impressing results with the high number of correct mappings.

Keywords: DBpedia · Wikipedia · Ontology · Mapping

1 Introduction

Nowadays, Linked Open Data (LOD) has grown rapidly to become large open datasets defined by RDF standards. Thanks to development of Data Web, the information on LOD is increasingly deeper, larger and easier to link in multi-domains, constituting the Linked Open Data cloud [2]. Currently, DBpedia is one of the central interlinking hubs of Linked Data [4]. This project exploits the Wikipedia content to publish in RDF data. Wikipedia is the most used encyclopedia and steadily maintained by thousands of active contributors. DBpedia project also develops and maintains a shared ontology that is populated using a rule-based semi-automatic approach that relies on Wikipedia templates. These templates are particular pages created to be included into other pages. Infoboxes

© Springer International Publishing AG, part of Springer Nature 2018
N. T. Nguyen et al. (Eds.): ACIIDS 2018, LNAI 10752, pp. 127–136, 2018.
https://doi.org/10.1007/978-3-319-75420-8_12

are a particular subset of templates. Each infobox is a set of attribute-value pairs that represents a summary of a Wikipedia article. Recently, contributors around the world join in together to map manually infoboxes and their attributes of Wikipedia articles to the classes and properties of the DBpedia ontology, respectively. This matching process is divided into two different steps. In detail, the infobox first is mapped to the corresponding class in the DBpedia ontology; then, infobox attributes are mapped to the properties owned by that class. Finally, the results are made available as Linked Data, and via DBpedia's main SPARQL endpoint. However, the number of infoboxes and attributes are so large for all Wikipedia articles in different languages. Therefore, this task is time-consuming and labor intensive. According to a statistic, DBpedia has extracted information in 128 languages, but there are 32 languages that have mappings.

Thus, in this paper, we focus on mapping infobox attributes with case study in Vietnamese Wikipedia. In particular, we propose a method to matching automatically with instance – based approach. This task will help to reduce time of mapping and building new DBpedia chapters for own languages that not yet mapped.

Organization: The remainder of the paper is structured as follows: Sect. 2 discusses some of the technologies used in actualizing the system and previous works done on the same field of study; Sect. 3 shows the problem definition; Sect. 4 describes mapping system; Sect. 5 represents experiments and evaluation; finally, a conclusion is given in Sect. 6 with some further suggested studies.

2   Related Work

Also related to our research, there are projects aiming to extract properties from some structured parts of the page different from infoboxes. For example, Yago exploits categories [10]. However, such approach is feasible only for a small number of attributes (for example the Wikipedia page Barack Obama is included in the 1961 births category, from which it can be inferred that Obama's birth year is 1961). Eytan et al. [1] present Ziggurat, an automatic system for aligning Wikipedia infoboxes, creating new infoboxes as necessary, filling in missing information, and detecting inconsistencies between parallel articles. Ziggurat uses self-supervised learning to allow the content in one language to benefit from parallel content in others. Experiments demonstrate the method's feasibility, even in the absence of dictionaries.

Nguyen et al. [7] propose WikiMatch, an approach for the infobox alignment task that uses different sources of similarity. The evaluation is provided on a subset of Wikipedia infoboxes in English, Portuguese and Vietnamese. More recently, Rinser et al. [9] and Airpedia [8] propose a three-stage general approach to infobox alignment between different versions of Wikipedia in different languages. First, it aligns entities using inter-language links; then it uses an instance-based approach to match infoboxes in different languages; finally, it aligns infobox attributes, again using an instance-based approach. The research in [3] considers the problem of automatic property alignment between two various DBpedia dataset by specifically object property alignment between English and Korean is their case study; they only focused on triples which must have URI-type object, and evaluated in limited dataset using DBpedia gold standard to estimate.

3   Problem Definition

In DBpedia, a real-world entity is represented by multiple instances [6] and each instance is described in a Wikipedia article in specific language l. Among the different components, we are interested in its infobox I, which is a set $A = a_{l_1}, a_{l_2}, \ldots, a_{l_n}$ of attribute with their corresponding values as (a_{l_i}, v_a) . This infobox summarizes important characteristics about an entity in that article. Figure 1 shows some attribute/value pairs in the English and Vietnamese infoboxes of entity about "Hồ Chí Minh" - the first president of Vietnam.

On the other hand, a DBpedia property r is a relation that describes a particular characteristic of an object. It has a domain and a range. The domain is the set of objects where such property can be applied. For instance, `birthPlace` is a property of `Person`, therefore `Person` is its domain. The range is the set of possible values of the property. It can be a scalar (*date, integer, string*, etc.) or an object (`Person`, `Place`, etc.). For example, the range of `age` is *integer* and the range of `birthPlace` is `Place`. Our system is to aim to achieve correct mappings between attributes and properties automatically. Thus, the problem can be classified as schema/ontology matching in which we are interested in equivalence relations attributes and properties. Formally, given an infobox I and an attribute a_i contained in I, our system will map this attribute to a corresponding relation r in the DBpedia ontology.

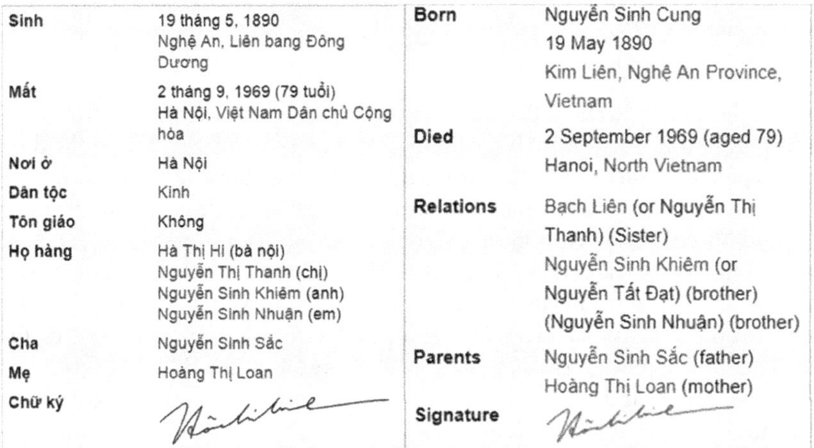

Sinh	19 tháng 5, 1890 Nghệ An, Liên bang Đông Dương	**Born**	Nguyễn Sinh Cung 19 May 1890 Kim Liên, Nghệ An Province, Vietnam
Mất	2 tháng 9, 1969 (79 tuổi) Hà Nội, Việt Nam Dân chủ Cộng hòa	**Died**	2 September 1969 (aged 79) Hanoi, North Vietnam
Nơi ở	Hà Nội		
Dân tộc	Kinh	**Relations**	Bạch Liên (or Nguyễn Thị Thanh) (Sister)
Tôn giáo	Không		Nguyễn Sinh Khiêm (or
Họ hàng	Hà Thị Hi (bà nội) Nguyễn Thị Thanh (chị) Nguyễn Sinh Khiêm (anh) Nguyễn Sinh Nhuận (em)		Nguyễn Tất Đạt) (brother) (Nguyễn Sinh Nhuận) (brother)
Cha	Nguyễn Sinh Sắc	**Parents**	Nguyễn Sinh Sắc (father)
Mẹ	Hoàng Thị Loan		Hoàng Thị Loan (mother)
Chữ ký		**Signature**	

Fig. 1. A part of infobox about Hanoi in English and Vietnamese Wikipedia

4   Mapping System

Our system is built upon the architecture as shown in Fig. 2. It consists of four components, data parser, candidate pair generator, similarity function generator and mapping filter. Data parser and similarity function generator are more important than others.

Workflow is in detail as following: Given two inputs and, our system first collects data needed for the matching between DBepdia and Wikipedia. Then, it creates an instance matrix based on retrieved data. This matrix is used to select the possibly candidate pairs between attributes and properties. Each pair is computed the similarity function basing on their values. The matching scores combine many similarities to choose the best case. Finally, mapping filter considers the returned pairs and produces the final mappings. Next, we describe the details of some important components.

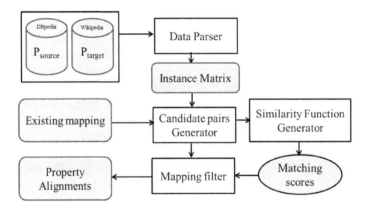

Fig. 2. The overview architecture of mapping system

4.1   Data Processing

One of our challenges is the difference between type of data retrieved from DBpedia and Wikipedia, especially Vietnamese Wikipedia. While, DBpedia data is strongly typed, Wikipedia articles are html pages that have no structure. Values of Wikipedia attributes often include data mixed among date, number and text and depend on editors with different editions. In addition, an article can be formatted by many ways depend on language. Hence, this module is built to retrieve and standardize the data needed for the mapping from Wikipedia attributes into DBpedia properties. As mentioned above, DBpedia uses an ontology in OWL format. This ontology describes classes, properties and all their characters. Each property has range and domain. Whereas, we divide properties into two kinds as following:

Object Properties. Two entities with different classes are related by a property. For instance, a person is connected to place by `birthPlace` property. The value

Table 1. Value/attribute in Infobox of Wikipedia article about "Hồ Chí Minh".

Atribute	Value
tên (name)	Hồ Chí Minh
hình (image)	Ho Chi Minh 1946.jpg
cỡ hình (image size)	320px
ngày sinh (birthdate)	{ngày sinh|1890|5|19}
nơi sinh (birth place)	[[Nghệ An]], [[Liên bang Đông Dương]]

v of an attribute a_i in an infobox I may contain one or more hyperlinks to other Wikipedia entities.

Data Properties. The relation connects instances of classes to literals of XML (scalar values). For example `birthDate` connects a `Person` to a date. In DBpedia, each RDF object has three main type including: date (*date, gYear-Month, gYear*), number (*double, float, nonNegativeInteger, positiveInteger, integer*), text (*string*). Besides something described above, we also consider some features of Vietnamese Wikipedia attributes obtained. For example, a Vietnamese Wikipedia uses date type in different formats such as "23 tháng 12 năm 2016" (dd/mm/yy), "23-12-2016" or "23/12/2016". In addition, number data of an entity maybe has different or approximate values in diverse editions. For an instance, population of Hanoi (the capital of Vietnam) is "7.588.150" and "7,587,800" in Vietnamese and English editions, respectively. They are also different in format.

To handle this problem, we set up a set of elements for each value of an attribute including: numbers, date, links, text tokens. All values of attributes are changed into these four sets for treating in next steps. Let's consider the attributes and their value of article about "Hồ Chí Minh" in Table 1. Value of "nơi sinh" (birth place) attribute will be changed two sets: one set of links {"Nghệ_An", "Liên_bang_Đông_Dương"} and one set of text token {Nghệ, An, ",", Liên, bang, Đông, Dương}. Value of "ngày sinh" (birth date) attribute is converted to three sets: one set of date {1890-5-19}, one number set {19, 5, 1890} and one set of token {19, "tháng", 5, 1890}. Converting to special set makes data become explicit and easy for performing task.

4.2 Candidate Pair Generating

The mission of this components is to find out pairs for matching. However, it is impractical to perform all pairwise comparisons, especially when repositories are large. Therefore, we limit the comparisons to only the instances of an entity, which are called candidates. To do this, we built a matrix based on cross-language link that aggregates the instances (rows) in the different editions of Wikipedia (columns). For example, the "Hồ Chí Minh" article in Vietnamese has infobox that consists name, birth date, birth place and etc. This data is also concluded in other Wikipedia editions. Building a matrix of instances that presents the same entity in different languages. In particular, we use pivot languages

Table 2. A part of instance matrix.

vi	en	de	nl
Hồ Chí Minh	Ho Chi Minh	Hồ Chí Minh	Hồ Chí Minh
Đảng Cộng sản Việt Nam	Communist Party of Vietnam	Kommunistische Partei Vietnam	Communistische Partij van Vietnam
Trường Đại học Bách Khoa Hà Nội	Hanoi University of Science and Technology	null	null

such as English, Germany, Netherland and Vietnamese. Thanks to DBpedia dumps, related resources are retrieved via cross-language links from an entity of Wikipedia article. An example about matrix of "Hồ Chí Minh" instance is given in Table 2.

We denote by $P_{l_1}, P_{l_2}, P_{l_3}, \ldots$ Wikipedia pages of a set of languages $L = l_1, l_2, \ldots, l_m$ respectively as column, and a page P as row. After that, properties r are collected to reduce the comparisons. These properties is matched with attributes in turn into $|r_{p_{l_i}}| \times |a_I|$ pairs and it is clear that this number is less than $|r_{P_{source}}| \times |a_{P_{target}}|$.

4.3   Mapping Extraction

Similarity Function. This section will be present how to determine whether an attribute a_I in infobox I can be match with a property r in DBpedia ontology. To extract mapping, we need to compute the similarity for every pairs (a_I, r). For each of the candidate pairs obtained in the previous step, the similarity sim_{l_i} is measured as follow:

$$sim_{l_i}(a_I, r_{l_i}) = \delta(f(a_I, P_l), g(r_{l_i}, P_{l_i})) \tag{1}$$

where, δ is inner function, it is used to calculate the similarity between the values of a_I and r for each pivot language. In total, this step produces a set of matches M for each infobox I. Then, pairs are chosen to become correct mapping with the highest $sim_{l_i}(a_I, r)$. The detail is described in next sections.

Matching Scores. Inner similar function $\delta(f(a_I, P_l), g(r_{l_i}, P_{l_i}))$ has value in [0,1]. And, $f(a_I, P_l)$ is used to address the value of Wikipedia attribute a_I in target language l, $g(r_{l_i}, P_{l_i})$ is used to extract the value of property r_{l_i} of DBpedia in l_i. We denote their definite values by $eval_W$ and $eval_D$, respectively. As mentioned above, we distinguish two kinds of property to compute the similarity. Thus, we apply these functions for each type of properties.

Object Property. When a range of property r is an object, the value of $eval_D$ consists one or more links connected to one ore more DBpedia resources. Using instance matrix, we can find out corresponding Wikipedia pages via cross-language link. Then, we will collect a set $eval_{D_l}$, whereas, each $eval_{D_{l_i}}$ can

Table 3. Comparing the value between a property and of an attribute "Hồ Chí Minh" entity.

Value of Property	Value of Atribute
dbr:Kim-Liên	[[Nghệ An]]
dbr:French-Indochina	[[Liên bang Đông Dương]]
dbr:Nghệ-An-Province	

be used to find in a set of links $eval_W$. Thus, the value of $\delta(eval_W, eval_D)$ is the reciprocal ratio $\frac{1}{ca}$, where ca is a cardinal number of the set $eval_W$. Let's consider the following example in Table 3. It shows the value of property "birthPlace" in English DBpedia and the value of the attribute "nơi sinh" in Vietnamese Wikipedia of "Hồ Chí Minh" entity. For DBpedia data, we have a set of $eval_D = \{$KimLiên,French Indochina, Nghệ An province$\}$ and $eval_{D_l} = \{$Kim Liên,Nam Đàn, Liên bang Đông Dương, Nghệ An$\}$ from instance matrix. And for Wikipedia, we have a set $eval_W = \{$Nghệ An, Liên bang Đông Dương$\}$. Therefore, the value of $\delta(eval_W, eval_D)$ is returned by $\frac{1}{2}$.

Datatype Property. When range of property is not an object, we only concern about data type. Score function works in flexible ways depending on the property type. For date data, there are three kinds of date (date with day, month and year; gYearMonth with month and year; gYear with year). Therefore, we compute the function for each case. The value of δ is calculated by searching the day, the month and the year of $eval_D$ and assign to each part of date that appear in $eval_W$ within a weight by $\frac{1}{3}$. Similarly, each part of gYearMonth and gYear are assigned a value of $\frac{1}{2}$ and 1, respectively.

For number, we need to handle about their unit because infobox attributes often have no unit or unclear unit. The number values often occur in attribute such as "population","total area" and "density" For example, the unit of property "areaTotal" is m^2 but the unit of attribute "diện tích" (total area) can be km^2 or m^2. Thus, we have to change them into a uniform unit to compare. Meanwhile the values of an attribute in different Wikipedia editions have changed over time. Therefore, the approximate comparison is used in these cases. In particular, we set a tolerance $t > 0$ and calculate $\varepsilon = \frac{|eval_D - ca|}{eval_D}$, with ca is the number of elements in the set $eval_W$. If $\varepsilon < t$, the function $\delta(eval_W, eval_D)$ has the value 1 and otherwise 0.

For text data, string kernel method [5] is used to compare the similarity between two strings. In this mapping case study, sub-sequences are n-grams that have converted into several sets of token from data parser step, whereas $n = \min(|eval_D, eval_W^*|)$ with $eval_W^*$ is a set of token which has been translated to target language. The similarity function is measured as follow:

$$\delta(eval_W, eval_D) = \frac{\hat{K}(eval_D, eval_W^*)}{n - i - 1} \qquad (2)$$

where K is a kernel of length n and it is defined by Lodhi in [5].

Matching Filter. This component filters the best pairs to construct the final property alignment. A pair (a_I, r_{l_i}) is correct mapping if its matching score $mScore(a_I, r_{l_i})$ satisfies the conditional statement of Eq. 3:

$$mScore(a_I, r_{l_i}) = \max(sim_{l_i}(a_{I_j}, r_{l_i}) \ for \ each \ i = 1 \ldots m, j = 1 \ldots n \quad (3)$$

where a_{I_i} is an element of attribute set A_I. This filtering strategy is specially effective on highly when many candidates are very similar but only the most similar one is the expected. On top of that, this component uses a majority voting method to give the final answers. Above, we have described the detail of system. The next section reports the experiments and the results.

5   Experiments and Evaluation

Experiments have been carried on Vietnamese Wikipedia, using existing DBpedia editions in three pivot languages (English, German and Dutch) as training data. Vietnamese data is a set of attributes extracted from Wikipedia infoboxes. Choosing an infobox for mapping is based on two following principles: the attribute number of test data and infobox are equal and all articles chosen have to link to the articles in pivot languages. However, the value of infobox attributes are often incomplete even null in Wikipedia articles. Thus, we have to create a dataset for each infobox in Vietnamese Wikipedia so that the number of attributes had value as much as possible. Most frequent infoboxes are mapped first. As a result, although the number of mappings may be small, a large number of articles can be added to the ontology. Fortunately, mapping statistic of DBpedia enlisted all infoboxes with occurrences in order from largest to smallest. In our experiment, we have created data test basing on the top 10 infoboxes with 1556 attributes. The evaluation has been performed on the Vietnamese mappings. In order to evaluate our approach for automatically mapping, we firstly explain how to estimate our performance; and then show the statistics of the experiment datasets. Similarly to infobox, each attribute has an occurrence index. The number presents its important role. In fact, while some attributes are occurred many times and others are appeared rarely. Therefore, the occurrence is one of the most criteria to evaluate the mapping results. To evaluate, we use Eqs. 4 and 5 to compute the proportion of mapped attributes (D) and the percentage of mapped attributes based on occurrences (E).

$$D = \frac{|N|}{|M|} \quad (4)$$

$$E = \frac{\sum_N c_i}{\sum_M c_i} \quad (5)$$

Where M is the total number of attributes in infobox I, N is the number of attributes mapped correct and c_i is the occurrence of an attribute a_I.

Table 4. The results of attribute mapping.

Occurrences	Infobox	M	N(D)	E
795025	Bảng phân loại (Categories)	167	62(37.1%)	98.8%
120715	Thông tin khu dân cư (Infobox setlement)	419	139(33.2%)	64.5%
11197	Thông tin hành tinh (Infobox planet)	87	35(40.2%)	45.1%
9742	Thông tin đơn vị hành chính Việt Nam (Infobox administrative divisions of Vietnam)	71	31(43.7%)	78.4%
9737	Thông tin nhân vật hoàng gia (Infobox royalty)	158	33(20.9%)	45%
2219	Thông tin tiểu sử bóng đá (Infobox football biography)	250	223(89.2%)	95.5%
1924	Thông tin nhạc sĩ (Infobox musical artist)	112	58(51.8%)	60.9%
1761	Thông tin đĩa đơn (Infobox single)	42	30(71.4%)	90.6%
1655	Thông tin phim (Infobox film)	53	33(62.3%)	85.9%
1631	Thông tin viên chức (Infobox officeholder)	197	100(50.8%)	65.7%

Let's consider the results of our method to mapping automatically as shown in Table 4 with the top 10 infoboxes. For "Thông tin đơn vị hành chính Việt Nam" (Infobox administrative divisions of Vietnam), there are 31 attributes which are mapped correctly ($N = 31$), approximately 44% (D). The remaining attributes, which have not mapped yet, amounted to 56%. However, most of them belong to attributes that have lower occurrences. Moreover, the relation between attribute and property does not exist or that attribute is too specific for only Vietnamese Wikipedia so that it is difficult to find out a corresponding property in DBpedia. For an instance, the attributes "người sáng lập" (founder), "cỡ bản đồ" (map size), "nhãn bản đồ" (map label) and etc. have occurrences that are less than or equal 1. Besides, attribute "xã" (commune) or "phường" (ward) could not match with any exist property in DBpedia. Hence, if this evaluation is based on occurrences, our method give the better value D with 98.8%. The result of our case study illustrated that using the instance-based approach is useful for mapping automatically. Therefore, this make encourage the deployment of DBpedia chapters effectively.

6   Conclusion and Future Work

In this paper, we introduced an approach to mapping Wikipedia attributes into corresponding properties in DBpedia ontology automatically basing on their values. We also presented a case study on Vietnamese Wikipedia that illustrated the benefits of automatic mapping by our approach. Our work is the first research in Vietnam about this area. Thus, this contribution is valuable to evolve into the development of linked data in Vietnam and fast deployment Vietnamese DBpedia chapter in the context of its mapping community is still weak. We developed a system to generate correct mapping. Experimental results demonstrated the effectiveness with high accuracy. However, there are a number of problems that we intend to pursue in future work. To further improve the effectiveness of algorithm, we will evaluate our method with deeper analyses, including comparison to more baselines. We also aim to improve our method by optimizing the matching scores or considering the label of attributes and properties.

References

1. Adar, E., Skinner, M., Weld, D.S.: Information arbitrage across multi-lingual Wikipedia. In: Proceedings of the Second ACM International Conference on Web Search and Data Mining, (WSDM), pp. 94–103. ACM (2009)
2. Bizer, C., Lehmann, J., Kobilarov, G., Auer, S., Becker, C., Cyganiak, R., Hellamnn, S.: DBpedia - a crystallization point for the web of data. Web Semant. Sci. Serv. Agents World Wide Web **7**(3), 154–165 (2009)
3. Kim, E.-K., Choi, K.-S.: Cross-lingual property alignment for DBpedia ontology using triple conceptualization. In: Proceedings of the 13th International Semantic Web Conference: Poster (2014)
4. Kobilarov, G., Bizer, C., Auer, S., Lehmann, J.: DBpedia - a linked data hub and data source for web applications and enterprises. In: The 18th International World Wide Web Conference (2009)
5. Lodhi, H., Saunders, C., Shawe-Taylor, J., Cristianini, N., Watkins, C.: Text classification using string kernels. Mach. Learn. Res. **2**, 419–444 (2002)
6. Nguyen, T.N., Takeda, H., Nguyen, K., Ichise, R., Cao, T.D.: Type prediction for entities in DBpedia by aggregating multilingual resources. In: Proceedings of the 15th International Semantic Web Conference: Poster (2016)
7. Nguyen, T., Moreira, V., Nguyen, H., Freire, J.: Multilingual schema matching for Wikipedia infoboxes. In Proceedings of the VLDB Endowment, pp. 133–144. ACM (2011)
8. Palmero Aprosio, A., Giuliano, C., Lavelli, A.: Towards an automatic creation of localized versions of DBpedia. In: Alani, H., Kagal, L., Fokoue, A., Groth, P., Biemann, C., Parreira, J.X., Aroyo, L., Noy, N., Welty, C., Janowicz, K. (eds.) ISWC 2013. LNCS, vol. 8218, pp. 494–509. Springer, Heidelberg (2013). https://doi.org/10.1007/978-3-642-41335-3_31
9. Rinser, D., Lange, D., Naumann, F.: Cross-lingual entity matching and infobox alignment in Wikipedia. Web Semant. Sci. Serv. Agents World Wide Web **38**(6), 887–907 (2013)
10. Suchanek, F.M., Kasneci, G., Weikum, G.: Yago: a core of semantic knowledge. In: Proceedings of the 16th International Conference on World Wide Web, pp. 697–706. CiteSeerX (2007)

A Fast Algorithm for Posterior Inference with Latent Dirichlet Allocation

Bui Thi-Thanh-Xuan[1,2]([⊠]), Vu Van-Tu[1], Atsuhiro Takasu[3], and Khoat Than[1]

[1] Hanoi University of Science and Technology, Hanoi, Vietnam
thanhxuan1581@gmail.com, vutu201130@gmail.com, khoattq@soict.hust.edu.vn
[2] University of Information and Communication Technology,
Thai Nguyen, Vietnam
[3] National Institute of Informatics, Tokyo, Japan
takasu@nii.ac.jp

Abstract. Latent Dirichlet Allocation (LDA) [1], among various forms of topic models, is an important probabilistic generative model for analyzing large collections of text corpora. The problem of posterior inference for individual texts is very important in streaming environments, but is often intractable in the worst case. To avoid directly solving this problem which is NP-hard, some proposed existing methods for posterior inference are approximate but do not have any guarantee on neither quality nor convergence rate. Based on the idea of Online Frank-Wolfe algorithm by Hazan [2] and improvement of Online Maximum a Posteriori Estimation algorithm (OPE) by Than [3,4], we propose a new effective algorithm (so-called NewOPE) solving posterior inference in topic models by combining Bernoulli distribution, stochastic bounds, and approximation function. Our algorithm has more attractive properties than existing inference approaches, including theoretical guarantees on quality and fast convergence rate. It not only maintains the key advantages of OPE but often outperforms OPE and existing algorithms before. Our new algorithm has been employed to develop two effective methods for learning topic models from massive/streaming text collections. Experimental results show that our approach is more efficient and robust than the state-of-the-art methods.

Keywords: Topic models · OPE · Stochastic inference · NewOPE
MAP estimation

1 Introduction

Latent Dirichlet analysis is a flexible latent variable framework for modeling high-dimensional sparse count data. Latent Dirichlet Allocation (LDA) [1] has found successful applications in a wide range of areas including text modeling [5], bioinformatics [6], history [7,8], politics [5,9], psychology [10]. Estimation of posterior distributions for individual documents is one of the core issues in LDA.

© Springer International Publishing AG, part of Springer Nature 2018
N. T. Nguyen et al. (Eds.): ACIIDS 2018, LNAI 10752, pp. 137–146, 2018.
https://doi.org/10.1007/978-3-319-75420-8_13

Recently, this estimation problem is considered by many researchers, and many methods such as Variational Bayes (VB) [1], Collapsed Variational Bayes (CVB) [11], CVB0 [12], Collapsed Gibbs Sampling (CGS) [7,13], and OPE [3] have been proposed and applied. We also see that the quality of LDA in practice is determined by the quality of the inference method being employed. However, almost mentioned methods do not have a theoretical guarantee of quality or convergence rate, except OPE algorithm.

Our first contribution is that we propose a new algorithm which is called NewOPE, for doing posterior inference of topic mixture in LDA. The posterior inference problem is in fact non-convex and is NP-hard [14]. Than and Doan [4] proved that OPE converges at a rate of $\mathcal{O}(1/T)$, which surpasses the best rate of existing stochastic algorithms for non-convex problems [15,16], where T is the number of iterations. One main drawback of OPE is that there is no guarantee for OPE escaping from saddle points of the inference problems [17][1].

Similar to OPE, NewOPE is stochastic in nature and theoretically converges to a local maximal/stationary point of the inference problem. We approximate the objective function by a combination of the upper and lower bounds using Bernoulli distribution. The usage of both bounds is stochastic in nature and helps us reduce the possibility of getting stuck at a local stationary point. Thus, it is an efficient approach for escaping saddle points in non-convex optimization. So, our new variant is appropriate and effective more than the original OPE. Existing methods become less relevant in high dimensional non-convex optimization.

Using NewOPE as core routines to do inference, our second contribution is that we obtain two effective methods for learning LDA (so-called ML-NewOPE, Online-NewOPE) from massive/streaming text collections. From extensive experiments on two large corpora New York Time and Pubmed, we find that our methods can reach state-of-the-art performance in both predictiveness and model quality.

Organization: The rest of this paper is organized as follows. We introduce an overview of posterior inference with LDA in Sect. 2. In Sect. 3, a new algorithm NewOPE for posterior inference is proposed in detail and is applied to online learning LDA. In Sect. 4, we give some results test with large datasets. Finally, we conclude the paper in Sect. 5.

Notation: Throughout the paper, we use the following conventions and notations. The unit simplex in the n-dimensional Euclidean space is denoted as $\Delta_n = \{ \boldsymbol{x} \in \mathbb{R}^n : \boldsymbol{x} \geq 0, \sum_{k=1}^n x_k = 1 \}$, and its interior is denoted as $\overline{\Delta}_n$. We will work with text collections with V dimensions (dictionary size). Each document $\boldsymbol{d}$ will be represented as a frequency vector, $\boldsymbol{d} = (d_1, \ldots, d_V)^T$ where d_j represents the frequency of term j in $\mathbf{d}$. Denote n_d as the length of $\boldsymbol{d}$, i.e., $n_d = \sum_j d_j$.

---

[1] A saddle point is critical point but not always a (local) maximal point. Further, the inference might have exponentially large number of saddle points.

2   Related Work

LDA [1] assumes that a corpus is composed from K topics $\boldsymbol{\beta} = (\boldsymbol{\beta}_1, \ldots, \boldsymbol{\beta}_K)$. Each document $\boldsymbol{d}$ is a mixture of those topics and is assumed to arise from the following generative process. For the nth word of $\boldsymbol{d}$:

- draw topic index $z_{dn} | \boldsymbol{\theta}_d \sim Multinomial(\boldsymbol{\theta}_d)$
- draw word $w_{dn} | z_{dn}, \boldsymbol{\beta} \sim Multinomial(\boldsymbol{\beta}_{z_{dn}})$.

We consider the MAP estimation of topic mixture for a given document d

$$\boldsymbol{\theta}^* = \arg \max_{\boldsymbol{\theta} \in \Delta_K} Pr(\boldsymbol{d} | \boldsymbol{\theta}, \boldsymbol{\beta}) Pr(\boldsymbol{\theta} | \alpha) \tag{1}$$

For a given document $\boldsymbol{d}$, the probability that a term j appears in $\boldsymbol{d}$ can be expressed as

$Pr(w = j | d) = \sum_{k=1}^{K} Pr(w = j | z = k). Pr(z = k | \boldsymbol{d})$ or $Pr(w = j | d) = \sum_{k=1}^{K} \beta_{kj} \theta_k$.

Hence the log likelihood of $\boldsymbol{d}$ is

$\log Pr(\boldsymbol{d} | \boldsymbol{\theta}, \boldsymbol{\beta}) = \log \prod_j Pr(w = j | d)^{d_j} = \sum_j d_j \log Pr(w = j | d)$

$= \sum_j d_j \log \sum_{k=1}^{K} \theta_k \beta_{kj}$

Remember that the density of the K-dimensional Dirichlet distribution with the parameter α is $P(\boldsymbol{\theta} | \alpha) \propto \prod_{k=1}^{K} \theta_k^{\alpha - 1}$. Therefore, problem (1) is equivalent to the following:

$$\boldsymbol{\theta}^* = \arg \max_{\boldsymbol{\theta} \in \Delta_K} \sum_j d_j \log \sum_{k=1}^{K} \theta_k \beta_{kj} + (\alpha - 1) \sum_{k=1}^{K} \log \theta_k \tag{2}$$

where $f(\boldsymbol{\theta}) = \sum_j d_j \log \sum_{k=1}^{K} \theta_k \beta_{kj} + (\alpha - 1) \sum_{k=1}^{K} \log \theta_k$ is the objective function of (2).

In the case of $\alpha \geq 1$, one can easily show that the problem (2) is concave, therefore it can be solved in polynomial time. Unfortunately, in practice of LDA, the parameter α is often small, says $\alpha < 1$, causing (2) to be non-concave. That is the reason for why (2) is intractable in the worst case. Sontag and Roy [14] showed that problem (2) is NP-hard in the worst case when $\alpha < 1$. Many "batch" posterior inference algorithms have been proposed, including variational Bayes (VB), collapsed variational Bayesian inference (CVB), CVB0, and collapsed Gibbs sampling (CGS). VB, CVB, and CVB0 try to estimate the distribution by maximizing a lower bound of the likelihood $p(d | \beta, \alpha)$, whereas CGS [7,13] tries to estimate $p(z | d, \beta, \alpha)$. However, those "batch" algorithms are not practical for large scale data analysis because they often require many sweeps through all documents in the corpus.

Based on Online Frank-Wolfe algorithm which is proposed by Hazan [2] solving effective convex optimization using stochastic approximation, Than and Doan [3,4] proposed Online Maximum a Posteriori Estimation (OPE) algorithm for doing inference of topic mixtures for documents. Details of OPE are presented in Algorithm 1.

Algorithm 1. OPE: Online maximum a posteriori estimation

Input: document $\mathbf{d}$ and model $\{\beta, \alpha\}$

Output: $\boldsymbol{\theta}^*$ that maximizes $f(\boldsymbol{\theta}) = \sum_j d_j \log \sum_{k=1}^{K} \theta_k \beta_{kj} + (\alpha - 1) \sum_{k=1}^{K} \log \theta_k$

    Initialize $\boldsymbol{\theta}_1$ arbitrary in Δ_K

    for $t = 1, 2, ...\infty$ **do**

        Pick f_t uniformly from $\{\sum_j d_j \log \sum_{k=1}^{K} \theta_j \beta_{kj}; (\alpha - 1) \sum_{k=1}^{K} \log \theta_k\}$

        $F_t := \frac{2}{t} \sum_{h=1}^{t} f_h$

        $e_t := \arg\max_{x \in \Delta_K} < F_t^{'}(\boldsymbol{\theta}_t), x >$

        $\boldsymbol{\theta}_{t+1} := \boldsymbol{\theta}_t + \frac{e_t - \boldsymbol{\theta}_t}{t}$

    end for

The main idea of OPE is to construct a stochastic sequence $F_t(\theta)$ that approximates for $f(\theta)$ by using uniform distribution, so that the (2) becomes easy to solve.

3   Proposed Method

We continue to consider the MAP problem (2) with the objective function

$$f(\boldsymbol{\theta}) = \sum_j d_j \log \sum_{k=1}^{K} \theta_k \beta_{kj} + (\alpha - 1) \sum_{k=1}^{K} \log \theta_k$$

We find out that OPE is a good algorithm for posterior inference, then we proceed to improve OPE by randomization to get better algorithm. To avoid solving directly non-convex optimization problem (2), we need to construct an approximation function which is easy to maximize and that approximates well for $f(\theta)$. We use Bernoulli distribution to construct the approximation of the objective function $f(\theta)$. Pick f_h has Bernoulli distribution with probability p from $\{g_1(\boldsymbol{\theta}); g_2(\boldsymbol{\theta})\}$ where $\Pr(f_h = g_1(\theta)) = p$, $\Pr(f_h = g_2(\theta)) = 1 - p$ and approximation $F_t(\theta)$ as a form $F_t(\theta) = \frac{2}{t} \sum_{h=1}^{t} f_h$. We see that $F_t(\theta)$ is a stochastic approximation which is easy to maximize and differentiable. We also see that

$$g_1(\boldsymbol{\theta}) = \sum_j d_j \log \sum_{k=1}^{K} \theta_k \beta_{kj} < 0; \quad g_2(\boldsymbol{\theta}) = (\alpha - 1) \sum_{k=1}^{K} \log \theta_k > 0$$

Hence, if we choose $f_1 = g_1$ then $F_1(\theta) < f(\theta)$, which leads $F_t(\boldsymbol{\theta})$ is a lower bound for $f(\boldsymbol{\theta})$. In contrast, if we choose $f_1 = g_2$ then $F_1(\theta) > f(\theta)$, and $F_t(\boldsymbol{\theta})$ is a upper bound for $f(\boldsymbol{\theta})$.

Although, OPE is a good candidate for solving a posterior inference in topic models, but we want to enhance OPE in different ways. It makes sense that two stochastic approximating sequences from above and below are better than one. So we construct two sequences that both approximating to $f(\theta)$, one begins with g_1 called the sequence $\{L_t\}$, another begins with g_2 called the sequence

$\{U_t\}$: Setting $f_1^l := g_1(\boldsymbol{\theta})$, $f_1^u := g_2(\boldsymbol{\theta})$. Pick f_t^l, f_t^u as Bernoulli distribution from $\{g_1(\boldsymbol{\theta}); g_2(\boldsymbol{\theta})\}$ with probability p, we have $L_t := \frac{2}{t}\sum_{h=1}^{t} f_h^l$ and $U_t := \frac{2}{t}\sum_{h=1}^{t} f_h^u$.

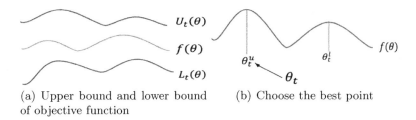

(a) Upper bound and lower bound of objective function   (b) Choose the best point

Fig. 1. The ideas of NewOPE

The idea of NewOPE is based on the greedy approach. At each iteration we always compare two values of $\{f(\boldsymbol{\theta}_t^u), f(\boldsymbol{\theta}_t^l)\}$ and take the point that makes the value of f highest as possible. NewOPE works differently from OPE. OPE just constructs one sequence $\{\boldsymbol{\theta}_t\}$ while NewOPE creates three sequences $\{\boldsymbol{\theta}_t^u\}, \{\boldsymbol{\theta}_t^l\}$ and $\{\boldsymbol{\theta}_t\}$ depending on each other. The structure of sequence $\{\boldsymbol{\theta}_t\}$ really changes, but OPE's good properties are remained in NewOPE (Fig. 1). Comparing with other inference approaches (including VB, CVB, CVB0 and CGS), NewOPE has many preferable properties as summarized

---

Algorithm 2. NewOPE

Input: document **d** and model $\{\beta, \alpha\}$
Output: $\boldsymbol{\theta}$ that maximizes $f(\boldsymbol{\theta}) = g_1(\boldsymbol{\theta}) + g_2(\boldsymbol{\theta})$

　Initialize $\boldsymbol{\theta}_1$ arbitrary in Δ_K
　$f_1^l := g_1(\boldsymbol{\theta})$, $f_1^u := g_2(\boldsymbol{\theta})$
　for $t = 2, 3, ...\infty$ **do**
　　Pick f_t^l as Bernoulli distribution from $\{g_1(\boldsymbol{\theta}) ; g_2(\boldsymbol{\theta})\}$ with probability p
　　$L_t := \frac{2}{t}\sum_{h=1}^{t} f_h^l$
　　$e_t^l := \arg\max_{x \in \Delta_K} < L_t'(\boldsymbol{\theta}_t), x >$
　　$\boldsymbol{\theta}_{t+1}^l := \boldsymbol{\theta}_t + \frac{e_t^l - \boldsymbol{\theta}_t}{t}$

　　Pick f_t^u as Bernoulli distribution from $\{g_1(\boldsymbol{\theta}) ; g_2(\boldsymbol{\theta})\}$ with probability p
　　$U_t := \frac{2}{t}\sum_{h=1}^{t} f_h^u$
　　$e_t^u := \arg\max_{x \in \Delta_K} < U_t'(\boldsymbol{\theta}_t), x >$
　　$\boldsymbol{\theta}_{t+1}^u := \boldsymbol{\theta}_t + \frac{e_t^u - \boldsymbol{\theta}_t}{t}$

　　$\boldsymbol{\theta}_{t+1} := \arg\max_{\boldsymbol{\theta} \in \{\boldsymbol{\theta}_{t+1}^u, \boldsymbol{\theta}_{t+1}^l\}} f(\boldsymbol{\theta})$
　end for

- NewOPE explicitly has a theoretical guarantee on fast convergence rate. This is the most notable property of NewOPE, for which existing inference methods often do not have, except OPE.
- Unlike CVB and CVB0 [12], NewOPE does not change the global variables when doing inference for individual documents. So, NewOPE is more beneficial than CVB and CVB0.

Our algorithm not only retains the good characteristics of OPE but also makes it better and more effective when solving the problem of inference with LDA. We focus on overcoming the disadvantages of ineffective batch methods, exploiting stochastic approximation, and putting probability distributions into our new algorithm.

We have seen many attractive properties of NewOPE that other methods do not have. We further show in this section the simplicity of using NewOPE for designing fast learning algorithms for topic models. More specifically, we obtain two algorithms: Online-NewOPE which learns LDA from large corpora in an online fashion, and ML-NewOPE which enables us to learn LDA from either large corpora or data streams based on ML-OPE and Online-OPE in [4].

4   Empirical Evaluation

This section is devoted to investigating practical behaviors of NewOPE, and how useful it is when NewOPE is employed to design new algorithms for learning topic models at large scales. To this end, we take the following methods, datasets, and performance measures into investigation.

Inference methods: Variational Bayes (VB) [1], Collapsed variational Bayes (CVB0) [12], Collapsed Gibbs sampling (CGS) [7], Online MAP estimation (OPE) [4], New Online MAP estimation (NewOPE).

Large-scale learning methods: ML-NewOPE, Online-NewOPE, ML-OPE, Online-OPE [4], Online-CGS [7], Online-CVB0 [12], Online-VB [18].

To avoid randomness, the learning methods for each dataset is run five times and reported its average results.

- Model parameters: The number of topics $K = 100$, the hyper-parameters $\alpha = \frac{1}{K}$ and $\eta = \frac{1}{K}$. These parameters are commonly used in topic models.
- Inference parameters: The number of iterations was chosen as $T = 50$.
- Learning parameters: $\kappa = 0.9$, $\tau = 1$ adapted best for existing inference methods.

Datasets: We used the two large corpora: Dataset PubMed consists of 330,000 articles from the PubMed central and dataset New York Times (NYT) consists of 300,000 news[2].

---

[2] The data sets were taken fromhttp://archive.ics.uci.edu/ml/datasets.

Measures: We used two measures *Log Predictive Probability* (LPP) [7] and *NPMI* [19]. Predictive probability measures the predictiveness and generalization of a model to new data, while NPMI evaluates semantics quality of an individual topic.

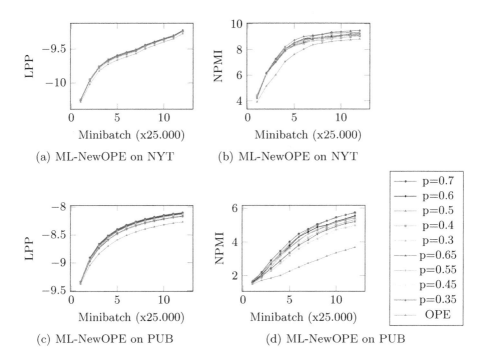

(a) ML-NewOPE on NYT   (b) ML-NewOPE on NYT

(c) ML-NewOPE on PUB   (d) ML-NewOPE on PUB

Fig. 2. ML-NewOPE with different parameter p

In Figs. 2 and 3, we find that the NewOPE algorithm which uses Bernoulli distribution instead of uniform better than OPE, especially when probability p is greater than 0.5. From Figs. 4 and 5, NewOPE is better than VB, CVB and CGS on two datasets and with two measures Predictive Probability and NPMI, especially when the probability p is greater than 0.5 such as 0.6, 0.65 or 0.7 in our experiments. This explains the contribution of prior/likelihood to solving the inference problem. In terms of semantic quality measured by NPMI, Fig. 5 shows the results from our experiments. It is easy to observe that Online-NewOPE often learns models with a good semantic and NewOPE makes ML-NewOPE and Online-NewOPE work more efficient. NewOPE demonstrates our idea of using two stochastic sequences $\{U_t(\theta), L_t(\theta)\}$ to approximate an objective function $f(\theta)$. The idea of increasing the randomness and greedy of the algorithm is exploited here. Firstly, two stochastic sequences of function $U_t(\theta), L_t(\theta)$ are used to raise our participants and information relevant to objective function $f(\theta)$. Hence, at the next iteration, we have more choices in θ_t. Secondly, choosing θ_t

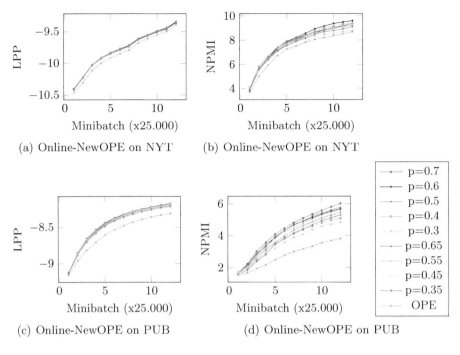

Fig. 3. Online-NewOPE with different parameter p

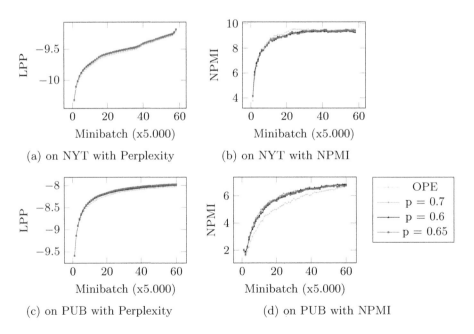

Fig. 4. ML-NewOPE compares with ML-OPE

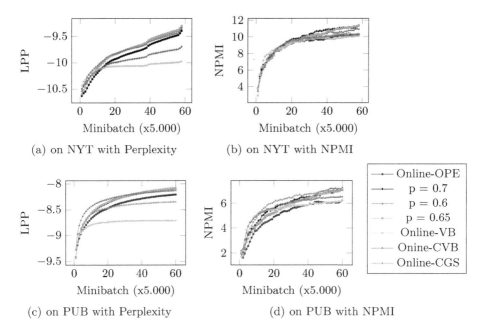

Fig. 5. Online-NewOPE compares with Online-OPE, Online-VB, Online-CVB0 and Online-CGS

from $\{\boldsymbol{\theta}_t^u, \boldsymbol{\theta}_t^l\}$ that makes the value of $f(\boldsymbol{\theta})$ higher in each iteration comes from idea of greedy algorithms. There are many ways of choices here, but we design a best way to create $\boldsymbol{\theta}_t$ from $\{\boldsymbol{\theta}_t^u, \boldsymbol{\theta}_t^l\}$. This approach is simple and there is no need for extra parameter.

5   Conclusion

We have discussed how posterior inference for individual texts in topic models can be done efficiently. Using Bernoulli distribution and stochastic approximation, we now provide NewOPE algorithm to deal well with this problem. By exploiting NewOPE carefully, we have arrived at two efficient methods for learning LDA from data streams or large corpora. As a result, they are good candidates to help us deal with text streams and big data.

References

1. Blei, D.M., Ng, A.Y., Jordan, M.I.: Latent Dirichlet allocation. J. Mach. Learn. Res. **3**(Jan), 993–1022 (2003)
2. Hazan, E., Kale, S.: Projection-free online learning. In: Proceedings of the 29th International Conference on Machine Learning, ICML 2012, 26 June – 1 July 2012, Edinburgh, Scotland, UK (2012)

3. Than, K., Doan, T.: Dual online inference for latent Dirichlet allocation. In: ACML (2014)
4. Than, K., Doan, T.: Guaranteed inference in topic models. arXiv preprint arXiv:1512.03308 (2015)
5. Blei, D.M.: Probabilistic topic models. Commun. ACM **55**(4), 77–84 (2012)
6. Falush, D., Stephens, M., Pritchard, J.K.: Inference of population structure using multilocus genotype data: linked loci and correlated allele frequencies. Genetics **164**(4), 1567 (2003)
7. Hoffman, M., Blei, D.M., Mimno, D.M.: Sparse stochastic inference for latent Dirichlet allocation. In: Proceedings of the 29th International Conference on Machine Learning (ICML 2012), pp. 1599–1606. ACM (2012)
8. Yao, L., Mimno, D., McCallum, A.: Efficient methods for topic model inference on streaming document collections. In: Proceedings of the 15th ACM SIGKDD International Conference on Knowledge Discovery and Data Mining, pp. 937–946. ACM (2009)
9. Grimmer, J.: A Bayesian hierarchical topic model for political texts: measuring expressed agendas in senate press releases. Polit. Anal. **18**(1), 1–35 (2010)
10. Schwartz, H.A., Eichstaedt, J.C., Dziurzynski, L., Kern, M.L., Blanco, E., Kosinski, M., Stillwell, D., Seligman, M.E., Ungar, L.H.: Toward personality insights from language exploration in social media. In: AAAI Spring Symposium: Analyzing Microtext (2013)
11. Teh, Y.W., Kurihara, K., Welling, M.: Collapsed variational inference for HDP. In: Advances in Neural Information Processing Systems, pp. 1481–1488 (2007)
12. Asuncion, A., Welling, M., Smyth, P., Teh, Y.W.: On smoothing and inference for topic models. In: Proceedings of the Twenty-Fifth Conference on Uncertainty in Artificial Intelligence, pp. 27–34. AUAI Press (2009)
13. Griffiths, T.L., Steyvers, M.: Finding scientific topics. Proc. Nat. Acad. Sci. **101**(suppl 1), 5228–5235 (2004)
14. Sontag, D., Roy, D.: Complexity of inference in latent Dirichlet allocation. In: Neural Information Processing System (NIPS) (2011)
15. Dang, C.D., Lan, G.: Stochastic block mirror descent methods for nonsmooth and stochastic optimization. SIAM J. Optim. **25**(2), 856–881 (2015)
16. Ghadimi, S., Lan, G., Zhang, H.: Mini-batch stochastic approximation methods for nonconvex stochastic composite optimization. Math. Program. **155**, 267–305 (2016)
17. Dauphin, Y.N., Pascanu, R., Gulcehre, C., Cho, K., Ganguli, S., Bengio, Y.: Identifying and attacking the saddle point problem in high-dimensional non-convex optimization. In: Advances in Neural Information Processing Systems, pp. 2933–2941 (2014)
18. Hoffman, M.D., Blei, D.M., Wang, C., Paisley, J.W.: Stochastic variational inference. J. Mach. Learn. Res. **14**(1), 1303–1347 (2013)
19. Lau, J.H., Newman, D., Baldwin, T.: Machine reading tea leaves: automatically evaluating topic coherence and topic model quality. In: EACL, pp. 530–539 (2014)

Design Thinking Based R&D, Development Technique, and Project Based Learning

A Development of Participatory Sensing System for Foreign Visitors in PBL

Shuta Nakamae[1(✉)], Wataru Sakamoto[1], Tetsuya Negishi[1], Shuhei Goto[1], Buntarou Shizuki[1], Chiemi Watanabe[2], and Toshiyuki Amagasa[1]

[1] University of Tsukuba, Tsukuba, Japan
{nakamae,shizuki}@iplab.cs.tsukuba.ac.jp,
{sakamoto,gotou,amagasa}@kde.cs.tsukuba.ac.jp,
negishi@mma.cs.tsukuba.ac.jp
[2] Advanced Institute of Industrial Technology, Tokyo, Japan
chiemi@acm.org

Abstract. In this project, we design a system to realize user participatory sensing. This system collects, organizes and visualizes data such as problems of foreign visitors in a city, thereby allowing city officials use that information for making decisions on various city policies. Furthermore, foreigners can be supported by the volunteers from city officials through exchanges of posts and comments in the application. One major problem is that many volunteers are not good at English. To alleviate this language barrier, we adopt a machine translation service and verified whether the service can serve as a communication method between application users.

Keywords: Machine translation · Information visualization
Big data · Tourism

1 Introduction

Nowadays in Japan, inbound tourism has been gaining much public attention towards the 2020 Summer Olympics which will be held in Tokyo. Moreover, the growth in the number of foreign visitors to Japan, 24,039,700 in 2016, is accelerating [2]. Furthermore, Tsukuba City, located in Ibaraki Prefecture and known as a city of science in Japan has been hosting many international events such as international conferences; thus there have been many foreign visitors. However, according to our interview with some city officials, it turned out that they do not have any statistics about foreign visitors (e.g., number, nationality, moving path).

One promising method to collect such statistics would be user participatory sensing [4]. It utilizes smartphones owned by ordinary users as sensing equipment to collect various kinds of information.

Meanwhile, Tsukuba City takes part in the BigClouT project [1]. The purpose of BigClouT is to return high-value-added information extracted from data

© Springer International Publishing AG, part of Springer Nature 2018
N. T. Nguyen et al. (Eds.): ACIIDS 2018, LNAI 10752, pp. 149–158, 2018.
https://doi.org/10.1007/978-3-319-75420-8_14

in cities to citizens and to make efficient use of it in various areas. In this project, which is a part of BigClouT project, we develop a participatory sensing system whereby foreign visitors can post any problems experienced during their stays in the city. By using the system, foreign visitors can post their problems (e.g., how to go to a certain destination or where to buy a specific item) via their smartphone, and volunteers can respond to it. In addition, when foreign visitors post their problems, the system collects various kinds of information that could be used to extract information about them, such as users' demographics and geographic locations. By utilizing these data, the city is able to acquire information of foreigners that the city could not know and can be utilized for making decisions on various city policies. To assess the feasibility of the system, we conducted an experiment. The main contributions of this paper can be summarized as the following: (1) the participatory sensing system for foreign visitors, which was developed in a project-based learning (PBL) course; and (2) many aspects of software development in a team learned during the development.

2   Related Work

While many systems for tourists were proposed (e.g., [3,5,7]), most of them focus on recommendation and navigation for the tourists.

There are numerous systems which focus on participatory sensing. For example, MinaQn [9] is a web-based participatory sensing platform, with which city officials can easily create a question and publish it on the city website, and citizens can answer the question. The answer consists of GPS information and text data. By collecting these answers, the city officials can collect qualitative and quantitative data, which can be utilized in policy decisions. In contrast to MinaQn, where city officials ask questions to citizens, the user voluntarily posts a problem in our case.

Lokemon [8] is another example of participatory sensing application, which uses monsters. Usually, participatory sensing suffers from privacy protection, incentives for posting and the quality of data. To solve these problems, "getting into a monster" type participatory sensing was proposed in Lokemon. Each monster is related to a specific place, and a user at that place (region) can post by getting into the monster (Lokemon). For example, when a user wants to post that the bus stop is crowded, the user can post the information by getting into the Lokemon related to the region; remote users can chat with the monster in order to know whether the bus stop is crowded. In our case, the user can post problems anonymously, and the incentive for posting is the responses from the volunteers.

3   System Overview

Our system consists of three applications: *client application for visitors*, *client application for volunteers* and *data visualization application*. Figure 1 shows the outline of this system. The data visualization application organizes and visualizes the data collected from foreign visitors and thus allows city officials to recognize the behavior of the foreign visitors in the city and use it for policy decisions or other uses. By organizing and visualizing the collected information from the foreign visitors, city officials can recognize the behavior of foreigners in the city by the data visualization application and use them for policy decisions or other uses.

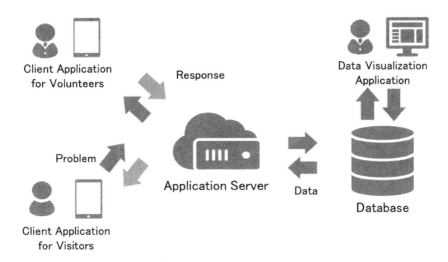

Fig. 1. Outline of the system.

3.1   Client Application for Visitors

The *client application for visitors* is an application for foreign visitors to post problems and to receive supports (i.e., responses) from volunteers for each problem. The visitor can post a problem by writing the contents of the problem and attaching photos, which is not mandatory, with her/his smartphone. For each problem, GPS information and user information are included. The GPS information is automatically acquired from the smartphone (confirmation is required when the visitor first uses the application). Before using the application, the visitor has to register information (shown below) within a range where individuals cannot be identified.

- Country of residence
- Age group (10's, 20's, 30's, 40's, 50's – 90's)
- Gender

When a volunteer sends a response to the visitor, the visitor receives a push notification and can check the detail. The visitor can also see problems posted by other visitors by choosing the middle tab on the bottom of the screen, shown in Fig. 2.

Client Application for Visitors   Client Application for Volunteers

Fig. 2. Screenshot of the client application for visitors and the client application for volunteers.

3.2   Client Application for Volunteers

The *client application for volunteers* is an application for volunteers in a city, typically city officials, to support foreign visitors by responding to their problems. A volunteer chooses one post from the list shown in Fig. 2 and responds to it. The volunteer can also send additional comments to the visitors when she/he receives appreciations or additional questions from the visitors.

One major problem in designing our system was how to alleviate language barriers; i.e., many Japanese city officials are not good at English whereas English is needed to communicate with foreign visitors. Our system is aimed at foreign visitors, which would post problems in English. Therefore, the responses to them must also be in English. Therefore, it is convenient if a Japanese user who cannot read/write in English can respond to those problems. Using a machine translation such as Google Translate[1] would be a solution. However, Japanese has a grammatical structure different from English [6]; thus the accuracy of the translation between these two languages is lower to other pairs of languages that have similar grammatical structures.

Therefore, we conducted a preliminary experiment in the development team to verify whether it is possible to understand the problem (i.e., post) and whether the right answer will be sent back to the users by using machine translation. One member participated as a foreign visitor who came to the city and had problems; three members participated as volunteers. First, the visitor made five English problems; Table 1 shows some examples. Then, these problems were translated into Japanese by using Google Translation and presented to the volunteers. The three volunteers created responses to each of the problems in Japanese. Then, those responses were translated into English by Google Translation, and

---

[1] https://translate.google.co.jp/.

presented to the visitor. Table 2 shows the responses which correspond to the problems in Table 1.

As these tables show, the accuracy of the translation was so high that conversation can be realized and thus problems can be solved with machine translation service. However, when a sentence included a proper noun, its translation occasionally did not make sense. For example, in Problem 2 of Table 1, the name of the spot for dogs was translated to "Tsukuba Dogwand Land", which does not make sense at all. On the other hand, in Problem 3, the name of a barbeque spot was translated to "Tsukuba Fureai no Sato", a Romanization of the Japanese, which makes sense. Since the translations of proper nouns in responses are vital on such occasions, we have to ensure the translations of the proper nouns in the future implementation. As a solution, we plan to establish rules [10] when using proper nouns to ensure that incorrect translation do not occur.

Table 1. Examples of the problems.

No.	Original problem (English)
1	I am from Egypt and I am Muslim. I can't eat pork. Can you tell me where there is halal service?
2	I don't know how to go to the park where there are dogs. Can you tell me the name of the park? And also, I want to use the bus. Can you tell me how to use the bus to the park?
3	I want to stay at a cottage in Tsukuba. Is there any place where I can stay? I'd love it if there's a supermarket near the place, where I can buy meat for BBQ

Table 2. Examples of the responses.

No.	Response
1	There is a dining room corresponding to Halal in Tsukuba University
	I will introduce the Haralmeet Restaurant in Tsukuba https://www2.kek.jp/rso/Living/pdf/vegetariandish.pdf
2	If it is a park where you can walk, "Penny Rain Tsukuba" in "Aeon Tsukuba" is given! If there is a dog facility, it will be "Tsukuba Dogwand Land"!
	Probably I think that Tsukuba is the land where you want to go. If you go by bus, please get off at Tsukuba Yamaguchi bus stop from Tsukuba center. I think that we can get there in ten minutes' walk from there
3	How about "Tsukuba Fureai no Sato"? There is no supermarket nearby, but you can enjoy BBQ in a very good place!

4 Experiment

We conducted an experiment with foreign students in University of Tsukuba. The purpose of this experiment is to test the developed client application for visitors, collect opinions from foreigners, and to examine what burdens and problems exist when volunteers solve problems by using the client application for volunteers. Additionally, we conducted an interview about the system and about opinions and information on various fields in Japan as foreigners.

4.1 Participants

We recruited 9 volunteer foreign students (7 males and 2 females) in University of Tsukuba in this experiment. 6 students were 20's and three were 30's, and there were 7 nationalities. All the 9 students participated in this experiment as foreign visitors. The members of the development team participated as volunteers.

4.2 Experiment Procedure

Visitors. In this experiment, the participants used the client application for visitors for 5 days. On the first day, we asked the participants to install the client application for visitors on their personal smartphones. After the installation, we had a short training session to make them familiar with the application. Then, we asked the participants to post more than three problems every day. Every night, we sent them an e-mail asking to fill a questionnaire of Google Forms, which asked them the understandability of responses, and bug reports of the application; the understandability was rated on a 5-point Likert scale (1: Couldn't understand at all, 5: Could completely understand). On the last day, we took an interview with each participant, which took about 20 min per participant. The questions of the interview are shown below.

- What kind of incentives would make foreigners want to use this application?
- Were there any differences in the English of the responses?
- How did you feel when using the application?
- What kind of problems/bugs did you find when using the application?
- What kind of functions do you want to have?
- What kind of problems did you have when you first came to Japan?
- How do you solve problems in your daily lives?
- Do you use the information (such as the homepage) published by the city? If you do, is it useful?
- What kind of problems did you have, when you visited other countries as a tourist? What did you do at that time?

We also investigated the most appropriate format of responses to the problem in the client by grouping the participants into two groups: *QA method* and *chat method*. The chat method allows the visitor and the volunteer to send messages to each other using ordinary chat-style interface. On the other hand, in the QA method, only one response can be added to a problem, and no additional comments from visitors can be made.

We also divided the participants into two groups by the response method. One group responded to the posted problem directly in English. Another group translated the posted problem into Japanese by Google Translation, answered it in Japanese, and sent the response in English by translating it. On the third day, we swapped the two groups for counter balance.

Volunteers. The volunteers responded to a post by using the client application for volunteers, every time a post was given by the participant. Whenever there was a new post, the volunteers received a notification via chat bot, and one volunteer was assigned to the post; the volunteer responded to it when he could respond to it. If it was difficult to answer the post, the volunteer gathered information on the internet and responded to it. It took 3 to 15 min in average to respond, depending on the content of the posts.

4.3 Results

In the experiment, 71 problems ($M = 7.89$, $SD = 4.77$) were posted. The average rate of understandability of responses was 4.8 for directly responded English, and 4.7 for the machine translation. This suggests that machine translation can be used in responses. We obtained 7 bug reports.

We obtained 16 opinions in the interview. Below are some of them.

- "I want a copy-and-paste function of responses."
- "I want to be able to save drafts, so I can use the application offline."
- "I want to use Google MAP features."
- "I want to send local photos."
- "I want to use Emoji[2]."
- "Automatic recommendation function would be great."
- "I'd be happy if there was a page for local information."
- "Some of the response was different from what I was expecting."
- "I wanted to see who was helping me. I think I would feel more reliability to the responses."
- "I would prefer responses which are based on experiences, not just pasting a link of a website."
- "It would be better if the visitors could also send responses to others, and could also see other people's problems."
- "Although I felt some difference of nuance in the English response, I could understand all the responses."

During the experiment, the volunteers corrected the mistranslation of the proper nouns manually when responding. Therefore, we need to implement a function to edit both original and translated sentences. Moreover, while we collected many opinions from the participants as a foreigner, all the participants were residents of the city for a long period. Therefore, we need to be careful in referring to these opinions for redesigning our system since some opinions would be peculiar to residents.

5 System Development in PBL

Our development team had four members. In this section, we describe how the team developed our system in PBL, and how we will continue our PBL.

---

[2] Ideograms and smiley faces.

5.1   Scrum Software Development

Our team employed Scrum[3] as the framework for software development. We divided the entire period into sprints of two weeks. Since the members were able to gather three days in a week and one whole day was necessary for the meetings of Scrum (i.e., planning meeting, daily scrum meeting, review and retrospective), we decided the length of a sprint to be two weeks to ensure five days for development.

In this project, we slightly changed the roles of the client and team in planning meetings to encourage the team to propose ideas in designing our system. Normally, a client proposes the specifications of the functions the client wants in detail. By contrast, in this project, the client presented the rough requirements and our team proposed functions from the requirements. Additionally, we created and clarified the *completion definitions* of each product backlog by using Google Docs. Using Google Docs allowed us to share the contents with the client and all the member of the team easily.

During the sprint, we recorded the logs of daily scrum meetings with Slack so that the members could share their progress and problems with each other even when a member could not attend a meeting. Moreover, we carefully broke down a development item to small tasks, each of which one member could complete in one day. This approach was effective in making clear for each member what to do in the day. At a retrospective, we also tried to change existing methods and way of thinking, to improve the team development for the next sprint. As one of the results of our retrospectives, we decided to add a "must, should, option" criteria, which is similar to the MoSCoW method, to each of the completion definitions. Sometimes, the members did not complete the completion definitions or did too much and thus failed to fulfill the requirements of the client. This was because the importance of each completion definition was vague for the members. By defining the "must, should, option" criteria before starting the development, the importance became clear and thus such failure diminished.

5.2   Small Experiments and Tests During the Development

We also carried out small experiments and tests during the development of the system to conduct our project steadily. For example, we conducted a small experiment in the team before the experiment described in Sect. 4. In this small experiment, all team members participated as both the volunteer and the visitor. From this experiment, we found issues of the applications and the server (e.g., fixing a bug), and thus were able to improve the system before conducting the experiment. Moreover, when developing a new function, our team wrote the test code for the function beforehand. By doing this, we could develop function separately and join them easily.

---

[3] A framework designed for small teams to complete duration cycles (sprints).

5.3   Future Development in This PBL

Based on the results of the experiment shown in Sect. 4, we held meetings within our team and with the client to decide the development policy. For the language, since serious problems did not occur, we decided to proceed with the implementation as it is. Although, as described in Sect. 4, the volunteers had to edit the proper nouns manually because the machine translation did not work properly. Thus, we have to implement a function that makes this process easier.

The client application for volunteers which we used in the experiment was simple because it only assumed to be used within the team. However, we are planning to conduct an experiment in Tsukuba City and city officials will use the application. Thus, we need to design and develop a simple and easy application for anyone to use. Also, we are designing the data visualization application. To do this, we are planning to conduct interviews with the city officials to investigate how and what kind of information they would want to examine. Moreover, we will conduct quantitative evaluations in various aspects such as the usability of the application in the experiment.

In this paper, we focused on the machine translation between Japanese and English as a communication method between application users. We are now implementing translations between other languages in the application.

6   Conclusions and Future Work

Although there are many foreigners who visit Tsukuba City, the city did not have much data and information of the foreign visitors. Therefore, we designed a system which collects, organizes and visualizes data such as problems of foreign visitors in a city, thereby allowing city officials use that information for making decisions on various city operation and/or policies. Furthermore, foreign visitors can be supported by the volunteers from city officials through exchanges of posts and comments in the application. We verified whether machine translation service could serve as a communication method between application users.

In our experiment, we only evaluated how well the foreign visitors understood the response. In the future, it is necessary to develop a framework for evaluating how precisely they could understand the response.

Acknowledgement. This work has been supported by "BigCLouT: Big data meeting Cloud and IoT for empowering the citizen ClouT in smart cities" (NICT).

References

1. BigClouT: Big data meeting Cloud and IoT for empowering the citizen clout in smart cities. http://bigclout.eu/. Accessed 14 Oct 2017
2. Trends in number of foreign visitors to Japan and number of Japanese traveling abroad. https://www.jnto.go.jp/jpn/statistics/marketingdata_outbound.pdf. Accessed 13 Oct 2017 (in Japanese)

3. Baraglia, R., Frattari, C., Muntean, C.I., Nardini, F.M., Silvestri, F.: RecTour: a recommender system for tourists. In: Proceedings of the 2012 IEEE/WIC/ACM International Joint Conferences on Web Intelligence and Intelligent Agent Technology, WI-IAT 2012, vol. 3, pp. 92–96. IEEE Computer Society, Washington, DC, USA (2012). http://dx.doi.org/10.1109/WI-IAT.2012.88

4. Burke, J., Estrin, D., Hansen, M., Parker, A., Ramanathan, N., Reddy, S., Srivastava, M.B.: Participatory sensing. In: Workshop on World-Sensor-Web (WSW 2006): Mobile Device Centric Sensor Networks and Applications, pp. 117–134 (2006)

5. Echtibi, A., Zemerly, M.J., Berri, J.: Murshid: a mobile tourist companion. In: Proceedings of the 1st International Workshop on Context-Aware Middleware and Services: Affiliated with the 4th International Conference on Communication System Software and Middleware, CAMS 2009, pp. 6–11. ACM, New York, NY, USA (2009). http://doi.acm.org/10.1145/1554233.1554236

6. Ohara, K.: Looking at Japanese from machine translation. Soc. Tech. Jpn. Educ. J. 5–8 (2001). (in Japanese)

7. Orso, V., Varotto, A., Rodaro, S., Spagnolli, A., Jacucci, G., Andolina, S., Leino, J., Gamberini, L.: A two-step, user-centered approach to personalized tourist recommendations. In: Proceedings of the 12th Biannual Conference on Italian SIGCHI Chapter, CHItaly 2017, pp. 7:1–7:5. ACM, New York, NY, USA (2017). http://doi.acm.org/10.1145/3125571.3125594

8. Sakamura, M.: Poster: research on participatory sensing and human behavior change with information presentation using monsters. In: Proceedings of the 14th Annual International Conference on Mobile Systems, Applications, and Services Companion, MobiSys 2016 Companion, pp. 76–76. ACM, New York, NY, USA (2016). http://doi.acm.org/10.1145/2938559.2948848

9. Sakamura, M., Ito, T., Tokuda, H., Yonezawa, T., Nakazawa, J.: MinaQn: web-based participatory sensing platform for citizen-centric urban development. In: Adjunct Proceedings of the 2015 ACM International Joint Conference on Pervasive and Ubiquitous Computing and Proceedings of the 2015 ACM International Symposium on Wearable Computers, UbiComp 2015/ISWC 2015 Adjunct, pp. 1607–1614. ACM, New York, NY, USA (2015). http://doi.acm.org/10.1145/2800835.2801632

10. Yamashita, N., Sakamoto, T., Nomura, S., Ishida, T., Hayashi, Y., Ogura, K., Isahara, H.: Analyzing user adaptation toward machine translation systems. J. Inform. Process. **47**(4), 1276–1286 (2006). (in Japanese)

A Practice for Training IT Engineers by Combining Two Different Types of PBL

Kazuhiko Sato[(✉)], Yosuke Kobayashi, Takeshi Shibata,
Hidetsugu Suto, Shinya Watanabe, Shun Hattori, and Sato Saga

College of Information and Systems, Muroran Institute of Technology,
Muroran, Hokkaido 050-8585, Japan
kazu@mmm.muroran-it.ac.jp

Abstract. PBLs signifies both problem-based learning and project-based learning; we clearly distinguish between the two types of PBLs. The former PBL involves finding ways to solve problems under unfavorable conditions. In the curriculum at our institute, problem-based learning has been conducted since 2006. By contrast, project-based learning involves a student team undertaking an actual project, devised by a company. We did not adopt project-based learning in our curriculum. In 2016, our institute joined the Education Network for Practical Information Technology project of Japan's Ministry of Education, Culture, Sports, Science and Technology. In that project, around 40 universities are engaged in developing a cooperative network. Each university develops practical education using PBL; its knowledge and resources obtained are shared among other universities in the network. For our part of the project, we added 3 practical exercises based on project-based learning. By combining the two different types of PBL, our curriculum provides exercises that introduce better problem-solving skills, instill a challenging spirit, and provide practical experience for students. In this paper, we report our exercises and practical results for 2017.

Keywords: Problem-based learning · Project-based learning
Proactive learning · Self-directed learning
Engineering education · enPiT

1 Introduction

At our institute, an original style of practical programming exercise using problem-based learning has been employed since 2006. The main purpose of that exercise is to provide students with experience in finding ways to solve problems under unfavorable conditions. With our approach, teachers do not provide students with considerable information or techniques for solving exercise-related tasks. Students have to do all their studying by themselves, and they meet the challenges of various tasks using the knowledge acquired in this way.

The exercise is undertaken in the latter half of the second grade in our curriculum. At that point, most students lack the knowledge and experience to deal with the tasks given by teachers. The students possess only basic programming skills in C language

© Springer International Publishing AG, part of Springer Nature 2018
N. T. Nguyen et al. (Eds.): ACIIDS 2018, LNAI 10752, pp. 159–168, 2018.
https://doi.org/10.1007/978-3-319-75420-8_15

and basic computer literacy. The task given in the exercise is to develop virtual reality (VR) software for educational use. The software is to be executed in an original VR environment of our institute. A basic framework written in Java and OpenGL is prepared and given to the students for developing the VR software. The framework can be used as it is and does not require enhancement; it serves as an example of the completed VR software. In the exercise, the framework is given to the students as both a task and learning material. Students begin by understanding the VR environment, VR framework, and new programming languages: Java and OpenGL. They also have to design their original educational VR software. It is a very high-load exercise. After we introduced the practice in 2006, it has undergone improvement every year. The details and results of the practice appear elsewhere [1, 2]. Our VR environment and basic structure of our framework for VR software are detailed in a previous study [3].

From the results of over 10 years' application, we are confident that our approach provides students with proactive learning habits and encourages a challenging spirit in dealing with unknown problems. However, some disadvantages with our exercise have become evident. First, the students lack the time to undertake the high-load exercises. Before designing and implementing the VR software, the students have to learn Java language and OpenGL. They also need to understand the VR environment and acquire basic knowledge of 3-D geometry. The students have to use most of the 15 weeks allotted time for this exercise to acquire that knowledge; thus, they have insufficient time for developing the software. Accordingly, the quality of the students' activities is very good, but the quality of the final product is not so high. The second disadvantage is a lack of experience in designing VR software and lack of skills in team development.

To address those disadvantages, we have introduced an additional practical exercise in the third grade. The purpose of that exercise is to learn techniques for designing software and writing technical documents. In addition, that exercise includes problem based and project-based learning (together abbreviated as PBLs) in designing software based on requests from virtual clients. In some reports, these two PBLs are regarded as the same. However, we clearly distinguish between these two types PBLs in the present study. In our curriculum, we have attempted to give students practical experience using these two type of PBL.

In this paper, we present an outline of our current exercises. We also detail the new practical exercises we introduced in 2017. Those new exercises were implemented as part of the Education Network for Practical Information Technology (enPiT) project of Japan's Ministry of Education, Culture, Sports, Science and Technology.

2 Outline of Current PBL Exercises

2.1 Overview of Exercises in Our Curriculum

The current flow of exercises in our curriculum appears in Fig. 1. In that curriculum, one or more exercises appear in each semester. In the first grade, basic exercises, such as computer literacy and basic C language programming, are covered. In the second semester of the second grade, after students have learned about algorithms, they

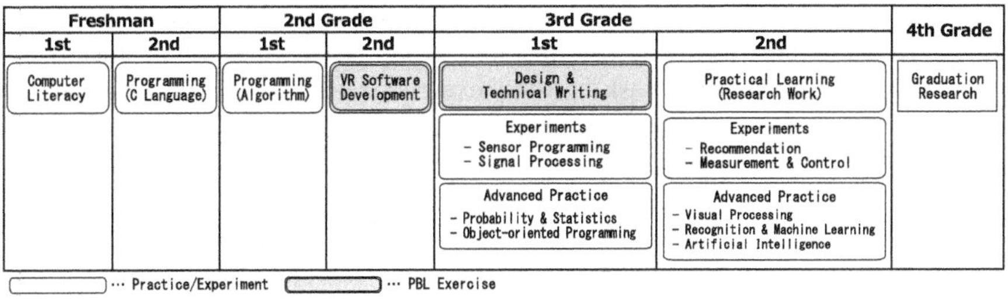

Freshman		2nd Grade		3rd Grade		4th Grade
1st	2nd	1st	2nd	1st	2nd	
Computer Literacy	Programming (C Language)	Programming (Algorithm)	VR Software Development	Design & Technical Writing	Practical Learning (Research Work)	Graduation Research
				Experiments – Sensor Programming – Signal Processing	Experiments – Recommendation – Measurement & Control	
				Advanced Practice – Probability & Statistics – Object-oriented Programming	Advanced Practice – Visual Processing – Recognition & Machine Learning – Artificial Intelligence	

⬚ ··· Practice/Experiment   ⬚ ··· PBL Exercise

Fig. 1. Current flow of exercises in our curriculum

undertake the PBL exercise of VR software development. Many exercises appear in the third grade. In the first semester, the PBL exercise of software design and technical writing is introduced. In parallel with that, an experiment is conducted on sensor programming and signal processing. Practical learning in laboratories begins in the second semester. Experiments on information recommendation as well as measurement and control are also undertaken in that semester. Some advanced practices, which are paired with lectures, also appear as elective subjects in the third grade.

2.2   PBL1: Exercise of VR Software Development

This exercise is called Information Engineering PBL: System Development Exercise. In the exercise, students collaborate in teams of five or six, and they attempt to develop software for educational use. Initially, a VR software framework, which is a package of base programs and sample data, is distributed to each group. The base program includes minimal functions for a viewer to visualize sample data as 3-D graphics; it uses a 3-D image drawing library called OpenGL. The base program is distributed to students as a finished product: it can be used as it is without enhancement. Students learn functions in their programs by executing and operating them, and they come to understand the programs' structure. After that, it is our aim that students develop the programs: by incorporating their own ideas into the programs, they create original educational software. The exercise assumed that the software developed would be used in teaching various subjects at our institute. Thus, we requested the actual lecturers responsible for those subjects to provide sample data. Starting with the exercise in 2010, we incorporated into the exercise VR equipment that had recently been introduced at our institute. The structure of the VR software that operates on the VR equipment is complex. The difficulty of the task was greater than it had been previously; hitherto, software development had been undertaken on ordinary PCs in the computer room.

In general programming exercises, the aim is mastery of programming techniques, and evaluation is based on the deliverables. By contrast, with our exercise, we evaluate how each student contributed and worked in conjunction with the group; we did not base our evaluation on deliverables. In addition, our evaluation included assessing how subjective learning was conducted toward achieving the task. Such an approach to evaluation demands accurately understanding the progress with individual students' work processes and the progress of their group as a whole. To achieve this understanding, we made three forms of visualization (group progress, group characteristics, and degree of comprehension task) of the exercise (step 1) [1]. Furthermore, we established a training-support environment that facilitated smooth learning and work progresses (step 2) [2]. For step 3, we set the Do Not Provide exercise, which encouraged students to learn subjectively by linking steps 1 and 2 [3].

With the Do Not Provide exercise, students have to undertake their tasks in an environment that is similar to a situation where they lack complete knowledge. Providing knowledge and skills beforehand or during the exercise is intentionally kept to a minimum. Descriptions of the task are also minimized. Students begin by understanding the contents of the task assigned to them. Thereafter, students have to select and learn the knowledge and skills they believe necessary to complete the task. Our exercise requires that students develop methods to undertake the task by themselves through trial and error.

The Do Not Provide exercise is a method that provides a sense of accomplishment for students who acted on their own initiative. In the exercise, the teachers do not actively give any information to the students, such as explanations of VR or object-oriented language. However, the environment is one in which students can obtain what they what if they act proactively. For students who are unable to act on their own initiative, the teachers provide certain "opportunities," which prompt the students to act. These may be hints in the form of comments about the program source or reference documents on Web pages for the exercises. By creating a situation in which students can notice such opportunities, we give a sense of accomplishment to students who are unable to act independently. We aim to improve students' motivation to undertake the next work challenge by giving them a sense of accomplishment, whereby the task was completed through their own discoveries or awareness.

Figure 2 shows the exercise being undertaken in 2016. For the exercise, we use a special room called the Virtual Reality Factory. That room has development areas with VR systems and a theater with a 3-D projector that is directly connected to the VR system. Students can demonstrate their VR software in the theater, and they can watch the results as a 3-D show using 3-D glasses.

2.3 PBL2: Exercise of Software Design and Technical Writing

This exercise is called Information Engineering PBL: Expression Techniques. In the exercise, students learn idea-making methods, such as brainstorming, software-modeling techniques, writing techniques for design documentation, and presentation techniques. Students are given a project task to design software based on requests from virtual clients set by teacher. The students create a first proposal in 4 weeks, which is reviewed by their fellow students; while next 8 weeks, the document undergoes a number of revisions before the teacher makes the final evaluation. With this exercise, the emphasis is on giving feedback on student activities as well as allowing the students to make frequent checks and provide advice.

Fig. 2. PBL exercise in 2016

3 New Practical Exercises for enPiT Project

3.1 Outline of enPiT and Our Additional PBL Exercises Based on It

In 2017, we introduced new practical exercises. The exercises are implemented as part of the enPiT project [4, 5], which is divided into four subject fields. Each field is organized by a number of institutes called Cooperating Universities (CUs). Over 40 CUs are engaged in enPiT. Our institute participates as a CU in the field of business system design (BizSysD) [6], which is covered by 10 CUs. Each CU offers courses for training and educating its own students as IT engineers. In addition, CUs collaborate with one another at various levels, such as sharing students, teachers, exercises, and teaching materials.

Each CU of enPiT accepts participation by other universities and colleges. Such schools are called Participating Universities (PUs). PUs may participate in different ways. The most common way is for PU students to take enPiT programs at a CU. PU students participate directly in exercises held in the form of training camps or they participate remotely. After completing all the programs, PU students can obtain certificates of completion in the same way as CU students. Another way of participation is in which only PU teachers take part in enPiT activities of a CU. This includes such participation as exercise instructors or lecturers taking part in FD activities.

When our institute first became involved in the enPiT program, we began by improving our curriculum. The new exercise flow appears in Fig. 3. For enPiT, we added two PBL exercises: Basic Business System Design Exercise (Basic BizSysD) and User-Centered Design Exercise (UCD). Both exercises are offered as training camps during the summer vacation for third grade students. We have also included a new specific PBL task about business system design in an existing experiment: Information Systems Exercise. This task is only for enPiT students. We offer this experiment in the second semester of the third grade.

3.2 PBL3: Basic BizSysD

This exercise is undertaken as a training camp in the summer vacation. The exercise involves design and implementing applications using sensor programming with Arduino [7] and Grove Starter Kit [8]. Arduino is an open-source hardware, and Grove Starter Kit is a sensor module set: it includes 12 Grove modules, which cover most functions needed by beginners with Arduino. After simple preliminary exercises related

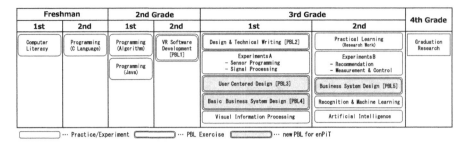

Fig. 3. Our exercise flow since 2017

to specifications and equipment use, students undertake a practical task in pairs. With this task, students receive a request from a virtual customer, which is a company in the Hokkaido city of Muroran. Using their own knowledge and techniques, the students have to design an application to comply with the request. Students need to develop a prototype within 2 days. The task is concluded when each student pair presents its deliverables to the virtual customer. In addition to students and teachers from our institute, students and faculty members from other universities and representatives from cooperating companies participate in the final presentation. Basic BizSysD students receive evaluations and comments from all attending individuals. By communicating with such people as other students and faculty from other universities and business-people, Basic BizSysD students are exposed to a broad range of perspectives. Our aim is that the students acquire practical skills in the process of mastering new technologies within a short period.

3.3   PBL4: UCD

This exercise is undertaken in the same summer camp as Basic BizSysD. Students are divided into groups of four or five. With UCD, fieldwork is conducted by targeting specific areas in Muroran. By means of a field survey, students have to identify problems in the target area. Then, using their imagination, students have to produce solutions to the problems, and they make a model or diorama to demonstrate their solution. The students use various means, such as designing techniques, sketches, miniature gardening, and picture-story shows, to present their ideas. Students also employ the technique of the customer journey map to visualize their activities in the field survey. We also encourage students to use what we term the quick-and-rough method (Q&R) to create ideas and evaluate them in a short time. With that method, they are able to enhance quality by repeatedly identifying problems. Q&R is a technique that emphasizes rapid trial and error rather than pondering a problem; it is based on the approach used with design in the arts. This exercise is thus closer to artistic design than information engineering or software engineering design. It is one of our initiatives to give students completely new ideas and perspectives.

3.4   PBL5: enPiT Special Task in Information Systems Experiment

This task is undertaken as completely project-based learning to solve a real problem from a real participating company. The main purpose of BizSysD is to acquire skills for designing a business system. This task also involves solving regional problems, such as low-quality support for inbound tourists. In this experiment, students collaborate with Hokkaido companies to produce solutions for regional problems. This experiment began in October 2017.

4   Report for 2017

The 2017 summer camp was held from September 4 to 9; 17 students from our institute and seven PU students participated. Chitose Institute of Science and Technology was our first PU.

4.1   Report for Basic BizSysD

In the first three days of the camp, Basic BizSysD was undertaken on the Muroran Institute of Technology campus. The theme of the exercise was this: present the Muroran night view interactively. Muroran is famous for its beautiful night scenery of a factory area around Muroran Bay. The students had to design exhibits to attract tourists' attention. It was assumed that the exhibits would appear at nearby New Chitose Airport, which receives many visitors. We divided the students into 12 teams and allocated PU students to different teams, where possible. Each team then produced its ideas using Arduino and Grove. The time flow appears in Fig. 4; photographs of the exercise appear in Fig. 5. Two teachers and six teaching assistant students supported the group work.

Sep.4				Sep.5			Sep.6			
Guidance	Short Lecture	Idea Making (1)	Training of Arduino & Grove	Night Tour	Idea Making (2)	Mid Presentation	Development (1)	Development (2)	Finishing	Final Presentation

Fig. 4. Time flow of Basic BizSysD in 2017

Fig. 5. Photographs of Basic BizSysD in 2017

In the day one, students attended lectures about the Internet of Things, and the physical programming using Arduino and Grove. Thereafter, students produced their initial ideas using an affinity diagram [9] or idea cards; they summarized their ideas on one idea sketch. That sketch served as the basis for further work. An afternoon task involved training in physical programming. At the end of day one, for information gathering, students saw the Muroran night view. The next day, students improved their initial ideas based on the experiences of day one. Secondaly, they heard a short lecture about the design of presentations. The lecture included viewpoints and techniques for designing business presentations. After the lecture, the students began developing their prototypes. In the afternoon on the last day, a final presentation was held. In addition to our faculty, professors from other CUs and a cooperating member of local IT companies participated; they gave various comments and advice to students. There were comments such as insufficient estimate or vision of financial resources, because they had little experience. The team names and titles of their deliverables appear in Table 1.

Table 1. Teams and deliverables

No.	Team name	Title of deliverables
01	Team Hako	Memorable night view
02	Toast	Night view spot guide "Navi-pyon"
03	Three arrows	Night viewer APP
04	Vier	Night viewer 360
05	Team Hakucho	Attractive night view display
06	B29	Romantic cafeteria
07	Initials O of the 3rd	My best view
08	Usagi-chan team	Night view on the wall
09	FPSP	MURO view
10	Tanbo no Ta	Project of "muroran attracted the world"
11	Team fighters	Night view spot guidance system
12	Oideyo Doubutu no Mori	The moving diorama - enjoy everyone! -

4.2 Report for UCD

The UCD was undertaken in the last 3 days of the summer camp at a vacant store located in the Nakajima shopping district in central Muroran. We did the exercise with the cooperation of a consortium of Nakajima businesses and could use the vacant store without charge. The students were divided into five groups; the groupings were different from those of previous exercises.

The exercise theme was designing future services for the Nakajima shopping district. "Future" meant after 5–10 years. The students had to identify the present merits of the Nakajima shopping district: they had to consider new services—not improvements to the current situation—to attract more people to the district. The time flow for the 3 days appears in Fig. 6; photographs of the exercise appear in Fig. 7.

Sep.7			Sep.8	Sep.9		
Guidance & Sketch-Dojo & Interview	Field Survey(1) & Making Customer Journey Map	Idea Making & Field Survey(2)	Prototyping (Diorama Making)	Making Picture-story Show	Final Presentaion & Reflecting

Fig. 6. Time flow of UCD in 2017

Fig. 7. Photographs of UCD in 2017

First, following guidance, sketching practice called Sketch Dojo was undertaken. Sketchbooks were the main equipment used in the exercise: students drew such items as scenery and people interviewed in their sketchbooks to serve as memos. After that practice, consortium members introduced the Nakajima shopping district and its current problems; students asked them for more details. Students then surveyed the district and interviewed passersby who used it on a daily basis. After returning to their base, students made customer journey maps to visualize people's activities. Each group developed one or two hypotheses about the district's problems based on its map. On day two, an additional field survey was undertaken to verify their hypotheses. Each group then had to make a diorama, showing a future solution to the problems it had identified. On the last day, each group presented slides in the form of a picture-story show using its diorama. In the final presentation, students presented their ideas using slides and a demonstration. It was attended by many people, including four members of other CUs, one city official, chairperson of the city council, and two newspaper reporters.

4.3 Considerations

After the summer camp, we obtained free comments about the two exercises from students. Many students stated that they were satisfied with our PBLs; however, one important comment was that they lacked sufficient time for development. With those exercises, considerable time was spent in guidance, explaining techniques, and training. In future, we intend to increase the time for development, e.g., creating preparation time before the camp as a proactive student-learning period. In addition to comments about insufficient development time, we heard that there was misunderstanding regarding the techniques used, such as Q&R. Particularly with UCD, it was necessary to repeat cycles of outputting, evaluation and improvement in a short time based on Q&R. Some students encountered difficulties working with short development cycles:

they believed that development required greater time with opportunities for reflection. In future, we intend to fully explain the methods used.

It was clear that the above improvements are necessary. However, the overall practice was evidently successful.

5 Conclusion

Since 2006, we have conducted VR software development exercises using PBL, originally as problem-based learning, not project-based learning. This exercise is undertaken at a quiet time of year; by imposing a difficult task, it encourages proactive learning habits and a challenging spirit among students dealing with unfamiliar problems. The purpose of the exercise was not learning software or user-centered design. The exercise did not originally include project-based learning for solving practical projects from a company. Thus, we added three new exercises in 2017. Following the two exercises conducted as a summer camp, there was evidently insufficient development time and a lack of understanding about the purpose and effects of the methods used in the exercise, such as Q&R. In future, we intend to address those problems.

Acknowledgments. This research was supported by the project Education Network for Practical Information Technology (enPiT), Ministry of Education, Culture, Sports, Science and Technology in Japan. We thank the Edanz Group (www.edanzediting.com/ac) for editing a draft of this manuscript.

References

1. Kazuhiko, S., Kentaro, K., Yoshifumi, O., Sato, S.: Implementation and evaluation of visualized software development exercising to increase students' motivation. Comput. Educ. **31**, 94–99 (2011)
2. Kazuhiko, S., Kentaro, K., Wataru, T., Yoshifumi, O., Yasuo, K., Sato, S.: Practice of exercises which "Do Not Provide Students a Lot" for achieving proactive learning. Eng. Educ. **61**(3), 56–61 (2012)
3. Kazuhiko, S., Kentaro, K., Yoshifumi, O., Sato, S.: A virtual reality atelier for VR software development and its application to problem-based learning. Jpn. J. Educ. Technol. **35**(4), 389–398 (2012)
4. Homepage about enPiT in the Ministry of Education, Culture, Sports, Science and Technology of Japan. http://www.mext.go.jp/a_menu/koutou/kaikaku/enpit/index.htm. Accessed 12 Dec 2017
5. enPiT Homepage. http://www.enpit.jp/. Accessed 12 Dec 2017
6. enPiT Homepage of Business System Design Field. http://bizsysd.enpit.jp/. Accessed 12 Dec 2017
7. Arduino Homepage. https://www.arduino.cc/. Accessed 12 Dec 2017
8. Grove Starter Kit for Arduino. https://www.seeedstudio.com/Grove-Starter-Kit-V3-p-1855.html. Accessed 12 Dec 2017
9. Kawakita, J.: The KJ Method: Chaos Speaks for Itself. Chuo Koron-sha, Japan (1986)

Utilizing Tablets in an Ideathon for University Undergraduates

Keiichi Endo$^{(\boxtimes)}$, Takuya Fujihashi, and Shinya Kobayashi

Graduate School of Science and Engineering, Ehime University,
Matsuyama 790-8577, Japan
endo@cs.ehime-u.ac.jp

Abstract. This paper discusses utilizing tablets during an ideathon for university undergraduates. When holding an ideathon, we had to meet some requirements such as appropriate and prompt team organization. Tablets were distributed to students and faculty members in order to meet those requirements. Tablets were used for organizing teams, answering a survey, sharing files, and using the Internet. As a result, we received positive evaluations from students in the survey conducted after the ideathon.

Keywords: Education · Tablet · Group activity · Team organization

1 Introduction

Recently, ideathons have garnered much attention and are being held at many corporations and universities. An ideathon is a neologism combining "idea" and "marathon," and is an event where participants cooperate to produce new ideas [1].

In March 2017 at Ehime University, an ideathon for undergraduates was held as part of their education. We had to meet some requirements such as appropriate and prompt team organization. Android tablets (ASUS ZenPad 3 8.0) were distributed to students and faculty members in order to meet those requirements.

This paper first gives an overview of the ideathon. Next, we enumerate some requirements we had to meet when holding an ideathon. Then, we explain in detail how the tablets were used. Afterwards, we discuss the results from the survey conducted after completion of the ideathon. Finally, we summarize the paper and describe future challenges.

2 Overview of the Ideathon

In this section, we give an overview of the ideathon held at Ehime University from March 27 to March 30, 2017. It was held for ascending juniors in this school. The ideathon was mainly aimed at letting students acquire tips on idea creation

© Springer International Publishing AG, part of Springer Nature 2018
N. T. Nguyen et al. (Eds.): ACIIDS 2018, LNAI 10752, pp. 169–176, 2018.
https://doi.org/10.1007/978-3-319-75420-8_16

Fig. 1. Brainstorming.

and improve their communication skills required for group activities. Teaching using nearly the same content was carried out twice, once in the first round (March 27–28) and then again in the second round (March 29–30). There were 25 students in the first round and 28 in the second round. For this ideathon, visits from other universities and corporations were permitted as part of faculty development. There were 10 visitors (from five universities) in the first round and 14 visitors (from six universities and three corporations) in the second round.

The theme for this ideathon was "the ideal university life." Before attending, students were instructed to think about the challenges they face directly in their daily lives as university students (troubles, troubling experiences, things they wish were different and so on).

The ideathon was held as follows.

Brainstorming: In a team composed of three or four students, they were asked to write down on sticky notes as many of the challenges they face in university life as they could think of. The sticky notes were placed in the center of the desk so all students in the team could see (Fig. 1). The students were instructed to think about challenges while referring to the sticky notes written by their team members.

KJ Method: Groups were created by gathering together similar challenges that were written on the sticky notes [2,3]. Each group was added names that made it easy to understand how the challenges were classified (Fig. 2).

Speedstorming: Students were arranged in pairs in two rows and asked to exchange ideas verbally for 5 min in their pairs (Fig. 3). They were asked to stay aware of expanding their ideas with reference to others' ideas. Then, they had 1 min to organize information, such as writing ideas that they heard or thought of on pieces of notepaper. They switched pairs and exchanged ideas four times in total.

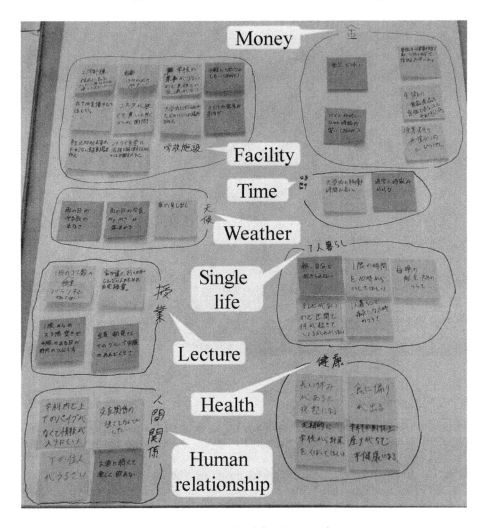

Fig. 2. KJ method (in Japanese).

Idea sketch: Students expressed problem-solving ideas on a piece of paper in the form of some pictures and short sentences. They were instructed to create at least three idea sketches each. Then, everyone's idea sketches were arranged on the desk and students and faculty members viewed and evaluated them by marking which ideas they thought were good. Referencing these evaluations, faculty members chose which ideas were to form the basis of future team activities. Considering diversity, they chose the same number of ideas as the number of teams that would be organized. Eight teams were organized in this ideathon.

Team reorganization: Students were asked to indicate their top three choices to discuss in future activities from the eight ideas selected by faculty members.

Fig. 3. Speedstorming.

They also answered questions to quantify individual characteristics based on the theory of Five Factors & Stress (FFS) [4]. Teams were then reorganized based on those answers.

Elevator pitch: Students studied ideas based on the idea sketches assigned to their teams and gave brief 30-s presentations on the features of the proposed applications. Then, faculty members and students had time for free discussion to talk about questions and opinions with each team taking a turn.

Presentation: After studying the idea further, students were asked to make 5-min presentations on the proposed applications. After each team's presentation, there was time for questions and answers. After all teams had finished their presentations, there was again time for free discussion.

3   Requirements in Holding an Ideathon

When we held the ideathon, we had to meet the following requirements:

(R1) It was required to organize teams appropriately and promptly. If we organized teams considering only students' interest, each team would be composed of students with similar characteristics. Inaga et al. [5] showed that a team composed of students with various characteristics would perform better in

team activities. However, it would take too long time if we manually organized teams considering both interest and characteristics of the students.

(R2) It was required to promptly summarize the results of the survey about
the ideathon answered by the students after the ideathon. This was because
we needed the results of the survey in a faculty development session after
the ideathon to exchange opinions among visitors (faculty members from
other universities and staff from corporations) and faculty members of our
university.

(R3) It was required to enable every student to see the slide-projection screen
clearly. We used a projector for description and instruction; however, it was
difficult to seat all the 25–28 students (and 10–14 visitors) in the place where
it is easy to see the only screen.

(R4) It was required to enable faculty members to check the progress of preparation for the presentation. By frequently checking the students' presentation
file, they can give advice to the students having difficulty summarizing the
idea. It would be difficult to advise students appropriately by only observing the display of the laptop used by the students. It would be also difficult
for faculty members (especially those from other universities) to make the
students stop the work and see the whole presentation.

(R5) It was required to enable students to use the Internet for investigation. For
example, they might need to check the novelty of the idea they were considering. They could use the laptop for the purpose; however, they could use only
one laptop in each team. Most of them had their own smartphones; however,
the display of a smartphone was too small for the purpose of investigation.

4   Utilizing Tablets

During this ideathon, tablets were distributed to students and faculty members
in order to meet the requirements stated in the previous section. Those tablets
were utilized for various purposes. The main uses of the tablets were as follows.

Team organization: To meet the requirement (R1), tablets were used for team
reorganization after the idea sketches were created (Fig. 4). Specifically, students were asked to enter their top three ideas of interest from the eight
faculty-selected ideas into the tablet. Additionally, they answered 25 questions to quantify individual characteristics in terms of condensability, accept-
ability, diffusibility, maintainability, and skill. Then, team organization was
performed automatically using a program that organizes teams by considering students' interests as much as possible, and ensuring that team members'
individual characteristics are as varied as possible. Performing this team organization manually would probably take several hours. With the program, team
organization was completed in a few minutes and it was possible to begin the
next task immediately.

Survey: To meet the requirement (R2), a survey about the ideathon was conducted on the tablets after the ideathon was completed. Since answers could

Fig. 4. Students answering questions for team organization using tablets.

be collected instantly using a program, the results of the survey could be distributed as materials at the subsequent faculty development session to exchange opinions among visitors and faculty members of our university.

File sharing: To meet the requirement (R3), we utilized the online storage service Dropbox [6] to share presentation files used for instruction with the students. In this way, students sitting in seats where it was difficult to see the slide-projection screen could check the slide in detail using the tablets. It was also possible to review slides later. Furthermore, by using Dropbox to share presentation files created in each team, visitors as well as the instructing professors could check progress at any time using the tablets and were able to give advice as necessary. Thus, the requirement (R4) was satisfied.

Internet use: To meet the requirement (R5), we permitted the students to use the tablets for the purpose of investigation. While students were considering ideas, they were able to check existing applications on the Internet.

5   Survey Results and Discussion

Figure 5 summarizes results of the survey conducted after completion of the ideathon. The number of students that answered the survey was 53.

With the team reorganization performed after the idea-sketch creation, approximately half of the students were compelled to discuss ideas they did not want to investigate as shown in Fig. 5 (1). However, in the end, 89% of the students became interested in the ideas that their team considered (2), and 75% of the students responded that they were able to actively produce ideas (3).

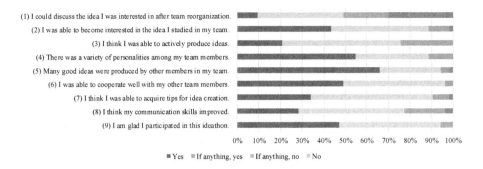

Fig. 5. Results of the survey after the ideathon.

Therefore, even though students were assigned ideas different from the ones they had wanted, this was not considered to be much of a problem.

Of the students, 89% responded that there was a variety of personalities in their team (4). It is thought that it was because teams were organized in consideration of not only assigning ideas as requested by students at the time of team reorganization but also maximizing variation of individual characteristics of team members. As a result, 94% of the students responded that many good ideas were produced by other members in their team (5). Despite having a variety of personalities within their teams, 96% of the students responded that they were able to cooperate well with their team members (6).

Ninety-one percent of the students said they thought they were able to acquire tips on idea creation by participating in the ideathon (7). Seventy-seven percent of the students said they thought their communication skills improved (8). Ideathons are considered to be effective for improving these skills. Ninety-four percent of the students said they were glad they participated in this ideathon (9); therefore, this ideathon can be considered a meaningful experience for students.

In the free descriptions of the ideathon's positive points, we found favorable opinions regarding tablet use, such as "the tablet was very useful" and "the tablet was useful for looking things up." Another opinion also expressed was that it was a good idea to use the tablets for team organization.

Additionally, many students said they felt they received helpful comments from the visitors from other universities and corporations. For example, one student reflected that "many opinions from those visiting the university had a different perspective from our own, and those opinions were helpful." It can be said that including participants from various positions in the discussion, such as faculty members from other universities and staff from corporations, proved effective in developing ideas. External faculty and staff members could check progress at any time during the presentation preparation period, and this was considered to be effective in terms of encouraging comments from them.

6   Conclusion

In this study, we evaluated an ideathon for university students. Tablets were used in this ideathon for organizing teams, answering a survey, sharing files, and using the Internet. As a result, we received positive evaluations from students, including a response in the survey conducted after the ideathon's completion that 94% of the students were glad they participated in the ideathon.

Although Wi-Fi was used in this ideathon, it cost a lot of money to install wireless access points and routers to enable approximately 50 tablets to be used simultaneously. In the future, we would like to investigate using an ad hoc network to directly connect tablets by Bluetooth. We expect that this would decrease implementation costs, enable a greater amount of students to participate, and make it difficult to be affected by network equipment failure. We previously verified the practicality of educational applications using Bluetooth communication in terms of communication delay [7]. Currently, we are also investigating the development of an application for tablets that support problem and idea organization through the KJ Method, and planning to use the application in an ideathon.

Acknowledgments. This work was supported by JSPS KAKENHI Grant Number JP15K16105.

References

1. Takagi, S.: An introduction to the economic analysis of open data. Rev. Socionetw. Strat. **8**(2), 119–128 (2014)
2. Kawakita, J.: The Original KJ Method (Revised Edition). Kawakita Research Institute (1991)
3. Scupin, R.: The KJ method: a technique for analyzing data derived from Japanese ethnology. Hum. Organ. **56**(2), 233–237 (1997)
4. Kobayashi, K., Furuno, T.: Soshiki Senzairyoku: Sono Katsuyo No Genri Gensoku - Five Factors & Stress. PRESIDENT Inc. (2008). (in Japanese)
5. Inaga, S., Washizaki, H., Yamada, Y., Kakehi, K., Fukazawa, Y., Yamato, S., Okubo, M., Kume, T., Tamaki, M.: Relationship between variations of personal characteristics and educational effectiveness in group assignment on software intensive systems development. In: Proceedings of the 8th International Technology, Education and Development Conference, INTED 2014 (2014)
6. Dropbox Inc.: Dropbox (2017). https://www.dropbox.com/
7. Endo, K., Onoyama, A., Okano, D., Higami, Y., Kobayashi, S.: Comparative evaluation of bluetooth and Wi-Fi direct for tablet-oriented educational applications. In: Nguyen, N.T., Tojo, S., Nguyen, L.M., Trawiński, B. (eds.) ACIIDS 2017. LNCS (LNAI), vol. 10191, pp. 345–354. Springer, Cham (2017). https://doi.org/10.1007/978-3-319-54472-4_33

Intelligent and Contextual Systems

An Approach to Representing Traffic State on Urban Roads Used by Various Types of Vehicles

Tha Thi Bui[1,2], Trung Vinh Tran[3], Linh Hong Thi Le[4],
Ha Hong Thi Duong[5], and Phuoc Vinh Tran[4(✉)]

[1] Ho Chi Minh City Vocational College, Ho Chi Minh City, Vietnam
bththa@gmail.com
[2] University of Information Technology (UIT), Ho Chi Minh City, Vietnam
[3] Fayetteville State University, 1200 Murchison Road,
Fayetteville, NC 28301, USA
ttranl@uncfsu.edu
[4] Thudaumot University (TDMU), Binhduong, Vietnam
panda.honglinh@gmail.com, phuoctv@tdmu.edu.vn,
Phuoc.gis@gmail.com
[5] Ho Chi Minh City Open University, Ho Chi Minh City, Vietnam
hadth.178i@ou.edu.vn, dthongha@gmail.com

Abstract. The priority-based traffic light controlling system at an intersection dedicates right-of-way to the highest priority direction. The system is applied for urban traffic system in Vietnam, where various types of vehicles run in close proximity to one another, needs a traffic model suitable to determine the priority of directions going through intersection. The paper has studied current traffic models, microscopic and macroscopic traffic flow models as well as three-phase traffic model, to propose spatio-temporal traffic statistics model. The paper considers red light as a traffic obstacle causing moving jam behind traffic light. The proposed spatio-temporal traffic statistics model studies traffic flow on a road segment bounded by two intersections, where various types of vehicles travel. The spatio-temporal traffic statistics model represents the road surface area occupied by vehicles at locations and over time as spatio-temporal histograms. The expectation of histogram is a factor determining the priority of direction going through intersection.

Keywords: Traffic · Traffic model · Traffic statistics · Traffic light
Traffic flow

1 Introduction

The priority-based system controlling traffic light at an intersection dedicates right-of-way to the highest priority direction. The priority of direction going through intersection depends on the vehicle crowd rate behind stop line and the appearance of prior vehicles according to the ground transportation code as fire truck, police car, and

© Springer International Publishing AG, part of Springer Nature 2018
N. T. Nguyen et al. (Eds.): ACIIDS 2018, LNAI 10752, pp. 179–188, 2018.
https://doi.org/10.1007/978-3-319-75420-8_17

ambulance. Therefore, in the priority-based traffic light controlling system, the vehicle crowd rate is an important factor to determine the priority of a direction going through intersection [1]. Meanwhile, the current traffic flow models and three-phase traffic model are considered with the traffic variables, flow rate and density, which are calculated with the number of vehicles counted by instruments installed along road. In fact, it is impossible to study the traffic state in Vietnamese cities with flow rate and density because the movement of various types of vehicles in close proximity to one another results in the impossibility of counting vehicles. Accordingly, it is necessary to constitute a traffic model suitable for the traffic state to estimate the priority of direction going through intersection.

This study proposes the spatio-temporal traffic statistics model characterized by a histogram representing road-surface area occupied by vehicles at locations along road center line behind stop line. The histogram changes over time. The studied object of this model is road-surface area occupied by vehicles at locations and over time. The histogram shapes cumulative and/or normal distribution function. The traffic variables of the spatio-temporal statistics model include normalized road-surface area occupied by vehicles, the expectation and variance of the histogram. The model studies the traffic state on urban road segment between two intersections installed traffic lights, where various types of vehicles such as cars, trucks, buses, motorbikes, bicycles, peddlers, and so on, travel in close proximity to one another.

The paper is structured as follows. The next section presents microscopic and macroscopic traffic flow models on space-time map, where microscopic model studies the movement of two consecutive vehicles, macroscopic model develops from microscopic model to study the relationship among flow rate, density, and speed of group of vehicles existing within a spatio-temporal domain; three-phase traffic model applies macroscopic traffic variables to study the movement at traffic bottleneck. In the third section, the spatio-temporal traffic statistics model is proposed to represent the road-surface area occupied by vehicles at locations and over time. In the fourth section, an illustration of the spatio-temporal traffic statistics model at an intersection of Dien Bien Phu Street in Ho Chi Minh city, Vietnam, in estimating vehicle crowd behind traffic light of an intersection. The fifth section summarizes the paper.

2 Traffic Flow Models

2.1 Microscopic Traffic Flow Model

Microscopic traffic flow model studies the movement of each vehicle on freeway. Vehicle along with driver is considered as a moving object, where diver controls vehicle to form the behavior of the moving object. According to temporal GIS, the movement of a moving object is represented by space-time trajectory on space-time map, where spatial locations are indicated on the axis X and temporal positions are indicated on the time axis T (Fig. 1) [2–5]. The microscopic traffic flow model studies the effect of a moving car on the vehicle following it. In that, the behavior of the leading vehicle (leader) affects the movement of the following vehicle (follower) and the follower is compelled to respond to the behavior of the leader [6–8]. As the example at the Fig. 1a, when the leader L decelerates at (x_1^L, t_1), the follower F is taken a delay time τ to respond to the leader's behavior. The delay time results in the decrease of distance between two vehicles; the smaller the speed of leader, the smaller the distance between two vehicles; the bigger the delay time, the smaller the distance between two vehicles.

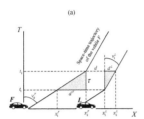
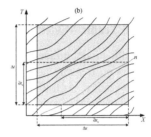

Fig. 1. Space-time maps represent traffic flow models: (a) traffic flow model at microscopic level studying two consecutive vehicles; (b) traffic flow model at macroscopic studying several moving vehicles within a space-time domain.

2.2 Macroscopic Traffic Flow Model

Macroscopic traffic flow models study the movement of several vehicles travelling on freeway during a time interval Δt on a road segment Δx (Fig. 1b). The movement of vehicles is represented as space-time trajectories on space-time map [4]. The domain $\Delta x \times \Delta t$ on space-time map indicates the existence of the vehicles during Δt on road segment Δx, e.g. the vehicle n exists during ∂t_n to travel a distance ∂x_n belonging to the domain $\Delta x \times \Delta t$ (Fig. 1b). The model is characterized by three traffic variables, flow rate q, density ρ, and speed v. In that, the flow rate is the number of vehicles passing a line across road during a unit of time, hour, minute, or second; the density is the number of vehicles exists on a unit of length of road at a time instant; the speed is the mean speed of all vehicles within domain $\Delta x \times \Delta t$. The values of the variables are calculated from the number of vehicles counted by technical instruments [9, 10].

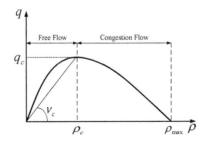

Fig. 2. The relationship between flow rate q, density ρ, and speed v in macroscopic traffic flow model.

The relationship among traffic variables flow rate q, density ρ, and speed v is shown in Fig. 2. The left side of ρ_c is considered as free flow because the traffic density is low, vehicles can accelerate freely and flow rate may be increased. The right side of ρ_c is considered as congestion flow because when the traffic density increases gradually, the flow rate reduces, and vehicles are compelled to decrease speed; when the density ρ reaches maximal density ρ_{max}, vehicles travel hardly and the flow rate reaches zero [5, 9, 10].

2.3   Three-Phase Traffic Model

The three-phase traffic model utilizes the results of macroscopic traffic flow to study the speed of traffic flow on freeway at a road segment shaped a traffic bottleneck due to works, accidents, obstacles. At traffic bottleneck, the phase of moving jam (J) appears because of very high density of vehicles which are moving very hard or even not being able to move. The moving jam affects the movement of vehicles upstream and downstream from the bottleneck. On the upstream road segment from the bottleneck, vehicles decelerate gradually from speed of free flow (F) to lower speed of synchronized flow (S), which is flow of vehicles moving at the same speed, then moving jam. In contrast, on the road segment downstream from traffic bottleneck, vehicles fast accelerate from the speed of moving jam to the speed of free flow and then the high speed of synchronized flow [11].

3   Spatio-Temporal Traffic Statistics Model

The state of traffic light at intersection affects the traffic state on the road segment behind the stop line of intersection. The red light is considered as a traffic obstacle and three-phase traffic model is suitable for studying the traffic state behind and in front of traffic light. The red light results in a moving jam spreading from stop line towards upstream, vehicles stop more and more behind stop line. The green light results in a free flow in front of stop line. On road segment between two intersections of traffic lights, upstream green light results in free flow running into the segment and

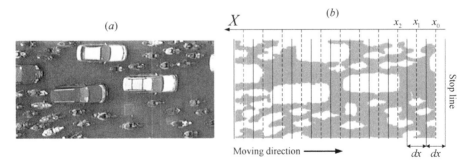

Fig. 3. A frame of video clip (Fig. 3a) is digitized to estimate band-vehicle-occupied area along road (Fig. 3b)

downstream red light results in a moving jam behind stop line. For the priority-based system controlling traffic light, the vehicle crowd rate behind stop line indicates the priority of direction going through intersection.

This spatio-temporal traffic statistics model estimates crowd rate behind stop line of urban road used by various types of vehicles. The model applies the spatial statistics in GIS for road segment joining intersection to represent the vehicle-occupied area as spatial histogram changing over time, where locations on road are indicated by the X axis toward upstream, which is stretched road center line. The model defines the following variables:

(1) vehicle-occupied area is the projection of a vehicle on road surface;
(2) road band is a band perpendicular to road center line;
(3) band-vehicle-occupied area is the sum of vehicle-occupied areas within a road band;
(4) normalized-band-vehicle-occupied area is the ratio of the band-vehicle-occupied area to the band area.

Mathematically, the studied road segment is divided into road bands of dx width (Fig. 3b). Let $s_{i,j}$ be the band-vehicle-occupied area within the road band at the location $x_j \,|\, j = 1, 2, ..$ and at the instant $t_i \,|\, i = 1, 2, ..$; in other word, $s_{i,j}$ is a variable depending on location and time. The state of moving jam appears when several road band areas are fast occupied by vehicles. The normalized-band-vehicle-occupied areas $r_{i,j} = s_{i,j}/w.dx \,|\, \forall i, j$ represented as bars perpendicular to the X axis constitute spatial histogram of normalized-band-vehicle-occupied area at different instants $t_i \,|\, i = 1, 2, ...$ At an instant t_i during the red period of downstream traffic light, the histogram of $r_{i,j}$ shapes cumulative distribution functions (Figs. 5 and 6). When the upstream green light switches on, a new distribution of normal shape appears from upstream intersection (Fig. 5e and f). The variables of expectation μ and variance σ^2 of the distribution of normalized-band-vehicle-occupied area at $t_i \,|\, i = 1, 2, ..$ are utilized to estimate the crowd rate of vehicles in the priority-based traffic light controlling system [1]. Let L be the length of the road segment, the normalized area occupied by vehicles on the whole road segment at an instant t_i is $R_i = \sum_j r_{i,j}/w.L$

4   Spatio-Temporal Traffic Statistics Model for Urban Roads

The paper studies traffic flows on a segment of Dien Bien Phu Street in Ho Chi Minh City, Vietnam, bounded by the upstream intersection Bahuyenthanhquan and the downstream intersection Truongdinh. The video clips of traffic state of the segment are recorded by instrument installed on flyer. Red traffic lights at intersections are considered as traffic bottlenecks because they stop all vehicles behind stop lines. Hence, three-phase traffic model may be applied to study the traffic state on the road segment, where traffic flow toward downstream intersection to go through the downstream intersection is affected by both upstream and downstream light.

After upstream light switches on green light, vehicles depart strongly from the upstream stop line to increase speed as a free flow, then they fast transfer to synchronized flow straight to downstream intersection. The synchronized flow is upholded to go through the downstream intersection if the downstream light is green, on the contrary, the flow transfers to moving jam if the downstream light is red. The synchronized flow constitutes a wave of vehicles running into the road segment whenever the upstream light switches on green light. The road-surface area occupied by vehicles on the segment increases fast if the downstream light is red and the upstream light is green (Fig. 4).

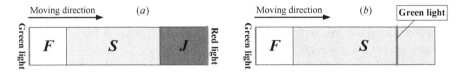

Fig. 4. The traffic state behind downstream traffic light is similar to three-phase flow: (a) moving jam constitutes behind red light; (b) synchronized flow going through green light. (Color figure online)

The specification of the traffic flow on the road segment is analyzed from data of video clips collected by flycams. The techniques of image processing are applied to convert regular-time-spaced frames of video clips to binary images, white for vehicle-occupied areas and grey for the free road surface. The segment is divided into bands perpendicular to road center line to estimate band-vehicle-occupied area. The traffic state on the segment at an instant is mathematically represented as spatial histogram of vehicle-occupied areas along the center line of the segment (Figs. 5 and 6). The specifications of spatial histograms changing over time are utilized to estimate vehicle crowd rate behind downstream stop line.

The Fig. 5 represents the traffic states behind traffic light of the downstream intersection during some traffic light periods. The traffic state is analyzed at the instants $t_0, t_1, t_2, t_3, t_4, t_5, t_6, t_7, t_8, t_9$, where t_0, t_8 are the end instants of green periods, t_1, t_9 are the beginning instants of red periods, t_4 is the end instant of red periods. The distribution of vehicle-occupied areas at instants shows:

– At t_0, the end instant of green period, the band-vehicle-occupied areas form thin distribution along Dienbienphu street from downstream intersection and wave of synchronized flow of vehicles results from upstream green light.
– During $[t_1, t_4]$, the duration of red period of downstream light, vehicles push each other in possible gaps of road surface behind stop line, band-vehicle-occupied areas shapes approximately a cumulative distribution sof which expectation increases over time. At t_4, part of new wave appears because of upstream green light.
– During $[t_5, t_8]$, the duration of green period, vehicle flow is transferred from moving jam to free phase, the crowd of vehicle behind stop line reduces and the expectation of cumulative distribution decreases over time. Meanwhile, the new wave grows gradually and advances towards stop line. At t_8, there is hardly any vehicle behind stop line because almost vehicles passed stop line, the expectation of cumulative distribution is near zero.

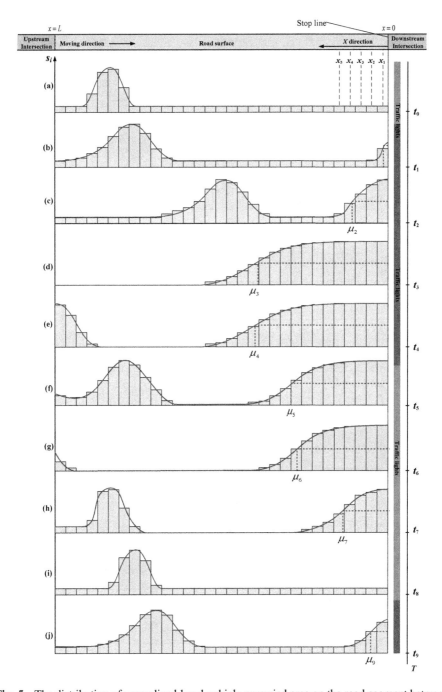

Fig. 5. The distribution of normalized-band-vehicle-occupied area on the road segment between two intersections.

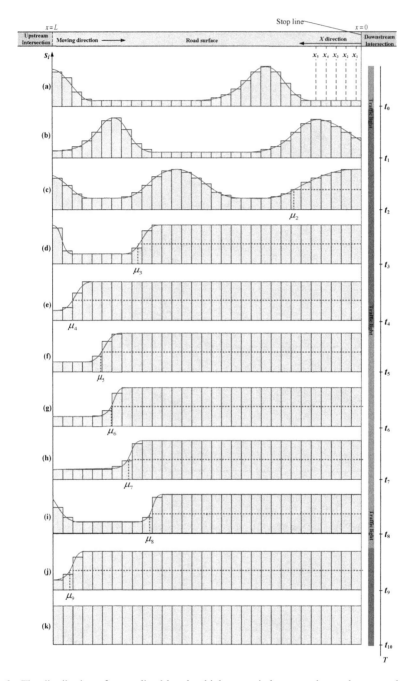

Fig. 6. The distribution of normalized-band-vehicle-occupied area on the road segment between two intersections.

– At t_9, vehicles congregate again behind stop line to form a new cumulative distribution while the next wave grows and advances towards stop line to increase the expectation.

The Fig. 6 represents the traffic states on the segment of Dienbienphu Street behind downstream traffic light during some traffic light periods at peak hours.

– At t_0, the end of green period, a big wave reaches downstream stop line while next wave appears in front of upstream intersection.
– During $[t_1, t_4]$, the duration of red period, vehicles jostle possibly in gaps of road surface behind downstream stop line. The crowd of vehicles expands fast from the stop line toward upstream. The expansion of the vehicle crowd results in the rapid increase of the expectation of cumulative distribution over time. At t_4, almost road surface is occupied by vehicles and the expectation is nearly equal to the length of the road segment.
– During $[t_5, t_8]$, the duration of green period, vehicles behind stop line are transferred from moving jam to free phase, the vehicle crowd reduces and the expectation of cumulative distribution decreases over time. However, at t_8, the end of green period, the expectation of cumulative distribution is not nearly equal to zero because the vehicle is still crowded.
– At t_9, the begin of next red period, the mergence of next waves to the available vehicle crowd behind stop line increases fast the size of vehicle crowd as well as the expectation.
– At t_{10}, the vehicle crowd grows to maximum, the expectation is equal to the length of the segment, no vehicle can move into the segment.

5   Conclusion

The paper studied microscopic and macroscopic traffic flow models and three-phase traffic model in order to propose spatio-temporal statistics model studying the traffic state behind traffic light at an intersection. The study considers red traffic light as a traffic obstacle causing moving jam spread behind stop line and green light as the disappearance of the obstacle. The vehicle flow on the road segment between downstream and upstream traffic light is affected by traffic lights at downstream and upstream intersections. The paper applies three-phase traffic model to study the traffic state on the road segment.

The study has proposed spatio-temporal traffic statistics model to estimate vehicle crowd rate behind traffic light in order to determine the priority of directions going through intersection. The model analyzes vehicle-occupied area as a traffic variable depending on space and time. The model applies the concept of spatial statistics in GIS to represent vehicle-occupied area as spatial histograms at different instants. The spatial histogram at an instant is cumulative distribution along road center line. The expectation and variance of the spatial cumulative distribution of vehicle-occupied area are variables depending on time, which are used to determine vehicle crowd rate behind stop line.

References

1. Tran, P.V., Bui, T.T., Tran, D., Pham, P.Q., Van Thi Tran, A.: Approach to Priority-Based Controlling Traffic Lights. In: Nguyen, N.T., Trawiński, B., Fujita, H., Hong, T.-P. (eds.) ACIIDS 2016. LNCS (LNAI), vol. 9622, pp. 745–754. Springer, Heidelberg (2016). https://doi.org/10.1007/978-3-662-49390-8_72
2. Hagerstrand, T.: What about people in regional science? Presented at the Ninth European Congress of Regional Science Association (1970)
3. Andrienko, N., Andrienko, G., Pelekis, N., Spaccapietra, S.: Basic concepts of movement data. In: Giannotti, F., Pedreschi, D. (eds.) Mobility, Data Mining and Privacy, Geographic Knowledge Discovery, pp. 15–38. Springer, Heidelberg (2008). https://doi.org/10.1007/978-3-540-75177-9_2
4. Tran, P.V., Nguyen, H.T.: Visualization Cube for Tracking Moving Object. Presented at the Computer Science and Information Technology, Information and Electronics Engineering (2011)
5. Darbha, S., Rajagopal, K.R., Tyagi, V.: A review of mathematical models for the flow of traffic and some recent results. Nonlinear Anal. **69**, 950–970 (2008)
6. Tang, T.Q., Li, J.G., Huang, H.J., Yang, X.B.: A car-following model with real-time road conditions and numerical tests. Measurement **48**, 63–76 (2013)
7. Li, Y., Sun, D.: Microscopic car-following model for the traffic flow: the state of the art. J. Control Theory **10**, 133–143 (2012)
8. Zheng, J., Suzuki, K., Fujita, M.: Evaluation of car-following models using trajectory data from real traffic. In: 8th International Conference on Traffic and Transportation Studies, Changsha, China, pp. 356–366 (2012)
9. Maerivoet, S., Moor, B.D.: Traffic Flow Theory. arXiv preprint arXiv:physics/0507126 (2005)
10. Immers, L.H., Logghe, S.: Traffic Flow Theory. Faculty of Engineering, Department of Civil Engineering, Section Traffic and Infrastructure, Kasteelpark Arenberg, vol. 40 (2002)
11. Kerner, B.S.: Introduction to Modern Traffic Flow Theory and Control. Springer, Heidelberg (2009). https://doi.org/10.1007/978-3-642-02605-8

Agent-Based Model of Ancient Siege Tactics

Ondrej Dolezal[1], Petr Kakrda[1], and Richard Cimler[2(✉)]

[1] Faculty of Informatics and Management, University of Hradec Kralove,
Hradec Králové, Czech Republic
[2] Faculty of Science, University of Hradec Kralove, Hradec Králové, Czech Republic
richard.cimler@uhk.cz

Abstract. *Introduction.* The use of an agent-based simulation of a problem of the ancient Celtic *oppidum* siege is presented. Computer models enable creating simulations of different strategic situations and testing various scenarios, which is very useful for studying complex systems.

Aim. The aim of the model is to study the military tactics of the Celtic population. The simulations are focused on different defense tactics of the Celtic *oppidum*. The succession rate of attack and the number of casualties is studied with various scenarios with varying ratios of the defenders and attackers, areas of attack, and combination of weapons.

Methods. The environment of this agent-based model of Celtic *oppidum* siege is based on archaeological findings from the long gone *oppidum*, Stare Hradisko, located in the Czech Republic. The agents operating in this model represent either residents of the *oppidum* who are defending their homes, or the attackers.

Results. The tool created enables studying different scenarios of attack on a Celtic *oppidum*. Simulations can be used for testing attacks on different locations if the GIS data of the terrain and fortification are provided.

Keywords: Agent-based · Simulation · Military tactics · Celts
Archeology · Computer model

1 Introduction

The increase in the computational power of computers and the development of different modeling techniques has made it possible to simulate various complex systems [5,18]. Computer simulations help in many fields of study [12,20]. One of them is archaeological research. This paper is a part of the research into the Celtic *oppidum* named Stare Hradisko, using agent-based modeling [4,7,13–15]. The research is focused on different scenarios which could lead to the collapse of the settlement. This particular paper describes a model of the defense of Stare Hradisko and deals with advanced modeling techniques in military history research.

One of the questions in the field of military history is whether the development of new technologies (such as ranged weapons) influenced the development

© Springer International Publishing AG, part of Springer Nature 2018
N. T. Nguyen et al. (Eds.): ACIIDS 2018, LNAI 10752, pp. 189–199, 2018.
https://doi.org/10.1007/978-3-319-75420-8_18

and use of different tactics, or vice versa. Despite studying many materials and hypotheses, historians have not come to a definitive conclusion. However, when addressing this issue, it is particularly useful to use agent-based modeling (ABM) when we consider that fighting or conflict can be perceived as a complex system (including technology, tactics, social behavior, etc.). The advantage of ABM is its ability to incorporate both qualitative (tactics, behavior, experience of soldiers, etc.) and quantitative (number of soldiers, rate of loss, etc.) data. In addition, ABM can combine historical research in a small scale (for example, the soldiers) and in a wider (tactical) scale.

The structure of this paper is as follows. The model background is in Sect. 2. Related work can be found in Sect. 3. The methodology is described in Sect. 4. In Sect. 5 there is information about the siege model created. A description of the experiments is in Sect. 6.

2   Model Background

The Celts were a nation of tribes that inhabited a vast territory of Europe, including the area of today's Bohemia and Moravia, approximately around the middle of the 2nd century. The Celts began to build so-called *oppida* in that territory. An *oppidum* is a generally fortified settlement that can serve as an administrative center and center of craft production or trade. The usual size of an *oppidum* is quite varied (from just several hectares to 100 or more). For these settlements, there is a typical complex ground plan, which usually includes several forecourts and an acropolis (fortified *oppidum* center). In the fortified areas, the presence of streets and various buildings with residential, economic (granary, granary, furnaces, etc.) or production functions (forging and ceramics workshops, coinage, etc.) is documented.

Archaeologists often argue about the importance of the military and defense purpose of *oppida*. *Oppida* were fortified and built in strategically advantageous locations (such as elevated areas surrounded by steep slopes or the river). However, the real question is how effective was the defense of these centers, such as the number of defenders needed to successfully defend the *oppida*, how many attackers could be deflected, what difference could the artificial obstacles (ditches, clay embankments) make, etc. From sources, for example [16], we know that the *oppida* were being besieged to starve their defenders, as well as attacked by direct attacks. One way to explore the defensive capabilities of a specific *oppidum* can be to model and perform simulations using agent modeling.

The proposed model is to illustrate the *oppidum* Stare Hradisko, which originated in the middle of the 2nd century BC. Its extinction dates to 100 years later. The remains of the *oppidum* are in Moravia on the eastern edge of the Drahanska Highlands.

3   Related Work

There are several computer models of military tactics. In the following, some of the most interesting are described.

The model of [3] focuses on testing hypotheses about possible 18th-century fighting on the Iberian Peninsula, while helping archaeologists choose the best battlefield strategy.

The model of [8] covers the area of medieval military engagements. Due to the nature of the battles at that time (mixed groups or individuals in combat without mutual coordination), the use of ABM is advantageous because it emphasizes the properties and activities of each soldier (agent). The result of a typical medieval struggle depends more on the emergence of behavior that arises from a collection of decisions and implementations rather than the expected result of a planned tactic.

Another model, [17], is focused on the study of infantry tactics in the War of the Spanish Succession (1714). The purpose of the model is to test the effectiveness of the tactics used during this period by simulating different battles.

4   Methodology

The aim of this research is to simulate the expected course of different types of attack on the Stare Hradisko *oppidum* during its existence (2nd – 1st century BC). Although not much evidence of the siege of the *oppidum* has been documented, it can be assumed that the potential attackers would be warriors from some Celtic or Germanic tribes [2]. The purpose of the model is to try to determine the defensibility of an *oppidum*. The possibilities of defense are determined by the position of the *oppidum* in a given location, by building artificial obstacles, by the number of defenders, as well as the interaction of all these elements. The model depicts the losses of the attackers when approaching the *oppidum* fortification due to the shooting of the defenders (bows, slings, and spears). The impact of the terrain and defensive obstacles in increasing the losses of the attackers is monitored.

Another parameter to be watched is the number and mutual ratio of the defenders and the invaders. The goal is to find out how many defenders are needed to effectively defend the *oppidum*. Another objective is to find out how important are natural and artificial obstacles in defending an *oppidum*. Dug ditches and clay embankments are considered artificial obstacles. Sloping terrain, most often in the form of a steep slope or hillside, is considered a natural obstacle. The primary task of both types of obstacles is to make it difficult and slower for the attacking enemy, thus giving the defenders more time to aim accurately.

Only attacking the fortifications (walls, gate) will be simulated, not the subsequent fighting within the fortifications. The fortifications of an *oppidum* are divided into several wide areas based on the cardinal direction (north, north-west, south, south-west, east): the simulation always focuses on the attack on one particular part. It is possible to simulate three types of attack (attacks with the help of ladders, attacks on the gate in a given part, and attacks against a walled fortification). It is also possible to select one of the three phases of the fortifications, which are based on archaeological research and differ in size and number of artificial obstacles. The model is primarily focused on the *oppidum*

Stare Hradisko, but if the relevant GIS data were provided, it would be possible to perform simulations for other *oppida* or fortified housing estates. The NetLogo version 5 software tool was chosen for the implementation platform to create the model. (More detailed information about this tool can be obtained from [19].)

4.1   Agent-Based Modeling

For the creation of this simulation, an agent-based approach has been chosen. This approach gives us the ability to model each attacker and defender as an autonomous unit. In this simulation, each shot from a long range weapon is also taken into account and the probability of a hit is computed based on the shooter's and target's elevation and proximity. The agents are operating in the given environment. The environment can be loaded from GIS data. The effect of the terrain, such as a slope or obstacles, can be integrated into the simulation.

4.2   Combat Logic

Each agent has its target. If the target is far away, and the agent is equipped with long range weapon, it is possible to shoot at the target. If the target is nearby, close combat is initiated.

The long range weapons are divided into two basic types: direct and indirect. The shots of direct ranged weapon go directly to the target (a fighter) and the shot to the target flies over an almost straight track. Indirect shooting is constituted by shooting not at individuals but at a location. Indirect shooting is traditionally used against close enemy formations where the probability of hitting the enemy is much higher. Direct shooting has a higher probability of hitting a target, which increases as the target approaches more closely. On the other hand, indirect shooting may be used at longer range, without significant reduction in accuracy (hundreds of meters), whereas direct firing is only effective at a certain distance (tens of meters) [9,10]. Ranged weapons are primarily used by defenders who are trying to repel enemy attacks. Each defender may have one of three long-distance weapons (bow, slingshot or javelin). Selected invaders can also shoot at the defenders on the fortifications and provide cover for the attackers. However, these attacking shooters cannot directly engage in the hand combat attack. The attackers can only have a bow or a slingshot (due to the short range of the javelin they would have to approach the fortifications and would be too vulnerable to the defenders). The aim of the shooters is to remove as many defenders as possible.

4.3   Friction

The difficulty in passing through the terrain is represented by the so-called friction factor, which is based on the degree of the slope. A friction factor (friction parameter) expresses the cost of movement through an individual field. The base unit is 1. If a patch has a friction factor of 5, then overcoming it is as demanding as overcoming 5 patches with values of 1 each. The friction factor varies, depending on the angle between the direction of motion of the agent and the direction of the slope. Moving speed of agents in the simulation is calculated based on the friction factor (Fig. 1). The calculation itself is based on the formula, which is described in detail in [11], $h = g^{\cos \alpha}$,

where

h ... A calculated friction value that takes the climb direction into account;
g ... The basic value of friction;
α ... The angle between the direction of motion of the agent and the direction of ascending slope.

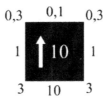

Fig. 1. Different friction values.

5   Siege Model

The model created will be described in this section. Agents are divided into attackers and defenders. They move in a given area which is loaded from GIS data. The aim of the attackers is to get into the fortication. To this end, they will choose one or more tactics to employ. The aim of the defenders is to prevent the attackers from getting into the *oppidum*.

5.1   Area

Area information is loaded from GIS data. Most important is the slope of the terrain, and also the location of the fortification. From archaeological research it is known that during the lifetime of the *oppidum*, there were different types of fortification. During the setup of the model it is possible to choose which type of fortification will be used.

5.2   Attackers and Defenders

There are two types of combat units in the simulation: attackers and defenders. Defenders mostly use long range weapons. They fight in close combat only when the attackers have breached the gate or climbed the wall. Attackers either engage in close combat after reaching the defenders, or long range shooting at the defenders from a distance.

5.3   Siege Tactics

One of the possible attack types can be chosen. There are currently three types of attack: attacks using ladders, attacks on a gate, and attacks on a fortification. In the current version of the model it is not possible to combine attack types in one simulation run. It is also necessary to set from where the attack will be conducted.

5.3.1   Attacks Using Ladders

The attack itself can take several forms. The fastest and the least coordinated is a direct attack on the fortifications, when the attackers try to reach the fortifications as quickly as possible and use ladders to climb to the top of the walls. If they succeed, they can fight with the defenders in close combat (swords, spears, etc.), which is usually more advantageous for the attackers if they can outnumber the defenders.

5.3.2   Attack on the Gate

Another type of attack is on the biggest weakness of each fortified site. This element is the gate that leads to the inner parts of the *oppidum* and which is usually placed at several points of the fortifications. The vulnerability of the gate arises from its construction and function. A wooden door is easier to damage than a solid stone wall. However, the builders of an *oppidum* realized this vulnerability and tried to eliminate it as much as possible by building some adjoining walls. Access to the gate is therefore often flanked by a wall of the fortifications, one on each side, so that the defenders can more easily shoot at the attackers. A wooden structure is also usually placed above the gates, allowing the defenders to be positioned directly above the entrance to the *oppidum*. These factors force the attackers to move slowly and cover themselves.

5.3.3   Attack on the Fortifications

Another type of attack involves the disruption and subsequent destruction of a stone brim, which is a wall of dry-laid stones, usually reinforced with an internal wooden beam construction. The warriors approach the wall and gradually disassemble it (with tools such as pickaxes), which can cause a complete or at least a partial collapse of the damaged wall [1]. The defenders, of course, try to prevent such damage to the walls, again by shooting bows and slinging or throwing spears.

6   Experiments

Several experiments have been performed to verify the functionality of the model. Each unique simulation setting was repeated 100 times to achieve more accurate results. The values generated for each setting combination were then averaged. The common goal of all experiments is to determine the influence of selected factors on the success or failure of the attack on *oppidum*. There are many parameters which can be set during setup of the model. Simulation allow to test different scenarios made by experts in a field of history to verify their ideas. In this paper we focus on two possible scenarios to present capabilities of the model.

6.1 Influence of the Ratio Between the Attackers and Defenders

This experiment aims to determine the role played, in the defense of the fortifications, by the ratio of the number of defenders to the number of attackers. The value sought is that ratio of the defenders to the attackers for which the attackers are guaranteed to be successful.

Not all possible combinations (weapons, types of attack, place, numbers of attackers and defenders) were tested in this particular experiment. To set aside the impact of different terrains and obstacles, only a certain part of the surroundings of the *oppidum* were used. An attack using leaders was chosen for this experiment. The defenders only had bows, a choice that was made with a regard to the higher number of such available resources compared to other types of weapons. Attackers with long range weapons were excluded. The numbers were set so that the defenders did not outnumber the attackers. The maximum and minimum were chosen with the assumption that hundreds of attackers could take part in the same fight. The attack was from the south-west, and the defense of the *oppidum* had only one ditch in that area.

Table 1. Results of experiment 1

Ratio	Success rate (%)
5,26	0
5,88	30
6,06	60
6,25	81
6,67	95
6,9	99
7,14	100

At a ratio of 6,90, the attackers eliminated all defenders in 99% of the cases; at a ratio of 7,14, the success rate increased to 100%, see the Table 1. From the results can be seen that model is sensitive to the initial values. The ratio sought is therefore approximately 7:1 attackers to defenders. Of course, this is a theoretical value. In fact, the outcome of a combat can not be simply determined. However, an overall picture can be made based on this experiment.

6.2 Impact of Terrain and Obstacles

This experiment explored how terrain and artificial obstacles affect the outcome of the attack. The experiment compared the defenses of five regions of the *oppidum*, with different terrains and levels of obstacles. In the case of the Southwest, the different stages of the fortification were also compared.

An attack using leaders was selected again, because the influence of the slope and obstacles (the attackers were spread out over a large area) can be easily observed. Different stages of the fortifications were placed, only in the southwest area, as in the other parts of the *oppidum*, the number of ditches and clay embankments was significantly reduced. The number of attackers was set to 500 men (20 groups); the number of defenders to 50 (ratio 10:1) or 100 (ratio 5:1). These numbers were chosen due to the results of the previous experiment. The degree of defensibility of an area is determined by the casualties that the defenders can inflict on the attackers. The defenders had only one type of weapon (bow), and attackers with shooting weapons were again excluded.

The results of the experiment at a ratio of 10:1 (50 defenders) are shown in Table 2. The least casualties for the attackers were observed in the south area.

Table 2. Attackers' causalities based on the direction of the attack. 50 defenders (ratio 10:1). All attacks were successful.

Direction	Level of fortification	Attackers' casualties
South-West	3-multiple	317
South-West	2-single	227
South-West	1-single	186
North-West	1-single	127
North	1-single	124
East	1-single	101
South	1-single	92

Table 3. Causalities of attacker based on the direction of the attack. 100 defenders (ratio 1:5)

Direction	Fortification	Attackers' casualties	Defense success rate (%)
South-West	3-multiple	451	100
South-West	2-single	460	100
South-West	1-single	429	97
North-West	1-single	353	85
North	1-single	256	2
East	1-single	227	1
South	1-single	196	0

The results of the experiment at a ratio of 5:1 (100 defenders) are in the Table 3. Key values are in the last column. Defense success rate is a percentage of simulations (100 runs has been simulated) that ended with the defeat of the attackers. In this case, the number of attacker's losses is affected by the fact that the defenders in some cases repelled the attack without eliminating all of them.

7   Discussion

The fortifications in the south-west are located in flat terrain, and practically protected throughout the entire length of their artificial barriers. The '1-single' phase includes a trench 2 meters deep, at the '2-single' stage this trench is 2, 5 m deep. The '3-multiple' phase includes a multiple set of trenches and clay embankments of different depths and heights (three ditches and two embankments). The north-west area is located in more rugged terrain. The fortifications in the remaining three areas are situated at the top of steep slopes without any ditches or embankments.

Surprisingly, the results may suggest that artificial obstacles strengthen the defense much more than the steep slope itself. The attackers are forced to overcome (and therefore to stop for a while) the sides of the trench, which are within reach of direct defensive shooting. Direct shooting is more accurate than shooting at short distances, and the defenders are able to cause great losses in a short period of time. The steep slope slows down the attackers when approaching the fortifications, but for most of the time, the attackers are within reach of only indirect shooting. However, the time (distance) during which the accuracy of the shooting is high is relatively short, as it is not prolonged by forced stops once the walls of the trench are overcome. So if we accept the results of the experiment as significant, the question arises as to how can the defenders strengthen the defense, for example, on the southern side of the *oppidum*. The answer to this question can be provided by archaeological research [6], according to which the southern, eastern and northern side of the *oppidum*, could be terraces with palisades (wooden walls that can be built without great expense). The results of the experiment may contribute to the confirmation of the above-mentioned archaeological research, as evidence of the presence of this fortification element is not unequivocally substantiated.

8   Conclusion

Agent-based modeling gives us the ability of studying many different complex systems. In this paper, a historical military model has been presented. The aim of this research was to create a model of the besieging of a Celtic *oppidum*, which can be used by archaeological experts to verify different scenarios.

The experiments performed showed the potential usefulness of this model for archaeological and historical research. The goal set for this research has therefore been achieved within the scope that was planned. The model can be further refined and expanded with the help of additional archaeological and historical sources.

Acknowledgment. The support of the Specific Research Project at FIM UHK is gratefully acknowledged. Thanks for consultation and GIS data to Alzbeta Frank Danielisova.

References

1. Anglim, S., Jestice, P., Rice, R., Rusch, S., Serrati, J.: Fighting Techniques of the Ancient World. Thomas Dunne Books, New York City (2007). ISBN 0312309325
2. Bouzek, J.: Keltove ceskych zemi v evropskem kontextu, 2nd edn. Triton, Praha (2009)
3. Campillo, X.R., Cela, J.M., Cardona, F.X.H.: Simulating archaeologists? using agent-based modelling to improve battlefield excavations. J. Archaeol. Sci. **39**(2), 347–356 (2012)
4. Cimler, R., Doležal, O., Kühnová, J., Pavlík, J.: Herding algorithm in a large scale multi-agent simulation. In: Jezic, G., Chen-Burger, Y.-H.J., Howlett, R.J., Jain, L.C. (eds.) Agent and Multi-Agent Systems: Technology and Applications. SIST, vol. 58, pp. 83–94. Springer, Cham (2016). https://doi.org/10.1007/978-3-319-39883-9_7
5. Cimler, R., Husáková, M., Koláčková, M.: Exploration of autoimmune diseases using multi-agent systems. In: Nguyen, N.-T., Iliadis, L., Manolopoulos, Y., Trawiński, B. (eds.) ICCCI 2016. LNCS (LNAI), vol. 9876, pp. 282–291. Springer, Cham (2016). https://doi.org/10.1007/978-3-319-45246-3_27
6. Čižmář, M.: Keltské oppidum Staré Hradisko. Archeologické centrum (2005)
7. Danielisová, A., Olševičová, K., Cimler, R., Machálek, T.: Understanding the iron age economy: sustainability of agricultural practices under stable population growth. In: Wurzer, G., Kowarik, K., Reschreiter, H. (eds.) Agent-based Modeling and Simulation in Archaeology. AGIS, pp. 183–216. Springer, Cham (2015). https://doi.org/10.1007/978-3-319-00008-4_9
8. Denti, A.M.E., Omicini, A.: Basi: multi-agent based simulation for medieval battles. http://ceur-ws.org/Vol-741/ID11_MolesiniDentiOmicini.pdf (2011)
9. Gabriel, R.A.: Ancient World (Soldiers' lives through history). Greenwood Publishing Group, Westport (2007). ISBN 0313333483
10. Gabriel, R.A., Metz, K.S.: From Sumer to Rome: The Military Capabilities of Ancient Armies. Praeger, Westport (1991). ISBN 0313276455
11. John, J.: Možnosti a limity počítačové rekonstrukce minulých cest na příkladu čertovy louky v krkonoších. Acta FF 4 (2010)
12. Macal, C.: Everything you need to know about agent-based modelling and simulation. https://doi.org/10.1057/jos.2016.7 (2016)
13. Machálek, T., Cimler, R., Olševičová, K., Danielisová, A.: Fuzzy methods in land use modeling for archaeology. In: Proceedings of Mathematical Methods in Economics (2013)
14. Machálek, T., Olševičová, K., Cimler, R.: Modelling population dynamics for archaeological simulations. In: Mathematical Methods in Economy, pp. 536–539 (2012). http://mme2012.opf.slu.cz/proceedings/proceedings.php
15. Olševičová, K., Cimler, R., Machálek, T.: Agent-based model of celtic population growth: NetLogo and Python. In: Nguyen, N., Trawiński, B., Katarzyniak, R., Jo, G.S. (eds.) Advanced Methods for Computational Collective Intelligence. SCI, pp. 135–143. Springer, Heidelberg (2013). https://doi.org/10.1007/978-3-642-34300-1_13

16. Riggsby, A.M.: Caesar in Gaul and Rome: War in Words. University of Texas Press, Austin (2010)
17. Rubio-Campillo, X., Cela, J.M., Cardona, F.X.H.: The development of new infantry tactics during the early eighteenth century: a computer simulation approach to modern military history. J. Simul. **7**(3), 170–182 (2013)
18. Tomaskova, H., Kuhnova, J., Cimler, R., Dolezal, O., Kuca, K.: Prediction of population with Alzheimers disease in the European union using a system dynamics model. Neuropsychiatric Dis. Treat. **12**, 1589 (2016)
19. Wilensky, U.: Netlogo user manual version 4.1. 3. Center for Connected Learning and Computer-Based Modelling. Northwestern University, Evanston, IL. http://ccl.northwestern.edu/netlogo/. Accessed July 2014
20. Wurzer, G., Kowarik, K., Reschreiter, H.: Agent-based Modeling and Simulation in Archaeology. Springer, Cham (2015). ISBN 978-3-319-00007-7

Weights Ordering During Training of Contextual Neural Networks with Generalized Error Backpropagation: Importance and Selection of Sorting Algorithms

Maciej Huk$^{(\boxtimes)}$

Department of Information Systems,
Wroclaw University of Science and Technology, Wroclaw, Poland
maciej.huk@pwr.edu.pl

Abstract. Contextual neural networks which are using neurons with conditional aggregation functions were found to be efficient and useful generalizations of classical multilayer perceptron. They allow to generate neural classification models with good generalization and low activity of connections between neurons in hidden layers. Their properties suggest also that usage of contextual neurons with conditional signals aggregation can cause similar effects as dropout technique in convolutional deep neural networks. The key factor to build such solutions is achieving self-consistency between continuous values of weights of neurons' connections and their mutually related non-continuous aggregation priorities. This allows to optimize neuron inputs aggregation priorities by simultaneous gradient-based optimization of connections' weights with generalized BP algorithm. But such method additionally needs to perform sorting of neuron inputs by its weights after each given number of training epochs. Thus within this text we compare efficiency of training of contextual neural networks with selected sorting algorithms. On this basis we discuss the theoretical properties of analyzed training algorithm which are related not only to characteristics of used weights sorting methods but also to application of self-consistency to selection of neural scan-paths in contextual neural networks.

Keywords: Classification · Self-consistency · Sorting · Aggregation functions

1 Introduction

Contextual neural networks are generalizations of known neural networks architectures which are using neurons with multi-step conditional aggregation functions [1]. Those models were used with success to solve both benchmark as well as real-life classification problems [2, 3]. In [4, 5] they were shown to be very good tools for fingerprints detection for crime-related analyses. They were also successfully used for spectrum prediction in cognitive radio and for research related to measuring awareness of computational systems [6].

© Springer International Publishing AG, part of Springer Nature 2018
N. T. Nguyen et al. (Eds.): ACIIDS 2018, LNAI 10752, pp. 200–211, 2018.
https://doi.org/10.1007/978-3-319-75420-8_19

Contextual neural networks can form classifiers with better generalization properties than their non-contextual versions such as MLP [3]. But what is more important they have also ability to considerably limit activity of connections between neurons without decreasing accuracy of their output values – both during and after training [2]. This can be used to decrease time and energy costs of running trained neural networks, especially in highly constrained applications. Limiting activity of given connections is done adaptively to processed data, thus for different input vectors given neuron can use different subsets of inputs and does not read signals from inputs not needed to calculate output value for given input vector. Finally, this changes the character of the neural network from black-box to grey-box model [3]. This is because by analyzing activity of inputs of neurons in the first hidden layer of contextual neural network one can check which data attributes are needed to calculate output values of the network for each input vector. Finally, with this technique data attributes can be ordered by their importance found by the contextual neural network for solving given problem.

In detail, neurons used to build contextual neural networks aggregate signals not in one but in multiple steps. Each aggregation step is used to read-in given subset of inputs and to decide if already processed information is enough to calculate the output value of the neuron with acceptable accuracy. The composition and order of groups of inputs, adequate for problem solved by the neural network, are selected independently for each neuron during training of the model. Given ordered list of groups of inputs is called a "scan-path", because multi-step conditional aggregation functions are realizations of Starks' scan-path theory [7]. Such functions in following steps aggregate signals from different subsets of inputs until neuron activation cumulated from previous groups of inputs is lower than given constant threshold. Examples of those functions are Sigma-if, CFA, OCFA and RFA functions [1].

It is also worth to notice that contextual neural networks with multi-step conditional aggregation functions in comparison to their non-contextual versions need only a small number of additional parameters, and values of those extra parameters can be easily selected with use of simple rules prior to the training [2]. This is the effect of special construction of aggregation functions of their neurons which allows to use generalized backpropagation algorithm (GBP) and self-consistency paradigm to setup their complex behavior and represent it within values of connection weights [8]. This is considerable improvement in comparison to multi-parameter aggregation functions of other contextual neurons: Clusteron, Sigma-Pi [9] or Spratling-Hayes neuron [10].

In this paper we find that elements of the generalized error backpropagation algorithm connected with sorting of data, can be further improved to increase the efficiency of the training of contextual neural networks. Thus the rest of the paper is organized as follows. In Sect. 2, brief description of the generalized backpropagation algorithm is given. Then Sect. 3 presents detailed discussion of possible improvements of the phase of GBP algorithm used to sort neuron's inputs. This is next used in Sect. 4 to experimentally test which sorting algorithms are best suited to be used within GBP while training contextual neural networks to solve selected UCI machine learning benchmark classification problems. Finally in Sect. 5, we discuss how obtained results and analyzed sorting methods can be used to characterize previously not studied theoretical properties of the GBP algorithm and contextual neural networks, especially the influence of self-consistency on neural scan-paths during GBP.

2 Generalized Backpropagation Algorithm

The generalized error backpropagation algorithm (GBP) is a modification of classical error backpropagation method extended to be able to train contextual neural networks which are using neurons with multi-step conditional aggregation functions [1, 2]. This is done by incorporating self-consistency paradigm known from physics [8]. It allows to use gradient based method to optimize simultaneously continuous (connection weights) and non-continuous, non-differentiable parameters of neuron's aggregation functions. The key to such abilities of the GBP method is maintaining mutual dependency of those both groups of parameters. It is realized by defining non-continuous parameters as function Ω of continuous weights. For neurons with multi-step aggregation function the Ω relation consists of two operations: sorting N neuron connection weights and then dividing ordered inputs into list of K equally-sized groups. N/K inputs with highest weights go to the first group, next N/K inputs with highest weights go to the second group, etc. Then the basic scan-path can be defined as the list of groups from first to last. It can be used within aggregation function, where groups of inputs are read-in one after another until given condition is met. Without details of aggregation the GBP algorithm can be represented as on the Fig. 1.

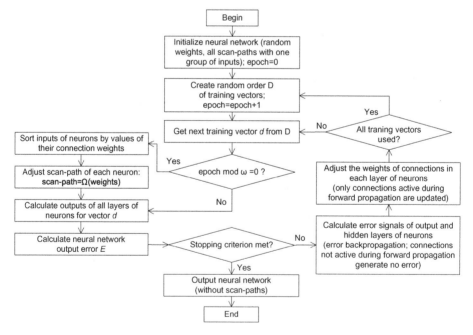

Fig. 1. General flowchart of the generalized error backpropagation algorithm (GBP) for given scan-path creation function Ω. Scan-paths update interval ω controls the strength of self-consistency between connections weights and parameters of neuron aggregation functions. After the training, temporary neuron scan-paths are discarded – they can be re-created later from weights with function Ω.

It is important to notice how interval ω controls the strength of self-consistency between connections weights and parameters of neuron aggregation functions. At the beginning and during the first ω epochs of GBP algorithm all neurons have all inputs assigned to the first group of their scan-paths. This is because at the beginning of learning we do not want to make any assumptions about the importance of the neurons inputs. Then, after each ω epochs scan-paths are updated with Ω function in accordance to actual connection weights. Thus the selection of interval ω can considerably influence the training process. If $\omega = \infty$ (or number of groups $K = 1$) the GBP algorithm behaves exactly as the classical error backpropagation method, and the contextual neural network with neurons using conditional multi-step aggregation functions behaves like MLP. On the other side, when $K > 1$ and the value of ω is close to one the error space of the neural network is frequently reorganized due to the updates of scan-paths of neurons after each ω epochs. This can make the training process to be ineffective. But it was shown experimentally that when $K > 1$ and $5 < \omega << \infty$, the output error of the neural network can temporarily increase after the update of the scan-path, but during following epochs error can drop down again, together with the activity of neural network connections.

Due to the above, the models built with GBP algorithm can have better generalization properties and lower average connections activity than their non-contextual versions trained with classical error backpropagation. One of the reasons for such effect is the following: cyclic reorganizations of the scan-paths of the neurons during the GBP together with evolving of decision space of each neuron during conditional multi-step signals aggregation, form mechanism analogous to the dropout technique extensively used in deep learning solutions. But there are also fundamental differences. Dropout is not dependent on the processed data, and the decision spaces of neurons of contextual neural network change according to actually processed data vector and during GBP take into account what the model has learned till given epoch. Dropout decreases the internal activity of the network connections only during the training process, and contextual neural networks limit it both during and after the training. This makes conditional multi-step aggregation functions valuable both for basic feedforward neural networks like MLP as well as for convolutional neural networks.

Presented description of the training process of contextual neural networks indicates that frequent updates of scan-paths of the neuron's aggregation functions are connected with high number of sorting of neuron inputs by their weights. For high number of neurons this can form considerable computational cost, thus it is worth to check which sorting algorithm is most suitable for use within the GPB method.

3 Sorting Weights of Neurons' Inputs Within GBP Algorithm

As it was indicated in the previous section, probably the most important element of the GBP algorithm is the update of neuron' scan-path realized with use of inputs sorting phase. This is because it allows to merge gradient-based search done by the error backpropagation algorithm with the self-consistency paradigm. But one can also notice that the selection of the sorting algorithm for GBP method is not straight-forward. This is because at least two factors can influence the efficiency of each sorting of inputs.

The first factor is the number of inputs of given hidden neuron – the size of the data to be sorted can considerably influence efficiency of given sorting method. And the second is the characteristic of the data to be sorted – which is hard to model in the general case of scan-path changes during the GBP process for different problems. This is because, depending on the data to be processed, values of parameters of GBP and on the phase of the training process, connection's weights can change a lot or almost not change between epochs of subsequent scan-paths updates (after each ω epochs). And different sorting algorithms can behave differently for sorted or partially sorted lists of data. In such cases their computational efficiency can be even much lower or much higher from their average efficiency.

To check the influence of the sorting method on the efficiency of the GBP algorithm, it will be helpful to analyze in detail how the sorting method is used by the training algorithm. At the beginning of the GBP the initial, random values are assigned to the connections weights of neuron inputs of given indexes. At the same time all inputs are assigned to the first aggregation group – this is to ensure that all inputs are used during first cycles of error backpropagation algorithm. Let's assume that after such initialization and first omega epochs of the training, connection weights w of neuron inputs of given indexes I as well as their assignment to scan-path groups G are as presented below:

```
W= [0.3,0.9,0.1,0.4,0.2,0.6,0.8] //connection weights
I= [  1,   2,   3,   4,   5,   6,   7] //indexes of inputs
G= [  1,   1,   1,   1,   1,   1,   1] //assignments to groups
```

Then given sorting method creates ordered list of connection weights, reordering accordingly also indexes of neuron inputs - without changing of the initial relation of given weight of its neuron input. After sorting of weights and reordering of connection indexes, neuron inputs are divided into given number of groups K. If possible, all groups have the same number of elements. But when the number of neuron inputs N can't be divided into K groups of equal size, the eventual rest of connections are added to the group of highest number. Then, during following epoch, neuron first aggregates signals from inputs belonging to group 1. Next, accordingly to details of its aggregation function, it conditionally aggregates signals from the remaining groups.

```
W= [0.9,0.8,0.6,0.4,0.3,0.2,0.1] //connection weights
I= [  2,   7,   6,   4,   1,   5,   3] //indexes of inputs
G= [  1,   1,   2,   2,   3,   3,   3] //assignment to 3 groups
```

Such weights sorting procedure and reassignment of neuron's input connections to aggregation groups is performed after every omega epochs of training, to keep mutual relation between values of weights and aggregation priorities after changes of weights values done in the effect of presentation of training vectors.

It is important to note, that for every sorting algorithm, given initial neural network is transformed exactly into the same trained neural model. This is because each sorting algorithm for the same input data outputs identical sorted list of values. But depending on the number of inputs of network' hidden neurons, different sorting methods will present different efficiency. However, the reader can notice that average sorting efficiency not necessarily must apply to data sorted during the training with generalized backpropagation. This is because GBP training procedure changes weights and their assignments to aggregation groups in non-random way. If the changes of values of weights between epochs are very small, then subsequent runs of sorting algorithm for given neuron will process list of data which is almost perfectly ordered. In such case superior efficiency would be achieved by algorithms that are optimal for sorted data – e.g. insertion-sort algorithm. In the opposite case, if the changes of values of weights between epochs are big enough to considerably change initial assignments to aggregation groups, then in most cases sorting algorithm for given neuron will process data which is not ordered. In such case better efficiency would be achieved by method such as quick-sort algorithm.

Finally, in this paper we want to check experimentally the efficiency of different sorting algorithms under conditions occurring during training of contextual neural networks with GBP algorithm. It will help to formulate guidelines how one should select the sorting algorithm for given contextual neural network. But it is more important, that it will help us to answer important question about characteristic of GBP method: how much the changes of weights between subsequent sorts influence the aggregation priorities of hidden neuron inputs? It can be, that after each sorting of weights the aggregation priorities of neuron inputs and the related error space of neural network change drastically. This would make the training more like random process. In the opposite case, the sorting would not change the initial aggregation priorities at all, what could mean that usage of the self-consistency paradigm has no influence on the training results. Is any of the above dominating the training? Does it depend on the frequency of scan-paths updates? Answering to the above questions will greatly increase our understanding of properties of the generalized backpropagation algorithm.

4   Results of Experiments

To answer the questions stated in the previous section, a set of experiments was performed for six different serial versions of sorting algorithms such as: Quick-sort (recursive), Merge-sort (recursive), Insertion-sort, Bubble-sort, Optimized Bubble-sort and Two-Way Optimized Bubble-sort [11–13]. Quick-sort, Insertion-sort and Merge-sort were used in their basic versions, without any improvements. In particular this means that Quick-sort was using middle element as a pivot and no specialized mechanism was used for sorting short sub-partitions [12]. By analogy the Insertion-sort was not performing initial placement of the smallest element at the beginning nor shifting larger elements prior to the insertion instead of carrying out many exchanges [11]. Decision to not use abovementioned optimizations was done because they were designed for the average case of unsorted data and we expect that this can be not valid in the case of partially sorted, GBP related data. Thus using simple algorithms should ease the interpretation of the results and help formulating possible sorting optimizations dedicated for use with GBP algorithm.

During the experiments contextual neural networks were trained with generalized backpropagation algorithm to solve example UCI ML benchmark problems such as Iris, Sonar, Soybean, Lung Cancer and Wine. All neurons within the neural network were using bipolar sigmoid or leaky rectifier activation function and Sigma-if aggregation function [2, 3]. Aggregation functions of all hidden neurons were dividing their inputs into K groups and were using aggregation threshold $\varphi^* = 0.6$. Continuous data attributes were normalized to range <0, 1>, and nominal attributes were represented with single binary input for every value (one-hot encoding). Resulting architectures of neural networks used during the experiments are presented in Table 1.

Table 1. Architectures of contextual neural networks used during the experiments for selected benchmark problems from UCI ML repository.

Training data set	Number of inputs of neural network	Number of hidden neurons (N)	Number of classes	Number of connections between neurons	Number of groups of inputs of neurons (K)
Iris	4	10	3	70	3
Wine	13	10	3	160	3
Hypo	35	10	5	400	14
Sonar	60	10	2	620	10
Soybean	134	20	19	3060	22
LungC	244	3	3	741	3

The numbers of neurons in hidden layer and the number of groups K were based on results of previous experiments and selected to achieve suboptimal classification accuracies of resulting models with considerable reduction of connections activity between neurons.

The values of parameters of the GBP training algorithm were as follows: constant training step $\alpha = 0.1$ (for networks with sigmoidal activation function) or $\alpha = 0.01$ (for networks with leaky rectifier activation function), interval of aggregation groups update $\omega = 25$, stopping criterion: perfect classification of training data or no classification accuracy improvement for more than 300 or 1300 subsequent training epochs. No momentum was used.

For each set of values of above parameters, training was performed within Borland Delphi 7 environment for set of ten different values of seed of built in pseudo-random generator used to initialize neural networks and to select sequences of training vectors during the training. The seed values were: 0, 1013420422, 550932796, 729462840, 923512745, 327401735, 423664106, 834385610, 245071639, 619168307. Additionally each training was repeated ten times to ensure that resulting neural models for given seed value and given set of values of other parameters are identical, and to increase precision of measurements of computational cost. Cross-validation was not used because analysis of the abilities of generalized backpropagation algorithm to build contextual neural networks of good classification properties was not the goal of performed experiments. Thus each training for given problem was done with use of the whole given training data set.

During each experiment following properties were measured:

- cumulated cost of execution of sorting method during the training,
- number of training epochs before the stopping criterion was met,
- average classification error of the resulting model,
- average, minimal and maximal activity of internal connections of the neural network.

Activity of the connections was measured for the neural network with the highest average classification accuracy observed during the GBP process. Computational cost of each sorting was measured with the use of main processor' integrated Time-Stamp-Counter (TSC) 64 bit register. Cumulated cost of execution of sorting method during given training was calculated by summing up numbers of ticks of the processor clock counted by TSC counter during each sorting (for each hidden neuron and each epoch). Then, the average cost of single sorting was calculated by dividing the cumulated cost of sorting by the number of calls of the sorting method. To minimize influence of the operating system on the measurement precision, TSC register was used with highest priority execution mode of the simulation program.

Basic results of experiments designed as described above are presented in Table 2. They include average classification errors, average number of epochs of training and average connections activity of contextual neural networks. Stopping criterion of the GBP was: network error equal 0 or no decrease of the error during last 1300 epochs.

Table 2. Average values of classification error, epochs of training as well as minimal, maximal and average activities of hidden neurons connections of contextual neural networks solving example UCI benchmark classification problems. Neurons with Sigma-if aggregation function as well as bipolar sigmoid (BS) and Leaky Rectifier (LR) activation functions were used.

Training data set	Activation function	Classification error [%]	Training epochs [1]	Activity (avg.) [%]	Activity (max) [%]	Activity (min) [%]
Iris	BS	3.9	149.1	86.0	87.5	85.7
Wine	BS	4.3	194.6	91.7	91.7	91.7
Hypo	BS	1.3	1763.3	25.7	73.1	17.5
Sonar	BS	0.8	2037.8	15.4	35.6	9.7
Soybean	BS	2.0	2286.8	22.9	36.0	15.9
LungC	BS	0.3	326.5	26.4	40.2	24.2
Iris	LR	5.0	748.7	57.1	57.1	57.1
Wine	LR	3.4	497.8	66.1	68.2	65.6
Hypo	LR	2.4	2285.7	19.8	58.9	17.5
Sonar	LR	6.9	725.7	19.7	39.2	11.5
Soybean	LR	2.1	2465.2	18.5	30.3	15.2
LungC	LR	2.2	1703.2	26.9	41.7	24.2

Achieved considerable decrease of connections activity proves contextual behavior of neural networks trained with GBP. It can be also seen that contextual neural networks of bigger structures (for Hypo, Sonar, Soybean and LungC problems) can achieve lower activity of connections than smaller networks (here for Iris and Wine problems). This was also expected because was observed in earlier experiments [3]. Thus performed simulations can be used to search for potential dependencies between properties of contextual neural networks trained with the GBP algorithm and computational efficiency of selected sorting methods used for scan-paths updates in GBP.

Within Table 3 we present measured relative average computational cost of six selected sorting methods during their use for neurons' scan-path update by GBP algorithm. Measures were taken by repeating GBP training for each of the sorting methods for the same set of benchmark problems, neural networks architectures and related parameters (including seed of pseudo-random generator, initial values of connections weights, etc.). For given set of initial values of parameters the result of the training was model identical for each of the sorting methods. This is because all used sorting methods give the same output for the same input data, and switching between different sorting methods can't change the result of the GBP. But each of the sorting methods can need different number of operations to sort given input data. This causes differences between computational costs of considered sorting methods during the GBP. To ease the analysis of results all obtained values of computational costs are presented in relation to the cost measured for the Insertion-sort method. For the same reason, within Table 3 the length of scan-path of hidden neurons is additionally presented. Scan-path length is defined here as the number of values which have to be sorted during single update of the scan-path of given neuron by the GBP algorithm.

Within presented results it can be noticed that the factor with strongest influence on the efficiency of scan-path updates is the length of scan-paths. This causes that for neural networks with hidden neurons having low number of inputs (data for Iris and Wine benchmarks) the most efficient sorting method is Insertion-sort. With the increase of the length of the scan-path above certain length the most efficient sorting method becomes Quick-sort. Based on the results from Table 3 this bound can be estimated by interpolation but it can be also presented graphically as on the Fig. 2. This shows that for considered contextual neural networks Quick-sort within GBP should be preferred for scan-paths longer than 35 elements, irrespective of the type of activation function, and Insertion-sort should be used for shorter scan-paths.

The above observation of effectiveness of sorting algorithms is considerably different than when Quick-sort is compared with Insertion-sort for random distribution of values to be sorted. In [12] as well as in well-known programming libraries the threshold, below which Insertion-sort is used instead of Quick-sort, is much lower than 35. For example, Bentley and McIlroy set this threshold to 7 [12]. GNU C Library (glibc), GNU ISO C++ Library and C++ Standard Library (STL) as this threshold use value 5. And in FORTRAN 90/95 one can find it to be equal 10. Such low values of this threshold are used because mentioned solutions are implemented to maximize average sorting efficiency for all possible sets of input data. Thus within conducted experiments Insertion-sort presents computational cost lower than average and closer to its best case of computational complexity $\Theta(n)$, where n is the length of scan-path. In the same case cost of Quick-sort is higher than average and closer to its worst case complexity of $\Theta(n^2)$.

Table 3. Average relative cost of connections sorting per neuron per epoch during training of contextual neural networks with GBP for different sorting algorithms. Sigma-if aggregation and example UCI problems were used for two activation functions: bipolar sigmoid (BS) and Leaky Rectifier (LR). Each result is average of 100 measurements. Scan-path length is the number of inputs of hidden neuron. Stopping: 300 epochs without improvement of classification accuracy.

Training data set	Activation function	Scanpath length	Merge sort [%]	Quick sort [%]	Insert sort [%]	Bub. sort two-way [%]	Bub. sort optimized [%]	Bubble sort [%]
Iris	BS	4	153.4	131.2	100	112.8	111.0	97.2
Wine	BS	13	180.8	137.1	100	172.4	188.4	242.1
Hypo	BS	35	154.0	98.1	100	265.9	271.3	280.6
Sonar	BS	60	103.1	83.1	100	250.7	272.9	297.7
Soybean	BS	134	51.2	44.0	100	242.8	252.6	293.0
LungC	BS	244	50.7	35.9	100	270.2	285.6	284.6
Iris	LR	4	186.3	135.8	100	119.8	117.4	101.6
Wine	LR	13	184.4	125.1	100	153.8	166.7	169.2
Hypo	LR	35	149.5	101.6	100	220.9	228.3	253.3
Sonar	LR	60	98.3	78.7	100	243.8	268.1	282.4
Soybean	LR	134	53.9	44.0	100	244.2	254.2	260.3
LungC	LR	244	42.0	47.15	100	225.2	237.9	236.0

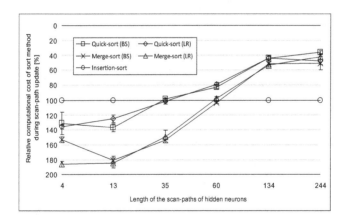

Fig. 2. Average computational cost of sorting during updates of scan-paths of contextual neural network by the GBP algorithm as a function of scan-paths length for selected sort algorithms and two activation functions of neurons: bipolar sigmoid (BS) and leaky rectifier (LR). Results are presented in relation to the measurements for Insertion-sort method. Subsequent, growing lengths of scan-paths correspond to the following UCI ML benchmark data sets: Iris, Wine, Hypothyroid, Sonar, Soybean and Lung Cancer. Scan-paths update interval was equal 25.

The interpretation of obtained results is straightforward - GBP algorithm can keep weights of inputs of given hidden neuron partially ordered between subsequent updates of its scan-path. Thus the scan-paths do not change in a random manner. Moreover, the average percentage of the scan-path which is changed during scan-paths update can be different for different activation functions of the neurons. We observe that it is higher for bipolar sigmoid and training step $\alpha = 0.1$. than for leaky rectifier function and $\alpha = 0.01$. Thus on average the scan-paths of neurons are not constant during the training of considered contextual neural networks with the GBP algorithm. Finally, if the scan-paths are changing during the training and those changes are not random, then self-consistency effect has considerable influence on the training. This is important, both practical and theoretical, result.

Using the outcomes of performed experiments it can be also checked how the frequency of scan-paths updates influences the efficiency of sorting methods. For Wine problem it is visualized on Fig. 3. For scan-paths interval ω from 5 to 60 epochs mutual relations of efficiency between sorting methods do not change. Thus, excluding cases when ω is higher than number of epochs of given training, earlier conclusions on functioning of self-consistency and GBP method stay valid for different ω values.

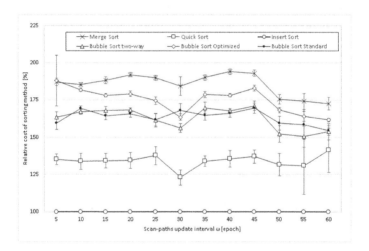

Fig. 3. Average computational cost of sorting during updates of scan-paths of contextual neural network by the GBP algorithm for Wine data set as a function of scan-paths update interval ω. Results are presented in relation to the measurements for Insertion-sort method. Activation function: bipolar sigmoid. Stopping: 300 epochs without error decrease. Training step $\alpha = 0.1$.

5 Conclusions

Measurements and analyses presented in this paper are unique in their nature because basic BP algorithm is not related with sorting methods. We have shown that self-consistency within GBP has considerable influence on the evolution of scan-paths of contextual neurons during training. In the effect, for scan-paths shorter than 35 elements Insertion-sort makes GBP method less computationally expensive than other considered sorting algorithms. Above this threshold Quick-sort was better choice and such result was noted for wide range of scan-paths update interval ω.

Moreover, measured values of this Insertion-sort threshold are considerably higher than analogous values used by hybrid sorting methods implemented within standard programming libraries. This is because those implementations of sorting algorithms are optimized for achieving maximal average efficiency – and the scan-path related data of neurons is preserved partially sorted during most of the epochs of their training with GBP. Thus in cases when computational efficiency of GBP would be important, one should consider using dedicated sorting method for scan-paths updates.

Presented experiments open new, previously not explored valley in the field of research on artificial neural networks. Thus possible further work includes analyzing relations between computational efficiency of GBP and such factors as different sorting methods (e.g. Shell-sort, partial interval sorting), various multi-step aggregation functions as well as parameters of neural networks and training methods. We expect that such research will allow to train contextual neural networks more efficiently and that it will extend our actual understanding of the nature of those models.

References

1. Huk, M.: Notes on the generalized backpropagation algorithm for contextual neural net-works with conditional aggregation functions. J. Intell. Fuzzy Syst. **32**, 1365–1376 (2017)
2. Huk, M.: Backpropagation generalized delta rule for the selective attention Sigma-if artificial neural network. Int. J. Appl. Math. Comput. Sci. **22**, 449–459 (2012)
3. Huk, M.: Learning distributed selective attention strategies with the Sigma-if neural network. In: Akbar, M., Hussain, D. (eds.) Advances in Computer Science and IT, pp. 209–232. InTech, Vukovar (2009)
4. Szczepanik, M., Jóźwiak, I.: Data management for fingerprint recognition algorithm based on characteristic points' groups. In: Pechenizkiy, M., Wojciechowski, M. (eds.) New Trends in Databases and Information Systems. Advances in Intelligent Systems and Computing, vol. 185, pp. 425–432. Springer, Heidelberg (2013). https://doi.org/10.1007/978-3-642-32518-2_40
5. Szczepanik, M., Jóźwiak, I.: Reliability and error probability for multimodal biometric system. In: Korbicz, J., Kowal, M. (eds.) Intelligent Systems in Technical and Medical Diagnostics. Advances in Intelligent Systems and Computing, vol. 230, pp. 325–332. Springer, Heidelberg (2014). https://doi.org/10.1007/978-3-642-39881-0_27
6. Huk, M.: Measuring the effectiveness of hidden context usage by machine learning methods under conditions of increased entropy of noise. In: 3rd IEEE International Conference on Cybernetics, pp. 1–6. IEEE Press (2017)
7. Privitera, C.M., Azzariti, M., Stark, L.W.: Locating regions-of-interest for the Mars Rover expedition. Int. J. Remote Sens. **21**, 3327–3347 (2000)
8. Raczkowski, D., Canning, A.: Thomas-Fermi charge mixing for obtaining self-consistency in density functional calculations. Phys. Rev. B **64**, 121101–121105 (2001)
9. Mel, B.W.: The Clusteron: toward a simple abstraction for a complex neuron. In: Advances in Neural Information Processing Systems, vol. 4, pp. 35–42. Morgan Kaufmann (1992)
10. Spratling, M.W., Hayes, G.: Learning synaptic clusters for nonlinear dendritic processing. Neural Process. Lett. **11**, 17–27 (2000)
11. Knuth, D.: The Art of Computer Programming: Sorting and Searching, vol. 3. Addison Wesley, Boston (1998)
12. Bentley, J.L., McIlroy, M.D.: Engineering a sort function. Softw.: Pract. Exp. **23**, 1249–1265 (1993)
13. Astrachan, O.: Bubble sort: an archaeological algorithmic analysis. SIGCSE Bull. **35**, 1–5 (2003)

Implementing Contextual Neural Networks in Distributed Machine Learning Framework

Bartosz Jerzy Janusz[✉] and Krzysztof Wołk

Wroclaw University of Science and Technology, Wroclaw, Poland
{210004,220999}@student.pwr.edu.pl

Abstract. Contextual neural networks are generalization of multilayer neural networks. They possess interesting property of automatic selection of data attributes needed for correct processing of given input vectors. To achieve that they are using neurons with conditional, multistep aggregation functions and error generalized error backpropagation algorithm based on self-consistency paradigm. According to the literature of the subject, currently there are no implementations of those models in high-performance machine learning platforms like Mahout or MLlib. In this paper we present initial results of implementation of contextual neural networks in distributed machine learning framework called H2O. The motivation behind this work is the need to analyze properties of contextual neural networks and conditional multi-step aggregation functions while solving large classification problems.

Keywords: Classification · Self-consistency · Multi-step aggregation functions
Generalized error backpropagation

1 Introduction

H2O is open source, distributed, parallel and scalable platform implementing core machine learning methods. It allows to build machine learning (ML) models on big data and has proven its industry-grade reliability [1–4] while solving real life problems [5–8]. H2O includes such algorithms as e.g. deep learning of multilayer neural networks, Gradient Boosting and Random Forest for creation tree-based models, ensemble learning, Principal Component Analysis, and k-Means clustering [9, 10]. It performs batch training of ML models, but can also make online predictions with use of Storm, but cannot train the models online. H2O realizes distributed computation with "distributed fork-join", which parallelizes jobs across many data nodes for efficient in-memory computation with a divide-and-conquer technique, and then combines the results. Final models can be exported as JAVA classes for further use.

H2O can be compared with other state-of-the-art ML frameworks such as Mahout, MLlib, Oryx, SAMOA, SINGA as well as WEKA, and among those is recognized for its usability, speed and extensibility [2, 4]. H2O includes interfaces to Java, R, Python and Scala and easily integrates with Spark. It provides also H2O Flow, an interactive, web-based notebook for manipulating data and building ML models using a hybrid of point-and-click and command-line approach. Basic start of H2O is as simple as:

© Springer International Publishing AG, part of Springer Nature 2018
N. T. Nguyen et al. (Eds.): ACIIDS 2018, LNAI 10752, pp. 212–223, 2018.
https://doi.org/10.1007/978-3-319-75420-8_20

unpacking its actual version to the chosen directory, loading included H2O server with "java –jar h20.jar" command and visiting local address http://localhost:54321 with favorite web browser for accessing H2O Flow web GUI. This allows H2O use on single workstation as well as in multi-node cloud environments, not only by programmers but also by researchers [8].

Although H2O implementations of ML algorithms are highly optimized for maximum performance, such approach causes that developers of H2O are highly concentrated only on basic ML models and methods. In the effect, a number of ML tools available in other frameworks can't be found in H2O. Thus one should also not expect from H2O realization of newest or highly specialized ML methods. This is why in this paper we present first approach to extend H2O Deep Learning functionality with capability of training contextual neural networks (CxNN) with Generalized Error Backpropagation algorithm [11–13].

In the effect, the rest of the paper is organized as follows. In Sect. 2, we briefly present contextual neural networks and the description of the generalized backpropagation algorithm. Then Sect. 3 includes detailed discussion of performed modifications of H2O architecture and shows related changes of H2O Flow web interface. Presented software is next used in Sect. 4 to solve selected UCI ML benchmark classification problems to experimentally test if properties of contextual neural networks built with H2O are as expected. Finally in Sect. 5, we discuss obtained results and possible areas of further development and research related with H2O and contextual neural networks.

2 Contextual Neural Networks and Generalized Error Backpropagation Algorithm

Considered contextual neural networks (CxNN's) were previously applied with success to solve benchmark as well as real-life classification problems [12, 13]. They were used for spectrum prediction in cognitive radio [14] and during research related to measuring awareness of computational systems through context injection [8, 15]. It was also shown, that they can be very good tools for fingerprints detection for crime-related analyses [16], can be helpful for solving such important problems as e.g. rehabilitation and elderly abuse prevention [17, 18]. Moreover it was shown, that they possess three important properties. The first property is that they are not black-boxes. The way how they operate allows to observe which input attributes are more important for classification of given data vectors. This can be very useful for many data-analytics-related tasks. The second property is that their neurons try to solve problems with use of as low number of input signals as possible, separately for each given data vector. This is done both during and after training of the neural network, practically without limiting model classification accuracy – and can considerably decrease costs of its use for data processing (especially time and energy costs). It was reported, that for many problems, above effect can limit connections activity of hidden neurons more than ten times in relation to analogous neural networks which process data with all connections. And the third important property of contextual neural networks is that the described behavior of their neurons during training causes further effects similar to outcomes of using dropout [13, 19–22]. The major difference between both techniques is that dropout is not related to the data processed by the neural network, nor to the knowledge which the neural network already possess. And finally the decrease of neural network internal activity caused by dropout is not preserved after the training process – what is not the case for contextual neural networks. All the above makes contextual neural networks to be good candidates for implementation within H2O framework.

In detail, contextual neural networks described above are direct generalizations of known neural networks and can have well known architectures (e.g. MLP), but are using neurons with multi-step conditional aggregation functions. As the name suggests, such functions aggregate input signals of the neuron not in one but in multiple steps [12, 13]. Each step of aggregation is used to read-in given subset of inputs and to decide if already accumulated information is enough to calculate the output value of the neuron with needed precision. Typically, the steps of aggregation are realized until the neuron activation, cumulated from groups of inputs processed in previous steps, is lower than some constant aggregation threshold φ^*. Examples of those functions are Sigma-if, CFA and OCFA functions [11]. It is worth to note, that given ordered list of K groups of neuron inputs is sometimes called "scan-path". This is because multi-step conditional aggregation functions are closely related with Starks' scan-path theory [23].

Unlike other neural networks built with Sigma-Pi, Clusteron or Spratling-Hayes neurons [24, 25], contextual models with conditional multi-step aggregation almost do not need separate parameters to describe the composition and priorities of groups of inputs of each neuron [13]. Except two mentioned parameters for defining number of groups of inputs K and aggregation threshold φ^*, which can be common for all neurons within the neural network, all information about scan-paths is stored within connection weights. This improvement can be further exploited by applying self-consistency paradigm to train contextual neural networks with gradient algorithm – Generalized Error Backpropagation (GBP) [12, 13].

The GBP algorithm is a generalization of classical error backpropagation method extended to be able to train contextual neural networks which are using neurons with multi-step conditional aggregation functions [12]. To do that, GBP is applying self-consistency paradigm known from physics to use gradient based method to optimize both output error of the network and non-continuous, non-differentiable scan-paths of aggregation functions of the neurons [26]. The algorithm is maintaining mutual dependency between connection weights of given neuron and virtual, non-continuous parameters defining priorities and grouping of neuron inputs. Those virtual parameters describing scan-paths are calculated from connection weights with Ω function and there is no need to remember their values after the training. But during the GBP scan-paths are updated not more often than once per ω epochs to control the strength of the dependency between connection weights and scan-paths. Thus scan-paths must be preserved between epochs of their calculation in structures separated from connection weights.

For neurons with multi-step aggregation the Ω function consists of two operations: sorting N connection weights of the neuron and then dividing ordered inputs into list of K equally-sized groups. N/K inputs with highest weights go to the first group, next N/K inputs with weights highest among other connections go to the second group, etc. Then the basic scan-path is defined as the list of groups from first to last. Finally scan-path is used by the aggregation function in a way that, groups of inputs are read-in one after another until given condition is met. Without details of specific aggregation functions, the GBP algorithm can be represented as on the Fig. 1.

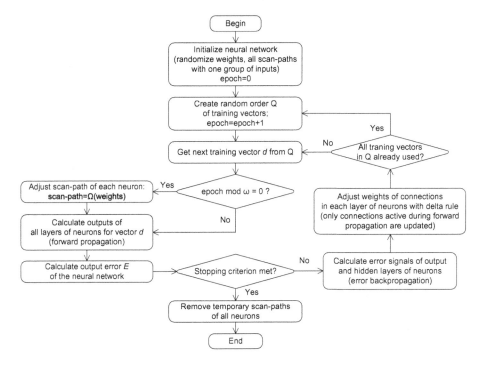

Fig. 1. Diagram of the generalized error backpropagation algorithm for given scan-path creation function Ω. Scan-paths update interval ω controls the strength of self-consistency between connections weights and virtual parameters of neuron aggregation functions. After the training, temporary scan-paths are discarded – they can be re-created later from weights with function Ω.

At the beginning and during the first ω epochs of GBP all neurons have all inputs assigned to the first group of their scan-paths. This is because at the beginning of training algorithm should not make any assumptions about the importance of the neuron inputs. Then, after each ω epochs scan-paths are updated with Ω function in accordance to actual connection weights.

3   Modification of the H2O.ai Framework

The previous section indicates that for scan-paths update interval $\omega = \infty$ the GBP algorithm behaves exactly as the classical error backpropagation method (BP). By analogy, contextual neural network with neurons using conditional multi-step aggregation for number of groups of inputs $K = 1$ behave exactly like multilayer perceptron network (MLP). This suggests that it should be possible to implement GBP method and contextual neural networks as direct extensions of actual realization of BP and MLP found in H2O. Unfortunately, analysis of highly optimized H2O code for distributed training of MLP neural networks with BP algorithm extended with deep learning techniques, showed that needed modifications are not straightforward.

3.1   Adding Aggregation Functions and Conditional Backpropagation

Actual, third version of H2O server implements different activation functions of the neurons not as methods of the "Neurons" class but as specializations of "fprop" and "bprop" methods of activation related child classes of the class "Neurons". This creates dedicated neuron classes for each activation function, such as "Tanh" neuron, "Maxout" neuron, etc. Such construction is reasonable under the assumption that low number of different aggregation functions is considered, especially for existing case of only one: linear combinations of weights. But our goal was to extend H2O with several considered aggregation functions (e.g. Sigma-if, CFA, OCFA, Random, etc.). Continuing original approach would then imply creating huge number of specialized neuron classes or sub-classes, such as "Tanh_SigmaIf", "Tanh_CFA", "Tanh_OCFA",…, "Maxout_SigmaIf", etc. This would lead to unnecessary code duplication and problematic maintenance of the source. On the other hand, we wanted to keep our changes of the H2O code structure to be as small and centralized as possible. This lead us to use mixed approach.

Having in mind the above observation that specialized neuron classes of H2O, named as activation functions, in fact realize whole neuron transfer functions, we extended their set with single "Ext" neuron class. Inside we encapsulated possibility to use both different activation and aggregation functions, leaving the original H2O code intact. Additionally, within this class we exchanged the basic error backpropagation method "bprop" with implementation of generalized error backpropagation named "conditionalBprop". The structures and methods related to the latter are added within new "NeuronConnectionGroups" and "LayerConnectionGroups" classes. Finally, it is worth to note, that we equipped "Ext" class not only with new aggregation and activation methods (Sigma-if, CFA, OCFA, Bipolar Sigmoid, Leaky Rectifier) but also with their forms existing in original H2O code (weighted linear aggregation, hyperbolic tangent, etc.). This allows us to check how they work with other added activation and aggregation functions and/or with generalized backpropagation algorithm. It is also worth to note that after defining required aggregation methods within "Ext" class the H2O framework automatically takes care about distributing calculations among available nodes, because this is done at the level where internal mechanisms of the trained model are not important.

3.2   Measuring Connections Activity in Distributed, Multi-threaded Environment

Special attention we had to pay during implementation of structures and methods for analyzing connections activity. This is due two reasons. First, because such kind of analysis is characteristic for contextual neural networks and it was completely absent within H2O. And the second reason is that H2O is by default distributed, multi-threaded environment. Even when run on single multi-core processor, it automatically adapts number of threads and their load to maximize efficiency of its tasks. In the effect, the measurements of activity of hidden connections of contextual neural network must be realized in a Map/Reduce manner. Additionally, each thread in each computational node within H2O cluster can train different neural networks during cross-validation procedure, as well as can process different portions of the training data. This is why it was decided that in presented version of the software, the possibility of hidden connections activity analysis will be limited to cases when the H2O cluster processes (trains or uses) neural network for one data set. Thus we assume that measurements of connections activity are not being done when cross-validation training is executed.

It is also worth to notice, that even without cross-validation, when multiple nodes and threads are available, H2O trains neural network in distributed way. In such case H2O independently maintains several local copies of the model and uses scheduled synchronization of their structures to create single, so called, shared model. At the end of the training process, shared model is returned as the result of the training process. Thus during the training we measure activity of connections only of the local neural networks – once after each epoch. In detail, when shared model is populated to all computational nodes, all threads are using this model to process disjoint portions of the given set of data vectors. For each data vector neurons of the local neural networks calculate their outputs and during this we are counting active inputs of the neurons. By summing up activities of all neurons of given neural network we get its number of active connections for given data vector. Finally, by summing up the activities for data vectors processed by all nodes and dividing this by the number of data vectors we get the final result – average activity of hidden connections of given model for given data set. This can be further presented also as the average percentage of active hidden connections of the local neural network (designated as avg_hca). By analogy, after the training we measure activity of hidden connections of the final, shared model of the neural network.

We have implemented the structures for aggregating data about activities of hidden connections of neural network within added "LayerConnectionsGroups" class. Measurement for given data vector is done within the call of forward propagation method "fprop". For simplicity we have modified the "fprop" function to return value being the number of active connections for processed data vector. Results of all "fprop" calls are then aggregated in 64 bit counters and averaged for all data vectors. Then such results for each epoch of training are saved to text file dedicated for given training. Such approach requires to perform measurements of activity of hidden connections with H2O parameter "train_samples_per_iteration" set to zero. This guarantees that the given set of data vectors will be processed by the neural network exactly one time during given epoch, regardless of the number of computing nodes.

3.3 Modifying the H2O Flow Web-GUI

Finally we have modified the original H2O Flow web application to be able to use new functionalities added to the H2O server. Its changes were limited to original DeepLearningModel.java and DeepLearningV3.java class files and were related to defining new parameters of H2O API as well as showing or hiding needed options within the web-GUI interface. Especially, we defined new "DeepLearningParameters.ExtActivation" parameter connected with the selection of our "Ext" transfer function computation with the GBP algorithm. We also decided to not change the name of original "activation" parameter to keep compatibility of modified API with the original one. In the effect the example screenshot of the modified H2O Flow application can look as on the Fig. 2.

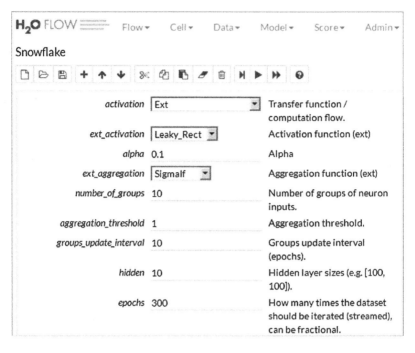

Fig. 2. Example fragment of the modified H2O Flow application with settings of characteristic parameters of contextual neural network prior to its training

One can notice, that to signal the true meaning of the "activation" parameter we have changed its description to "Transfer function/computation flow". After selecting its new "Ext" value, additional parameters are shown. Especially, the activation and aggregation functions can be selected by changing values of "ext_activation" and "ext_aggregation" parameters, respectively. Fields "number_of_groups", "aggregation_threshold" and "groups_update_interval" are related to parameters K, φ^* and ω, which in our implementation of GBP are common for all neurons within the neural network.

4   Results of Experiments

To verify correctness and analyze the properties of the presented modification of the H2O software we have run several tests. Here we present the most important results obtained for selected classification benchmark problems from UCI Machine Learning repository: Sonar, Crx and KDDCup (1999) [27]. Sonar and Crx are small data sets but are popular within the ML related literature. On the other hand, the KDDCup is one of the biggest UCI ML data sets (over 700 MB of training data). Analyzed properties of contextual neural networks were average classification accuracy as well as average activity of hidden connections (avg_hca). The software was modified version of the H2O Server 3.11 and H2O Flow 0.52. The detailed experiments setup was as follows.

In all experiments we have used default settings of the H2O Flow, except the following changes. Cross-validation and adaptive rate of training step α were not used. Constant training step α was 0.1. Parameters score_training_samples and number of training samples per MapReduce iteration (train_samples_per_iteration) were set to zero, what means "all vectors in the training data set". After initial tests, during described experiments for Sonar and Crx problems, score_interval, maximal number of training epochs, aggregation threshold and interval of groups update were 0.1 s, 600, 0.6 and 25, respectively. For KDDCup10 (around 10% of full KDDCup data) and KDDCup data sets values of the same parameters were 1 s, 30, 0.001 and 3 respectively. Values of the other parameters, related to the processed data sets are presented in Table 1.

Table 1. Architectures of contextual neural networks used during the experiments for selected benchmark problems from UCI ML repository.

Training data set	Number of inputs of neural network	Number of hidden neurons	Number of connections between neurons	Number of groups of neuron inputs (K)	Number of classes (network outputs)	Number of data vectors
Crx	60 (9)	10	400	10	5	690
Sonar	60 (0)	10	620	10	2	208
KDDCup10	119 (0)	100	14200	14	23	494022
KDDCup	124 (2)	100	14700	14	23	4898431

It is worth to notice that H2O by default sets the neural network architecture with use of extended "one-hot encoding" of inputs. Continuous attributes are represented by single input, and nominal attributes are represented with single binary input for every value. In addition to this standard method H2O performs also analysis of interactions between data attributes and in the effect extends the list of network inputs with additional elements. Counts of those extra inputs for considered data sets are given in Table 1 in round brackets. How interactions-based inputs are created is described in DataInfo class within H2O source code.

The results of experiments for Sonar and Crx data sets are presented on Figs. 3 and 4. They show how for both problems the average classification error decreases with subsequent epochs of GBP training of contextual neural networks. But it can be also noticed that in both cases with gradual decrease of classification error also the average activity of hidden connections semi-logarithmically drops down from 100% to 14% for Sonar and to 20% for the Crx problem.

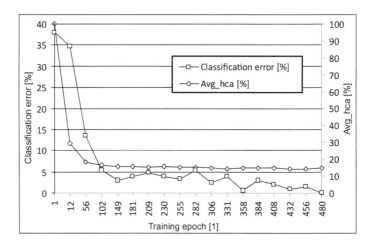

Fig. 3. Average classification error and hidden connections activity of contextual neural network solving Sonar problem, during its training with GBP algorithm implemented in H2O framework. Number of hidden neurons = 10, number of neuron inputs groups $K = 10$, aggregation threshold $\varphi^* = 0.6$, interval of groups update $\omega = 25$.

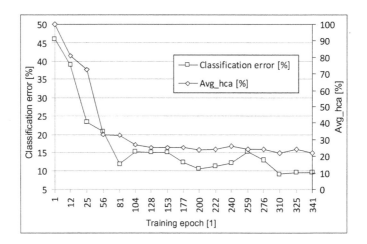

Fig. 4. Average classification error and hidden connections activity of contextual neural network solving Crx problem, during its training with GBP algorithm implemented in H2O framework. Number of hidden neurons = 10, number of neuron inputs groups $K = 10$, aggregation threshold $\varphi^* = 0.6$, interval of groups update $\omega = 25$.

The constant 100% activity level is the maximum of possible avg_hca results – e.g. for MLP neural networks or CxNN with number of neuron inputs groups K set to 1. During performed experiments at the beginning of GBP algorithm each CxNN had all inputs of all neurons assigned to the same, first group thus such models were behaving like the MLP. This caused that the initial avg_hca in both cases is 100%.

What is also interesting, the observed final classification accuracies of trained contextual models for Sonar and Crx problems measured for training data are comparable to the generalization (accuracy measured for test data) reported for MLP of analogous architectures. Most probably this is caused by the dynamic changes of contextual neural network architecture during the training which can prevent overfitting of constructed models.

After the above experiments two additional tests were performed for KDDCup 1999 and its 10% version dataset (designated in this paper as KDDCup10) to check the described solution against largest (in terms of training examples) problem served by UCI ML repository.

Then the results were compared with outcomes of training of two MLP neural networks of the same architecture – one originally implemented in H2O 3.10.0.3 software (MLP1), and the second one constructed as contextual neural network with number of groups $K = 1$ (MLP2). The results are given in Table 2.

Table 2. Hidden connections activities (avg_hca) and classification accuracies of the training data (train_acc) of contextual neural networks solving KDDCup 1999 benchmark problem from UCI ML repository. MLP1- original H2O implementation of Multi-Layer Perceptron, MLP2 – contextual neural network with $K = 1$, CxNN - contextual neural network for $K = 14$.

Training data set	MLP 1 avg_hca [%]	MLP 2 avg_hca [%]	CxNN avg_hca [%]	MLP 1 train_acc $[10^{-3}\%]$	MLP 2 train_acc $[10^{-3}\%]$	CxNN train_acc $[10^{-3}\%]$
KDDCup10	100 ± 0	100 ± 0	22.6 ± 0.3	1.55 ± 0.04	1.60 ± 0.07	1.5 ± 0.08
KDDCup	100 ± 0	100 ± 0	22.5 ± 0.2	0.8 ± 0.4	0.9 ± 0.4	1.1 ± 0.3

Obtained measurements show that implemented contextual neural network works as intended. As expected, for both data sets the activity of hidden connections of CxNN for $K = 1$ is the same as for MLP model and equal 100%. Moreover, when the number of groups of neuron inputs is increased to $K = 14$ the avg_hca decreases almost five times in comparison to MLP. This is achieved without considerable decrease of classification accuracy. Such behavior is characteristic for CxNN models and proves correctness of GBP implementation. It can be also noticed that for both data sets the avg_hca of CxNN models is the same. This is also correct because both considered data sets represent the same classification problem. In such case, for given set of values of parameters, independently from the number of training vectors both CxNN need similar average number of active hidden connections to solve the problem.

5   Conclusions

The above text describes the first approach to implement contextual neural networks with multi-step aggregation functions in scalable, distributed ML framework. Presented results for selected classification problems from UCI ML repository show that the mechanisms added to H2O software are working as intended, including Generalized Error Backpropagation method and CxNN model which is direct generalization of the

MLP neural network. The value of constructed software lies in the fact that it can be used for further research on contextual neural networks and conditional multi-step aggregation functions. It allows to perform experiments with large data sets and easily distribute calculations among many nodes within available H2O clusters.

Further research in the presented area can include detailed analysis of computational efficiency and optimization of the modified H2O.ai code. It would be also valuable to test CxNN behavior with different aggregation functions and other large classification benchmark data sets, such as Poker Hand, Susy and HIGGS. But especially valuable would be to try CxNN to solve "URL Reputation" classification problem – which has not only large number of training vectors but also over three millions of attributes. This would be interesting test of abilities of contextual neural networks to dynamically select attributes needed for correct data processing.

References

1. Grolinger, K., Capretz, M.A.M., Seewald, L.: Energy consumption prediction with big data: balancing prediction accuracy and computational resources. In: 2016 IEEE International Congress on Big Data (BigData Congress), pp. 1–8 (2016)
2. Ng, S.S.Y., Zhu, W., Tang, W.W.S., Wan, L.C.H., Wat, A.Y.W.: An independent study of two deep learning platforms - H2O and SINGA. In: 2016 IEEE International Conference on Industrial Engineering and Engineering Management (IEEM), pp 1–5. IEEE Press, Bali (2016)
3. Niu, F., Recht, B., Christopher, R., Wright, S.J.: HOGWILD!: a lock-free approach to parallelizing stochastic gradient descent. In: Advances in Neural Information Processing Systems, pp. 693–701 (2011)
4. Richter, A.N., Khoshgoftaar, T.M., Landset, S., Hasanin, T.: A multi-dimensional comparison of toolkits for machine learning with big data. In: 2015 IEEE International Conference on Information Reuse and Integration, pp. 1–8. IEEE, San Francisco (2015)
5. Suleiman, D., Al-Naymat, G.: SMS spam detection using H2O framework. Procedia Comput. Sci. **113**, 154–161 (2017)
6. Domingos, S.L., Carvalho, R.N., Carvalho, R.S., Ramos, G.N.: Identifying IT purchases anomalies in the Brazilian government procurement system using deep learning. In: 15th IEEE International Conference on Machine Learning and Applications (ICMLA) (2016)
7. Al Najada, H., Mahgoub, I.: Big vehicular traffic data mining: towards accident and congestion prevention. In: International Wireless Communications and Mobile Computing Conference, pp. 256–261 (2016)
8. Huk, M.: Measuring the effectiveness of hidden context usage by machine learning methods under conditions of increased entropy of noise. In: 3rd IEEE International Conference on Cybernetics, pp. 1–6. IEEE Press (2017)
9. Liang, M., Trejo, C., Muthu, L., Ngo, L.B., Luckow, A., Apon, A.W.: Evaluating R-based big data analytic frameworks. In: 2015 IEEE International Conference on Cluster Computing (CLUSTER), pp. 1–2. IEEE, Chicago (2015)
10. Cook, D.: Practical Machine Learning with H2O Powerful, Scalable Techniques for Deep Learning and AI. O'Reilly Media, Newton (2016)
11. Huk, M.: Notes on the generalized backpropagation algorithm for contextual neural networks with conditional aggregation functions. J. Intell. Fuzzy Syst. **32**, 1365–1376 (2017)

12. Huk, M.: Backpropagation generalized delta rule for the selective attention Sigma-if artificial neural network. Int. J. Appl. Math. Comput. Sci. **22**, 449–459 (2012)
13. Huk, M.: Learning distributed selective attention strategies with the Sigma-if neural network. In: Akbar, M., Hussain, D. (eds.) Advances in Computer Science and IT, pp. 209–232. InTech, Vukovar (2009)
14. Huk, M., Pietraszko, J.: Contextual neural-network based spectrum prediction for cognitive radio. In: 4th International Conference on Future Generation Communication Technology (FGCT 2015), pp. 1–5. IEEE Computer Society, London (2015)
15. Huk, M.: Context injection as a tool for measuring context usage in machine learning. In: Nguyen, N.T., Tojo, S., Nguyen, L.M., Trawiński, B. (eds.) ACIIDS 2017. LNCS (LNAI), vol. 10191, pp. 697–708. Springer, Cham (2017). https://doi.org/10.1007/978-3-319-54472-4_65
16. Szczepanik, M., Jóźwiak, I.: Data management for fingerprint recognition algorithm based on characteristic points' groups. In: Pechenizkiy, M., Wojciechowski, M. (eds.) New Trends in Databases and Information Systems. Advances in Intelligent Systems and Computing, vol. 185, pp. 425–432. Springer, Heidelberg (2013). https://doi.org/10.1007/978-3-642-32518-2_40
17. Huk, M.: Using context-aware environment for elderly abuse prevention. In: Nguyen, N.T., Trawiński, B., Fujita, H., Hong, T.-P. (eds.) ACIIDS 2016. LNCS (LNAI), vol. 9622, pp. 567–574. Springer, Heidelberg (2016). https://doi.org/10.1007/978-3-662-49390-8_55
18. Huk, M.: Context-related data processing with artificial neural networks for higher reliability of telerehabilitation systems. In: 17th International Conference on E-health Networking, Application & Services (HealthCom), pp. 217–221. IEEE Computer Society, Boston (2015)
19. Huk, M., Kwasnicka, H.: The concept and properties of sigma-if neural network. In: Ribeiro, B., Albrecht, R.F., Dobnikar, A., Pearson, D.W., Steele, N.C. (eds.) Adaptive and Natural Computing Algorithms, pp. 13–17. Springer, Vienna (2005). https://doi.org/10.1007/3-211-27389-1_4
20. Huk, M.: Sigma-if neural network as the use of selective attention technique in classification and knowledge discovery problems solving. Ann. UMCS Sectio AI – Inf. **4**(2), 121–131 (2006)
21. Huk, M.: Manifestation of selective attention in Sigma-if neural network. In: 2nd International Symposium Advances in Artificial Intelligence and Applications, International Multiconference on Computer Science and Information Technology IMCSIT/AAIA 2007, vol. 2, pp. 225–236 (2007)
22. Srivastava, N., Hinton, G., Krizhevsky, A., Sutskever, I., Salakhutdinov, R.: Dropout: a simple way to prevent neural networks from overfitting. J. Mach. Learn. Res. **15**, 1929–1958 (2014)
23. Privitera, C.M., Azzariti, M., Stark, L.W.: Locating regions-of-interest for the Mars Rover expedition. Int. J. Remote Sens. **21**, 3327–3347 (2000)
24. Mel, B.W.: The Clusteron: toward a simple abstraction for a complex neuron. In: Advances in Neural Information Processing Systems, vol. 4, pp. 35–42. Morgan Kaufmann (1992)
25. Spratling, M.W., Hayes, G.: Learning synaptic clusters for nonlinear dendritic processing. Neural Process. Lett. **11**, 17–27 (2000)
26. Raczkowski, D., Canning, A.: Thomas-Fermi charge mixing for obtaining self-consistency in density functional calculations. Phys. Rev. B **64**, 121101–121105 (2001)
27. UCI Machine Learning Repository. http://archive.ics.uci.edu/ml

DeepEnergy: Prediction of Appliances Energy with Long-Short Term Memory Recurrent Neural Network

Erdenebileg Batbaatar[1] , Hyun Woo Park[1] , Dingkun Li[1] ,
Meijing Li[2] , and Keun Ho Ryu[1(✉)]

[1] Database and Bioinformatics Laboratory, School of Electrical and Computer
Engineering, Chungbuk National University, Cheongju, South Korea
{eegii,hwpark,jerryli,khryu}@dblab.chungbuk.ac.kr
[2] College of Information Engineering, Shanghai Maritime University,
213, 1550 Haigang Avenue, Pudong New Area, Shanghai,
People's Republic of China
mjli@shmtu.edu.cn

Abstract. Our world is becoming more interconnected and intelligent, huge amount of data has been generated newly. Home appliances' energy usage is the basis of home energy management and highly depends on weather condition and environment. Using weather in context, it is theorized that usage of home energy would be higher in cold days. Time series and contextual data collected from sensors can be monitored and controlled in home appliances network. The aim of this work is to propose a deep neural network architecture and apply it to a contextual and multivariate time series data. Long short-term memory (LSTM) models are powerful neural networks based on past behaviours in long sequences. LSTM networks have been demonstrated to be particularly useful for learning sequences containing longer-term patterns of unknown length, due to their ability to maintain long-term memory. In this work, we incorporate contextual features into the LSTM model because of ability of keeping context of data for a long-time, and for analysing it we integrated two different datasets; the first dataset contains measurements about house temperature and humidity measured over a period of 4.5 months by a 10 min intervals using a ZigBee wireless sensor network. The second dataset contains measurements about individual household electric power consumptions gathered over a period of 47 months. From the wireless network, the data from the kitchen, laundry and living room were ranked the highest in importance for the energy prediction.

Keywords: Internet of things · Neural network · Recurrent neural network
Long-short term memory · Time series data · Appliances energy

1 Introduction

The Internet of Things (IoT) is a complex paradigm where billions of devices are connected to a network [1]. These connected devices form an intelligent system of systems that share the data without human-to-computer or human-to-human interaction.

© Springer International Publishing AG, part of Springer Nature 2018
N. T. Nguyen et al. (Eds.): ACIIDS 2018, LNAI 10752, pp. 224–234, 2018.
https://doi.org/10.1007/978-3-319-75420-8_21

Deep Learning 'mimics the brain functionality' with the help of robust neural network algorithms. The wide range of deep learning applications include image recognition, computer vision, speech recognition, pattern recognition and behaviour recognition. In the world of IoT, the datasets are high-dimensional, temporal and multi-modal. The ability to predict future energy requirements is a critical component of a variety of applications that seek to conserve or improve management of energy resources. Utilities, for example, use forecasting of future demand to determine how to manage energy generation. With the recent widespread deployment of the smart grid, forecasting at the scale of individual homes and even appliances becomes necessary for enabling effective demand response systems and user-side energy management.

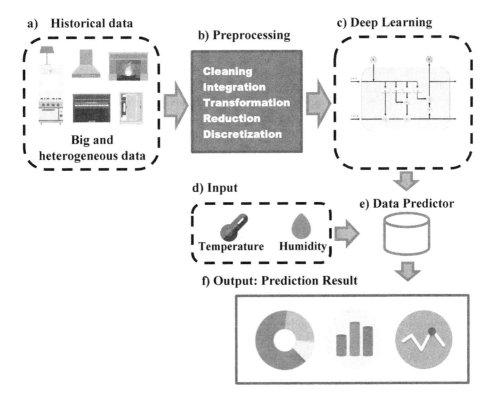

Fig. 1. (a) Historical data. The big and heterogeneous data is stored in the database and divided into 60% training data, 20% validation data, 20% test data, (b) Preprocessing of dataset that contains cleaning, integration, transformation, reduction, discretization steps, (c) Deep Learning models. LSTM model structure has been presented, (d) Input data as home appliance energy data, (e) Data Predictor. We have developed predictive model, (f) Output: Prediction result. It shows experimental results as graphs.

Understanding the underlying causes that impact home energy consumption is effective for home energy management. Usage of home energy is correlated with weather condition and a set of insights into key factors influencing energy usage and significant associations between energy usage and key features such as hour of a day, day of the week, use of other appliances in the home, and user-supplied annotations of activities such as working or cooking [2]. Also other factors such as income groups, occupant types (age, number of home occupants, and occupation), and home types (rentals and user-owned) make a system context-rich [3]. A contextual system that is able to accurately estimate energy requirements at the scale of individual appliances in the home, for example, can enable feedback and recommendations to support load shifting to reduce peak demand or help homeowners better understand how they can modify usage to minimize electricity bills [4].

As indicated in [5], the electricity consumption in domestic building is explained by two main factors: the type and number of electrical appliances and the use of the appliances by the occupants. Naturally, both factors are interrelated. Prediction models of electrical energy consumption in buildings can be useful for a number of applications: to determine adequate sizing of photovoltaics and energy storage to diminish power flow into the grid [6], to detect abnormal energy use patterns [7], to be part of an energy management system for load control [8–10], to model predictive control applications. Figure 1 shows the pipeline of our predictive model. In this work, the prediction was carried out using different data sources and environmental parameters (indoor and outdoor conditions). Specifically, data from a nearby airport weather station, temperature and humidity in different rooms in the house from a wireless sensor network and one sub-metered electrical energy consumption (lights) have been used to predict the energy use by appliances. Four regression models have been tested, namely (a) multiple linear regression model (lm), (b) support vector machine with radial basis function kernel (SVM radial), (c) random forest (RF) and (d) gradient boosting machines (GBM) with different combinations of predictors.

Big heterogeneous data. Big data is currently defined using three data characteristics: volume, variety and velocity. In this research, we have collected big and heterogeneous data, and their characteristics can be summarized as follows:

- **Appliances energy prediction data.** We have collected the dataset (https://archive. ics.uci.edu/ml/datasets/Appliances+energy+prediction) that comprise of measurements of house temperature and humidity conditions with a 10-min interval for a period of 4.5 months. The house temperature and humidity conditions were originally monitored with a ZigBee wireless sensor network. Each wireless node transmitted the temperature and humidity conditions around 3.3 min. Then, the wireless data was averaged for 10 min periods. The energy data was logged every 10 min with m-bus energy meters. On the other hand, Weather data from the nearest airport weather station (Chievres Airport, Belgium) was downloaded from a public data set from Reliable Prognosis (rp5.ru), and merged together with the experimental data sets using the date and time column. Two random variables have been included in the data set for testing the regression models and to filter out non-predictive attributes (parameters).
- **Individual household electric power consumption dataset.** This household electricity consumption dataset (https://archive.ics.uci.edu/ml/datasets/individual +household+electric+power+consumption) contains 260,640 measurements gathered between January 2007 and June 2007 (6 months). It is a subset of a larger,

original archive that contains 2,075,259 measurements gathered between December 2006 and November 2010 (47 months). (global_active_power*1000/60 - sub_metering_1 - sub_metering_2 - sub_metering_3) represents the active energy consumed every minute (in watt-hour) in the household by electrical equipment not measured in sub-meters 1, 2 and 3. The dataset contains some missing values in the measurements (nearly 1.25% of the rows). All calendar timestamps are present in the dataset but for some timestamps, the measurement values are missing: a missing value is represented by the absence of value between two consecutive semi-colon attribute separators. For instance, the dataset shows missing values on April 28, 2007.

Data features and importance. Since the dataset contains several features or parameters and considering that the airport weather station is not at the same location as the house, it is desirable to find out which parameters are the most important and which ones do not improve the prediction of the appliances' energy consumption. Table 1 shows the list of appliances in each room or house zone.

Table 1. List of appliances in each room or house zone

Room	Equipment
Laundry	Small fridge, upright freezer, wine cellar for 160 bottles, washing machine, dryer, internet router, internet hub, network attached storage
Kitchen	Fridge, induction cooktop, kitchen hood, microwave, oven, dishwasher, coffee machine
Dining	WIFI booster, ZigBee coordinator, electrical blinds
Living	TV 138 cm, hard drive enclosure, DVD player, cable box, laptop, Ink-jet printer, electric blinds
Office	2 desktop computers, 3 computer screens, 1 router, 1 laptop, 1 copier-printer, electric blinds
Ironing	Alarm clock, radio, Iron, electric blind
Game	93 cm TV, Internet router, DVD player, PlayStation
Bathroom	2 electric toothbrushes, hair dryer
etc.	...

Research objectives and methodology outline. The purpose of this work is to understand the relationships between appliances' energy consumption and different predictors. Also to discuss the performance of different models (recurrent neural network, long short-term memory) to predict energy consumption. Furthermore, to rank the influence of predictors/parameters in the prediction.

The problem of appliance usage prediction through consumption data is new. Although, the problem of the user behaviour prediction in a home automation system using a Bayesian network for a single appliance is dealt with in [11], a general model for appliance prediction is still lacking.

Data integration. In order to train our deep model, we have integrated 2 different datasets from different sources, particularly, appliances energy prediction (AEP) dataset and individual household electric power consumption (IHEPC) dataset (Fig. 2). Both of which contain household electric and energy consumption.

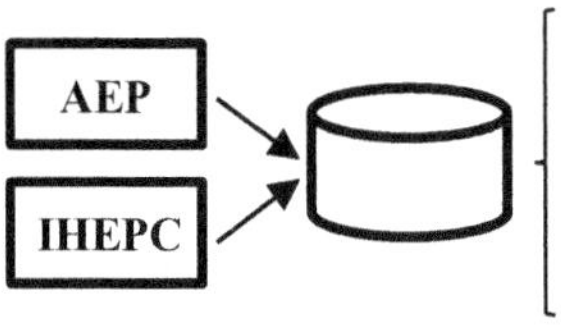

№	date	light	visibility	voltage	temperature	humidity	...
1	January 11	30	63	234.840	19.2	47.59	
2	January 12	30	59.16	233.630	19.2	46.69	
3	January 13	30	55.33	233.290	19.23	46.06	
4	January 14	30	53.35				
5	January 15	30	51.5	233.740	20.2	46.33	
6	January 16	30	47.66	235.680	21	46.02	
19736	May 27	10	46.6	239	25.5	42.97	

Fig. 2. Data integration and feature extraction

AEP dataset is collected from January to May, IHEPC is collected for 47 months. Household energy is highly based on weather condition. Only the data during the period between January and May (available in years 2007, 2008, 2009, 2010) are selected from IHEPC where the mathematical average of each value was calculated and integrated with AEP. The period is special case of weather that is becoming from cold-time to warm-time between January to May.

Feature selection. AEP contains data such as household global minute-averaged active and reactive power, voltage, global intensity, energy sub-metering of kitchen, laundry room and electric water-heater and air-conditioner for each day. IHEPC contains data such as temperature and humidity in kitchen, living room, laundry room, office room, bathroom, outside room, etc. for each day. After integration, we selected all single attributes and merged them day by day for training our model.

The content of this paper is organized as follows. Section 2 presents related works with recurrent neural network and energy consumption management. Section 3 presents the methods recurrent neural network and LSTM models, Sect. 4 presents experimental results and Sect. 5 presents conclusion of this work.

2   Related Work

The appliance energy prediction dataset was released in February 2017 where data-driven predictive models [12] for the energy use of appliances were discussed. Generally, the approaches can be divided into two types. The first type of models is based on the physical principles that map out the thermal dynamics and energy behaviour at the building level. These include models of space heating systems, natural ventilation, air conditioning, photovoltaic systems, occupants' behaviour, indoor and outdoor climate, price-responsiveness, and so on. The second type is based on statistical methods. These methods are used to predict building energy consumption by correlating it with influencing variables such as weather and energy prices. Interested readers are referred to [13, 14] for a more comprehensive discussion of the modelling and prediction of energy consumption in buildings, and to recent reviews such as

[15, 16]. Moreover, to account for the evolution of future buildings energy systems, there are also hybrid approaches which combine some of the above modelling methods to optimize predictive performance, such as semi-parametric regression models [17], exponential smoothing [18] and seasonal time series models [19–22]. On the other hand, it is worth noting that some of the most widely used machine learning methods for energy prediction are Artificial Neural Networks (ANNs) and Support Vector Machines (SVMs) [17, 23–25].

Various network architectures, such as three, four and five-layers, a number of recurrent types, and a number of feedforward ones have been investigated aiming at finding the one that could result in the best overall performance. Also a number of different network sizes and learning parameters have been tried [26].

3   Methods

3.1   Recurrent Neural Network

LSTMs were originally introduced in [27], following a long line of research into recurrent neural network (RNN) for sequence learning. RNN (Fig. 3) deals with sequence problems because their connections form a directed cycle. In other words, they can retain state from one iteration to the next by using their own output as input for the next step. In programming terms this is like running a fixed program with certain inputs and some internal variables. The simplest recurrent neural network can be viewed as a fully connected neural network if we unroll the time axes. It is possible to adapt Backpropagation algorithm to train a recurrent network, by "unfolding" [28] the network through time and constraining some of the connections to always hold the same weights. Recurrent neural network can take a contextual information both in input and output layers and modify behavior of RNN.

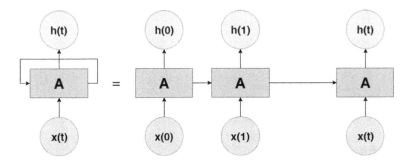

Fig. 3.   Recurrent neural network

3.2   Long Short-Term Memory for Sequence Modeling

Long short-term memory (LSTM) is special recurrent neural network, capable of learning long-term dependencies. Appliances energy patterns have a high degree of temporal and spatial correlation. One problem that arises from the unfolding of an RNN is that the gradient of some of the weights starts to become too small or too large if the network is unfolded for too many time steps. This is called the vanishing gradient problem. A type of network architecture that solves this problem is the LSTM. In a typical implementation, the hidden layer is replaced by a complex block of computing units composed by gates that trap the error in the block, forming a so-called "error carrousel". Figure 4 shows a local architecture of LSTM neural network. There are three types of gates within a unit [29]:

- **Forget gate:** conditionally decides what information to throw away from the block.
- **Input gate:** conditionally decides which values from the input to update the memory state.
- **Output gate:** conditionally decides what to output based on input and the memory of the block.

The gates decide what information to throw away and what information to keep, that depends on context of data for capturing long-range dependencies in sequences. We input context to first layer of LSTM, model keeps the context for long-time in memory.

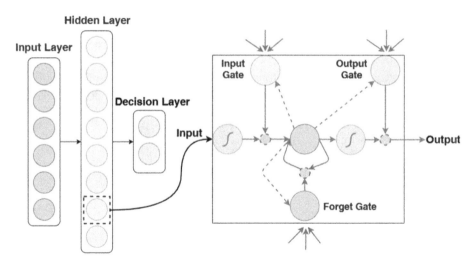

Fig. 4. Long short-term memory

4   Experimental Result

In this section, we present extensive experimental results and evaluate our system for the prediction of home appliance energy. In our LSTM model for energy appliance prediction with contextual data, one sequence was defined as a sequential collection of the daily dataset of any single day in a time period (4.5 months). Our model is composed of a single input layer with a number of memory cells as that of the sequence learning features one sequence may hold, followed by multiple LSTM layers and a dense layer and a single output layer with the number of memory cells as that of the categories of the sequence performance.

Experimental setup. One sequence contains 35 attribute values. We got 19,735 sequences from 2016/01/11 to 2016/05/27. We have divided the dataset into 3 parts such as 60% training data, 20% validation data and 20% test data. We used 11,841 sequences for training purpose and 3,947 sequences for validation and 3,947 for testing.

Parameter setting. The learning architecture of our system contained three LSTM layers, with 50 neurons in the first hidden layer and 1 neuron in the output layer for prediction appliances energy. The input shape will be 1-time step with 8 features. We used the Mean Absolute Error (MAE), Mean Absolute Percentage Error (MAPE) and Root Mean Square Error (RMSE) loss function and the efficient Adam version of stochastic gradient descent. The model was fit for 100 training epochs with a batch size of 72.

Feature setting. IHEPC dataset contains 7 attributes that are measurements of electric power consumption. AEP dataset contains 28 attributes representing appliances' energy uses. We have divided all features into 3 parts (Table 2) by their importance for energy prediction.

Table 2. Feature setting for training

Name	Features	Number of attributes
P1	IHEPC + AEP (others)	15
P2	AEP (Appliances, light, kitchen, living room, laundry)	8
P3	AEP (office, bathroom, outside of building, ironing room, teenager room, parents room)	12

In Table 3, we compared combination of the features, and observed that the error rate (RMSE, MAPE, MAE) of P2 is generally greater than other 2 combinations and combination of P2 and P3 is greater than other combinations. Our proposed LSTM model is always achieves the best RMSE, MAE and MAPR across the two datasets of home appliance energy use.

Table 3. Result of home appliances energy prediction

Features combination	MAPE	MAE	RMSE
P1 + P2 + P3	**1.841**	**1.156**	**1.546**
P1	**2.214**	1.879	2.137
P2	2.346	**1.458**	**1.894**
P3	2.297	1.654	1.987
P1 + P2	1.986	**1.337**	1.745
P1 + P3	2.012	1.406	1.732
P2 + P3	**1.942**	1.421	**1.648**

Training LSTMs for the two datasets were implemented by using the Keras (https://keras.io/) package in conjunction with the Tensorflow library (https://www.tensorflow.org/).

5   Conclusion

In this paper, we proposed a deep neural network architecture and applied contextual features to long short-term-memory, special kind of recurrent neural network. LSTM is powerful to keep context for a long time. We used a time series context-data taken at 10 min interval for a period of about 4.5 months. The dataset represented house temperature and humidity conditions which were monitored with a ZigBee wireless sensor network. This dataset was merged to another individual household electric power consumption dataset that contains measurements gathered for 47 months. We divided all dataset into 3 combinations of features and compared the error rate. We can see that P2 is a collection of highest importance for energy prediction and generally achieves the best result across the datasets. And we implemented the LSTM with Keras deep learning framework and measured the value 0.1546 of root-mean-square-error (RMSE), 1.841 of mean-absolute-percentage-error (MAPE) and 1.156 of mean-absolute-error (MAE).

Future work. The period is special case of weather that is becoming from cold-time to warm-time between January to May. Our work is extendable with another approach for during the whole period of a year. Our model will be compared with other approaches and non-contextual systems.

Acknowledgement. This work was supported by Basic Science Research Program through the National Research Foundation of Korea (NRF) funded by the Ministry of Science, ICT & Future Planning (No. 2017R1A2B4010826) in the Republic of Korea and also was supported by the National Natural Science Foundation of China (61702324) in People's Republic of China.

References

1. Atzori, L., Iera, A., Morabito, G.: The internet of things: a survey. Comput. Netw. **54**(15), 2787–2805 (2010)
2. Rollins, S., Banerjee, N.: Using rule mining to understand appliance energy consumption patterns. In: 2014 IEEE International Conference on Pervasive Computing and Communications (PerCom), pp. 29–37. IEEE (2014)
3. Irwin, G., Banerjee, N., Hurst, A., Rollins, S.: Contextual insights into home energy relationships. In: 2015 IEEE International Conference on Pervasive Computing and Communication Workshops (PerCom Workshops), pp. 305–310. IEEE (2015)
4. Han, D.M., Lim, J.H.: Smart home energy management system using IEEE 802.15. 4 and zigbee. IEEE Trans. Consum. Electron. **56**(3) (2010)
5. Firth, S., Lomas, K., Wright, A., Wall, R.: Identifying trends in the use of domestic appliances from household electricity consumption measurements. Energy Build. **40**(5), 926–936 (2008)

6. Spertino, F., Di Leo, P., Cocina, V.: Which are the constraints to the photovoltaic grid-parity in the main European markets? Sol. Energy **105**, 390–400 (2014)
7. Seem, J.E.: Using intelligent data analysis to detect abnormal energy consumption in buildings. Energy Build. **39**(1), 52–58 (2007)
8. Barbato, A., Capone, A., Rodolfi, M., Tagliaferri, D.: Forecasting the usage of household appliances through power meter sensors for demand management in the smart grid. In: 2011 IEEE International Conference on Smart Grid Communications (SmartGridComm), pp. 404–409. IEEE (2011)
9. Zhao, P., Suryanarayanan, S., Simões, M.G.: An energy management system for building structures using a multi-agent decision-making control methodology. IEEE Trans. Ind. Appl. **49**(1), 322–330 (2013)
10. Castillo-Cagigal, M., Caamano-Martín, E., Matallanas, E., Masa-Bote, D., Gutiérrez, A., Monasterio-Huelin, F., Jiménez-Leube, J.: PV self-consumption optimization with storage and active DSM for the residential sector. Sol. Energy **85**(9), 2338–2348 (2011)
11. Hawarah, L., Ploix, S., Jacomino, M.: User behavior prediction in energy consumption in housing using Bayesian networks. In: Rutkowski, L., Scherer, R., Tadeusiewicz, R., Zadeh, L.A., Zurada, J.M. (eds.) ICAISC 2010. LNCS (LNAI), vol. 6113, pp. 372–379. Springer, Heidelberg (2010). https://doi.org/10.1007/978-3-642-13208-7_47
12. Candanedo, L.M., Feldheim, V., Deramaix, D.: Data driven prediction models of energy use of appliances in a low-energy house. Energy Build. **140**, 81–97 (2017)
13. Krarti, M.: Energy Audit of Building Systems: An Engineering Approach. CRC Press, Boca Raton (2016)
14. Dounis, A.I.: Artificial intelligence for energy conservation in buildings. Adv. Build. Energy Res. **4**(1), 267–299 (2010)
15. Foucquier, A., Robert, S., Suard, F., Stéphan, L., Jay, A.: State of the art in building modelling and energy performances prediction: a review. Renew. Sustain. Energy Rev. **23**, 272–288 (2013)
16. Zhao, H.X., Magoulès, F.: A review on the prediction of building energy consumption. Renew. Sustain. Energy Rev. **16**(6), 3586–3592 (2012)
17. Fan, S., Hyndman, R.J.: Short-term load forecasting based on a semi-parametric additive model. IEEE Trans. Power Syst. **27**(1), 134–141 (2012)
18. Taylor, J.W.: Exponentially weighted methods for forecasting intraday time series with multiple seasonal cycles. Int. J. Forecast. **26**(4), 627–646 (2010)
19. De Livera, A.M., Hyndman, R.J., Snyder, R.D.: Forecasting time series with complex seasonal patterns using exponential smoothing. J. Am. Stat. Assoc. **106**(496), 1513–1527 (2011)
20. Aydinalp-Koksal, M., Ugursal, V.I.: Comparison of neural network, conditional demand analysis, and engineering approaches for modeling end-use energy consumption in the residential sector. Appl. Energy **85**(4), 271–296 (2008)
21. Hurtado, L.A., Nguyen, P.H., Kling, W.L.: Smart grid and smart building inter-operation using agent-based particle swarm optimization. Sustain. Energy Grids Netw. **2**, 32–40 (2015)
22. Xuemei, L., Lixing, D., Jinhu, L., Gang, X., Jibin, L.: A novel hybrid approach of KPCA and SVM for building cooling load prediction. In: Third International Conference on Knowledge Discovery and Data Mining, WKDD 2010, pp. 522–526. IEEE (2010)
23. Wong, S.L., Wan, K.K., Lam, T.N.: Artificial neural networks for energy analysis of office buildings with daylighting. Appl. Energy **87**(2), 551–557 (2010)
24. Kalogirou, S.A.: Artificial neural networks in energy applications in buildings. Int. J. Low-Carbon Technol. **1**(3), 201–216 (2006)

25. Yang, J., Rivard, H., Zmeureanu, R.: On-line building energy prediction using adaptive artificial neural networks. Energy Build. **37**(12), 1250–1259 (2005)
26. Mocanu, E., Nguyen, P.H., Gibescu, M., Kling, W.L.: Deep learning for estimating building energy consumption. Sustain. Energy Grids Netw. **6**, 91–99 (2016)
27. Hochreiter, S., Schmidhuber, J.: Long short-term memory. Neural Comput. **9**(8), 1735–1780 (1997)
28. LeCun, Y., Bengio, Y., Hinton, G.: Deep learning. Nature **521**(7553), 436–444 (2015)
29. Hochreiter, S., Schmidhuber, J.: Long short-term memory. Neural Comput. **9**(8), 1735–1780 (1997)
30. Bengio, Y., Simard, P., Frasconi, P.: Learning long-term dependencies with gradient descent is difficult. IEEE Trans. Neural Networks **5**(2), 157–166 (1994)
31. Bengio, Y.: Learning deep architectures for AI. Foundations and trends®. Mach. Learn. **2**(1), 1–127 (2009)
32. Bengio, Y., Lamblin, P., Popovici, D., Larochelle, H.: Greedy layer-wise training of deep networks. In: Advances in Neural Information Processing Systems, pp. 153–160 (2007)
33. Basu, K., Hawarah, L., Arghira, N., Joumaa, H., Ploix, S.: A prediction system for home appliance usage. Energy Build. **67**, 668–679 (2013)
34. Li, D., Park, H.W., Ishag, M.I.M., Batbaatar, E., Ryu, K.H.: Design and partial implementation of health care system for disease detection and behavior analysis by using DM techniques. In: 2016 IEEE 14th International Conference on Dependable, Autonomic and Secure Computing, 14th International Conference on Pervasive Intelligence and Computing, 2nd International Conference on Big Data Intelligence and Computing and Cyber Science and Technology Congress (DASC/PiCom/DataCom/CyberSciTech), pp. 781–786. IEEE, August 2016

Efficient Ensemble Methods for Classification on Clear Cell Renal Cell Carcinoma Clinical Dataset

Kwang Ho Park[1] , Musa Ibrahim M. Ishag[1] , Kwang Sun Ryu[1] ,
Meijing Li[2] , and Keun Ho Ryu[1(✉)]

[1] Database and Bioinformatics Laboratory, School of Electrical and Computer Engineering,
Chungbuk National University, Cheongju, South Korea
`{khblack,ibrahim,ksryu,khryu}@dblab.chungbuk.ac.kr`
[2] College of Information Engineering, Shanghai Maritime University,
213, 1550 Haigang Avenue, Pudong New Area, Shanghai, People's Republic of China
`mjli@shmtu.edu.cn`

Abstract. Kidneys play an important role in human body. In essence, a kidney maintains homeostasis and removes harmful materials by making and ejecting a form of urine. Especially 2–3% of humans who have malignancies, also suffered a clear cell renal cell carcinoma (ccRCC) which is one kind of kidney diseases. When diagnosed early, this renal cell carcinoma can be easily treated with some incision surgical method. Nonetheless, some patients who cannot undergo incision surgery need a customized medical service. The ensemble method is usually used to improve the classification performance by combining classifier. For this reason, in this paper, we suggest an implementation of classification algorithm on clinical data to find important clinical factors for ccRCC using an ensemble method and compare the results with a recent work in the literature. The experimental results showed that classification with ensemble methods improved the classification result, especially bagging method.

Keywords: Clear cell renal cell carcinoma · Feature selection · Classification
Ensemble method · C4.5 · Support vector machine · Artificial neural network
Bayesian network

1 Introduction

Renal cell carcinoma (RCC), which is one of kidney diseases, occupies to 2–3% of human tumor type disease [1]. This RCC can be categorized into three subtypes: clear cell renal cell carcinoma (ccRCC), chromophobe renal cell carcinoma (chRCC) and papillary renal cell carcinoma (pRCC). Among these subtypes, ccRCC is the most hazardous and common type of RCC. Also, ccRCC incidence occurs mainly in the clear cell which are located close to many blood streams. Because of this reason, cancer cell readily transfer to other organs than the two other types of RCC [2].

Nowadays, an increasing demand is arising from various medical disciplines to use medical decision support systems in order to improve patients' personalized medical

© Springer International Publishing AG, part of Springer Nature 2018
N. T. Nguyen et al. (Eds.): ACIIDS 2018, LNAI 10752, pp. 235–242, 2018.
https://doi.org/10.1007/978-3-319-75420-8_22

care and reduce diagnostic errors [3]. Most of all classifiers have been broadly adopted for medical decision support system [4, 5]. Jung et al. [6] applied C4.5, CART, and CHID on breast cancer patients' data. Baxt [7] in their paper concluded artificial neural network(ANN) can achieve good performance on clinical data. Akay et al. [3] used support vector machine (SVM) on Wisconsin breast cancer data set for breast cancer diagnosis with feature selection and could get reasonable accuracy on their data. Jaya-surya et al. [8] used two classification algorithms which are Bayesian network and support vector machine on lung cancer patients' clinical dataset and they suggested Bayesian network is more better than support vector machine for lung cancer data. Also in our previous papers, [9, 10] we performed four classification algorithms, which are Bayesian network, C4.5, support vector machine and artificial neural network, with and without feature selection algorithms. And we found Bayesian network was the most suitable for classifying ccRCC clinical data with and without feature selection methods.

To improve our previous work [9, 10], we tried to adopt an ensemble method on our classification. There are two typical approaches in this ensemble method. One is boosting and the other is bagging. We applied both of these approaches to our work and compared the results with our previous work. The performance evaluation was conducted on ccRCC clinical data, which was obtained from the Cancer Genome Atlas (TCGA) [11]. In the results, we found bagging usually improves all of four classification algorithms performance, especially artificial neural network. And Bayesian network still has a good performance on our data. Performed experiments were also used to check if techniques such as Bagging and Boosting are efficient both for contextual (e.g. decision trees, Bayesian networks) and non-contextual systems (e.g. MLP, SVM).

2   Related Work

2.1   Feature Selection

Feature selection is an essential preprocessing step in data mining for finding optimal subset from original dataset. Highly dimensional data may cause some issues such as computational complexity and reducing classification accuracy. Although, feature selection seems to lose information of the original dataset, this process eliminates redundant and irrelevant features and improves performance of classifiers or other data mining analysis results.

These feature selection algorithms can be broadly grouped into three approaches, which are filter, wrapper and embedded approach. The filter approaches do not require any classification algorithm. These approaches are faster than wrapper approach and they are suited for high dimensional data. However, in those cases feature selection method cannot corporate with the classification models. The wrapper approaches are included in a classification model, which is used to select some features. Therefore, it is more accurate than filter approach. Nevertheless, if the underlying training data is too small, it might lead to overfitting problem. Also, when the data size is too large, it increases the computation complexity. Embedded approaches are merged form of filter and wrapper approaches. Therefore, they can cover high dimensional data as well.

However, the range in which these approaches can be applied in limited cases than filter or wrapper approaches [12].

2.2 Classification Algorithms

C4.5 algorithm is based on Interactive Dichotomizer 3 algorithm, which is one of decision tree-based contextual classification algorithms, and was developed to solve the limitation of difficulty in classifying continuous attributes. To resolve this problem, C4.5 algorithm uses gain ratio [13, 14].

Support vector machine (SVM) separates training dataset by binary label with a maximal margin hyper-plane using structure risk minimization technique. As using maximal margin hyper-plane it can prevent model over-fitting problem. Also, even if SVM is non-contextual in nature, SVM can be used linearly and nonlinearly separable data by mapping them into a high-dimensional space [15].

Artificial Neural Network (ANN) is based on perceptron, which is one of binary classifier, and this ANN consists of several layers which are divided into input layer, hidden layer and output layer. Calculation of ANN is occurred by several layers' perceptron in hidden layer. Therefore, between each layer, there are a number of calculations. and typically all inputs of neural network are used for processing input vectors [16]. This makes ANN MLP models to be non-contextual.

Bayesian Network (BN) consists of directed acyclic graph (DAG) and property between each variable. Because each of DAG edges is consists of conditional property of each node, when given other attributes, using conditional probability, BN can calculate unknown variable's class and represents typical contextual system [17].

2.3 Bagging

Bootstrap aggregating, as known as bagging, suggested to improve the classification result by combining classifications of randomly generated training set and aggregated classifiers. The algorithm takes D number of training data set and N number of bootstrap samples as input then constructs an ensemble classifier, which is the combined form of the several classifiers trained from the multiple bootstrap samples. D' is obtained by repeatedly sampling instances from a dataset according to probability distribution. Because of the sampling with replacement, some instances may appear several times in the same training set, while others may not. Through this procedure, there are N number of bootstrap samples, D_1, D_2, ..., D_N are generated, and classification algorithm C^* is trained by each bootstrap sample D_i. Finally a combined classification algorithm C^* is built by C_1, C_2, ..., C_i, and this C^* classify instance x by counting votes [18].

2.4 Boosting

Boosting is one kind of ensemble methods to enhance the performance of classification algorithms that construct multiple classifiers and vote on their algorithms. Differently from bagging method, boosting gives a weight on each instance that could change at the end of each boosting round. There are several boosting methods, the most commonly used is Adaboost. This adaboost takes as input training data D, containing m instances and iteration parameter N and then outputs a combined classifier. Initially, all of the instances are equally assigned the same weight. Then, this boosting constructs classifier by modifying the weights of training instances based on result of previous classifier. After reaching an optimal classifier model, the algorithm calculates the rate of the incorrectly classified instances. If weighted error is larger than 0.5, training set D_i will be set to a bootstrap sample with weight 1 for every instance. On the other hand, the weight of correctly classified instances will be updated by a factor inversely proportional to error. In other words, if the classifier finds some object hard to classify, that object should be assigned a larger weight at next round. And if the object is easily classified, then that object should be assigned smaller weight at next round. Finally, these classifiers are combined into one classifier by voting scheme [19].

3   Materials and Methods

3.1   Dataset

For this experiment, we obtained data from The Cancer Genome Atlas (TCGA) Data Portal, which is a platform for researchers who want to analyze cancer related dataset [11]. This TCGA has many kinds of cancer related data: clinical data, image data, genomic data, high level sequencing data. In this paper, we just used clear cell renal cell carcinoma, as known as one of kidney cancer, patients' data.

3.2   Experimental Workflow

Our experimental process was formulated as shown in Fig. 1 below. Firstly, collect ccRCC patients' clinical data from TCGA website. Then we adopted preprocess step on the original clinical data to clean the data and reduce the noise. This step resulted in a preprocessed data, which has 439 objects with 24 attributes for downstream analysis. It assumes two distinct values; namely, "Dead", and "Alive". Thus, rendering the task to a binary classification problem. Then we used feature selection methods, which are information gain (IG), symmetrical uncertainty (SU), ReliefF (RF), gain ratio (GR), chi-squared (Chi), OneR (OR) and pearson correlation (PE), to narrow down the attribute number before classification. Using these selected features, we applied two ensemble methods, which are adaboost and bagging with majority vote as voting method. The two ensemble methods were composed of four classification algorithms: SVM, ANN, BN, C4.5 with 10-fold cross validation method. Finally, we calculated accuracy, sensitivity and specificity on each experiment and compared the results.

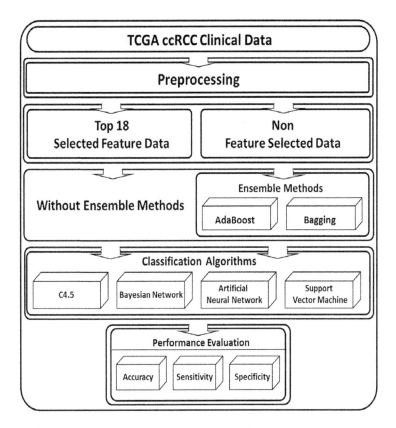

Fig. 1. Workflow of the experiment

4   Results of Experiments

Before applying the classifiers, best features were selected by adopting the seven-feature selection methods mentioned in the work follow section. Afterwards, the top features selected by each method were merged and the best 18 features, which yield high classification accuracy, were finally selected for the classification task. The top 18 features were namely; Tumor Status, AJCC tumor pathologic pm, AJCC tumor pathologic tumor stage, Initial pathologic dx year, Tumor grade, AJCC tumor pathologic pt, Prospective collection, Retrospective collection, Platelet count, Lymph des examined, Hemoglobin level, Serum calcium level, Tobacco smoking history indicator, Ldh level, Laterality, AJCC tumor pathologic pn, white cell count.

Next, we put both feature-selected data and non-feature selected data into four classification algorithms. And we try to adopt ensemble method on these data to compare the results of using the ensemble technique with and without feature selection. In the case of ANN experiment, we set backpropagation as training algorithm and used sigmoid function as an activation function. The number of hidden layers was set to 20 and the epoch to 500 for each experiment. Also learning rate and momentum rate for backpropagation were set to 0.3 and 0.2. Then we got the results shown in Tables 1 and 2.

Table 1. Performance of four classification methods on non-feature selected ccRCC clinical dataset. (Acc = Accuracy, Sen = Sensitivity, Spe = Specificity)

	Without Ensemble			Adaboost			Bagging		
	Acc	Sen	Spe	Acc	Sen	Spe	Acc	Sen	Spe
C4.5	77.9%	84.9%	64.2%	78.1%	84.5%	65.5%	81.3%	89.3%	65.5%
SVM	82.7%	89.7%	68.9%	81.3%	87.3%	69.6%	81.5%	89.7%	65.5%
ANN	77.4%	85.9%	60.8%	77.4%	85.9%	60.8%	81.5%	86.9%	70.9%
BAY	83.6%	89.0%	73.0%	79.7%	85.2%	68.9%	83.1%	88.3%	73.0%

Table 2. Performance of four classification methods on feature selected ccRCC clinical dataset. (Acc = Accuracy, Sen = Sensitivity, Spe = Specificity)

	Without Ensemble			Adaboost			Bagging		
	Acc	Sen	Spe	Acc	Sen	Spe	Acc	Sen	Spe
C4.5	78.4%	86.6%	62.2%	76.8%	82.8%	64.9%	82.2%	89.3%	68.2%
SVM	83.4%	89.7%	70.9%	82.9%	89.0%	70.9%	81.8%	89.3%	66.9%
ANN	79.7%	85.2%	68.9%	79.7%	85.2%	68.9%	81.1%	87.3%	68.9%
BAY	83.8%	89.0%	62.2%	80.2%	85.2%	70.3%	83.6%	89.3%	72.3%

In Table 1, we could find that using ensemble methods on ccRCC clinical data improve C4.5 and ANN classification results. Also same as in Table 2. Comparing the two ensemble methods, which are adaboost and bagging method, using bagging method as ensemble of the classification methods, results in better values across all performance metrics in comparison to adaboost.

5   Conclusion

In our previous work, we have tried adopting four classification algorithms with feature selection methods on this TCGA ccRCC patients' dataset [9, 10]. And this time we used the ensemble method to improve our classification results by applying ensemble method. However, in some case of adaboost's results, there are little decrease of classification measurement than results of bagging. Comparing this with the result of bagging, there is decreased classification measurement but not much than adaboost. Rather, in case of artificial neural network and C4.5, both accuracy and sensitivity are increased, when using bagging as ensemble method. Among all of these results, the most acceptable method for classifying ccRCC patients' is feature selected Bayesian network with bagging. The results also show that bagging and boosting can increase accuracy, specificity and sensitivity of classification with both non-contextual and contextual methods. This makes them tools to be considered also with other contextual systems.

Acknowledgement. This research was supported by Basic Science Research Program through the National Research Foundation of Korea (NRF) funded by the Ministry of Science, ICT & Future Planning (No. 2017R1A2B4010826), by the MSIT (Ministry of Science and ICT), Korea, under the ITRC (Information Technology Research Center) support program (IITP-2017-2013-0-00881) supervised by the IITP (Institute for Information & communications Technology Promotion), supported by the KIAT (Korea Institute for Advancement of Technology) grant funded by the Korea Government (MOTIE: Ministry of Trade Industry and Energy). (No. N0002429) in Republic of Korea and also supported by the National Natural Science Foundation of China (61702324) in People's Republic of China.

References

1. Koelzer, V.H., Rothschild, S.I., Zihler, D., Wicki, A., Willi, B., Willi, N., Voegeli, M., Cathomas, G., Zippelius, A., Mertz, K.D.: Systemic inflammation in a melanoma patient treated with immune checkpoint inhibitors - an autopsy study. J. Immuno Ther. Cancer **4**(1), 13 (2016)
2. Siegel, R., Ma, J., Zou, Z., Jemal, A.: Cancer statistics, 2014. Cancer J. Clin. **64**(1), 9–29 (2014)
3. Akay, M.F.: Support vector machines combined with feature selection for breast cancer diagnosis. Expert Syst. Appl. **36**(2), 3240–3247 (2009)
4. Lim, K.H., Ryu, K.S., Park, S.H., Shon, H.S., Ryu, K.H.: Short-term mortality prediction of recurrence patients with ST-segment elevation myocardial infarction. J. Korea Soc. Comput. Inf. **17**(10), 145–154 (2012)
5. Choi, N.H., Piao, Y., Li, M., Ryu, K.H.: Comparison of combination of feature selection methods and classification methods for multiclass cancer classification from RNA-seq gene expression data. In: The 7th International Conference FITAT 2014 and 4th International Symposium ISPM 2014, vol. 1, no. 1, pp. 79–81 (2014)
6. Jung, Y.G., Lee, S.H., Sung, H.J.: Effective diagnostic method of breast cancer data using decision tree. J. Inst. Internet Broadcast. Commun. **10**(5), 57–62 (2010)
7. Baxt, W.G.: Application of artificial neural networks to clinical medicine. Lancet **346**(8983), 1135–1138 (1995)
8. Jayasurya, K., Fung, G., Yu, S., Dehing-Oberije, C., De Ruysscher, D., Hope, A., Dekker, A.L.A.J.: Comparison of Bayesian network and support vector machine models for two-year survival prediction in lung cancer patients treated with radiotherapy. Med. Phys. **37**(4), 1401–1407 (2010)
9. Park, K.H., Ryu, K.S., Shon, H.S., Ryu, K.H.: Performance evaluation of contemporary classification algorithms for high risk patients with clear cell renal carcinoma. In: TSDAA 2015, vol. 1, no. 1 (2015)
10. Park, K.H., Ryu, K.S., Ryu, K.H.: Determining minimum feature number of classification on clear cell renal cell carcinoma clinical dataset. Int. Conf. Mach. Learn. Cybern. **1**(1), 894–898 (2016)
11. The Cancer Genome Atlas Homepage. http://cancergenome.nih.gov
12. Namsrai, E., Munkhdalai, T., Namsrai, O.E.: Comparison of feature selection techniques for efficient mining of arrhythmia. In: The 4th International Conference on Frontiers of Information Technology, Applications and Tools, vol. 1, no. 1, pp. 144–148 (2011)
13. Quinlan, J.R.: Programs for Machine Learning, pp. 1–149. Morgan Kaufmann Publishers Inc., San Francisco (1993)
14. Quinlan, J.R.: Induction of decision trees. Mach. Learn. **1**(1), 81–106 (1986)

15. Li, D.C., Liu, C.W.: A class possibility based kernel to increase classification accuracy for small data set using support vector machines. Expert Syst. Appl. **37**(4), 3104–3110 (2010)
16. Abraham, A.: Artificial neural networks. In: Handbook of Measuring System Design, pp. 901–908 (2005)
17. Heckerman, D.: Bayesian Networks for Data Mining, pp. 79–119. Kluwer Academic Publishers, Dordrecht (1997)
18. Breiman, L.: Bagging predictors. Mach. Learn. **24**(2), 123–140 (1996)
19. Freund, Y., Schapire, R.E.: Experiments with a newboosting algorithm. In: International Conference on Machine Learning, pp. 148–156 (1996)

Intelligent Systems and Algorithms in Information Sciences

Solution of Dual Fuzzy Equations Using a New Iterative Method

Sina Razvarz[1], Raheleh Jafari[2(✉)] , Ole-Christoffer Granmo[2], and Alexander Gegov[3]

[1] Departamento de Control Automatico, CINVESTAV-IPN,
National Polytechnic Institute, Mexico City, Mexico
srazvarz@yahoo.com
[2] Department of Information and Communication Technology,
Agder University College, 4876 Grimstad, Norway
Jafari3339@yahoo.com, ole.granmo@uia.no
[3] School of Computing, University of Portsmouth, Buckingham Building,
Portsmouth PO13HE, UK
alexander.gegov@port.ac.uk

Abstract. In this paper, a new hybrid scheme based on learning algorithm of fuzzy neural network (FNN) is offered in order to extract the approximate solution of fully fuzzy dual polynomials (FFDPs). Our FNN in this paper is a five-layer feed-back FNN with the identity activation function. The input-output relation of each unit is defined by the extension principle of Zadeh. The output from this neural network, which is also a fuzzy number, is numerically compared with the target output. The comparison of the feed-back FNN method with the feed-forward FNN method shows that the less error is observed in the feed-back FNN method. An example based on applications are given to illustrate the concepts, which are discussed in this paper.

Keywords: Fully fuzzy dual polynomials · Fuzzy neural network
Approximate solution

1 Introduction

Artificial neural networks (ANNs) are mathematical or computational models based on biological neural networks. They make effort to imitate the information presentation, processing scheme and discrimination capability of natural neurons in the human brain. ANNs are a prominent component of artificial intelligence, which emulate the learning procedure of the human brain for extracting patterns from historical data. Neural networks can be categorized as feed-forward and feed-back ones. The primary futileness of feed-forward neural networks is that the weight updating does not employ any information on the local data structure, also the function approximation is impressionable to the training data [1]. However, feed-back neural networks have impressive representation abilities so that can successfully overcome the futileness of feed-forward neural networks. In recent years, there have been a wide spread of studies in the field of neural networks [2–6]. Ishibuchi et al. [7] designed a FNN with triangular fuzzy

© Springer International Publishing AG, part of Springer Nature 2018
N. T. Nguyen et al. (Eds.): ACIIDS 2018, LNAI 10752, pp. 245–255, 2018.
https://doi.org/10.1007/978-3-319-75420-8_23

weights. Abbasbandy et al. [8] investigated the solution of polynomials like $a_1 x + \ldots + a_n x^n = a_0$ where $x \in \Re$ and $a_0, a_1, \ldots, a_n$ are fuzzy numbers by a learning algorithm of FNN. Jafarian et al. [9] proposed a new learning algorithm for solving fuzzy polynomials. Friedman et al. [10] represented a new model for solving a fuzzy $n \times n$ linear system with crisp coefficient matrix and a fuzzy vector in the right-hand side. Also they investigated the duality of the fuzzy linear systems like $Ax = Bx + y$ where A and B are two real $n \times n$ matrices and the unknown x and y are two vectors with n fuzzy numbers components [11]. In [12] a FNN model is utilized in order to extract the coefficients of fuzzy polynomial regression.

Up to now, however, none has been reported on applications of the FNNs to solve FFDP. This kind of polynomial has been widely studied due to its promising potential for applications in different fields such as engineering, physics, economics and optimal control theory. Thus, in this paper the FNN is a first and important step for solving these polynomials. In [13], the authors have been proposed Newton's method for solving fuzzy nonlinear equations. Dehghan et al. [14] introduced a numerical method to solve a system of linear fuzzy equation. The solution of fuzzy polynomial equation based on the ranking method has been investigated by [15]. In [16] the ranking technique is implemented in order to obtain the real roots of dual fuzzy polynomial equation. Muzzioli et al. [17] applied nonlinear programming method for the solution of fuzzy linear system. Amirfakhrian in [18] presented a numerical iterative method to find the roots of an algebraic fuzzy equation of degree n with fuzzy coefficients. In [19] the fully fuzzy system of linear equations with an arbitrary fuzzy coefficient is investigated. Ezzati [20] developed a new method for solving an arbitrary general fuzzy linear system by using the embedding approach. In [21] solving fully fuzzy system of linear equations by using multi objective linear programming and the embedding approach is discussed. Waziri et al. [22] applied a new approach for solving dual fuzzy nonlinear equations by using Broyden's and Newton's methods. Also, in [23] the Adomian decomposition method for solving these polynomials is introduced. In [24] the exponent to production technique is illustrated in order to generate an analytical and approximated solution of fully fuzzy quadratic equation. Babbar et al. [25] have applied a new approach to find the nonnegative solution of a fully fuzzy linear system, where the elements of the coefficient matrix are defined as arbitrary triangular fuzzy numbers. More information on fuzzy polynomials can be found in [26, 27].

The objective of this paper is to design a new model based on FNNs for approximate solution of FFDPs. In this work, a model of feed-back FNN equivalent to dual fuzzy polynomial of the form $a_1 x + \ldots + a_n x^n = b_1 x + \ldots + b_n x^n + d$ is built, where a_j, b_j, d and x are fuzzy numbers (for $j = 1, \ldots, n$). The input-output relation of each unit of the designed neural network is defined by the extension principle of Zadeh [28]. The proposed feed-back FNN is able to estimate the fuzzy solution related to FFDP to any level of preciseness. In continues, by comparing our results with the results obtained by using feed-forward FNN, it can be observed that the feed-back method yields faster convergence rate and less computational complexity in the adjusting the weights. This paper is started with a brief description of FFDPs in Sect. 2. In this section, the feed-back FNN and feed-forward FNN are introduced. Furthermore, by using the learning algorithm which is derived from the cost function, we will be capable of finding a fuzzy solution associated with the FFDP. An example is likewise presented in Sect. 3. Section 4 finishes the paper with conclusions.

2 Fully Fuzzy Dual Polynomials

In this part the interest is vested in finding a solution for the following FFDP

$$p_1 y + \ldots + p_n y^n = q_1 y + \ldots + q_n y^n + \Phi \tag{1}$$

where $p_i, q_i, \Phi, y \in E (i = 1, \ldots, n)$. In order to extract an estimated solution associated with the FFDP two models of feed-back FNN and feed-forward FNN equivalent to Eq. (1) are presented in Figs. 1 and 2 respectively.

Generally, for an arbitrary fuzzy number $a \in E$, there exists no element $b \in E$ such that, $a + b = 0$. In fact, for all non-crisp fuzzy number $a \in E$ we have $a + (-a) \neq 0$. Hence, Eq. (8) cannot be equivalently substituted by $(p_1 - q_1) y + \ldots + (p_n - q_n) y^n = \Phi$, which had been investigated.

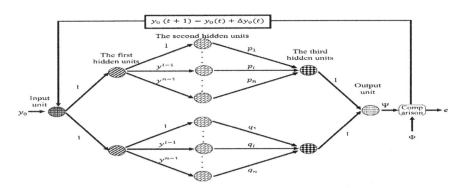

Fig. 1. Feed-back FNN for resolving dual fuzzy polynomials

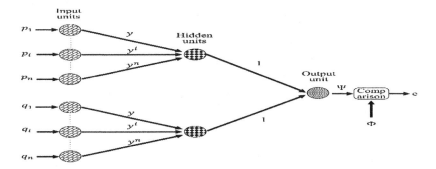

Fig. 2. Feed-forward FNN for resolving dual fuzzy polynomials

2.1   Computation of Fuzzy Output in Feed-Back FNNs

In the presented neural network training data are written as $\{(P, Q); \Phi\}$ where $P = (p_1, \ldots, p_n)$ and $Q = (q_1, \ldots, q_n)$. The α-level sets associated with the fuzzy coefficients p_i as well as q_i are nonnegative, i.e., $0 \leq \underline{p}_i^{\alpha} \leq \bar{p}_i^{\alpha}$ and $0 \leq \underline{q}_i^{\alpha} \leq \bar{q}_i^{\alpha}$ for all i's. We have

- Input unit

$$[y_0]^{\alpha} = (\underline{y}_0^{\alpha}, \bar{y}_0^{\alpha}) \tag{2}$$

- The first hidden units

$$[Z_{11}]^{\alpha} = \left(\underline{y}_0^{\alpha}, \bar{y}_0^{\alpha}\right), [Z_{12}]^{\alpha} = (\underline{y}_0^{\alpha}, \bar{y}_0^{\alpha}). \tag{3}$$

- The second hidden units

$$[Z_{21}]^{\alpha} = \left(\sum_{i \in M_{21}} \left(\underline{y}_0^{\alpha}\right)^i + \sum_{i \in C_{21}} \left(\bar{y}_0^{\alpha}\right)^i + \sum_{i \in N_{21}} \left(\underline{y}_0^{\alpha}\right)^i, \sum_{i \in M_{22}} \left(\bar{y}_0^{\alpha}\right)^i \right.$$
$$\left. + \sum_{i \in C_{22}} \left(\underline{y}_0^{\alpha}\right)^i + \sum_{i \in N_{22}} \left(\bar{y}_0^{\alpha}\right)^i \right) \tag{4}$$

$$[Z_{22}]^{\alpha} = \left(\sum_{i \in M_{21}} \left(\underline{y}_0^{\alpha}\right)^i + \sum_{i \in C_{21}} \left(\bar{y}_0^{\alpha}\right)^i + \sum_{i \in N_{21}} \left(\underline{y}_0^{\alpha}\right)^i, \sum_{i \in M_{22}} \left(\bar{y}_0^{\alpha}\right)^i \right.$$
$$\left. + \sum_{i \in C_{22}} \left(\underline{y}_0^{\alpha}\right)^i + \sum_{i \in N22} \left(\bar{y}_0^{\alpha}\right)^i \right) \tag{5}$$

Where $M_{21} = \left\{i | \underline{y}_0^{\alpha} \geq 0\right\}, C_{21} = \left\{i | \underline{y}_0^{\alpha} < 0, \ i \ is \ even \ number\right\}$, $N_{21} = \left\{i | \underline{y}_0^{\alpha} < 0, \right.$ $i \ is \ odd \ number\}$, $M_{22} = \{i | \bar{y}_0^{\alpha} \geq 0\}, C_{22} = \{i | \bar{y}_0^{\alpha} < 0, \ i \ is \ even \ number\}$, $N_{22} = \left\{i | \bar{y}_0^{\alpha} < 0, \ i \ is \ odd \ number\right\}$, $M_{21} \cup C_{21} \cup N_{21} = \{1, \ldots, n\}$ and $M_{22} \cup C_{22} \cup N_{22} = \{1, \ldots, n\}$.

- The third hidden units

$$[Z_{31}]^{\alpha} = (\sum_{j \in M_{31}} \underline{Z}_{21}^{\alpha} \underline{p}_i^{\alpha} + \sum_{j \in C_{31}} \underline{Z}_{21}^{\alpha} \bar{p}_i^{\alpha}, \sum_{j \in M_{31}'} \bar{Z}_{21}^{\alpha} \bar{p}_i^{\alpha} + \sum_{j \in C_{31}'} \bar{Z}_{21}^{\alpha} \underline{P}_i^{\alpha}) \tag{6}$$

where $M_{31} = \{i | \underline{Z}_{21}^{\alpha} \geq 0\}$, $C_{31} = \{i | \underline{Z}_{21}^{\alpha} < 0\}$, $M_{31}' = \{i | \bar{Z}_{21}^{\alpha} \geq 0\}$, $C_{31}' = \left\{i | \bar{Z}_{21}^{\alpha} < 0\right\}$, $M_{31} \cup C_{31} = \{1, \ldots, n\}$ and $M_{31}' \cup C_{31}' = \{1, \ldots, n\}$.

$$[Z_{32}]^{\alpha} = (-\sum_{i \in M_{32}} \underline{Z}_{22}^{\alpha} \underline{q}_i^{\alpha} - \sum_{j \in C_{32}} \underline{Z}_{22}^{\alpha} \bar{q}_i^{\alpha}, - \sum_{j \in M_{32}'} \bar{Z}_{22}^{\alpha} \bar{q}_i^{\alpha} - \sum_{j \in C_{32}'} \bar{Z}_{22}^{\alpha} \underline{q}_i^{\alpha}) \tag{7}$$

where $M_{32} = \{i | \underline{Z}_{22}^{\alpha} \geq 0\}$, $C_{32} = \{j | [\underline{Z}_{22}^{\alpha} < 0\}$, $M_{32}' = \{i | \bar{Z}_{22}^{\alpha} \geq 0\}$, $C_{32}' = \{j | \bar{Z}_{22}^{\alpha} < 0\}$, $M_{32} \cup C_{32} = \{1, \ldots, n\}$ and $M_{32}' \cup C_{32}' = \{1, \ldots, n\}$.

- Output unit

$$[\Psi]^\alpha = (\underline{Z}^\alpha_{31} + \underline{Z}^\alpha_{32}, \bar{Z}^\alpha_{31} + \bar{Z}^\alpha_{32}) \tag{8}$$

The triangular fuzzy weight y_0 is indicated considering the three parameters mentioned as $y_0 = (y_0^1, y_0^2, y_0^3)$. Taking into account the fuzzy parameter y_0, its adjustment rule has been portrayed as below

$$y_0^r(t+1) = y_0^r(t) + \Delta y_0^r(t), \quad r = 1, 2, 3$$
$$\Delta y_0^r(t) = -\eta \frac{\partial e^\alpha}{\partial y_0^r} + \gamma \Delta y_0^r(t-1) \tag{9}$$

where t is referred as the number of adjustments, η is taken to be the learning constant, also γ is referred as a momentum constant. We calculate $\frac{\partial e^\alpha}{\partial y_0^r}$ as follows

$$\frac{\partial e^\alpha}{\partial y_0^r} = \frac{\partial \underline{e}^\alpha}{\partial y_0^r} + \frac{\partial \bar{e}^\alpha}{\partial y_0^r} \tag{10}$$

Hence complexities lies in the calculation of the derivatives $\frac{\partial \underline{e}^\alpha}{\partial y_0^r}$ and $\frac{\partial \bar{e}^\alpha}{\partial y_0^r}$. So we have

$$\frac{\partial \underline{e}^\alpha}{\partial y_0^r} = -\alpha(\underline{\Phi}^\alpha - \underline{\Psi}^\alpha)(\frac{\partial \underline{net}^\alpha_{31}}{\partial y_0^r} - \frac{\partial \underline{net}^\alpha_{32}}{\partial y_0^r}) \tag{11}$$

where

$$\frac{\partial \underline{net}^\alpha_{31}}{\partial y_0^r} = \sum_{j \varepsilon M_{31}} \underline{P}^\alpha_i \frac{\partial \underline{Z}^\alpha_{21}}{\partial y_0^r} + \sum_{j \varepsilon C_{31}} \bar{P}^\alpha_i \frac{\partial \underline{Z}^\alpha_{21}}{\partial y_0^r} \tag{12}$$

$$\frac{\partial \underline{net}^\alpha_{32}}{\partial y_0^r} = \sum_{j \varepsilon M_{32}} \underline{q}^\alpha_i \frac{\partial \underline{Z}^\alpha_{22}}{\partial y_0^r} + \sum_{j \varepsilon C_{32}} \bar{q}^\alpha_i \frac{\partial \underline{Z}^\alpha_{22}}{\partial y_0^r} \tag{13}$$

$$\frac{\partial \underline{Z}^\alpha_{21}}{\partial y_0^r} = \frac{\partial \underline{Z}^\alpha_{22}}{\partial y_0^r} = \sum_{j \varepsilon M_{21}} i(\underline{y}^\alpha_0)^{i-1} \frac{\partial \underline{y}^\alpha_0}{\partial y_0^r} + \sum_{j \varepsilon C_{21}} i(\bar{y}^\alpha_0)^{i-1} \frac{\partial \underline{y}^\alpha_0}{\partial y_0^r} + \sum_{j \varepsilon N_{21}} i(\underline{y}^\alpha_0)^{i-1} \frac{\partial \underline{y}^\alpha_0}{\partial y_0^r} \tag{14}$$

and

$$\frac{\partial \bar{e}^\alpha}{\partial y_0^r} = -\alpha(\bar{\Phi}^\alpha - \bar{\Psi}^\alpha)(\frac{\partial \overline{net}^\alpha_{31}}{\partial y_0^r} - \frac{\partial \overline{net}^\alpha_{32}}{\partial y_0^r}) \tag{15}$$

where

$$\frac{\partial \overline{net}^\alpha_{31}}{\partial y_0^r} = \sum_{j \varepsilon M'_{31}} \bar{P}^\alpha_i \frac{\partial \bar{Z}^\alpha_{21}}{\partial y_0^r} + \sum_{j \varepsilon C'_{31}} \underline{P}^\alpha_i \frac{\partial \bar{Z}^\alpha_{21}}{\partial y_0^r} \tag{16}$$

$$\frac{\partial \overline{net}_{32}^{\alpha}}{\partial y_0^r} = \sum_{j \varepsilon M_{32}'} \overline{q}_i^{\alpha} \frac{\partial \overline{Z}_{22}^{\alpha}}{\partial y_0^r} + \sum_{j \varepsilon C_{32}'} \underline{q}_i^{\alpha} \frac{\partial \overline{Z}_{22}^{\alpha}}{\partial y_0^r} \tag{17}$$

$$\frac{\partial \overline{Z}_{21}^{\alpha}}{\partial y_0^r} = \frac{\partial \overline{Z}_{22}^{\alpha}}{\partial y_0^r} = \sum_{j \varepsilon M_{22}} i(\overline{y}_0^{\alpha})^{i-1} \frac{\partial \overline{y}_0^{\alpha}}{\partial y_0^r} + \sum_{j \varepsilon C_{22}} i(\underline{y}_0^{\alpha})^{i-1} \frac{\partial y_{-0}^{\alpha}}{\partial y_0^r} + \sum_{j \varepsilon N_{22}} i(\overline{y}_0^{\alpha})^{i-1} \frac{\partial \overline{y}_0^{\alpha}}{\partial y_0^r} \tag{18}$$

In above relations the derivatives $\frac{\partial y_{-0}^{\alpha}}{\partial y_0^r}$ and $\frac{\partial \overline{y}_0^{\alpha}}{\partial y_0^r}$ can be summarized as follows

$$\frac{\partial y_{-0}^{\alpha}}{\partial y_0^r} = \begin{cases} 1 - \alpha, & r = 1 \\ \alpha, & r = 2, \\ 0, & r = 3 \end{cases} \qquad \frac{\partial \overline{y}_0^{\alpha}}{\partial y_0^r} = \begin{cases} 0, & r = 1 \\ \alpha, & r = 2 \\ 1 - \alpha, & r = 3 \end{cases} \tag{19}$$

2.2   Computation of Fuzzy Output in Feed-Forward FNNs

The α-level sets associated with the fuzzy inputs p_i's as well as q_i's are nonnegative, i.e., $0 \leq \underline{p}_i^{\alpha} \leq \overline{p}_i^{\alpha}$ and $0 \leq \underline{q}_i^{\alpha} \leq \overline{q}_i^{\alpha}$ for all i's. We have

- Input units

$$[p_i]^{\alpha} = \left(\underline{p}_i^{\alpha}, \overline{p}_i^{\alpha} \right), [q_i]^{\alpha} = \left(\underline{q}_i^{\alpha}, \overline{q}_i^{\alpha} \right), \qquad i = 1, 2, \ldots, n \tag{20}$$

- Hidden units

$$[Z_1]^{\alpha} = \left(\sum_{i \varepsilon M} \left(\underline{y}^{\alpha} \right)^i \underline{p}_i^{\alpha} + \sum_{i \varepsilon C} \left(\underline{y}^{\alpha} \right)^i \overline{p}_i^{\alpha}, \sum_{i \varepsilon M'} (\overline{y}^{\alpha})^i \overline{p}_i^{\alpha} + \sum_{i \varepsilon C'} (\overline{y}^{\alpha})^i \underline{p}_i^{\alpha} \right) \tag{21}$$

$$[Z_2]^{\alpha} = \left(-\sum_{i \varepsilon M} \left(\underline{y}^{\alpha} \right)^i \underline{q}_i^{\alpha} - \sum_{i \varepsilon C} \left(\underline{y}^{\alpha} \right)^i \overline{q}_i^{\alpha}, -\sum_{i \varepsilon M'} (\overline{y}^{\alpha})^i \overline{q}_i^{\alpha} - \sum_{i \varepsilon C'} (\overline{y}^{\alpha})^i \underline{q}_i^{\alpha} \right) \tag{22}$$

where $M = \left\{ i | (\underline{y}^{\alpha})^i \geq 0 \right\}$, $C = \left\{ i | (\underline{y}^{\alpha})^i < 0 \right\}$, $M' = \left\{ i | (\overline{y}^{\alpha})^i \geq 0 \right\}$, $C' = \left\{ i | (\overline{y}^{\alpha})^i < 0 \right\}$, $M \cup C = \{1, \ldots, n\}$ and $M' \cup C' = \{1, \ldots, n\}$.

- Output unit

$$[\Psi]^{\alpha} = \left(\underline{Z}_1^{\alpha} + \underline{Z}_2^{\alpha}, \overline{Z}_1^{\alpha} + \overline{Z}_2^{\alpha} \right) \tag{23}$$

Assume Φ to be the fuzzy target output in association with the fuzzy coefficient vectors (p_i, q_i). A cost function which is required to be minimized is stated at par with the α-level sets of the fuzzy output Ψ as well as the target output Φ as $e^{\alpha} = \underline{e}^{\alpha} + \overline{e}^{\alpha}$, where $\underline{e}^{\alpha} = \alpha \frac{(\underline{\Phi}^{\alpha} - \underline{\Psi}^{\alpha})^2}{2}$ and $\overline{e}^{\alpha} = \alpha \frac{(\overline{\Phi}^{\alpha} - \overline{\Psi}^{\alpha})^2}{2}$. The $\underline{e}^{\alpha}$ and $\overline{e}^{\alpha}$ are demonstrated as the squared errors for the lower limits as well as the upper limits associated with the α-level sets of the fuzzy output Ψ and target output Φ, respectively.

The triangular fuzzy weight y is indicated considering the three parameters mentioned as $y = (y^1, y^2, y^3)$. The weight is adjusted by the following rule [7]

$$
\begin{aligned}
y^r(t+1) &= y^r(t) + \Delta y^r(t), \quad r = 1, 2, 3 \\
\Delta y^r(t) &= -\eta \frac{\partial e^\alpha}{\partial y^r} + \gamma \Delta y^r(t-1)
\end{aligned}
\tag{24}
$$

where t is referred as the number of adjustments, η is taken to be as the learning constant, also γ is referred as a momentum constant. We calculate $\frac{\partial e^\alpha}{\partial y^r}$ as follows

$$
\frac{\partial e^\alpha}{\partial y^r} = \frac{\partial \underline{e}^\alpha}{\partial y^r} + \frac{\partial \overline{e}^\alpha}{\partial y^r}
\tag{25}
$$

Hence complexities lies in the calculation of the derivatives $\frac{\partial \underline{e}^\alpha}{\partial y^r}$ and $\frac{\partial \overline{e}^\alpha}{\partial y^r}$. So we have

$$
\frac{\partial \underline{e}^\alpha}{\partial y^r} = -\alpha(\underline{\Phi}^\alpha - \underline{\Psi}^\alpha)\left(\frac{\partial net_{21}^\alpha}{\partial y^r} - \frac{\partial net_{22}^\alpha}{\partial y^r}\right)
\tag{26}
$$

where

$$
\frac{\partial net_{21}^\alpha}{\partial y_0^r} = \sum_{i \in M} \underline{p}_i^\alpha i (\underline{y}^\alpha)^{i-1} \frac{\partial \underline{y}^\alpha}{\partial y^r} + \sum_{i \in C} \overline{p}_i^\alpha i (\underline{y}^\alpha)^{i-1} \frac{\partial \underline{y}^\alpha}{\partial y^r}
\tag{27}
$$

$$
\frac{\partial net_{22}^\alpha}{\partial y_0^r} = \sum_{i \in M} \underline{q}_i^\alpha i (\underline{y}^\alpha)^{i-1} \frac{\partial \underline{y}^\alpha}{\partial y^r} + \sum_{i \in C} \overline{q}_i^\alpha i (\underline{y}^\alpha)^{i-1} \frac{\partial \underline{y}^\alpha}{\partial y^r}
\tag{28}
$$

and

$$
\frac{\partial \overline{e}^\alpha}{\partial y^r} = -\alpha(\overline{\Phi}^\alpha - \overline{\Psi}^\alpha)\left(\frac{\partial \overline{net}_{21}^\alpha}{\partial y^r} - \frac{\partial \overline{net}_{22}^\alpha}{\partial y^r}\right)
\tag{29}
$$

Where

$$
\frac{\partial \overline{net}_{21}^\alpha}{\partial y^r} = \sum_{i \in M'} \overline{p}_i^\alpha i (\overline{y}^\alpha)^{i-1} \frac{\partial \overline{y}^\alpha}{\partial y^r} + \sum_{i \in C'} \underline{p}_i^\alpha i (\overline{y}^\alpha)^{i-1} \frac{\partial \overline{y}^\alpha}{\partial y^r}
\tag{30}
$$

$$
\frac{\partial \overline{net}_{22}^\alpha}{\partial y^r} = \sum_{i \in M'} \overline{q}_i^\alpha i (\overline{y}^\alpha)^{i-1} \frac{\partial \overline{y}^\alpha}{\partial y^r} + \sum_{i \in C'} \underline{q}_i^\alpha i (\overline{y}^\alpha)^{i-1} \frac{\partial \overline{y}^\alpha}{\partial y^r}
\tag{31}
$$

In above relations the derivatives $\frac{\partial y^\alpha}{\partial y^r}$ and $\frac{\partial \bar{y}^\alpha}{\partial y^r}$ can be summarized as follows

$$\frac{\partial y^\alpha}{\partial y^r} = \begin{cases} 1-\alpha, & r=1 \\ \alpha, & r=2 \\ 0, & r=3 \end{cases}, \quad \frac{\partial \bar{y}^\alpha}{\partial y^r} = \begin{cases} 0, & r=1 \\ \alpha, & r=2 \\ 1-\alpha, & r=3 \end{cases} \tag{32}$$

3 Numerical Examples

To show the behavior and properties of the proposed method, an examples is solved.

Example: A vertical propeller shaft AQ with a diameter of $d = 0.015$ is connected to the fixed base. The propeller shaft is made of steel with $G = 80 \times 10^9$ and the resultant torques in the points A, B, C and D are $T_1 = y, T_2 = y^2, T_3 = y$ and $T_4 = y^2$, respectively. The resultant torques T_1 and T_2 can cause a twisting in shaft which is equal to φ degree, see Fig. 4a. The resultant torques T_3 and T_4 can cause a twisting in shaft which is equal to $\varphi \ominus (5, 9, 14)$ degree, see Fig. 3b. According to the torsion equation we will have [29]:

$$\varphi = \frac{L_1 T_1}{JG} \oplus \frac{L_2 T_2}{JG} = \frac{L_3 T_3}{JG} \oplus \frac{L_4 T_4}{JG} \oplus (5, 9, 14) \tag{33}$$

where

$$J = \frac{\Pi}{2} d^4. \tag{34}$$

The length of the shafts are not exact, which satisfy the triangular function (1),

$$\begin{aligned} L_1 &= (7, 8, 11) = p_1 \\ L_2 &= (1, 2, 3) = p_2 \\ L_3 &= (1, 2, 3) = q_1 \\ L_4 &= (2, 3, 4) = q_2 \end{aligned} \tag{35}$$

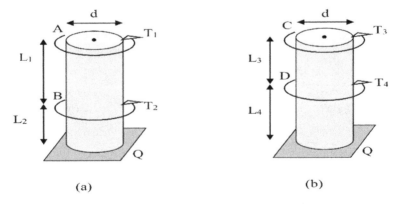

(a)   (b)

Fig. 3. The cylindrical force

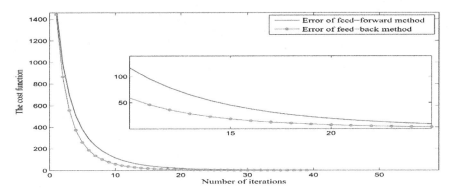

Fig. 4. The cost function associated with Example 2 considering the number of iterations with both techniques

Table 1. The estimated solutions at par with error analysis associated to Example

t	$y_0(t)$ by feed-forward FNN	Error	t	$y_0(t)$ by feed-back FNN	Error
1	(2.9685, 5.8452, 6.9025)	1443.3563	1	(2.8692, 5.7905, 6.8479)	1443.3563
2	(2.7245, 5.6413, 6.6585)	999.52256	2	(2.5892, 5.5812, 6.4475)	806.67385
3	(2.4998, 5.4025, 6.3011)	696.92256	3	(2.3098, 5.2560, 6.0313)	517.06672
4	(2.2458, 5.0255, 5.9915)	509.34256	4	(2.0514, 4.9047, 5.7065)	375.04018
5	(2.0150, 4.8961, 5.6162)	382.84254	5	(1.8256, 4.5973, 5.3560)	231.27915
⋮	⋮	⋮	⋮	⋮	⋮
55	(1.0099, 3.0043, 4.0067)	0.32515382	35	(1.0075, 3.0081, 4.0097)	0.6460055
56	(1.0081, 3.0036, 4.0052)	0.22514552	36	(1.0068, 3.0062, 4.0078)	0.4542216
57	(1.0075, 3.0029, 4.0041)	0.11254587	37	(1.0052, 3.0044, 4.0057)	0.3298205
58	(1.0061, 3.0020, 4.0030)	0.12552141	38	(1.0043, 3.0029, 4.0040)	0.1649085
59	(1.0048, 3.0011, 4.0022)	0.09254654	39	(1.0035, 3.0019, 4.0022)	0.0888326

We use feed-back FNN and feed-forward FNN shown in Figs. 1 and 2 to approximate the solution y. The exact solution is termed as = (1, 3, 4). The training starts with $y(0) = (3, 6, 7), \eta = 4 \times 10^{-4}$ as well as $\gamma = 5 \times 10^{-4}$. Table 1 displays the estimated solution considering the number of iterations. The preciseness of the computed solution $y_0(t)$ is portrayed in Fig. 4, where t is taken to be the number of iterations.

4 Concluding Remarks

This paper describes the design and training of a FNN which is used for solving FFDP. To obtain a solution of a FFDP, the adjustable parameter of FNN is systematically adjusted by using a learning algorithm that is based on the gradient descent method. The effectiveness of the derived learning algorithm is demonstrated by computer simulation on numerical examples. We proposed two examples based on applications. The comparison of the feed-back FNN method with the feed-forward FNN method shows that the feed-back FNN method is better or at least more suitable than the feed-forward FNN method. The reason behind it is that, the speed of convergence is increased which depends on the number of computations.

References

1. Jin, L., Gupta, M.M.: Stable dynamic backpropagation learning in recurrent neural networks. IEEE Trans. Neural Networks **10**(6), 1321–1334 (1999)
2. Jafari, R., Yu, W., Li, X.: Numerical solution of fuzzy equations with Z-numbers using neural networks. Intell. Autom. Soft Comput. 1–7 (2017)
3. Jafari, R., Yu, W., Li, X., Razvarz, S.: Numerical solution of fuzzy differential equations with Z-numbers using bernstein neural networks. Int. J. Comput. Intell. Syst. **10**(1), 1226–1237 (2017)
4. Jafari, R., Yu, W.: Fuzzy differential equation for nonlinear system modeling with Bernstein neural networks. IEEE Access (2017). https://doi.org/10.1109/access.2017.2647920
5. Jafari, R., Yu, W.: Uncertainty nonlinear systems modeling with fuzzy equations. Math. Probl. Eng. (2017). https://doi.org/10.1155/2017/8594738
6. Jafarian, A., Jafari, R., Mohamed Al Qurashi, M., Baleanud, D.: A novel computational approach to approximate fuzzy interpolation polynomials. Springer Plus **5**, 14–28 (2016)
7. Ishibuchi, H., Kwon, K., Tanaka, H.: A learning of fuzzy neural networks with triangular fuzzy weights. Fuzzy Sets Syst. **71**(3), 277–293 (1995)
8. Abbasbandy, S., Otadi, M.: Numerical solution of fuzzy polynomials by fuzzy neural network. Appl. Math. Comput. **181**(2), 1084–1089 (2006)
9. Jafarian, A., Measoomynia, S.: Solving fuzzy polynomials using neural nets with a new learning algorithm. Aust. J. Basic Appl. Sci. **5**(9), 2295–2301 (2011)
10. Friedman, M., Ming, M., Kandel, A.: Fuzzy linear systems. Fuzzy Sets Syst. **96**(1), 201–209 (1998)
11. Friedman, M., Ma, M., Kandel, A.: Duality in fuzzy linear systems. Fuzzy Sets Syst. **109**(1), 55–58 (2000)
12. Otadi, M.: Fully fuzzy polynomial regression with fuzzy neural networks. Neurocomputing **142**, 486–493 (2014)
13. Abbasbandy, S., Asady, B.: Newton's method for solving fuzzy nonlinear equations. Appl. Math. Comput. **159**(2), 356–379 (2004)
14. Dehghan, M., Hashemi, B.: Iterative solution of fuzzy linear systems. Appl. Math. Comput. **175**(1), 645–674 (2006)
15. Rouhparvar, H.: Solving fuzzy polynomial equation by ranking method. In: First Joint Congress on Fuzzy and Intelligent Systems. Ferdowsi University of Mashhad, Iran (2007)
16. Rahman, N.A., Abdullah, L.: An interval type-2 dual fuzzy polynomial equations and ranking method of fuzzy numbers. Int. J. Math. Comput. Phys. Electr. Comput. Eng. **8**(1), 92–99 (2014)
17. Muzzioli, S., Reynaerts, H.: The solution of fuzzy linear systems by non-linear programming: a financial application. Eur. J. Oper. Res. **177**(2), 1218–1231 (2007)
18. Amirfakhrian, M.: Numerical solution of algebraic fuzzy equations with crisp variable by Gauss-Newton method. Appl. Math. Model. **32**(9), 1859–1868 (2008)
19. Kumar, A., Bansal, A., Babbar, N.: Solution of fully fuzzy linear system with arbitrary coefficients. Int. J. Appl. Math. Comput. **3**(3), 232–237 (2011)

20. Ezzati, R.: Solving fuzzy linear systems. Soft. Comput. **15**(1), 193–197 (2011)
21. Allahviranloo, T., Mikaeilvand, N.: Non zero solutions of the fully fuzzy linear systems. Appl. Comput. Math. **10**(2), 271–282 (2011)
22. Waziri, M.Y., Majid, Z.A.: A new approach for solving dual fuzzy nonlinear equations using Broyden's and Newton's methods. Adv. Fuzzy Syst. **2012**(1), 1–5 (2012)
23. Otadi, M., Mosleh, M.: Solution of fuzzy polynomial equations by modified Adomian decomposition method. Soft. Comput. **15**(1), 187–192 (2011)
24. Allahviranloo, T., Gerami Moazam, L.: The solution of fully fuzzy quadratic equation based on optimization theory. Sci. World J. **2014**(1), 1–6 (2014)
25. Babbar, N., Kumar, A., Bansal, A.: Solving fully fuzzy linear system with arbitrary triangular fuzzy numbers (m, α, β). Soft Comput. **17**(4), 691–702 (2013)
26. Oh, S.K., Pedrycz, W., Roh, S.B.: Genetically optimized fuzzy polynomial neural networks with fuzzy set-based polynomial neurons. Inf. Sci. **176**(23), 3490–3519 (2006)
27. Wang, C.C., Tsai, C.F.: Fuzzy processing using polynomial bidirectional hetero-associative network. Inf. Sci. **125**(1–4), 167–179 (2000)
28. Zadeh, L.A.: Toward a generalized theory of uncertainty (GTU) an outline. Inf. Sci. **172**(1–2), 1–40 (2005)
29. Beer, F.P., Johnston, E.R.: Mechanics of Materials, 2nd edn. Mcgraw-Hill, New York (1992)

Transformation of Extended Entity Relationship Model into Ontology

Zdenka Telnarova$^{(\boxtimes)}$

Department of Informatics and Computers,
University of Ostrava, 30. dubna 22, Ostrava 1, Czech Republic
zdenka.telnarova@osu.cz

Abstract. The paper presents several possibilities of database transformation into ontology and focuses on EER model transformation into the OWL model. Section 2 is devoted to two possible approaches with references to relevant literature, including a description of disadvantages of these approaches. Section 3 brings a set of rules which consider various situations and EER constructs taking into account transformation into OWL. It concerns transformation of entities, strong and weak entity, relationship transformation with various cardinalities, including relationships with attributes. In addition, the paper presents transformation of a composite attribute, either concerning attribute of an entity or relationship. Section 4 focuses on a practical demonstration of several described situations.

Keywords: EER model · OWL · Conversion EER into OWL

1 Introduction

According to [1], we can integrate databases into Semantic Web through two methods. The most explored method is to map database into domain ontology [2]. The second method is to make Semantic Web representations for example RDF or OWL ontologies [3].

There are five approaches how to transform a database schema into RDF [4], resp. OWL:

– Approach that uses a relational schema analysis.
– Approach that uses tuples' analysis.
– Approach that uses HTML pages.
– Approach that works with Entity Relationship (ER) or Extended Entity Relationship models (EER).
– Approach that uses Structure Query Language (SQL).

In this paper, we will focus only on the approach based on EER.

2 Approaches Based on Extended Entity Relationship Model (EER)

This approach is based on the idea that databases as well as ontologies are created according to a conceptual model, which is the most abstract form of modelling reality. A conceptual model preserves more semantics than for example a database schema.

© Springer International Publishing AG, part of Springer Nature 2018
N. T. Nguyen et al. (Eds.): ACIIDS 2018, LNAI 10752, pp. 256–264, 2018.
https://doi.org/10.1007/978-3-319-75420-8_24

The task is to transform one conceptual model (in our case EER) to another one (in our case OWL).

Literature asserts different approaches how to transform EER onto OWL. Some of them are described below.

2.1 Upadhyaya Approach

The goal is to create domain ontology (OWL) using Extended Entity Relationalship diagrams [5]. This approach allows using automatic process and shows the differences and the similarities between EER and OWL [1].

When we transform an EER diagram to relational data schema partial semantic information is lost. For example, information about inheritance, composite attributes, complex attributes, N:M cardinality in binary relationships, self-relationships, etc. It is one of the reasons why using EER is more accurate than using, for example, a relational schema or SQL approach.

The reality that both EER and OWL present conceptual models causes that this attitude is different from others. The Extended Entity Relationalship model is important in this situation because both EER and ontology models can be accounted a semantic models [7]. This approach requests an expert that can help with finding relevant semantic information.

Disadvantages
The main problem is that we do not have an EER model in an existing database. For example, EER was not developed during the process of the design of database scheme.

Another problem is that even if we have an EER model, most changes will be done on the scheme of an existing database and no changes were done in EER model.

This approach requires the domain expert who is responsible for the quality of generated ontology.

There is also difficulty in representation of entities, resp. attributes:

- Complex attributes are not considered.
- The primary key attribute is represented like data type property which has functional and inverse functional characteristics. Cardinality is set to one and this makes a lot of problems [5].

2.2 Xu *et al.* Approach

Xu et al. use formal rules for translating EER model into ontology [6]. It disposes of an instrument that makes analyses of the EER schema script coded in XML. XML code is automatically produced by CASE tool. After that mapping rules are used to translate the EER schema to OWL.

Disadvantages of this approach
This approach is based on the hypothesis that we have only single-valued attributes and there are all mandatory. This is a very strong rule which is not compatible with the reality.

Only simple-attribute keys are considered. It is not possible to use the composed key. Complex attributes available in the EER model are not considered as well.

3   Conversion of EER Model into OWL

In this chapter, we provide general rules for generation an OWL model from existing EER model.

3.1   General Rule

The conversion rules can be divided into three parts:

– Entities are transformed into OWL classes.
– Relationships are represented either by a pair of object properties or by two pairs of inverse object properties.
– Attributes are transformed into datatype properties or into a class which has datatype properties.

There is a significant moment that differs identifying of concepts in ontology and databases. In databases, any EER component is identified by its unique name, whereas in ontologies any OWL component is identified by its URI.

3.2   Creation of Classes

This operation consisted of four main steps:

– Strong as well as weak entities are converted into an *owl: class*.
– Composite attributes are converted into an *owl: class*.
– N-ary relationships for n > 2 are converted into an *owl: class*.
– Many To Many relationships with additional attributes are converted into an *owl: class*.

3.3   Creation of Datatypes

All entity or relationship attributes are transformed into datatype properties. There is a lot of types of attributes that an entity or relationship can have. For example, simple attribute, composite attribute, multivalued attribute, complex attribute, etc. Therefore, we need separate ways how to map these attributes into OWL datatypes properties. EER do not support constraint such as NOT NULL or UNIQUE. For that reason we proceed with attributes as described:

– *Owl*: *Cardinality* that can treat the same as to primary key, alternate key or the discriminator attribute of a weak entity. This restriction is usable for any simple attribute as well as for all attributes which is part of a composite attribute.

- Simple attributes will be transfered into the functor *owl*: *Functional*. This functor assures restriction that each simple attribute takes only one value.
- Complex attribute is either a composite or multivalued attribute. There are two possibilities how to convert a composite attribute to OWL datatype property. The first is to convert only the components and ignore the composite attribute. For example, we have composite attribute Address which consists of city, street, house_number, post_code, etc. In this case, simple attributes: city, street, house_number, post_code, etc. are mapped and attribute Address is ignored. The second is to convert the composite attribute into datatype property and after that convert its elements into a subproperty of a consistent datatype property. All datatype properties should have *Functional* restriction. As for the multivalued attributes, they are represented without *Functional* restriction.
- Relationship attributes: in case the relationship is transformed into a class all relationship attributes are assigned to the corresponding class. This conversion is used only for a relationship with cardinality N:M. For other types of cardinalities (1:N and 1:1), the relationship did not create a class:
    For relationship with cardinality 1:N, the attributes will be allocated to the class corresponding to the entity with cardinality N.
    For relationship with cardinality 1:1, the attributes will be allocated to whichever class entering the relationship.

3.4   Object Properties Creation

Object properties are created for relationships in EER. Each relationship in EER is transformed into a pair of object properties in OWL. Each object property representing relationship which is part of definition of the class has its inverse property in the corresponding class. There are several types of relationships:

Binary relationship with cardinality 1:N or 1:1

Each relationship will obtain a pair of object properties and there is an inversion of them. It means that one object property is the inverse to the other. Domain is represented by the class which corresponds to the entity created from this entity. Range is represented by the class which makes reference to the other entity in the relationship.

    1:1 relationship: object properties of both are Functional.
    1:N relationship: only side with cardinality 1 is Functional.

Self-relationship

Self-relationship can be defined as a binary relationship in which the domain as well as the range refers to the same class. To create object properties concerning with mentioned relationship it is necessary to use the class corresponding to the entity with Self Relationship for the domain as well as for the range.

Relationships between weak and strong entity

Any instance of a weak entity has to have a corresponding instance of a strong entity. It is a so-called mandatory relationship. This type of a relationship means that there is no possibility for the existence of an instance of a weak entity without the existence of the instance of a strong entity.

At the same time, an instance of a weak entity is not possible to identify without an instance of a strong entity (identifying relationship). To model this situation in OWL, it is necessary to map this relationship by an object with cardinality of one. The domain is the class refers to the weak entity and the range is the class corresponding to the strong entity.

Binary relationships with cardinality M:N

To transform binary relationship with cardinality N:M creation of a single object property for each relationship without inversion is needed. One of them will be assigned to the class refers to the entity with cardinality M as its domain and the range will be the class with cardinality N. The second will be assigned to the class refers to the entity with cardinality N as its domain and the range will be the class with cardinality M.

Relationships that are representing with class

There are several situations when to transform relationship with classes:

– Binary Relationships (M:N) with additional attributes

Relationship that has attributes is also converted into an OWL class. Attributes are converted into the datatype property of a corresponding OWL class.

– Representing a Relationship for a Composite Attribute

Composite Attribute is mapped into a separate class. To have connection with the original entity, it is necessary to connect a new class (based on the composite attribute) with the class created according to the original entity. Domain is the class which was created from original entity and the class of the composite attribute is its range.

– Representing Inheritance

EER model also enriches the model with ISA hierarchy. It means generalisation and specification is possible, general attributes from the supertype are inherited by all corresponding subclasses. Mapping entities from ISA hierarchy (supertype/subtype) into OWL classes, we will create two classes (class/subclass) and relationships between them called subsumption.

4 Examples

Example 1

Let us have an EER diagram which uses strong and weak entities, binary relationship with cardinality 1:N, and composite attribute (Fig. 1).

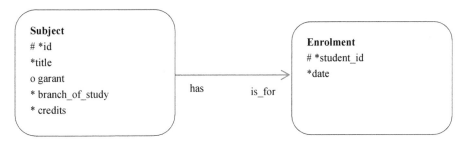

Fig. 1. EER for Example 1

where:

Subject is a strong entity
Enrolment is a weak entity
Binary relationship (Subject → Enrolment) has cardinality 1:N
garant is a composite attribute that consists of simple attributes: first_name, last_-name, department

OWL representation of this EER will have three classes. One for the strong entity, one for the weak entity and one for the composite attribute (Fig. 2).

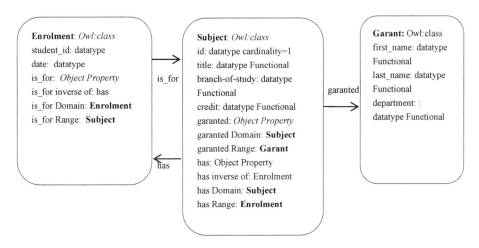

Fig. 2. OWL for Example 1

Example 2

Let us have relationship (1:N) with an attribute. For example, Teacher belongs_to Department. Relationship belongs_to has attribute start_date (Fig. 3).

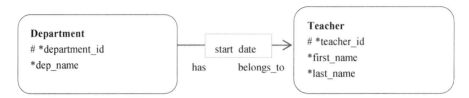

Fig. 3. EER for Example 2

OWL representation of this EER counts with relationships attributes as attributes of a class with cardinality N (Fig. 4).

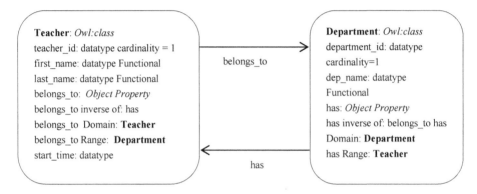

Fig. 4. OWL for Example 2

Example 3

Let us have relationship (N:M) with an attribute. For example, Teacher teaches Subjects with attributes year and semester (Fig. 5).

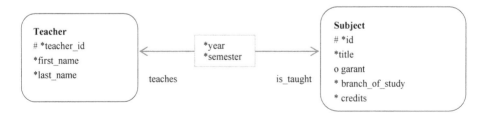

Fig. 5. EER for Example 3

This situation generates another class (**Schedule**) which represents N:M relationship. Attributes of the relationship are the attributes of new class Schedule (Fig. 6).

Example 4

This example shows Self Relationship where a teacher is managed by a boss who is also a teacher, for example, head of department (Figs. 7 and 8).

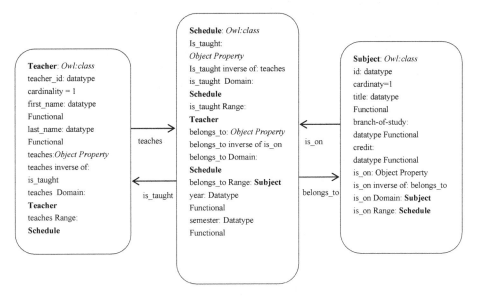

Fig. 6. OWL for Example 3

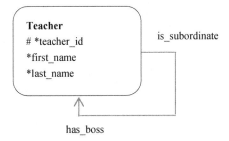

Fig. 7. EER for Example 4

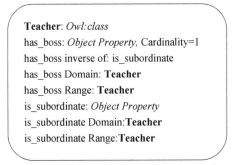

Fig. 8. OWL for Example 4

5   Conclusion

This paper analyses various situations which can be proposed within the EER model and how to map the proposed constructs to the OWL model. Entities and relationships in cardinalities 1:1, 1:N and N:M are mentioned as well as identifying and non-identifying type of a relationship, composite attribute (either of an entity or relationship), and Self Relationship. The paper does not mention: multivalued attributes, complex attributes, inheritance, and n-ary relationships. These topics will be addressed in another work. Practical demonstrations in a form of four examples document the identifying type of a relationship, composite attribute, cardinality 1:N and N:M, including relationship attributes and Self Relationship.

References

1. Alalwan, N., Zedan, H., Siewe, F.: Generating OWL ontology for database integration. In: Advances in Semantic Processing, SEMAPRO 2009, pp. 22–31, October 2009
2. Tirmizi, S.H., Sequeda, J., Miranker, D.: Translating SQL applications to the semantic web. In: Bhowmick, S.S., Küng, J., Wagner, R. (eds.) DEXA 2008. LNCS, vol. 5181, pp. 450–464. Springer, Heidelberg (2008). https://doi.org/10.1007/978-3-540-85654-2_40
3. Žáček, M.: Ontology or formal ontology. In: Simos, T., Tsitouras, C. (eds.) Proceedings of the International Conference on Numerical Analysis and Applied Mathematics 2016 (ICNAAM-2016), p. 070012. American Institute of Physics, Melville (2017). https://doi.org/10.1063/1.4992234. ISBN 978-0-7354-1538-6. ISSN 0094-243X. WOS 000410159800084
4. Lukasová, A., Žáček, M., Vajgl, M.: Reasoning in formal systems of extended RDF networks. In: Nguyen, N.T., Tojo, S., Nguyen, L.M., Trawiński, B. (eds.) ACIIDS 2017. LNCS (LNAI), vol. 10192, pp. 371–381. Springer, Cham (2017). https://doi.org/10.1007/978-3-319-54430-4_36. ISBN 978-3-319-54429-8. ISSN 0302-9743
5. Upadhyaya, S., Kumar, P.: ERONTO: a tool for extracting ontologies from extended E/R diagrams. In: Proceedings of SAC 2005 ACM Symposium on Applied Computing, Santa Fe, New Mexico, USA, March 2005
6. Xu, Z., Cao, X., Dong, Y., Su, W.: Formal approach and automated tool for translating ER schemata into OWL ontologies. In: Dai, H., Srikant, R., Zhang, C. (eds.) PAKDD 2004. LNCS (LNAI), vol. 3056, pp. 464–475. Springer, Heidelberg (2004). https://doi.org/10.1007/978-3-540-24775-3_57
7. Sugumaran, V., Storey, V.C.: Ontologies for conceptual modelling: their creation, use, and management. Data Knowl. Eng. **42**, 251–271 (2002)

Pathfinding in a Dynamically Changing Environment

Eva Volna$^{(\boxtimes)}$ and Martin Kotyrba$^{(\boxtimes)}$

Department of Informatics and Computers,
University of Ostrava, 30. dubna 22, 70103 Ostrava, Czech Republic
{eva.volna,martin.kotyrba}@osu.cz

Abstract. This paper is focused on a proposal of an algorithm for dynamic path planning which uses the input speed of the vehicle as a dynamic value based on which paths are searched in a graph. The proposed algorithm AV* stems from algorithm A*, works with proposed weight functions of evaluation function and, once again, uses information about nodes which have already been calculated in order to find out a faster update of a new path. This algorithm has been subjected to tests, whose results were compared with the results of algorithm A*. The proposed algorithm AV* complies with the condition of pathfinding between two points in a finite time based on the criterion of the fastest path using the maximum of the speed limit. From the point of view of dynamic optimization, all tests achieved a higher speed when the algorithm was run again at a different speed. This speed-up corresponds to approximately 25% and is linearly dependent on the amount of searched nodes.

Keywords: Dynamic pathfinding · A* algorithm · AV* algorithm

1 Introduction

Path planning is one of the basic tasks when navigating robots. Path planning is a complex process based on the use of information about an environment. The information is based on the map layout of the environment, but it can also come directly from sensors placed on robots, i.e. visual sensors (cameras), laser or sonar. Overview of algorithms for path planning used in robotics can be found in [1]. Path planning falls into Artificial Intelligence, Image Processing, and Algorithmization. Algorithms for path planning and robotic navigation differentiate depending on what type of work environment, i.e. the static or dynamic algorithms.

There are a number of algorithms and approaches to planning a path from a start point to a final point to plan a path in a static environment, with the resultant proposed trajectory, which is not in collision with obstacles. Gavrilut et al. [3] propose a path planning algorithm based on cellular neural network. This algorithm uses image processing techniques to convert camera output to a sequence of motions. The advantage of this method lies in its easy parallelization. Although this method has been implemented, no comparative tests with existing approaches are recorded. Kroumov and Yu [7] present their model of mobile robots' navigation, which is based on path planning using neural networks. Their methods use the potential field for path planning and

© Springer International Publishing AG, part of Springer Nature 2018
N. T. Nguyen et al. (Eds.): ACIIDS 2018, LNAI 10752, pp. 265–274, 2018.
https://doi.org/10.1007/978-3-319-75420-8_25

propose their own planning algorithm that is easily parallelizable. The inputs of the networks are the coordinates of the points of the path. The output neuron is described by the expression, which is called a repulsive penalty function and has a role of repulsive potential. The algorithm also solves the deadlock in a local minimum, however, simulations and tests are proposed for static environments only. Overview of other algorithms for path planning is given in [8]. Here are described algorithms covering the following areas: discrete planning, logic-based planning methods, motion planning, sampling-based motion planning, combinatorial motion planning, decision-theoretic planning, planning under sensing uncertainty, and planning under differential constraints.

On the other hand, path planning algorithms in a dynamic environment are more complex with regard to the amount of requirements placed on the resulting path, such as real-time requirement, obstacle prediction efficiency, etc. An overview of dynamic path planning algorithms used in robotics can be found in [10]. Authors' own work presents an algorithm capable of negotiating obstacles in static, partially dynamic and dynamic environments. The advantage of dynamic path planning algorithms lies in their ability to reuse information processed from previous searches to find a new solution faster. These algorithms include, for example, LPA*, Dynamic A* (D*) and later D* Lite, etc. Even though path planning algorithms have been developed for over two decades, there is still a space for improvement of existing solutions, either to accelerate the overall algorithm progress or to implement these algorithms in conjunction with the given hardware.

The research of path planning algorithms often focuses on finding the shortest path to reduce time for calculation, number of nodes, use as few sources as possible, etc. By focusing on these factors, a great deal of generality of design and use is guaranteed. This is well-founded, as path mapping algorithms are useful for various graph simulations, computer games, or robotic navigation. Areas of real and virtual applications are so different that they cannot be treated equally. Virtual applications have their boundaries and complexity defined, while real-world applications have to work not only with the degree of coincidence but also with a large increase in complexity with a number of factors that affect the calculation. Applications in the real world also work with the orientation of the path towards the current orientation of the robot. The path labeled as the shortest can take longer than the other way, which includes more number of nodes, depending on how many times the robot is forced to change the direction. Changing the direction affects the speed of the robot, especially in the case of axial movement, when it is almost forced to stop due to a change of direction. A general solution is thus necessary to adjust not only to find the shortest way, but also on finding the "easiest" way. For example, publications deal with this issue [4, 11], where it was found that the way found with the least number of directional changes was on average about 20% longer than the shortest route. Although the path planning area is considered to be explored, new hybrid algorithms based on existing algorithms are created, e.g. algorithms PRM, D*, D* Lite [3, 10] and existing approaches are optimized [2, 9]. Most of the proposed solutions are tested only in the form of program simulation, which are quite different from real-world use. Another relevant literature is rather focused on creating an overview, categorizing algorithms and comparing their effectiveness. From the conclusions of the current state analysis, we can say that the algorithms used for dynamic robotic path planning do not reflect the acceleration of robots but only reflect their instantaneous speed. They also do not define time requirements for way calculation and its update in the case of an obstacle, which in their use in the real world is a problem. It is usually calculated with zero speed when changing the direction of movement of the robot. The robot must stop, turn and accelerate at every change of direction, which is time consuming. This solution is presented in [9].

The research goal of this paper is to propose a path planning algorithm in a dynamically changing environment. It will be taken into account the requirement of real time processing in association with the movement of the agent. The theoretically proposed algorithm will be supported by computer simulation models.

2 Algorithm Proposal

The state of the art shows that many algorithms of dynamic pathplanning are based on the A* algorithm. This algorithm belongs to the category of algorithms working with visibility of graphs [4, 5, 9]. Therefore, this algorithm was also chosen as the basis for the pathfinding algorithm in a dynamically changing environment. To ensure a dynamic pathfinding, the criterion of finding the fastest route is chosen for the A* algorithm with respect to the maximum available speed. This speed is specific to each object and represents a changeable component for repeated searches.

The course of the algorithm A* is controlled by means of weights g_{cost} and h_{cost}. These weights are used to calculate the function value f_{cost} (1):

$$f_{cost} = (g_{cost} + h_{cost}) \tag{1}$$

where

g_{cost} indicates the distance from a particular node to the start node;
h_{cost} indicates the distance from a particular node to the stop node;
f_{cost} is the evaluation function.

2.1 Decreasing the Number of Turns

Decreasing the number of turns is one of many criteria of static and dynamic pathplanning. Paper [11] defines the path as "the simplest" in the case it contains a minimum number of direction changes. The core of the proposed algorithm for dynamic planning uses the principle of A* algorithm controlled by a heuristic function f_{cost} (1). The value of the heuristic function was added with a coefficient t_{cost}, which enables to prefer path alternatives with the lowest number of direction changes (2). This principle differs from implementation used in literature [11], where breadth-first-search is the subject.

$$f_{cost} = (g_{cost} + h_{cost} + t_{cost}) \tag{2}$$

Value t_{cost} in formula (2) is determined based on the information about the direction for each expanded node, which is recorded in a form of direction vector v. This vector is then used to calculate value f_{cost} successor by calculating the vector product of vector v with direction vector u of the currently expanded node (3). With this optimization, algorithm A* prefers nodes which do not change direction.

$$t_{cost} = \cos \alpha = \frac{u \cdot v}{|u| \cdot |v|} \tag{3}$$

2.2 Implementation of a Dynamic Part of Algorithm A*

The dynamic part of algorithm A* represented changeable speed of the vehicle. In real world, there are curves not influencing the speed of a vehicle. In such a case, it is necessary to consider physical and kinematic characteristics of a given vehicle which enable to identify such curves. The following modification of the algorithm focuses on this issue. In order to calculate values of function t_{cost}, parameters such as vehicle weight and speed as well as curve radius are used.

Weight. The influence of weight on the path trajectory can be imagined as a ball on a rope tied to a pole. If the ball is rotated around the axis created by the center of the pole, the ball is affected by a relatively small centripetal force. In the case that the ball is replaced by a bowling ball, its rotation needs a far more force to achieve the same speed. The same principle applies to the movement of an agent.

Speed. If the agent is controlled slowly in a circle, tires do not have problems with generating enough friction to keep it in a circular direction. However, if it speeds up, there is moment when the tires are not able to generate enough friction and skidding occurs. The maximum speed in a curve can be achieved using a general equation to calculate a centripetal force (4).

$$F_d = \frac{m \times v^2}{r} \tag{4}$$

where r is the curve radius, v is the agent speed, and m is its weight.

Curve radius. It influences agent movement similarly to its speed. If the curve radius decreases while the speed is the same, there is a certain point of tires skidding on road surface.

2.3 Speed Modification

Weight function t_{cost} ensures finding the fastest path by decreasing frequent speed drop. During the algorithm, the value of function t_{cost} is calculated and stored into individual nodes. The created map keep information on path surface, roughness, and radii which do not have to be calculated when finding a path, see Fig. 1.

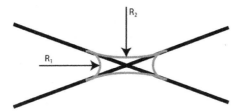

Fig. 1. Radii of non-linear paths

Map preprocessing is done during its uploading, before launching the algorithm respectively. During this process, the graph environment is supplied with data about all points of path crossings and individual curves. Calculating the value of function t_{cost} uses the relationship for centripetal force (4). The input value is the maximum vehicle speed v_{max}. During searching, all suitable alternatives of non-linear paths are tested as well as the maximum speed for radius r_{vmax}. The proposed algorithm AV* works in the following steps [6]:

1. For each expanded node:
   (a) Test if the node has two predecessors.
   (b) If yes, calculate the maximum speed in a curve r_{vmax}.
      (i) If $r_{vmax} < v_{max}$: set t_{cost} to $v_{max} - r_{vmax}$.
2. Calculate the value of the heuristic function (2).
3. Continue in algorithm A*.

3   Experimental Verification

The testing was realized on Lenovo IC 510S, 90GB007TCK. For the verification purposes, algorithm A* was compared with the proposed AV* in the following configurations:

1. Classical implementation of A* without any modifications or optimizations.
2. The proposed algorithm AV* (with modified weights to stem from the speed of the navigated agent (vehicle).
3. The proposed algorithm AV* with a pre-set speed. This algorithm is launched every time after the second iteration in order to analyze dynamic behavior of the proposed algorithm.

Input for each test is the graph environment containing nodes and edges. The memory stores the following information about it:

- Number of nodes and their positions.
- Number of paths and their node reference numbers.
- Friction coefficient for a given path.
- Maximum speeds for individual nodes stored in the course of searching.

Output of each test is:

- Number of nodes composing the found path.
- Length of the found path: real number in meters (m).
- Time to go through the found path: integers in seconds (s).
- Time of the calculation: integers in microseconds (μs).
- Average memory workload: integers in bits (bit).

Both tested algorithms A* and AV* are functionally incompatible. The same view cannot be applied on results from algorithm A*, which is focused on finding the shortest path. However, algorithm AV* is to find the shortest path in time. The basic version of algorithm A* is here only for purposes of evaluating the influence of the implemented modifications on the original algorithm. Each test is done fifty times and the stated results are averaged.

3.1   Test Task 1

Figure 2 depicts a testing environment with 25 nodes, where the start (blue point) and end (yellow point) are defined on nodes with several possible subsequent paths. This experiment verified that reducing the number of turns during searching can result in decreasing the time to go through the path. The experimental results are stated in Table 1.

Table 1. Experimental results for the 1st testing environment

Criterion	A* $v_{max} = 100$ km/h	AV* $v_{max} = 100$ km/h	A* $v_{max} = 20$ km/h	AV* $v_{max} = 20$ km/h
Number of nodes	8	5	8	8
Path length (m)	300	308	300	300
Time to do the path (s)	14	13	50	50
Time of calculation (μs)	1227	1744	1250	1432
Memory (bit)	1025	1501	990	1496

Both algorithms found a path and their results differed in all values, see Fig. 2. At $v_{max} = 100$ km/h, the path found by algorithm A* contains a higher number of nodes and is 8 m shorter. Considering the speed at which it can be done, its estimation is set to 14 s, which makes 1 s more than for the AV* algorithm. The characteristics of both paths is also significantly different. At $v_{max} = 20$ km/h, both algorithms found the same path.

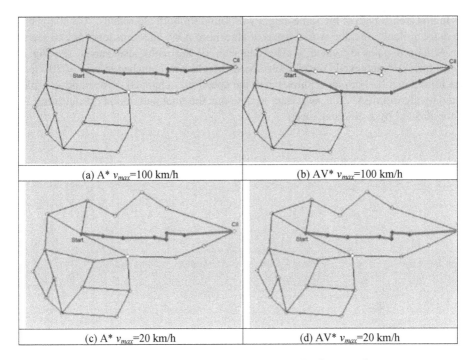

(a) A* v_{max}=100 km/h	(b) AV* v_{max}=100 km/h
(c) A* v_{max}=20 km/h	(d) AV* v_{max}=20 km/h

Fig. 2. Found paths for testing task 1. (Color figure online)

3.2 Test Task 2

Figure 3 depicts a testing environment with 129 nodes. This test uses a graph with a higher density of nodes. It primarily monitors saving in memory sources and using of previously acquired information. Experimental results are stated in Table 2. Decreasing the time to go through the path is a result of implementation of the maximum speed as a dynamic part.

Table 2. Experimental results for the 2nd testing environment

Criterion	A* $v_{max} = 100$ km/h	AV* $v_{max} = 100$ km/h	A* $v_{max} = 20$ km/h	AV* $v_{max} = 20$ km/h
Number of nodes	11	10	11	11
Path length (m)	380	391	380	380
Time to do the path (s)	19	17	64	64
Time of calculation (μs)	3356	16038	3412	11878
Memory (bit)	3341	4512	3410	4650

In this area, a path of the same length was found for both algorithms. As anticipated, the more difficult area, the worse speed of algorithm AV* due to its tendency to search non-penalized paths by value T. The character of the paths also varies, see Fig. 3. Algorithm AV* achieved considerable speed-up when re-launched at $v_{max} = 20$ km/h. The fond paths by both algorithms when the speed was changed correspond to a path found by algorithm A*. It is necessary to note that the final path found by algorithm A* is not affected by a different speed.

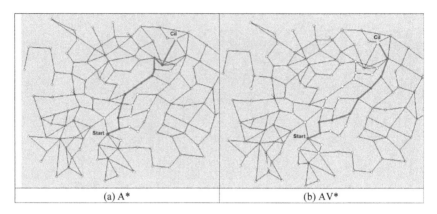

Fig. 3. Found paths for test task 2.

3.3   Test Task 3

Figure 4 depicts a testing environment with 129 nodes. This test also uses a graph with a higher density of nodes with a difference that the starting and final node are placed close to each other. The objective is to verify that the algorithms do not search though a high number of nodes. Experimental results are stated in Table 3. In this experiment, there were changes in the dynamic part of the algorithm, which modifies re-use of previously acquired information.

Table 3. Experimental results for the 3rd testing environment

Criterion	A* $v_{max} = 100$ km/h	AV* $v_{max} = 100$ km/h	A* $v_{max} = 20$ km/h	AV* $v_{max} = 20$ km/h
Number of nodes	14	15	14	14
Path length (m)	515	533	515	531
Time to do the path (s)	26	24	86	80
Time of calculation (µs)	13277	40286	13145	25456
Memory (bit)	3402	4863	3402	5143

Increasing distance between the start and finish in algorithm AV* resulted in multiplied time requirements for the calculation. The character of the found paths is different, see Fig. 4.

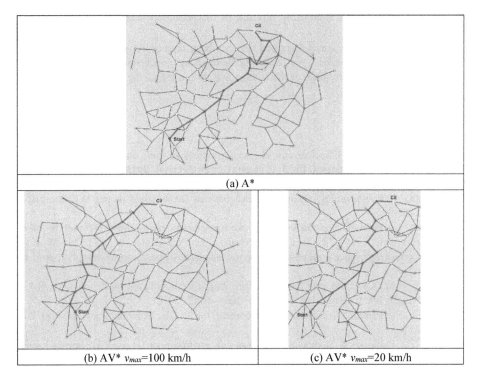

Fig. 4. Found paths for test task 3.

4   Conclusions

All tested algorithms succeeded in finding a path in all performed tests. The proposed algorithm AV* differs from algorithm A* by the amount of searched nodes due to repetitive search with a different value t_{cost}. The number of searched nodes depends on environment complexity and the position of the start and finish. There is a correlation between the number of found nodes and the character of the final path as the number of the found nodes serves for geographical orientation of the algorithm. It affect the algorithm run only from the point of view of memory and time of the calculation. On the contrary, the real length has direct influence on the character of the found path and it is crucial for both compared algorithms as it influences the choice of expansion in the next step. The time to go through the path represents the main criterion for algorithm AV* as it is proposed to its minimization. All tested algorithms succeeded in finding a path shorter in time.

Based on the results, it is possible to claim that the algorithm complies with the condition of finding a path between the start and finish in a finite time based on the criterion of the fastest alternative using the maximum speed limit possible. Considering dynamic optimization, all tests achieved a speed increase when the algorithm was repeated at a different speed. This speed-up corresponds to approximately 25% and it is linearly dependent on the increasing amount of the searched nodes.

Acknowledgments. The research described here has been financially supported by University of Ostrava grant SGS07/PřF/2017. Any opinions, findings and conclusions or recommendations expressed in this material are those of the authors and do not reflect the views of the sponsors.

References

1. Algfoor, Z.A., Sunar, M.S., Kolivand, H.: A comprehensive study on pathfinding techniques for robotics and video games. Int. J. Comput. Games Technol. **2015**, 11 (2015). Article ID 736138
2. El Khaili, M.: Path planning in a dynamic environment. Int. J. Adv. Comput. Sci. Appl. **1**(5), 86–92 (2014)
3. Gavrilut, I., Tiponut, V., Gacsádi, A.: An integrated environment for mobile robot navigation based on CNN images processing. In: Proceedings of the 11th WSEAS International Conference on SYSTEMS, Agios Nicolaos, Crete, Greece, vol. 5117, pp. 81–86 (2007)
4. Jiang, B., Liu, X.: Computing the fewest-turn map directions based on the connectivity of natural roads. Int. J. Geogr. Inf. Sci. **25**(7), 1069–1082 (2011)
5. El Khaili, M.: Planning in a dynamic environment. Int. J. Adv. Comput. Sci. Appl. **5**(8), 86–92 (2014)
6. Kosík, D.: Pathfinding in dynamically changing environment (in Czech). Master thesis, University of Ostrava, Czech Republic (2017)
7. Kroumov, V., Yu, J.: Neural networks based path planning and navigation of mobile robots. In: Recent Advances in Mobile Robotics, pp. 174–190. INTECH Open Access Publisher (2011)
8. Lavalle, S.M.: Planning Algorithms, 1st edn. Cambridge University Press, New York (2006)
9. Narayanan, V., Phillips, M., Likhachev, M.: Anytime safe interval path planning for dynamic environments. In: Proceedings of the 2012 IEEE/RSJ International Conference on Intelligent Robots and Systems (IROS), Vilamoura-Algarve, Portugal, pp. 4708–4715 (2012)
10. Raja, P., Pugazhenthi, S.: Path planning for a mobile robot in dynamic environments. Int. J. Phys. Sci. **6**(20), 4721–4731 (2011)
11. Zhou, Y., Wang, W., He, D., Wang, Z.: A fewest-turn-and-shortest path algorithm based on breadth-first search. Geo-Spatial Inf. Sci. **17**(4), 201–207 (2014)

Ontology Languages for Semantic Web from a Bit Higher Level of Generality

Martin Žáček(✉) , Alena Lukasová, and Marek Vajgl

Department of Computers and Informatics,
University of Ostrava, Ostrava, Czech Republic
{martin.zacek,alena.lukasova,marek.vajgl}@osu.cz

Abstract. An article tries to see a group of semantic web representation and communication means like description logics, semantic networks, RDF model and language, OWL language, RDF CFL etc. taking a part at the development of formal ontologies from a slightly higher (meta) level. A focus has been oriented towards properties that semantic web language ought to have to fulfill according to a semantic web idea. Consequently the article tries to draw by the help of the means above the trip towards a proper semantic-web ontology language fulfilling the main demands of expressivity and also easy-to-use conditions. Even if the development of a semantic web language was until now much more complicated with a taking part of a lot of people or companies the authors of the article try to contribute to an analyzing of the actual state by drawing it as a cloud of demands, streams, events or condition's fulfilling that at the end would have brought into the world an easy-to-usable semantic web ontology language having all the demands fulfilled. The article moreover shows that a language RDF CFL belongs nowadays to the semantic web family as a unique well usable tool of reasoning within semantic web knowledge bases.

Keywords: RDF model · Ontology · CFL · Clausal formal logic
Semantic web · Knowledge base · OWL language

1 Introduction

A vision of semantic web is an idea that in the near future of the web development information and knowledge would have its explicit meaning, making them usable not only for people but also for machines.

The development of a semantic web language with a taking part of a lot of people or companies was until now much more complicated that our article wants and could to capture. But the most important parts of the development form something like clouds of demands solving, streams or conditions and events that at the end would have brought into the world an easy-to-usable semantic web ontology language having fulfilled all the demands. A focus has been oriented before all towards properties that semantic web language ought to have to fulfil according to a semantic web idea and what kinds of formal means contribute to the fulfilling.

© Springer International Publishing AG, part of Springer Nature 2018
N. T. Nguyen et al. (Eds.): ACIIDS 2018, LNAI 10752, pp. 275–284, 2018.
https://doi.org/10.1007/978-3-319-75420-8_26

The article shows also a role of the language RDF CFL that has not been usually accompanied into the family of web languages, but nowadays it seems clear that the language serves a unique usable tool for reasoning within semantic web knowledge bases.

Formal tools (languages) at the Fig. 1 represented by graph nodes are here in the graph connected by oriented arcs expressing properties overtaken from the subject node to the object node. The graph shows also how semantics of information or knowledge has taken stepwise its dominant position within language syntax. At the beginning, it represents only a simple semantic network, but nowadays RDF [2] modelling principle with its clear and precise expression of a language meaning stays high in a scale of usability.

2   From a Natural Language Description Towards a Formal Domain Ontology

The concept of ontology has been derived from the Greek as a branch of metaphysics. It means ontology to be a study about which entities are there in the universe. Artificial intelligence (AI) sees the domain ontology as a framework fulfilling a role of a working model for entities, their hierarchical ordering and their interactions in some of particular domain of discourse. The way to general principles of building domain ontologies we can see today as a composition of various directions each of them preferred its own goals or preference. Nevertheless they have formed together a unique stream with many interesting branches like it is shown in the graph at the Fig. 1. The graph shows in a form of a associative network [3] formal languages, natural language, paradigms and linguistic viewpoints systems (graph nodes) in mutual relationships represented by arcs.

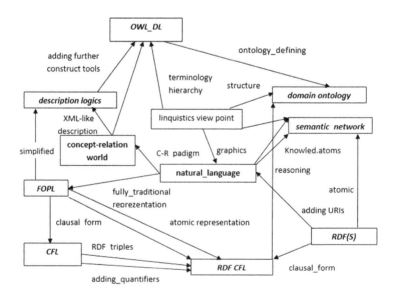

Fig. 1. Formal tools taking a part in a building of semantic web ontology language.

2.1 First Order Predicate Logics

At the beginning the full description of all the important facts in the language of first order predicate logics (FOPL) within a domain have been taken by specialists of AI as a best correct picture with a maximal expressivity and factual reliability of representation a domain in question. Both together, standard syntax and the model-theory (MT) semantics of the FOPL language have served to a (people) user complete information. But computers could have got a meaning of information only with a help of IT experts. The full traditional knowledge representation in the FOPL language has brought it's contribute to the goal by the high quality of expressivity and possibilities of precise descriptions. Later new languages could to be compared with the FOPL as with a symbol of quality.

It has become clear that some simplification of the knowledge representation process has to take a place. But at first it has been a necessity to begin at the conceptual level of knowledge representation with a new paradigm of seeing a domain to be described. It has to be the concept – relationship (C–R) paradigm that makes it possible to manipulate formally only with sets of entities not with singular entities themselves like in the FOPL

2.2 C–R Modelling Paradigm and Description Logics

Idea of semantic web adds to the human recognizing the world a new demand: besides of human readers and users it must be recognized and used knowledge and information in the web also by computers. The demand lays also a new condition on a web language: to be readable and recognizable to people as well as to computers. Solving the problem has been found in a new style of modelling by means of RDF model [5] supported on the C–R paradigm.

In description logic an ontology developer defines important notions of domain ontologies, especially classes or relations of them in terms of concepts and roles. The concepts of the DL language are represented by unary and roles by binary predicates. Further constructors in a quantified form give in the DL possibilities to restrict concepts or roles to their various specifications. Some incorrectness in the DL definitions [1] has brought a high expectation in general usability and simplifications until a believing that a decidable version of the simplified FOPL language has been found (see [6, 9, 11]).

At the beginning of formal modeling a conceptualization generally begins with a special seeing a structure of a domain by a special paradigm. In the case of description logics (DL) s the concept – relation C–R paradigm probably at the first time has taken its leading place. Until now there has been only a natural demand to have a simplified language with sharing a high expressivity like in the FOPL. Both demands have lead towards the concept-relationship construction of description logics as a simplified version of the FOPL. It has appeared there that a demand of building a new language on the base of atomic elements recognized also by human minds has been fulfilled.

2.3   Associative (Semantic) Networks

A bit more semantics [3] into syntactic forms of representation (for example in the FOPL) have brought some specialists in English linguistic [4] as authors of an idea of semantic networks.

Semantic networks in linguistics has been used for a support of (people, not computers) better understanding of information content. Graphic form [12] expressing for example a simple sentence like "cat is a beast and catches the mouse" is shown at the Fig. 2.

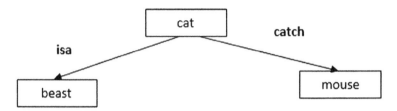

Fig. 2. The semantic networks

The possibility of a graphic drawing of knowledge has brought a new direction of portioning information into bags of linked atomic statements according real connections of their meanings. But it has also come a time to solve a further problem – adding a correct meaning to a concrete knowledge direct into the language of formal representation.

3   RDF Representation in Web Languages

RDF model [2] follows the approach of semantic networks - seeing and writing statements about the world by rewriting the arguments from natural language into a graphic easy-to-understand form and also in a form readable by computers. Taking the world of discourse under the general C–R paradigm led the formal representation towards a crucial simplified form of elementary statement about the world. So, ontology [10] of the world to be represented by formal means has been built as a set of concepts and their mutual relations similar as it is recognized in human minds.

3.1   Triples and URIs in the RDF

Natural language is indeed the suitable for communication between people, but not suitable for processing by computers. In order to translate a statement into such a suitable form minimal two conditions are to be fulfilled:

- to introduce a system of unique identifiers for the subject, predicate and object, which could become an automated process,
- to build a machine processing capable language suitable for writing statements and mutual exchange them between computers.

As the tool of mutual communication or as the tool of general data interchange on the web a standard conceptual model RDF and its communication language has been developed (W3C document).

Information about an entity is in RDF model expressed by means of its subject - predicate - object triple statement with URIs pointed to their resources. RDF identifies each entity on the web using URIs leading towards a resource explaining properties of the entity by means of a document at the web. RDF provides a way to write simple statements about resources, using named properties and values. This approach describes in the form of a single fact a concrete meaning of an entity described and used within the web.

Originaly RDF triples was designed mainly to provide metadata of web documents. For example the statement:

title of the data_x is "Using RDF triple statements",

contains the three parts. (see Table 1).

Comparing with the description logics language RDF needs only binary predicate representation of all the elementary atoms of the language. For example a representation of the concept "red" by a unary predicate $red(x)$ must be transformed to binary predicate like $isa(x, red)$ at the RDF representation.

So **red**(ball) in DL has its corresponding expression for example **isa**(ball, red)

RDF in general is a method of conceptual data modelling. Generally URI's are used for the subject and predicate. The subject denotes the resource and the predicate expresses the relationship between the subject and the object. The object is pointed either as another URI or it is a literal such as a number or a string.

RDF as well as DL provides a tool to define new vocabularies, which could be used in practical tasks. "Domain Ontology" means the special vocabulary of the domain in question.

Now, RDF triple can be used to publish information on the various resources in the web. Moreover it is applicable to provide also information about objects that are on the web somehow identifiable, but not through the web directly accessible. RDF extends the linking structure of the web to use URIs to name the relationship between things. It has an ability to combine data even if the underlying schemas differ. Using this simple model, it allows structured and semi-structured data to be mixed, exposed, and shared across different applications. A resource (source) can be considered all entities identifiable on the web, for example persons or companies or specific parts of a document.

The linking structure of RDF statements as well as in the case of semantic networks gives a possibility to form a directed, labeled graph, where the edges represent the named link between two resources, represented by the graph nodes.

The graph view is the easiest possible mental model for RDF and serves a means for natural language understanding. It is often used in easy-to-understand visual explanations (Fig. 3).

Adding quantifiers similarly as in CFL language [8] makes RDF capable to express general and existential formulas that are usually necessary components of knowledge bases for reasoning of further knowledge.

title

Fig. 3. The semantic networks of example.

Besides the graphic representation is RDF usually based on XML syntax (RDF/XML).

3.2 Making RDF Expressivity Higher by Means of Special Constructors

RDF has only a capability of building an instance of a class. In the frame of RDF vocabulary is there the "rdf:type" term used to specify descriptions of instances of the class in subject.

Xclass   rdf:type teacher.

RDF provides a set of basic terms of a standard RDF vocabulary that have the purpose to be generally reused for describing data. Now it is possible to add descriptions to the data.

3.3 Making RDF Expressivity Higher by Means of RDFSchema

The basic RDF [5] does not provide means for the definition of new classes and properties. For this purpose is here an extension RDF Schema (RDFS). So, RDFS is a semantic extension of RDF.

RDFS has terms for creating classes. The vocabulary RDFS gives a possibility of subject specification by means of rdfs:subClassOf constructor:

teacher   rdfs:subClassOf employee

Graph at the Fig. 1 is a semantic network with labeled nodes by formal languages or approaches taking a part in a whole evolution process of domain ontology building. A label of the graph arc represents in a RDF triple a predicate item expressing adding a property or making an action by the subject to the object of the triple.

Table 1. RDF triples (without URIs) corresponding to the graph at the semantic network of the Fig. 1

subject	predicate	object
data_x	**title**	"Using RDF triples statements..."
linguistic_theory	**using_terminology_hierarchy**	natural language
FOPL	**clausal form**	CFL
natural_language	**concept-relation (CR) paradigm**	description_logics
...	...	...

The graph could become a real RDF graph if completed with an explanation of RDF triples (see Table 1) and corresponding URIs of each subject, predicate and object for each the RDF triple.

4 RDF CFL Knowledge Bases in Ontologies

4.1 Necessity of RDF Model with Quantifiers

The language RDF CFL uses both properties together, RDF triples knowledge atoms and clausal forms of formulas easy to usable in inferring logical consequents as results of resolution reasoning [7].

Adding quantifiers similarly as in CFL language [8] makes RDF as well as RDF CFL capable to express general and existential formulas that are usually necessary components of knowledge bases for inference of further knowledge.

4.2 From CFL to RDF CFL

The author of the CFL language [8] introduced a special form and notation for predicate clauses. He also modified the resolution rule for using in clausal forms. The same uses RDF CFL but with a further restriction: Atoms within clauses must be RDF triples represented in the form of simple binary predicate formulas. It means not only a necessity of a transformation of predicate formulas into clausal form but moreover the rewriting all the n-ary predicates, $n \neq 2$, into compositions of binary predicates with a corresponding meaning.

RDF CFL uses the well-known resolution rule to obtain logical consequents from a knowledge base. The rule holds at the original RDF CFL formal system at the text version as well as in the graph-based RDF CFL [7].

Resolution inference rule works by the help of two partial rules - the substitution rule and the cut rule.

RDF CFL Cut rule with a special notation for the graph version:

If the knowledge base designed for derivation contains a pair of clauses sharing the same atom (vector), one in the antecedent (with a dashed line in the graph of the atom), one in the consequent (with a full line in the graph of the atom), then, as a result a new clause is derived from this pair of clauses. This clause cuts the shared atom and connects it together (drawing one common network) with the remaining atoms of their antecedents (consequents) by conjunction (disjunction) connectivity.

The substitution rule for both versions says that a uniform substitution of a term of the language into a clause to the position of a universal variable returns a new holding clause.

4.3 Special Axioms in Domain Ontologies

Using the "if – then" clausal form notation gives as well as in the CFL a possibility of using the resolution reasoning rule for deriving logical consequents of a set of special axioms holding in the frame of the corresponding domain ontology.

5   OWL Ontology Language for Semantic Web

More complex descriptions of web entities then RDF(S) serves using OWL, as another vocabulary that provides a further set of terms. By using both of them it is possible to start making more detailed descriptions of data.

OWL is a language for expressing ontologies as sets of precise descriptive statements about some part of the world (domain of interest). In order to precisely describe a domain of interest, it is necessary to create a terminology of the domain - a set of central terms and fix their meanings. The meaning of a term can be characterized by stating how this term is interrelated to the other terms. A terminology, providing a vocabulary together with such interrelation information constitutes an essential part of a typical OWL document.

6   Reasoning within Semantic Web

6.1   Reasoning on the Base of Description Logics

The investigation of algorithms for reasoning services and their complexity was the main focus of the DL research community in the last decades. Despite the high complexity, highly optimized DL reasoning systems were implemented especially those based on the tableau method published in [8] (FACT or RACER systems) corresponding to that one of first order predicate logics.

Experience with the reasoning by means of the tableau algorithm based on the so called decidable DL formalism defined in [8] has shown some incorrectness as a consequent of the unprecise defined language semantics. This has been the reason of a new definition and presentation of the description logics DL1 [8]. It has been shown the correspondence of the DL1 with the partially decidable FOPL, namely in its semantic correctness and completeness.

6.2   Properties that Ought to Share an Ontology Language

To have a high level tool of knowledge representation and manipulation on the Web site it is important to have a formal language (together with a formal system) on the base of the first order predicate logic (FOPL) that

- uses a relative simple language syntax, allowing a machine readability and handling,
- is shared by a wide community,
- has a property, knowledge represented in the language be placed into some external contexts, in order to capture better its semantics,
- a language of knowledge representation should be a web language that ought to follow the RDF(S) (Resource Description Framework (Schema)) model.
- The formal language with the characteristics above naturally ought to provide a mechanism of automated deduction in the frame of semantically correct and complete formal system.

Recently all the requirements listed above in the computer science have been working partly in some way and have brought many good results.

7 Conclusion

The Web Ontology Language OWL extends RDF and RDFS, so it fulfils the first four conditions of semantic web language listed in the paragraph 6.2. Its primary aim is to bring the expressive and reasoning power of description logic to the semantic web. The first level above RDF required for the semantic web is an ontology language what can formally describe the meaning of terminology used in web documents. OWL language fulfils this demand. It is a mark-up language designed for creating ontologies that extends RDFS vocabulary on elements related to classes and their properties. OWL contains a broad set of elements through which it is possible to create ontologies from the simplest to the most sophisticated structures built on an unlimited use of RDF constructs.

Reasoning in OWL belongs to those problems until now waiting to a corrected solving. Tableau algorithm built on incorrect definition of universal quantification semantics in DL could return false decision if used its incorrectly obtained negation. Alternative tableau algorithm in DL1 is a simple copy of that one of the FOPL and does not belong to easy to usable methods.

RDF CFL fulfils all the conditions discussed above. Using the "if – then" clausal form is a natural way of expressing knowledge in its connection with a world around similarly like in human minds. The notation of the RDF CFL uses the well-known resolution rule to obtain logical consequents from a set of special axioms of a domain in question that holds at the original RDF CFL formal system at the text version as well as in the graph-based RDF CFL.

Generally, ontologies play an important role in many fields and applications or knowledge systems. In the frame of fulfilling the idea of semantic web a variety of new applications based on ontologies has been and will be created.

Acknowledgments. The research described here has been financially supported by University of Ostrava grant SGS10/PřF/2017. Any opinions, findings and conclusions or recommendations expressed in this material are those of the authors and do not reflect the views of the sponsors.

References

1. Baader, F. (ed.): The Description Logic Handbook: Theory, Implementation, and Applications. Cambridge University Press, New York (2003). ISBN 0-521-78176-0
2. Hjelm, J.: Creating the Semantic Web with RDF: Professional. Wiley, New York (2001)
3. Lukasová, A.: Knowledge representation in associative networks. In: Proceedings of the Znalosti (2001). (in Czech)
4. Lukasová, A., Žáček, M.: English grammatical rules representation by a meta-language based on RDF model and predicate clausal form. Information **19**, 4009–4015 (2016). ISSN 1343-4500
5. Lukasová, A., Vajgl, M., Žáček, M.: Knowledge represented using RDF semantic network in the concept of semantic web. In: International Conference of Numerical Analysis and Applied Mathematics 2015, ICNAAM 2015: Proceedings of the International Conference on Numerical Analysis and Applied Mathematics 2015 (ICNAAM-2015), 23 September 2015. American Institute of Physics Inc., Rhodes (2016). https://doi.org/10.1063/1.4951895. ISBN 978-0-7354-1392-4

6. Lukasová, A.: Formal Logics in Artificial Intelligence (Formální logika v umělé inteligenci). Computer Press (2003)
7. Lukasová, A., Žáček, M., Vajgl, M.: Carstairs-McCarthy's morphological rules of English language in RDFCFL graphs. In: Ntalianis, K., Croitoru, A. (eds.) APSAC 2017. LNEE, vol. 428, pp. 169–174. Springer, Cham (2018). https://doi.org/10.1007/978-3-319-53934-8_20
8. Richards, T.: Clausal Form Logic. An Introduction to the Logic of Computer Reasoning. Addison – Wesley, Boston (1989)
9. Vajgl, M., Lukasová, A., Žáček, M.: Knowledge bases built on web languages from the point of view of predicate logics. In: 1st International Conference on Applied Mathematics and Computer Science, ICAMCS 2017: AIP Conference Proceedings 19 January 2017. American Institute of Physics Inc., Rome (2017). https://doi.org/10.1063/1.4981998. ISBN 978-073541506-5
10. Žáček, M.: Ontology or formal ontology. In: International Conference of Numerical Analysis and Applied Mathematics 2016, ICNAAM 2016: AIP Conference Proceedings 19 September 2016. American Institute of Physics Inc., Rhodes (2017). https://doi.org/10.1063/1.4992234. ISBN 978-073541538-6
11. Žáček, M., Lukasová, A., Miarka, R.: Modeling knowledge base and derivation without predefined structure by Graph-based Clausal Form Logic. In: The 2013 International Conference on Advanced ICT (Information and Communication Technology) for Education: Proceedings of the 2013 International Conference on Advanced ICT and Education 20 September 2013, Sanya, China, AISR 2013, pp. 546–549. Atlantis Press, France (2013). ISBN 978-90786-77-79-6. WOS 000327670900110
12. Žáček, M., Miarka, R., Sýkora, O.: Visualization of semantic data. In: Silhavy, R., Senkerik, R., Oplatkova, Z.K., Prokopova, Z., Silhavy, P. (eds.) Artificial Intelligence Perspectives and Applications. AISC, vol. 347, pp. 277–285. Springer, Cham (2015). https://doi.org/10.1007/978-3-319-18476-0_28

Intelligent Applications of Internet of Thing and Data Analysis Technologies

A New Method for Establishing and Managing Group Key Against Network Attacks

Dao Truong Nguyen$^{(\boxtimes)}$ and My Tu Le

Academy of Cryptography Techniques, Hanoi, Vietnam
truongnguyendao@gmail.com, tulemy@hotmail.com

Abstract. The one-way function tree is an efficient group key management scheme. Many methods have been proposed to improve it against different types of internal attack. However, these previous works have not considered external attacks such as man in the middle. This is a very dangerous attack because it can intervene in the key establishment and key agreement process. This paper proposes a new method to prevent man in the middle attack by combining the OFT scheme with digital signatures scheme which is used to authenticate the participants. Beside, the proposed method can also prevent the collusion attack.

Keywords: One-way function tree (OFT) · Replay attack
Collusion attack · Man in the middle (MITM)

1 Introduction

One-way function tree (OFT) [3,4] is a centralized group key management scheme. It is based on the Logical Key Hierarchy (LKH) scheme [5,6] and works as a key managing structure. It adds one-way functions to the LHK. The total communication overhead of the group manager when a member is eliminated is reduced to $log_2 n$, where n is the total number of members in the group. It is reduced by a half comparing to LKH.

Beside, OFT has some vulnerabilities. First is the vulnerability under a sophisticated membership-evolving attack, such as collusion attack that Horng pointed out [1]. Next is the vulnerability under an external attack and man in the middle attack is an example [2]. In order to improve OFT scheme's invulnerability, Ku and Chen researched these attacks and proposed other collusion attack methods [7]. Xu et al. also showed the ability to attack between two members of the system [8]. However, these researches and proposed improvements only focused on internal attacks or collusion attack in specific despite the fact that these networks are increasingly exposed to public external networks.

In this paper, we propose a security enhanced solution for the OFT scheme by using digital signatures [13] in establishing, managing group key in order to assure security against external attacks like MITM attacks [2].

© Springer International Publishing AG, part of Springer Nature 2018
N. T. Nguyen et al. (Eds.): ACIIDS 2018, LNAI 10752, pp. 287–296, 2018.
https://doi.org/10.1007/978-3-319-75420-8_27

2   Related Works

2.1   Previous Works

In their research [7], Ku and Chen showed collusion attacks based on OFT scheme's vulnerabilities that Horng pointed out in [6]. Then, these authors improved the OFT scheme. In their improved schemes, they updated every key on the path from evicted node to the root node. However, total time cost for that action is approximately $h^2 + h$, where h is the height of the key tree. In order to reduce communication overhead of group manager, Xu et al. [8] studied the vulnerabilities to collusion attacks between two arbitrary members. They discovered that with two arbitrary members, it is not always able to find unknown key's information. Therefore, group manager does not need to update all secret keys every time a member is eliminated. In [9], Liu and Yang synthesized previous research's results and analyzed Xu's main idea. Liu gave an anti-example to prove that Xu's conditions for collusion attack are not really necessary. Afterward, Liu proposed to improve OFT to HOFT (self-homomorphic one-way function tree). The new scheme is proved to be immune to collusion attack by Liu. In [10], the authors developed OFT into two schemes, ROFT (Repeated one-way function tree) and NOFT (Node one-way function tree) with minimum cost increased.

However, all these previous works only focused on prevent collusion attacks or in another word, internal attacks. In today connected Internet of Things (IoT) world, prevention against internal threats only is not enough. The key management scheme must be immune to man in the middle attack [2]. Therefore, in this paper we proposed to improve the OFT scheme invulnerability against these attacks. Beside, our proposed scheme is also effective against internal attack and meet system's attacks immunity requirements.

2.2   The OFT Scheme

The OFT scheme is a group key management scheme, proposed by Sherman and his colleages [3]. It is based on combination of Logical Key Hierarchy (LKH) scheme and using one-way function in key management [4,5]. In OFT scheme, there is a centralized group manager who is in charge of updating, storing, and distributing keys. OFT scheme's management structure is a binary key tree. Each internal node i in the tree contains: a node key k_i, a blinded node key y_i, where blinded node key y_i is output of one-way function that takes node key k_i as input. The blinded node key y_i is used to compute the node keys of the upper nodes in the key tree. Key of the root is the group key. The blinded node key y_i is calculated by formula $y_i = g(k_i)$, which $g()$ is a one-way function. k_i is used to encrypt the updated key's information when re-keying. Every member of the group stores the blinded node key of the siblings of the nodes in the path from its leaf node to root. Therefore, every members can use its leaf node's key and blinded node key to compute other node's keys in the path from bottom up.

In Fig. 1, node i is an internal node of the key tree. The left children tree is $2i$, the right children tree is $2i+1$. Corresponding to each node i is the sub-group

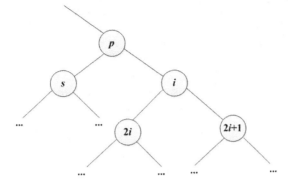

Fig. 1. The tree structure of one way function key

G_i, in which consists of m_i members. Members in the sub-tree (sub-group) G_i can compute the sub-group key at node i by formula $k_i = f(g(k_{2i}), g(k_{2i+1}))$, where f, g are cryptographic functions that immune to cryptographic attacks.

In this paper, we use a one-way function, g, and a trapdoor one-way function, f, that symmetry over g (which mean $f : K \times K \longrightarrow K, f(k_1, g(k_2)) = f(g(k_1), k_2)$, where K is the key space). They can calculate y_i as following: $y_i = g(k_i)$. These members store y_s of node s (sibling of node i). Thus, they can compute the sub-group key at node p (parent node of node i) by formula $k_p = f(y_s, y_i)$. Similarly, every member can calculate all the node key (sub-group key) in the path from that node to the root and obtain the root key which is the group key. Updating (adding, evicting) group's members will adjust the tree's structure (demonstrated in Sect. 3) and update corresponding key (described in Sect. 4).

3   Binary Tree Structure for Group

3.1   Building Binary Tree for Group

The binary tree's structure of group G which consists of n members is built based on following rules:

1. Subgroup $G_{0,1}$ is subgroup G with height 0, noted as $G_{0,1}(0, n)$.
2. If subgroup $G_{h,j}(h, m_{h,j})$ with height h has more than 1 member $(m_{h,j} > 1)$, then this subgroup is divided into two $h+1$ height subgroups, with the number of left and right members are $m_{h+1,2 \times j-1}$ and $m_{h+1,2 \times j}$ respectively. Where j is the jth subgroup from left to right that has same height.

In group binary tree's structure of group G, if leaf nodes are noted as $G_{h,j}(h, m_{h,j}), m_{h,j} = 1, (1 \le j \le n)$ and other nodes noted as $G_{i,j}(i, m_{i,j}), (0 \le i \le h-1, 1 \le j \le 2^i)$ then the total number of members in group is computed by formula $m_{i,j} = m_{i+1,2 \times j-1} + m_{i+1,2 \times j}, (0 \le i \le h-1, 1 \le j \le 2^i)$.

For example, the group has n members, denoted as $G = \{A_1, A_2, \ldots, A_n\}$. At each node $I = (h_I, m)$, of the tree, where h_I is the height of node I, m is the number of members in group. The set of members which belong to subgroup at node I is $S = \{A_1, A_2, \ldots, A_m\}$. The structure of group's binary tree is illustrated in Fig. 2.

The height of node i, h_i is the total number of nodes in the path from root to that node. In Fig. 2, $h_1, h_2, h_3, h_4, h_5, h_6, h_7, h_8$ are all equals to 3, where $h_1, \ldots, h_8$ are heights corresponding to members $A_1, \ldots, A_8$ respectively. Therefore, the height of the binary tree $h(G)$ or h is $h = \max\{h_i\}$, $i = 1 \ldots n$, $\log_2 n \le h \le n$.

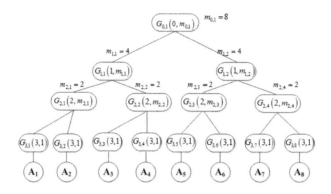

Fig. 2. Binary tree for group G with 8 members

Algorithm 1. Building tree structure for group G

Input: there are n members
Output: Binary tree T for group G.

```
Tree GroupTreeBuid()
{ Int n1= ⌊log₂n⌋;
Int n2 = n − n1;
Tree T1= BuidBinaryTree(n1);
Tree T2 = BuidBinaryTree(n2);
Return CombineTree(T1, T2); }
```

3.2   Adding New Member to Group

When a new member is added to the group, the group manager will choose a leaf node, j, which is nearest to the root. Node j will be modified, the existing member of node j will be moved to node $2j$ (left children of j), the new member will become node $2j + 1$ (right children of j). Algorithm 2 demonstrates how to restructure the group.

3.3 Evict a Member from Group

When a member attached to leaf node, j, is evicted from group, if node s (sibling of node j) is also a leaf node then the leaf node s will be moved to the node p (which is parent node of j and s). If node s is the root of another sub tree (sub group), s will be moved to node p to form a new sub tree that has root at node s (closer to the root). The rebuilding group's structure is followed the Algorithm 3.

Algorithm 2. Rebuild the group binary tree when a new member is added

---

Input: Group management tree T, new member A_{n+1}
Output: Updated group management tree T.

Tree ***AddMemtoTree***(T, A_{n+1})
$\{\ x = \text{SelectLeafNode}(T);$
$x_{old} = \text{Member}(x);$
$(x_{left}, x_{right}) = \text{Split}(x);$
$x_{left} \longleftarrow x_{old};\ x_{right} \longleftarrow A_{n+1};$
Return $(T);\}$

Algorithm 3. Rebuild the group binary tree when a member is evicted

---

Input: OFT tree T, member A_j;
Output: Updated OFT tree$T$;

Tree RemoveMemfromTree(T, A_j)
$\{\ y = \text{LeafNode}(A_j);$
$p = \text{ParentNode}(y);$
$s = \text{siblingNode}(y);$
if s is LeafNode then Member$(p) \longleftarrow$ Member(s)
Else $p = s$;
Return $T;\}$

---

4   Proposed Scheme

In this section we use some symbols $sign()$ as signature and $ver()$ as signature verification function, $E_K()$ as encryption function by key K.

4.1 Establishing Key Between Two Members

Suppose that two members A and B exchange information with each other, the key will be established as demonstrate in Algorithm 4.

Algorithm 4. Establishing secure key between two members.

Phase	A(*Sender*)	B(*Receiver*)
Phase I	$k_A = random(); y_A = g(k_A);$ $s_A = sign_A(y_A);$ **A** send (y_A, s_A) to **B**	$k_B = random(); y_B = g(k_B);$ $s_B = sign_B(y_B);$ **B** send (y_B, s_B) to **A**
Phase II	$if(ver(s_B, y_B) = true)then$ $k_{AB} = f(g(k_A), g(k_B))$	$if(ver(s_A, y_A) = true)then$ $k_{AB} = f(g(k_B), g(k_A))$

4.2 Establishing Key Between Two Groups

The establishing group key k for group G has been discussed in Sect. 3. In many previously proposed OFT schemes [3, 4, 7–12], there is no authentication in establishing key process. Today, due to the needs of connecting to outside networks, it is very necessary to authenticate the group members in order to avoid handshakes attacks such as MITM. Our scheme proposed to include the authentication by using digital signatures [13]. Detailed of establishing secure group key is demonstrated in Algorithm 5. In this algorithm, A_1, B_1 are group manager of group **A** and group **B** respectively. **A** has group key k_A and **B** has group key k_B.

Algorithm 5. Establishing secure key between two groups.

Phase	A(*Left SubGroup*)	B(*Right SubGroup*)
Phase I	A_1 (Left subgroup manager) $y_A = g(k_A); s_A = sign_{A_1}(y_A);$ A_1 send (y_A, s_A) to B_1	B_1 (Right subgroup manager) $y_B = g(k_B); s_B = sign_{B_1}(y_B);$ B_1 send (y_B, s_B) to A_1
Phase II	$A_i(i = 1, \ldots, m)$ (members in Group) $if(ver(s_B, y_B) = true)then$ $k_{AB} = f(k_A, y_B)$	$B_j(j = 1, \ldots k)$ (members in Group) $if(ver(s_A, y_A) = true)then$ $k_{AB} = f(k_B, y_A)$

4.3 Updating Key When a New Member is Added to Group

The updating key process when a new member is added to group is executed in two periods:

1. First period (old key is still valid): The most important thing of the key establishment and management protocol is to maintain normal operation of the system, all current members must operate normally with their distributed key. The key establishing process when a new member joins to the group is executed as Algorithm 6. The new one will register his signature with group manager. So, if faulty member wants to join into the group, he will not be authenticated and then he will be rejected by Algorithm 6.
2. Second period (old key is expired): After the old group key expired, if there is a new member added then two tasks must be done: First is reconstructing the group management tree as *Rebuild the group binary tree when a new member joins to group* (Sect. 3). Second is re-establishing the group key as Algorithm 7. Figure 3 describes an example that simulates the updating key process when a new member is added to group.

Algorithm 6. Updating key process when a new member is added and the old group key is still valid.

Phase	G_{old}(**Old Group**)	B(**New member**)
Initiation	G_{old} have the old group key, $k_{G_{old}}$	B has a new key, k_B
Phase I	A_1 (Old group manager)	B (New member)
	$y_{G_{old}} = g(k_{G_{old}}); s_{G_{old}} = sign_{A_1}(y_{G_{old}});$ A_1 send $(y_{G_{old}}, s_{G_{old}})$ to B	$y_B = g(k_B); s_B = sign_B(y_B);$ B send (y_B, s_B) to A_1
Phase II	A_1	B
	$if(ver(s_B, y_B) = true)then$ $k_{G_{new}} = f(k_{G_{old}}, y_B)$	$if(ver(s_{G_{old}}, y_{G_{old}}) = true)then$ $k_{G_{new}} = f(k_B, y_{G_{old}})$

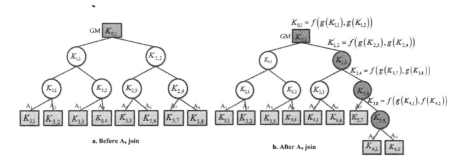

Fig. 3. Updating key when a new member is added

Algorithm 7. Updating key process when a new member is added and the old group key is expired.

Input: OFT T, new member A_{n+1}.
Output: Updated OFT T.

1. $x = \text{SelectLeaftoAdd}(T); K_{i,j} = Key(x);$
2. $\text{OldMember} = \text{Member}(x); (a, b) = \text{Split}(x); a = \text{OldMember}; b = A_{n+1};$
3. $k_b = \text{random}(); K'_{i,j} = f(g(k_a), g(k_b)); (K'_{3,8}$ in Fig. 3)
4. Group manager (GM) encrypt new key $K'_{i,j}$ with old key $K_{i,j}$ then send unicast to A_n: $GM \longrightarrow A_n : \{ E_{K_{i,j}}(K'_{i,j})\}$
5. While parentNode(b) is not NULL do
$\{ p = \text{parentNode}(b); k_p = f(g(k_{p_{left}}), g(k_{p_{right}})) ; Node(b) = Node(p); \}$
In Fig. 3 includes: $K'_{3,8} = f(g(K_{4,1}), g(K_{4,2})); K'_{2,4} = f(g(K_{3,7}), g(K'_{3,8})); K'_{1,2} = f(g(K_{2,3}), g(K'_{2,4})); K'_{0,1} = f(g(K_{1,1}), g(K'_{1,2})).$
6. GM encrypt these new keys with the corresponding old key then send multicast the updated key to other corresponding members, then send unicast to new node $GM \rightrightarrows \{A_m\} :\{E_{K_{i,j}}(K'_{i,j}, ...)\}; GM \longrightarrow \{A_{n+1}\} : \{E_{K_{A_{n+1}}}(K'_{i,j}, ...)\}.$
7. GM sends reminding messages to all sibling members of that node in the path from new node to the root. The message reminds them that there is a new member the group and they must update all their keys.

8. *GM* sends unicast the parent node key of the new node to its siblings. As illustrated in Fig. 3: $GM \longrightarrow \{A_8\} : \{E_{K_{A_8}}(K'_{0,1}, K'_{1,2}, K'_{2,4}, K'_{3,8})\}$

4.4   Updating Key When a Member is Evicted from Group

The updating key process when a member is evicted from group is executed as Algorithm 8. An example simulates the updating key process when a member is evicted from group is described in Fig. 4.

Algorithm 8. Updating key when a member is evicted from group.

Input: OFT T, Evicted member A_j.
Output: Updated OFT T.

---

1. Rebuild the group management tree as Algorithm 3.
2. Group manager (GM): A_s =sibling(A_j); K'_{A_s} = random(); (In Fig. 4, $K'_{2,4}$ is assigned to A_7.)
3. b = Node(A_s);
While parentNode(b) is not NULL do
$\{p = \text{parentNode}(b); k_p = f(g(k_{p_{left}}), g(k_{p_{right}})); b = p; \}$. In Fig. 4, $K'_{1,2} = f(g(K_{2,3}), g(K'_{2,4})); K'_{0,1} = f(g(K_{1,1}), g(K'_{1,2}))$.
4. *GM* encrypts these new keys by the corresponding old key then sends multicast the updated key to other corresponding member before sending unicast to the node that just moved to parent node: $GM \rightrightarrows \{A\} : \{E_{K_{i,j}}(K'_{i,j}, \ldots)\}$; $GM \longrightarrow \{A_s\} : \{E_{K_s}(K'_{i,j}, \ldots)\}$
5. *GM* sends reminding messages to all sibling members of that node in the path from new node to the root. The message reminds them that there is an evicted member and they must update all their keys.
6. *GM* sends unicast the parent node key of the new node to its siblings. As illustrated in Fig. 4: $GM \longrightarrow \{A_7\}:\{E_{K_{A_7}}(K'_{0,1}, K'_{1,2})\}$.

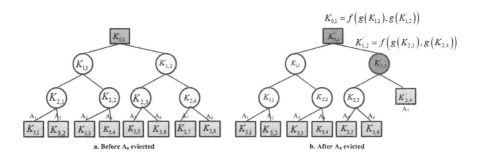

Fig. 4. Updating key when a member is evicted from group

4.5   Comparison of Proposed Scheme and Others

Table 1 shows the detailed comparison between the proposed scheme and other previous works, where AC is collusion attack, RA is replay attack, $MITM$ is man in the middle attack, h is the height of key tree, L is the key size, C_E is the computational cost to encrypt, C_D is the computational cost to decrypt, C_H is the computational cost to calculate hash function, C_f is the computational cost to calculate one-way trap-door function, C_M is the computational cost to calculate the modulo operation, C_{sign} is the computational cost to sign the digital signatures, C_{ver} is the computational cost check the signatures, S is the number of nodes from evicted or added node to the root.

Table 1. Comparison of computational cost between our scheme and others

Schemes		Against MITM	Against AC	Against RA	GM Com. cost (join; evict)	GM's computational cost (join; evict)	Total computational cost of members (join; evict)
Our scheme	Group key is valid	Yes	Yes	Yes	1; $(h+1)*L$	$2*C_H + C_{sign} + C_{ver} + C_E + C_D$; $(h+1)*C_E + (h-1)*C_H$	$2*C_H + C_{sign} + C_{ver} + C_E + C_D$; $C_D + 2h*C_H$
	Group key is expired				$(2h+1)*L$; $(h+1)*L$	$(2h+1)*C_E + (h-1)*C_H$; $(h+1)*C_E + (h-1)*C_H$	$2C_D + (h-1)*C_H$; $C_D + 2h*C_H$
ROFT, NOFT [10]		No	Yes	Yes	$(2h+1)*L$; $(h+1)*L$	$(2h+1)*C_E + (2h-1)*C_H$; $(h+1)*C_E + (h-1)*C_H$	$2C_D + (2h-1)*C_H$; $C_D + 2h*C_H$
OFT [3,4]		No	No	No	$(2h+1)*L$; $(h+1)*L$	$(2h+1)*C_E + (h-1)*C_H$; $(h+1)*C_E + h*C_H$	$2C_D + h*C_H$; $C_D + h*C_H$
Ku and Chen [7]		No	Yes	No	$(2h+1)*L$; $(h+1)*L$	$(2h+1)*C_E + (h-1)*C_H$; $(h^2+h+1)*C_E + (h^2+h)*C_H$	$2C_D + h*C_H$; $h*C_D + (1/2)h^2 * C_H$
Xu et al. [8]		No	Yes	No	$(2h+1)*L$; $(h+1)*L$	$(2h+1)*C_E + (h-1)*C_H$; $(h+1)*C_E + (h-2)*C_H$	$2C_D + h*C_H$; $C_D + h*C_H$
HOFT [9]		No	Yes	No	$(2h+1)*L$; $(h+1)*L$	$(2h+1)*C_E + (2S*h+1)*C_M + (h+S*h+1)*C_f$; $(h+1)*C_E + (h+2)*C_M + (h-1)C_f$	$(1+h)*C_H + 2h*C_M + (h+S*h)*C_f$; $C_D + (h+1)*C_M + h*C_f$

The proposed OFT guarantee to immune against internal attack such as collusion attacks and external attacks such as MITM. Furthermore, the computational cost of the proposed scheme when the old group key (session key) is still valid is always much lower comparing to previous schemes. However, in the case of expired old group key, the total computational cost of adding or evicting member is increased because of signing and authenticating signatures. Specifically, when evicting member from group, the computational cost and communication cost are equivalent because it is compulsory to reconstruct the managing group key tree right after the elimination of that member, then all informations related to that members must be updated.

5   Conclusions

The establishing and managing group key is the most important factor to ensure the continuity of the communication when a member is evicted from or added to the group. This paper proposes the improved OFT scheme used for managing group key that has outstanding advantages compared to previous schemes. Firstly, the proposed scheme is immune to external attacks like MITM. It can also be invulnerable against internal attack such as collusion attack because it reconstructs the managing key tree and refreshes all information of previous old keys.

References

1. Horng, G.: Cryptanalysis of a key management scheme for secure multicast communications. IEICE Trans. Commun. **E85–B**(5), 1050–1051 (2002)
2. Maynard, P., McLaughlin, K., Haberler, B.: Towards understanding man-in-the-middle attacks on IEC 60870-5-104 SCADA networks. In: Proceedings of the 2nd International Symposium for ICS and SCADA Cyber Security Research (2014)
3. Sherman, A.-T., McGrew, D.-A.: Key establishment in large dynamic groups using one-way function trees. IEEE Trans. Softw. Eng. **29**(5), 444–458 (2003)
4. Balenson, D., McGrew, D., Sherman, A.: Key Management For Large Dynamic Groups: One-Way Function Trees and Amortized Initialization. Internet Research Task Force (2000)
5. Wallner, D.-M., Harder, E.-J., Agee, R.-C.: Key Management for Multicast: Issues and Architectures. Internet Engineering Task Force (1998)
6. Wong, C.-K., Gouda, M., Lam, S.-S.: Secure group communication using key graphs. IEEE/ACM Trans. Netw. **8**(1), 16–30 (2000)
7. Ku, W.-C., Chen, S.-M.: An improved key management scheme for large dynamic groups using one-way function trees. In: Proceedings International Conference Parallel Processing Workshops, Kaohsiung, Taiwan, pp. 391–396 (2003)
8. Xu, X., Wang, L., Youssef, A., Zhu, B.: Preventing collusion attacks on the one-way function tree (OFT) scheme. In: Katz, J., Yung, M. (eds.) ACNS 2007. LNCS, vol. 4521, pp. 177–193. Springer, Heidelberg (2007). https://doi.org/10.1007/978-3-540-72738-5_12
9. Liu, J., Yang, B.: Collusion-resistant multicast key distribution based on homomorphic one-way function trees. IEEE Trans. Inf. Forensics Secur. **6**(3), 980–991 (2011)
10. Sun, Y., Chen, M., Bacchus, A., Lin, X.: Towards collusion-attack-resilient group key management using one-way function tree. Comput. Netw. **104**, 16–26 (2016)
11. Zhang, Y., Zheng, Z., Szalachowski, P., Wang, Q.: Collusion-resilient broadcast encryption based on dual-evolving one-way function trees. Secur. Commun. Netw. **9**, 3633–3645 (2016)
12. Jusoh, N.-A., Seman, K., Nawawi, N.-M., Sayuti, M.-S.-M.: The improvement of key management based on logical key hierarchy by implementing Diffie Hellman algorithm. J. Emerg. Trends Comput. Inf. Sci. **3**(3) (2012). ISSN 2079-8407
13. Ali, A.-I.: Comparison and evaluation of digital signature schemes employed in NDN network. Int. J. Embed. Syst. Appl. (IJESA) **5**(2), 15–29 (2015)

Detection of the Bee Queen Presence
Using Sound Analysis

Tymoteusz Cejrowski[1], Julian Szymański[1(✉)] ⓘ, Higinio Mora[2],
and David Gil[2]

[1] Faculty of Electronics, Telecommunications and Informatics,
Gdańsk University of Technology, Gdańsk, Poland
`tymoteusz.cejrowski@gmail.com`, `julian.szymanski@eti.pg.gda.pl`
[2] Department of Computer Science Technology and Computation,
University of Alicante, Alicante, Spain
`{hmora,david.gil}@ua.es`

Abstract. This work describes the system and methods of data analysis we use for beehive monitoring. We present overview of the hardware infrastructures used in hive monitoring systems and we describe algorithms used for analysis of this kind of data. Based on acquired signals we construct the application that is capable to detect an absence of honey bee queen. We describe our method of signal analysis and present results that allow us to drown conclusions on honey bee behaviour.

Keywords: Honey bees · Hive monitoring · Signal analysis

1 Introduction

Honeybee (Apis mellifera) is probably one of the most important insects in the agriculture. This extremely valuable insects are treated as the key factor in plants pollination [1,2]. It is crucial that the number of bees increases. Their work is considered as the guiding light for the hard-workers and is appreciated for ages. People and especially beekeepers should provide special care for this insects. Unfortunately today's beekeeping is facing many issues which cause the number of insects to decrease [3]. Disease or uncontrolled swarming can be the cause of bee distinction. Also insufficient care of the beekeeper is very often the main factor of collapsing the whole bee's families.

One of the important tasks for beekeeper is to check whether the queen bee is healthy and capable to lay eggs. This is done by opening the hive and inspecting the hive frames. If there is no eggs or larvae, the queen bee might be dead, ill or just not present in the hive. In such situations life of the whole swarm could be in danger and an immediate action is required. Whenever there is no reproduction because of the death or disability of the queen bee there is also no more younger bees that could replace the older ones [4]. Usually the bee worker lives only up to 40 days and after that her work should be overtaken by another bee [5].

ⓒ Springer International Publishing AG, part of Springer Nature 2018
N. T. Nguyen et al. (Eds.): ACIIDS 2018, LNAI 10752, pp. 297–306, 2018.
https://doi.org/10.1007/978-3-319-75420-8_28

The lack of healthy queen bee is extremely unfavorable and should be detected as soon as it is possible. But on the other side, daily hive inspections and checking whenever queen bee is present can be harmful for the whole bee family. Frequent disturbances can be a stressful factor and introduce the anxiety to the swarm. To avoid such situations it is necessary to use a non-invasive method that is able to detect lack of the queen bee.

In this paper we present the remote, non-invasive system that monitors and analyzes the honey bees behavior according to different conditions. For our study we create the situation where the queen bee is not present inside the hive. We monitor that situation with the set of the sensors and based on it we create the classifier that indicates whether the bee family becomes aware of the missing queen. The system presented in this paper can possibly detect different illnesses of bee colony such as presence of Varroa Destructor or predict bee swarming but that analysis is a plan for the future work.

2   Related Works

In recent years the interest in bees and their habits is increasing rapidly. Such situation cause growth the number of systems which are capable of collecting data from the hives.

For example, the commercial Arnia[1] system is designed for collecting weight, temperature and humidity measurements. The device is also equipped with microphone to obtain sound samples from inside the beehive. All data is transferred remotely to the server and then presented to the user. There are many similar projects with the same core objectives. For example, projects presented in [6] and [7] differ only in the set of sensors. Some of them like the BuzzBox[2] additionally provides open access to recorded data.

There are also number of scientific projects which are focused on detecting particular situations inside the beehive. For example, problem of bee swarming (when the insects leave the hive because of newborn queen) has been studied in [8]. The proposed solution uses cyclic temperature measurements and pattern recognition techniques which are based for a predictive algorithm. System is able to detect the preswarming moment by evaluating the increase in temperature inside the hive. Authors found five patterns that may occur during the year. Anomalies, which are accompanied by elevated temperatures within the hive, and hence the inability to classify data from a particular moment may be a sign of the incoming swarming.

Ferrari's work about the bee swarming prediction [9] describes wireless network of sensors that collect sounds, temperature and humidity values from the hive during the swarming periods. Based on empirical graphs observations some patterns were specified and determined.

---

[1] Arnia system: www.arnia.co.uk, access 10 Sep 2017.
[2] BuzzBox: www.osbeehives.com, access 10 Sep 2017.

There are also some projects focused on bee's diseases detection. Project which was developed at Edith Cowan University in Australia [10] aims to completely eliminate external parasitic mite Varroa Destructor from the Australian continent. Device collects sound samples and converts them to feature vector, at the end the data is classified using SVM or LDA algorithms. Mite detection such as Varroa Destructor can also be solved using image analysis processing. In Larissa Chazette's work [11] camera-equipped system has been developed which recognizes Varroa Destructor infected bees by using Convolutional Neural Networks (CNN). In Schurischuster's work [12] a Raspberry Pi 3 based device is able to collects high resolution and well-exposed pictures of bees entering the hive. The combination of [11,12] systems could lead to even better results.

3   System Design

Our system is composed from three parts: server, client and embedded module. The wireless network of embedded devices was made according to master-slave architecture. The *endpoints* which are mounted directly inside the hive are responsible for collecting measurements and passing them to *accesspoint*. The *accesspoint* uploads raw data to the server where is processed.

Analysis of bees behavior can only be performed by usage of sensors that provide case-essential data. Significant type of data can be specified basing on the related work. Bees like most of the insects produce sound during the flight. The bee worker emits sounds at 250 Hz on the air. But bees can be also considered as one super-organism where their sounds accumulates to one, common buzz. Seemingly irrelevant noise emanating from the center of the hive can be a source of valuable information. For example, when bees are preparing for swarming, they also change their extremely ordered behavior. Some of the insects start becoming restless, bring excitement to the hive and finally change nature of the common buzz. Without doubt sound is one of the most important factors in bee analysis. In presented work, sound samples are collected by specially designed microphone. The band-pass filters have been selected so that the microphone will be sensitive for bee's sounds (20–2000 Hz).

Bees also need proper level of temperature and humidity inside the hive. Without suitable conditions the colony can leave their current place of occupation [8]. Invalid humidity level is causing multiple bee diseases so it is also significant value for monitoring. Temperature and humidity are the most sensitive factors among the bees and these two values are monitored using an integrated sensor HDC1008.

In our system the microphone, temperature and humidity sensors was inserted into a specially designed bee hive frame in order to not disturb the bees. The designed frame is shown in the Fig. 1. Single data set contains one second sound sample and information about levels of temperature and humidity. Data is acquired every 15 min and then uploaded to the server.

Fig. 1. Bee hive frame used in experiment.

4   Data Processing

In our approach we divide the data into two sets: one derived from normal bees work and one from abnormal, where there is a lack of bee queen in the hive. Thanks to that it is possible to extract the pattern, which will allow us to differentiate particular behaviors. At the beginning of the analysis process the data is downloaded from server. This step must ensure data consistency in which the sound, temperature and humidity values must be available at a given moment. Then, the features are extracted from the available data and classification potential is tested. At the end the model is worked out and final classification is performed.

4.1   Feature Extraction - Linear Predictive Coding (LPC)

The process of the data analysis start from the transformation of the input data from the hive into a form that can be used as input for algorithms. Sound signal should be compressed into the finite element vector of the size significantly less than the original length of the sound sample vector. For this purpose Linear Predictive Coding was used.

LPC is a method used in a speech audio compression. This method assumes that signal is produced by a buzzer located at the end of the tube [13]. For the correct sound characterization it is important to determine the output signal parameters as the inverse of the impulse response of FIR filter that represents the vocal tract. In order to facilitate the use, the input is the Dirac delta function. Model can be represented as in (1)

$$H(z) = \frac{G}{1 - \sum_{k=1}^{m} a_k z^{-k}} \tag{1}$$

where G is the gain, m level of the model and a_k represents searched characteristic coefficients.

Using Z transform, and Levinson-Durbin algorithm a coefficient vector is obtained [14]. It is assumed that 10–14 LPC coefficients describe well the signal and further increasing this number results in an insignificant improvement in signal approximation. In this work the given sound sample with a dimension of 3000 was characterized using LPC algorithm by vector of size $N = 14$. The final data was extended by temperature and humidity values collected from the same moment of time as the sound sample.

4.2   Classification Potential - t-SNE

Having some set of a data it is crucial to determine if the classification and model extraction is possible at all. This process is a supportive step and performed only in the case of recognizing new features. Could be carried out by viewing the 2D or 3D points which corresponds to input data. If there is possibility for data separation it means that examined feature can be used in hive modeling. To get 2D points from multidimensional input vector it is necessary to use technique for dimensionality reduction such as t-SNE.

Algorithm was introduced by Laurens van der Maaten in 2008 and is the variation of previously existing algorithm called SNE developed by Geoffrey Hinton and Sam Roweis in 2002. t-SNE converts multidimensional set of data $\chi = \{x_1, x_2, \ldots, x_n\}$ to 2D or 3D vectors $\Upsilon = \{y_1, y_2, \ldots, y_n\}$. The basis of this algorithm is to compare the density distribution of multivariate variables with the distribution of their projection on a two or three-dimensional plane [15]. The difference between these two distributions is calculated by Kullback–Leibler divergence and minimized by gradient descent.

In our work the t-SNE algorithm was used to evaluate the potential of the hive classification according to presence of the queen bee inside the hive. We treat LPC coefficients vectors, extended with humidity and temperature values, as the input for t-SNE algorithm. We perform dimension reduction on vectors of size $n = 16$ to obtain 2D or 3D map of points that corresponds to state of the colony at given moment. Then we can decide if future classification is reasonable. It is desired to observe clusters of data representing samples taken during the normal work and a separate set representing anomalies. If so, we can perform the last step which was the classification according to previously chosen feature.

4.3   Learning - SVM

Information of the separation capacity is important for the next stage of the mathematical bees modeling. Input data in the form of n-dimensional vector is given as the input to the classifier. Presented system use SVM classifier developed by Vapnik [16] in 1963 with the modification [17] from 1992 introduced by Boser, Guyon and Vapnik himself.

The basic SVM classifier is capable of separating two sets that are linearly separated so that the hyperplane spreading the training data maximizes the value of the geometric margin. The output of SVM algorithm is the separating hyperplane which form is presented in (2)

$$y(\boldsymbol{x}) = \boldsymbol{w}^T \boldsymbol{x} + b = 0 \tag{2}$$

where $\boldsymbol{w} = [w_1, w_2, \ldots, w_N]^T$ is the N dimension weight vector and $\boldsymbol{x} = [x_1, x_2, x_N]^T$ is the input vector. As a matter of fact the input data rarely can be linearly divided into two separate sets. Separation of the data that can not be linearized is solved with the help of a so-called kernel trick [18]. It transforms nonlinearly input data so that they are likely to be linearly separable.

In presented work non-linear, Gaussian-kernel, SVM classifier was used to obtain a model of the hive in relation to the designated feature. We have used SVM method on two data sources: previously described $n = 16$ dimensional vectors and the output of t-SNE. Both approaches provide similar results. The SVM output is the hyperplane dividing learning set into two separate clusters, one indicating anomaly and second describing normal swarm behavior. Such model can be later evaluated on testing data and relevant conclusions regard to the behavior of bees can be learned.

5   Experimental Results

Bee colony was monitored in the period from February 2017 to August 2017. Unfortunately, the bees did not swarm in that time also they were not infected by any of the diseases. To test our classification system, it was decided to force a critical situation for the bees. Absence of a bee's queen in the hive was chosen and it was caused manually by a beekeeper. The aim of the experiment is to develop a hive model in which will be possible to observe and specify the patterns characteristic for bees living without the queen.

The experiment was carried out using embedded system described in Sect. 3. Bee hive frame was inserted into the hive as third frame from the entrance. The mother exchange process together with periods of downloading of sound data is presented in Fig. 2.

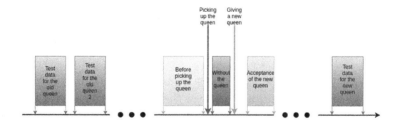

Fig. 2. Queen exchange workflow.

In the first step, two data sets previously designated as "Before picking up the queen" and "Without the queen" were used as input data. The audio samples

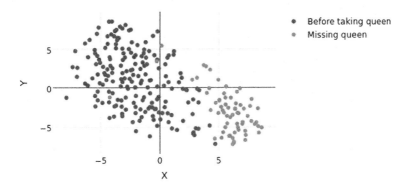

Fig. 3. Output of t-SNE algorithm with data "Before picking up the queen" and "Without the queen".

were compressed into a feature vector, which was the LPC coefficients. These vectors were also extended with temperatures and humidity values. At the end data was normalized. Such prepared vector was provided for the input of the t-SNE dimension reduction algorithm to identify the classification potential. Result was shown in Fig. 3.

Based on output of t-SNE algorithm it is clear that bees with and without the queen act differently. Thus detection of this two cases can be performed using classification method. For that purpose Support Vector Machine algorithm with C-classification was used. Cost was set to $C = 100$ and kernel was defined as $K(x_j, x) = exp\left(-\gamma \|x - x_j\|^2\right)$ with $\gamma = 0.4$. Result as shown in Fig. 4.

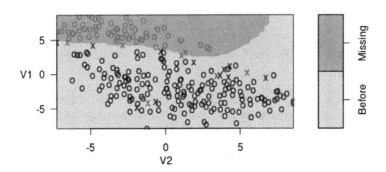

Fig. 4. SVM classification borders plot. Modeling the hive with absence queen bee as feature: queen inside the hive (Before) and Queen taken (Missing).

Such defined model was tested with test data presented in Fig. 2. Table 1 presents classification on the test data.

Data named as "Test old queen" and "Test old queen 2" indicates normal work of the swarm. Such situations was classified correctly and model was very

Table 1. Test data classification.

Name	Samples	With queen	Without queen	Error
Test old queen	92	90	2	2.17%
Test old queen 2	72	70	2	2.77%
Test new queen	130	98	32	75.38%

accurate. It is possible to determine the moment when the queen bee may suffer or even die for example from a pest attack. In such cases we should found samples which significantly differ from the others.

It was also expected that after some time the data and situation inside the hive should return to the situation before the mother's removal. The experiment showed that the colony did not return to the same state even 15 days after new queen bee insertion. To more precisely examine this situation it was decided to check classification potential between "Test data for the old queen bee" and "Test data for the new queen". Result as shown in Fig. 5.

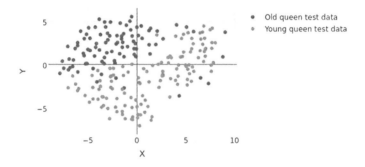

Fig. 5. Swarm classification with two different queen bees.

Output data is quite easy separable what indicates that different bee queens cause different behaviors across the swarm. For queen bee collapse detection the system should only analyze changes of the behavior patterns but expecting same behavior patters with fresh and old queen could be misleading. These observations was discussed with two independent beekeepers who confirmed that queen bee influences the behavior of the whole family. The new queen bee after introduction into the family makes the bees subordinate to her disposition and sound significantly changes. Model derived from "normal state" of two different bee queens has been tested on two extra test datasets.

Results presented in Table 2 show that it is necessary to calculate a new model for the newly introduced queen bee in order to detect next possible swarm collapse. New queen is significantly different from her predecessor, and thus the old model loses its usefulness.

Table 2. Classification of two test datasets from two different queen.

Name	Samples	Old queen	New queen	Error
Test data old queen	72	68	4	5.55%
Test data new queen	183	17	166	9.28%

6   Conclusion and Future Work

The experiment and its results confirm the validity of the proposed model. It has been proved that there is a pattern that characterizes the normal work of the swarm and it can be correctly identified using the system presented in this paper. The anomalies such as the exchange of the mother are distinguishable and extracted by the presented system.

The developed classification system can also be significantly improved. It is necessary to collect much more data from different anomalies occurring inside the hive in order to develop a global model of bees behavior. Detection of swarming or diseases with usage of the described system can be real. For that purpose we plan to use more sophisticated classifiers [19] ant their parallel implementation [20] that should allow us to process larger set of the data in more effective way. We can also obtain some improvement on the level of the data representation. Here, we plan to add to Linear Predictive Coding analysis of particular feature context in the similar way done in [21].

Our system is only a starting point for further work that is currently being conducted. The life and behavior of animals in particular bees can be a source of valuable information. Researchers at Nanchang University in China [22] have found that bees work harder the day before the expected rain. This observation is the basis for extending the system for predicting temperature and humidity based on data coming from the hive.

Acknowledgments. This work has been supported partially by COST project CA15118 "Mathematical and Computer Science Methods for Food Science and Industry" and founds of the Department of Computer Architecture, Faculty of Electronics, Telecommunications and Informatics, Gdańsk University of Technology.

References

1. Gill, R.: The value of honeybee pollination to society. Acta Hortic. **228** (1991)
2. Svensson, B.: The importance of honeybee-pollination for the quality and quantity of strawberries. Acta Hortic. **228**, 260–264 (1991)
3. Cox-Foster, D.L., Conlan, S., Holmes, E.C., Palacios, G., Evans, J.D., Moran, N.A., Quan, P.L., Briese, T., Hornig, M., Geiser, D.M., et al.: A metagenomic survey of microbes in honey bee colony collapse disorder. Science **318**(5848), 283–287 (2007)
4. Ratnieks, F.L.: Egg-laying, egg-removal, and ovary development by workers in queenright honey bee colonies. Behav. Ecol. Sociobiol. **32**(3), 191–198 (1993)

5. Tautz, J.: The Buzz About Bees. Springer Science, Berlin (2008). https://doi.org/10.1007/978-3-540-78729-7
6. Zacepins, A., Kviesis, A., Ahrendt, P., Richter, U., Tekin, S., Durgun, M.: Beekeeping in the future—smart apiary management. In: 2016 17th International Carpathian Control Conference (ICCC), pp. 808–812. IEEE (2016)
7. Strob, M., Kašparu, M.: Beehive electronic measuring system
8. Kridi, D.S., de Carvalho, C.G.N., Gomes, D.G.: A predictive algorithm for mitigate swarming bees through proactive monitoring via wireless sensor networks. In: Proceedings of the 11th ACM Symposium on Performance Evaluation of Wireless Ad Hoc, Sensor, and Ubiquitous Networks, pp. 41–47. ACM (2014)
9. Ferrari, S., Silva, M., Guarino, M., Berckmans, D.: Monitoring of swarming sounds in bee hives for early detection of the swarming period. Comput. Electron. Agric. **64**(1), 72–77 (2008)
10. Qandour, A., Ahmad, I., Habibi, D., Leppard, M.: Remote beehive monitoring using acoustic signals (2014)
11. Chazette, L., Becker, M., Szczerbicka, H.: Basic algorithms for bee hive monitoring and laser-based mite control. In: 2016 IEEE Symposium Series on Computational Intelligence (SSCI), pp. 1–8. IEEE (2016)
12. Schurischuster, S., Zambanini, S.: Sensor study for monitoring varroa mites on honey bees (apis mellifera)
13. Atal, B.S., Hanauer, S.L.: Speech analysis and synthesis by linear prediction of the speech wave. J. Acoust. Soc. Am. **50**(2B), 637–655 (1971)
14. Makhoul, J.: Linear prediction: a tutorial review. Proc. IEEE **63**(4), 561–580 (1975)
15. van der Maaten, L., Hinton, G.: Visualizing data using t-SNE. J. Mach. Learn. Res. **9**(Nov), 2579–2605 (2008)
16. Cortes, C., Vapnik, V.: Support-vector networks. Mach. Learn. **20**(3), 273–297 (1995)
17. Boser, B.E., Guyon, I.M., Vapnik, V.N.: A training algorithm for optimal margin classifiers. In: Proceedings of the Fifth Annual Workshop on Computational Learning Theory, pp. 144–152. ACM (1992)
18. Schölkopf, B.: The kernel trick for distances. In: Advances in Neural Information Processing Systems, pp. 301–307 (2001)
19. Draszawka, K., Szymański, J.: Thresholding strategies for large scale multi-label text classifier. In: 2013 the 6th International Conference on Human System Interaction (HSI), pp. 350–355. IEEE (2013)
20. Czarnul, P., Rościszewski, P., Matuszek, M., Szymański, J.: Simulation of parallel similarity measure computations for large data sets. In: 2015 IEEE 2nd International Conference on Cybernetics (CYBCONF), pp. 472–477. IEEE (2015)
21. Szymański, J.: Words context analysis for improvement of information retrieval. In: Nguyen, N.-T., Hoang, K., Jędrzejowicz, P. (eds.) ICCCI 2012. LNCS (LNAI), vol. 7653, pp. 318–325. Springer, Heidelberg (2012). https://doi.org/10.1007/978-3-642-34630-9_33
22. He, X.J., Tian, L.Q., Wu, X.B., Zeng, Z.J.: RFID monitoring indicates honeybees work harder before a rainy day. Insect Sci. **23**(1), 157–159 (2016)

A Design for an Anti-static Wrist Strap Wireless Monitoring System in a Smart Factory

Chia-Chi Chang, Shien-Yi Chen, Ching-Chuan Wei[(✉)], and Yung-Hung Hsu

Chaoyang University of Technology, Taichung 41349, Taiwan
{ccchang,ccwei}@cyut.edu.tw

Abstract. Electrostatic protection is a basic requirement for general electronics factories, so these factories require an anti-static wrist strap system. However, since the staff on specific production lines generally work with their hands, often anti-static wrist straps retain poor contact. As a result, it is quite possible to cause anti-static wrist strap monitors to misjudge. Once the current anti-static wrist strap status is lost, the whole production line is under high risk, and this may lead to numerous losses. This research has a twofold focus. The first goal is to design anti-static wrist strap sensing hardware that will integrate with a specific ion-fan and infrared sensors. After the sensed data are collected, the second goal is to transmit that data to the back-end database via wireless communication, in order to verify whether the specified working guidelines have been followed. Finally, experimental results show that the new system's false alarm rate is 0.23%, which is an improvement over the false alarm rate of traditional systems, and related research results can be a useful reference for constructing smart factories.

Keywords: Static electrical shielding · Statically electrical wristband
Wireless communication · Intelligent factory

1 Introduction

On the production line of 3C products, static electricity is the focus. Most companies use anti-static wrist strap monitors to monitor whether staff members have put their anti-static wristbands on. The anti-static wrist strap monitors [1] on the market today usually are paired up with a warning light. The red warning light goes off with a siren whenever a staff member removes the wristband. However, the anti-static wrist strap monitor is unstable and occasionally, misjudgments occur, in which the wristbands are not able to detect the connection correctly. Therefore, staff members are unwilling to wear the wristbands and are searching for systematic flaws [5].

Anti-static wrist strap monitors have a lot of flaws [4]. For example, if the static wristband is removed from the anti-static wrist strap monitor, the monitor is not able to detect whether the staff member is wearing the wristband properly, or it is just simply not in working time. Staff members can take advantage of this flaw for their own benefit, and managers cannot determine whether staff members have followed work regulations. Conclusively, the qualities of the product cannot be ensured. To avoid such and other similar issues, this research designs and implements a monitoring system wristband in intelligent factories.

© Springer International Publishing AG, part of Springer Nature 2018
N. T. Nguyen et al. (Eds.): ACIIDS 2018, LNAI 10752, pp. 307–314, 2018.
https://doi.org/10.1007/978-3-319-75420-8_29

The goal of this paper is to use an ATmega microcontroller as the core in intelligent factories for monitoring system wristbands. The following data would be required: an original warning light, power supplies for both a monitor and specific ion-fan, an infrared sensor, and a control warning light. After that, the system must be able to transmit the collected data to the back-end server through Wi-Fi for processing. Last, our study compares an original monitoring system wristband with monitoring system wristbands in intelligent factories through experiments that record error rates of both under certain limb movements. The newly designed product would be improved and verified to complete the development of intelligent factories for monitoring system wristbands.

2   System Architecture and Components

2.1   System Architecture

This paper proposes several goals for improvement: (1) An ATmega chip that reads the signal from the monitoring wristband and controls the warning light, so that it will prevent a malfunction of the wristband when it is removed. (2) Use of a variable resistance (VR) so that we not only can adjust and decrease the sensibility of the wristband, but also can determine the buffering time of the wristband, thereby, improving the situation in which bad connections result in misjudgment of the monitoring system and trigger the warning light. (3) Monitoring the conditions of the power supplies in the monitoring wristband and affiliate infrared sensor, so to monitor the working seats of the staff members. And transmitting the data from both monitoring systems to the back-end server for the manager's convenience to confirm if there is any improper action, such as cutting off the wristband power supply. (4) Installing a specific ion-fan to improve protective measuring. This prevents static electricity from damaging the elements when bad connections occur. (5) Collecting data from the wristbands, the monitoring power supplies, specific ion-fans, and the infrared sensors, and transmitting them to the back-end server through Wi-Fi for the manager's use. Figure 1 demonstrates the flow chart of the static electric system.

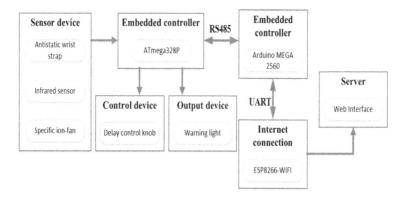

Fig. 1. Anti-static wrist strap system block diagram

2.2 ATmega Series Chips

ATmega series chips are AVR processors produced by the Atmel Corporation. The Atmel AVR series employs an 8–32 bit reduced instruction set computing (RISC) system, a microcontroller, which has several powerful characteristics, including a built-in RC oscillator, and a 0–20 MHz oscillator that can be plugged in. The Atmel AVR series possesses a 40 mA current output identical to PIC I/O with a 10–20 mA sink current and the ability to simplify a peripheral hardware circuit, such as a built-in EEPROM, UART, AADC, SPI, TWI, etc.

2.3 Wireless Transmission Chips

ESP8266 is a Wi-Fi chip produced by Espressif Systems (Shanghai) Pte. Ltd [4]. It contains only thirty-two pins with a low manufacturing cost of 90 NT dollars. In fact, this chip is developed for fourteen different types of circuit boards, ESP-01 to ESP-12F. This paper uses ESP-12F for the ESP86266 as a newer version of the circuit board. It has stronger signals and possesses extra features, such as UART, SPI, and ADC. Moreover, it has an embedded PCB antenna with the advantages of both Station (STA) and Access Point (AP).

2.4 MAX485 Model

MAX485 is a low power consumption package with 8 pins that allow both-way conversions between UART signals and RS-485 signals [6]. It has an operating voltage of 5 V and up to a 10 M transmission rate. It supports UART input voltage, ranging −0.5 V– (VCC + 0.5) V, and RS-485 logic voltage, ranging −8 V–+12.5 V logic level. Data transmission distance can reach 1200 m. Normally, more than one anti-static wristband is used in factories; it is common to use dozens of them. Also, every anti-static wristband has a certain distance from one another. Thus, this paper opts for long distances for the convenience of connecting up to thirty-two sets of MAX485 as a tool for data exchange.

3 Hardware Circuit Design

3.1 Anti-static Wristband Sensory Circuit Board

The circuit board consists of two parts, an anti-static wristband sensor and Wi-Fi transmission. The anti-static wristband sensor uses an ATmega 328P [2] microcontroller as its core. To prevent connecting errors, an XH2.5 m/m foolproof connector is used for detecting the wristband, specific ion-fan, and infrared sensor. A 7447 BCD decoder is used for controlling the seven-segment display and decreasing the microcontroller pin usage. An ADC is utilized to read the variable resistance voltages, and the voltage change is utilized to adjust the delay time of the warning light. The above hardware elements integrate into a complete anti-static wristband circuit board (Fig. 2).

Fig. 2. Anti-static wristband system block

3.2   Wi-Fi Controlling Circuit Board

A Wi-Fi Controlling circuit board uses an ATmega 2560 microcontroller [3] as the core. The ATmega 2560 is not installed directly on the circuit board but the ATmega 2560 chip development board and Arduino MEGA 2560 are combined with the Wi-Fi transmission circuit board. The Wi-Fi transmission circuit board integrates with the ESP 8266 drive circuit, which allows the board to choose between the USB Type B from the board itself and the Arduino MEGA 2560 for the power supply. It utilizes both pins and Arduino MEGA 2560 to communicate with ESP8266-12F and MAX485, respectively. The above hardware elements integrate into a complete Wi-Fi circuit board system (Fig. 3).

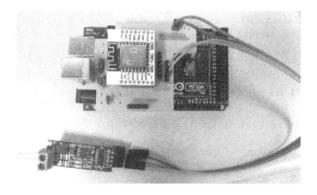

Fig. 3. Anti-static wristband system product

4   System Verification and Experimental Results

4.1   Environmental Setup and System Verification

The environmental setup mainly focuses on the simulation of the factory's layouts. The table area is supplied with a specific ion-fan, anti-static wristband, wristband monitoring machine, warning light set, and the system's core - wristband sensory board. The wristband sensory board connects with the Wi-Fi transmission circuit board though MAX485. Normally, for one production line, there are multiple wristband sensory boards connected to one Wi-Fi transmission circuit board through RS-485. Hence, only one Wi-Fi transmission circuit board is necessary (Fig. 4).

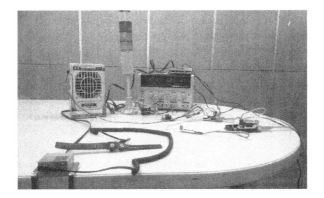

Fig. 4. Anti-static wristband environmental setup

The wristband sensory board would wait for the Wi-Fi circuit board to respond. Whenever it responded, the wristband sensory board would transmit the data from the wristbands and other sensory devices to the Wi-Fi circuit board through RS-485. The Wi-Fi circuit board would transmit the data to the back-end server according to the

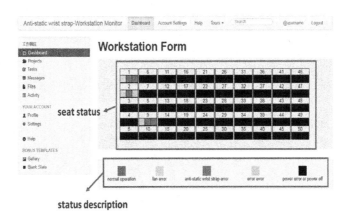

Fig. 5. Workstation display interface of wristband conditions

change of previously transmitted data. Eventually, the server would display the data on a website interface (Fig. 5).

4.2   Experimental Methods and Results

To verify the improvement in the wristbands, this paper compares the error rates between the anti-static wrist strap monitoring system and the anti-static wrist strap wireless monitoring system. Using a manual mode to simulate the working motions of staff members by swinging both arms, it consists of six main types of movements: left and right, up and down, upper left and lower right, lower left and upper right, and clockwise and counterclockwise. By using these movements, one can examine the error rates of the anti-static wristband monitoring program.

The experiment is processed under simulation of the six main types of movements with both wristbands. Every movement is executed 100 times. According to the OP light from the machine and the warning light from the control system, the error rates for both are recorded. The experiment lasts ten days with 12000 data values for comparison.

Since the warning light in the anti-static wrist strap monitoring system is directly controlled by the machine, the error rates of the OP light and the system light are identical. According to Table 1, the high error rates of the machine easily trigger the warning lights which influences the work flow.

Table 1. Anti-static wrist strap monitoring system – error rates

Light type	Moving type					
	Left and right	Up and down	Upper left and right lower	Lower left and right upper	Forward	Reverse
OP light	12.3%	15.3%	16.2%	5.4%	10.7%	8.2%
System light	12.3%	15.3%	16.2%	5.4%	10.7%	8.2%

The anti-static wristband system, discussed in this paper, has a different design. The machine does not directly control the warning light. Instead, the wristband sensory board bridges the two. Thus, the machine and the system have different detections. The system has a delaying time of 0–3 s. This paper chooses 2 s as the median for experimentation. According to Table 2, the two-second delay time resolves the instantaneous, bad connection issue that results in misjudgment of the warning light.

Table 2. Anti-static wrist strap wireless monitoring system – error rate

Light type	Moving types					
	Left and right	Up and down	Upper left and right lower	Lower left and right upper	Forward	Reverse
OP light	9.8%	11.4%	12.7%	5%	9.2%	7.6%
System light	0.1%	0.6%	0.4%	0.1%	0.1%	0%

After comparing the results of the two different systems, the average error rate of the anti-static wrist strap monitoring system is 11.35% and the average error rate of the anti-static wrist strap wireless monitoring system, which is designed and discussed in the paper, is 0.23% (Fig. 6).

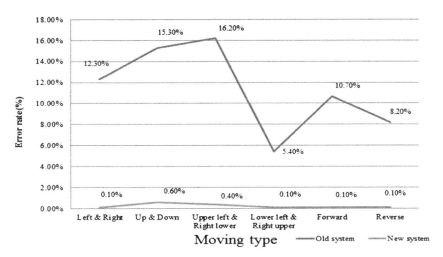

Fig. 6. The comparison of two systems

5 Conclusion

This paper successfully designs an intelligent factory of monitoring systems and anti-static wristbands by using ATmega 328P and Arduino MEGA 2560 to collect and transmit data, respectively. Through RS-485, sensing data transmission is carried out, finally transmitting the received data back to the server through Wi-Fi. The system uses adjustable delay time to determine the wristbands' conditions and effectively reduce the system's error rate. Manually simulating the working movements of staff members at workstations allows us to determine the different error rates according to the OP light and warning light from the system. The experiment resulted in an error rate of 0.23% for the intelligent factory with a monitoring system and anti-static wristbands, compared to an error rate of 11.35% for the anti-static wrist strap monitoring system. The comparison allows us to conclude that the use of a monitoring system and anti-static wristbands in an intelligent factory is superior to the anti-static wrist strap monitoring system.

Acknowledgment. This work was founded in part by Ministry of Science and Technology of Taiwan under Grant MOST 105-2221-E-324-016 and 105-2622-E-324-003-CC2.

References

1. Anti-static products and coatings. http://www.explainthatstuff.com/howantistatic coatingswork.html
2. Atmel ATmega328P. http://www.microchip.com/wwwproducts/en/ATmega328P
3. Atmel ATmega2560. http://www.microchip.com/wwwproducts/en/ATmega2560

4. Lin, N.-C.: Wearable and wireless human body electrostatic monitoring system. In: 2015 International Conference on Intelligent Information Hiding and Multimedia Signal Processing (IIH-MSP), pp. 23–25. IEEE, Adelaide, SA, Australia (2015)
5. Tang, M., Jin, Y., Yao, L.: WiFi-ZigBee coexistence based on collision avoidance for wireless body area network. In: 2017 3rd International Conference on Big Data Computing and Communications (BIGCOM), pp. 10–11. IEEE, Chengdu, China (2017)
6. RS-485 Reference Guide. http://www.ti.com/lit/sg/slyt484a/slyt484a.pdf
7. Cho, S., Lee, D., Ali, I., Kim, S., Pu, Y., Yoo, S., Lee, K.: Highly reliable automotive integrated protection circuit for human body model ESD of +6 kV, over voltage, and reverse voltage. Electron. Lett. **53**(13), 843–845 (2017)

Enhanced Passcode Recognition Based on Press Force and Time Interval

Hua-Yuan Shih[1], Song Guo[2], Rung-Ching Chen[1(✉)],
and Chen-Yeng Peng[1]

[1] Department of Information Management,
Chaoyang University of Technology, Taichung, Taiwan
shih.huayuan@gmail.com, rungching@gmail.com,
davud8407@gmail.com
[2] Department of Computer Science and Engineering,
The University of Aizu, Aizuwakamatsu, Japan
sguo.uaizu@gmail.com

Abstract. Mobile devices applied to many businesses applications. Constructing a useful security system for a mobile device is necessary. As to lots of personal information on the mobile device, it must protect completely. Now, the mobile device has many protection mechanisms. However, even we have these mechanisms that still cannot protect someone who wants to use your passcode to intrude private information. The 3D-Touch features use people's habit and action. According to these functions, we construct them to become a decision tree. In this paper, we will use 3D-Touch techniques to propose an enhanced passcode mechanisms for protection mobile device system. Experimental results show that the method is feasible and efficient.

Keywords: 3D-touch · Decision tree · Machine learning · Identity recognition

1 Introduction

More and more apps or services use on mobile devices. Since their inception less than three years ago, almost 30% of adults in the US have their tablet computer [1] and about half of American adults have their smartphone [2]. The mobile device lets our life and job become convenient. For example, people used a mobile device to keep emotion with friends [3], sent mail on the mobile device, and accessed cloud driver and home security [4]. All of the services use passwords to protect their system, but the password cannot prevent someone who wants to invade your mobile device. Now, we have the password, graphic locking, and Touch ID. Those mechanisms are protection our device; only the owner has the password, graphic locking or 3D-touch. However, if the proprietor lets someone get your passcode who can get into your device without your agreement. For example, you lend your phone, and he/she remainders your passcode. If any close friend they want, they always can open your mobile phone and access. People don't know about the security of the mobile device. Kruger and Kearney explain the following factors which show the results from addressing awareness levels

© Springer International Publishing AG, part of Springer Nature 2018
N. T. Nguyen et al. (Eds.): ACIIDS 2018, LNAI 10752, pp. 315–323, 2018.
https://doi.org/10.1007/978-3-319-75420-8_30

in an organization [5]: (1) knowledge: what people know; (2) attitude: what people think; (3) behavior: what people do.

However, most of the people don't have knowledge of information security with a mobile device. People just use and set the password, but they don't know how to protect their mobile device in the right way. This phenomenon exists on using mobile devices. Mobile devices also have critical applications such as computer connection, surveillance, health care and robotics interactive [6]. Now, some mobile devices have the 3D-touch. In this paper, we will propose a new idea based on the human behavior inertia. The system can monitor the features of inputted when people input the password. By using the 3D-touch, we can get the time of inputted password and force of inputted password. The system will use a decision tree to judge whether the user is the device's owner or not which according to the press time and force. Recall, Precious and F1 values are used to evaluate the system effectively.

2 Related Works

Mobile devices stores lots of user's information such as e-mail, bank account, files, money, friends, and house [7, 8]. When people can access your mobile device, you will leak your personal information. People change customs from cash to electronic money of mobile devices [9, 10]. If a user lost her/his mobile device, the consequences could be disastrous.

Most of the users don't have the knowledge about the information security. Users usually set a short password or simple graphic locking. Even user set a complex password and graphic locking. Some close friends when you unlock your mobile device. They may see your password or graphic locking. If they get the password or graphic lock, they also can unlock your mobile devices.

Touch-ID is one kind of Biometrics [11], it has better safety than a password or graphic locking. But, even if it was an advanced technology when users sleep, other people can use your finger to unlock your mobile device, too.

In this paper, we propose a method when users are not the mobile device owners, they can't unlock the mobile devices just using a password or graphic locking even they have passcodes.

In the 1960s, the psychologist Gunnar Johansson performed a series of famous experiments in which he attached lights to people's limbs and recorded videos of them performing different activities, such as walking, running, and dancing [12]. Observers can usually carry out this task easy, and they could sometimes determine gender and even recognize specific individuals in this way. The motion signature is a perceptible element of human action [13]. In analogy with handwrite names, people have characteristic motion signatures that can individualize their features. When people use the different postures, they are different people. They generate different force and different interval time when they touch the mobile device screen, too.

In this study, according to spite of the above, we will record the time of interval and the force of touch when people unlock password. Time and pressing force will be the features for recognitions.

Decision Tree can be converted to rules or intuitively visualized [14]. They have usually used in a business setting and data mining software [15].

Decision Tree works by dividing data sets into smaller and more homogeneous subsets. Therefore, it works in a gap and conquers approach. This recursive division of data sets means the subsets on lower levels have few instances [16]. Therefore, Decision Trees tend to over fit in the lower level of the tree. Decision tree normally has a better outperformance than other learning algorithms such as Neural Networks [17], Support Vector Machines [18].

In addition, domain knowledge improves the performance of Machine Learning algorithms. Therefore, utilizing domain knowledge would improve the accuracy of Decision Trees. The most common methods for creating Decision Tree are those that create Decision Tree from a set of examples (data records). We refer to these methods as data-based Decision Tree methods.

A Decision Tree is a useful tool for guiding a decision process as long as no changes occur in the dataset. However, it is hard to manipulate or restructure Decision Trees because a Decision Tree is a procedural knowledge representation, which imposes an evaluation order on the attributes. In contrast, rule-based Decision Tree methods handle manipulations in the data, through the rules induced from the data not the Decision Tree itself. A declarative representation, such as a set of decision rules is much easier to modify and adapt to different situations than a procedural one. The easiness is due to the absence of constraints on the order of evaluation the rules [19].

Mobile devices develop many mechanisms, but most people set password easily. The easy password is to log in easily. Many people set '12345' or 'password' to be the password because the password remembered easily, so mobile device develops graphic locking. People accord 9 points to draw the graphic to set the password. The graphics are the problem, smooth graphics are remembered easily by users.

In recently, mobile devices have fingerprint recognition. The fingerprint recognition looks safety, but people may lose some detail. If you drink, sleep, the intruders also can unlock your mobile device. In the criminal identification, a signature is one kind of way to know. We can know the real writer from the signature force and speed. Then, in this study, we will according to the human behavior designs the system.

In this study, we use force recognition based on Decision Tree. People register the mobile device. Mobile device collect people's press strength and interval time to create a Decision Tree. This system matches time and force to know device owner or not.

3   Methodology

The system has register user interface, login user interface and the statistic of result user interface. By this process, the user input three times of password. Then, the system records those data which are used to build the Decision Tree. The register data include user touch time and force of each button. The system will calculate the average time of three times and average force of three times of input. Also, the system calculates force interval of three times. These data will be used to compare with user login data when users login system. When users finished register process, they can use login user interface. The system will use the Decision Tree to identify the user. First, it compares the average system time and total time of user login and uses each interval time to analyze conditions. Next, the system will compare average force and force of user login in the decision tree. The analyzing conditions are each interval force that system already calculates the force of three times.

When the user login is successful, the system will use the user's login data to re-build a Decision Tree. The system is shown in Fig. 1. The system has a simple and easy user interface to get the result of preciseness. In this system, we limit the number of passwords is six.

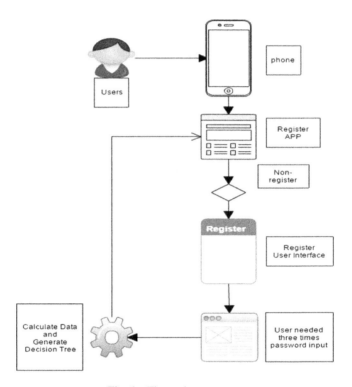

Fig. 1. The register process

The buttons can extract user's force when the user presses buttons. When users input the password which has six buttons, the system will calculate the interval time. When finish register process, the system calculates the average duration of three times, average force of three times, interval time of three times and interval of the strength of three times which are shown in Table 1.

Due to, the user needs to input six numbers of a password. So, it has five interval time. The three times of interval time shows on first three rows in which stand for by Time 1, Time 2 and Time 3. Then, use Time 1, Time 2 and Time 3 to find the Average Time. And Time 1, Time 2 and Time3, they need to subtraction each other. Thus, the system will use the max interval time from Time 1, Time 2 and Time 3, and it becomes the Time Interval. The time Interval is found from Formula 1.

$$TI_i = \max\{|T_{i,j+1} - AT_{i,j}|, j = 1, 2, 3\} \tag{1}$$

Where TIi is Time interval of the i times press; i is the index of times that values are {1, 2, 3, 4, 5, 6}; j is corresponding to Time j.

Table 1. Record register data of a user

	1	2	3	4	5	6
Time 1	0.26553	0.35173	0.26493	0.31681	0.24995	
Time 2	0.29859	0.28445	0.23219	0.31679	0.26709	
Time 3	0.31874	0.29975	0.31581	0.3323	0.2333	
Force 1	0.0225	0.035	0.0375	0.055	0.0125	0.0125
Force 2	0.035	0.1	0.055	0.015	0.025	0.0175
Force 3	0.025	0.0375	0.04	0.025	0.0375	0.035
Average time	0.29429	0.31198	0.2709	0.32198	0.25014	
Time interval	0.05321	0.06727	0.08361	0.01555	0.03371	
Average force	0.03166	0.07916	0.05	0.01833	0.02916	0.0233
Force interval	0.0125	0.065	0.0175	0.04	0.025	0.0225

Due to the system let user input six numbers of force so that system can collect six forces, too. Then use Force 1, Force 2 and Force 3 to calculate the verage Force. And subtract each other to get the Force Interval. By the same way, the system finds the Force Interval from the three times force values shown in Formula 2.

$$FI_i = \max\{|F_{i,j+1} - AF_{i,j}|, j = 1, 2, 3\} \tag{2}$$

Where FI_i is Force interval of the i times press; i is the index of times that values are {1, 2, 3, 4, 5, 6}; j is corresponding to Force j. The Decision Tree will use the records and statistics data to set up.

When user finished register process, they can log in the system. Users don't need to input the user name but in order to do the experiments; the system joins the user's name and password in the user interface. The login of user's interface also has a button to label whether the user is mobile device owner or not. The "Myself Login" button doesn't be used to judgment and just be used to calculate the precious. So, it won't affect this experiment result. Figure 2 show the login of the user interface.

If user login success, the system will start to update the record data and modify Decision Tree. It is different to tradition Decision tree. When users login is successful, the system will get their input time, input force, interval time and interval energy. The system will modify record data.

Fig. 2. User login interface

The Interval Time and Interval Force show in Table 2. Then, the system needs to update the record data and Decision Tree. So, it needs to re-calculate Average Time and Average Force. The system uses Average Time and Average Force from and Time and Force form. Then Interval Time and Interval Force, they also need to update the Average Time and Force dynamically. When those records data update finished, the system rebuilds the Decision Tree.

Table 2. Login data

	1	2	3	4	5	6
Time	0.24855	0.34869	0.2676	0.23331	0.18343	
Force	0.025	0.0675	0.065	0.0375	0.015	0.035
Interval time	0.04573	0.0367	0.0032	0.0886	0.0667	
Interval force	0.00666	0.01166	0.015	0.01916	0.01416	0.01166

The system calculates the Recall, Precious, and F1. These three values are general to estimate the accuracy of the scheme [20].

The system will use confuse matrix to find the precious, recall and F1 values to evaluate the system effectively. Table 3 shows the confusion matrix where "real events" is the real operator who enters the system and "decision event" is our decision tree evaluation results. "0" is not the owner and 1 is the owner who uses the mobile device. Then, the precision, recall, and F1 are defined as to Formulas 3, 4 and 5 listed as follows.

$$Recall = d/(c+d) \tag{3}$$

$$Precision = d/(b+d) \tag{4}$$

$$F1 = 2 * (Recall * Precision)/(Recall + Precision) \tag{5}$$

Table 3. The confusion matrix

Decision events

		0	1
Real events	0	a	c
	1	b	d

4 Experiments

In this experiment, five users are asked to test the system accuracy. Each user inputs his password 50 times and randomly selects the other users, not the five users, to input 50 times. So, each account has 100 times of input data. The results show in Table 4. According to Table 4, we can find the Average Recall is around 96%, Average Precision is about 81%. The values of Precision can use to locate the login of users are the mobile devices owners correctly. The values of Recall can use to observe mobile device owner and the other users are identified right or not. F1 shows the total accuracy based on recall and precious.

Table 4. Experiment results

	Recall	Preciseness	F1
User 1	0.9772	0.7001	0.8129
User 2	0.966	0.895	0.9281
User 3	0.9525	0.7716	0.8488
User 4	0.9289	0.7457	0.8189
User 5	0.9871	0.9667	0.9766
Average	0.96234	0.81582	0.87706

Figure 3 shows the decision tree creation accuracy. The system initial states, the F1 values of five users, are very high or very low because the accuracy depends on users first times login success that is warm start problem. Then, the Decision Tree just starts to learn users' habits, so the accuracy has higher changes during the period. After a period, the system efficiency will become more stable. Recall, Preciseness and F1 values are between 80% and 100% respectively. So, according to all the results, our proposed method is useful for users' identification.

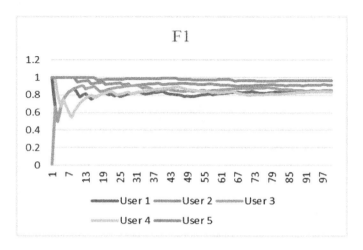

Fig. 3. The F1 values

5   Conclusions

In this paper, we have proposed a method which based on user press password to identify the legal users of mobile devices. The system was implemented on iPhone 6 cell phone. The features are users press time interval and press force on the screen. The decision tree is used to construct a learning system for each user. Experiments results indicated our system is useful. In the future, we will use the different algorithm to improve the preciseness, and it can be utilized in various application or APP.

Acknowledgements. This work was supported by Ministry of Science and Technology, Taiwan, R.O.C. (Grant No. MOST-104-2221-E-324-019-MY2; MOST-106-2221-E-324; MOST-106-2218-E-324-002).

References

1. Rainie, L.: Tablet and e-book reader ownership nearly double over the holiday gift-giving period. Pew Internet. Pew Research Center (2012)
2. Smith, A.: Nearly half of American adults are smartphone owners. Pew Internet. Pew Research Center (2012)
3. Hooten, E.R., Hayes, S.T., Adams, J.A.: Communicative modalities for mobile device interaction. Hum. Comput. Stud. **71**, 998–1002 (2013)
4. http://www.tahrd.ntnu.edu.tw/files/archive/157_9684612a.pdf

5. Kruger, H., Kearney, W.: A prototype for assessing information security awareness. Comput. Secur. **25**, 223–229 (2006)
6. Stirling, D., Pan, Z., Ros, M.: Recognizing human motions through mixture modeling of inertial data Matthew field. Pattern Recognit. **48**, 2394–2406 (2015)
7. Green, N.: Who's watching whom? Monitoring and accountability in mobile relations. In: Brown, B., Green, N., Harper, R. (eds.) Wireless World. Computer Supported Cooperative Work, pp. 32–45. Springer, London (2002). https://doi.org/10.1007/978-1-4471-0665-4_3
8. Jones, O., Williams, M., Fleuriot, C.: "A new sense of place?" The implications of mobile wearable ICT devices for the geographies of urban childhood. Child. Geograph. **1**(2), 165–180 (2003)
9. Fox S.: Online Banking. Report of Pew Internet (2013). http://www.pewinternet.org/Reports/2013/online-banking.aspx
10. Kennedy, N., Sifiso, N.: On privacy calculus and underlying consumer concerns influencing mobile banking subscriptions. In: Information Security for South Africa (ISSA), pp. 1–9, August 2012
11. Moço, N.F., Correia, P.L., Soares, L.D.: Smartphone-based palmprint recognition system. In: Telecommunications (ICT), pp. 457–461, May 2014
12. Johansson, G.: Visual motion perception. Sci. Am. **232**(6), 76–88 (1975)
13. Alex, M., Vasilescu, O.: Human motion signatures: analysis, synthesis, recognition. Department of Computer Science, University of Toronto, vol. 3, pp. 456–460. IEEE, August 2002
14. Lqbal, M.R.A., Rahman, S., Nabil S.I.: Knowledge based decision tree construction with feature importance domain knowledge. In: Electrical Computer Engineering (ICECE), pp. 659–662, December 2012
15. Clayton, A.: A Practical Guide to Knowledge Acquisition. Addison-Wesley, Reading (1991)
16. Mitchell, T.M.: Machine Learning. McGraw-Hill, New York (1997)
17. Mitchell T.M.: Artificial Neural Networks, pp. 81–126. McGraw-Hill Science/Engineering/Math (1997)
18. Vapnik, V.N.: Statistical Learning Theory. Wiley, New York (1998)
19. Imam, I.F., Michalski, R.S.: Should decision trees be learned from examples or from decision rules? In: Komorowski, J., Raś, Z.W. (eds.) ISMIS 1993. LNCS, vol. 689, pp. 395–404. Springer, Heidelberg (1993). https://doi.org/10.1007/3-540-56804-2_37
20. Yukun, C., Yunfeng, L.: An intelligent fuzzy-based recommendation system for consumer electronic products. Expert Syst. Appl. **33**, 230–240 (2007)

A Low-Costed Positioning System Based on Wearable Devices for Elders and Children in a Local Area

Chuan-Bi Lin[(✉)], Yung-Fa Huang[(✉)], Long-Xin Chen,
Yu-Chiang Chang, Z-Ming Hong, and Jong-Shin Chen

Department of Information and Communication Engineering, Chaoyang
University of Technology, Taichung 41349, Taiwan (R.O.C.)
{cblin, s10530603, s10630609, s10630614}@gm.cyut.edu.tw,
yfahuang@mail.cyut.edu.tw, jschen26@cyut.edu.tw

Abstract. Internet of Thing (IoT) technology is an attractive issue because it has been applied widely to a number of aspects, such as smart home, indoor positioning, and autonomous produce and control. Currently, the cares of elders and children existed some problems of management for several years. Therefore, the IoT is expanded to position and monitor walks of elders and children in a local area. However, the proposed positioning or monitoring systems have high costs and power consumption. In addition, most proposed systems are demanded to match up a smartphone, but the smartphones bring other problems for elders and children. Therefore, a wearable device positioning system, in this study, is proposed. This study main aims at addressing the challenges for those proposed systems to develop a low cost and power consumption system by wearable devices and raspberry pi boards through Bluetooth low energy (BLE) transmission. Furthermore, the text marquee and Google Cloud Message push-notification are added to the wearable device positioning system to prevent dangers occurred in the bathrooms and outdoor areas.

Keywords: Wearable device · Positioning system · Bluetooth low energy
Text marquee · Push-notification

1 Introduction

Internet of Thing (IoT) technology has grown rapidly, as it can be applied to a number of aspects including smart home, positioning, monitoring, and autonomous produce and control, etc. [1, 2]. Currently, because the long-term cares for elders become an important issue, the IoT is applied for positioning and monitoring their statuses. Therefore, many IoT systems are proposed for long-term care institutions. According to our concepts, some sensing technologies are used to position and track the elders indoor, as ZigBee [3], Bluetooth Low Energy (BLE) [4], Wi-Fi [5], Radio Frequency IDentification (RFID) [6], and Infrared [7].

Although the proposed RFID techniques, as LANDMARC [8], have been used on the aspect of tracking positions for indoor areas, they are too high cost due the high dense deployments of the tags, locaters, and readers. Besides, a number of nodes, as

© Springer International Publishing AG, part of Springer Nature 2018
N. T. Nguyen et al. (Eds.): ACIIDS 2018, LNAI 10752, pp. 324–332, 2018.
https://doi.org/10.1007/978-3-319-75420-8_31

Zigbee and BLE, is demanded for tracking positions. The power consumption of WiFi technology is higher than those of other proposed techniques. Therefore, the high-costed problem is addressed for the above-mentioned technologies of positioning systems.

To solve the high-costed problem, a novel positioning system is proposed, called iBeacon [9]. The iBeacon is based on BLE [10, 11], and has lower cost for the high dense deployments of iBeacon nodes. The elders have to take smart phones for the iBeacon positioning system. The smart phone can receive a signal form iBeacon nodes and get the information of positions from the databases of computers or servers according to setting information of each iBeacon node, where presenting their locations in the databases of computers. However, taking smart phones is a big problem for elders in their daily routine because the smart phones are lost easily by the elders. The kindergartens exist the same problems for the children as those of long-term care institutions.

Therefore, in this study, a low cost tracking system is proposed. In the proposed system, a wearable device, a raspberry pi board [12] where model B is used, and a website are just demanded, where the wearable device is placed on the hand of an elder or a child. We can track their positions through the proposed approach without any mobile devices and the high dense deployments of iBeacon nodes and RFID requirement. Besides, we use the text marquee and a Google Cloud Messaging (GCM) push-notification [13] to assistant the caregivers to prevent dangers occurred in the bathrooms and outdoor areas.

2 Wearable Device Positioning System Design

The proposed wearable device positioning system has three components: wearable devices (watches), access networks, and a positioning website. The proposed approach aims at addressing the challenges in those long-term care institutions and kindergartens to develop a low-cost wearable device positioning system to understand the environmental, location, and activity variables when elders and children leave for other areas or buildings and close the raspberry pi board (model B) with customized positioning information to caregivers to prevent them to get lost. An overview of the system design is shown in Fig. 1. A wearable device is placed on the hand of a child or an elder to provide the unique MAC address as his/her identity. The device has a Bluetooth transmitter and limited computation power to perform complex data processing. To process data, wearable devices broadcast the sensed data to a sink for processing through a wireless network. The sink, in the form of a raspberry pi board, will pre-process the collected data for compression and encoding and forward the processed data to a database through a wireless and wired access network. Data will be stored in the database and processed to the positioning website for caregivers.

2.1 Wearable Device

We adopt the existing wearable devices. The wearable device includes a microcontroller, nanosecond pulse near-field sensing, 3-axis accelerometer, Bluetooth transmitter, and a rechargeable lithium-ion battery, as shown in Fig. 2(a). The wearable device can be placed on the hand of an elder or a child to broadcast the packet information including Mac address, iBeacon data, Tx Power and RSSI (Received Signal Strength Indicator) parameters, as shown on Table 1. Information from wearable devices is collected and stored on flash memory and also transmitted via a Bluetooth transmitter to a raspberry pi board, as shown in Fig. 2(b).

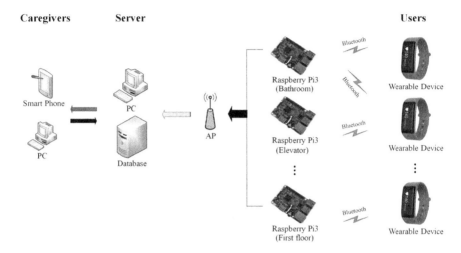

Fig. 1. Wearable device positioning system

Table 1. Packet information of wearable device

Mac	uuid	Major	Minor	Tx Power	RSSI
cc:5d:6c:da:9e:b7	14031900000201060303f0ff08ff0002	1539	512	−96	−60

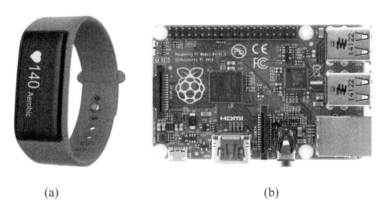

(a)   (b)

Fig. 2. (a) Wearable Device (b) Raspberry Pi Board (Model B)

2.2 Access Networks

We adopt the infrastructure of the existing networks. The access network is used to collect data from wearable devices where raspberry pi boards are used as sink nodes. The radio frequency of wearable devices is based on BLE (2.4 GHz). Wearable devices broadcast the collected data. The raspberry pi boards receive this data and store it on a secure digital (SD) card. The data in a card (and the saved information) can be transmitted through a secure wireless/wired local area network (WiFi/Ethernet). The server or pc receives this data and uses UI (User Interface) to present the positions of elders and children in a local area. Besides, the caregivers can use smart phones or tablets to access the UI via 4G or WiFi.

2.3 Positioning Website

Data from a sink will be send to the database of a server or pc through a wireless and wired access network. A positioning website is made according to these data where the website presents the photos, names, time, and locations of elders and children in a local area. The unique MAC addresses of wearable devices and raspberry pi boards are collocated with names of elders and buildings or areas, respectively. Besides, the text marquee and a GCM push-notification is added to the positioning website to prevent dangers occurred in the bathroom and outdoor areas.

3 Experimental Results

In this study, the proposed wearable device positioning system was tested on senior center and campus at Chaoyang University of Technology, as shown in Figs. 3 and 4. The following activities were recorded from the elders with wearable devices strapped on the hands of these subjects: sitting in the room, going to the bathroom, leaving from center, walking and standing on the campus.

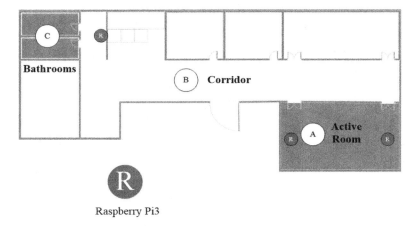

Fig. 3. Senior center plan blueprint

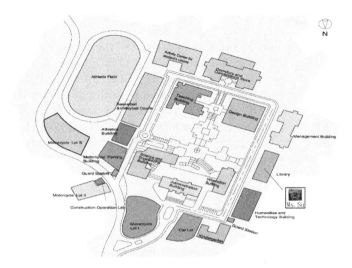

Fig. 4. Campus plan blueprint

Figure 5 shows the positions of elders with wearable devices through the positioning website page, while standing on and going to any area or building of campus. To prevent danger occurred in the bathroom and outside of active room, we apply a text marquee and a GCM push-notification to the proposed system. If an elderly people spends a longer time than usual time while staying at the restrooms or outdoor areas, the text marquee with red color is immediately appeared on the upper left of a positioning website, and an email and a GCM push-notification is also sent by the system to alert staff regarding the potential danger, as shown in Fig. 6(a) and (b).

Statistics Total: 18 Attendance: 16 Absence: 2

#	Name	Location	Time	#	Name	Location	Time	#	Name	Location	Time
	Ms. Su	Absence			Ms. Chen	Library			Ms. Syue	Management Building	
	Ms. Lin	Indoor			Ms. Yang	Library			Ms. Dai	Bathroom	
	Ms. Zeng	Indoor			Ms. Liu	Indoor			Ms. Wen	Indoor	
	Ms. Chen	Outdoor			Ms. Bian	Bathroom			Ms. Lin	Indoor	
	Ms. Huang	Absence			Mr. Liu	Indoor			Mr. Yang	Indoor	
	Mr. Sie	Indoor			Mr. Chen	Indoor			Mr. Chen	Indoor	

Fig. 5. Positioning website

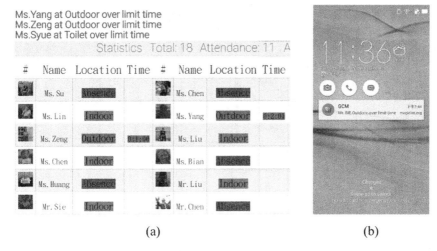

(a)   (b)

Fig. 6. (a) Text marquee on website (b) FCM push-notification

On the other hand, to observe the accurate data of positions between elders and buildings from the measured wearable device data, the collected data are processed using dynamic walk analysis through relative majority decision. We demonstrate the advantages of using dynamic walk analysis (DWA) with the kinematic data collected when subjects perform the walk for a short period of time and using RSSI (Received Signal Strength Indicator) to differentiate past, in and leave movements.

4 Dynamic Walk Analysis

When a child or an elder goes through the building installed Raspberry Pi Boards, we cannot know whether he/she enter the building or not. To increase the accuracy of judgments, the DWA is adopted to the proposed system. The DWA main collects the experimental data and analyses three kinds of walking statuses: in, leave, and past. We put two raspberry pi boards, called Ri and Rd, in the gate and central position of a room, respectively, as shown in Fig. 7. Where the red, black, and white arrow lines are used to present past, in, and leave walks, respectively. Here, we test 30 times at a speed 0.5–1.0 m/s for each walk and simulate them according to receiving the strengths of RSSI from two raspberry pi boards. Then, we create three kinds of referenced models through simulations for three kinds of walks: in, leave, and past. Therefore, when the elders and children are in, leave, or past the buildings, we can adopt the two Raspberry Pi boards and three referenced models to increase the accuracy of judgments.

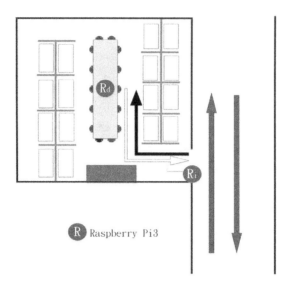

Fig. 7. Experimental field for dynamic walk analysis

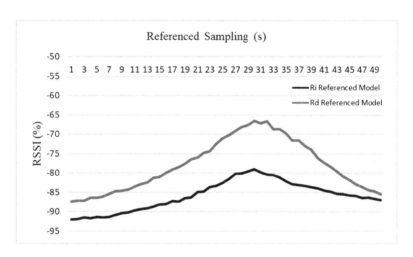

Fig. 8. Referenced sample of "Past Walk"

Figure 8 shows the Ri and Rd referenced models of "Past Walk", when an elder or a child goes through the building according to receiving signal strengths from Ri and Rd boards. Rd has stronger signal than Ri because the wearable device of elder is closed to Rd board when going through the room. On the other hand, we also create the Ri and Rd referenced models of "In Walk" and "Leave Walk", as shown in Figs. 9 and 10, respectively. For instance, Ri has stronger signal than Rd at the beginning, but Rd becomes stronger than Ri with leaving steps, as shown in Fig. 10.

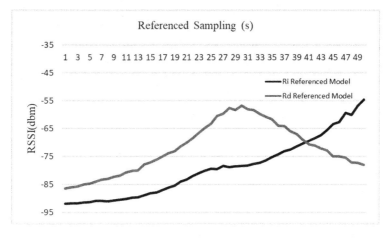

Fig. 9. Referenced sample of "In Walk"

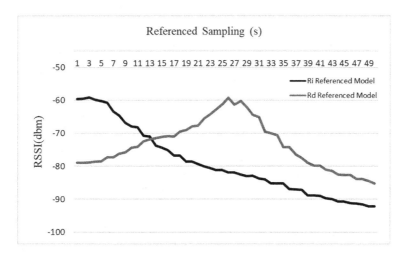

Fig. 10. Referenced sample of "Leave Walk"

5 Conclusions

In this paper, an approach for the positioning of the movements of children and elders in a local area is presented. The approach is based on wearable devices and raspberry pi boards (model B) supported by wireless transmissions to forward collected data to storage for location analysis. The wireless transmissions are based on IEEE 802.15.1 and IEEE 802.11g, also known as BLE and Wireless Local Area Network, respectively. When a child or an elder with a wearable device is in a building or an area installed a raspberry pi board, we can know obviously his/her location from the monitoring web. In addition, we adopt the experiment of Dynamic Walk Analysis

(DWA) to analyze collected data from two raspberry pi boards to increase the accuracy of estimating locations. The result is expected to be spread to kindergartens and long-term care institutions to enhance the care safeties of children and elders, and to decrease the costs of caregivers and time. Furthermore, it is expected that more physical signals of elders and children from the wearable devices, and we can prevent more dangers via the data in the future.

References

1. Bhilare, R., Mali, S., Dutkiewicz, E.: IoT based smart home with real time E-metering using E-controller. In: Annual IEEE India Conference (INDICON 2015), pp. 1–6, New Delhi (2015)
2. Zhang, L., Liu, J., Jiang, H.: Energy-efficient location tracking with smartphones for IoT. In: IEEE SENSORS 2012, pp. 1–4, Taipei (2012)
3. Fang, S.-H., Wang, C.-H., Huang, T.-Y., Yang, C.-H., Chen, Y.-S.: An enhanced ZigBee indoor positioning system with an ensemble approach. IEEE Commun. Lett. 16(4), 564–567 (2012)
4. Collotta, M., Pau, G.: A novel energy management approach for smart homes using bluetooth low energy. IEEE J. Sel. Areas Commun. 33(12), 2988–2996 (2015)
5. Bahl, P., Padmanabhan, V.N.: RADAR: an in-building RF-based user location and tracking system. In: Proceedings IEEE INFOCOM 2000, Conference on Computer Communications, Nineteenth Annual Joint Conference of the IEEE Computer and Communications Societies (Cat. No. 00CH37064), pp. 775–784, Tel Aviv (2000)
6. D3.7 A Structured Collection on Information and Literature on Technological and Usability Aspects of Radio Frequency Identification (RFID). http://www.fidis.net/resources/fidis-deliverables/hightechid/int-d3700/. Accessed 18 Nov 2016
7. Want, R., Hopper, A., Falcao, V., Gibbons, J.: The active badge location system. ACM Trans. Inform. Syst. (TOIS) 10(1), 91–102 (1992)
8. Ni, L.M., Liu, Y., Lau, Y.C., Patil, A.P.: LANDMARC: indoor location sensing using active RFID*. In: Proceedings of IEEE International Conference on Pervasive Computing and Communications, (PerCom 2003), pp. 407–415 (2003)
9. Burzacca, P., Mircoli, M., Mitolo, S., Polzonetti, A.: iBeacon technology that will make possible Internet of Things. In: International Conference on Software Intelligence Technologies and Applications & International Conference on Frontiers of Internet of Things, pp. 159–165, Hsinchu (2014)
10. Ji, M., Kim, J., Jeon, J., Cho, Y.: Analysis of positioning accuracy corresponding to the number of BLE beacons in indoor positioning system. In: 17th International Conference on Advanced Communication Technology (ICACT 2015), pp. 92–95, Seoul (2015)
11. Aman, M.S., Jiang, H., Quint, C., Yelamarthi, K., Abdelgawad, A.: Reliability evaluation of iBeacon for micro-localization. In: IEEE 7th Annual Ubiquitous Computing, Electronics & Mobile Communication Conference (UEMCON 2016), pp. 1–5, New York (2016)
12. Meet the New Raspberry Pi 3—A 64-bit Pi with Built-in Wireless and Bluetooth LE. https://makezine.com/2016/02/28/meet-the-new-raspberry-pi-3/. Accessed 18 Jan 2017
13. Yilmaz, Y.S., Aydin, B.I., Demirbas, M.: Google Cloud Messaging (GCM): An evaluation. In: IEEE Global Communications Conference, pp. 2807–2812, Austin (2014)

Intelligent Systems and Methods in Biomedicine

PCA-KNN for Detection of NS1 from SERS Salivary Spectra

N. H. Othman[1], Khuan Y. Lee[1(✉)] ⓘ, A. R. M. Radzol[1], W. Mansor[1],
P. S. Wong[2], and I. Looi[3]

[1] Faculty of Electrical Engineering and Computational Intelligence EK,
Pharmaceutical & Lifesciences Communities of Research,
Universiti Teknologi MARA, Shah Alam, Malaysia
leeyootkhuan@salam.uitm.edu.my
[2] Infectious Unit, Pulau Penang General Hospital, George Town, Malaysia
[3] Seberang Prai General Hospital, Seberang Perai, Malaysia

Abstract. K-Nearest Neighbor (kNN) has shown its strong capability in pattern recognition, classification and machine learning applications. In this paper, kNN was used to distinguish between Non-structural protein 1 (NS1) positive and NS1 negative dengue patients from salivary Raman spectra. The presence of NS1 was detected in the saliva of dengue infected subjects. It was found Raman active, producing a molecular Raman fingerprint. Surface Enhanced Raman Spectroscopic (SERS) technique was adopted in obtaining the NS1 Raman spectra dataset. Performance of kNN with different K-values, optimized with Scree, Cumulative Percentage Variance (CPV) and Eigenvalue One Criterion (EOC) stopping criteria, was investigated and compared in term of sensitivity, specificity, accuracy and kappa. The best performance is found with the use of CPV stopping criteria and a K-value of 5, which attained an accuracy of 84.5% and kappa of 0.69.

Keywords: NS1 · SERS · PCA · kNN

1 Introduction

Raman Spectroscopy (RS) and Surface Enhanced Raman Spectroscopy (SERS) are emerging as useful clinical non-invasive and label free techniques to extract useful information from body fluids [1]. RS and SERS are laser based analytical techniques that capture the inelastic scattering of photons occurring from interaction of samples with monochromatic light [2]. The advantages of RS and SERS are as follows: the analysis is simple and fast; the quantity of samples required is small; the samples do not require special preparation; the tests are non-destructive and easily reproducible [3]. Owing to this, RS and SERS are increasingly gaining acceptance in disease detection applications, that includes oral cancer [1], breast cancer [4], acute myocardial infraction [5], colon cancer [6] and nasopharyngeal cancer [7]. SERS is more popular in practice than RS because it is able to amplify weak Raman signal to a usable range. SERS has been reported to have boosted the Raman signal by 10^{13} to 10^{15} times [4].

© Springer International Publishing AG, part of Springer Nature 2018
N. T. Nguyen et al. (Eds.): ACIIDS 2018, LNAI 10752, pp. 335–346, 2018.
https://doi.org/10.1007/978-3-319-75420-8_32

For the past few years, the usage of saliva for diagnostic applications has been on an increasing trend [8]. In comparison to blood collection, specially trained specialists are not required for saliva collection. In addition, saliva is readily available and its collection is simple, non-invasive, painless, and inexpensive [1, 3]. Furthermore, saliva contains abundance of human metabolites and proteins of significant to biological functions [4, 5]. With the wide spectrum of valuable molecular information, it presents itself a preferred clinical diagnostic medium to other body fluids, especially as an alternative to blood [1]. Raman spectra of saliva are used in disease diagnosis such as acute myocardial infraction [5], oral cancer [1] and breast cancer [4].

Non-structural protein 1 (NS1) is one of non-structural protein in the genome of flaviviruses such as Dengue virus, Japanese Encephalitis virus, Murray Valley Encephalitis virus, Tick borne Encephalitis virus, West Nile Encephalitis virus and Yellow Fever virus [9]. Its presence was detected in the saliva of dengue infected subjects [10]. However, detection of NS1 in saliva is lower in performance than NS1 in blood. NS1 is found Raman active, hence producing a molecular Raman fingerprint [11]. Integrating the advantages of SERS and signal processing techniques, detection of NS1 in saliva is potentially promising.

In this work, K-nearest neighbor (kNN) classification algorithm is integrated with Principal Component Analysis (PCA) to classify the positive and negative dengue cases based on salivary Raman spectra. The effect of nearest neighbor value (K-value) and PCA stopping criteria on performance of the PCA-kNN classifier is investigated. Section 2 details on the theoretical background of PCA, kNN and performance indicating coefficients. Section 3 elaborates on the methodology. Section 4 discusses the performance of kNN classifier with different nearest neighbor values and stopping criteria, in terms of sensitivity, specificity accuracy and kappa coefficients.

2 Theoretical Background

2.1 Principal Component Analysis (PCA)

PCA is the most famous and probably the oldest multivariate analysis techniques introduced by Pearson (1901) [12]. This technique is not widely used until the electronic computer was invented.

PCA is a variable reduction algorithm, which reduces the dimensionality of dataset with a large number of variables, while maintaining as much as possible the characteristic of the dataset. Variance of the dataset is ranked as principal components (PCs) based on the covariance eigenvalues, which measures the variability in the distribution of the dataset. The average squared deviation of each sample from its mean describes the variance [12] as follows,

$$Variance, \ s^2 \ = \ \frac{\sum(sample - mean)^2}{Total \ sample - 1} \tag{1}$$

Variables with significant variance are captured as highly ranked PCs, while variables with less variance are ranked lower. Retaining only the highly ranked PCs maintains essence of the original dataset. This means lower ranked PCs can be obsoleted to reduce data dimension. The number of PCs to be kept is central to reporting originality of data. Overestimation includes unimportant characteristic, while underestimation removes important characteristic. PCA stopping criteria serve as guidelines to determine the cutoff between variance of different significance. In our study, three stopping criteria are considered, (i) Eigenvalue-One Criterion (EOC), (ii) Cattell's Scree test and (iii) Cumulative Percent of Variance (CPV).

2.2   K-Nearest Neighbor (KNN)

The concept of K-Nearest Neighbor (kNN) algorithm is first developed by Fix and Hodges in year 1970 [13]. It is a method used to perform the classification of objects based on learning data that are located closest to the object [14, 15]. It is also known as the lazy learner algorithm [17], being one of the simplest classifiers at no sacrifice to accuracy [16]. Good performance has been reported by researchers using kNN algorithm on different experiments and dataset [15].

The kNN classifier is a non-parametric learning algorithm for classification and regression problems. It determines the class for the query data based on the class of the majority of the K nearest neighbors. This method is suitable if the information on the data distribution is insufficient [15]. Unlike other classifying algorithms such as ANN, SVM, LDA, kNN requires no training stage, since all the data can be used as training data to classify an unknown samples [14, 18].

Implementation of kNN involves two steps: first is to select the optimum classifier value (K-value), then second to calculate the distance between a query sample to its K-nearest neighbors. The query sample is classified as label of the majority K- nearest neighbors with similar features. In kNN, data space (X) is described by an n-dimensional vector as in Eq. (2). Each element X_i has m features as described in Eq. (3).

$$X = \{X_1, X_2, \ldots, X_n\} \tag{2}$$

$$X_i = (X_{i1}, X_{i2}, \ldots, X_{im}) \tag{3}$$

To determine the class of an element, X_i, its distance from the neighbouring elements is measured. Then, it is assigned to the class of the K-nearest neighbours specified by a distance rule. The optimum K-value is highly dependent on the characteristic of the data. In general, a larger K-value reduces the classification noise, but blurs the inter-class boundaries, while a smaller one otherwise. An optimum K-value can be determined through parameter estimation approaches such as cross validation [14]. Classification performance of kNN also depends on the distance rule. There exist many types of distance rule to compute the distance between the query sample and its neighbors [13]; examples are Euclidean, Cityblock, Cosine and Correlation distance rules.

Euclidean distance rule is the most preferred and widely used [14, 18–20]. Equation (4) defines the Euclidean distance between the query sample and its neighbor.

$$d = \sqrt{(x_{i1} - x_1')^2 + (x_{i2} - x_2')^2 + \ldots + (x_{if} - x_f')^2} \tag{4}$$

2.3   Performance Indicator

Sensitivity, specificity and accuracy are indicators used to evaluate the performance of a classifier. In this study, sensitivity refers to the ability of the kNN classifier to correctly classify positive dengue patients. Specificity is measure for the ability of the classifier to identify negative samples correctly, while accuracy reflects on the classifier ability in providing correct classification for both cases [21]. The classifier performance is benchmarked against NS1- Enzyme-Linked Immunosorbent Assay (ELISA) test, a WHO recommended test. Equations (5), (6) and (7) describe accuracy, sensitivity, and specificity, respectively.

$$Accuracy = \frac{TP + TN}{TP + FN + TN + FP} \tag{5}$$

$$Sensitivity = \frac{TP}{TP + FN} \tag{6}$$

$$Specificity = \frac{TN}{TN + FP} \tag{7}$$

where TP is true positive cases, TN is true negative cases, FN is false negative cases and FP is false positive cases

Kappa is a measure that reflects agreement between researchers, clinicians or classifiers for the same or different examination tools in disease diagnosis [22]. In addition to accuracy (observed agreement), it includes the probability of random agreement (chance agreement) between the raters. The following Eq. (8) calculates the kappa coefficient,

$$Kappa = \frac{Observed\,agreement \; - \; Chance\,agreement}{1 \; - \; Chance\,agreement} \tag{8}$$

where the chance agreement, p_e is defined as,

$$p_e = \left(\frac{TP + FP}{TP + FN + TN + FP}\right)\left(\frac{TP + FN}{TP + FN + TN + FP}\right)$$
$$+ \left(\frac{FN + TN}{TP + FN + TN + FP}\right)\left(\frac{FP + TN}{TP + FN + TN + FP}\right) \tag{9}$$

Kappa value spreads over a range [−1 to 1]. A value of less than '0' indicates no agreement; that between '0.01' to '0.20' as none to slight agreement; that between '0.21' to '0.40' as fair agreement; that between '0.41'–'0.60' as moderate agreement; that between '0.61' to '0.80' as substantial agreement and that between '0.81'–'1.00' as almost perfect agreement [23]. In this study, Kappa value is used to measure classification agreement between the NS1-ELISA test and the kNN classifier models.

3   Methodology

This section describes the database and classification procedure for this study. Dataset for this study is obtained from UiTM-NMRR 12868-NS1-DENV database. It consists of Raman spectra from 284 saliva samples of suspected dengue patients and control

Table 1. NS1-ELISA dataset

Dataset	Patients (positive)	Patients (negative)	Healthy volunteer (negative-control)
NS1-ELISA	142	82	60

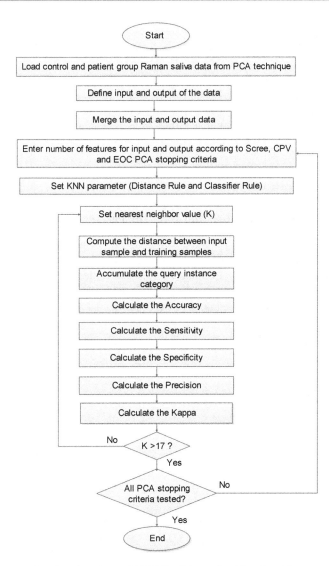

Fig. 1. Flowchart of kNN-Euclidean classifier algorithm

group. Out of this, 142 of the patient samples are diagnosed as dengue positive and 82 are diagnosed as dengue negative, using NS1-ELISA technique. The remaining 60 are samples collected from healthy volunteers. Table 1 summarizes the dataset.

In this study, PCA was used as the feature extraction algorithm and kNN was used as the classification algorithm. Prior to PCA, the Raman spectra were pre-processed to remove the spurious noise, background noise, fluorescent effect, signal drifts and spikes from the cosmic ray in the spectra using background subtraction, baseline removal, smoothing and normalization methods. Then, the pre-processed spectra were analyzed with PCA. The number of PCs retained are estimated based on the PCA stopping criteria, as mentioned in Sect. 2.1.

The retained PCs were served as inputs to the kNN algorithm. Euclidean distance metric was chosen as the distance rule, while nearest neighbor the classifier rule, with nearest neighbor value (K-value) of 1, 3, 5, 7, 9, 11, 13, 15 and 17. The first step computes the Euclidean distance between the input samples and training samples. Then, the classifier divides the test samples into positive or negative classes, accordingly. For every combination of K-values and stopping criteria, the classifier performance was evaluated in term of accuracy, precision, specificity, sensitivity and kappa. The algorithm, developed in Matlab R2014a, is depicted in flowchart, as shown in Fig. 1.

4 Result and Discussion

4.1 PCA for Feature Extraction and Data Dimensional Reduction

This section presents the finding obtained from PCA, the feature extraction stage. PCA produces PCs to input to kNN classifier. Figure 2 shows the score plot of the two most significant principal components, PC1 and PC2. Negative dengue cases are represented by (blue 'o') and positive dengue cases are represented by (red 'x'). From the plot, it is observed that all of the positive samples of PC1 are located on left side while negative

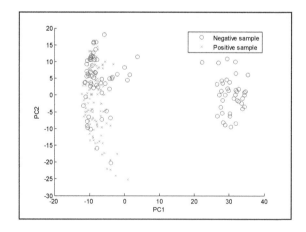

Fig. 2. 2D scatter plot of PC1 and PC2 (Color figure online)

samples of PC1 are distributed on both sides; illustrating the dataset is not linearly separable. If only these two PCs are used as classifier inputs, poor classification performance is expected. Thus, it is necessary to include more PCs to improve the performance.

Three PCA stopping criteria mentioned in Sect. 2.1 are used to estimate the optimum number of PCs. The number of PCs retained by the different PCA stopping criteria, are summarized in Table 2.

Table 2. Number of retained PCs, cumulative percentage variance and percentage of data reduction according to Scree test, CPV and EOC criteria

Stopping criteria	No of retained PCs	Cumulative percentage variance (%)	Data reduction (%)
Scree test	7	46.94	99.61
CPV	95	90.18	94.73
EOC	126	94.49	93.00

Table 2 shows PCs of 7, 95 and 126 being retained according to Scree test, CPV and EOC, respectively. Scree test achieved the most in data reduction (99.61%). However, it only embodies 46.94% of cumulative variance within the dataset. More cumulative variance is achieved with CPV (90.18%) and EOC (94.49%) with acceptably high reduction in data dimension of more than 93%. Using the number of retained PCs proposed by the criteria, the performance of kNN classifier is evaluated.

4.2   Performance of kNN Classifier

This section discusses the performance of kNN classifier for the different PCA stopping criteria and different K-values in term of sensitivity, specificity, accuracy, and kappa value. The results are presented in Tables 3, 4, 5 and 6, respectively.

Table 3 shows the sensitivity performance. Overall, it is observed that, the sensitivity performance ranges from 71.4% to 95.2%. Better sensitivity is achieved with CPV and EOC relative to Scree test. The highest sensitivity performance of 95.2% is achieved with both CPV and EOC criteria but at different K-values. For all the criteria, it is observed that as K-value increases, the sensitivity increases to its maximum performance and then settles at a slightly lower sensitivity.

Table 4 shows the specificity performance, which reports achievement from 52.4% to 83.3%. Similar to sensitivity, better specificity is found with more retained PCs as suggested by CPV and EOC criteria. As K-value increases, the specificity performance increases to it maximum value, then decreases to a lower value, in the case of CPV and EOC criteria. However, a fluctuating trend is observed in the case of EOC criterion. The highest specificity of 83.3% is obtained with EOC criterion.

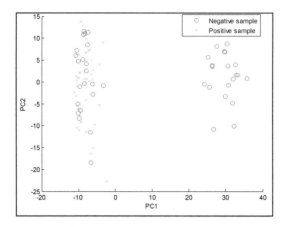

Fig. 3. Score plot of PC1 and PC2 of test samples. (Color figure online)

Figure 3 shows the score plot distribution (PC1 and PC2) of the test sample, benchmarked against the NS1-ELISA test. From the plot, there are 22 negative samples (blue 'o') located on the left side, mixing with positive samples. These samples are easily misclassified as positive samples by the classifier. Figure 4(a) and (b) show classification by the kNN algorithm for Scree test and CPV, respectively. It shows that most of the samples located on the left side of the plot are classified as positive samples, by Scree test criterion. Only 9 samples are classified as negative samples. It agrees with the low specificity performance in Table 4. As more PCs are included as classifier inputs based on CPV (95 PCs), more samples are classified as negative samples as shown in Fig. 4(b). This increases the specificity performance as shown in Table 4.

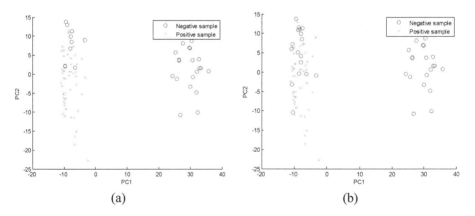

(a)   (b)

Fig. 4. Classification based on kNN classifier with (a) Scree test criterion (lowest accuracy), (b) classification based on kNN classifier with CPV criterion (highest accuracy) (Color figure online)

Table 3. Sensitivity Performance for Different PCA-Euclidean-kNN.

Nearest neighbor value (K)	PCA stopping criteria		
	Scree (7 PCs)	CPV (95 PCs)	EOC (126 PCs))
K = 1	78.6	76.2	*71.4
K = 3	85.7	83.3	88.1
K = 5	83.3	90.5	88.1
K = 7	83.3	90.5	88.1
K = 9	83.3	92.9	*95.2
K = 11	83.3	*95.2	*95.2
K = 13	83.3	*95.2	92.9
K = 15	83.3	92.9	92.9
K = 17	81.0	92.9	92.9

Table 4. Specificity performance for different PCA-Euclidean-kNN.

Nearest neighbor value (K)	PCA stopping criteria		
	Scree (7 PCs)	CPV (95 PCs)	EOC (126 PCs)
K = 1	59.5	76.2	*83.3
K = 3	61.9	81.0	78.6
K = 5	61.9	78.6	76.2
K = 7	61.9	71.4	71.4
K = 9	66.7	66.7	69.0
K = 11	59.5	69.0	71.4
K = 13	59.5	69.0	71.4
K = 15	54.8	64.3	66.7
K = 17	*52.4	64.3	61.9

Table 5. Accuracy performance for different PCA-Euclidean-kNN.

Nearest neighbor value (K)	PCA Stopping Criteria		
	Scree (7 PCs)	CPV (95 PCs)	EOC (126 PCs)
K = 1	69.0	76.2	77.4
K = 3	73.8	82.1	83.3
K = 5	72.6	*84.5	82.1
K = 7	72.6	81.0	79.8
K = 9	75.0	79.8	82.1
K = 11	71.4	82.1	83.3
K = 13	71.4	82.1	82.1
K = 15	69.0	78.6	79.8
K = 17	*66.7	78.6	77.4

Table 6. Kappa performance for different PCA-Euclidean-kNN.

Nearest neighbor value (K)	PCA stopping criteria		
	Scree (7 PCs)	CPV (95 PCs)	EOC (126 PCs)
K = 1	0.38	0.52	0.55
K = 3	0.48	0.64	0.67
K = 5	0.45	*0.69	0.64
K = 7	0.45	0.62	0.60
K = 9	0.5	0.60	0.64
K = 11	0.43	0.64	0. 67
K = 13	0.43	0.64	0.64
K = 15	0.38	0.57	0.60
K = 17	*0.33	0.57	0.55

With reference to Table 5, the accuracy performance achieved for the different PCA and K-values ranges from 66.7% to 84.5%. For Scree test criterion, the highest accuracy achieved is 75% with K-value of 9. Increasing the number of retained PCs to 95 as proposed by the CPV criterion, the accuracy increases to 84.5% when K-value is set to 5. Using the EOC criterion, the highest accuracy achieved is 83.3%, corresponding to K-value of 3 and 11. Thus the best accuracy is achieved with the CPV criterion.

It can be observed that the accuracy and kappa value are directly proportional. As the accuracy increases, the kappa value also increases. From Table 6, it is observed that the best kappa value achieved is 0.69 with the CPV criterion. This indicates a substantial agreement between classification by NS1-ELISA and kNN algorithm.

Based on the result obtained, it can be ascertained that, of the three criteria, the CPV criterion scores the highest performance in accuracy and kappa value, when K-value is 5. The Scree test gives a lower performance than the CPV and EOC criterion, which is in concordance to our previous findings using an ANN classifier [24]. The hypothesis is that, since the Scree test criterion retains only 7 PCs, inheriting only 46.94% of the total cumulative percentage variance of the dataset, it is insufficient to represent important attributes of the original signal for classification of infected salivary SERS spectra, as much as the CPV and EOC criterion.

5   Conclusion

Our work here investigates the performance of PCA-kNN classifier models in detection of NS1 from SERS spectra of saliva of dengue positive and negative samples. It is found that performance of the kNN classifier models is dependent on the nearest neighbor value and number of principal components. Results here show that by using suitable number of principal component and K-value, kNN is capable to detect NS1 from salivary SERS spectra of dengue infected saliva. It is found that the best K-value is 5, which yields an accuracy of 84.5% and kappa of 0.69, with the CPV-kNN classifier model.

Acknowledgment. The author would like to thank the Ministry of Education (MOE) of Malaysia, for providing the research funding 600-RMI/ERGS 5/3 (20/2013); the Research Management Institute, and the Faculty of Electrical Engineering, Universiti Teknologi MARA, Malaysia, for the support and assistance given to the authors in carrying out this research.

References

1. Cao, G., Chen, M., Chen, Y., Huang, Z., Lin, J., Lin, J., Xu, Z., Wu, S., Huang, W., Weng, G., Chen, G.: A potential method for non-invasive acute myocardial infarction detection based on saliva Raman spectroscopy and multivariate analysis. Laser Phys. Lett. **12**, 1–6 (2015). https://doi.org/10.1088/1612-2011/12/12/125702
2. Connolly, J.M., Davies, K., Kazakeviciute, A., Wheatley, A.M., Dockery, P., Keogh, I., Olivo, M.: Non-invasive and label-free detection of oral squamous cell carcinoma using saliva surface-enhanced Raman spectroscopy and multivariate analysis. Nano Med. Nanotechnol. Biol. Med. **12**, 1593–1601 (2016). https://doi.org/10.1016/j.nano.2016.02.021
3. Sun, S., Wang, X., Gao, X., Ren, L., Su, X., Bu, D., Ning, K.: Condensing Raman spectrum for single-cell phenotype analysis. BMC Bioinform. **16**, 1–7 (2015). https://doi.org/10.1186/1471-2105-16-S18-S15
4. Gonchukov, S., Sukhinina, A., Bakhmutov, D., Minaeva, S.: Raman spectroscopy of saliva as a perspective method for periodontitis diagnostics. Laser Phys. Lett. **9**, 73–77 (2012). https://doi.org/10.1002/lapl.201110095
5. Feng, S., Huang, S., Lin, D., Chen, G., Xu, Y., Li, Y., Huang, Z., Pan, J., Chen, R., Zeng, H.: Surface-enhanced Raman spectroscopy of saliva proteins for the noninvasive differentiation of benign and malignant breast tumors. Int. J. Nano Med. **10**, 537–547 (2015). https://doi.org/10.2147/IJN.S71811

6. Li, X., Yang, T., Li, S., Wang, D., Song, Y., Zhang, S.: Raman spectroscopy combined with principal component analysis and k nearest neighbour analysis for non-invasive detection of colon cancer. Laser Phys. **26**, 1–9 (2016). https://doi.org/10.1088/1054-660X/26/3/035702
7. Xu, Z., Ge, X., Huang, W., Lin, D., Wu, S., Lin, X., Wu, Q., Sun, L.: Nasopharyngeal carcinoma detection by tissue smears using surface-enhanced Raman spectroscopy. Spectrosc. Lett. **50**, 17–22 (2017). https://doi.org/10.1080/00387010.2016.1268164
8. Kaufman, E., Lamster, I.B.: The diagnostic applications of saliva—review. Crit. Rev. Oral Biol. Med. **13**, 197–212 (2002). https://doi.org/10.1177/154411130201300209
9. Lindenbach, B.D., Rice, C.M.: Flaviviridae: The Viruses and Their Replication. Fields Virology, 5th edn. Lippincott-Raven, Philadelphia (2007). https://doi.org/10.1016/0038-092x(88)90131-4
10. Andries, A.-C., Duong, V., Ong, S.: Evaluation of the performances of six commercial kits designed for dengue NS1 and anti-dengue IgM, IgG and IgA detection in urine and saliva clinical specimens. BMC Infect. Dis. **16**, 1–9 (2016). https://doi.org/10.1186/s12879-016-1551-x
11. Korhonen, E.M., Huhtamo, E., Virtala, A.M.K., Kantele, A., Vapalahti, O.: Approach to non-invasive sampling in dengue diagnostics: exploring virus and NS1 antigen detection in saliva and urine of travelers with dengue. J. Clin. Virol. **61**, 353–358 (2014). https://doi.org/10.1016/j.jcv.2014.08.021
12. Aparna, A., Vinod Kumar, R., Muralikrishna, V., Kumar, R.P.: Diagnostic utility of saliva as non-invasive alternative to serum in suspected dengue patients. IOSR J. Dent. Med. Sci. **14**, 26–32 (2015). https://doi.org/10.9790/0853-14962632
13. Andries, A.C., Duong, V., Ly, S., Cappelle, J., Kim, K.S., Lorn Try, P., Ros, S., Ong, S., Huy, R., Horwood, P., Flamand, M., Sakuntabhai, A., Tarantola, A., Buchy, P.: Value of routine dengue diagnostic tests in urine and saliva specimens. PLoS Negl. Trop. Dis. **9**, 1–30 (2015). https://doi.org/10.1371/journal.pntd.0004100
14. Radzol, A.R.M., Lee, Y.K., Mansor, W.: Raman molecular fingerprint of non-structural protein 1 in phosphate buffer saline with gold substrate. In: Proceedings of International Annual Conference on IEEE Engineering Medical Biology Society, pp. 1438–1441 (2013). https://doi.org/10.1109/embc.2013.6609781
15. Joliffer, I.T.: Principal Component Analysis. Springer Series of Statistic, 2nd edn. Springer, Heidelberg (2002). https://doi.org/10.2307/1270093
16. Fix, J., Hodges, L.: Discriminatory analysis, nonparametric discrimination: consistency properties. US Air Force School of Aviation Medicine. Technical report 4 (1951). https://doi.org/10.1007/978-3-642-38652-7
17. Nirmaladevi, M., Alias Balamurugan, S.A., Swathi, U.V.: An amalgam KNN to predict diabetes mellitus. In: Proceedings of IEEE International Conference on Emerging Trends in Computing, Communication and Nanotechnology, pp. 691–695 (2013). https://doi.org/10.1109/ice-ccn.2013.6528591
18. Shirvan, R., Tahami, E.: Voice analysis for detecting Parkinson's disease using genetic algorithm and KNN classification method. In: Proceedings of Biomedical Engineering, pp. 14–16 (2011). https://doi.org/10.1109/icbme.2011.6168572
19. Dudani, S.A.: The distance-weighted k-nearest-neighbor rule. In: Proceedings of IEEE Transactions on Systems, Man, and Cybernetics, pp. 325–327 (1976). https://doi.org/10.1109/tsmc.1976.5408784
20. Anchalia, P.P., Roy, K.: The k-nearest neighbor algorithm using MapReduce paradigm. In: Proceedings of International Conference on Intelligent Systems, Modelling and Simulation, pp. 513–518 (2015). https://doi.org/10.1109/isms.2014.94

21. Odajima, K., Pawlovsky, A.P.: A detailed description of the use of the kNN method for breast cancer diagnosis. In: Proceedings of International Conference on Biomedical Engineering and Informatics, pp. 688–692 (2014). https://doi.org/10.1109/bmei.2014.7002861

22. Saini, I., Singh, D., Khosla, A.: QRS detection using K-Nearest Neighbor algorithm (KNN) and evaluation on standard ECG databases. J. Adv. Res. **4**, 331–344 (2013). https://doi.org/10.1016/j.jare.2012.05.007

23. Yuan, W., Juan, L., Zhou, H.-B.: An improved KNN method and its application to tumor diagnosis. In: Proceedings of International Conference on Machine Learning and Cybernetics, pp. 2836–2841 (2004). https://doi.org/10.1109/icmlc.2004.1378515

24. Othman, N.H., Lee, K.Y., Radzol, A.R.M., Mansor, W.: PCA-SCG-ANN for detection of non-structural protein 1 from SERS salivary spectra. In: Nguyen, N.T., Tojo, S., Nguyen, L. M., Trawiński, B. (eds.) ACIIDS 2017. LNCS (LNAI), vol. 10192, pp. 424–433. Springer, Cham (2017). https://doi.org/10.1007/978-3-319-54430-4_41

Dynamic Modeling of the Czech Republic Population with a Focus on Alzheimer's Disease Patients

Hana Tomaskova[1(✉)], Petra Maresova[1], Daniel Jun[2], Martin Augustynek[3],
Jan Honegr[2], and Blanka Klimova[1]

[1] Faculty of Informatics and Management, University of Hradec Kralove,
Hradec Kralove, Czech Republic
hana.tomaskova@uhk.cz
[2] Biomedical Research Centre, University Hospital Hradec Kralove,
Hradec Kralove, Czech Republic
[3] Faculty of Electrical Engineering and Computer Science,
Technical University of Ostrava, Poruba, Czech Republic

Abstract. This article focuses on the Czech population modeling with
the main aim to predict the Alzheimer's disease patients numbers. We
choose a system dynamics as a modeling tool for this phenomenon. The
article describes Alzheimer's disease and its causes, the expected development of the population by 2100, the AD patient's development, and
last but not least the theory of system thinking. The outputs are based
on the statistical data analysis and created by a dynamic model, which
can simulate the development of the Czech Republic population and the
AD patients.

Keywords: System dynamics · Alzheimer's disease · Prediction
Czech Republic · Simulation · Population modeling

1 Introduction

This article deals with the dynamic modeling of the Czech Republic population
but specifically focuses on Alzheimer's disease patients. Alzheimer's disease is
a genetically complex, slowly progressive, irreversible neurodegenerative brain
disease [9]. Symptoms usually develop slowly, worsen over time, and become
so severe that they interfere or even make it impossible to carry out everyday
activities [1]. The authors of papers [6,10,13–20,28] focus on the Alzheimer's
disease economic and financial implications. Obviously, in the last decades the
number of illnesses has increased and long ago it is not a disease affecting only
the oldest people. The age limit and the number of illnesses vary every year.
Based on available data and prevalence studies, dynamic model allow to predicts
an estimate of the AD patients number across age groups by 2100. The system
modeling principles and system dynamics will be used to the problem simulation.
Dynamic modeling, and in particular system dynamics, are the appropriate tools
for these simulations, as seen, for example, in [5,8,11,23,25–27,29].

© Springer International Publishing AG, part of Springer Nature 2018
N. T. Nguyen et al. (Eds.): ACIIDS 2018, LNAI 10752, pp. 347–356, 2018.
https://doi.org/10.1007/978-3-319-75420-8_33

2 System Dynamics

System dynamics is part of system engineering and in the paper [24] was described that system dynamics is used to understand how the system evolves over time, and to compare system outputs with different input conditions and different system settings. System Dynamics by Sterman [23] uses for their models a combination of linear and non-linear differential equations. Basic principles have been developed to study managerial dynamic decision making using management principles [7]. System dynamics is based on the feedback's idea, which is illustrated by computer models. Modeling is one of the most important parts of system dynamics. Thanks to the special software, the computer can simulate the real system behavior. Forrester [7] set the system dynamics basic principles: The first principle is to assert that if the fluxes accumulate in the layers then the dynamic behavior occurs in the given environment. The second principle is the fact that the levels and flows of the system create mutual feedback loops. Feedback transmits information. Defines a situation where the output of a particular system affects its backward input. The third principle of modeling is the rule that feedback loops are linked by nonlinear linkages.

3 Alzheimer's Disease

As the Czech Alzheimer Society [22] says, Alzheimer's disease disrupts part of the brain and causes the decline of cognitive functions - thinking, memory, judgment. It is the most common cause of dementia, which gradually leads to dependence of the patient on the daily assistance of another person. Nevsimalova [21] report that AD is a common disease with a prevalence of 1% of the population, causes at least 60% dementia and the fourth or fifth most frequent cause of death. Holmerova [12] report that the Alzheimer's disease manifestation and progress are connected with: worsening of brain cell metabolism and degeneration, acetylcholine deficiency, oxidative stress, amyloid formation and deposition, decrease in estrogen in postmenopausal women. But women also play a significant role in having a higher age. In total, by Alzheimer's disease in the beginning of the new millennium has suffered up to 25 million people in the world and approximately 70–90 thousand in the Czech Republic. These numbers, however, as Zgola [30] said, unfortunately increase in the context of aging populations.

Table 1. Numbers and costs for AD patients in the Czech Republic by VZP [3]

	Number of patients	Cost of care	Average per patient
Year 2013	29 352	684 702 000 CZK	23 327 CZK
Year 2014	32 037	743 189 000 CZK	23 197 CZK
Year 2015	35 862	891 100 000 CZK	24 848 CZK
Comparison 13/15	+22.2%	+30.1%	6.5%

The General Health Insurance Company (VZP) records not only the numbers of patients, but also the cost associated with this disease. A simple overview of the results for the years 2013 to 2015 is given in the Table 1.

3.1   Age and Gender

Age is one of the basic factors. People less than 60 years old rarely have this disease. For people over 85, every fifth human suffers from the disease. In this respect, women are threatened by somewhat more than men, and this is supported by Nevsimalova [21], and adds that women with a low level of education are even more at risk.

The Czech Alzheimer's Society annual report [2] from 2015 states that the correct diagnosis has about 20–30% of people with dementia in the Czech Republic. Therefore, the numbers of people with different types of dementia can only be estimated. For the calculation, this association uses the 2009 Alzheimer Europe prevalence study. Based on this study, it is estimated that the number of people with dementia in 2014 (December 31) will be 152.7 thousand.

4   Population of Czech Republic

According to the Czech Statistical Office (CZSO) [4] forecast, by 2050, the Czech Republic population will decrease slightly due to reduced birth rates. The CZSO further expects the Czech Republic to remain an immigrant country. The population will, therefore, be affected by global migration. Thanks to technological advances in the field of science, there is a higher expectation of longevity for men and women, so the average age of the population will also be affected. Very important information is the assumption that the population will become very old, especially due to the above-mentioned facts, according to the CZSO it is necessary to expect that the proportion of persons over the age of 65 should be closer to one-third, which is double the current state. The biggest growth rate is assumed for the highest age, the number of people over 85 should be back up again by 2050. It is therefore likely to be estimated that this assumption will affect the model final phase, the number of people suffering from Alzheimer's disease, whose key factor is age.

5   Dynamic Model

A system dynamics model contains the basic elements associated to some subsystems. The main subsystem focuses on the general population. The elements are the different population age cohorts in the Czech Republic for 2013–2100. The relationships between the elements are given by aging, the growth rates of the population, the death rates, etc. The second subsystem is related to the population with AD. Its basic elements are the different population age cohorts with Alzheimer disease. Its relationships are the incidence rates, the death rates due to AD, etc.

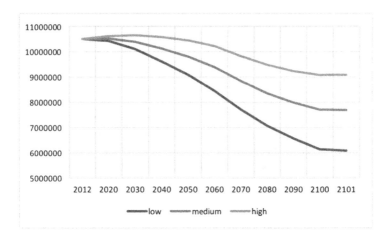

Fig. 1. Czech population prediction, CZSO

The model's basic relationships were used according to an article [27] for current CZSO data. The models shown in Figs. 2 and 3 were modified (split) into two sexes, again with CZSO data.

The STELLA iseesystems.com modeling tool has been chosen for problem modeling. It is a user-friendly drag-and-drop environment containing predefined building elements. These elements can be divided into the following four basic shapes:

– The warehouse (rectangle) can have four forms of use, namely "conveyor", "stock", "queue" and "oven". The naming fully corresponds to their use. For the conveyor, the maximum quantity is determined. The stock stores the elements received, the queue sorts the elements with the FIFO style, and the oven holds the elements to the time before they move as the "baked" state.
– Flows (double arrow with valve) have three forms: unidirectional one-way flow, two-way bi-directional flow and unit converter.
– Converters (circles) represent the reference quantity, different rates, etc.
– The connector (simple arrow) represents the two elements relationship, which is transferred to a mathematical representation.

5.1   Dynamic Model of Population of Czech Republic

The population model of the Czech Republic had to be taken from the broader perspective. The main reason is that in the Alzheimer's disease case it is necessary to know the patient age. As has been said, the most important risk factor is undoubtedly age, so it was necessary to know the exact number of people of the given age, especially men and women, because of another risk factor of the gender in question, which also differed in the chances of illness, that life expectancy and other population factors differ in both sexes. Thus, the basic part of the system is the Czech Republic population model, which simulates the citizen's growth by 2100, recording the individual ages and their aging. Based on these groups, it is then possible to focus on age-specific individuals who play a role in the model's second part.

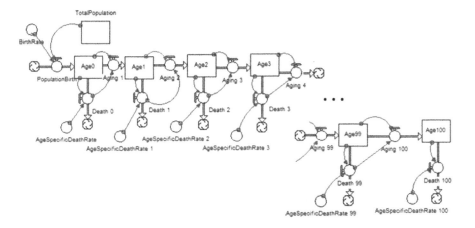

Fig. 2. Model without gender specification

The population model contains (twice - by gender) 101 elements labeled as a stock and representing age cohorts "0 years" to "100 years". For each age cohort, the death probability at the given age is determined. This probability was estimated based on the death probability in previous years. These features were created based on available population data in the CZSO databases.

The population model shown in Fig. 2 is a non-sex model view. In the model, the individual age cohorts, that is, the population size at that age is represented by a "stock". The population growth rate is shown as well as the mortality rates in the converter element. The transition to another age cohort, as well as the turn to the "dead" state, is shown by a double arrow. A slim arrow shows the relationship or linkage during the calculations.

5.2 Dynamic Model of AD Patients in Czech Republic

The model second part focuses on the AD, it is less extensive than the first, mainly because it monitors the disease in patients from the age of 60. The chance of becoming ill in younger patients is so small that it can be neglected. Each level monitors the age-specific patients number and their aging with the disease. The second model part was divided into two gender's parts because it can be expected from the scientific texts, that the female patient's number will be higher not only because of the higher expected older women population but also because of the significantly higher disease risk.

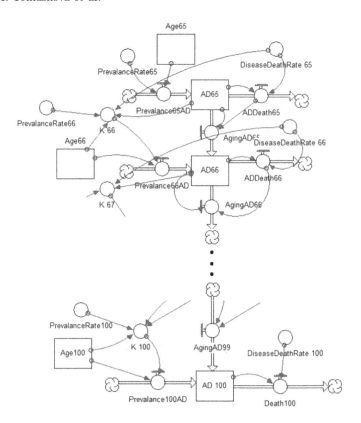

Fig. 3. Model AD patients without gender specification

6   Results

The first part model result is the Czech population development built on the basis of the CZSO projection. The model, as well as projection, records the population decline. The results achieved correspond to the projection calculation CZSO. Thanks to this basic verified model, it is possible to work with the follow-up model of AD patients.

All outputs from the STELLA listed in the Figs. 4, 5 and 6 indicate 0 as the simulation start representing the year 2013, and the simulation end as the 87th simulation step representing the year 2100.

Despite a decrease in the total population cohorts of the oldest population grows and life expectancy is extended.

In Fig. 5 it is possible to see the peculiar prognosis of AD patients for both men and women in the Czech Republic. Line number 1 indicates a man and at its peak does not reach 115,000 patients. While line number 2 represents women - patients and the peak of this prediction is below 200,000 patients.

It is interesting to compare the course of the overall poll of the Czech Republic and patients with the AD, shown in Fig. 6. This outlook shows that despite the

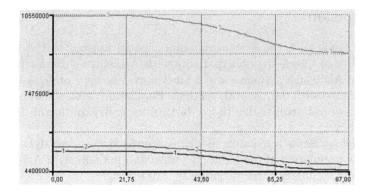

Fig. 4. Prognosis of CR population (3), women (2) and men (1)

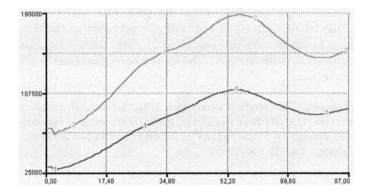

Fig. 5. Prognosis of AD patients, women (2) and men (1)

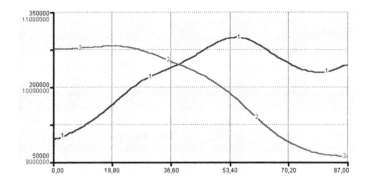

Fig. 6. Forecast population size (2) and patient counts (1) in CR

decrease in the overall population due to lower birth rates, the number of AD patients increases due to the population aging and the longer life expectancy of women.

7 Conclusion

The article deals with the system dynamics model applied to the prediction of the number of Czech Republic patients with Alzheimer's disease by gender. Dementia in Alzheimer's disease is the most prevalent type of dementia and is highly associated with the age of the patient. Population aging will be an ongoing problem for several decades due to low birth rates and a continuous increase in life expectancy.

The article presents a dynamic modeling approach to system dynamics. The model of the population of the Czech Republic and patients with AD is divided by gender and estimates these variables by 2100.

All models are based on the same mathematical formulas described in the text above. The results of modeling the whole population can be seen in Fig. 4 and in Figs. 5 and 6 for the number of people with AD.

The modeling Czech population and its prediction and comparison with CZSO verify modeling methods in the field of worldwide accepted predictions. The differences in the results are negligible and are caused by rounding, and internal algorithms of modeling SW. The results correspond to the CZSO Czech population prediction.

Acknowledgement. This study was supported by the research project The Czech Science Foundation (GACR) 2017 No. 15330/16/AGR Investment evaluation of medical device development run at the Faculty of Informatics and Management, University of Hradec Kralove, Czech Republic.

References

1. Alzheimer's Disease & Dementia. http://www.alz.org/alzheimers_disease_what_is_alzheimers.asp
2. Czech Alzheimer Society, the dementia report 2015. http://www.alzheimer.cz/res/archive/002/000331.pdf?seek=1452679851
3. General Health Insurance Company of Czech Republic, VZP CR. https://www.vzp.cz/o-nas/aktuality/analyza-nemocnych-s-alzheimerem-pribylo-za-tri-roky-o-22-naklady-se-blizi-miliarde
4. The official website of the Czech statistical office CZSO. https://www.czso.cz/csu/czso/projekce-obyvatelstva-ceske-republiky-do-roku-2100-n-fu4s64b8h4
5. Bonabeau, E.: Agent-based modeling: methods and techniques for simulating human systems. Proc. Nat. Acad. Sci. U.S.A. **99**(Suppl. 3), 7280–7287 (2002)
6. Carrillo, M.C.: Leveraging global resources to end the Alzheimer's pandemic. Alzheimer's Dement. J. Alzheimer's Assoc. **9**(4), 363–365 (2013)
7. Forrester, J.W.: Industrial dynamics. J. Oper. Res. Soc. **48**(10), 1037–1041 (1997)
8. Grimm, V., Berger, U., Bastiansen, F., Eliassen, S., Ginot, V., Giske, J., Goss-Custard, J., Grand, T., Heinz, S.K., Huse, G., et al.: A standard protocol for describing individual-based and agent-based models. Ecol. Model. **198**(1), 115–126 (2006)
9. Hampel, H., Prvulovic, D., Teipel, S., Jessen, F., Luckhaus, C., Frolich, L., Riepe, M.W., Dodel, R., Leyhe, T., Bertram, L., et al.: The future of Alzheimer's disease: the next 10 years. Prog. Neurobiol. **95**(4), 718–728 (2011)

10. Handels, R.L., Wolfs, C.A., Aalten, P., Joore, M.A., Verhey, F.R., Severens, J.L.: Diagnosing Alzheimer's disease: a systematic review of economic evaluations. Alzheimer's Dement. **10**(2), 225–237 (2014)
11. Hirsch, G., Homer, J., McDonnell, G., Milstein, B.: Achieving health care reform in the United States: toward a whole-system understanding. In: 23rd International Conference of the System Dynamics Society, pp. 17–21 (2005)
12. Holmerova, I., Janeckova, H., Niklova, D.: To Help Caregiving Families. 10 edn. Česká alzheimerovská společnost, Praha (2014). Technical report. ISBN 978-80-86541-33-4
13. Klimova, B., Maresova, P., Valis, M., Hort, J., Kuca, K.: Alzheimer's disease and language impairments: social intervention and medical treatment. Clin. Interv. Aging **10**, 1401 (2015)
14. Marešová, P., Klimova, B., Kuča, K.: Alzheimers disease: cost cuts call for novel drugs development and national strategy. Ceska a Slovenska farmacie: casopis Ceske farmaceuticke spolecnosti a Slovenske farmaceuticke spolecnosti **64**(1–2), 25–30 (2015)
15. Maresova, P., Klimova, B.: Supporting technologies for old people with dementia: a review. IFAC-PapersOnLine **48**(4), 129–134 (2015)
16. Maresova, P., Mohelská, H., Dolejs, J., Kuca, K.: Socio-economic aspects of Alzheimer's disease. Curr. Alzheimer Res. **12**(9), 903–911 (2015)
17. Marešová, P., Mohelská, H., Kuča, K.: Economics aspects of ageing population. Procedia Econ. Financ. **23**, 534–538 (2015)
18. Maresova, P., Tomaskova, H., Kuca, K.: The use of simulation modelling in the analysis of the economic aspects of diseases in old age. In: Bilgin, M.H., Danis, H., Demir, E., Can, U. (eds.) Business Challenges in the Changing Economic Landscape - Vol. 1. ESBE, vol. 2/1, pp. 369–377. Springer, Cham (2016). https://doi.org/10.1007/978-3-319-22596-8_26
19. Mohelska, H., Maresova, P.: Economic and managerial aspects of Alzheimer's disease in the Czech Republic. Procedia Econ. Financ. **23**, 521–524 (2015)
20. Mohelska, H., Maresova, P., Valis, M., Kuca, K.: Alzheimer's disease and its treatment costs: case study in the Czech Republic. Neuropsychiatr. Dis. Treat. **11**, 2349 (2015)
21. Nevsimalova, S., Ruzicka, E., Tichy, J.: Neurologie. 1 edn, 368 p. Galén, Praha (2002). Technical report. ISBN 80-7262-160-2
22. Czech Alzheimer Society: Alzheimers Disease. http://www.alzheimer.cz/alzheimerova-choroba/
23. Sterman, J.D.: Business Dynamics: Systems Thinking and Modeling for a Complex World, vol. 19. Irwin/McGraw-Hill, Boston (2000)
24. Tako, A.A., Robinson, S.: The application of discrete event simulation and system dynamics in the logistics and supply chain context. Decis. Support Syst. **52**(4), 802–815 (2012)
25. Tomaskova, H., Kuhnova, J., Kuca, K.: Economic model of Alzheimer's disease. In: Proceedings of the 25th International Business Information Management Association Conference-Innovation Vision 2020: From Regional Development Sustainability to Global Economic Growth, pp. 7–8 (2015)
26. Tomaskova, H., Kuhnova, J., Kuca, K.: Ageing and Alzheimer disease - system dynamics model prediction [Alzheimerova choroba a stárnutí populace - predikce s pomocí systémového modelování]. Ceska Slov. Farm. **65**, 99–103 (2016)
27. Tomaskova, H., Kuhnova, J., Cimler, R., Dolezal, O., Kuca, K.: Prediction of population with Alzheimer's disease in the European union using a system dynamics model. Neuropsychiatr. Dis. Treat. **12**, 1589 (2016)

28. Wimo, A., Jönsson, L., Gustavsson, A., McDaid, D., Ersek, K., Georges, J., Gulacsi, L., Karpati, K., Kenigsberg, P., Valtonen, H.: The economic impact of dementia in europe in 2008-cost estimates from the eurocode project. Int. J. Geriatr. Psychiatry **26**(8), 825–832 (2011)
29. Wolstenholme, E., Monk, D., Smith, G., McKelvie, D.: Using system dynamics to influence and interpret health and social care policy in the UK. In: system dynamics conference, Oxford, UK (2004)
30. Zgola, J.M.: Successful Care for a Person with Dementia. 1 edn. 232 p. Grada, Praha (2003). Technical report (2003). ISBN 80-247-0183-9

Detection and Dynamical Tracking of Temperature Facial Distribution Caused by Alcohol Intoxication with Using of Modified OTSU Regional Segmentation

Jan Kubicek[1(✉)], Marek Penhaker[1], Martin Augustynek[1],
Martin Cerny[1], David Oczka[1], and Petra Maresova[2]

[1] VSB–Technical University of Ostrava, FEECS, K450, 17. listopadu 15,
708 33 Ostrava–Poruba, Czech Republic
{jan.kubicek,marek.penhaker,martin.augustynek,
martin.cerny}@vsb.cz, David-oczka@seznam.cz
[2] Faculty of Informatics and Management, University of Hradec Kralove,
Rokitanskeho 62, 500 03 Hradec Králové, Czech Republic
petra.maresova@uhk.cz

Abstract. The alcohol intoxication is a significant issue influencing many social areas. There are commonly utilized standards for the alcohol measurement like is the breath and blood analyzers. Nevertheless such methods require direct contact, and agreement of tested person. New potential trend of the alcohol measurement reflect a fact that the alcohol cause thermal facial changes which can be observable from the IR image records. In this regard, we have performed analysis dealing with the alcohol effect modeling via multiregional segmentation algorithm based on the modified OTSU method. This method autonomously extracts areas exhibiting significant temperature variations whilst person drinking. By this way we have tracked the alcohol effect over the time which corresponded with the gradual alcohol consumption. Significant part of the analysis deals with comparison and verification of the facial temperature distribution caused by the alcohol against breath analysis.

Keywords: Facial temperature distribution · Image segmentation
OTSU method · Alcohol consumption · Correlation analysis

1 Introduction

At lower doses, the alcohol can serve as a stimulant inducing positive feeling of euphoria and talkativeness. Nevertheless excessive alcohol drinking can lead up to drowsiness, respiratory depression – breathing is getting slower, shallow or completely stops. After swallowing a drink, the alcohol is rapidly absorbed into the blood (20% by the stomach and 80% by the small intestine) [1, 2].

The BAC (blood alcohol concentration) rises, and the feeling of drunkenness occurs, when alcohol is drunk faster than the liver can break it down. The BAC does not exactly correlate with the symptoms of drunkenness and different people have

© Springer International Publishing AG, part of Springer Nature 2018
N. T. Nguyen et al. (Eds.): ACIIDS 2018, LNAI 10752, pp. 357–366, 2018.
https://doi.org/10.1007/978-3-319-75420-8_34

variable symptoms even after drinking the same amount of the alcohol. The BAC level and corresponding consequent reactions are influenced by:

- The liver ability to metabolise alcohol. This fact is influenced by various genetic differences exhibiting in the liver enzymes that break the alcohol down.
- The presence or absence of food in the stomach.
- The alcohol concentration in the beverage – high concentration (spirits) is rapidly absorbed.
- How quickly alcohol is consumed.
- Fatter and more muscular people contain more fat and muscles to absorb the alcohol.
- Other attributes like is age, sex, or ethnicity (e.g. woman have a higher BAC after drinking a same amount of alcohol than men which is caused by metabolic differences and absorption). Especially significant is a fact that men have more fluid in their bodies than woman to distribute the alcohol. Some of the ethnic populations have different levels of a liver enzyme which is responsible for the alcohol breakdown.
- Alcohol inclination – how frequently a person drinks alcohol. A regular drinker can tolerate a higher dose of alcohol in a comparison with a person drinking occasionally [3, 4].

2   Related Work

The IR imaging can allow for the detection, and tracking of the temperature distribution. In the context of the alcohol detection the thermal distribution well reflect the alcohol intoxication as a representative feature for the dynamical variations measurement. This area is relatively new; anyway the recent literature contains several contributions regarding the temperature distribution on the drunken people [3, 4].

In [5] a distinguishable space between the drunk and sober people is done on the base of the discriminant analysis. Simultaneously, the authors devoted to tracking the temperature variations between different facial areas to recognize the most significant areas in the drunken people. In [6] authors analyzed the blood vessels activity, on the base of the anisotropic diffusion, and the Top-Hat transformation, which can be a reliable indicator of the drunkenness. In [7] the eye temperature distribution (relative iris and sclera temperature) of the alcohol intoxicated people was analysed. In [8] the authors employed the neural networks with the target of the classification the facial areas being the most differentiable whilst the person drinking. In [9] they make use of different technics such as Discrete Wavelet Transform, Discrete Cosine Transform and the Support Vector Machine. In [10] the Local Difference Patterns were applied on the forehead to obtain discriminant alcohol features.

3 Problem Definition

On the base of the review outputs, and our own measurements, we have found out two significant areas well reflecting the alcohol intoxication in a context of the facial temperature. (1) The forehead area contains a less proportion of the higher temperature when person is sober. During the alcohol drinking, this area is getting increasingly expanded in a sense of the area and pixel's values distribution. (2) The nose area is relatively mostly cold when person is sober. Whilst drinking, the alcohol this area is getting hotter in a sense of the cold temperature disappearing. Other words speaking the cold area is gradually shrunk. Such temperature phenomenon from areas (1) and (2) may be segmented, and extracted by the multiregional segmentation whilst drinking. The Fig. 1 reports the IR monochromatic images of the nose (a–c), and the forehead (d–f) whilst person is drinking.

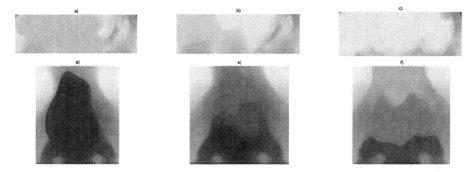

Fig. 1. Comparison of the sober person (a, d), and individual stages of the drinking of the 40% alcohol: (b, e) – after 80 ml, and (c, f) – after 120 ml.

The main idea is an extraction and classification of the IR pixels belonging to the facial temperature map showing the most significant differences whilst alcohol consumption. In this regards the main attention has been paid to the nose and forehead areas well reflecting the dynamical process of the drinking. In such segmentation task, the modified OTSU method has been adopted. The multiregional OTSU segmentation allows for the differentiation of the temperature map into a predefined number of the classes consequently allow for the dynamical alcohol modeling.

4 IR Data and Measurement

We have performed a set of measurements on the 20 volunteers. For each person, six IR image records have been taken. Firstly, we have taken the sober state, it serves as reference image. Rest five images have been taken whilst gradual drinking. Since the temperature measurement may be influenced by various surrounding factors, the measurements have been done within standardized conditions. All the persons have been measured indoor within predefined stabilized conditions: atmospheric pressure 1002.5 hPa, temperature 20 °C and humidity 50%.

After arrival to place of the measurement, the tested individuals were kept in calm state during 30 min to prevent the outside temperature influence. Also, all the tested persons were asked about using medicaments may influence the temperature distribution (e.g. pills for thermoregulation). Individual using such medicaments have been excluded from the measurement.

During the alcohol measurement, the individuals were given a 38% spirit. Time gap between individual alcohol consumption was 10, 20 and 30 min. For each individual, six IR images we stored ((1)–(6)) in the following structure:

(1) Sober state.
(2) After 40 ml of alcohol.
(3) After 80 ml of alcohol.
(4) After 120 ml of alcohol.
(5) After 160 ml of alcohol.
(6) After 200 ml of alcohol.

In order to ensure the alcohol distribution in the human body, the IR image has been taken in 10, 20 and 30 min time after drinking. The forehead and nose areas were simultaneously manually extracted, and consequently converted into monochromatic format linearly mapping the original temperature values ($0 \approx 22.6$ °C; $255 \approx 37.4$ °C).

As part of the measurement, other parameters which may be important, in a context of the alcohol intoxication, have been measured: the systolic and diastolic blood pressure and the heartbeat. The person distance from the IR camera was 2 m, and the emissivity was set on 0.98. The IR scanning background was the white stucco plaster. For the alcohol objectification, the breath analyzer Dräger Alcotest® 7510 has been used which is, according to the Czech Metrology Institute, perceived as a determined measurer with the standard deviation 1.55%. For the blood pressure measurement, the Tonometr Sencor SBP 690 has been used. It achieves an accuracy of the blood pressure ±3 mmHg, and heartbeat ±5%.

The IR camera FLIR T640 has been used for the measurement. FLIR T640 thermal detector resolution is 640×480 px, and thermal sensitivity (NETD) <0.035 °C. For the measurement, two types of the detectors have been used: FOL41 (f = 41.3 mm, 15 °C) and FOL25 (f = 24.6 mm, 25 °C).

5 Proposal of Modified OTSU Regional Segmentation

In this method, image having L gray levels is being classified into two classes C_0 and C_1 (binary classification). By this way we can suppose that C_0 contains an area of gray levels: $\{0, 1, 2, \ldots, k\}$, and C_1 is defined for the grey level area: $\{k + 1, k + 2 \ldots, L - 1\}$. In the principle, we are seeking for the thresholding separating temperature distribution regarding to statistical variance. Between class variance (BCV) should be maximized. The BCV is given by ϑ_B:

$$\vartheta_B = \frac{1}{\omega_0(1 - \omega_0)} (\omega_0 \mu_T - \mu_k)^2 \tag{1}$$

where ω_0 denotes to the cumulative moments, μ_T is the grey level mean of the IR image and μ_T stands for the first-order cumulative of the histogram up to the gray level k.

We can suppose that the BCV is represented by a differentiable form with respect to the gray level, and let the first BCV derivate is zero.

$$f(k) = k - \frac{\mu_0 + \mu_1}{2} \tag{2}$$

Where μ_0 and μ_1 stand for the mean gray levels for the classes C_0, respectively C_1. The Bisection has been adopted to solve the BCV-derivative equation. The Bisection method seeks for a root. Such interval is called active interval. The active interval is bisected again. It is an iterative method being performed until no further bisection needs to be performed. In our case, the Bisection starts by: $k_0 = 0$ and $k_2 = L - 1$. We work with 256 gray levels, it means that $L = 256$. In this optimization procedure, we use $\log_2 L$ iterations in order to find the root. Among the generated midpoints, the threshold is selected as the minimal absolute value of the BCV-derivative function.

6   Testing and Evaluation

A current state of the alcohol intoxication is routinely assessed by many ways utilizing different physical principles e.g. the blood analysis. Those principles are able to detect a current state of the alcohol, not dynamical variations. We have analyzed the facial temperature not directly representing the alcohol intoxication, but effect closely related to the alcohol consumption. By a set of the IR facial measurements, we can track the dynamical temperature variations in the nose and forehead. The dynamical variations are well represented by the multiregional segmentation allowing for a differentiation of the temperature areas into predefined segmentation classes. On the forehead, a class representing the highest temperature is being consecutively expanded contrarily a class representing the coldest area on the nose is being shrunk whilst the alcohol drinking.

In order to observe, and differentiate the temperature variations in the time delay, we have taken the IR images after 10, 20 and 30 min after drinking. Simultaneously, we have measured the alcohol via the breath analyzer to perform the correlation analysis among the temperature – facial features against the breath. The Fig. 2 brings a comparison of the forehead area segmentation by the modified OTSU method where six classes are selected. The yellow class label represents the highest temperature which is being expanded whilst the person is drinking. The Fig. 3 represents a same situation of the nose area whilst drinking. In this case the lowest facial temperature is represented by the blue label color being disappeared whilst drinking from the sober state up to the maximum alcohol consumption.

Fig. 2. Segmentation results of the hottest forehead area (yellow) from the sober state (left), after 80 ml of the 40% alcohol (middle), and after 120 ml of the 40% alcohol (right). (Color figure online)

Fig. 3. Segmentation results of the coldest nose area (blue) from the sober state (left), after 80 ml of the 40% alcohol (middle), and after 120 ml of the 40% alcohol (right). (Color figure online)

Through individual columns we can observe the segmentation model of the nose and forehead exhibit dynamical variation whilst drinking. In the case of the forehead the hottest temperature area is getting expanded, contrarily in the case of the nose the coldest area is getting shrunk whilst alcohol drinking. Those temperature-alcohol features can be objectivized against the whole nose and forehead area. For each tested person, the nose and forehead area was measured. Respective regional segmentation model was expressed in a form of the area size. Parameter $R_{n,a}$, respective $R_{f,a}$ evaluate ratio of the alcohol feature (1), respectively (2). Where a represents an amount of the alcohol. First row indicates average results of 14 men, while second row indicate average of 6 women. Average results of twenty measured persons are summarized in the Tables 1 and 2.

Table 1. Ratio for the nose coldest temperature area during the alcohol drinking from segmentation model.

$R_{n,0}[\%]$	$R_{n,40}[\%]$	$R_{n,80}[\%]$	$R_{n,120}[\%]$	$R_{n,160}[\%]$	$R_{n,200}[\%]$
85	76	66	45	32	15
89	70	64	42	27	20

Table 2. Ratio for the forehead hottest temperature area during the alcohol drinking from segmentation model.

$R_{f,0}[\%]$	$R_{f,40}[\%]$	$R_{f,80}[\%]$	$R_{f,120}[\%]$	$R_{f,160}[\%]$	$R_{f,200}[\%]$
15	22	42	55	77	90
19	26	40	59	79	91

6.1 Correlation Analysis of the Temperature Distribution Model

We have proposed a mathematical model which is able to extract and consequently track representative temperature-alcohol features well reflecting the alcohol state based on the facial temperature distribution. In many studies, it has been proved and justified that alcohol causes temperature variations. Nevertheless, it hasn't been proved whether there is a relationship between temperature distribution and real content of the consumed alcohol. This fact is substantially important regarding objectification of the segmentation model. In an ideal case, we suppose that there is a relationship between consumed alcohol measured by the breath analyzer and the facial temperature variation. In this regard we have computed the Pearson correlation coefficient expressing a level of the linear dependence in the range: $[-1; 1]$. We take advantage a fact that we are working with the increasing linear dependence in the case of the forehead, it means that increasing alcohol volume causes higher temperature. This fact is expressed by expanded forehead area. The original interval can be rewritten to the range: $[0; 1]$. In a case of zero we have no linear correlation between consumed alcohol, and segmentation model, contrarily 1 indicate a full linear correlation. The correlation analysis has been performed for the IR images taken 10 min after drinking (Fig. 4), after 20 min (Fig. 5) and 30 min (Fig. 6). The correlation results are summarized in the Table 3.

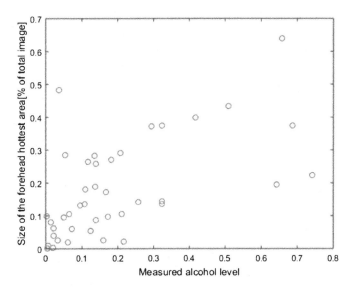

Fig. 4. Correlation analysis for the forehead area where images have been taken after 10 min from drinking.

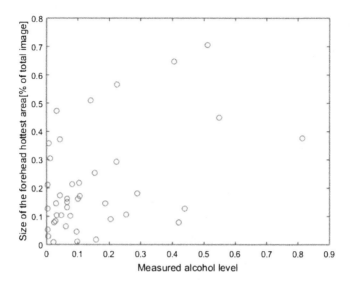

Fig. 5. Correlation analysis for the forehead area where images have been taken after 20 min from drinking.

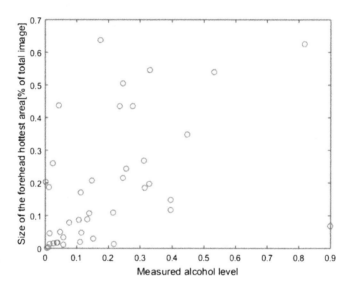

Fig. 6. Correlation analysis for the forehead area where images have been taken after 30 min from drinking.

Table 3. Results of correlation analysis for forehead area model with the breath analysis.

	Modified OTSU Corr	OTSU Corr
T_{10}	0.412	0.361
T_{20}	0.589	0.577
T_{30}	0.891	0.691

7 Conclusion

IR facial temperature modelling surely has ambitions to be considered for building a predictive model for segmentation, and classification of a current state of the alcohol amount. Nevertheless, we have to take into account some limitations. This approach is based on the facial temperature maps which represent implication of the alcohol.

The multiregional segmentation allows for dynamical tracking of the alcohol manifestation in a form of the temperature maps. The segmentation extract areas well reflecting individual stages of the alcohol intoxication, thus we can objectivise a relationship between a level of the intoxication and temperature distribution. On the other hand we have to be aware that we do not measure alcohol content, but temperature variations, as a product of the alcohol drinking.

We consider important the correlation analysis which reflects linear relationship of the proposed model with the breath analyser. We should know that the temperature variations reflect only one alcohol effect (facial temperature). Nevertheless, there are other side effects which may be more or less linked with the alcohol drinking. In this regard, it seems to be important a time delay of taking the IR images from the alcohol consumption. Judging by the correlation results (Table. 3) with the breath analysers, the best correlation is achieved on time 30 min (0.891). We have also compared our results with a standard version of the OTSU method. Regarding correlation analysis, the modified OTSU method produces robust results. The proposed segmentation model also objectively show trend of the nose and forehead temperature features (Tables 1 and 2). Important aspect is also a correlation of the blood pressure and heartbeat whilst alcohol drinking. In the future research we are going to focus on the sensitivity of the segmentation model regarding those physiological parameters which exhibit dynamical variations whilst drinking.

Acknowledgment. The work and the contributions were supported by the project SV4507741/2101, 'Biomedicínské inženýrské systémy XIII'. This study was supported by the research project The Czech Science Foundation (GACR) No. 17-03037S, Investment evaluation of medical device development.

References

1. Guzanová, A., Ižaríková, G., Brezinová, J., Živčák, J., Draganovská, D., Hudák, R.: Influence of build orientation, heat treatment, and laser power on the hardness of Ti6Al4V manufactured using the DMLS process. Metals **7**(8), 1–17 (2017). ISSN 2075-4701
2. Ťúková, V., Poláček, I., Tóth, T., Živčák, J., Ižaríková, G., Kovačevič, M., Somoš, A., Hudák, R.: The manufacturing precision of dental crowns by two different methods is comparable. Lekar a Technika **46**(4), 102–106 (2016)
3. Jadhav, K.S., Magistretti, P.J., Halfon, O., Augsburger, M., Boutrel, B.: A preclinical model for identifying rats at risk of alcohol use disorder. Sci. Rep. **7**(1), Article no. 9454 (2017)
4. Bodnárová, S., Hudák, R., Živčák, J.: Príprava a návrh biologického testovania biokompatibility keramických skáfoldov na báze hydroxyapatitu a trikalcium fosfátu. In: Novus Scientia 2017, pp. 17–21. TU, Košice (2017). ISBN 978-80-553-3080-8

5. Koukiou, G., Anastassopoulos, V.: Drunk person identification using thermal infrared images. Int. J. Electron. Secur. Digit. Forensics **4**(4), 229–243 (2012)
6. Koukiou, G., Anastassopoulos, V.: Facial blood vessels activity in drunk persons using thermal infrared. In: 4th International Conference on Imaging for Crime Detection and Prevention, London, UK, November 2011, pp. 1–4. Kingston University (2011)
7. Koukiou, G., Anastassopoulos, V.: Drunk person screening using eye thermal signatures. J. Forensic Sci. **61**(1), 259–264 (2016)
8. Koukiou, G., Anastassopoulos, V.: Neural Networks for identifying drunk persons using thermal infrared imagery. Forensic Sci. Int. **252**, 69–76 (2015)
9. Xie, Z., Jiang, P., Xiong, Y., Li, K.: Drunk identification using far infrared imagery based on DCT features in DWT domain. In: Proceedings of SPIE 10157, Infrared Technology and Applications, and Robot Sensing and Advanced Control, 101571F, 25 October 2016. https://doi.org/10.1117/12.2246469
10. Koukiou, G., Anastassopoulos, V.: Drunk person identification using local difference patterns. In: 2016 IEEE International Conference on Imaging Systems and Techniques (IST) (2016)

Modelling and Objectification of Secondary X ray Irradiation on Skiagraphy Images in Clinical Conditions

Jan Kubicek$^{(\boxtimes)}$ ⓘ, Martin Augustynek ⓘ, Andrea Vodakova ⓘ,
and Marek Penhaker ⓘ

VSB–Technical University of Ostrava, FEECS, K450,
17. listopadu 15, 708 33 Ostrava–Poruba, Czech Republic
{jan.kubicek,martin.augustynek,andrea.vodakova.st,
marek.penhaker}@vsb.cz

Abstract. Secondary, also called scattered irradiation is directly linked with each clinical X ray examination. In the principle, the secondary irradiation spreads itself to all directions. Thus, besides the scanned skiagraphy image, other stored images in surrounding are affected as well. More frequently are skiagraphy images affected, worse image features the skiagraphy image has. In this paper, we have proposed the multiregional segmentation model which is able to classify and extract areas corresponding with the secondary X-ray deterioration. On the base of the segmentation model, we have defined four significantly important features well reflecting the secondary irradiation through tested distances of the X-ray source and the skiagraphy records. Those features allow for modelling of the secondary irradiation effect in a dependence of respective distance. Such predictive model is well usable in the clinical conditions for an assessment of the deterioration level of the skiagraphy image. The prediction model also shows significant differences between mobile and fixed X ray in a sense of the amount of the scattered irradiation.

Keywords: X-ray · Secondary radiation · Skiagraphy image
Multiregional segmentation

1 Introduction

X ray has an ability to penetrate through any matter which becomes it more or less weak. This fact is depended on the quality of the irradiation, and also on the features of the irradiated matter. During the penetration of the X ray through the matter it goes to the X ray scattering, and consequently to the deflection from the primary beam [1–3].

The secondary irradiation has a negative influence on the patient, also on the clinical staff, and on the imaging quality [11, 12]. An amount of the secondary irradiation and the image quality is determined by the primary irradiation. Bigger area of the beam is set by the collimator; less of the secondary irradiation in the irradiated object is created [4–6].

© Springer International Publishing AG, part of Springer Nature 2018
N. T. Nguyen et al. (Eds.): ACIIDS 2018, LNAI 10752, pp. 367–375, 2018.
https://doi.org/10.1007/978-3-319-75420-8_35

If the scattered irradiation is caught by the image detector it has a negative influence on the resulting image. It may have a result impairing of the image quality. The scattered irradiation negatively influence the image contrast of the X ray image, thus significantly deteriorates its quality [13–15]. It is supposed that as more the scattered irradiation is created worse image quality. This fact is insoluble problem, in the digital radiology decreased contrast may be compensated by the supplementary image processing [7–10].

2   Measurement Results on Fixed and Mobile X ray Device

By performing a set of the experiments, a latent image was created on the skiagraphy record. Experiments were performed by irradiation of the skiagraphy cassettes from three distances: 1, 1.5, and 2 m, and a level of the image deterioration was analyzed. Consequently, a cassette with the skiagraphy image was inserted to the digitizer, and by using the red laser the electrical analogue signal was registered. This signal was converted into the digital image. The clinical workplace for digitizing image records is depicted on the Fig. 1.

Fig. 1. Digitizer with cassette and PC with the RIS system for the digital image processing and sending to the PACS (picture archiving and communication system) system.

The digital X ray device Siemens Mobilett was used for the experimental measurements. The mobile skiagraphy is used in the situations where patient cannot be transported to the fixed X ray device, also it is used on the operating rooms, or on the emergency units. The mobile X ray device used for the measurements is reported in the Fig. 2.

Image Preprocessing
Digital images of the scattered irradiation from the experiments were acquired in the DICOM format. Consequently, those images were anonymized in the information system (RIS) – all the text and scales were removed. Finally, the images were converted into .png format. Before the modeling procedure, the images have been preprocessed. In this step, the brightness transformation was applied to improve image contrast. Note that we processed images having the resolution: 1000×830 px. All the images were consecutively measured in the distances 1, 1.5 and 2 m (Fig. 3) to observe a level of the scattered irradiation effect.

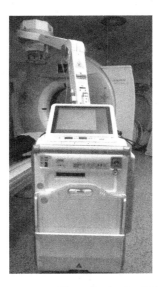

Fig. 2. Mobile X ray device Siemens Mobilett Mira in the University hospital of Ostrava used for the experiment realization.

Fig. 3. Resulting digital images representing the secondary irradiation from the radiation source: 1 m (left), 1.5 m (middle), and 2 m (right).

3   Multiregional Image Segmentation of Scattered Irradiation

For the modeling of the scattered irradiation the Otsu multiregional segmentation was used. This method performs a separation the image regions into predefined number of classes representing different image features. This method defines individual image regions based on the statistical variance. It is supposed that the variance inside the class should be minimized to prevent deviating pixels having significantly different image features that other pixels lying close to the class centroid.

The between class variance is given on the base of the background weight (W), and the average background intensity (μ) separately for foreground (f) and background (b). The image having dimension $M \times N$ is represented by individual pixels with different gray levels L. A number of pixels in given gray level i is denoted as n_i.

$$W_f \;=\; \sum_{i=1}^{L} \frac{n_i}{N} \qquad (1)$$

$$W_b \;=\; \sum_{i=1}^{k} \frac{n_i}{N} \qquad (2)$$

$$\mu_f \;=\; \sum_{i=k+1}^{L} \frac{n_i \cdot i}{N - N_k} \qquad (3)$$

$$\mu_b \;=\; \sum_{i=1}^{k} \frac{n_i \cdot i}{N_k} \qquad (4)$$

Between class variance σ^2 is consequently given by equation:

$$\sigma_B^2 \;=\; W_b \cdot W_f \cdot \left(\mu_b - \mu_f\right)^2 \qquad (5)$$

The image segmentation was performed for the 8 thresholding levels (Fig. 4). The thresholding levels <5 represent the region of interest, and they are assigned 1 in the binary segmentation. Other classes represent the plumb object, and they were given 0 in the binary segmentation (Fig. 5). Such binary segmented image was finally multiplied by the native image so that the pixels of the secondary irradiation were perceived.

Fig. 4. Original image taken from 1 m from the irradiation source (left), and the multiregional Otsu segmentation with 8 classes (right).

The Fig. 6 reports the histogram of the original image in comparison with the Otsu regional segmentation. We can observe two peaks representing two major image areas. Less dominating peak not appearing in the segmented histogram represents the plumb object manifestation. The dominant peak reliably represents main area of the scattered irradiation.

Fig. 5. Resulting segmented image after binarization procedure.

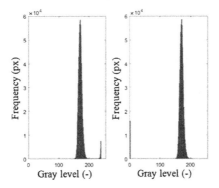

Fig. 6. Comparison of the histogram analysis: original image (left) and segmented image (right).

4 Analysis of Model and Statistical Evaluation

The resulting histogram (Fig. 6) was fitted by the continuous curve. This curve approximates an area of the scattered irradiation. This curve is also characterized by two dominating peaks representing a finite image interval of the scattered irradiation manifestation. The peak maximum defines the maximum intensity of the irradiation. We have defined four image features well representing the secondary irradiation manifestation in the dependence of measured distance from the irradiation source. (1) amount of the pixels under the histogram, (2) gray level intensity of the maximum, (3) lower limitation of the band, and (4) upper limitation of the band (Tables 1 and 2).

Table 1. Extraction of the scattered irradiation features for the fixed X ray based on the histogram analysis.

	Distance (m)	Feature (1)	Feature (2)	Feature (3)	Feature (4)
1	1	821745	169	144	191
2	1	821184	168	139	189
3	1.5	823439	192	171	211
4	1.5	823118	193	174	215
5	2	821361	215	199	221
6	2	821433	213	191	219

Table 2. Extraction of the scattered irradiation features for the mobile X ray based on the histogram analysis.

	Distance (m)	Feature (1)	Feature (2)	Feature (3)	Feature (4)
1	1	823115	82	61	111
2	1	821666	92	72	129
3	1.5	820991	135	115	126
4	1.5	823388	131	112	160
5	2	821420	169	148	203
6	2	823049	168	151	199

Statistical Analysis – Descriptive Parameters

On the base of the descriptive statistical analysis we have described individual scattered irradiation features in the dependence of the distance from the irradiation source. The resulting data do not meet the normal (Gaussian) distribution. All the deviating values were kept for the purpose of the analysis. The data file contains 10 values for each distance (1 and 2 m) of the image from the irradiation source (Table 3). Figure 7 brings a comparison of the box plots for fixed and mobile X ray for individual analyzed distances from the radiation source (1, 1.5 and 2 m).

Table 3. Statistical indicators of number of pixels under the histogram curve for 1 and 2 m distance on the fixed and mobile X ray.

	1 m		2 m	
	Fixed X ray	Mobile X ray	Fixed X ray	Mobile X ray
Number of measurements	10	10	10	10
Minimum	815000	810000	815400	820000
Lower quartile	823200	821100	823100	823300
Average	820000	820000	820000	820000
Median	800999	820000	814500	823900
Upper quartile	839000	833000	838000	821000
Maximum	844000	839000	844000	842000
Variance	130000	380000	150000	760000
Standard deviation	365.78	619.59	381.55	872.16

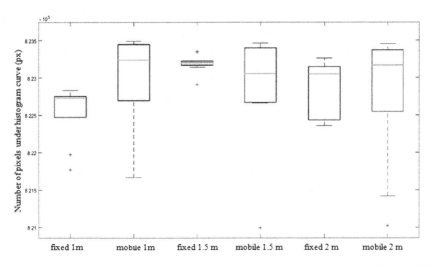

Fig. 7. Comparison of the histogram analysis: original image (left) and segmented image (right).

Important part of the analysis is the prediction model which is able to estimate general scattered irradiation features representing the image effect in the dependence on the particular the irradiation source distance. On the base of the histogram manifestation two histogram limitations were selected: lower and upper limitations. By using these extremes we define a finite interval for each distance.

Interval for the fixed X ray

- Distance 1 m: $\langle 142; 194 \rangle$ gray levels
- Distance 1.5 m: $\langle 177; 218 \rangle$ gray levels
- Distance 2 m: $\langle 196; 226 \rangle$ gray levels

Interval for the mobile X ray

- Distance 1 m: $\langle 73; 133 \rangle$ gray levels
- Distance 1.5 m: $\langle 117; 182 \rangle$ gray levels
- Distance 2 m: $\langle 151; 204 \rangle$ gray levels

5 Conclusion

The X ray imaging is one of the most frequently used clinical alternative. Unfortunately each such examination is closely connected with the secondary irradiation which is perceived as unwanted side effect may significantly deteriorate quality of the stored skiagraphy images. Particular level of such effect represents one of the major aspects influencing the image quality. Unfortunately, there is a lack of the clinical SW methods which would assess this phenomenon. Based on the multiregional analysis we have extracted area of the scattered irradiation on the mobile, and fixed X ray device in the clinical conditions of the radiology ward of the University hospital in Ostrava. The segmentation procedure defines a finite interval of the histogram manifestation which is described by four significantly differentiable features objectively reflecting the image effect in the dependence of the irradiation source distance. One of the main applicable outputs of the analysis is a prediction model allowing for estimation of a finite band of the scattered irradiation on the skiagraphy images. In this regard, we have found out two important outputs. Higher distance is, higher brightness values represent the scattered irradiation. Also, in a comparison with the fixed X ray the mobile X ray less influences the quality of the analyzed images because its brightness interval is represented by lower pixel's values.

Acknowledgment.   The work and the contributions were supported by the project SV4507741/2101, 'Biomedicínské inženýrské systémy XIII'. This study was supported by the research project The Czech Science Foundation (GACR) No. 17-03037S, Investment evaluation of medical device development.

References

1. Boydaş, E., Yılmaz, D., Cömert, E.: Characterization of the effective atomic number for first row transition elements by the ratio of coherent to compton scattering intensities obtained by wavelength dispersive X-ray fluorescence. Instrum. Sci. Technol. **44**(6), 642–650 (2016)
2. Gabor, M., Litoborski, M.: Dose measurement verification in solid state phantom in place of field connection for non-standard radiotherapy conditions. Rep. Pract. Oncol. Radiother. **13**(5), 247–256 (2008)
3. Garg, A., Rajkhowa, M.P., Dawar, S.: Real time remote diagnosis and distant education using CollabDDS. In: ACM International Conference Proceeding Series, Part F127653, pp. 6–10 (2017)
4. Giordano, A.V., Arrigoni, F., Bruno, F., Carducci, S., Varrassi, M., Zugaro, L., Barile, A., Masciocchi, C.: Interventional radiology management of a ruptured lumbar artery pseudoaneurysm after cryoablation and vertebroplasty of a lumbar metastasis. Cardiovasc. Intervent. Radiol. **40**(5), 776–779 (2017)
5. Ha, J.Y., Jeon, K.N., Bae, K., Choi, B.H.: Effect of Bone Reading CT software on radiologist performance in detecting bone metastases from breast cancer. Br. J. Radiol. **90**(1072) (2017). Article no. 20160809
6. Kubicek, J., Valosek, J., Penhaker, M., Bryjova, I., Grepl, J.: Extraction of blood vessels using multilevel thresholding with color coding. In: Sulaiman, H.A., Othman, M.A., Othman, M.F.I., Rahim, Y.A., Pee, N.C. (eds.) Advanced Computer and Communication Engineering Technology. LNEE, vol. 362, pp. 397–406. Springer, Cham (2016). https://doi.org/10.1007/978-3-319-24584-3_33
7. Mikulka, J., Kabrda, M., Gescheidtová, E., Peřina, V.: Classification of jawbone cysts via orthopantomogram processing. In: 2012 35th International Conference on Telecommunications and Signal Processing, TSP 2012 - Proceedings, pp. 499–502 (2012). Article no. 6256344
8. Mikulka, J., Gescheidtová, E., Kabrda, M., Peřina, V.: Classification of jaw bone cysts and necrosis via the processing of orthopantomograms. Radioengineering **22**(1), 114–122 (2013)

9. Mori, S., Amano, S., Furukawa, T., Shirai, T., Noda, K.: Effect of secondary particles on image quality of dynamic flat panels in carbon ion scanning beam treatment. Br. J. Radiol. **88**(1047) (2015). Article no. 20140567

10. Oczka, D., Penhaker, M., Knybel, L., Kubicek, J., Selamat, A.: Design and implementation of an algorithm for system of exposure limit in radiology. Stud. Comput. Intell. **642**, 433–443 (2016)

11. Pooler, B.D., Lubner, M.G., Kim, D.H., Chen, O.T., Li, K., Chen, G.-H., Pickhardt, P.J.: Prospective evaluation of reduced dose computed tomography for the detection of low-contrast liver lesions: direct comparison with concurrent standard dose imaging. Eur. Radiol. **27**(5), 2055–2066 (2017)

12. Robinson, J.W., Ryan, J.T., McEntee, M.F., Lewis, S.J., Evanoff, M.G., Rainford, L.A., Brennan, P.C.: Grey-scale inversion improves detection of lung nodules. Br. J. Radiol. **86**(1021) (2013). Article no. 27961545

13. Smirg, O., Liberda, O., Smekal, Z.: Automated detection of important facial curves in radiographs. In: 2013 36th International Conference on Telecommunications and Signal Processing, TSP 2013, pp. 778–782 (2013). Article no. 6614044

14. Uzunoğlu, Z., Yilmaz, D., Şahin, Y.: Quantitative x-ray spectrometric analysis with peak to Compton ratios. Radiat. Phys. Chem. **112**, 189–194 (2015)

15. Bodnárová, S., Hudák, R., Živčák, J.: Príprava a návrh biologického testovania biokompatibility keramických skáfoldov na báze hydroxyapatitu a trikalcium fosfátu. In: Novus Scientia 2017, pp. 17–21. TU, Košice (2017). ISBN 978-80-553-3080-8

Framework for Effective Image Processing to Enhance Tuberculosis Diagnosis

Tsion Samuel[1], Dawit Assefa[1,2], and Ondrej Krejcar[2(✉)]

[1] Division of Biomedical Computing, Center of Biomedical Engineering,
Addis Ababa Institute of Technology, Addis Ababa University,
Addis Ababa, Ethiopia
tsisami@gmail.com, dawit.assefa@aau.edu.et
[2] Faculty of Informatics and Management,
Center for Basic and Applied Research,
University of Hradec Kralove, Hradec Králové, Czech Republic
Ondrej.Krejcar@uhk.cz

Abstract. Tuberculosis (TB) is one of the disease causing microorganisms which is infectious with a high morbidity and mortality around the world. According to the latest World Health Organization (WHO) report, there were an estimated 10.4 million new TB cases in 2015 and more than 1.8 million deaths were attributed to the disease the same year. A person with TB causing bacteria need to know the presence of the bacteria at its latent (passive) stage so that she/he can take medication only for a couple of weeks to a couple of months (daily, bi-weekly or weekly depending on the treatment regimen of choice) to be free of TB. A person who is diagnosed with an active form of TB has to take medication for a minimum of 6 to 9 months. Negligence to take this medication properly will make the TB causing bacteria drug resistant. Diagnosing TB causing bacteria in its different stages is an effective mechanism which can enhance the performance of present detection schemes available in different clinics. In that regard, this paper presents a methodology for use in enhancing TB diagnosis specificity based on image processing of lung images acquired using normal x-ray as well as those from a sputum smear microscopy. A general framework is designed preceded by a step to extract useful imaging features for use in robust characterization of latent, active, drug resistant and TB free samples.

Keywords: Tuberculosis · Drug resistant TB · X-ray · Classifiers
Image processing

1 Introduction

Microbial infections including bacterial and viral infections are potentially life-threatening and require rapid identification and characterization of the microorganisms. Rapid and accurate detection of microorganisms from food, environment and clinical samples requires characterization to the species or subspecies level as well as determining their antibiotic susceptibility. These steps are important for selecting appropriate antibiotics, reducing morbidity and mortality and reducing health care costs [1].

© Springer International Publishing AG, part of Springer Nature 2018
N. T. Nguyen et al. (Eds.): ACIIDS 2018, LNAI 10752, pp. 376–384, 2018.
https://doi.org/10.1007/978-3-319-75420-8_36

Tuberculosis (TB) is one of disease causing microorganisms which is infectious with a high morbidity and mortality around the world. The TB causing Mycobacterium can be categorized into three different states as latent (passive), active and multi drug resistant. People with latent TB infection have TB germs in their bodies, but they are not sick because the germs are not active. These people do not have symptoms of TB disease, and they cannot spread the germs to others [2]. However, they may develop TB disease in the future. They are often prescribed treatment to prevent them from developing TB disease. Patients with active pulmonary TB are the source of Mycobacterium TB (MTB). In more than 90% of persons infected with MTB, the pathogen is contained as asymptomatic latent infection. Recent studies raise the possibility that some persons acquire and eliminate acute infection with MTB. The risk of active disease is estimated to be approximately 5% in the 18 months after initial infection and then approximately 5% for the remaining lifetime. An estimated 2 billion persons worldwide have latent infection and are at risk for reactivation. Contained latent infection reduces the risk of reinfection on repeated exposure, whereas active TB is associated with an increased risk of a second episode of TB on re-exposure [3]. Symptomatic active form of TB is the most crucial factor for transmission of infection. Even though competent immune system will limit the multiplication, some bacilli may remain dormant latently. Relatively, small proportion of exposed people develop TB disease at any course of their life while the probability of developing TB is much higher among people infected with HIV.

2 Related Works

There are several TB diagnosing devices that are different in their accuracy, sensitivity, the way they diagnose and their price. Below is a list of present TB diagnosing techniques used in different clinics with their merits and demerits.

A technique developed by Kamble et al. [4] uses TB screening based on Computer-aided Diagnostic (CAD) system using Matlab based chest x-ray image processing. Two major approaches are followed in the CAD system: finding the location of lesions and quantifying the features of the radiography images to check for any manifestation of TB patterns inside the lung. For a given CXR image as an input, the system first applies pre-processing to reduce noise and other artefacts aiming for image enhancement. The system then applies an active contour method to effectively segment out the lung region. This is followed by statistical feature extraction on the segmented lung region and these features are used as input to a classifier to dictate classification of the image as either normal or abnormal. Results showed that normal chest radiographs assume lower values in features such as variance, third moment and entropy, but higher mean.

Another study by Kandaswamya et al. [5] tried to demonstrate that acoustic signals generated by the lungs during inspiration and expiration give important information about the condition of the respiratory system. They used signal analysis based on joint time-frequency localized transformation to implement spectral analysis. The mutually exclusive time and frequency domain representations (e.g. the signal time series or its complex Fourier transform) are not highly successful in the diagnostic classification of lung sounds. Hence, they used wavelets for joint time and frequency characterization of the signals. Figure 1 below shows a schematic of the classification technique which was based on the wavelet transform.

Fig. 1. Schematic of the classification technique proposed in [5].

The process of converting a signal from the time domain to the frequency domain is achieved conventionally using the Fourier transform (FT). FT determines only the frequency components of a signal, but not their location in time. This could be adequate to analyse signals which are stationary in which case their frequency characteristics is not changing with time. Most signals are non-stationary and their frequency components show temporal variation (e.g. most medical signals). In order to overcome the FT drawback, the short time Fourier Transform (STFT) was proposed [6].

Cicero et al. [7] in their paper used different color specification to segment bacilli based on the color ratio method using color information from pixels that belong to the bacilli and neighbor. The input for the segmentation was selected by combining color information extracted from the microscopic images represented in RGB, HIS, YBCr and Lab color spaces. The authors compared two classification schemes for use in bacillus segmentation, namely, SVM and ANN and color ratio (CR) method is used during post processing to identify the bacilli.

Rani et al. [8] in their research proposed that the automation of microscopy may help to increase the yield of screening, as machines can screen a greater number of fields and can detect many TB cases in their early stages. Schematic of a fully automated TB detection system based on microscopic sputum smear images is shown in Fig. 2.

Fig. 2. Schematic of the automated TB detection system proposed in [8].

Manuel et al. [9] proposed a method for identification of TB bacteria using shape and color. In their study specimens taken from patients were stained with flourochrome to specifically interact with acid fast cell wall and give different red color from surrounding cell. Flourochrome can make non fluoresce molecules to fluoresce. The authors compared different autofocus algorithms for use in fast autofocus measures during automation of fluorescence stained mycobacterium and proposed to use the variance of the logarithm of the upper part of the image histogram computed which correspond to tiny bacteria.

MTBs when they appear in an image, they fluoresce in the range between green and yellow up to white [9].

The authors [12] presented their experimental results using two sets of performance metrics: Recall/Precision and Sensitivity/Specificity. Fluorescence images of sputum smear slides were taken using Cell Scope, which has a 0.4NA objective and an 8-bit monochrome CMOS camera. The algorithm first identifies potential TB objects and characterizes each candidate using Hu moments, geometric and photometric features, and histograms of oriented gradients. The candidate objects were then classified using an Intersection Kernel Support Vector Machine (IKSVM), achieving average precision of 89.2% ± 2.1% for object classification. At the slide level, the algorithm performed as well as human readers.

Table 1 below summarizes merits and demerits of the different TB diagnosing techniques in terms of their accuracy, sensitivity, price and ease of use.

Table 1. Comparison of different TB diagnosing techniques.

Diagnosing mechanism	Pros	Cons
Sputum smear microscopy	Simple; easy to interpret; less expensive; good specificity	It takes long time; 50–60% sensitivity; delay in result
Chest x ray	Quick; no need of specimen	Can't differentiate earlier scar; difficult to interpret; expensive; lack specificity
Fluorescence microscopy	70% sensitivity; accurate	Very expensive; difficult to interpret
Sound test	Less contamination; quick	Lack sensitivity; noise disturbance

3   Development of a New TB Diagnosis Tool

3.1   Criteria

Goldberger principle [10] listed a number of attributes that new tests for TB diagnosing might be expected to demonstrate improvements, which are adopted as criteria to develop a new TB diagnosing tool in the current study. These performance attributes include: (i) Sensitivity: more than 70%; (ii) Specificity: more than 98%; (iii) Simplicity: easily administered by a technician requiring little or no training; (iv) Equipment and reagents required; (v) Stability: a shelf life of two or more years; and (vi) Cost: no more expensive than current test.

3.2   Methodology

3.2.1   Effective Image Processing of Chest X-ray Images Using CAD

The main interest here is to enhance the capacity of X-ray diagnosis by computer aided detection (CAD) with high specificity. The specificity of screening TB is affected by image quality and the radiologist's level of expertise. CAD technology can improve the performance of radiologists, by increasing specificity to rates comparable to those obtained by double reading, in a cost-effective manner. The radiologist first reviews the exam, then activates the CAD software and re-evaluates the CAD-marked areas of concern before issuing the final report. In this paper, we present an overview of digital image processing and pattern analysis techniques to address several issues related to TB diagnosis using a CAD system. The CAD system follows four major stages to process the images to automate TB diagnosis: image preprocessing, image segmentation, features extraction and selection and image classification. The preprocessing involves: lung shape analysis, clavicle detection, and texture analysis. The CAD system will be adapted to identify the following physical features of interest of TB in X-ray images:

- *Fibrosis* - tissue deep in the lungs becomes thick, stiff and scarred by TB. The scarring is called fibrosis. As the lung tissue becomes scarred, it interferes with a person's ability to breathe.
- *Pleural changes* – the pleural cavity contains a relatively small amount of fluid, approximately 10 mL on each side. Pleural fluid volume is maintained by a balance between fluid production and removal, and changes in the rates of either can result in the presence of excess fluid, traditionally known as a pleural effusion.
- *Cavity* - is a gas-filled area in the surface of the lung as a result of TB infection.
- *Lung field zones* - the lungs are assessed and described by dividing into upper, middle and lower zones, the structure of the zones is altered upon infection by TB.
- *Loss of lung volume* - the average total lung capacity of an adult person is about 6 L of air and any severe loss of lung volume indicates abnormality.
- *Scar tissue in upper lobes* - pulmonary fibrosis is a respiratory disease in which scars are formed in the lung tissues and bacterial infections like TB may cause fibrotic changes in both lungs upper or lower lobes and other microscopic injuries to the lung.
- *Consolidations* - this refers to alveolar spaces that contain liquid instead of gas.
- *Lymphadenopathy* - it is a chronic lymph node enlargement.
- *Granulomas* – the granuloma contains mostly blood-derived macrophages, epithelioid cells differentiated macrophages.
- *Calcification* - in active pulmonary TB, infiltrates or consolidations and cavities are often seen in the upper lungs with or without mediastinal or hilar lymphadenopathy. Conversely calcified nodular lesions (calcified granuloma) pose a very low risk for future progression to active TB.

The image of the lung captured by the x-ray will be taken to identify the different states of the TB causing bacteria using image processing. First the image needs to be pre-processed by histogram equalization and filtering in order to improve the quality of the image and reduce undesired portions from the background of the images. Then lung segmentation will be modelled as an optimization problem that takes properties of lung boundaries, regions, and shapes into account. The key functions of the segmentation are clustering pixels having same intensity value from the whole image regions, separating regions or objects of desired part of the original image and hiding the undesired region or surfaces. Active contour method can be used as image segmentation tool. Then the input data will be transformed into a reduced set of features.

Table 2. Different classifiers built in Matlab.

Classifier	Prediction speed	Memory usage	Interpretability
Decision tree	Fast	Small	Easy
Discriminant analysis	Fast	Small for linear, large for quadratic	Easy
Logistic regression	Fast	Medium	Easy
SVM	Medium for linear, slow for others	Medium for linear. All others: medium for multiclass, large for binary	Easy for Linear SVM. Hard for all other kernel types
Nearest neighbor classifiers	Slow for cubic. Medium for others	Medium	Hard
Ensemble classifiers	Fast to medium depending on choice of algorithm	Low to high depending on choice of algorithm	Hard

After the feature extraction stage, the features have to be analysed to identify the active, latent, drug resistant or old healed TB. Then classification will proceed using the best classification method and the classification scheme is evaluated for its effectiveness by comparing its performance against the available ground truth information. Table 2 compares different classification schemes built in Matlab with respect to their computational cost as well as interpretability. Figure 3 depicts a rough flow chart of the complete X-ray image processing scheme.

In this study, a rigorous comparison will be carried out to check which of the above listed classification schemes do their purpose based on quantitative matrices which will be developed for assessment of each. But generally, based on what has been reported in various literatures, also decision tree, discriminant analysis and logistic regression are

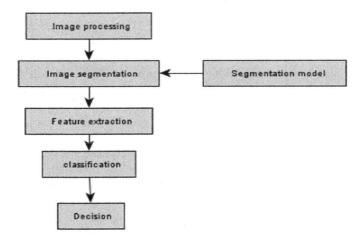

Fig. 3. Flow chart of the X-ray image processing scheme.

efficient in terms of time and memory requirement, all face difficulty handling large number of categorical features and highly biased training set, while SVM with medium prediction speed is highly accurate owing to its ability to model complex nonlinear decision tree boundaries and it is much less prone to over fitting than other methods. To identify the TB state, SVM could then be preferred [11].

At the first stage the x-ray images will be processed to remove undesired portion from their background aiming acceptable contrast enhancement and elimination of specific artefacts. A segmentation scheme will then be developed based on features that can uniquely characterize TB lung cavities. The application of an active contour method in effective segmentation of the lung cavities will thoroughly be investigated and compared against similar other segmentation schemes. For use in describing the different states of TB, intensity histogram, gradient magnitude histogram and shape descriptors/features will be extracted and a classification scheme will be developed accordingly based on both SVM and Artificial Neural Network (ANN) to pick the best classification strategy before final decision is made.

3.2.2   Robust Image Processing for Use in Analysing Images Generated on Sputum Smear Microscopy

The sputum smear images are captured by attaching a digital imager or a camera to the eyepiece of the microscope. These images can then be stored in the computer and can be processed in real-time or offline mode using image processing techniques. Before performing the image acquisition, auto focusing has to be done if the microscope is not a digital one (such as using the method used by Rani et al. [8]). In our work, the basic stages used in analysing the resulting image data comprise of pre-processing of the images to filter noises and other artefacts, followed by segmentation of TB bacilli and final classification of the bacilli. Filtering of the noise and other possible artefacts will be carried out by applying a 3×3 wiener spatial domain filter applied on each colour channel.

MTBs and similar microorganisms have acid-fast cell wall, which makes the cells impervious to acid–alcohol mixture [11]. Therefore, acid-fast staining technique is used for detection of acid-fast bacilli (AFB). Random forest (RF)-based method for the automated pixel segmentation and identification of TB bacilli in microscopic images is used to obtain stained sputum smears by using a light-field microscope. A data set need to be collected from different slides taken from various patients for experimentation. In each training image, the pixels belonging to regions of TB bacilli will manually be labelled by medical technician. To minimize the number of pixels manually marked incorrectly in each image, noisy data elimination will be performed on the RGB images based on their colour distributions. To achieve RF based supervised learning algorithm for pixel segmentation, a training procedure must be employed on different three class pixels. The first class pixels manually marked as the latent state, the second class pixels represented as active TB and the third as drug resistant TB. Therefore, each pixel in the stained images in the test set is automatically labelled by using RF-based supervised learning algorithm as latent, active and drug resistant TB. The TB bacilli pixels are then grouped into the regions by using connected component analysis. Each region is then rotated, resized and centrally positioned within 30×30 bounding box, respectively, in order to utilize appearance-based TB bacilli identification algorithms. As a result of the pixel segmentation, the bounding box can include different pixels for different states. Once the image is segmented, only the region of pixels given same bacilli colours is retained. Subsequently, appearance-based TB bacilli identification process is then performed for determining the state. Finally, the segmented and positioned region (pixels) into the bounding box is classified as either TB active, latent or drug resistant bacilli or not by using the proposed RF-based learning algorithm. The overall flowchart of the proposed algorithm is the one shown in Fig. 4.

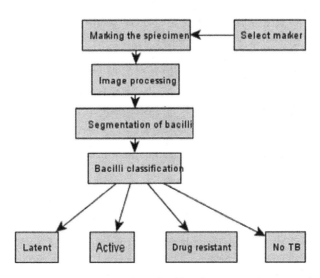

Fig. 4. Proposed image processing algorithm for smear microscopy images.

4   Conclusions

Few apps were developed in the literature to diagnose TB and none were able to differentiate the different states of TB: active, latent, and drug resistant. Differentiating the different states of TB could greatly enhance the quality of care and treatments. Increasing the sensitivity of TB diagnosing devices using a CAD system could increase the effectiveness of the diagnosis of TB causing bacteria before any associated effects on other organs inside the body. In that regard, the current study highlighted a framework for a more comprehensive diagnosis strategy based on a CAD system applied on both X-ray as well as sputum smear microscopic images. The authors of this paper mainly wanted to convince readers that there is still an immense need for the development of a comprehensive TB assessment tool inside clinics and the two methods which are proposed in this work, which are complementary to each other, could offer some promises, subject to further investigations.

Acknowledgement. This work was supported by internal student's project at FIM, University of Hradec Kralove, Czech Republic (under ID: UHK-FIM-SP-2018).

References

1. Rasha, B., Mona, S.: Molecular characterization of clinical isolates with elevated resistance to carbapenems. Open Microbiol. **11**(1), 152–159 (2017)
2. Clinical Disease Control: Tubeculosis Elimination Division of Tuberculosis, pp. 3–4, July 2017
3. Alimudin, Z., Mark, R., Richard, H., Fordhan, C.: Current concepts: tuberculosis. New Engl. J. Med. **3**(10), 2–3 (2013)
4. Kamble, P.A., Anagire, V.V., Chamataguodar, S.N.: CXR tuberculosis detection using Matlab image proccesing. IRJET **3**(06), 2342–2344 (2016)
5. Kandaswamy, A., Kumar, C.S., Ramanathan, R.P., Malmurugana, N.: Neural classification of lung sounds using wavelet cofficent. Comput. Biol. Med. **34**(6), 523–537 (2004)
6. Saed, S., Aouni, L., Mac, T.: Gear fault diagnosis using time frequency methods, pp. 7–27. National Research Councile, December 2015
7. Costa Filho, C.F.F., Levy, P.C., Xavier, C. de M., Fujimoto, L.B.M., Costa, M.G.F.: Automatic identification of tuberculosis mycobacterium. Res. Biomed. Eng. **31**(1), 33–43 (2015)
8. Rani, O., Biju, S., Gagan, S., Jeny, R.: A review of automatic methods based on image processing techniques for tuberculosis detection from microscopic sputum smear images. J. Med. Syst. **40**, 1–13 (2015)
9. Manuel, G.F., Fillip, S., Gabriel, C.: Identification of TB bacteria based on shape and color. Real Time Imaging **10**(4), 251–262 (2004)
10. Product Development Guidelines New York: WHO Tuberculosis Diagnostics Workshop, Ohio, Cleveland, 27 July 1997
11. Ayas, S., Ekinci, M.: Random Forest-based Tuberculosis Bacteria Classification in Images of ZN-stained Sputum Smear Samples. Sig. Image Video Process. **8**(Suppl. 1), 49–61 (2014)
12. Chang, J., Arbeláez, P., Switz, N., Reber, C.: Automated tuberculosis diagnosis using flourescence image from a mobile microscope. Med. Image Comput. Assist. Interv. **15**(3), 345–352 (2012)

Multiregional Soft Segmentation Driven by Modified ABC Algorithm and Completed by Spatial Aggregation: Volumetric, Spatial Modelling and Features Extraction of Articular Cartilage Early Loss

Jan Kubicek$^{(\boxtimes)}$ ⓘ, Marek Penhaker ⓘ, Martin Augustynek ⓘ,
Martin Cerny ⓘ, and David Oczka ⓘ

FEECS, VSB–Technical University of Ostrava, K450 17. listopadu 15,
708 33 Ostrava–Poruba, Czech Republic
{jan.kubicek, marek.penhaker, martin.augustynek,
martin.cerny}@vsb.cz, David-oczka@seznam.cz

Abstract. In a clinical practise of the orthopaedics and medical imaging systems, the early cartilage loss, and cartilage lesions are challenging tasks. Due to an insufficient contrast, such pathologies are badly observable by naked eyes. Furthermore, objectification and quantification of those pathological findings are usually only subjectively estimated without the SW support. We propose a multiregional segmentation model based on the histogram classification with using of a sequence of triangular fuzzy functions where each such function represents specific knee area. To ensure a robustness of the model, respective fuzzy class location is driven by the ABC (Artificial Bee Colony) genetic algorithm respecting statistical features of the physiological cartilage. In the second step of the algorithm, a spatial aggregation is applied in order to consider spatial relationships in every region to prevent the image noise deterioration. Such multiregional segmentation model allows for an extraction of significant features well corresponding with the early cartilage loss like is the cartilage volume.

Keywords: Articular cartilage · Knee · MRI
Multiregional image segmentation · ABC · Genetic algorithm · Fitness function

1 Introduction

In a clinical practice of the orthopedics, the articular cartilage assessment is one of the major tasks which are routinely done [1]. Clinical assessment of the articular cartilage is especially important when the cartilage loss is present. Latent stages of the cartilage loss may lead up to the complete cartilage dysfunction [2]. In this regard, a prevention of such pathologies is substantially important. Correct assessment of the early cartilage loss is a crucial task for the clinical practice. Unfortunately, the early cartilage loss is badly observable from the native image records, thus physicians have a complicated role to correctly distinguish and discover respective pathological state of the cartilage [3–5].

© Springer International Publishing AG, part of Springer Nature 2018
N. T. Nguyen et al. (Eds.): ACIIDS 2018, LNAI 10752, pp. 385–394, 2018.
https://doi.org/10.1007/978-3-319-75420-8_37

2 Related Work

Judging by the recent literature, the articular cartilage segmentation can be done by the three essential ways. Firstly, it is manual segmentation. This approach is based on the manual drawing of the cartilage outline. This way is significantly affected by the subjective error, and presently it is utilized as a gold standard for the segmentation performance evaluation [13]. Semiautomatic segmentation is intended for a partial reduction of the user interaction when the cartilage is being segmented. This process requires a certain level of the user supervision to focus the segmentation process in the particular direction. This process usually requires an initialization in a form of specification of the image parameters. Particularly, the following methods have been adopted for the semiautomatic segmentation: thresholding [6, 7], watershed [8], edge detection [9], energy-minimization [10], Live Wire [11], graph-cuts [7] and active contours [10]. In the ideal case of the cartilage modeling, it should be perceived balance among the algorithm performance, preciseness, robustness and level of the user interaction. The main present target of the cartilage segmentation is a development of fully autonomous systems serving for a precise morphological extraction of the cartilage even in cases when data are corrupted by the image noise, and in the same time the human's interaction should be minimized. The automatic cartilage segmentation includes the methods: texture analysis [12] and supervised learning (Neural networks kNN) [13].

3 Problem Definition

Recent literature is often focused only to the physiological cartilage modelling and extraction, and do not reflect other related fact including pathological disorders. Other words speaking particular segmentation model may behave differently in environment where the pathological changes are present. This issue can be effectively figured out by using the multiregional segmentation procedure separating knee area into a finite number of the image regions. Thus, we suppose that the cartilage model will be effectively separated from the early cartilage loss even in cases when contrast of the healthy cartilage missing. The Fig. 1 reflect situation of the early cartilage loss which is shown under week contrast. For this reason we have proposed the soft multiregional segmentation model driven by the ABC algorithm and completed by statistical local aggregation.

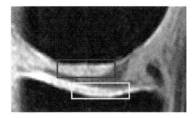

Fig. 1. RoI (Region of Interest) of the articular cartilage. Healthy cartilage is shown by the dark white color. Red area indicates healthy tibial cartilage, while blue area indicates the femoral cartilage. After the blue area, the white spectrum is getting weaker – it is clinically indicated as the early cartilage loss. (Color figure online)

4 Proposal of Brightness Cartilage Segmentation

The proposed multiregional segmentation method for the cartilage segmentation should be robust even in the noisy environment. Therefore, the segmentation strategy utilizes two stages of the pixels classification. Firstly, the brightness spectrum is decomposed into a finite number of the segmentation classes specified by the user. On the base of the experiments we use six segmentation classes well reflecting the cartilage mor-phological structure. In this step, the histogram is approximated by finite number of the triangular fuzzy classes (each segmentation region is given by one class), thus each pixel r is assigned to the each region by a certain level of the membership $\mu(I(r))$ (Fig. 2). In this regards, we can speak about the fuzzy descriptor which transforms the original brightness values $(I(r))$ to a space of the fuzzy membership functions. The Eq. (1) indicates the fuzzy membership vector for L classes.

$$\mu(I(r)) = [\mu_1(I(r))\ \mu_2(I(r))\ldots\mu_L(I(r))] \tag{1}$$

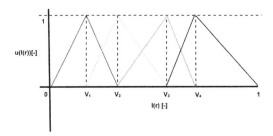

Fig. 2. Sequence of the fuzzy triangular classes defining the multiregional segmentation model where four classes are used.

Such segmentation model is determined by a sequence of the vertexes (V) speci-fying individual regions. The following rules are used for building of the soft bright-ness model:

- complete partitioning: $\forall x, \exists \mu_l(x), 1 \leq l \leq L, \mu_l(x) > 0,$
- consistency: if $\mu_l(x_0) = 1$, then $\mu_k(x_0) = 0, \forall k \neq l,$
- normality: $\max(\mu_l(x)) = 1,$
- and intersection of the adjacent fuzzy classes: $\mu_l(x_0) = \mu_{l+1}(x_0) = 0.5.$

Triangular function appears as effective alternative. It is assumed that the vertex is placed in the histogram peak, other words speaking it reflects the maximal probability of the region. Furthermore, the vertex is only one criterial point which is needed to be defined. Thus, the main issue of the model is a determination of the vector: $V(V_1, V_2, \ldots, V_n)$. In the simplest case, the clustering analysis (e.g. K-means or FCM) may be adopted where V would be given as centroid of the respective cluster. Unfortunately, such approach sometimes badly converges to correct centroid in a dependence of initial selection of the centroids, furthermore it does not reflect features of the image regions such is the cartilage. Therefore, we are using a genetic optimization procedure which is based on the ABC (Arteficial Bee Colony) algorithm.

Genetic optimization of the soft brightness model

Here, we figure out a problem to find out an optimal distribution of the triangular classes well approximating the healthy cartilage region. For this task, the ABC algorithm is adopted. Firstly, the initial population is defined. In comparison of the standard ABC, we use the C-means clustering for a definition of the one initial solution of the optimization problem. Other solutions are defined on the base of the random numbers. This process is described by the Eq. 2.

$$\overline{X}_i = \left(R_{i1,1}V_{i,1}, R_{i2,2}V_{i,2}, \ldots, R_{in,n}V_{i,n}\right) \tag{2}$$

where $\overline{X}_i$ depicts i^{th} solution of the optimization problem, $R_{i,n}$ represents n^{th} random number and $V_{i,n}$ stands for n^{th} centroid from FCM method. The ABC algorithm is composed from the three stages including EB (Employed Bees), OB (Onlooker Bees) and SC (Scout Bees). It is supposed that: $EB = OB = SN$, where SN stands for a number of the sources (number of initial solutions).

Phase of Employed Bees

In this phase, the artificial EB are looking for new food sources (possible solutions). In a practical point of the view, for each defined solution $\overline{X}_i$, the new adjacent solution $\overline{V}_i$ is defined (Eq. (3)).

$$V_{ik} = X_{ik} + \phi_{ik} \times \left(X_{ik} - X_{jk}\right) \tag{3}$$

where X_{jk} represents a random candidate solution ($i \neq j$), k represents the random index and the function ϕ_{ik} represents the random numbers generator: $\phi_{ik} \in [-0.1; 0.1]$. In this phase, $\overline{X}_i$ and $\overline{V}_i$ are being compared on the base of their fitness functions. If $fit_{X_i} > fit_{V_i}$ then a new $\overline{V}_i$ is generated. This process can be repeated n-times, where n specifies limitation of the selection (l_v). After taking n-selections of the $\overline{V}_i$, the $\overline{X}_i$ is perceived as exhausted, and it is erased from the memory. If $fit_{X_i} < fit_{V_i}$ then the $\overline{V}_i$ is stored in the memory as possible solution of the optimization problem.

Phase of Onlooker Bees

In this phase, the onlooker bees are looking for the food sources containing as much of the nectar as it is possible. Practically, the probabilistic selection is figured out (Eq. 4). It is supposed that a higher value of P_i indicates better solution. Eventually, when the $\overline{X}_i$ having the $\max(P_i)$ is selected, the adjacent $\overline{V}_i$ is generated. As same as in the EB phase, a higher value of the fitness function indicates better result which is taken.

$$P_i = \frac{fit_i}{\sum_{j=1}^{SN} fit_j} \tag{4}$$

In the case that $OB \neq EB$ (due to exhausted sources $\overline{X}_{i,out}$), the new $\overline{X}_i$ solutions are generated and process is repeated.

Phase Scout Bees

The scout bees are looking for the exhausted sources ($\overline{X}_{i,out}$). In their surroundings, the new solutions $\overline{X}_i$ and $\overline{V}_i$ are done and the evaluation process is repeated. This three stage process (EB, OB, SB) is repeated within a predefined number of iterations (IT). The Table 1 summarizes parameters of the ABC for the cartilage modeling.

Table 1. Parameters of the ABC algorithm for the soft brightness model.

SN	l_v	IT
150	10	100

Fitness function

Lastly, the fitness function is introduced. The fitness function assesses a quality of the individual solution of the optimization process. In this regard, the entropy of each region is taken into account. The entropy calculates the pixels probability distribution of each segmented region. If the histogram area is concentrated (do not contain significantly deviating pixel values) it has a higher value of the entropy because the pixels appear in this area with a higher probability. Contrarily, the significantly different pixel values which can represent noise and artefacts may have a lower probability. Thus, the fitness function maximizes the entropy function. In our algorithm, a contribution of the Kapur entropy (H_m) is calculated for thresholding t_{m1} and t_{m2} (Eq. 5).

$$H_m = -\sum_{i=t_{m1}}^{t_{m2}} \frac{p_i}{\omega_m} ln\left(\frac{p_i}{\omega_m}\right), \omega_m = \sum_{i=t_{m1}}^{t_{m2}} p_i, p_i = h(i)/\sum_{i=0}^{L-1} h(i) \qquad (5)$$

Fitness function of the ith solution is figured out by the Eq. 6 (n stands for a number of the segmentation classes).

$$fit_{X_i} = \sum_{m=1}^{n} H_m \qquad (6)$$

5 Spatial Cartilage Segmentation

In this section, the spatial local aggregation is introduced. The spatial aggregation takes into account the spatial relationships inside the individual regions, and allows for modification of originally assigned membership value. A great benefit of this procedure is that it works in the space of the fuzzy membership functions and not in the brightness space. It means that the original pixel value is not modified, and clinical information is perceived. Local aggregation may be applied in a form of the statistical operators taking into account the spatial relationships. In our method, the mean (*AvgAg*) and median aggregation (*MedAg*) are employed. The local aggregation works on a principle of the cyclic convolution. It means that iteratively goes through the individual regions, and in each horizontal and vertical shift it performs the local aggregation within the predefined size of the convolution window. In our algorithm, the window size 3 (3×3) is adopted. This procedure is especially beneficial in situations when the image noise having significantly different brightness spectrum is present. Such pixels probably have high membership level in different region but lay inside the cartilage region. If we had not used the local aggregation, such pixels would represent so called blind spots may significantly deteriorate the segmentation model. Such situation is depicted on the Fig. 3 where the central noise pixel has membership level 0 (a) of the cartilage region, after the median local aggregation (median is computed from the surrounding pixels) same pixel has membership level 0.9 (b).

0.9	0.8	0.9		0.9	0.8	0.9
0.7	0	0.9		0.7	0.9	0.9
0.7	1	0.9		0.7	1	0.9

a)   b)

Fig. 3. Principle of *MedAg:* (a) original pixel surrounding 3×3 and (b) median aggregation.

6   Testing and Quantitative Comparison

In this section, the testing and verification of the proposed method is carried out. We have tested the proposed multiregional segmentation method on a sample of the 150 MR (magnetic resonance) image records of the articular cartilage exhibiting the early loss of the cartilage which is badly clinically assessable due to insufficient contrast between the cartilage and spots where the cartilage is interrupted or completely missing. In this regard it is important to mention that the cartilage is relatively small in a comparison with adjacent knee structures (bones and other tissues), thus the RoI (Region of Interest), completed by the cubic interpolation of third order has been applied to increase an area where the early cartilage loss is presented (Fig. 4). This spatial model allows for the extraction of the physiological cartilage (red areas). Those areas are divided into each other by interruption clearly indicating the early cartilage loss (Fig. 4).

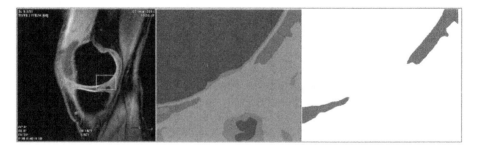

Fig. 4. Native MR data of the knee area with RoI (left), multiregional (six classes) segmentation model (middle) and final extraction of the cartilage (right). (Color figure online)

Such model may be applied either for the single MR slice (one image), or whole image series to make 3D volumetric model (Fig. 5) which allows for a volumetric modeling of the cartilage. The Fig. 5 shows the final 3D model of the articular cartilage which reliably reflects the early cartilage loss (blue RoI) and cartilage lesions (red RoI).

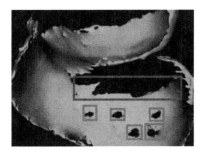

Fig. 5. 3D (volumetric) multiregional segmentation model of the articular cartilage indicating the cartilage lesions (red RoI) and the early cartilage loss is clearly represented in the blue RoI. (Color figure online)

2D and 3D cartilage models reliably reflect manifestation, morphological structure and pathological processes like is the early cartilage loss or the cartilage lesions which deteriorate the cartilage structure. Nevertheless, the main output of the models is features extraction. We needed to specify significant cartilage features well reflecting the cartilage deterioration from the healthy cartilage. If cartilage structure is impaired by the pathologies, the cartilage volume is being changes in a comparison with the healthy cartilage. Therefore, we have performed the cartilage volume analysis from the cartilage model. The model results have been compared with three independent clinical experts who subjectively estimated volume of the cartilage. Each clinical expert has performed three-times manual segmentation of the cartilage on its base the volume has been calculated. The volumetric results (cartilage volume calculated from the segmentation model V_{model} and by the clinical experts V_{exp}) of 150 patients with the early cartilage loss are summarized in the Table 2.

Table 2. Volumetric parameters from the multiregional soft segmentation model.

	$\widetilde{V}_{model}$ [ml]	$\widetilde{V}_{exp}$ [ml]	$\overline{V}_{model}$ [ml]	$\overline{V}_{exp}$ [ml]	σ_{model} [ml]	σ_{exp} [ml]
Men	4.611	4.599	4.599	4.467	0.871	0.991
Woman	3.981	3.756	3.911	3.691	0.661	0.456

Eventually, the verification procedure is carried out. In this phase, we have compared our model against representative state of the art segmentation methods. We have consciously selected alternative segmentation method utilizing completely different segmentation strategy. (1) Multiregional segmentation is represented by the OTSU method, (2) the probabilistic segmentation is represented by MASP (Maximal Spatial

Probability) and clustering is represented by the FCM (Fuzzy C-means). The following measures are considered for the objective comparison.

- Variation of Information (*VI*): this parameter measures a distance between two segmentation classes C_1 and C_2 in a sense of the average conditional entropy which is given by the Eq. 7:

$$VI(C_1, C_2) = \mathcal{H}(C_1) + \mathcal{H}(C_2) - 2I(C_1, C_2) \tag{7}$$

- Rand Index (*RI*): this parameter measures a level of the similarity between two regions. *RI* compares a compatibility of assignment between two pairs of the elements in two clusters. The *RI* is given by the Eq. 8.

$$RI(C_1, C_2) = \frac{2(n_{11} + n_{00})}{N(N-1)} \tag{8}$$

where N denotes a total number of the pixels, n_{11} stands for a number of the pairs which are in the same cluster C_1 and C_2 and n_{00} is a number of the pairs which are assigned to the different clusters. The *RI* has values from the: $[0; 1]$, where 0 indicates that clusters are completely different, while 1 indicates that two clusters are completely identical.

- Mean Squared Error (*MSE*): it is an estimator measuring the average of the error squares between two segmentation results. The *MSE* represents a risk function which corresponds with the expected value of the squared or quadratic error loss.

The Table 3 summarizes results of the verification procedure. The best results for each test are highlighted. For the *RI* better result is given by higher number, contrarily for the *VI* and *MSE* better results are given by smaller numbers. The results represent averaged values from 150 MR knee image records.

Table 3. Objective comparison of the soft segmentation algorithm.

	MedAg	AvgAg	FCM	OTSU-N	MASP
VI	2.551	2.566	2.788	2.598	3.121
RI	0.811	0.801	0.7881	0.671	0.611
MSE	31.144	31.011	31.133	32.221	34.188

7   Conclusion

We have presented the multiregional soft segmentation method driven by the ABC algorithm for the cartilage segmentation and the early cartilage loss extraction. Segmentation method is robust even in the noisy environment. The image noise is a substantial component which is present in each image data from the MR. Such noise significantly affects the native image data, clinical information is modified and segmentation models are limited in a sense of the feature extraction. The proposed method utilizes the brightness classification model approximating the image histogram by a finite sequence of the triangular functions which are driven by the genetic (ABC) algorithm. Second segmentation strategy takes into account spatial relationships with surrounding pixels, thus original membership functions may be modified.

Based on the experimental testing, we are using six segmentation classes for the knee area segmentation. We have tested the soft segmentation for the 2D MR image records of the patients suffering from the early cartilage loss. The segmentation model reliably reflects an area of the physiological cartilage and significant interruption representing the early cartilage loss (Fig. 4). Second part of the testing deals with the volumetric modelling when the soft segmentation has been applied in a multiple form to individual slices. In a result we had obtained a set of the segmentation results which have been interpolated to the final model. The Fig. 5 reflects the volumetric model well reflect the cartilage lesions and the early cartilage loss. Based on the model, we have computed the volumetric features (cartilage volume) which have been compared against the expert segmentation. The Table 2 reports statistical insignificant differences of the model against the expert's segmentation. Eventually, the proposed model has been compared against the state of the art methods where the median aggregation (*MedAg*) appears as the best compromise. The median aggregation has the biggest *RI* index (0.811), contrarily the lowest *VI* index (2.551).

Acknowledgment. The work and the contributions were supported by the project SV4507741/2101, 'Biomedicínské inženýrské systémy XIII'. This study was supported by the research project The Czech Science Foundation (GACR) No. 17-03037S, Investment evaluation of medical device development.

References

1. Guzanová, A., Ižaríková, G., Brezinová, J., Živčák, J., Draganovská, D., Hudák, R.: Influence of build orientation, heat treatment, and laser power on the hardness of Ti6Al4V manufactured using the DMLS process. Metals **7**(8), 1–17 (2017). ISSN 2075-4701
2. Linka, K., Itskov, M., Truhn, D., Nebelung, S., Thüring, J.: T2 MR imaging vs. computational modeling of human articular cartilage tissue functionality. J. Mech. Behav. Biomed. Mater. **74**, 477–487 (2017)
3. Ťúková, V., Poláček, I., Tóth, T., Živčák, J., Ižaríková, G., Kovačevič, M., Somoš, A., Hudák, R.: The manufacturing precision of dental crowns by two different methods is comparable. Lekar a Technika **46**(4), 102–106 (2017)
4. Bodnárová, S., Hudák, R., Živčák, J.: Príprava a návrh biologického testovania biokompatibility keramických skáfoldov na báze hydroxyapatitu a trikalcium fosfátu. In: Novus Scientia 2017, pp. 17–21. TU, Košice (2017). ISBN 978-80-553-3080-8
5. Nebelung, S., Sondern, B., Oehrl, S., Tingart, M., Rath, B., Pufe, T., Raith, S., Fischer, H., Kuhl, C., Jahr, H., Truhn, D.: Functional MR imaging mapping of human articular cartilage response to loading. Radiology **282**(2), 464–474 (2017)
6. Kumarv, A., Jayanthy, A.K.: Classification of MRI images in 2D coronal view and measurement of articular cartilage thickness for early detection of knee osteoarthritis. In: 2016 IEEE International Conference on Recent Trends in Electronics, Information and Communication Technology, RTEICT 2016 - Proceedings, pp. 1907–1911 (2017). Article no. 7808167

7. Mallikarjuna Swamy, M.S., Holi, M.S.: Knee joint cartilage visualization and quantification in normal and osteoarthritis. In: International Conference on Systems in Medicine and Biology, ICSMB 2010 - Proceedings, pp. 138–142 (2010). Article no. 5735360

8. Fripp, J., Crozier, S., Warfield, S.K., Ourselin, S.: Automatic segmentation and quantitative analysis of the articular cartilages from magnetic resonance images of the knee. IEEE Trans. Med. Imaging **29**(1), 55–64 (2010). Article no. 5071225

9. Wang, P., He, X., Lyu, Y., Li, Y.-M., Qiu, M.-G., Liu, S.-J.: Automatic segmentation of articular cartilages using multi-feature SVM and elastic region growing. Jilin Daxue Xuebao (Gongxueban)/J. Jilin Univ. (Eng. Technol. Ed.) **46**(5), 1688–1696 (2016)

10. Kubicek, J., Vicianova, V., Penhaker, M., Augustynek, M.: Time deformable segmentation model based on the active contour driven by Gaussian energy distribution: extraction and modeling of early articular cartilage pathological interuptions. Frontiers Artif. Intell. Appl. **297**, 242–255 (2017)

11. Gougoutas, A.J., Wheaton, A.J., Borthakur, A., Shapiro, E.M., Kneeland, J.B., Udupa, J.K., Reddy, R.: Cartilage volume quantification via Live Wire segmentation. Acad. Radiol. **11**(12), 1389–1395 (2004)

12. Dodin, P., Pelletier, J.P., Martel-Pelletier, J., Abram, F.: Automatic human knee cartilage segmentation from 3D magnetic resonance images. IEEE Trans. Bio-Med. Eng. **57**(11) (2010)

13. Xia, Y., Manjon, J.V., Engstrom, C., Crozier, S., Salvado, O., Fripp, J.: Automated cartilage segmentation from 3D MR images of hip joint using an ensemble of neural networks. In: Proceedings - International Symposium on Biomedical Imaging, pp. 1070–1073 (2017). Article no. 7950701

Pseudo-Relevance Feedback
for Information Retrieval in Medicine
Using Genetic Algorithms

Lanh Nguyen[1(✉)] and Tru Cao[1,2]

[1] Faculty of Computer Science and Engineering,
Ho Chi Minh City University of Technology, Ho Chi Minh City, Vietnam
nguyenlanh2580@gmail.com, tru@cse.hcmut.edu.vn
[2] John von Neumann Institute, Vietnam National University at Ho Chi Minh City,
Ho Chi Minh City, Vietnam

Abstract. Pseudo-Relevance Feedback is one of the methods for improving search engine results. By automatically extracting information from a previous search result, a new query is posed as an expansion of the original query, and then it is searched again. In this paper, we apply a genetic algorithm to improve the Pseudo-Relevance Feedback method in searching medical texts. First, a set of candidate terms is constructed by extracting keywords from the documents returned from the initial search using the original query. Then, the seed terms are selected from the candidate term set using our proposed genetic algorithm, to be merged with the original query to create a new query. The new query is searched again, returning a final ranked list of documents. Experimental results on the TREC 2014 CDS dataset show that the proposed method outperforms the baseline method that does not use a genetic algorithm for Pseudo-Relevance Feedback.

Keywords: Medical case report · Clinical question type
Query expansion · Candidate terms · Jaccard similarity coefficient

1 Introduction

Information retrieval focuses on organization of a collection, and storage and communication of different kinds of data structures (e.g. texts, images and sounds) [1]. The purpose of an information retrieval system is to provide users with those documents that satisfy their information need without taking much time. There are search engines that may be available to the public, such as Educational Resources Information Center[1] system for education research and information, or FinAstronomy[2] and Infotopia[3] systems in the domains of science and biology, etc.

---

[1] http://eric.ed.gov.
[2] http://www.findastronomy.com.
[3] http://www.infotopia.info.

© Springer International Publishing AG, part of Springer Nature 2018
N. T. Nguyen et al. (Eds.): ACIIDS 2018, LNAI 10752, pp. 395–404, 2018.
https://doi.org/10.1007/978-3-319-75420-8_38

In the medical domain, relevant biomedical documents to a patient case report are searched by search engines [2]. For example, a case report may describe information such as a patient's current symptoms, the patient's medical history, the patient's diagnosis, or the steps taken by a physician to treat the patient. The challenges of medical information retrieval include the followings [3]:

- Highly specialized information.
- Different kinds of medical information.
- Medical documents in multiple languages.

Today, users can access online search engines to search for medical information. For instance, PubMed was developed by the National Center for Biotechnology Information (NCBI) at the National Library of Medicine. PubMed consists of more than 26 million citations for biomedical literature. PubMed is a free resource and also provides access via websites.

Entrez[4] is the online search engine developed by the NCBI to search on PubMed. This search engine requires a user to enter search terms that are used to search for relevant documents. Therefore, physicians can easily seek out information about how to best care for their patients to make a clinical decision.

Usually, a user's query is not expressive enough to represent the actual user's information need. Therefore, the initial list of search results may not be satisfactory. There are some methods to improve the search effectiveness due to incomplete representation of a query.

The relevance feedback method performs the interaction between the search system and users, in which a user gets improved retrieval performance thanks to their feedback to the system. The idea behind relevance feedback is using the user intervention and feedback to the initial search results to pose a revised query. When this automatic technique works without the user interaction, it is called Pseudo-Relevance Feedback (PRF), also known as Blind Relevance Feedback [4,5].

Query expansion is often combined with PRF where the original query is expanded with some terms automatically extracted from selected documents in the initial search results [5–7]. Further, genetic algorithms (GA) have also been applied to extract those added terms, which are crucial to improve the search performance [8]. However, to the best of our knowledge, GA-based PRF has not been employed for medical information retrieval.

Therefore, in this paper, we propose a genetic algorithm used with PRF for searching medical texts. The main role of GA is to select the seed terms in the top-ranked documents resulted from the initial search, based on their similarity with all of their context terms. A new query is then generated by adding the seed terms to the original query. Finally, the new query is searched again to return the final ranked list of documents.

The remainder of the paper is organized as follows. Section 2 defines the problem of discourse and describes the system architecture, baseline method, and proposed method. Section 3 presents details of the dataset, evaluation methods, and the main experiments. Finally, the conclusion and future work are given in Sect. 4.

---

[4] https://www.ncbi.nlm.nih.gov/gquery/.

2   Proposed Method

2.1   Problem Definition

To approach and extract the information from biomedical literatures, some modern search engines have been developed. In some cases, the needs of physicians may be found in a document or collection of documents. However, when the medical information become large and overlap, finding the relevant information for patient care becomes a significant challenge.

The focus of the TREC 2014 Clinical Decision Support Track is to retrieve relevant biomedical articles for answering generic clinical questions about medical records [9]. A detailed description of its input and output is presented below.

Input. The input data are divided into two parts: the first part is a document collection and the second is a query, also called as a topic. The document collection for the track is an open access subset of PubMed Central, an online repository of free available full-text biomedical literature. A database is created by indexing the text of the articles in the collection.

Each topic consists of a medical case report and one of the three generic clinical question types as follows:

- Medical case report: A case report typically describes a challenging medical case and is represented in the free-text format.
- Clinical question type: The three most common generic clinical question types are *diagnosis*, *test*, and *treatment*, which account for a majority (52.72%) of the clinical questions posed by primary care physicians [10]. Each case report has one associated clinical question type. Table 1 describes the meanings of these clinical question types.

Table 1. Description of the clinical question types.

Question type	Description
Diagnosis	Question about determining the diagnosis of the patient
Test	Question about suggesting relevant interventions for diagnosing the patient
Treatment	Question about suggesting the best treatment plan for the condition exhibited by the patient

For each of the clinical question types, the resulting documents should be relevant to it. A topic with the *diagnosis* label, for example, requires the system to retrieve those documents that a physician would find useful for determining the diagnosis for the patient described in its case report. Meanwhile, for the *test* label, the search results should suggest required medical tests to be conducted for the patient. Finally, for the *treatment* label, the retrieved documents should suggest to a physician appropriate treatment plans for the condition exhibited by the described patient.

Output. The expected search result is a ranked list of retrieved documents based on their relevance to the query. As in [9], our method evaluates results in 1,000 top-ranked documents for each topic.

2.2   System Architecture

To solve the problem defined above, we first present a general architecture for full-text medical information retrieval systems. Moreover, we briefly describe the baseline method used for each module [11]. This architecture has six main modules, as illustrated in Fig. 1 and presented below.

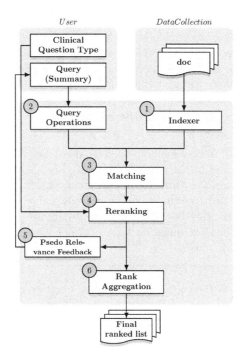

Fig. 1. System architecture for medical information retrieval.

Module 1. Indexing is the process of tagging search terms or phrases to each document to facilitate faster search and retrieval. As in [11], the raw documents is indexed by using Lucene[5].

Module 2. The query operations include two stages: query preprocessing and query expansion. The preprocessing stage may include the steps such as spelling checking, identification of words, phrases, sentences, stop words elimination, stemming, etc. Next, query expansion is the process of reformulating the original query such as finding synonyms or various morphological forms of words, in

---
[5] https://lucene.apache.org/.

order to better represent the meaning of the query. As in [11], the second stage expands queries by using the Medical Subject Headings (MeSH)[6] thesaurus.

Module 3. This module matches documents and query terms. In this module, [11] employed the Vector Space Model (VSM) [12], Language Model (LM) [13, 14], Best Matching (BM) with the two variants BM25L and BM25$^+$ in [15,16].

Module 4. Based on the clinical question type of a query, this module reranks the search results. The method in [11] counted the number of occurrences of candidate terms by using the MeSH thesaurus for each question type. If a document contains more candidate terms corresponding to the query's question type, it is more relevant.

Module 5. The PRF module is for simulating the user behaviour to select seed terms for query expansion from initial search results. In this work, we apply GA with PRF to improve the system's performance.

Module 6. The rank aggregation module is to combine multiple ranked document lists into a single ranked list to improve the rankings produced by individual systems. The Reciprocal Rank Fusion (RRF) algorithm in [17] is used for this module.

2.3 Genetic Algorithm in Pseudo-Relevance Feedback

For PRF, [18] shows that using a search process is more effective than simply using terms from initially retrieved documents. In this paper, we propose to use GA for this searching to select seed terms from a set of candidate terms in medical information retrieval.

Specifically, each document in the initial search results is represented by a string of 0's and 1's as a chromosome, in which each word is a gene. A set of chromosomes together with their associated fitness values is called a population. The population size (N) is the number of chromosomes in each generation.

Specifically, the steps using GA [8] to select the seed terms for PRF are as follows:

- **Step 1:** Generate an initial population of documents from the top of the search result list.
- **Step 2:** Encode retrieved documents into chromosomes in the binary format.
- **Step 3:** Create a new population by carrying out the genetic operations selection, crossover, and mutation on the previous population.
- **Step 4:** Verify the fitness of the new generation of individuals. If converged, stop. Otherwise go to Step 3.
- **Step 5:** Decode the optimized chromosomes to obtain the seed terms for PRF.

---

[6] https://www.nlm.nih.gov/mesh/.

Details of each step is presented below.

Initial Population. A candidate set is initialized with the top-30 ranked documents.

Chromosome Representation. Each chromosome encodes a binary string. The length of chromosomes depends on the size of the candidate set. When a keyword is present in a document, the corresponding bit is set to 1; otherwise, it is 0.

Fitness Function. As in [8], the fitness of each chromosome in a population is evaluated using the Jaccard similarity measure, which ranges between 0 and 1.

Selection Operation. As in [19], the selection process is to remove some bad (with low fitness) chromosomes. It is based on spinning the roulette wheel in N times.

Crossover Operation. As in [19], let P_c be the crossover probability. This probability gives us the expected number $P_c * N$ of chromosomes

Mutation Operation. Let P_m be the mutation probability. This probability gives us the expected number of $P_m * N$ of chromosomes. Every bit in all chromosomes of the whole population has an equal chance to undergo mutation, that is, change from 0 to 1 or vice versa. According to [8], typically P_c ranges between 0.7 and 0.9 and P_m ranges between 0.01 and 0.03.

3   Experiments

In this section, we evaluate and compare empirical performances of the baseline method and the proposed method on the TREC 2014 Clinical Decision Support (CDS) dataset [20].

3.1   Dataset

The focus of TREC 2014 CDS Track is retrieval of biomedical articles relevant to generic clinical question about medical records. The track dataset is divided into two separate parts: the first part includes medical documents such as full-text biomedical articles and the second part contains case reports as topics. Details are presented below.

Documents. The document dataset is an open access subset from PubMed Central (PMC)[7]. This set contains the abstracts, full texts, and other metadata of 733,138 articles (47.2 GB) in the biomedical domain. The articles are presented in the NXML format using the National Library of Medicine's Journal Archiving and Interchange Tag Set [9].

Each article in the collection is uniquely identified by the PubMed Central Identifier (PMCID) number, which is specified by the <*article-id*> element within its NXML file. The article is named using the same PMCID number.

---

[7] https://www.ncbi.nlm.nih.gov/pmc/.

Topics. The query set of the dataset includes 30 different topics in total, with 10 topics for each of the query types *diagnosis, test,* and *treatment.* Each topic consists of a summary which describes a patient's case report created by expert topic developers at the U.S National Library of Medicine maintaining actual medical records. Figure 2 shows examples of the three topics types used in the task.

```
<topic number="3" type="diagnosis">
     58-year-old female non-smoker with left lung mass on x-ray. Head
     CT shows a solitary right frontal lobe mass.
</topic>
```

```
<topic number="12" type="test">
     25-year-old woman with fatigue, hair loss, weight gain, and cold
     intolerance for 6 months.
</topic>
```

```
<topic number="26" type="treatment">
     Group traveling to the Amazon rainforest, including 3 pregnant
     women. All members' immunizations are up-to-date but they require
     malaria prophylaxis.
</topic>
```

Fig. 2. Three of 30 topics from the TREC 2014 CDS Track.

3.2   Evaluation Methods

In this section, we present some methods to evaluate information retrieval systems. According to [21], Precision and Recall are the basic measures expressed as percentages. Precision and Recall are set-based measures in unordered sets of documents. However, the quality of a search engine is also expressed via ranking of relevant documents retrieved. That is, more relevant documents are expected to be at higher positions in the result list. Therefore, other measures such as *P@k* and *R-precision* are introduced, as in [22].

3.3   Experimental Results

We conduct 4 experiments on the dataset presented in Sect. 3.1. The first two experiments (Method 1, Method 2) use the baseline method and the others (Method 3, Method 4) use the proposed method. Following the TREC standard, the 1,000 top-ranked documents are retrieved for each query in the evaluation. Also, in the experiments, we set $P_c = 0.783$, $P_m = 0.029$, and $N = 30$ (i.e., top-30 ranked documents).

As shown in Table 2, in the experiments we apply various algorithms such as BM25L, BM25$^+$, VSM, LM in the matching module. Method 1 uses BM25L model while Method 2 combines multiple models by using the RRF algorithm as introduced above. The first two experiments expand queries by using the MeSH thesaurus and PRF. The PRF module in Method 1 and Method 2 only extracts keywords from the top-3 documents as in [11]. In contrast, we combine GA and PRF in Method 3 and Method 4 that correspond to Method 1 and Method 2, respectively.

Table 2. The overall performance of different methods.

Method ID	Method description	R-prec	$P@10$
Method 1	BM25L, MeSH, PRF	0.1850	0.3533
Method 2	RRF: (BM25L, MeSH, PRF) (BM25$^+$, MeSH, PRF) (VSM, MeSH, PRF) (LM, MeSH, PRF)	0.1913	0.3667
Method 3	BM25L, MeSH, PRF, GA	**0.1950**	**0.3733**
Method 4	RRF: (BM25L, MeSH, PRF, GA) (BM25$^+$, MeSH, PRF, GA) (VSM, MeSH, PRF, GA) (LM, MeSH, PRF, GA)	**0.2036**	**0.3933**

As one can see, for R-prec, Method 3 outperforms Method 1 by 5.4% (0.1950 vs 0.1850), and Method 4 outperforms Method 2 by 6.4% (0.2036 vs 0.1913). Meanwhile, for $P@10$, the improvement of Method 3 over Method 1 is by 5.6% (0.3733 vs 0.3533), and the improvement of Method 4 over Method 2 is by 7.2% (0.3933 vs 0.3667).

In addition, Fig. 3 shows the R-prec measures for each clinical question types. As compared to Method 2, Method 4 outperforms it on all of the three query types. However, Method 3 is only better than Method 1 with the *test* query type. It could be due to different initial search results in different runs and on different question types, where Method 2 and Method 4 use a different matching model from that of Method 1 and Method 3.

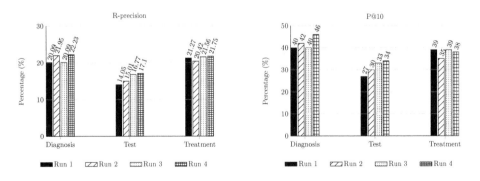

Fig. 3. The overall performance of different methods for each query types.

4   Conclusions

We have first proposed and presented the use of GA in PRF for medical information retrieval. First, relevant documents are retrieved and ranked for the initial search. Second, a set of candidate terms is extracted from the set of top-k retrieved documents. Third, the seed terms are selected from the candidate set by the GA and added to the original query. Finally, the system matches the new query and documents in the database, and returns the final ranked list.

We have experimented the proposed GA method on the TREC 2014 CDS dataset. Unlike traditional information retrieval datasets, here each query is associated with a medical question type. The results show that the proposed method improves the system performance, in comparison with the use of PRF without GA.

For the future work, we suggest using semantic relations between medical terms in query expansion and content matching. Besides, we are applying and adapting the proposed method to information retrieval on Vietnamese medical documents.

Acknowledgments. This work is funded by Vietnam National University at Ho Chi Minh City under the grant number B2016-42-01.

References

1. Chou, S., Chang, W., Cheng, C.Y., Jehng, J.C., Chang, C.: An information retrieval system for medical records & documents. In: 2008 30th Annual International Conference of the IEEE Engineering in Medicine and Biology Society, pp. 1474–1477 (2008). https://doi.org/10.1109/IEMBS.2008.4649446
2. Goeuriot, L., Jones, G.J.F., Kelly, L., Müller, H., Zobel, J.: Medical information retrieval: introduction to the special issue. Inf. Retr. J. **19**(1–2), 1–5 (2016)
3. Palotti, J., Hanbury, A., Müller, H., Kahn Jr., C.E.: How users search and what they search for in the medical domain - understanding laypeople and experts through query logs. Inf. Retr. J. **19**(1–2), 189–224 (2016)
4. Cao, G., Nie, J., Gao, J., Robertson, S.: Selecting good expansion terms for pseudo-relevance feedback. In: Proceedings of the 31st Annual International ACM SIGIR Conference on Research and Development in Information Retrieval, SIGIR 2008, Singapore, 20–24 July 2008, pp. 243–250 (2008). https://doi.org/10.1145/1390334.1390377
5. Lv, Y., Zhai, C., Chen, W.: A boosting approach to improving pseudo-relevance feedback. In: Proceeding of the 34th International ACM SIGIR Conference on Research and Development in Information Retrieval, SIGIR 2011, Beijing, China, 25–29 July 2011, pp. 165–174 (2011). https://doi.org/10.1145/2009916.2009942
6. Cao, G., Nie, J.-Y., Gao, J., Robertson, S.: Selecting good expansion terms for pseudo-relevance feedback. In: Proceedings of the 31st Annual International ACM SIGIR Conference on Research and Development in Information Retrieval, SIGIR 2008, pp. 243–250. ACM, New York (2008)
7. Vargas, S., Santos, R.L.T., Macdonald, C., Ounis, I.: Selecting effective expansion terms for diversity. In: Open Research Areas in Information Retrieval, OAIR 2013, Lisbon, Portugal, 15–17 May 2013, pp. 69–76 (2013)

8. Chen, H.: Machine learning for information retrieval: Neural networks, symbolic learning, and genetic algorithms. JASIS **46**(3), 194–216 (1995)

9. Simpson, M.S., Voorhees, E., Hersh, W.: Overview of the TREC 2014 clinical decision support track. In: Proceedings of the 23rd Text Retrieval Conference (TREC), Gaithersburg, MD, USA (2014)

10. Del Fiol, G., Workman, T.E., Gorman, P.N.: Clinical questions raised by clinicians at the point of care: a systematic review. JAMA Intern. Med. **174**(5), 710–718 (2014)

11. Mourão, A., Martins, F., Magalhães, J.: NovaSearch at TREC 2014 clinical decision support track. In: Proceedings of the Twenty-Third Text REtrieval Conference, TREC 2014, Gaithersburg, Maryland, USA, 19–21 November 2014

12. Singh, J.N., Dwivedi, S.K.: Analysis of vector space model in information retrieval. In: Proceedings Published by International Journal of Computer Applications® (IJCA), vol. 2, pp. 14–18 (2012)

13. Trotman, A., Puurula, A., Burgess, B.: Improvements to BM25 and language models examined. In: Proceedings of the 2014 Australasian Document Computing Symposium, ADCS 2014, Melbourne, VIC, Australia, 27–28 November 2014, p. 58 (2014). https://doi.org/10.1145/2682862.2682863

14. Banerjee, P., Han, H.: Language modeling approaches to information retrieval. JCSE **3**(3), 143–164 (2009)

15. Lv, Y., Zhai, C.: Lower-bounding term frequency normalization. In: Proceedings of the 20th ACM Conference on Information and Knowledge Management, CIKM 2011, Glasgow, United Kingdom, 24–28 October 2011, pp. 7–16 (2011)

16. Lv, Y., Zhai, C.: When documents are very long, BM25 fails! In: Proceeding of the 34th International ACM SIGIR Conference on Research and Development in Information Retrieval, SIGIR 2011, Beijing, China, 25–29 July 2011, pp. 1103–1104 (2011). https://doi.org/10.1145/2009916.2010070

17. Cormack, G.V., Clarke, C.L.A., Büttcher, S.: Reciprocal rank fusion outperforms condorcet and individual rank learning methods. In: Proceedings of the 32nd Annual International ACM SIGIR Conference on Research and Development in Information Retrieval, SIGIR 2009, Boston, MA, USA, 19–23 July 2009, pp. 758–759 (2009). https://doi.org/10.1145/1571941.1572114

18. Carpineto, C., Romano, G.: A survey of automatic query expansion in information retrieval. ACM Comput. Surv. **44**(1), Article ID 1–1150 (2012). https://doi.org/10.1145/2071389.2071390

19. Gen, M., Liu, B.: A genetic algorithm for optimal capacity expansion. J. Oper. Res. Soc. Jpn. **40**, 1–9 (1997)

20. Roberts, K., Simpson, M.S., Demner-Fushman, D., Voorhees, E.M., Hersh, W.R.: State-of-the-art in biomedical literature retrieval for clinical cases: a survey of the TREC 2014 CDS track. Inf. Retr. J. **19**(1–2), 113–148 (2016)

21. Zuva, K., Zuva, T.: Evaluation of information retrieval systems. Int. J. Comput. Sci. Inf. Technol. (IJCSIT) **4**, 35–43 (2012)

22. Mogotsi, I.C., Manning, C.D., Raghavan, P., Schütze, H.: Introduction to Information Retrieval. Cambridge University Press, Cambridge (2008). 482 p. ISBN: 978-0-521-86571-5. Inf. Retr. **13**(2), 192–195 (2010)

Analysis and Modelling of Heel Bone Fracture with Using of Active Contour Driven by Gaussian Energy and Features Extraction

Jan Kubicek[1]([⊠]) [iD], Marek Penhaker[1] [iD], David Oczka[1] [iD],
Martin Augustynek[1] [iD], Martin Cerny[1] [iD], and Petra Maresova[2]

[1] VSB–Technical University of Ostrava, FEECS, K450,
17. listopadu 15, 708 33 Ostrava–Poruba, Czech Republic
{jan.kubicek,marek.penhaker,
martin.augustynek,martin.cerny}@vsb.cz,
David-oczka@seznam.cz
[2] Faculty of Informatics and Management, University of Hradec Kralove,
Rokitanskeho 62, 500 03 Hradec Králové, Czech Republic
petra.maresova@uhk.cz

Abstract. In a clinical practice of the traumatology, the heel bone fracture is a routine procedure which is commonly diagnostic via the X-ray imaging. A reliable indicator of the particular heel bone fracture is the periosteal callus having dynamical geometrical features over the time. Unfortunately, the clinical imaging techniques allow for only the basic visualization without further post processing. Thus, such kind of the diagnosis may be affected by the subjective error depending on the particular experience of the clinical physician. In cooperation with the clinical experts, we have developed a mathematical model which is based on the active contours utilizing the Gaussian fitting energy to autonomously specify an area of the periosteal callus. Consequently, the periosteal callus features are figured out to objectively measure and track the heel bone fracture.

Keywords: Heel bone · Periosteal callus · Image segmentation
Dynamical tracking · Active contour · Gaussian energy · Features extraction

1 Introduction

Periosteal callus is a key factor for starting the secondary reparation of the bone fracture. In the case of damaging periosteal vascular supplementing, it must not to be damaged endosteal supplementing during the treatment [1–3]. Callus evolution goes through four phases:

- Proliferation phase (0–7 days).
- Differentiation phase (8–21 days).
- Ossification phase (since 4th week).
- Modeling and remodeling phase (8–12 weeks).

© Springer International Publishing AG, part of Springer Nature 2018
N. T. Nguyen et al. (Eds.): ACIIDS 2018, LNAI 10752, pp. 405–414, 2018.
https://doi.org/10.1007/978-3-319-75420-8_39

The Fig. 1 reports a typical example of the fracture bone where periosteal callus is manifested (arrows). It is apparent by the naked eyes that the periosteal callus is badly recognizable due to weak contract from the bone structure. Recognition of the periosteal callus is subjective, with inter-physician variability of 20–25%. Image processing algorithms have potential to render callus measurement objective and thereby reduce observer error. However, previous studies which measured callus with image processing protocols did not document the accuracy and objectivity of their methods. Badly recognizable object are commonly pre-processed to increase its contrast and distinguish their borders for consequent modelling. Anyway, such procedure is not recommended is this case due to modification of the clinical information. Any image pre-processing procedure may influence character and distribution of the pixels, thus the original clinical information may be modified or lost [4, 5].

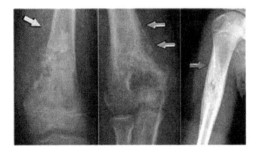

Fig. 1. X-ray imaging of the bones where the periosteal callus is manifested (arrows) [4]. (Color figure online)

2   Relevance of the Problem: Clinical Assessment of Periosteal Callus

The periosteal callus is a key indicator of the secondary reparation of the heel bone and well reflects dynamical process of the heel bone recovering [6]. In this context, the feature well representing the healing period is a crucial for a proper clinical diagnosis. Clinical experts specify two major features which are being tracked whilst recovering period. (1) Size of the periosteal callus area is an important indicator. It is supposed that at the beginning of the healing period where the bone fracture is opened, the periosteal callus is minimal, and over the time is getting more and more expanded. (2) When the heel bone fracture is diagnosed on the CT (Computer Tomography) allowing for the 3D visualization, the periosteal callus volume is estimated [7]. Finally, the determining features (1, 2) are compared in a dependence of their dynamical evolution over the time: from beginning of the treatment up to the final recovering where it is supposed that the periosteal callus comes to the heel bone structure. Such tracking procedure should be ideally done autonomously without any intervention to prevent subjective errors which may significantly influence diagnosis. In a present time, a contour of the periosteal callus is usually drawn to specify an area of the manifestation [8].

Consequently, stated features (1, 2) are estimated via a number of pixels belonging inside of such area. On the base of those facts, autonomous modelling allowing for representative features extraction have great ambitions to overcome subjective diagnosis and significantly improve the clinical diagnosis [9, 10].

3 Materials and Methods

Since the heel bone fracture, including the periosteal callus, is observable due to clinical imaging methods, the image segmentation is employed. In our work, the concept of the segmentation algorithm utilizes the active contour driven by the Gaussian energy fitting. In this method, an area of the periosteal callus is approximated by smooth continuous curve dynamically approximating an area of the periosteal callus. In predefined number of the discrete steps, the callus area is being gradually approximated by the Gaussian energy fitting. In this mathematical procedure, the active contour curve well spread itself when pixels distribution reliably reflects the Gaussian distribution otherwise the active contour is significantly limited or completely terminated. Eventually, we obtain a smooth curve approximating an area of the periosteal callus. Such area allows for calculation either area of the callus (number of pixels) or volume (number of voxels) quantifying the periosteal callus and heel bone fracture grade. Such model represents the dynamical features of the heel bone fracture depending on objective features extraction of the periosteal callus over the time.

3.1 Active Contour Segmentation Driven by Gaussian Energy

The segmentation algorithm for the periosteal callus is based on the time-deformable model. An active curve gradually approximates area of the periosteal callus within predefined number of the iteration steps defined by the user [9, 10]. A contrast of the different tissues such is BP (bone-periosteal callus) indicates situation when distribution of the pixels is rapidly changed, thus original Gaussian energy is substantially modified. Such spots cause termination of the curve. Other words speaking, an interface between the heel bone and the periosteal callus reliably serves as indicator for the callus definition.

It is a region-based model where the local image intensities are described by the Gaussian distributions with different means and variances. This model is based on the energy functional defined by the Eq. 1.

$$E^{LGDF} = \int_{\Omega} E_X^{LGDF} dx = \int_{\Omega} \left(\sum_{i=1}^{N} \int_{\Omega_i} -\omega(x-y) \log p_{i,x}(I(y)) dy \right) dx \quad (1)$$

The first parameter $p_{i,x}(I(y))$ represents a probability density in the region Ω_i, $\omega(x-y)$ is a non-negative weighting function, and $-log$ demonstrates conversion from maximization to minimization. In order to perform minimization procedure of the E_X^{LGDF} the double integral for the energy functional is defined. We suppose that the image domain (periosteal callus) can be separated into two different regions (binary segmentation) belonging inside and outside of the level set ϕ. The energy is expressed on the base of the Heaviside function H in the Eq. 2.

$$
\begin{aligned}
E_X^{LGDF} &\left(\phi,\ u_1(x),\ u_2(x),\ \sigma_1(x)^2,\ \sigma_2(x)^2\right) \\
&= -\int \omega(x-y) \log p_{1,x}(I(y)) M_1(\phi(y)) dy \\
&\quad -\int \omega(x-y) \log p_{2,x}(I(y)) M_2(\phi(y)) dy
\end{aligned}
\tag{2}
$$

where $M_1(\phi(y)) = H(\phi(y))$ and $M_2(\phi(y)) = 1 - H(\phi(y))$. Eventually, the energy can be defined by the Eq. 3.

$$
E^{LGDF}\left(\phi,\ u_1,\ u_2,\ \sigma_1^2,\ \sigma_2^2\right) = \int_\Omega E_X^{LGDF}\left(\phi,\ u_1(x),\ u_2(x),\ \sigma_1(x)^2,\ \sigma_2(x)^2\right) dx
\tag{3}
$$

In order to achieve precise evolution, the regularization procedure of the level set function is proposed. Such procedure is based on the penalizing of its deviation from the signed distance function (Eq. 4).

$$
P(\phi) = \int \frac{1}{2} (|\nabla\phi|(x) - 1)^2 dx
\tag{4}
$$

The penalizing procedure of the zero level set is defined in the Eq. 5.

$$
L(\phi) = \int |\nabla H(\phi(x))| dx
\tag{5}
$$

The entire energy functional is defined in the Eq. 6.

$$
F\left(\phi,\ u_1,\ u_2,\ \sigma_1^2,\ \sigma_2^2\right) = E^{LGDF}\left(\phi,\ u_1,\ u_2,\ \sigma_1^2,\ \sigma_2^2\right) + vL(\phi) + \mu P(\phi)
\tag{6}
$$

The parameters $v,\ \mu > 0$ represent the weight constants practically the Heaviside function may be approximated by the smooth form (Eq. 7).

$$
H_\varepsilon = \frac{1}{2}\left[1 + \frac{2}{\pi} \arctan\left(\frac{x}{\varepsilon}\right)\right]
\tag{7}
$$

An approximated form of the energy functional is defined in the Eq. 8.

$$
F_\varepsilon\left(\phi,\ u_1,\ u_2,\ \sigma_1^2,\ \sigma_2^2\right) = E_\varepsilon^{LGDF}\left(\phi,\ u_1,\ u_2,\ \sigma_1^2,\ \sigma_2^2\right) + vL_\varepsilon(\phi) + \mu P(\phi)
\tag{8}
$$

The minimization procedure of the F_ε is done on the base of the gradient descent flow equation defined in the Eq. 9.

$$
\frac{\partial\phi}{\partial t} = -\delta_\varepsilon(\phi)(e_1 - e_2) + v\delta_\varepsilon(\phi)div\left(\frac{\nabla\phi}{|\nabla\phi|}\right) + \mu\left(\nabla^2(\phi) - div\left(\frac{\nabla\phi}{|\nabla\phi|}\right)\right)
\tag{9}
$$

where:

$$e_1(x) = \int_\Omega \omega(y - x) \left[\log(\sigma_1(y)) + \frac{(u_1(y) - I(x))^2}{2\sigma_1(y)^2} \right] dy \qquad (10)$$

and:

$$e_2(x) = \int_\Omega \omega(y - x) \left[\log(\sigma_2(y)) + \frac{(u_2(y) - I(x))^2}{2\sigma_2(y)^2} \right] dy \qquad (11)$$

There is a set of the controlling parameters driving a proper flow of the segmentation method:

- Δt – time step of the curve
- ε – geometry of the Dirac impulse (width)
- σ – definition of the kernel
- c_0 – constant that creates binary step function from level set function ϕ
- v – weighting constant from Eq. (6)
- μ – weighting constant from Eq. (6)
- n – number of iterations (we define 500 iterations)
- λ_1 – outer weight, constant that multiply the e_1 function
- λ_2 – inner weight, constant that multiply the e_2 function.

4   Results and Discussion

We have collected historical clinical data of 200 patients who have undergone a surgery of the heel bone. Those patients have been implemented special fixators to the heel bone helping during the recovering procedure. Immediately after the surgery, the patients underwent the first X-ray and CT imaging of the heel bone. The X-ray example is shown on the Fig. 2. It is observable that the contrast between the heel bone and the periosteal callus (red RoI) is relatively week. This phenomenon is done by the X-ray limitation in a sense of the tissue density. Dense tissues are commonly imagined under higher contrast. This fact is apparent on the heel bone, and the metal structure of the fixator (Fig. 2). Contrarily, the periosteal callus represents a connective tissue not being as dense as the surrounding bone. On the other hand even a lower level of density causes at least partial contrast which is observable and serves as boundaries for the active contour method.

Each X-ray examination has been verified by the parallel scanning of the CT imaging. Thus, we can make a simultaneous modeling of those structures for comparison. Nevertheless, the CT data are commonly affected by a specific noise (Beam hardening) which is present when the metal objects are imagined. Despite this unfavorable phenomenon, the CT data serve as a reference level for the X-ray heel bone modeling. After the first imaging from the surgery, the patients were regularly examined after each two weeks four times. Thus, we obtained 5 examinations of the X-ray

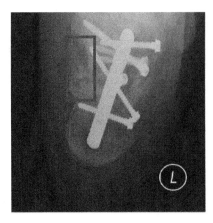

Fig. 2. X-ray imaging after the surgery of the heel bone fracture with implemented fixator. The red RoI indicates an area of the periosteal callus. (Color figure online)

and CT to make a mathematical model of the heel bone recovering period. The Fig. 3 shows a set of three native X-ray images of one patient where the periosteal callus area is indicated by the red RoI defined by the clinical expert from the traumatology department.

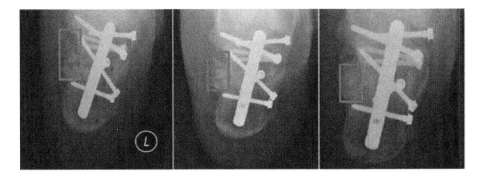

Fig. 3. Evolution of the heel bone fracture where the periosteal callus is indicated by red: X-ray image after surgery (left), after two week (middle) and after four weeks (right). (Color figure online)

The segmentation procedure based on the active contour iteratively approximates an area of the periosteal callus. By this way this process is perceived as an optimization procedure defining the callus model. The active contour is initialized by the initial contour which should roughly reflect an area of the callus. Based on the experimental observations, we are using the ellipse placed inside the periosteal callus. The Fig. 4 reports results of the active contour model (green contour) on a set of the three X-ray images from the Fig. 3.

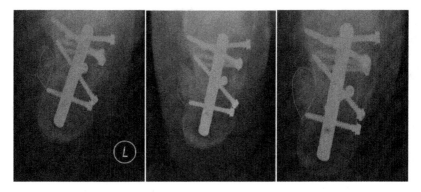

Fig. 4. Evolution of the heel bone fracture segmentation based on the active contour with 500 iterations where the periosteal callus is segmented by green contour: X-ray image after surgery (left), after two week (middle) and after four weeks (right). (Color figure online)

Based on the segmentation model we have perform an extraction of the periosteal callus in the binary form. Segmentation model by the active contours is accompanied by the energetic map reflecting energy of the active contour. Inside the contour (periosteal callus) the energy is negative contrarily outside (image background) the contour energy is positive. By this energy selection, the periosteal callus area is extracted, while other structures are perceived in the native X-ray image (Fig. 5). Final model of the periosteal callus is presented in the Fig. 6 where the image background is suppressed.

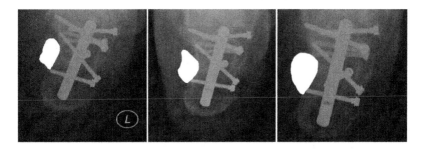

Fig. 5. Extraction of the periosteal callus area (white) by using the energy map where other structures are perceived.

The model allows for the features extraction reliably quantifying a respective stage of the bone healing. We analyzed data of 200 patients examined on the X-ray (spatial modeling) and CT (volumetric modeling). We have extracted a spatial area of the model (X-ray) and the volume area (CT). Those features have been confronted with two clinical experts from the traumatology department. We have tracked a proportional growth of the callus from operation up to the complete healing. The Table 1 reports average values of 200 patients. SG_i represents a proportional growth of the callus area

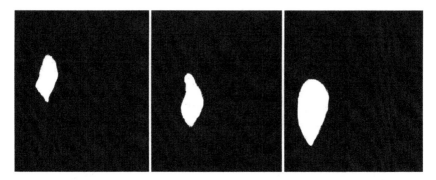

Fig. 6. Final model of the periosteal callus in the binary form. The periosteal callus model is indicated by the white area.

after (i = 1) two weeks, (i = 2) four weeks, (i = 3) six weeks and (i = 1) eight weeks. The parameter SG_{ei} represents the clinical expert opinion based on the manual tracing of the callus. The Table 2 reports same situation for the callus volume area (VG_i) from the CT. Based on the results we observe two important facts. (1) Dynamical proportional variation has decreasing tendency. At the beginning of the healing period, the callus exhibits the most significant dynamical changes of its geometrical structure. (2) We consider that the active contours results are robust because difference against the manual segmentation by the clinical experts is lower than 5% for all the cases.

Table 1. Extraction of spatial area based on the active contour model of the periosteal callus against estimation of the clinical experts.

SG_1 [%]	SG_{e1} [%]	SG_2 [%]	SG_{e2} [%]	SG_3 [%]	SG_{e3} [%]	SG_4 [%]	SG_{e4} [%]
33.5	31.2	22.1	21.4	20.5	18.6	13.6	12.5

Table 2. Extraction of volume area based on the active contour model of the periosteal callus against estimation of the clinical experts.

VG_1 [%]	VG_{e1} [%]	VG_2 [%]	VG_{e2} [%]	VG_3 [%]	VG_{e3} [%]	VG_4 [%]	VG_{e4} [%]
44.5	40.2	19.5	19.1	23.5	18.1	9.1	7.4

4.1   Verification and Quantitative Comparison

In the last part of the analysis, the segmentation model based on the active contour driven by the Gaussian energy has been verified. We have performed the verification against the Otsu-N method (multiregional segmentation) and FCM (clustering). We have done quantitative comparison including two verification metrics. Individual testing procedures are done in a comparison with the manual segmentation (ground truth) performed by the clinical expert from the traumatology. Results are stated in the Table 3. The best results are highlighted.

- Rand Index (*RI*): this parameter measures similarity between two data clusters. *RI* gives values in $[0; 1]$, where 0 indicates zero similarity while 1 indicates that the data clusters are completely identical. The *RI* index is given by the Eq. 12.

$$\text{RI}(C_1, \; C_2) \; = \; \frac{2(n_{11} + n_{00})}{N(N-1)} \tag{12}$$

Where N denotes a total number of the points, n_{11} denotes a number of the pairs that they are in the same cluster in C_1 and C_2 and n_{00} is a number of the pairs belonging to different clusters.

- Variation of information (*VI*): this parameter measures a distance between two data clusters in a sense of the conditional entropy. The *VI* parameter is given by the Eq. 13.

$$\text{VI}(C_1, \; C_2) \; = \; \mathcal{H}(C_1) \; + \; \mathcal{H}(C_2) \; - \; 2I(C_1, \; C_2) \tag{13}$$

where $\mathcal{H}(C_i)$ represents an entropy associated with the C_i cluster, and I represents a mutual information between C_1 and C_2 clusters. Lower *VI* indicates better results of the test.

The Table 3 brings a comparison among the proposed model based on the active contours, and the multiregional Otsu-N, and fuzzy C-means clustering. Based on the analyzed X-ray images, the active contour has the best results in *RI* index (0.992).

Table 3. Quantitative comparison of the periosteal callus model.

	Active contour	Otsu-N	FCM
RI	0.992	0.756	0.818
VI	2.411	3.556	2.410

5   Conclusion

In our analysis, we have proposed a mathematical model for the periosteal callus dynamical modelling based on the active contour model driven by the Gaussian energy. It is autonomous model gradually approximating an area of the periosteal callus (X-ray images), and the volume area (CT images). Periosteal callus represents a crucial indicator for the assessment of the heel bone fracture recovering period. It is supposed that the periosteal callus is not stable over that period, but it exhibits dynamical process.

Acknowledgment. The work and the contributions were supported by the project SV4507741/2101, 'Biomedicínské inženýrské systémy XIII'. This study was supported by the research project The Czech Science Foundation (GACR) No. 17-03037S, Investment evaluation of medical device development.

References

1. Guzanová, A., Ižaríková, G., Brezinová, J., Živčák, J., Draganovská, D., Hudák, R.: Influence of build orientation, heat treatment, and laser power on the hardness of Ti6Al4 V manufactured using the DMLS process. Metals **7**(8), 1–17 (2017). ISSN 2075-4701

2. Scott, A.T., Pacholke, D.A., Hamid, K.S.: Radiographic and CT assessment of reduction of calcaneus fractures using a limited sinus tarsi incision. Foot Ankle Int. **37**(9), 950–957 (2016)

3. Bodnárová, S., Hudák, R., Živčák, J.: Príprava a návrh biologického testovania biokompatibility keramických skáfoldov na báze hydroxyapatitu a trikalcium fosfátu. In: Novus Scientia 2017, pp. 17–21. TU, Košice (2017). ISBN 978-80-553-3080-8

4. Ťúková, V., Poláček, I., Tóth, T., Živčák, J., Ižaríková, G., Kovačevič, M., Somoš, A., Hudák, R.: The manufacturing precision of dental crowns by two different methods is comparable. Lekar a Technika **46**(4), 102–106 (2016)

5. Chou, S.H., Hwang, J., Ma, S.-L., Vokes, T.: Utility of heel dual-energy X-ray absorptiometry in diagnosing osteoporosis. J. Clin. Densitom. **17**(1), 16–24 (2014)

6. Nasr-Esfahani, E., Karimi, N., Jafari, M.H., Soroushmehr, S.M.R., Samavi, S., Nallamothu, B.K., Najarian, K.: Segmentation of vessels in angiograms using convolutional neural networks. Biomed. Signal Process. Control **40**, 240–251 (2018)

7. Núñez, J.A., Goring, A., Hesse, E., Thurner, P.J., Schneider, P., Clarkin, C.E.: Simultaneous visualisation of calcified bone microstructure and intracortical vasculature using synchrotron X-ray phase contrast-enhanced tomography. Sci. Rep. **7**(1), Article no. 13289 (2017)

8. Aira, J.R., Simon, P., Gutiérrez, S., Santoni, B.G., Frankle, M.A.: Morphometry of the human clavicle and intramedullary canal: a 3D, geometry-based quantification. J. Orthop. Res. **35**(10), 2191–2202 (2017)

9. Lee, H., Tajmir, S., Lee, J., Zissen, M., Yeshiwas, B.A., Alkasab, T.K., Choy, G., Do, S.: Fully automated deep learning system for bone age assessment. J. Digit. Imag. **30**(4), 427–441 (2017)

10. Mahfouz, M.R., Mustafa, A., Abdel Fatah, E.E., Herrmann, N.P., Langley, N.R.: Computerized reconstruction of fragmentary skeletal remains. Forensic Sci. Int. **275**, 212–223 (2017)

Intelligent Biomarkers of
Neurodegenerative Processes in Brain

Automatic Detection and Classification of Brain Hemorrhages

Anh-Cang Phan[1]([⊠]) ⓘ, Van-Quyen Vo[2] ⓘ,
and Thuong-Cang Phan[3] ⓘ

[1] Faculty of Information Technology,
Vinh Long University of Technology Education, Vinh Long City, Viet Nam
cangpa@vlute.edu.vn
[2] Center of Information Technology,
Can Tho University of Medical and Pharmacy, Can Tho City, Viet Nam
vovanquyen@ctump.edu.vn
[3] Faculty of Information Technology,
Can Tho University, Can Tho City, Viet Nam
ptcang@cit.ctu.edu.vn

Abstract. Computer-aided detection and diagnosis systems have been the focus by a great number of endeavor researchers, particularly in detecting, diagnosing, and classifying brain hemorrhages. In this paper, we propose a system, which can automatically identify and classify the existence of brain hemorrhages. Our proposed method emphasizes on analyzing brain hemorrhage regions from medical images. It includes six stages: determining Hounsfield units, processing image segmentation, extracting the brain hemorrhage regions, extracting features and classifying brain hemorrhages, estimating the timing of hemorrhage. Our experimental results show that the accuracy of detection of brain hemorrhages is 100% and the classification of brain hemorrhages achieves the accuracy of 95.3%. In addition, our method also determines the bleeding timing to assist doctors with timely treatment.

Keywords: Medical image · Brain CT/MRI · Brain hemorrhage

1 Introduction

During the patient treatment, subclinical outcomes play a very important role in assisting doctors to detect, diagnose and treat pathologies, especially pathologies related on the brain hemorrhage. The techniques of computed tomography [3] (CT), magnetic resonance imaging (MRI), and digital subtraction angiography (DSA) are very helpful in supporting to detect the brain hemorrhage regions. However, there are about a hundred slices (images) for each case of CT/MRI scan depending on the scan region and the thickness of the slices. Therefore, specialized physicians spend a lot of time looking at all images and finding abnormalities in each CT/MRI image to provide the best diagnosis and treatment. To overcome this disadvantage, we propose a system that can automatically detects, diagnoses and classifies brain hemorrhages in patients based on CT, MRI medical images.

© Springer International Publishing AG, part of Springer Nature 2018
N. T. Nguyen et al. (Eds.): ACIIDS 2018, LNAI 10752, pp. 417–427, 2018.
https://doi.org/10.1007/978-3-319-75420-8_40

In recent years, there are many researches in the field of medical images. Ahmed and Nordin [1] proposed highlighting the several features of medical images to improve the quality of the images and to display these images more clearly. Their method improved the quality of images without considering the automatic diagnosis and classification of diseases.

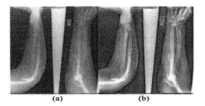 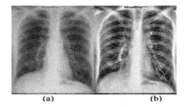

Fig. 1. Radiographic images (a) Before, (b) After treatment

Al-Ayyoub et al. [2] proposed another method. The authors likewise succeeded with automatically detecting and classification of brain hemorrhages. However, the authors used the Ostu method [4] to detect brain hemorrhage regions. Their method depended on a threshold of the grayscale [5]. To determine this threshold, they must convert the DICOM images to grayscale images. This method may affect image quality and threshold value. Besides, the classification of brain hemorrhages based on a number of features such as the hemorrhage dimension (in pixels), the focal point, the area and shape of the hemorrhage region. It is still limited in determining the duration of bleeding, and reference to the experienced doctors and medical imaging specialists is indispensable.

In order to overcome these limitations, we propose a method for determining the area of brain hemorrhage based on Hounsfield unit (HU) values [6] computed from the values of Rescale Intercept and Rescale Slope available in medical image files of DICOM format. The use of HU values is well suited for medical professionals in medical imaging. Furthermore, our method can accurately determine whether there is brain hemorrhage. In particular, we determine the duration of brain hemorrhage, which is critical to doctors' decision-making for a patient's treatment. This is a new idea of our approach compared to preceding research. Additionally, our method focused on four major types of brain hemorrhages: Epidural hematoma, subdural hematoma, subarachnoid hemorrhage and intracerebral hemorrhage [6, 8] as shown in Fig. 1. We also collect opinions from experienced doctors and specialists in the field of medical imaging in Can Tho University Hospital to help us identify and distinguish the 4 popular types of brain hemorrhages from our data sets (Fig. 2).

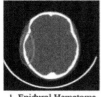 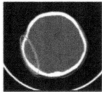 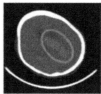 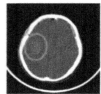

1. Epidural Hematoma   2. Subdural Hematoma   3. Subarachnoid Hemorrhage 4. Intracerebral Hemorrhage

Fig. 2. CT images of the four types of brain hemorrhages.

2 Related Work

We use the interactive learning machine by establishing an interactive tool between experts and doctors in order to create a training database. Experts and doctors will determine and classify the brain hemorrhages from the database. Next, we process DICOM [9] images by the algorithms: compute the HU values to extract the image segmentations and determine areas of brain hemorrhages; extract and store features of medical images in the database. In order to classify the brain hemorrhage of a new patient from a medical image, we determine the bleeding timing; detect and identify the region of the brain hemorrhage. The KNN algorithm [10] is used for this identification. The following works related to our proposed method.

2.1 Determination of the HU Value

Our method processes the source data on medical images from a DICOM standard [11] directly without transforming into JPG, BMP, PNG, etc. Therefore, we do not need to consider preprocessing these images. It aims at saving useful information from the DICOM images. Therefore, our method can save the useful information to determine the HU values for the detection of the brain hemorrhage and bleeding timing. This is a new idea and a contribution of our approach. To determine the HU value, we apply a linear transformation by the following equation:

$$HU = Pixel\_value * Rescale\_slope + Rescale\_intercept \ [23] \qquad (1)$$

where: *Pixel_value* is the value of each image point, the *Rescale_slope* and *Rescale_intercept* are the parameter values provided in DICOM images. To access these values, we use the dicominfo() function in Matlab [12, 13].

2.2 Image Segmentation for Detecting Brain Hemorrhages

After computing the HU value of each image point, we make an image segmentation based on Table 1. Figure 3 shows the result of the image segmentation based on HU values to detect a brain hemorrhage region. The HU values are in the range of 40 to 90 to coincide with the HU value of brain hemorrhage [14–16].

Table 1. The HU of common substances [21]

Substance	HU
Water	0 HU
Bone	1000 HU
Air	−1000 HU
Grey matter	35–40 HU
Blood	**40–90 HU**
Calcification	More than 120 HU

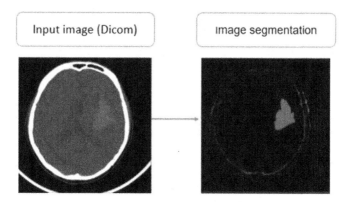

Fig. 3. Image segmentation from a DICOM image to detect a brain hemorrhage region

2.3   Extraction Area of Brain Hemorrhage

After the stage of image segmentation, some extracted features (areas) are close to the cortex since X-rays map them. Therefore, we removed these features such as the cortex and other features outside the brain hemorrhage area as shown in Fig. 4. The HU values are chosen in the range of 40 to 50 [17] that coincide with the HU value of brain hemorrhage.

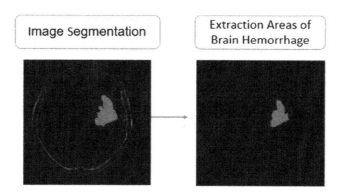

Fig. 4. An area of brain hemorrhage is extracted after image segmentation

2.4   Feature Extraction

After the above-mentioned stages, we extract features to serve for identification of brain hemorrhages. These features include:

− The image points with the HU values in range of 40 to 90 as represented in Table 1.

- The greatest, smallest, and average HU values of the image points in the hemorrhage area. These values are very important for the determination of bleeding timing. The determination of bleeding timing contributes to assisting doctors and experts in choosing the proper and timely treatment for patients.
- The center of hemorrhage: the determination of the center of the hemorrhage area helps to detect the hemorrhage position in the cortex. This makes the classification more accurate.
- The size of hemorrhage (computed by pixel): the size of hemorrhage area is also important since it helps figure out the size and the seriousness of the brain hemorrhage.

2.5   Detection and Classification of Brain Hemorrhages

After extracting and storing features from medical images in a testing data set, we compared those features with features in our training database. The Euclidean-based KNN algorithm [20] is used to detect and classify the brain hemorrhages. The results are represented in Fig. 5.

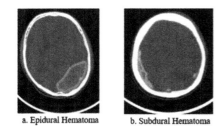 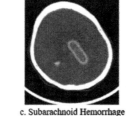 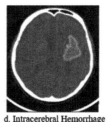

a. Epidural Hematoma   b. Subdural Hematoma   c. Subarachnoid Hemorrhage   d. Intracerebral Hemorrhage

Fig. 5. Results of detection and classification of brain hemorrhages

2.6   Determination of Timing of Brain Hemorrhages

The determination of the bleeding timing plays a vital role in deciding the proper therapeutics for patients. In this work, we estimate the bleeding timing based on the average HU value of the brain hemorrhage. The bleeding timing of the brain hemorrhage is divided into three levels [18, 19]:

- Hyper acute: the most recognizable hyper attenuating phenomenon (within 3 days) with the HU estimated to be about 50–60 HU compared to the normal brain's HU (18–30 HU).
- Acute: the dark level is relatively reduced from day 3 to day 21 (normally from day 3 to day 14) with the HU value less than 40.
- Chronic: homogeneity on the region of hemorrhage from the day of 14 to 21 makes the injured region of brain difficult to distinguish due to HU is a half of the normal region of brain.

3   Overview of Our Proposed Method

Our proposed method was described in Fig. 6 including the following stages:

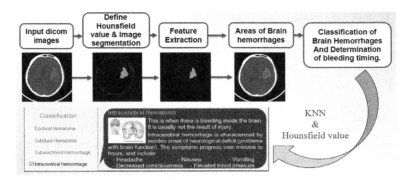

Fig. 6. Model of the system of automatic detection and classification of brain hemorrhages

– Stage 1: The determination of the HU value: the most valuable feature of the DICOM image data is ability to store a lot of the necessary information for the computation of the HU values. In Fig. 7, the two most important parameters are Rescale Intercept and Rescale Slope. The HU value is computed by Eq. (1).

```
Filename: 'D:\Brain_Hemorrhages\2-I150'
FileModDate: '07-Jan-2017 03:41:49'
Format: 'DICOM'
Width: 512
Height: 512
ColorType: 'grayscale'
Modality: 'CT'
Manufacturer: 'Philips'
InstitutionName: 'BV TRUONG DAI HOC Y DUOC CAN THO'
InstitutionAddress: 'CAN THO, VIETNAM'
StudyDescription: 'SO NAO'
Rows: 512
Columns: 512
RescaleIntercept: -1024
RescaleSlope: 1
...
```

Fig. 7. Information for the computation of the HU values in CT images by DICOM standard

– Stage 2: Image segmentation: after computing the HU values of image points, we detect areas of brain hemorrhages by image segmentation based on HU values, which are in the range from 40 to 90 (mentioned in Table 1).
– Stage 3: Determination of brain hemorrhage areas: we remove un-related image areas and image areas due to some effects in CT technique such as the cortex.
– Stage 4: Feature extraction: we extract some important features from the areas of brain hemorrhages using the Regionprops tool in Matlab as shown in Fig. 8.

Fig. 8. Orientation parameters using the Regionprops tool in Matlab

- Stage 5: Classification of brain hemorrhages: from extracted features, we apply the KNN algorithm [22] using the Euclidean distance to identify brain hemorrhages (Fig. 9).

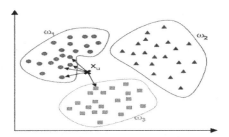

Fig. 9. Illustration of KNN algorithm

The Euclidean distance is computed by the Eq. (2) [20]:

$$d(p, q) = \sqrt{(p_1 - q_1)^2 + (p_2 - q_2)^2 + \dots + (p_n - q_n)^2} = \sqrt{\sum_{i=1}^{n} (p_i - q_i)^2} \tag{2}$$

Where: p and q are the two typical vectors; p_i and q_i are their elements (i = 1,..., n).

- Stage 6: Determination of timing of hemorrhage: the bleeding timing was considered to support doctors making timely treatment for patients. Based on the smallest, largest, average values of the HU values, we estimate the bleeding timing as presented in Sect. 2.6.

4   Experimental Results

Our proposed method is experimented on a system that automatically detects, diagnoses, and classifies brain hemorrhages. We collected 500 CT images by DICOM standard from the patients' skull in CanTho University Hospital. The training and testing data sets are selected at a 7:3 ratio [2]. It means that the training set of 350 CT images is classified into four types of brain hemorrhages based on the experience of doctors and specialists of Can Tho University Hospital. This data set includes 95 CT images of Epidural Hematoma, 85 CT images of Subdural Hematoma, 80 CT images of Subarachnoid Hemorrhage, and 90 CT images of Intracerebral Hemorrhage. The files of the training set are 179 Mb. The testing set of 150 images is classified into types of brain hemorrhages and normal brain (no bleeding), which includes 45 normal brain images, 25 images of Epidural Hematoma, 20 images of Subdural Hematoma, 30 images of Subarachnoid Hemorrhage, and 30 images of Intracerebral Hemorrhage.

Our system is installed on Matlab version R2015a using a computer with CPU i7, 16 Gb RAM, 500 Gb HDD, SSD 256 Gb, Windows 10 Pro 64 bit. The system interface automatically detects, diagnoses and classifies brain hemorrhages shown in Fig. 10.

The regions of brain hemorrhage are classified into four types of common brain hemorrhages: epidural hematoma, subdural hematoma, subarachnoid hemorrhage or intracranial hemorrhage.

In order to evaluate the accuracy of our proposed method, we test on a data set of 150 images using KNN algorithm with the parameter values of K in the range of 1 to 10. As a result, our method provides the best classification with K = 3. The results are presented in the confusion matrix in Table 2.

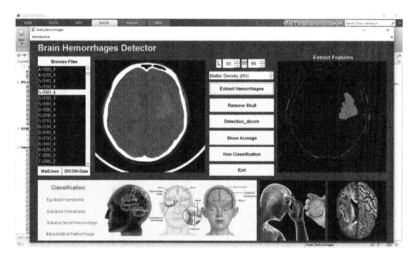

Fig. 10. Interfaces of the system "automatic detection and classification of brain hemorrhages"

From the confusion matrix, we can see that in 25 testing images of the Epidural Hematoma, 24 images are correctly identified (96%) and one image is incorrectly identified as Subdural Hematoma (5%). Similarly, in the case of testing 30 images of Subarachnoid Hemorrhage, 27 images are correctly identified (90.00%), one image of the Subdural Hematoma (3.33%) and two images of Intracranial Hemorrhage (6.67%).

Table 2. Table of the confusion matrix

4 types of brain hemorrhages	Epidural hematoma	Subdural hematoma	Subarachnoid hemorrhage	Intracerebral hemorrhage	No brain hemorrhage
Epidural hematoma (25)	**24 (96.00%)**	1 (4.00%)	0	0	0
Subdural hematoma (20)	1 (5.00%)	**19 (95.00%)**	1 (3.33%)	0	0
Subarachnoid hemorrhage (30)	0	1 (3.33%)	**27 (90.00%)**	2 (6.67%)	0
Intracerebral hemorrhage (30)	0	1 (3.33%)	2 (6.67%)	**28 (93.34%)**	0
No brain hemorrhage (45)	0	0	0	0	**45 (100%)**

From the confusion matrix shown in Table 2, the average accuracy is computed:

$$\frac{24 + 19 + 27 + 28 + 45}{150} * 100\% = 95.33\% \tag{3}$$

The result of our method achieves the average accuracy of 95.33% based on the HU values and the image processing techniques. It gives more accurate identification results than some other research as in [2] (92%). Their method performs well only on three of four kinds of lesions of brain hemorrhages. Additionally, our method also determines the timing of brain hemorrhages to help doctors making timely treatment like surgery, cerebrovascular intervention, or conservative treatment. It is very significant to determine the stage of cerebral palsy such as acute, subacute or chronic stages as described in Sect. 2.6. This also helps doctors making decisions faster and easier. Moreover, in preceding research, the presence of brain hemorrhage is detected achieving 100% using the Otsu algorithm [4]. Our method uses some available information stored in the DICOM images to determine the HU values with a linear transformation equation. They are our contributions in our research.

5 Conclusion

A computer-aided diagnosis system for automatically detection and classification of brain hemorrhages is proposed in our research based on the HU values. The HU values are used in identifying the brain hemorrhage and the bleeding timing. Our system can assist doctors in detecting and classifying the brain hemorrhages through medical images (CT/MRI) by DICOM standard, especially the bleeding timing. We focus on four types of brain hemorrhages namely Epidural Hematoma, Subdural Hematoma, Subarachnoid Hemorrhage and Intracerebral Hemorrhage. We also focus on directly processing CT images based on the DICOM standard. Our method of classification of brain hemorrhages obtains the accuracy of 95.33%.

According to Vietnamese stroke association's statistics, there are about 200.000 Vietnamese people suffer from strokes (or brain hemorrhages) every year. Where, 50% of the total sufferers die of this disease due to late diagnosis, and 50% remaining survivals in which 92% of the total number of them having sequelaes of moving, 68% having unserious sequelaes, 27% having serious sequelae, and 92% of the survivals in half paralysis who need to exercise at home for recovery. Therefore, our research provides a promising service for all neurosurgeons and patients of brain hemorrhages in near future.

References

1. Ahmed, H.S.S., Nordin, M.J.: Improving diagnostic viewing of medical images using enhancement algorithms. J. Comput. Sci. **7**(12), 1831–1838 (2011)
2. Al-Ayyoub, M., Alawad, D., Al-Darabsah, K., Aljarrah, I.: Automatic detection and classification of brain hemorrhages. WSEAS Trans. Comput. **12**(10) (2013)
3. CT scan – Mayo Clinic (2016). https://www.mayoclinic.org/. Accessed 20
4. Otsu, N.: A threshold selection method from gray-level histograms. Automatica **11**(285–296), 23–27 (1975)
5. Johnson, S.: Stephen Johnson on Digital Photography. O'Reilly, Sebastopol (2006)
6. Hounsfield, G.N.: Med. Phys. **7**, 283–290 (1980). PubMed
7. Hoa, P.N., Van Phuoc, L.: CT Chan thuong dau. Medical publishing House Ho Chi Minh city branch (2010)
8. Josephson, C.B., White, P.M., Krishan, A., Al-Shahi Salman, R.: Computed tomography angiography or magnetic resonance angiography for detection of intracranial vascular malformations in patients with intracerebral haemorrhage. The Cochrane Library (2014)
9. Clunie, D., Cordonnier, E.: Digital Imaging and Communications in Medicine (DICOM) – Application/DICOM MIME Sub-type Registration (2014)
10. Altman, N.S.: An introduction to kernel and nearest-neighbor nonparametric regression. Am. Stat. **46**(3), 175–185 (1992)
11. Mustra, M., Delac, K., Grgic, M.: Overview of the DICOM standard (PDF). In: 50th International Symposium ELMAR, Zadar, Croatia, pp. 39–44 (2008)
12. Ferreira, A.J.M.: MATLAB Codes for Finite Element Analysis. Springer, Netherlands (2009). https://doi.org/10.1007/978-1-4020-9200-8
13. Smith, S.T.: MATLAB Advanced GUI Development. Dog Ear Publishing, Indianapolis (2006)
14. Buzug, T.M.: Computed Tomography From Photon Statistics to Modern Cone-Beam CT. Springer, Heidelberg (2008). https://doi.org/10.1007/978-3-540-39408-2
15. Heymsfield, S.: Human Body Composition. Human Kinetics (2005)
16. Prokop, M.: Spiral and Multislice Computed Tomography of the Body (2003)
17. Gong, T., et al.: Classification of CT brain images of head Trauma. In: Rajapakse, J.C., Schmidt, B., Volkert, G. (eds.) PRIB 2007. LNCS, vol. 4774, pp. 401–408. Springer, Heidelberg (2007). https://doi.org/10.1007/978-3-540-75286-8_38
18. Brant, W.E., Helms, C.A.: Fundamentals of Diagnostic Radiology. Lippincott Williams & Wilkins, Philadelphia (2007)
19. Thust, S.C., Burke, C., Siddiqui, A.: Neuroimaging findings in sickle cell disease. Br. J. Radiol. (2013)

20. Deza, E., Deza, M.M.: Encyclopedia of Distances. Springer, Heidelberg (2009). https://doi. org/10.1007/978-3-642-00234-2
21. http://radclass.mudr.org/content/hounsfield-units-scale-hu-ct-numbers
22. https://en.wikipedia.org/wiki/K-nearest_neighbors_algorithm
23. http://www.idlcoyote.com/fileio_tips/hounsfield.html

Rough Set Data Mining Algorithms and Pursuit Eye Movement Measurements Help to Predict Symptom Development in Parkinson's Disease

Albert Śledzianowski[1], Artur Szymański[1], Stanisław Szlufik[2],
and Dariusz Koziorowski[2(✉)]

[1] Polish-Japanese Academy of Information Technology, 02-008 Warsaw, Poland
[2] Neurology, Faculty of Health Science, Medical University of Warsaw,
03-242 Warsaw, Poland
dkoziorowski@esculap.pl

Abstract. This article presents research on pursuit eye movements tests conducted on patients with Parkinson's disease in various stages of disease and phases of treatment. The aim of described experiment was to develop algorithms allowing for measurements of parameters of pursuit eye movement in order to reference calculated results to the previously collected neurological data of patients. An additional objective of the experiment was to develop an example of data-mining procedure, allowing for classification of neurological symptoms based on oculometric measurements. Definition of such correlation enables assignment of particular patient to a given neurological group on the base of parameters values of pursuit eye movements. By using created decision table, we have achieved good results of prediction of the neurological parameter UPDRS, with total accuracy of 93.3% and total coverage of 60%. This allows for better evaluation of stage of the disease and its progression. It also might provide additional tool in determining efficacy of different disease treatments.

Keywords: Neurodegenerative disease · Parkinson's disease
Pursuit eye movements · Eye tracking · Data mining · Machine learning

1 Introduction

Parkinson's disease is the second, most common neurodegenerative disease, still incurable. Total, global number of patients is difficult to obtain, but disease statistics in developed countries are about 1% for citizens over 60, with estimates of up to 4% of older adults [1]. Symptomatic treatment of Parkinson's disease (PD) is based on pharmacological methods, but in serious cases surgical treatments are applied to alleviate patients symptoms. Pharmacological methods involve use of L-Dopa which affect the dopaminergic system and have strong

© Springer International Publishing AG, part of Springer Nature 2018
N. T. Nguyen et al. (Eds.): ACIIDS 2018, LNAI 10752, pp. 428–435, 2018.
https://doi.org/10.1007/978-3-319-75420-8_41

side efects. The beneficial effect of this treatment lasts from a few to a dozen years, followed by movement fluctuations and/or involuntary movements (dyskinesias), which can completely disturb the daily functioning. Alternatives are non-pharmacological methods. One of the most popular non-pharmacological methods of symptomatic PD treatment is Deep Brain Stimulation (DBS) of Subthalamic Nucleus (STN). Patients undergoing surgical treatment requires constant monitoring of the severity of symptoms and depending on the results, pharmacological treatment modifications and adjustments of stimulation parameters. The overall mobility of patients is assessed on the basis of the Unified Parkinson's Disease Scale (UPDRS) and Hoehn and Yahr (H&Y) scale. Assessments are usually performed in a separated neurological examinations. Measurements are usually made in 3 different patient groups, tested 3 times every six months along with oculo-metric test for pursuit eye movement. Significant effect on the effectiveness of PD treatment by the DBS method, has the precision of implantation of the electrodes. Because STN is a small structure and is not visible on standard in vivo imaging, precise positioning of the electrodes is difficult. Verification of the position of the electrodes is made by intraoperative micro registration and postoperative magnetic resonance imaging (MRI), also using interactive 3D anatomy atlas.

Efficiency of treatments can also be analyzed using objective oculometric techniques such as eye tracking. The aim of this study was to develop and test algorithms, allowing for evaluation of the pursuit movements of the eye, which could be used as a simple tool for objective, automatic and quantitative assessment of the condition of the patient. This evaluation could be automated by dedicated software, using standard and publicly accessible devices like tablets or smartphones. This method will allow for better understanding of the pathophysiology of the disease. The research hypothesis assumes that presented methodology will allow for better monitoring of the progress of the disease.

2   Methods of the Experiment

The experiment was conducted in clinical conditions using head-mounted infrared eye-tracker JAZZ-Novo. The device has been created in cooperation with the Polish Institute of Biocybernetics and Biomedical Engineering and Ober Consulting Poland. The device measures eye movements, head speed and speed acceleration of the head in the horizontal and vertical axis, also plethysmography signals in two wavelengths of light. All measurements were conducted in frequency of 1000 Hz. Patients has been divided into two groups: those who were operated during the project (the DBS group) and postoperative patients (POP) with electrodes implanted before start of the project. Each group consisted of 10 subjects, with two type of treatments enabled: pharmacological (BMT - Best Medical Treatment) and DBS. Subjects were examined in three different sessions composed of possible variants of the treatment:

- Session 1 (S1): No treatment (BMT Off, DBS Off)
- Session 2 (S2): Patients were only on non-pharmacological treatment (BMT Off, DBS On)

- Session 3 (S3): Patients were only on pharmacological treatment (BMT On, DBS Off)
- Session 4 (S4): Both types of treatments were active (BMT On, DBS On)

During the experiment, patients followed moves of the light spot generated by tracking device, which stimulated pursuit eye movement. The spot was moved in horizontal axis in 3 different frequencies of 0.125, 0.25 and 0.50 Hz, preceded by a calibration mechanism. The patient's task was to track the movement of the spot with greatest possible accuracy and possible smallest delay.

3   Computational Basis

Two parameters, Gain and Accuracy, were calculated on the basis of registered motion of spot and patient's eyes. The Gain value was calculated for each of the 3 different amplitude frequencies, as the mean value of the control window defined at 20% of the sine length at the maximum amplitude points, then averages for each of the 3 different amplitude frequencies were calculated. Accuracy was measured as the absolute difference between the spot and the eye position, normalized to the absolute sinusoidal surface (the quotient of the sum of vectors of difference between coordinates of the spot and coordinates of the eye at the maximum amplitude points). Half of the first and last sine wave were excluded from the calculation as possibly loaded with artifacts.

The collected measurements were placed in the dataset and further used as decision table in Rough Set Exploration System (RSES) analysis. Rough set (RS) theory is founded on the assumption that every object associate some information which are characterized by the same information in view of the available information about them [5]. The indiscernibility relation or similarity relation is the mathematical basis of this theory, stating that a set of similar objects forms a basic granule of knowledge and any union of those elementary sets formulates a precise set - otherwise the set is rough [5]. Each RS has boundary-line of objects which cannot be certainly classified and any RS contains associated pair of precise sets - called the lower and the upper approximation [5]. The lower approximation consists of all objects certainty belong to the set and the upper approximation contains all objects which possible belong to the set. The difference between the upper and the lower approximation constitutes the boundary region of the RS and such approximations are basic operations used in this data-mining methodology [5].

4   Results

In order to find correlations, Rough Set Exploration System (RSES) has been used for data analysis. RSES is a tool set of methods coming from Rough Set Theory [3]. Data from recordings of different frequency have been compiled together as one decision table and additionally pivoted. The resulting dataset was then mapped to the neurological database of patients. Initial database contained more

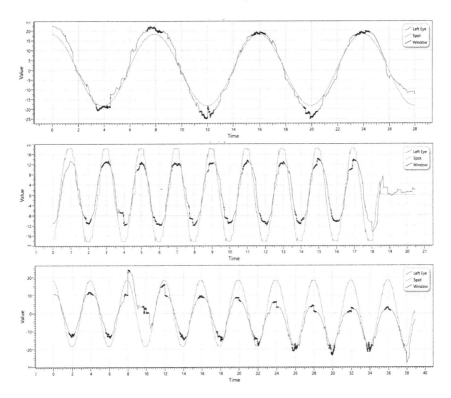

Fig. 1. Sample of the recording of patient's smooth pursuit eye movement against spot amplitude frequency, respectively of 0.125, 0.25 and 0.5 Hz. Blue line marks the eye movement of the patient, red line motion of the spot. Green dots mark the width of the window from which gain parameter was calculated. Horizontal axis represents time in seconds, vertical axis coordinates in the manufacturer's contractual units. (Color figure online)

than 300 attributes. Before RSES classification this amount has been limited in WEKA by the Attribute Selection filter with CfsSubset evaluator, using Best First selection algorithm [6]. The list of selected attributes contained (Fig. 1):

– UPDRS III result
– H&Y scale result
– Schwab&England scale result
– Slowness Indicator
– Arrhythmia Indicator
– Anticholinergic Indicator

Finally neurological attributes were reduced only to UPDRS (which was chosen as target scale for patients classification) and Arrhythmia, as this parameter was the only one showing differences in patients examined in oculomotor test. The dataset containing attributes presented below, have been used for initial input to the RSES.

Table 1. Sample of the initial dataset

Group	Session	PatId	DisDur	Gain125	Acc125	Gain250	Acc250	Gain500	Acc500	Arrh	UPDRS
DBS	S2	46	9	1.07	0.79	0.47	0.48	0.55	0.56	0	20
DBS	S3	46	9	0.92	0.28	0.68	0.64	0.62	0.63	0	25
DBS	S1	46	9	0.73	0.62	0.51	0.53	0.59	0.59	0	38
POP	S2	35	10.08	1.07	0.82	0.64	0.65	0.35	0.31	0	25
POP	S3	35	10.08	1.13	0.8	0.61	0.62	0.56	0.56	0	29
POP	S1	35	10.08	0.99	0.86	0.56	0.6	0.61	0.61	0	38

The entire dataset contained 11 attributes and 23 observations (experimental measurements). Attributes presented in Table 1 are defined as follow: Group - type of the group (DBS or POP), Session - type of the session (S1, S2, S3, S4) PatId - patient number, DisDur - duration of the disease, Arrh - indicator of Arrhythmia, UPDRS - value of patient's UDPRS III classification, Gain 125, 250, 500 - calculated Gain value for adequate frequency of spot amplitude, Accu 125, 250, 500 - calculated Accuracy value for adequate frequency of spot amplitude.

In the next phase of analysis, reduction of attributes has been performed along with the discretization of the selected attributes in method of global three-fold cross-validation. The resulting dataset contained reduced number of attributes and attributes values replaced by their range (Table 2).

We can see on sample below that attributes DisDur, Acc 250, Gain 500, Acc 500 were not considered as important by the selection algorithm. It is also interesting that the UPDRS were divided only into two different ranges and that the duration of the illness was considered not relevant. Use of auto-binner tool from KNIME package allowed to group UPDRS III numeric data in proper intervals of equal frequencies. It has been experimentally determined that the division of numerical values of UDPRS III into 5 bins gives the best predictive results. The decision tree has been obtained, by classifying dataset in WEKA using J48 classifier [6]. The visualization in Fig. 2 shows relationships between values of tree's attributes in the context of assigning to a particular UPDRS value group.

Next, the dataset previously binned in the KNIME was again analyzed in RSES, where the discretization method has been applied to conditional attributes. Discretization type was selected as local, with symbolic attributes, allowing for nominal values analysis. It gave better outputs in terms of sensitivity

Table 2. Sample of the reduced and discretized dataset

Group	Session	PatId	DisDur	Gain125	Acc125	Gain250	Acc250	Gain500	Acc500	UPDRS
DBS	S2	46	*	(1.055, inf)	(−inf, 0.81)	(−inf, 0.635)	*	*	*	(−inf, 31.5)
DBS	S3	46	*	(−inf, 1.055)	(−inf, 0.81)	(0.635, inf)	*	*	*	*(−inf, 31.5)
DBS	S1	46	*	(−inf, 1.055)	(−inf, 0.81)	(−inf, 0.635)	*	*	*	(31.5, inf)
POP	S2	35	*	(−Inf, 1.055)	(0.81, Inf)	(0.635, Inf)	*	*	*	(−Inf, 31.5)
POP	S3	35	*	(−Inf, 1.055)	(−Inf, 0.81)	(−Inf, 0.635)	*	*	*	(−Inf, 31.5)
POP	S1	35	*	(−Inf, 1.055)	(0.81, inf)	(−Inf, 0.635)	*	*	*	(31.5, Inf)

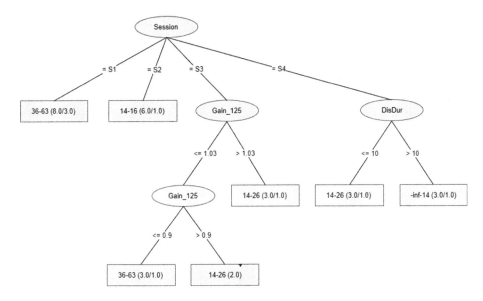

Fig. 2. Decision tree visualization for discretized dataset with attribute selection performed in RSES.

of discretization as in general local methods gives more number of cuts [4]. The final dataset has been presented on Table 3 presented below.

For such dataset, the value of the 5 folds cross-validation prediction of UPDRS attribute has been estimated with total accuracy of 93.3% and total coverage of 60%, what should be considered as very good result with such limited number of observation. We compared these results with other classifiers available in WEKA. For same dataset with 5 folds cross-validation, Mulitlayer Perceptron classifies correctly only 34.4% of instances (the result increases up to 48.2% using 10 folds cross-validation), Random Forest 51.7%, Naive Bayes 55.1% and Support Vector Machine (SMO) 58.6% [6].

What seems to be interesting at first glance when looking at RSES results, is the fact that attribute Group determining the date of the implantation of the electrode has been recognized as not significant. It is also interesting, that one parameter determining the outcome of the oculomotor test is enough for correct assignment to the session and to the group of results of the UPDRS neurological examination.

Table 3. Sample of the final decision table

Group	Session	PatId	DisDur	Gain125	Acc125	Gain250	Acc250	Gain500	Acc500	UPDRS
*	('S2', 'S4')	46	(−inf, 9.5)	(−inf, 1.095)	*	*	*	*	*	(14, 26)
*	('S1', 'S3')	46	(−inf, 9.5)	(−inf, 1.095)	*	*	*	*	*	(14, 26)
*	('S1', 'S3')	46	(−inf, 9.5)	(−inf, 1.095)	*	*	*	*	*	(36, 63)
*	('S2', 'S4')	35	(9.5, 14.55)	(−inf, 1.095)	*	*	*	*	*	(14, 26)
*	('S1', 'S3')	35	(9.5, 14.55)	(1.095, inf)	*	*	*	*	*	(26, 36)
*	('S1', 'S3')	35	(9.5, 14.55)	(−inf, 1.095)	*	*	*	*	*	(36, 63)

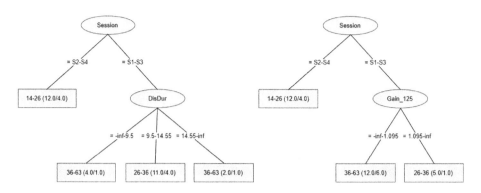

Fig. 3. Decision tree visualization for the most important decision attributes.

Table 4. My caption

UPDRS III	−inf–14	14–26	36–63	26–36
−inf–14	0	2	0	0
14–26	0	8	1	1
36–63	0	1	3	4
26–36	0	1	3	5

Basing on correlations contained in the decision table created by RSES, we can derive an equation that allows to allocate ranges of values of patient's attributes to the appropriate neurological group of UPDRS results. In example: (Session = 'S2', 'S4') & (Disease Duration = −inf, 9.5) & (Gain 125 = −inf, 1.095) = (UPDRS = 14, 26).

The decision trees derived from obtained results are presented in Fig. 3. Both decision trees has been obtained in WEKA using J48 classifier [6]. The strongest attribute determining the shape of the tree was the type of session, under which combinations of treatment methods has been defined. Next successively, the duration of the disease, obtained from database of neurological examinations and gain result obtained from measurements of pursuit eye movements in frequency of 0.125 Hz (Table 4).

5   Discussion and Conclusions

As can be seen in the results, duration of the disease and value of gain measurements, was both sensitive for patients examined in S1 and S3 sessions, respectively when there was no treatment or patients were on Best Pharmacological Treatment. It raises the question, why both attributes are irrelevant for patients with active Deep Brain Stimulation. Perhaps DBS may mask other methods of treating patients, because it is much more invasive. In terms of disease duration, as disease progresses the symptoms get worse as the UPDRS results are increasing. Seems that during DBS, UPDRS is reduced and the differences between new and older PD patients are getting smaller, therefore, the time of the disease becomes irrelevant. However this finding, suggests further researches in this area.

The analysis of data from only 20 patients has a high numerical sensitivity, which makes hard to make further generalization of these results. However, obtained very good results of the predictions are indicative and prove further experiments in presented area. On the other hand, when comparing the results of RSES classification with other methodologies, a significant difference in the results of correct classification can be noticed. This is due to the fact that RSES is good at dealing with predicates based on small datasets, as opposed to classifiers like Multilayer Perceptron, Decision Forests or Support Vector Machine which results were presented earlier in this text. Thus, from the point of view of number of correctly classified instances, RSES appears to be the optimal classifier for presented type of data containing limited number of distinct patients.

It is hoped that further development of methodology described in this text, may be helpful in determining PD symptoms. It is also hoped that such methodology can help in tracking disease progression, as in the future, applied algorithms can be used in widely available applications installed on non-medical devices like PCs, tablets or smartphones, which are widely available to every patient. Such a widespread availability of diagnostic method, should increase amount of examination data, which should lead to more accurate diagnosis of the patients. It might also help in catching errors in treatments and in applying appropriate corrections.

References

1. de Lau, L., Breteler, M.: Epidemiology of Parkinson's disease. Lancet Neurol. **5**(6), 525–535 (2006)
2. Przybyszewski, A.W., Kon, M., Szlufik, S., Szymanski, A., Habela, P., Koziorowski, D.M.: Multimodal learning and intelligent prediction of symptom development in individual Parkinson's patients. Sensors **16**, 1498 (2016)
3. Pawlak, Z.I.: Rough Sets. Springer, Dordrecht (1991). https://doi.org/10.1007/978-94-011-3534-4
4. Peters, J.F.: Transactions on Rough Sets III. Springer Science & Business Media, Heidelberg (2005). https://doi.org/10.1007/b136502
5. Pawlak, Z.: Rough sets and data mining. Institute of Theoretical and Applied Informatics, Polish Academy of Sciences. http://bcpw.bg.pw.edu.pl/Content/1884/RSDMEAK.pdf
6. Hall, M., Frank, E., Holmes, G., Pfahringer, B., Reutemann, P., Witten, I.H.: The WEKA data mining software an update. SIGKDD Explor. **11**(1), 10–18 (2009)

Rules Determine Therapy-Dependent Relationship in Symptoms Development of Parkinson's Disease Patients

Andrzej W. Przybyszewski[1]([✉]) [iD], Stanisław Szlufik[2] [iD],
Piotr Habela[1] [iD], and Dariusz M. Koziorowski[2] [iD]

[1] Polish-Japanese Institute of Information Technology, 02-008 Warsaw, Poland
{przy,piotr.habela}@pjwstk.pl
[2] Neurology Faculty of Health Science,
Medical University Warsaw, Warsaw, Poland
stanislaw.szlufik@gmail.com, dkoziorowski@esculap.pl

Abstract. We still do not have cure for neurodegenerative disorders (ND) such as Parkinson's disease (PD). Recent findings demonstrated that neurodegenerative processes related to ND have long periods without symptoms that effects lack in effective therapy when ND is diagnosed. Neurologists estimate PD progression on the basis of their tests: Hoehn and Yahr (H&Y) and Unified Parkinson's Disease Rating (UPDRS) scales, but results of these tests are partly school-dependent. We have previously proposed that eye movement tests can give objective and precise measure of the PD progression.

In this study, we have recorded reflexive saccades in patients with different disease stages and with different treatments. We put together patients' demographic data, results of neurological and eye movements' tests. In order to estimate effectiveness of different therapies we have placed data in information tables, discretized and used data mining (RS - rough set theory) and machine learning (ML). In end-effect we have obtained rules that determine longitudinal course of disease progression in different group of patients. By using of ML and RS rules obtained for the first visit of BMT/DBS/POP (only on medication/ recent DBS surgery/earlier DBS surgery) patients we have predicted UPDRS values in next year (two visits) with the global accuracy of 70% for both BMT visits; 56% for DBS, and 67, 79% for POP second and third visits. We have used rules obtained in BMT patients to predict UPDRS of DBS patients; for first session DBSW1: global accuracy was 64%, for second DBSW2: 85% and the third DBSW3: 74% but only for DBS patients during stimulation-ON. We could not predict UPDRS in DBS patients during stimulation-OFF visits and in all conditions of POP patients. We have compared rules in BMT patients with POP group and found many contradictory rules. It means that long-term brain electrical stimulation has changed brain mechanisms.

Keywords: Neurodegenerative disease · Rough set · Decision rules Granularity

© Springer International Publishing AG, part of Springer Nature 2018
N. T. Nguyen et al. (Eds.): ACIIDS 2018, LNAI 10752, pp. 436–445, 2018.
https://doi.org/10.1007/978-3-319-75420-8_42

1   Introduction

Parkinson (PD) is the second most popular neurodegenerative diseases (ND). ND are related to neuron death in many different brain's structures, starting in the case of PD from the substantia nigra (SN). As SN is responsible for the release of the transmitter dopamine (Dopa), which is involved in the movements control, PD patients have movements regulations instabilities and in addition often depression (Dopa is transmitter related to the reward) and/or also cognitive problems. As each patient has different disease symptoms, and rate of the progression, experienced neurologist has to find an optimal therapy. Finding an optimal one depends on results of doctor's tests, her/his experience and time. The neurologists get their knowledge based on their patient cases and they might use an intuitive prediction about results of different therapies for the particular patient's symptoms. As results of different doctor's tests are partly subjective, we propose to use eye movement measurements as the method that is objective, easy to apply and can be performed by technical staff. In addition, we would like to improve the diagnostic process by storing data from many visits and in additional to the statistical analysis that does not take into account individual patient's symptoms use AI methods. Our methods are based on the data mining and machine learning algorithms.

In this work, we have developed intelligent (AI) methods of symptoms and stages of the disease classification, which are following that found in the visual system of primates [1]. Classification of symptoms has many similarities to the recognition of the unknown complex objects. The recognition process is related to the logic system of the visual brain [1]. We have certain in-born mechanisms, related, for example, to anatomical structures of our brains, which coexist with the brain plasticity (learning mechanisms). Objects that we know help us in partial prediction of new object's properties, but often these properties have contradictory features. As we have demonstrated earlier [2], brain solves those problems by using approach similar to rough set theory. It means that the descending and ascending pathways interact in order to minimize the boundary region between known and actual object's properties. The classification is perfect (crisp) if this region is empty (see below).

The primary task of neurologists it is to estimate the disease stage because it determines different sets of treatments, in most cases at the beginning – dosage of medications. In PD, the 'Golden standards' are based on Hoehn and Yahr (H&Y) and the UPDRS (Unified Parkinson's Disease Rating) scales. As the H&Y scale has only five stages is not very precise but gives a general view of the disease severity. The UPDRS is result of many tests related to: patients mood, behavior, self–evaluation, motor symptoms and complications of therapy, etc. is more precise and it will be the purpose of study to use it as indicator of the PD progression.

We have estimated disease progression in different groups of patients that were under different therapies and they were tested during three every half-year visits. We hope that our method will lead to introduce more precise and more automatic follow ups in the perspective possibilities of the remote diagnosis and treatments.

2 Methods

We have analyzed tests from Parkinson Disease (PD) patients divided into three groups: (a) 23 patients that have therapy limited only to medication - group with the best medical treatment (named as BMT - group); (b) 24 patients on medications and with implanted (Institute of Neurology and Psychiatry Warsaw Medical University) Deep Brain Stimulation (DBS) electrodes in the STN (subthalamic nucleus). If the surgery was performed during our study these patients became part of the DBS – group; (c) if DBS surgery was performed earlier before our study we put these patients to POP-group (postoperative group). Patients from all groups were tested in at least two sessions: MedOn/Off sessions (ses. #3 with -, and ses. #1 without-medication). Patients from DBS and POP groups were additionally tested in StimOn/Off sessions (sess. on: #2/#4; off: #1/#3). All tests: neurological, neuropsychological, and EM tests were performed in dept. of Neurology, Brodno Hospital, Warsaw Medical University. In this study, we have only measured parameters of the fast, saccadic eye movements (EM) so-called reflexive saccades (RS). Detailed methodology was described earlier [2, 3]. In short, each patient was watching a computer screen while sitting in a stable position without head movements. After fixating in the middle of the screen, she/he has to follow horizontally moving randomly $10°$ to the right or $10°$ to the left dot. These tests were performed for each session (as described above). We have extracted the following parameters from saccades: the delay that is defined as a time difference between the beginning of the stimulus and eye movements; relative amplitude of the saccadic related to the light spot movements; max saccadic movement velocity; duration of the saccade as a time between the end and the beginning of each saccade. Eye movements were recorded by head-mounted saccadometer (Ober Consulting, Poland).

2.1 Theoretical Basis

Our data mining analysis is based on the Pawlak's rough set theory (Pawlak [4]). Our data in converted to the decision table where rows were related to different measurements (may be obtained from the same or different patients) and columns represent different attributes. An information system [4] a pair $S = (U, A)$, where U, A are nonempty finite sets called the universe of objects U and the set of attributes A. If $a \in A$ and $u \in U$ the value $a(u)$ is a unique element of V (where V is a value set).

We define as in [4] the *indiscernibility relation* of any subset B of A or $IND(B)$ as: $(x, y) \in IND(B)$ or $xI(B)y$ iff $a(x) = a(y)$ for every a $\in B$ where the value of $a(x) \in V$. It is an equivalence relation $[u]_B$ that we understand as a *B-elementary granule*. The family of $[u]_B$ gives the partition U/B containing u will be denoted by $B(u)$. The set $B \subset A$ of information system S is a reduct $IND(B) = IND(A)$ and no proper subset of B has this property [5]. In most cases, we are only interested in such reducts that are leading to expected rules (classifications). On the basis of the reduct we have generated rules using four different ML methods (RSES 2.2): exhaustive algorithm, genetic algorithm [6], covering algorithm, or LEM2 algorithm [7].

A *lower approximation* of set $X \subseteq U$ in relation to an attribute B is defined as $\underline{B}X = \{u \in U : [u]_B \subseteq X\}$. The *upper approximation* of X *is* defined as $\overline{B}X = \{u \in U : [u]_B \cap X \neq \phi\}$. The difference of $\overline{B}X$ *and* $\underline{B}X$ is the boundary region of X that we denote as $BN_B(X)$. If $BN_B(X)$ is empty then set than X is *exact* with respect to B; otherwise if $BN_B(X)$ is not empty and X is not *rough* with respect to B.

A decision table (training sample in ML) for S is the triplet: $S = (U, C, D)$ where: C, D are condition and decision attributes [8]. Each row of the information table gives a particular rule that connects condition and decision attributes for a single measurements of a particular patient. As there are many rows related to different patients and sessions, they gave many particular rules. Rough set approach allows generalizing these rules into universal hypotheses that may determine optimal treatment options for an individual PD patient. The decision attribute D is giving a particular object (patient's state) classification by an expert (neurologist). Therefore a decision table classifies data by *supervised learning (ML)* where teaching is related to decisions made by neurologist(s). Each row is an example of the teacher's decision.

Recent literature supports view that neurodegenerative processes start to accelerate a decade or two before the first PD symptoms. These processes are dependent on individual compensatory mechanisms therefore they are not exactly the same in different patients. In the consequence, there are often large inconsistencies between disease progression and symptoms between individual patients ("no two PDs are the same"). In addition, effects of similar treatments also often differ between patients. Therefore, our algorithms should be flexible enough to cover all these differences. Such methodology is related on the granular computation that might simulate interactions between neurologists and patients. Not so many experienced doctors have the ability to perceive patient's symptoms from various levels of abstraction (as defined above different granular levels). Granular computing is similar to the intelligent visual bran object classifications [2]. In this work, we have analyzed data from different groups of patients: one group (BMT patients) was used for the training (findings rules) and another groups (DBS, POP) were tested with BMT rules. The purpose was to find what are limits of rules' flexibilities that may describe symptoms development of patients with different treatments, and in different disease stages.

We have used the RSES 2.2 (Rough System Exploration Program) [9] with implementation of RS rules to process our data. We have demonstrated earlier that the RS method is superior to other classical methods [3].

3   Results

All 62 PD patients were divided into three groups as described in the Methods section: BMT group (only medication), DBS group (medication and STN stimulation, surgery during the study) and POP group (medication and STN stimulation, surgery before the study).

In 23 patients of BMT group the mean age was 57.8 ± 13 (SD) years; disease duration was 7.1 ± 3.5 years, UPDRS was 48.3 ± 17.9. In 15 patients of POP group the mean age was 63.1 ± 18.2 (SD) years and disease duration was 13.5 ± 3.6 years (stat. diff. from BMT p < 0.025, and from DBS-group: p < 0.015), UPDRS was 59.2 ± 24.5 (stat. diff. than BMT-group: p < 0.0001).

Our main interest was in 24 patients of DBS group, the mean age of 53.7 ± 9.3 years, disease duration was 10.25 ± 3.9 years; UPDRS was 62.1 ± 16.1 (stat. diff. than BMT-group: p < 0.0001).

These statistical data are related to the data obtained during the first session for each group: BMT W1 (visit one), DBS W1 (visit one) and POP W1 (visit one).

3.1   Comparing Longitudinal UPDRS Changes

The first plot from the right presents UPDRS of DBSW1 group. There are only two sessions as patients were before the surgery. In ses. #1 mean UPDRS was 62.2 ± 16.1, in the second (ses. #3) was 29.9 ± 13.3 strongly (p < 0.0001) different from ses. #1 and it represents effect of medication. Plot in the middle of Fig. 1 represents DBSW2 after the surgery and UPDRS in ses. #1 is larger than that before the surgery 65.3 ± 17.6 but there are not stat sig differences, however UPDRS in ses. #1 of DBSW3 is 68.7 ± 17.7 and stat diff (p < 0.03) then in W2. Effects of different therapies (session numbers) are significantly different in W1, W2, and W3, but not different between the same session numbers in different visit (with an exception of the ses. #3 in W1 as after the surgery the dosage of medication is reduced).

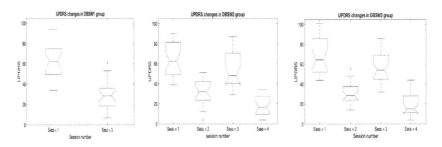

Fig. 1. The boxplot shows that the median of UPDRS in different sessions and visits; W1, W2, W3 of the DBS group. There are minimum and maximum values in each session. Since the notches in the box plot do not overlap, you can conclude, with 95% confidence that the true medians do differ.

In POP-group UPDRS values are similar. There is an increase of the UPDRS ses. #1 from W1: 63.1 ± 18.2 to W2: 68.9 ± 20.3 to W3: 74.2 ± 18.4 but there were smaller differences for ses. #4 (Med&DBSOn) W1: 21 ± 11.3 to W2: 23.3 ± 9.5 to W3: 23.8 ± 10.7. Therefore, we have assumed that groups DBS and POP are similar.

In BMT group UPDRS in ses. #1 was W1: 48.3 ± 17.9; W2: 57.3 ± 16.8 (p < 0.0005 diff. than W1); W3: 62.2 ± 18.2 (p < 0.05 diff. than W2). In ses. #3 UPDRS was W1: 23.6 ± 10.3; W2: 27.8 ± 10.8; W3: 25 ± 11.6 (no stat. diff. between visits for ses. #3).

3.2 Rough Set Rules and Machine Learning Approach for BMT Group

There were patient only on medication, and they have only two sessions (#1 - MedOff and #3 - MedOn) measured every half of the year.

We have divided UPDRS into 4 ranges: "(-Inf, 24.0)", "(24.0, 36.0)", "(36.0, 45.0)", "(45.0, Inf)" with help of the discretization RSES software and by using machine learning and rough set algorithms [6] we have obtained 70 rules from BMTW1 and use them to predict UPDRS W2 and W3.

Table 1. Discretized-table extract for BMT patients

P#	t_dur	S#	PDQ39	Epworth	SccLat	SccDur	SccAmp	SccVel	UPDRS
4	"(-Inf, 9.7)"	1	"(-Inf, 55.0)"	"(-Inf, 3.0)"	"(-Inf, 181.5)"	*	*	*	"(36.0, 45.0)"
4	"(-Inf, 9.7)"	3	"(-Inf, 55.0)"	"(-Inf, 3.0)"	"(-Inf, 181.5)"	*	*	*	"(-Inf, 24.0)"
5	"(9.7, Inf)"	1	"(-Inf, 55.0)"	"(3.0, Inf)"	"(181.5, 395.0)"	*	*	*	"(36.0, 45.0)"
5	"(9.7, Inf)"	3	"(-Inf, 55.0)"	"(3.0, Inf)"	"(181.5, 395.0)"	*	*	*	"(24.0, 36.0)"
7	"(-Inf, 9.7)"	1	"(55.0, Inf)"	"(3.0, Inf)"	"(181.5, 395.0)"	*	*	*	"(36.0, 45.0)"
7	"(-Inf, 9.7)"	3	"(55.0, Inf)"	"(3.0, Inf)"	"(181.5, 395.0)"	*	*	*	"(-Inf, 24.0)"

In this table (Table 1) are discretized results of three patients: 4, 5, and 7 in two sessions #1, #3 with parameters related to psychological testing (PDQ39 – quality of life, Epworth – quality of sleep), and parameters of saccades where only latency was significant (SccLat), and in the last column the decision attribute UPDRS.

Each row gives a particular rule and by using RSES we have obtained more general 70 rules like:

$$(S\# = 3) \,\&\, (PDQ39 = \text{``}(-Inf, 50.5)\text{''}) \Rightarrow (UPDRS = \text{``}(-Inf, 33.5)\text{''}\,[12]) \quad (1)$$

It states that if the session is #3 and PDQ39 is smaller than 50.5 then UPDRS will be smaller than 33.5. It was fullfield in 12 cases.

We have used these rules to predict UPDRS in BMTW2 and W3 using 6-fold cross validation we have obtained for both visits global accuracy 0.7, and the global coverage 1.0. It means that in BMT group patients we can have good expectation about symptoms changes during the disease progression.

3.3 Rough Set Rules and Machine Learning Approach for DBS Group

We have predicted UPDRS of DBSW3 by rules from DBSW2 (both groups have 4 sessions), and we have obtained the global accuracy 0.56 and global coverage 1.

In the next step, we have applied the same BMTW1 rules to the DBS group. It was successful for DBSW1 pre-operative patients as they were also in two sessions with a high dosage of medication. We have obtained the global accuracy 0.64 with the global coverage 0.5.

However, prediction of UPDRS from BMT group for DBSW2 and W3 groups was not very effective as in these groups we have 4 sessions and the BMT group has only 2 sessions. Therefore, we have divided DBSW2, W3 patients into two subgroup each:

one without stimulation (DBSOff) and another one with DBSOn. We could not predict UPDRS in DBS groups without stimulation (DBSOff) only with DBSOn.

UPDRS of DBSW2 (DBSOn only) were predicted from BMTW1 rules with good global accuracy 0.85, but the global coverage was 0.3 and some classes were not at all predicted (UPDRS larger than 63). We have obtained similar results for UPDRS of DBSW3 (DBSOn only) from BMTW1 rules; the global accuracy was 0.74 but the global coverage 0.56.

3.4   Rough Set Rules and Machine Learning Approach for POP Group

We have predicted UPDRS for POPW2 and POPW3 from rules obtained from POPW1. UPDRS in POPW2 was predicted with the global accuracy 0.67 and global coverage 1; UPDRS in POPW3 group was predicted from POPW1 group with the global accuracy 0.8 and global coverage 0.97. In the next step we have divided each od POPW1, W2, W3 group into two subgroups: on with DBSOff, another one with DBSOn (in the way as for DBS group). These subgroup have similar sessions (MedOff/On) to the BMT group. We could predict disease progression of W2, W3 on the basis of W1 for both subgroups with the global accuracy 0.5 (large differences in UPDRS) global coverage = 1 for DBSOff, accuracy 0.6 for DBSOn subgroups (Table 2).

In contrast to the DBS group we were not successful in using rules from the BMT patients to predict UPDRS of the POP group patients. In order to find possible reasons we have compared rules from the BMTW1 patients with rules of POPW3 group as UPDRS values for both groups were similar:

BMTW1 rules:

$$(Ses = 1) \& (PDQ39 = ``(-Inf, 50.5)")$$
$$\Rightarrow (UPDRS = ``(43.0, 63.0)"[6], ``(33.5, 43.0)"[2], ``(-Inf, 33.5)"[4]) \, 12 \qquad (2)$$

Table 2. Confusion matrix for UPDRS of POPW3-subgroup (StimOn) by rules obtained from POPW1-subgroup (StimOn).

Actual	Predicted				
	"(26.5, Inf)"	"(19.0, 26.5)"	"(-Inf, 13.5)"	"(13.5, 19)"	ACC
"(26.5, Inf)"	11. 0	3.0	1.0	0.0	0.73
"(19.0, 26.5)"	1.0	5.0	1.0	1.0	0.625
"(-Inf, 13.5)"	0.0	0.0	2.0	1.0	0.67
"(13.5, 19)"	0.0	0.0	4.0	0.0	0.0
TPR	0.92	0.625	0.25	0.0	

TPR: True positive rates for decision classes; ACC: Accuracy for decision classes: the global coverage was 1 and the global accuracy was 0.6, the coverage for all decision classes was 1.

POPW3 rules:

$$(Ses = 1) \,\&\, (PDQ39 = \text{``}(-Inf, 48.0)\text{''}) \,\&\, (RSLat = \text{``}(301.0, Inf)\text{''})$$
$$\Rightarrow (UPDRS = \text{``}(66.0, Inf)\text{''}\,[2])\,2 \tag{3}$$

Rules (2) and (3) contradictory. In both rules values of conditional attributes were the same: ses. #1, and PDQ39 was less than 50. These conditions implied in rule (2) that UPDRS in 6 cases was (43, 63), in 2 cases was (33.5, 43), and in 4 cases was less than 33.5. In contrast same values of attributes (with additional condition for the latency of the saccades) imply for POP group that UPDRS should be larger than 63.

BMTW1 rules:

$$(Ses = 3) \,\&\, (PDQ39 = \text{``}(-Inf, 50.5)\text{''}) \Rightarrow (UPDRS = \text{``}(-Inf, 33.5)\text{''}\,[12])\,12 \tag{4}$$

POPW3 rules:

$$(Ses = 3) \,\&\, (PDQ39 = \text{``}(-Inf, 48.0)\text{''}) \,\&\, (RSPeak = \text{``}(522.0, Inf)\text{''})$$
$$\Rightarrow (UPDRS = \text{``}(51.0, 66.0)\text{''}\,[2])\,2 \tag{5}$$

Above rules (4) and (5) are contradictory as the same values of conditional attributes give opposite results of the UPDRS, in Eq. (4) UPDRS is less than 33.5, whereas in Eq. (5) UPDRS should be between 51 and 66.

BMTW1 rules:

$$PDQ39 = \text{``}(-Inf, 50.5)\text{''}) \,\&\, (RSLat = \text{``}(264.0, Inf)\text{''}) \,\&\, (dur = \text{``}(-Inf, 5.65)\text{''})$$
$$\Rightarrow (UPDRS = \{\text{``}(-Inf, 33.5)\text{''}\,[4], \text{``}(33.5, 43.0)\text{''}\,[1]\})\,5 \tag{6}$$

POPW3 rules:

$$(PDQ39 = \text{``}(-Inf, 48.0)\text{''}) \,\&\, (RSLat = \text{``}(301.0, Inf)\text{''}) \,\&\, (RSPeak = \text{``}(403.5, 522.0)\text{''})$$
$$\Rightarrow (UPDRS = \text{``}(66.0, Inf)\text{''}\,[1])\,1 \tag{7}$$

In Eqs. (6) and (7) conditions attributes are PDQ39 with values below 50, and the latency of the reflex saccades that is longer than 264 ms. Both equations have additional also condition attributes, but their decision attribute has opposite values: in Eq. (6) below 43, but in Eq. 7 above 66. Therefore these rules are contradictory.

In other words, the long-term electrical brain stimulation may inverse relationship between such parameter as quality of life or effects of medication on the UPDRS.

4 Discussion

Neurological procedures are changing every several years (as e.g. UPDRS procedures were upgraded to MDS-UPDRS that is more precise in many tests) based on new norms and therapies (like e.g. new medications). New procedures improve PD statistical patient's treatments, but question is, if the actual procedures are optimal for a specific individual case? We still do not know how far our procedures are from the optimal one for a particular patient. In this study, we have verified how to use the data mining and machine learning methods in order to compare different neurological protocols and their effectiveness. Such comparisons might lead us, in the near future, to more optimal therapies for different groups of patients especially if will make all tests automatically (quantitatively objective). Then doctor's decisions will be compared with the AI system classifications (based on the supervised learning from many doctors decisions).

We have tested three groups of patients in the following stages: BMT group with mean disease duration 7 years, DBS – 10 years, and POP – 13.5 years; mean UPDRS in ses. #1: BMTW1: 48; DBSW1: 62; POPW1: 63; BMTW2: 57; DBSW2: 65; POPW2: 69; BMTW3: 62; DBSW3: 69; POPW3: 74. UPDRS of all patients increases with time with mean increase for half of the year for BMT: 7, for DBS: 3.5, for POP: 5.5. The major problem with prediction of UPDRS for DBS and POP groups on the basis of rules from BMT group was that there were significant difference in UPDRS. UPDRS for BMT and POP groups were more similar. UPDRS during electric brain stimulation (DBSOn) for DBS and POP groups were significant lower and it was realistic to predict their values from BMT rules. However, we were successful with our prediction for DBS group but not for POP group. In conclusion, it means that a long-term STN electric stimulation lowers UPDRS (improves symptoms) but it changes relationships psychology (like quality of life PDQ39), eye movement and UPDRS. We do not know what is exact reason for changes in these rules (relationships).

Acknowledgements. This work was partly supported by projects Dec-2011/03/B/ST6/03816, from the Polish National Science Centre.

References

1. Przybyszewski, A.W.: Logical rules of visual brain: from anatomy through neurophysiology to cognition. Cogn. Syst. Res. **11**, 53–66 (2010)
2. Przybyszewski, A.W.: The neurophysiological bases of cognitive computation using rough set theory. In: Peters, J.F., Skowron, A., Rybiński, H. (eds.) Transactions on Rough Sets IX. LNCS, vol. 5390, pp. 287–317. Springer, Heidelberg (2008). https://doi.org/10.1007/978-3-540-89876-4_16
3. Przybyszewski, A.W., Kon, M., Szlufik, S., Szymanski, A., Koziorowski, D.M.: Multimodal learning and intelligent prediction of symptom development in individual Parkinson's patients. Sensors **16**(9), 1498 (2016). https://doi.org/10.3390/s16091498
4. Pawlak, Z.: Rough Sets: Theoretical Aspects of Reasoning About Data. Kluwer, Dordrecht (1991)
5. Bazan, J.G., Nguyen, H.S., Nguyen, T.T., Skowron, A., Stepaniuk, J.: Desion rules synthesis for object classification. In: Orłowska, E. (ed.) Incomplete Information: Rough Set Analysis, pp. 23–57. Physica-Verlag, Heidelberg (1998)
6. Bazan, J., Nguyen, H.S., Nguyen, S.H., Synak, P., Wróblewski, J.: Rough set algorithms in classification problem. In: Polkowski, L., Tsumoto, S., Lin, T. (eds.) Rough Set Methods and Applications, pp. 49–88. Physica-Verlag, Heidelberg/New York (2000)

7. Grzymała-Busse, J.: A new version of the rule induction system LERS. Fundamenta Informaticae **31**(1), 27–39 (1997)
8. Bazan, J.G., Szczuka, M.: The rough set exploration system. In: Peters, James F., Skowron, A. (eds.) Transactions on Rough Sets III. LNCS, vol. 3400, pp. 37–56. Springer, Heidelberg (2005). https://doi.org/10.1007/11427834_2
9. Bazan, J.G., Szczuka, M.: RSES and RSESlib - a collection of tools for rough set computations. In: Ziarko, W., Yao, Y. (eds.) RSCTC 2000. LNCS (LNAI), vol. 2005, pp. 106–113. Springer, Heidelberg (2001). https://doi.org/10.1007/3-540-45554-X_12

Analysis of Image, Video and Motion Data in Life Sciences

Vertebra Fracture Classification from 3D CT Lumbar Spine Segmentation Masks Using a Convolutional Neural Network

Charmae B. Antonio, Louise Gillian C. Bautista, Alfonso B. Labao, and Prospero C. Naval Jr.[✉]

Computer Vision and Machine Intelligence Group,
Department of Computer Science, College of Engineering,
University of the Philippines Diliman, Quezon City, Philippines
pcnaval@dcs.upd.edu.ph

Abstract. Accurate and efficient identification of vertebra fractures in spinal images is of utmost importance in improving clinical tasks such as diagnosis, surgical planning, and post-operative assessment. Previous methods that tackle the problem of vertebra fracture identification rely on quantitative morphometry methods. Standard six-point morphometry involves manual identification of the vertebral bodies' corners and placement of points on identified corners. This task is time-consuming and requires effort from experts and technicians and prone to subjective errors in visual estimation in spinal images. In this paper, we propose an automated method to detect and classify vertebra fractures from 3D CT lumbar spine images. Fifteen 3D CT images with accompanying fracture labels for each of the five lumbar vertebra from the xVertSeg Challenge were utilized as data set. Each vertebra from the 3D image is processed into 100×50 2D 3-channel images composed of three grayscale images. The three grayscale images represent the vertebral slices in the sagittal, coronal, and transverse anatomical planes. These 100×50 2D images are fed into the 152 layer Residual Network. A total of 13,400 images were generated from the data pre-processing stage. 12,700 of which having varying classifications were used as training data, and 100 images for each of the seven vertebra fracture classifications were used as testing data. The network achieved 93.29% testing accuracy.

Keywords: Vertebra fracture detection
Convolutional neural network · ResNet

1 Introduction

Vertebra fractures are categorized into three major patterns, namely: flexion, extension, and rotation. This work focuses on flexion fracture patterns, which are further categorized into two: the compression fracture and the axial burst fracture. A compression fracture is characterized by a reduction in height of the

© Springer International Publishing AG, part of Springer Nature 2018
N. T. Nguyen et al. (Eds.): ACIIDS 2018, LNAI 10752, pp. 449–458, 2018.
https://doi.org/10.1007/978-3-319-75420-8_43

anterior portion of the vertebra, while an axial burst fracture is characterized by reduction of both anterior and posterior portions. These reductions in height are caused either by high-energy trauma such as various physical accidents, or by low-impact activities such as reaching or twisting. Furthermore, diseases or physiological abnormalities that may be inferred from having vertebra fractures are osteoporosis, tumors, and various conditions that weaken the bone.

A precise vertebra fracture identification method in spinal imaging is in high demand due to its importance in orthopedics, neurology, oncology, and many other medical fields. A common method in identifying vertebra fractures is the semi-quantitative (SQ) method proposed by Genant et al. [1]. This method introduced specific morphological cases and grades of vertebral body fractures and is accepted as the ground truth for the evaluation of vertebra fractures. By estimating the differences in the anterior, central and posterior heights of the vertebral body in sagittal radiographic images, the SQ method presents the following morphological cases of 3 vertebra fractures.

Morphological Cases

- Wedge: characterized by distinctive difference between the anterior and posterior vertebral height.
- Biconcavity: characterized by similar anterior and posterior vertebral height but smaller central vertebral height.
- Crush: characterized by the mean vertebral height lower than the statistical value for that vertebra or for adjacent vertebral bodies.

For morphological grades, mild grade is characterized by a height difference of 20–25%. Moderate grade is characterized by a height difference of 25–40%. Severe grade is characterized by a height difference of more than 40%.

For identification of these cases in CT Scans or MRI scans of bones, the usual recourse is through visual inspection of vertebral bodies, which requires the expertise of radiologists and physicians. The visual inspection of each vertebra is time consuming and prone to subjective errors.

For our proposed procedure, we automate the classification of 3D bone images according to the seven classes of the semi-quantitative method: (1) Normal, (2) Mild Wedge, (3) Moderate Wedge, (4) Severe Wedge, (5) Mild Biconcavity, (6) Moderate Biconcavity and (7) Moderate Crush. Automation is done using a deep residual network from [2], which achieved outstanding benchmark results in the ImageNet classification challenge. The deep residual network makes use of 'skip-connections' to allow information to freely pass between network layers. This facilitates training and allows the network to extend to very deep layers. Employing deeper layers allow the network to abstract better and generate features that contain high-value information.

We train the automated procedure using a total of 12,700 images, and then test using 700 images spread equally to each of the seven vertebra fracture classifications. The automated network procedure achieved 93.29% accurate classification.

2   Methodology

2.1   Data Acquisition

The data set used in this paper were obtained from the xVertSeg challenge, an open online computational challenge in the field of spine medical image analysis. The data set consists of 25 3D computed tomography (CT) lumbar spine images in MetaImage (MHD) format, which requires a pair of .mhd and .raw file per image. The fracture labels for the lumbar vertebra were given in a CSV file. Fracture labels used in the data set follow the semi-quantitative vertebra fracture classification system proposed by Genant et al.

This semi-quantitative system presents a total of 10 classifications, but we exclude the following classifications due to insufficient samples: (1) Severe Biconcavity, (2) Mild Crush, and (3) Severe Crush.

The data set split the 25 images into two: Data1 and Data2. Data1 consists of 15 CT images, and Data2 consists of 10 CT images. The images from Data1 have accompanying segmentation masks and fracture labels, while the images from Data2 do not. For the purpose of vertebra fracture classification, only the Data1 images were used.

The segmentation mask given assigns a corresponding pixel value for each of the five lumbar vertebra: 200 for the L1 vertebra, 210 for the L2 vertebra, 220 for the L3 vertebra, 230 for the L4 vertebra, and 240 for the L5 vertebra.

2.2   Data Preprocessing

The data to be fed to the deep residual network are 100×50 CT images that are created from 3 grayscale images corresponding to the sagittal, coronal, and transverse slices of the vertebra. Preprocessing for each of the 5 lumbar vertebra in a single CT image undergoes the following steps:

1. **Find the areas in the 3D image that are indicative of the vertebral shape.** The 3D image is repeatedly sliced into a 2D image in the sagittal plane for every 2 pixels. Each sliced plane is checked for the 5 lumbar vertebral bodies that are cleanly separable from their spinal parts.
2. **Find the z and y coordinates for the coronal and horizontal planes.** From the first sagittal slice of each vertebra deemed indicative by step 1, the contour of each of the five vertebral bodies are obtained using OpenCV's findContour function. From each of the contour coordinates, the leftmost coordinate marks the area where the coronal planes of the respective vertebra will be sliced from, and the topmost coordinate marks the area where the horizontal planes of the respective vertebra will be sliced from.
3. **Crop the vertebra in three planes.** Step 1 and step 2 results to a voxel coordinate for each of the lumbar vertebra. This voxel coordinate marks the probable corner of the vertebral body. From this coordinate, the coronal, and horizontal planes are each inspected. An algorithm checks if the slices contain enough data or pixels to be indicative of the vertebral shape. Once

the algorithm deems both planes indicative, it proceeds to crop the vertebral shapes using OpenCV's findContour function. The next iteration moves the voxel coordinate by 2 pixels in each plane.

4. **Create a single BGR image from the three grayscale images.** Step 3 generates sets of three 2D grayscale images, each associated to each other according the same voxel coordinate. These 3 grayscale images corresponds to the sagittal (yz), coronal (xz), and horizontal (xy) planar views. Using OpenCV's merge function, the 3 grayscale images are combined to form a 3-channel BGR image. A single instance of this 3D image is equivalent to one sample to be used either for training or testing (Fig. 1).

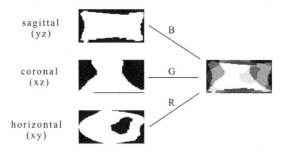

Fig. 1. Patches from the sagittal, axial, and horizontal planes are merged into a BGR image

For every vertebral body, the algorithm iterates through steps 2 and 3. The voxel coordinate is moved by 2 pixels to each of the planes, until it reaches the probable end of the vertebral body. The data is augmented through rotation of the 3D image through $-30°$ to $30°$, with an interval of $5°$. Each BGR image produced is vertically flipped to double the sample data. In total, this produces 13,400 vertebra samples, with the following classifications: Normal (4602), Mild Wedge (1384), Moderate Wedge (3222), Severe Wedge (1454), Mild Biconcavity (570), Moderate Biconcavity (1172), Moderate Crush (996).

3   Methods

After the data pre-processing step, we build a 3D convolutional network based on ResNet-152 which takes 3-channel images as input. A residual learning framework brings ease of training compared to other convolutional networks by adding a +x to the traditional networks' activation which is y = f(x), allowing the gradient to pass backwards directly. This formulation of y = f(x) + x can be realized in the shortcut connections of feed forward networks. The difference is that, ResNet's shortcuts perform identity mapping and adding its output to the outputs of the stacked layers [2]. Figure 2 shows a visualization of the shortcut operation in a feed forward network.

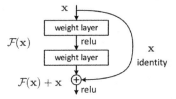

Fig. 2. Residual learning: a building block [2]

3.1   Layers and Architecture

A generic ResNet is composed of Batch Normalization, Rectified Linear Unit, Max Pooling, Convolutional Layers and Fully Connected. ResNet-152, or Deep Residual Network with 152 Layers is implemented for this research. The ResNet used in this research is largely based on the Residual Network implementation for ImageNet. The network accepts 3-channel images as input and are resized to 244×244, with its shorter side randomly sampled in [256, 480] for scale augmentation. Batch Normalization is done after each convolution and before activation. Training is done from scratch since the data on hand is 3-dimensional and therefore needed to be augmented to fit the requirements of 3-channel Residual Net. The learning rate is 0.0001. Stochastic Gradient Descent is used with mini-batch size of 256. The last layer is a single fully connected layer with an output vector of size 1000.

1. **Convolutions.** Convolution operations are applied to input volumes to form a 3-dimensional arrangement of neurons arranged by height, width and depth. Neurons are connected to a region of the input volume called a receptive field of size $N \times N$. Each neuron therefore has $N \times N \times 3$ weighted inputs. A convolution operation takes a kernel on an input image and transforms information encoded in pixels. A specified kernel size is convolved through the entire input image forming "activated regions" [3].
2. **Pooling.** Pooling reduces the amount of parameters and the computations needed to be executed in the network. This is also employed to reduce overfitting [3]. Max Pooling is used after the first batch of convolution operation. A 3×3 sliding window is used and for every channel the values will be down-sampled according to the results of the max pooling operation. Average Pooling is used at the end of all the residual blocks with sliding window of size 7.
3. **Batch Normalization.** Batch Normalization is an operation used in deep neural networks that allows the usage of higher learning rates. It is a step between each layer where the output of the previous layer is normalized first before the next layer [4]. Batch Normalization is used every after convolution operation in this research. The learning rate used is 0.0001.
4. **Rectified Linear Unit (ReLU).** ReLU is the most common activation function for the outputs of CNNs used in deep learning. It has an equation $max(0,x)$ [3]. All residual blocks in this research uses ReLU.

5. **Fully Connected Layer.** Neurons of the fully connected layer have full connections to all the activations from the previous layer [3]. It is used to construct the desired number of outputs and therefore usually placed on the last layers of the network. This research used only one Fully Connected layer after average pooling operation. It is a single vector with 1000 outputs.

ResNet-152 Architecture

Layer	Output Size	Num. of Outputs	Kernel Size	Stride	Padding
conv1	112x112	64	7	2	3
bn_conv1					
scale_conv1					
conv1_relu					
max pooling	56x56		3 (sliding win)	2	
res_2a	56x56	64	1	1	0
bn_2a					
scale_2a					
res_2a_relu					
res_2b	56x56	64	1	1	1
bn_2b					
scale_2b					
res_2b_relu					
res_2c	56x56	256	1	1	0
bn_2c					
scale_2c					
res_2c_relu					
res_3a	28x28	128	1	1	1
bn_3a					
scale_3a					
res_3a_relu					
res_3b	28x28	128	3	1	1
bn_3b					
scale_3b					
res_3b_relu					
res_3c	28x28	512	1	1	0
bn_3c					
scale_3c					
res_3c_relu					
res_4a	14x14	256	1	1	0
bn_4a					
scale_4a					
res_4a_relu					
res_4b	14x14	256	3	1	1
bn_4b					
scale_4b					
res_4b_relu					
res_4c	14x14	1024	1	1	0
bn_4c					
scale_4c					
res_4c_relu					
res_5a	7x7	512	1	1	0
bn_5a					
scale_5a					
res_5a_relu					
res_5b	7x7	512	3	1	1
bn_5b					
scale_5b					
res_5b_relu					
res_5c	7x7	2048	1	1	0
bn_5c					
scale_5c					
res_5c_relu					
average pool	1x1		7	1	
fully connected	1x1	1000			
softmax	1x1000				

Building block annotations: ResNet Building Block 1 (res_2) x3; ResNet Building Block 3 (res_3) x8; ResNet Building Block 4 (res_4) x36; ResNet Building Block 5 (res_5) x3

Fig. 3. ResNet with 152 layers is used to create models and classifiers for vertebra fracture classification problem. It is composed of 4 building blocks that are stacked together

6. **Residual Block.** This research uses the ResNet architecture that is composed of 4 building blocks. Each building block is composed of 4 groups of 4-layers of different operations stacked together. In Fig. 3 we refer to these 4 groups as, ResNet Building Block 1, ResNet Building Block 2, ResNet Building Block 3, and ResNet Building Block 4. After the first group, an identity mapping is applied where the output of the previous group is added to the new output. This identity mapping allows the "shortcut" connections. Each of the four ResNet Building Blocks is stacked on top of the same ResNet Building Block design 3, 8, 36, and 3 times, respectively. ResNet Building Block 1 is stacked together 3 times. This will attach to ResNet Building Block 2 which is stacked together 8 times. This is attached to ResNet Building Block 3 which is stacked together 36 times. This is attached to ResNet Building Block 4 which is stacked together 3 times. The number of outputs and the parameters used for each layers stacked and a visual representation of the ResNet design are shown in Fig. 3.

4   Results and Discussion

The network was trained up to 3500 iterations. Weights are saved every 100th interval from the 1st iteration to the 3500th iteration. A randomly selected 700 validation data was fed to every model saved. From the 3000th iteration, only the 3500th iteration was saved coming up with 29 weights saved, thus yielding 29 validation results. Each of the 7 classifications had 100 test samples each. The testing was done with batch size 100, therefore arriving with 7 test results.

Figure 4 suggests that ResNet-152 presents 93.29% overall classification accuracy for the validation data using the model trained at 3000 iterations. As the number of iterations increases, validation accuracy also increases.

Training (testing) accuracy is the accuracy when a model is applied on the training (testing) data. Figure 5 shows that training accuracy is generally higher than the testing accuracy with an average of 0.092 difference.

Figure 6 shows the predicted labels of vertebra fracture classification after 700 randomly selected and balanced test data was applied to the classifier. The classifier used was trained on 3000 iterations. The classifier was able to correctly predict 99 true normal 3D vertebra images. Among all the classes, normal class got the highest correct predictions followed by moderate crush and moderate wedge.

Figure 7 shows comparison between the predictions given by ResNet-152 trained on 3000 iterations. Normal class got the most correct predictions and the least number of predicted as false moderate wedge. It can also be inferred that 5 out of 7 classes tend to be falsely classified as normal. This includes mild wedge, moderate wedge, severe wedge, mild biconcavity, and moderate biconcavity. Such trend can be a result of the number of training data set available for normal 3D vertebra images amounting to 4 502, the largest among all the classes. In majority, all 3D vertebra images in the test dataset ended up being predicted correctly.

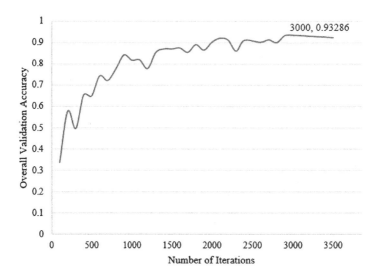

Fig. 4. Trend of the classification accuracy for 700 validation data fed to 29 models trained on different iterations from 100 to 3500

Fig. 5. Training accuracies and testing accuracies when 12700 training data and 700 validation data were fed to 29 models trained on different iterations from 100 to 3500

Vertebra Fracture Class	normal	mild wedge	moderate wedge	severe wedge	mild biconcavity	moderate biconcavity	moderate crush
normal	99	0	1	0	0	0	0
mild wedge	5	94	0	1	0	0	0
moderate wedge	4	0	96	0	0	0	0
severe wedge	6	0	0	94	0	0	0
mild biconcavity	10	3	4	0	83	0	0
moderate biconcavity	7	0	4	0	0	89	0
moderate crush	0	0	2	0	0	0	98

Fig. 6. 7-way classification confusion matrix of lumbar vertebra fracture classes

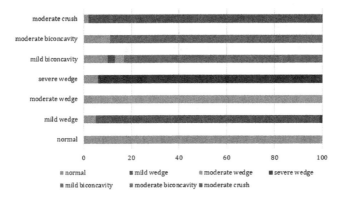

Fig. 7. Distribution of class predictions given by ResNet-152 trained on 3000 iterations

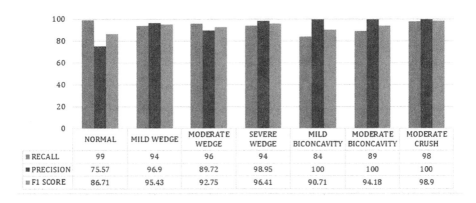

	NORMAL	MILD WEDGE	MODERATE WEDGE	SEVERE WEDGE	MILD BICONCAVITY	MODERATE BICONCAVITY	MODERATE CRUSH
■ RECALL	99	94	96	94	84	89	98
■ PRECISION	75.57	96.9	89.72	98.95	100	100	100
■ F1 SCORE	86.71	95.43	92.75	96.41	90.71	94.18	98.9

Fig. 8. Evaluation measures using recall, precision and f1-score per lumbar vertebra classification

Figure 8 shows the Recall, Precision and F1 Scores for all seven lumbar vertebra fracture classes. The ResNet-152 trained on 3000 iterations returned 100% predictions of Mild Biconcavity, Moderate Crush and Moderate Biconcavity. This means that out of all the 3D Vertebra Images predicted as moderate biconcavity, mild biconcavity, or moderate crush, 100% of it are truly moderate biconcavity, mild biconcavity, or moderate crush. It can be seen from Fig. 6 that there were a total of 98, 89, and 83 3D vertebra images classified as moderate crush, moderate biconcavity, and mild biconcavity consecutively and 98, 89 and 83 of them are truly moderate crush, moderate biconcavity, and mild biconcavity. Normal and Moderate Wedge scored lowest in precision. There were 131 and 101 3D vertebra images classified as normal and moderate wedge. 32 and 11 of these two were wrongly classified as such. Normal scored highest in recall measure with 99. It can be deduced that out of the all the 100 normal 3D vertebra Images, 99 of them were classified as normal. Mild biconcavity scored lowest in recall with 84.

Among the 700 testing data, the ResNet-152 model trained on 3000 iterations has identified a total number of 653 correct classifications. This shows that the automated model has an overall accuracy of 93.29%.

5   Conclusion

The research proposes an automated method to detect and classify vertebra fractures from 3D CT Lumbar Spine Images. This system is limited to 7 classifications only, namely the normal, mild-wedge, moderate-wedge, severe-wedge, mild-biconcavity, moderate-biconcavity, and moderate-crush. The techniques on handling and preparing data for the network is also included in this paper. The architecture followed the ResNet-152, or the Deep Residual Network with 152 layers. ResNet-152 resulted to 93.29% overall testing accuracy for the model trained at 3000 iterations. We have achieved these results by using extracting multiple 3D RGB images from 3D Greyscale CT Images of the lumbar spines.

References

1. Genant, H.K., van Wu, C.Y., Kuijk, C., Nevitt, M.C.: Vertebral fracture assessment using a semiquantitative technique. J. Bone Miner. Res. **8**(9), 1137–1148 (1993)
2. He, K., Zhang, X., Ren, S., Sun, J.: Deep residual learning for image recognition. ArXiv e-prints, December 2015
3. Karphaty, A.: Cs231n convolutional neural networks for visual recognition. ArXiv e-prints (2015)
4. Loffe, S., Szegedy, C.: Batch normalization: accelerating deep network training by reducing internal covariate shift. ArXiv e-prints 18, 1484–1496 (2015)

Evaluation of Similarity Measuring Method of Human Body Movement Based on 3D Chain Code

Truong Hong Ngan Pham[1(✉)], Teruhisa Hochin[2],
and Hiroki Nomiya[2]

[1] Graduate School of Science and Technology,
Kyoto Institute of Technology, Kyoto, Japan
pthngan@ctu.edu.vn
[2] Kyoto Institute of Technology, Kyoto, Japan
{hochin,nomiya}@kit.ac.jp

Abstract. This paper experimentally evaluates the method for measuring human body movement similarity considering the human body motion as a set of 3-dimensional curves representing the changes of direction. After the direction sequences describing the motion paths are derived, an improved method inspired by the dissimilarity measure of the 3D curves is used for measuring the similarity of human body movements. The tests for statistical significance is used in order to evaluate the efficiency of measure in the experiments. The experimental results show that the proposed method successfully measures the similarity of human body movements for the low complexity dataset and the similar motion types.

Keywords: Space harmony · 3D chain code · Spatial quantization
Labanotation · Dissimilarity measure of 3D curves

1 Introduction

The analysis of human body movement involves many areas and has a wide range of applications. In clinical applications, the performing human motion analysis goes from comprehension of the multiple mechanisms that translate muscular contractions about articulated joints into improving functional stability [16]. In the skills, the human body movement is analyzed. These include the analysis of the movement of craft works [7], the prototype Taekwondo training system using a hybrid sensing technique of a body sensor (accelerometer) and a visual sensor [8], and a method for deriving pause in traditional skills for obtaining fundamental movements [6].

To accomplish the analysis of human body motion for such different purposes, different approaches are used and developed to suit their specific requests: High-speed data acquisition for sports, or real time tracking for virtual reality applications restricting the delay time between data capture and simulation. Generally, those approaches include motion analysis data collection protocols [3], estimating precision [15], and modeling movements for reducing data [7].

© Springer International Publishing AG, part of Springer Nature 2018
N. T. Nguyen et al. (Eds.): ACIIDS 2018, LNAI 10752, pp. 459–471, 2018.
https://doi.org/10.1007/978-3-319-75420-8_44

The motion trajectory shapes of the human body parts in movements were concentrated on analyzing in order to distinguish the human body movement [12]. The motion trajectory shapes of the human body parts are 3D curves. The similarity values between human body movements are calculated by the 3D chain codes of these 3D curves without measuring limbs for calculating links and joints. The adjusting method for the movement speeds and the normalization method of human pose in motion data are introduced. The weight values of each human body parts in movement are estimated and are used to calculate similarity values.

This paper experimentally evaluates the method described above. The statistical significance is used to evaluate the efficiency of the measurement in the experiment. The results show that the proposed method successfully measures the similarity of human body movements for the low complexity dataset and the similar motion types.

The remaining part of this paper is as follows: Sect. 2 describes the related works used on this research. Section 3 briefly describes the similarity measure method for human body movement. Section 4 describes the experiment conducted to test the method described in Sect. 3. In Sect. 5, discussions are presented. Finally, the summary is described in Sect. 6.

2 Related Works

The similarity measuring method of human body movement [12] is based on the chain code method [1, 4, 5], the spatial quantization in human body movement [13], space harmony theory [9], the dissimilarity measure for 3D curves [2, 10], and the similarity measure [14]. Here, we briefly describe the chain code method, the space harmony theory, the definition of the partial common couple for 3D chain code, and the split value of the similarity measuring method because of space limitation. Please refer to the literature [12] for the other works.

2.1 The Chain Code Method

The chain code method is a popular and important technique in many fields: computer vision, image processing, pattern recognition, and automated cartography in geographic information systems. The first approach called Chain Coding to represent digital curves is proposed by Freeman [4]. The method quantizes and encodes the two-dimension (2D) figure. A uniform mesh superimposes the line. The represent points of the line drawing are mesh nodes, which are close to the line drawing in terms of the quantization schemes. The movable directions of a curve point are expressed by eight code values. A direction code represents a line segment in the approximation according to the segment direction. In short, a sequence of direction codes represents a 2D line drawing.

Freeman [5] proposed another chain code method extending the chain coding scheme to 3D line structures quantized on a cubic lattice in order to represent 3D digital curves in 1974. The movable directions of a curve point were extended from eight directions to 26 directions.

Bribiesca [1] presented a 3D chain code technique using five orthogonal direction changes for a digitalizing 3D continuous curve as a 3D discrete curve. Five possible direction changes representing any 3D discrete curve were proposed in the method. Three continuous straight-line segments constituted a chain code element.

2.2 Space Harmony Theory

Laban and Ullmann [9] proposed the movement theory and practice well-known as Space Harmony or Choreutic. Space Harmony includes many aspects. The similarity measuring method concentrates on Kinesphere and Labanotation.

The space around the human body whose limbs can reach without changing the position of a person is called Kinesphere. The most proximal skeletal joint, which is articulated, is sometimes considered as the center of the movement space.

Laban created a symbol system in order to represent the movable directions of dancer body, and known as a notation system or Labanotation. Labanotation has 26 directions, which are derived from the vertices of the Octahedron, the Icosahedron, and the Cube. The system is used for describing where a person is moving in space.

2.3 Definition of the Partial Common Couple for Chain Code

The similarity measuring method of human body movement [12] are based on the measure of shape dissimilarity for 3D curves [2]. In order to calculate the similarity value of two motions, the similarity value and the dissimilarity value of motion paths of corresponding marker pairs are needed to be calculated. As these motion paths of corresponding marker pair are represented by 3D chain codes, all partial common couples of corresponding marker pair chain codes can be found. The definitions of the Partial Common Couple [2] is presented as follows:

Definition 1. Subchain.

Let $A = a_1 a_2 \ldots a_m$ be a chain. A subchain S of A with initial index i is defined as:

$$S = a_i a_{i+1} \ldots a_n \text{ such that } 1 \leq i \leq m \text{ and } n \leq m$$

Definition 2. Maximum common subchain.

Let $A = a_1 a_2 \ldots a_m$ and $B = b_1 b_2 \ldots b_n$ be two chains. A maximum common sub-chain S of A and B with initial indices q and r respectively, satisfy the following conditions:

(i) $1 \leq L(S) \leq min(m, n)$
(ii) S is a subchain of A and B, with initial indices q and r respectively.
(iii) $a_{q-1} \neq b_{r-1}$ and $a_{q+L(S)} \neq b_{r+L(S)}$

where $L(S)$ is the length of the subchain, and $1 \leq q \leq m$ and $1 \leq r \leq n$

Definition 3. Left maximization.

Let $A = a_1 a_2 \ldots a_m$ and $B = b_1 b_2 \ldots b_n$ be two chains. Let S be a subchain of A and B, with initial index i in A and initial index j in B, such that:

$$A = a_1a_2\ldots a_ia_{i+1}\ldots a_{i+L(S)-1}\ldots a_m \; where, a_i = s_1, a_{i+1} = s_2, \ldots, a_{i+L(S)-1} = s_{L(S)}$$

$$B = b_1b_2\ldots b_jb_{j+1}\ldots b_{j+L(S)-1}\ldots b_n \; where, b_j = s_1, b_{j+1} = s_2, \ldots, b_{j+L(S)-1} = s_{L(S)}$$

The function lmax is defined as:

$$lmax(A, B, S) = a_ka_{k+1}\ldots a_ia_{i+1}\ldots a_{i+L(S)-1} = b_{k'}b_{k'+1}\ldots b_jb_{j+1}\ldots b_{j+L(S)-1} = u_{lmax}$$

where u_{lmax} is a maximum subchain of $A' = a_1a_2 \ldots a_{i+L(S)-1}$ and $B' = b_1b_2 \ldots b_{j+L(S)}$ -1.

Definition 4. Partial Common Couple.

Let $A = a_1a_2 \ldots a_m$ and $B = b_1b_2 \ldots b_n$ be two chains. Let P' be a common subchain of A and B with initial indices i'_1 and i'_2 respectively, and $L(P') = k'$, for $1 \leq k' \leq min(m, n)$.
$$P' = p'_1p'_2 \ldots p'_{k'}$$
Let $P = lmax(A, B, P')$, with left maximization index in A i_1 and left maximization index in B i_2.

$$P = p_1p_2\ldots p_k = a_{i_1}a_{i_1+1}\ldots a_{i_1+L(p)-1} = b_{i_2}b_{i_2+1}\ldots b_{i_2+L(p)-1}$$

Let S be a subchain of A with initial index 1 and $L(S) = i_1 + L(P) - 1$.

$$S = a_1a_2\ldots a_{i_1}a_{i_1+1}\ldots a_{i_1+L(p)-1} = a_1a_2\ldots p_1p_2\ldots p_k$$

Let T be a subchain of B with initial index 1 and $L(T) = i_2 + L(P) - 1$.

$$S = b_1b_2\ldots b_{i_2}b_{i_2+1}\ldots b_{i_2+L(p)-1} = b_1b_2\ldots p_1p_2\ldots p_k$$

For convenience, the couple (S, T) will be named partial common couple.

2.4 Split Value

In the similarity measuring method of human body movement [12], the active space of the human body is superimposed on by a rectangular lattice, and the kinespheres of the human body main joints are superimposed on by the cubic lattices. In the other words, these spaces are split into multiple sub-spaces. These sub-spaces are cubes, and they have the same size. The split value is the maximum number of the sub-spaces on a dimension of the active space or the kinesphere. In a rectangular lattice, the sub-space number of the largest dimension in three dimensions is the split value, and the sub-space number of the remaining dimensions are based on the ratios of these dimensions to the largest dimension. In a cubic lattice, the sub-space number of all dimensions are the same, and they are equal to the split value.

3 Method for Human Body Movement Similarity Analysis

The method for human body movement similarity analysis without identifying joints was proposed [12]. Each marker motion path is represented by a 3D curve. After extracting data from c3d file, the speeds of the movements are adjusted, and the movement is normalized. After that, the 3D coordinate data of markers continue to go through the four-step process as follows.

3.1 Calculating Motion Direction Sequences of Markers

The method uses 26 alphabet characters to represent 26 possible motion directions as shown in Fig. 1 and the minus sign ("−") for the unmoved state. By comparing the values of two consecutive data frames from the first frame to the last frame of data, the description strings of moving directions of markers are obtained. After getting a direction sequence, the minus sign "−" representing the stop value is eliminated from the results. These direction sequences of the whole space normalization are further converted to the chain codes. The weight values are calculated by the direction sequences of the subspace normalization. The detail is presented in the Sect. 3.4.

3.2 Converting Direction Sequences to 3D Chain Codes

The extending chain code inspired by Bribiesca's work [2] for representing a 3D motion path will be presented in this section. A chain code element is identified by three consecutive direction values of the direction sequence, say a, b and c, and is represented by three vectors, say u, v and w as shown in Fig. 2.

There are 130 relative positions of u, v, and w [12]. Total chain code element quantity is 130. The code value is calculated by Eq. (1):

$$code(\alpha, \beta, \gamma) = part(\alpha) \times 26 + part(\beta) \times 8 + part(\gamma)(equal(equal(part(\alpha), 4)$$
$$+ equal(part(\gamma), 4), 2) + greater(part(\beta), 0)) \times 7 \tag{1}$$

Here α, β, and γ are angles, part (N) is a function that returns the integer part of dividing N by 45, equal(A, B) is a function that returns 1 if A is equal to B, and returns 0 if A is not equal to B, greater(A, B) is a function that returns 1 if A is greater than B, and returns 0 if A is not greater than B.

3.3 Calculating Dissimilarity Values Between Motions of Corresponding
      Marker Pairs

Firstly, the common parts of sub-curves are found [12]. Then the dissimilarity of them are calculated. Next, the sub-curves are sorted in ascending order by the dissimilarity values. If overlapping occurs, the sub-curves with lower dissimilarity will be kept. The retained sub-curves are stored in a *selected list*. Finally, dissimilarity between motions of corresponding marker pairs is the sum of the dissimilarity values of all common couples and the different value of elements left without correspondence.

Finding Motion Similarities. 3D curves are represented by changes of direction. This means that a sub-curve that could be found in a 3D curve has the same representation regardless its position and orientation. The sub-curve's first element index within the 3D curve indicates the sub-curve position. The first element of the common sub-curve is needed to be defined by the first two elements preceding the sub-curve, which represent the two changes of direction. The sub-curve can add imaginary elements if the common sub-curve begins at indices one or two of the 3D curve.

Finding Dissimilarities. In case the longest common-sub-curves for two 3D curves overlap each other, the dissimilarity measure can be used as a selecting parameter for choosing the best matching. The method is based on the sub-curve length, the starting element index of the sub-curve, the element's number needed to define the first element of the sub-curve, and the accumulated direction, which is the final orientation of the sub- curve after affected by all its preceding chain elements.

After finding all the common couples between two 3D curves, Eq. (2) calculates the dissimilarity of sub-curves S and T. It is used as a sorting criterion. The common couples with lower dissimilarity values will have higher rank in the sorted list.

$$d(S,T) = \begin{bmatrix} 0, & when\ S = T \\ \frac{|L(S)-L(T)|}{max(L(S)-L(T))-1} + (min(m,n) - L(P) + \Delta(S,T)), & when\ S \neq T \end{bmatrix} \quad (2)$$

where m and n are the lengths of curve A and curve B, S and T are sub curves of A and B, P is the common part of S and T, $L(P)$, $L(S)$, $L(T)$ are the lengths of P, S, and T, $\frac{|L(S)-L(T)|}{max(L(S)-L(T))-1}$ is the displacement of the two common sub-curves within their respective curves, $min(m, n) - L(P)$ is the measure how large is the common-sub-curve (P) with respect to the 3D curve where it is contained, and $\Delta(S, T)$ is a pseudo-metric of accumulated direction.

The dissimilarity normalized to the range [0, 1], d'(S, T), is obtained by dividing d(S, T) by m + n.

The dissimilarity measure of motion for a corresponding marker pair between two human body movements whose curves are A and B is obtained by Eq. (3):

$$D(A,B) = \sum_{r=1}^{l} d'(s_r, t_r) + m + n - \left(\sum_{r=1}^{l} 2L(P'_r) \right) \quad (3)$$

where $d'(s_r, t_r)$ is the dissimilarity value of the rth sub-curves corresponds to the rth partial common sub-curve in the selected list, l is the total number of partial common sub-curves in the selected list, P'_r is the rth partial common sub-curve in the selected list, $L(P'_r)$ is the length of P'_r, and m and n are the lengths of curves A and B.

The dissimilarity normalized to the range [0, 1], D'(A, B), is obtained by dividing D(A, B) by m + n.

3.4   Calculating Dissimilarity Values Between Movements

These weights are determined by the direction sequences of subspace normalization as shown in Eq. (4):

$$w_k = \frac{min(L(A_k), L(B_k))}{\sum_{r=1}^{l} min(L(A_r), L(B_r))} \times l \tag{4}$$

where w_k is the weight of the kth marker in both two movements, $L(A_r)$ and $L(B_r)$ are the direction sequence lengths of the rth markers in two movements, and l is the number of markers.

Finally, the dissimilarity value of two movements whose curves are A and B is calculated by Eq. (5):

$$Dissimilarity(A, B) = \frac{\sum_{r=1}^{l} w_r \times D'_r}{l} \tag{5}$$

where w_r is the weight of the rth marker in movements, D'_r is the motion dissimilarity value of the rth corresponding marker pair, and l is the number of markers.

4 Experiments

4.1 Experimental Data

To evaluate the proposed method, the Basic Walking dataset from the Karlsruhe Institute of Technology whole-body human motion database (KIT-Database) is chosen [11]. The number of the selected subjects is eight. The selected subjects for the experiment include subject 4, subject 6, subject 7, subject 8, subject 9, subject 10, subject 11, and subject 12. These subjects are required to perform six kinds of actions. The selected actions include walking 8 steps in clockwise circle (a1-walking in clockwise circle), walking 8 steps in counter-clockwise circle (a2-walking in counter-clockwise circle), walking 4 steps straight backwards (a3-walking straight backwards), walking 4 steps straight forwards (a4-walking straight forwards), walking 4 steps and turn right (a5-walking and turn right), and walking 4 steps and turn left (a6-walking and turn left). However, these subjects did not perform all kinds of actions. Subject 4 did not do "walking and turn left", Subject 6 did not do "walking straight forwards", Subject 8 did not do "walking and turn left", and Subject 10 did not do "walking straight forwards". The total number of motions is 44. With each split value, each motion is measured each of the other motions exactly once. The total number of dissimilarity values per each split value is $C_{44}^2 = \frac{44!}{40! \times 2!} = 946$. The number of the selected markers corresponding to the main human body parts is 26.

4.2 Evaluation Method

Tests for statistical significance are used in order to prove that the method reflects the differences between the various groups of human body movements. The one-tail testing is chosen for calculating the statistic. The conditions were as follows:

- The hypotheses:
  - The null hypothesis (H_0): There is no difference in the dissimilarity values of movements.
  - The alternative hypothesis (H_a): The dissimilarity value of movements in the same action is lower than the dissimilarity values of motion in different types.
- The significance level is 0.05 ($\alpha = 0.05$).
- The dissimilarity values are split into two groups. The dissimilarity values between the motions in the same action are considered as a group, the other motions are in the other group.

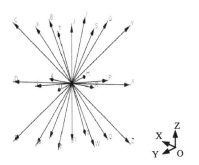

Fig. 1. Symbols represent for 26 directions

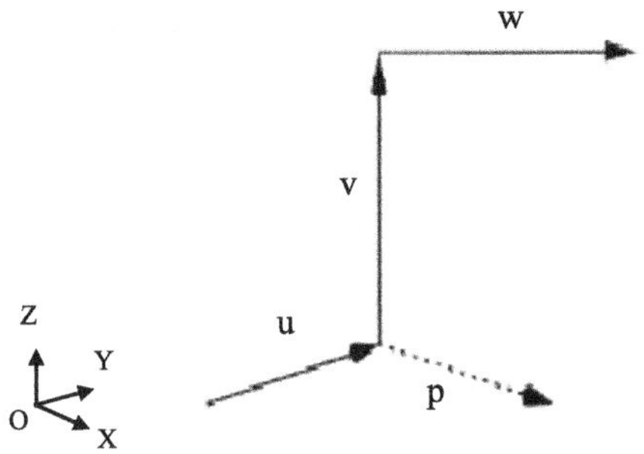

Fig. 2. Three vectors representing a chain code element

4.3 Condition for Selecting Split Values

Difference tests were conducted to find the suitable split values. The selected split value has to meet the following requirements:

1. The p-values of all actions are less than the significance value.
2. The smaller average of dissimilarity values in the same actions is better.
3. The shorter conducted time of experiments is better.

4.4 Experimental Method

As presenting in Sect. 3.4, the method uses the length of a chain code converting from subspace quantization type for calculating the weight values. In addition, the efficiency when using weight values should also be checked. Therefore, the following experiments are conducted:

- The experiment aims to find the split value for calculating the weight values. For conducting experiments, the tested split values are 4, 6, 8, 12, 16, 20, and 24.
- The experiment aims to find the split value for calculating dissimilarity values without using the weight values. For conducting experiments, the tested split values are 10, 20, 40, 60, 70, and 200.

- The experiment aims to find the split value for calculating dissimilarity values using the weight values. For conducting experiments, the tested split values are 10, 20, 40, 60, 70, and 200.

4.5   Experimental Results

The conducted times. The conducted time has a relationship with the Split Value. When the split value increases, the conducted time increases as well. Figure 3 provides a graphical description of the conducted time of the experiments using weight values of all action types.

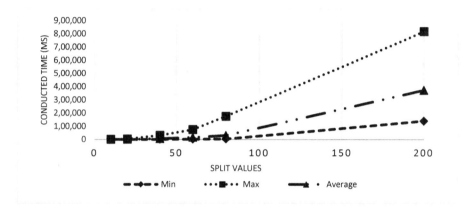

Fig. 3. The conducted times of the experiments using weight value of all action types

Finding Split Values of Subspaces Quantization Type. The split value 24 is the only split value, which is met the first requirement. Due to the length of the chain code is increased when the split value is increased; the conducted times for calculating dissimilarity values are also increased. With the split value is 24, the consuming time is smallest among the split values that the p-values of all actions are less than 0.05. As the result, the selected split value is 24. Figure 4 provides a graphical description of the averages of dissimilarity values.

Finding Split Values for Calculating Dissimilarity Values Without Using Weight Values. The p-values of all actions are less than 0.05. In this case, the averages of split values are considered. Clearly, the trends of the average of dissimilarity values can be split into two groups: the increasing trend group includes actions 1, 2, 5, and 6 and the fluctuation trend group includes actions 3 and 4.
According to the results as shown in Fig. 5, we have two options:

- Using multiple split values for actions. The split values, which have the smallest averages of dissimilarity values, are selected: the split value for actions 1 and 2 is 10, the split value of actions 3 and 4 is 40, and the split value of actions 5 and 6 is 10.

- Using a single selected split value for all actions. The selected split value must be ensured the balance between the conducted time and the magnitude of the average value. There are two smallest split values for two groups. If the selected split value is 10, the dissimilarity values of actions 1, 2, 5, and 6 are smallest. However, the dissimilarity values of actions 3 and 4 are high. If the selected split value is 40, the dissimilarity values of actions 3 and 4 are smallest but the other actions are higher. The execution time of comparing is also longer when the split value is 40. Therefore, the selected split value is 20. The average values of all actions are almost equally.

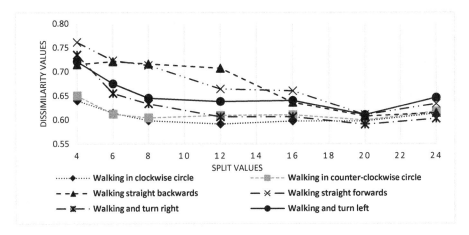

Fig. 4. The averages of dissimilarity values with subspace quantization type of all action types

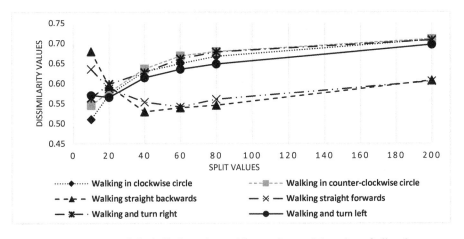

Fig. 5. The averages of dissimilarity values without using weight value of all action types

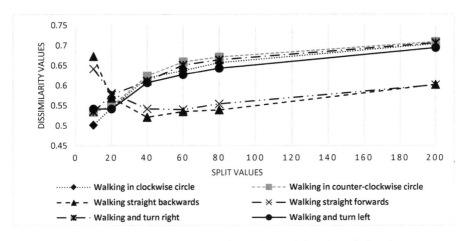

Fig. 6. The averages of dissimilarity values using weight value of all action types

Finding Split Values for Calculating Dissimilarity Values Using Weight Values.
The split value of subspace quantization type is 24, which is used to calculate the
weight values. The p-values of all actions are less than 0.05. In this case, the averages
of split values are considered. The split value, which has the smallest averages of
dissimilarity values, is selected. Similar to the above experiment, the trends of the
average of dissimilarity values can be split into two groups: the increasing trend group
includes actions 1, 2, 5, and 6, and the fluctuation trend group includes actions 3 and 4.

According to the results as shown in Fig. 6, we have two options:

- Using multiple split values for actions. The split value for action 1 and action 2 is
  10, the split value of action 3 and 4 is 40, and the split value of action 5 and 6 is 10.
- Using a single selected split value for all actions. With the same selection method
  described in Sect. 4.3, the selected split value must be ensured the balance between the
  execution time and the magnitude of the average value, the selected split value is 20.

Comparing Two Dissimilarity Measure Methods. The p-values of two methods are
less than 0.05. Therefore, they can be used to calculate the dissimilarity values between
human motions. The method using the weight values has the smaller averages. Only
one average value of 36 average values of the method using the weight values is larger
than the average values of the method without using the weight values. The results
prove that our proposed method matches in calculating the dissimilarity values of
human body movements.

5 Discussions

The split value of spatial quantization plays an important role in the similarity measuring method. This parameter affects the accuracy of the direction sequence. Besides,
the split parameter value affects the length of chain code. When the split value

increases, the length of the chain code also increases. Therefore, the processing time increases. The action types of a dataset affect the threshold value. The complexity of a dataset depends on the purpose of the application. The applications, where the subjects move in a narrow range or unmoved, have the simple and similar sample dataset. Conversely, the applications, in which the subjects move in a wide range, have complex sample dataset. The threshold value to determine whether two movements are the same depends on the data type and the purpose of the application.

The research aims to measure the similarity of human body movements. The experiment results with the low complexity dataset and the similar movement types proved that the proposed method achieved the requirements. The dissimilarity values in the same action are smaller than those are in the different actions.

6 Conclusions

A method for measuring the similarity of human body movements without representing human body was evaluated. The set of markers for comparison does not need to be specified. The roles of markers are expressed through weight values. It is, however, necessary to determine the appropriate split value for each type of action in order to achieve accurate result. The statistical significance result shows that the method is effective for evaluating the similarity of human body movement. The experimental results show that the combination of the weight value and the adjusting speed method increases the accuracy.

The method only works well on the good dataset, without errors. Due to the conducted time is not optimized; the method cannot be used in real time applications.

References

1. Bribiesca, E.: A chain code for representing 3D curves. Pattern Recogn. **33**(5), 755–765 (2000)
2. Bribiesca, E., Aguilar, W.: A measure of shape dissimilarity for 3D curves. Int. J. Contemp. Math. Sci. **1**(15), 727–751 (2006)
3. Davis, R.B., Davids, J.R., Gorton, G.E., Aiona, M., Scarborough, N., Oeffinger, D., Tylkowski, C., Bagley, A.: A minimum standardized gait analysis protocol: development and implementation by the Shriners Motion Analysis Laboratory network (SMALnet). In: Pediatric Gait: A New Millennium in Clinical Care and Motion Analysis Technology, pp. 1–7 (2000)
4. Freeman, H.: On the encoding of arbitrary geometric configurations. IRE Trans. Electron. Comput. **EC-10**(2), 260–268 (1961)
5. Freeman, H.: Computer processing of line-drawing images. ACM Comput. Surv. (CSUR) **6**(1), 57–97 (1974)
6. Hochin, T., Nomiya, H.: Deriving pauses for obtaining fundamental movements in traditional skills. In: Lee, R.Y. (ed.) Applied Computing and Information Technology. SCI, vol. 553, pp. 19–30. Springer, Cham (2014). https://doi.org/10.1007/978-3-319-05717-0_2
7. Kume, M., Yoshida, T.: Characteristics of technique or skill in traditional craft workers in Japan. In: Advances in Affective and Pleasurable Design, p. 140 (2012)

8. Kwon, D.Y.: A study on taekwondo training system using hybrid sensing technique. J. Korea Multimed. Soc. **16**(12), 1439–1445 (2013)
9. Laban, R., Ullmann, L.: Choreutics. Macdonald & Evans, London (1966)
10. Lin, D.: An information-theoretic definition of similarity. In: Proceedings of the Fifteenth International Conference on Machine Learning, pp. 296–304. Morgan Kaufmann Publishers Inc., Burlington (1998)
11. Mandery, C., Terlemez, Ö., Do, M., Vahrenkamp, N., Asfour, T.: The KIT whole-body human motion database. In: Proceedings of the International Conference on Advanced Robotics (ICAR), pp. 329–336. IEEE (2015)
12. Ngan, P.T.H., Hochin, T., Nomiya, H.: Similarity measure of human body movement through 3D chaincode. In: Proceedings of 18th IEEE/ACIS International Conference on Software Engineering, Artificial Intelligence, Networking and Parallel/Distributed Computing (SNPD), pp. 607–614 (2017)
13. Okada, Y., Etou, H., Niijima, K.: Intuitive interfaces for motion generation and search. In: Grieser, G., Tanaka, Y. (eds.) Intuitive Human Interfaces for Organizing and Accessing Intellectual Assets. LNCS (LNAI), vol. 3359, pp. 49–67. Springer, Heidelberg (2005). https://doi.org/10.1007/978-3-540-32279-5_4
14. Reyes, F.J.T.: Human motion: analysis of similarity and dissimilarity using orthogonal changes of direction on given trajectories. Dissertation, Department of Computer Science, University of Colorado Colorado Springs (2014)
15. Schmidt, J., Berg, D.R., Ploeg, H.-L., Ploeg, L.: Precision, repeatability and accuracy of Optotrak® optical motion tracking systems. Int. J. Exp. Comput. Biomech. **1**(1), 114–127 (2009)
16. Shull, P.B., Jirattigalachote, W., Hunt, M.A., Cutkosky, M.R., Delp, S.L.: Quantified self and human movement: a review on the clinical impact of wearable sensing and feedback for gait analysis and intervention. Gait Posture **40**(1), 11–19 (2014)

Analysis of Convolutional Neural Networks and Shape Features for Detection and Identification of Malaria Parasites on Thin Blood Smears

Kristofer Delas Peñas[1], Pilarita T. Rivera[2], and Prospero C. Naval Jr.[1(✉)]

[1] Department of Computer Science, College of Engineering,
University of the Philippines, Diliman, Philippines
pcnaval@dcs.upd.edu.ph
[2] Department of Parasitology, College of Public Health,
University of the Philippines, Manila, Philippines

Abstract. The gold standard for malaria diagnosis still remains to be microscopy. However, cases from remote areas needing immediate diagnosis and treatment can benefit from a faster diagnostic process. Several intelligent systems for malaria diagnosis have been proposed using different computer vision techniques. In this research, models using convolutional neural networks, and a model using extracted shape features are implemented and compared. The CNN models, one trained from scratch and the other utilizing transfer learning, with accuracies of 92.4% and 93.60%, both outperform the shape feature model in malaria parasite recognition.

1 Introduction

Malaria is considered to be endemic to 91 countries and territories. In 2015, the World Health Organization (WHO) estimated 212 million cases of malaria globally. Four hundred twenty-nine thousand of these cases resulted to death [1]. With these figures, WHO formed an initiative to actively monitor malaria cases and eventually reduce malaria-endemic countries to 35 by the year 2030.

A critical point in malaria mitigation is the remote location of majority of the cases and the unavailability of personnel trained on microscopy to analyze blood samples for diagnosis. With the emergence of intelligent systems, this can become manageable as diagnosis can be automated with photographs of the blood smears as input.

2 A Review of Malaria

Malaria parasites are protozoans belonging to the genus *Plasmodium*. This genus is further divided into two subgenera: *Plasmodium* and *Laverania*. *Plasmodium vivax*, *Plasmodium malariae*, and *Plasmodium ovale* all belong to the subgenus

© Springer International Publishing AG, part of Springer Nature 2018
N. T. Nguyen et al. (Eds.): ACIIDS 2018, LNAI 10752, pp. 472–481, 2018.
https://doi.org/10.1007/978-3-319-75420-8_45

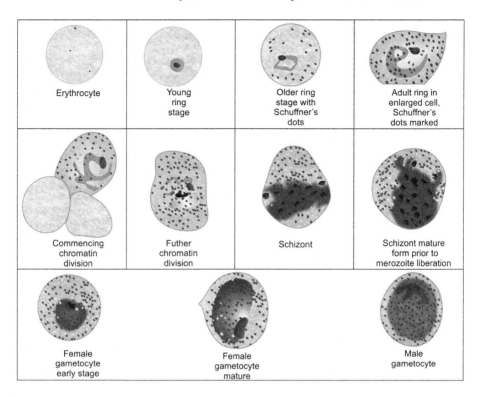

Erythrocyte	Young ring stage	Older ring stage with Schuffner's dots	Adult ring in enlarged cell, Schuffner's dots marked
Commencing chromatin division	Futher chromatin division	Schizont	Schizont mature form prior to merozoite liberation
Female gametocyte early stage	Female gametocyte mature		Male gametocyte

Fig. 1. *Plasmodium vivax* [2]

Plasmodium, while *Plasmodium falciparum* is classified under the subgenus *Laverania* due to its significant differences from the other three species [2]. Figures 1 and 2 show the erythrocytic schizogony of *P. vivax* and *P. falciparum*.

Malaria is often characterized by recurring fever. Aside from fever, a malaria-infected person exhibits the following symptoms:

– chills
– sweating
– headaches
– nausea and vomiting
– body aches

Further physical findings may include increased respiratory rate, enlarged spleen, mild jaundice, and enlargement of the liver [3].

Although other diagnostic methods like molecular biology methods (PCR), and serology and rapid diagnostic testing (RDT) exist, the golden standard in malaria diagnosis is still microscopy on blood smears. A thin smear is used in determining the infecting species. This type of smear is prepared from capillary blood drawn from a fingertip, applied to a slide, and stained with *Giemsa* to highlight the parasite [2]. Figure 3 shows an example of a thin smear.

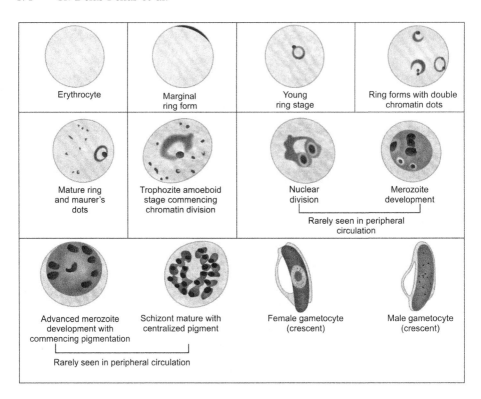

Erythrocyte	Marginal ring form	Young ring stage	Ring forms with double chromatin dots
Mature ring and maurer's dots	Trophozite amoeboid stage commencing chromatin division	Nuclear division	Merozoite development
		Rarely seen in peripheral circulation	
Advanced merozoite development with commencing pigmentation	Schizont mature with centralized pigment	Female gametocyte (crescent)	Male gametocyte (crescent)
Rarely seen in peripheral circulation			

Fig. 2. *Plasmodium falciparum* [2]

In blood smears, ring forms of all Plasmodium species appear with a distinguishable blue cytoplasm and detached nuclear dot(s). Ring forms of *P. vivax*, *P. ovale*, and *P. malariae* appear compact, while those of *P. falciparum* appear fine and with double chromatin [2].

Oval-shaped gametocytes are seen in *P. vivax*, *P. ovale*, and *P. malariae*, while *P. falciparum* appear crescent-shaped [2]. Enlarged red blood cells with *Schuffner's dots* are characteristic of *P. vivax*, while in *P. falciparum*, cells are normal in size but with *Maurer's dots* and basophilic stippling [2].

Treating malaria requires administration of anti-malarial drugs. These drugs are used to prevent relapse, to prevent transmission, and to initiate prophylaxis. The most common anti-malarial drugs are chloroquine, amodiaquin, quinine, pyrimethamine, doxycycline, and artemesinin [2].

The course of treatment is determined by the infecting species. *P. vivax*, *P. ovale*, and *P. malariae* are treated similarly. Chloroquine is administered to the patient over a period of three days. To prevent relaspe, primaquine is also given to the patient for fourteen days [2].

Treatment for *P. falciparum* is different. Chloroquine is also administered at the same dosage, but primaquine is only given once to prevent relapse. In cases of chloroquine resistance, artemesinin is given instead [2].

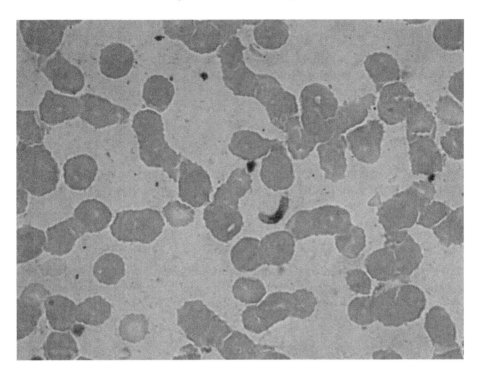

Fig. 3. A thin smear (Color figure online)

3 Malaria Intelligent Systems

Several research on malaria intelligent systems were developed in the past few years. The different approaches utilized by these studies yielded varying degrees of accuracy, but all can generally be regarded as effective identifiers of malaria.

Makkapati and Rao [4] used a segmentation scheme based on the HSV color space to separate the red blood cells (RBC) and parasites and addressed the problems due to variability and presence of artifacts in microscope images of blood samples. Alternative to HSV is the La*b* color space as presented in the work by Nanoti et al. [5]. Their work showed that shape and texture features can be effectively extracted in the a* and b* channels.

Pinkaew et al. [6] explored the use of support vector machines in malaria species classification. Their work used statistical measurements such as mean intensity, standard deviation, kurtosis, skewness and entropy to characterize regions of interest. Training on 40 P. falciparum and 25 P. vivax images, the study's SVM model with radial basis kernel gained an accuracy of 85.71% in identifying P. falciparum and 78.72% in identifying P. vivax.

Delas Peñas et al. [7] developed a convolutional neural network model utilizing transfer learning on Inceptionv3. The research obtained an accuracy of 92.4% in identifying malaria parasites.

4   Shape Features

Shape analysis is the automatic analysis and description of geometric shapes in digital images. Each region of interest in a digital image is characterized by a vector of values representing some measurements. Some of the measurements used in shape analysis are the following:

Area

In shape analysis of a digital image, the area is given by the number of pixels contained in the region. For a binary image, with 1 representing the foreground, the area is equal to the number of pixels whose value is 1 [8].

Convex Area

Convex Area is computed in the same manner as Area except a convex image is used instead of the original. A convex image specifies convex hulls containing parts of the region of interest, with all the holes filled in [8].

Eccentricity

The eccentricity of the region of interest in a digital image is the eccentricity of the ellipse with the same second moment as the region [8].

Equivalent Diameter

Equivalent diameter is the diameter of the circle with the same area as the region [8].

Euler Number

Euler number is the number of connected components minus the number of holes in the components [8].

Extent

Extent is the ratio of the computed area to the area of the bounding box of the region [8].

Major Axis Length

Major axis length is the length of the major axis of the ellipse with the same second moment as the region [8].

Minor Axis Length

Minor axis length is the length of the minor axis of the ellipse with the same second moment as the region [8].

Orientation

The orientation of the region of interest is the angle the major axis of the ellipse with the same second moment as the region makes with respect to the x-axis [8].

Perimeter

Perimeter is the distance around the boundary of the region of interest [8].

Solidity

Solidity is the ratio of the area and the computed convex area.

5   Convolutional Neural Networks

Convolutional Neural Networks or CNNs are very similar to ordinary artificial neural networks having trainable weights on sets of neurons given some inputs and expected outputs. One key difference is that in CNNs, the input are three-dimensional data (height, width and depth) such as images. This additional dimension (depth) would require a restructuring of neurons in ANNs and a huge additional computational complexity, giving rise to another key difference of CNNs and ANNs: the weight of the neurons are shared.

Different architectures over the few years of active research on CNNs were proposed. These architectures mainly differ on the number of layers for each type used, and the parameter values used for each. One of the most recent architectures is Google's Inception. Inception tries to minimize the set of parameters by factorizing convolutions with large filter size into a set of multiple convolutions with small filter sizes. For example, a convolution of 5×5 convolution is replaced by two 3×3 convolutions in Inception. Developers of Inception have shown that this replacement is indeed less computationally complex than the original convolution. With only around 13 million parameters, Inception (version 3) outperformed VGGNet, with only 3.58% top-1 error and 17.2% top-5 error [9].

Training a convolutional network depends greatly on the depth of the network and the hardware used for training. With CPUs alone, training an Inception network takes more than a month to converge. With GPUs, the training could be brought down to two weeks. This complexity in training time is one weak point of using CNNs. However, several research have shown that the first few layers of a network trained on a completely different task could be used for another. Through transfer learning, as researchers call it, only the last few fully-connected layers need to be trained for a particular task. Pre-trained models like Inception were made available online.

6 Methodology

The dataset used is composed of 363 images. The smears used for the images were obtained during a field study done in Palawan, Philippines conducted by researchers of the College of Public Health, University of the Philippines. Images were taken using a digital single-lens reflex camera attached to a microscope with a resolution of 2592 × 1944 pixels at 100X magnification using an oil immersion objective lens. Images of P. vivax (142 images) were included in the dataset to effectively discriminate between P. falciparum (221 images) from this species.

6.1 Data Preprocessing and Segmentation

As demonstrated by Makkapati and Rao [4], the HSV color space is a better fit than the RGB color space for the segmentation task to follow. In this study, the saturation channel was used since it gives the best contrast between parasite and non-parasite regions in the images.

A series of morphological transforms were done for data preprocessing to enhance the digital images before segmentation.

Each image underwent image opening and closing. This step ensures that salt-and-grain noise that would otherwise register as candidate regions were removed. After which, the mean and standard deviation of each image is computed and subtracted from the image. Lastly, binary thresholding was done, followed by connected components analysis to extract the regions of interest for classification. Figure 5 shows some regions of interest obtained from the dataset. Figure 6 summarizes the segmentation process.

After obtaining the regions of interest, shape features were extracted from each region. The shape features examined were: Area, Convex Area, Eccentricity, Equivalent Diameter, Euler Number, Extent, Filled Area, Major Axis Length, Minor Axis Length, Orientation, Perimeter, and Solidity.

6.2 Identifying *P. falciparum* Gametocytes

For this task, an artificial neural network that uses the shape features extracted was trained. Another model that uses transfer learning on Inceptionv3 was also built (Fig. 4).

6.3 Detection and Identification of *Plasmodium* Species

For the second task, two CNN models were built – one from scratch and one using transfer learning [7]. Both models use the Inceptionv3 architecture. With the obtained regions of interest working as masks to the original RGB color images, the models were trained and evaluated.

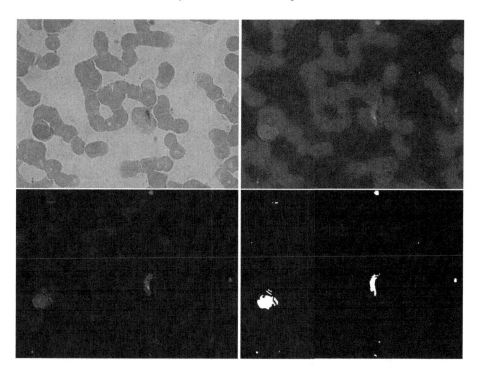

Fig. 4. Preprocessing steps (top to bottom, left to right: original image, image opening and closing on the saturation channel, subtracting the mean and standard deviation values, thresholding)

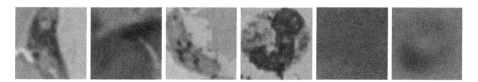

Fig. 5. Segmented (magnified) regions of interest for classification

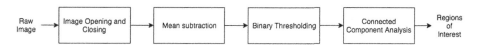

Fig. 6. Preprocessing steps to extract regions of interest

7   Comparative Results

7.1   Identifying *P. falciparum* Gametocytes

Table 1 shows the comparison of the performance of the two models built for this task. As seen from the high accuracy and sensitivity values, CNN clearly outperforms the ANN with shape features. Both models, however, obtained an accuracy significantly higher than that of Pinkaew et al. (85.71%) [6].

Another point of comparison worth noting is the time spent to extract features used and train the two models. Using the same hardware, the ANN trained for 5 h while the CNN was built in under 30 min.

Table 1. Performance comparison of shape features and pre-trained Inceptionv3 CNN model on identifying *P. falciparum* gametocytes

Performance measure	ANN - shape features	CNN - Inceptionv3
Accuracy	92.26%	**97.99%**
Specificity	**99%**	96.82%
Sensitivity	78%	**98.62%**

7.2   Detection and Identification of *Plasmodium* Species

Table 2 shows the comparison of the performance of the two CNN models built. As the figures show, although the CNN trained from scratch detects *Plasmodium* parasites better than the CNN from transfer learning, the latter obtained higher sensitivity and accuracy in identifying *P. vivax*. Both models, however, obtained an accuracy significantly higher than that of Pinkaew et al. in identifying *P. falciparum* and *P. vivax* (85.72% and 78.72%, respectively) [6].

Table 2. Performance comparison of CNN models trained from scratch and using transfer learning

Performance measure	CNN - transfer learning	CNN - trained from scratch
Accuracy (detection)	92.46%	**93.60%**
Accuracy (*P. falciparum*)	89.64%	**89.71%**
Accuracy (*P. vivax*)	**88.35%**	88.14%
Specificity	84.7%	**89.67%**
Sensitivity	**95.2%**	95.06%

8   Conclusion

The shape features examined in this study, as the results show, are good attributes to consider in the automation of the diagnostic process for malaria. However, the CNN model outperformed shape feature analysis. This implies that even without manually engineering features, problems like malaria diagnosis could still be done. However, as is the case for most CNN models, the time requirement for training and testing grew exponentially.

Comparing the CNN models (trained from scratch and transfer learning), given a decision bias on more effective parasite detection, training the model from scratch proves to be the better choice, but at the expense of longer training time and slightly lower accuracy on *P. vivax* detection compared to the other CNN model.

References

1. World Health Organization: World malaria report 2016 (2016)
2. Paniker, C.J.: Paniker's Textbook of Medical Parasitology, 7th edn. Jaypee Brothers Medical Publishers (P) Ltd., New Delhi (2013)
3. Centers for Disease Control Prevention: About Malaria (2015)
4. Makkapati, V., Rao, R.: Segmentation of malaria parasites in peripheral blood smear images, pp. 1361–1364 (2009)
5. Nanoti, A., Jain, S., Gupta, C., Vyas, G.: Detection of malaria parasite species and life cycle stages using microscopic images of thin blood smear. In: 2016 International Conference on Inventive Computation Technologies (ICICT), vol. 1, pp. 1–6 (2016)
6. Pinkaew, A., Limpiti, T., Trirat, A.: Automated classification of malaria parasite species on thick blood film using support vector machine. In: 2015 8th Biomedical Engineering International Conference (BMEiCON), pp. 1–5 (2015)
7. Delas Peñas, K.E., Rivera, P.T., Naval, P.C.: Malaria parasite detection and species identification on thin blood smears using a convolutional neural network. In: 2017 IEEE/ACM International Conference on Connected Health: Applications, Systems and Engineering Technologies (CHASE), pp. 1–6, July 2017
8. MathWorks: Measure properties of image regions (2014)
9. Szegedy, C., Liu, W., Jia, Y., Sermanet, P., Reed, S., Anguelov, D., Erhan, D., Vanhoucke, V., Rabinovich, A.: Going deeper with convolutions. In: 2015 IEEE Conference on Computer Vision and Pattern Recognition (CVPR), pp. 1–9 (2015)

Optical Flow for Collision Avoidance in Autonomous Cars

Damian Pęszor$^{(\boxtimes)}$, Marcin Paszkuta, Marzena Wojciechowska,
and Konrad Wojciechowski

Research and Development Center in Bytom of Polish-Japanese Academy
of Information Technology, Warsaw, Poland
damian.peszor@pja.edu.pl, marcin.paszkuta@polsl.pl,
mwojciechowska@pjwstk.edu.pl, konrad.wojciechowski@pj.edu.pl
http://bytom.pja.edu.pl/

Abstract. Autonomous cars, robotic platforms and other devices capable of unassisted movement are becoming widely considered as superior to human-based control in many areas. Such platforms, however, often are constructed using expensive equipment. We investigate the possibility of using simple setup consisting of an embedded computer module, such as smartphone, single camera and inertial measurement unit along with the concept of the optical flow to detect possible collisions given real-time, onboard processing as an alternative to compound systems based on radar and lidar devices. While most optical flow algorithms are not applicable for real-time processing, our findings prove that those which sacrifice accuracy to gain speed can still be upgraded while remaining accurate enough for the field of collision avoidance. We propose modifications to further enhance optical flow for given context. Our findings prove that using proposed setup consisting of both hardware and software allows for omitting expensive sensors in the field of collision avoidance.

Keywords: Optical flow · Collision detection · Autonomous

1 Introduction

Number of injured people in traffic accidents is staggering. In European Union alone, over 1 400 000 were injured in 2014 with over 800 000 fatalities [1]. Many of these accidents are a result of human error which could possibly be avoided if additional artificial systems would analyse the situation in front of the car and either signal the problem in advance, take some measure of control over the vehicle, or generally, take complete control as in case of autonomous cars. However, such autonomous (or at least automated) systems use expensive equipment, such as high-class radars or laser scanners to obtain the information about distance to obstacles in the immediate surroundings of the vehicle [2]. Since such equipment is not easily available for most car owners, it is worthwhile to investigate whether cheaper equipment can be used to obtain similar information that could be used

© Springer International Publishing AG, part of Springer Nature 2018
N. T. Nguyen et al. (Eds.): ACIIDS 2018, LNAI 10752, pp. 482–491, 2018.
https://doi.org/10.1007/978-3-319-75420-8_46

to avoid collision between autonomous or automated platform and obstacles in form of other vehicles on the road or its surroundings.

Now a days, visible light cameras are both easily accessible and cheap. Attaching such devices to car is simple, as it requires only to be directed so that the camera would register what lies in front of the vehicle. They can also be mounted inside of the cabin. Many cars are already equipped with some in order to record any events that might be important for insurance purposes. As such, cameras are very useful in order to provide some additional degree of safety in case of automated systems and are also an important part of reducing the costs of fully autonomous vehicles. With each passing year, smartphone devices offer capabilities which rival those of dedicated hardware platforms. Typical smartphone offers high-quality camera, inertial measurement unit and processing unit. Some of the devices are even capable of highly parallelized calculations using solutions primarily designed for Graphics Processing Units. As such, some models can already be used for task presented in this article, while the presence of devices with such capabilities can be expected in near future.

One of types of algorithms that can be used to obtain information about what happens on the scene is known as optical flow. Optical flow algorithms are widely used in Unmanned Aerial Vehicles, movement detection in alarm systems, robot navigation, [3–5] and similar areas. It can also provide important information that is needed to find obstacles on the motion trajectory of autonomous vehicle.

Calculated optical flow is a vector field which describes relative motion of the camera and obstacles in camera's field of vision. It is composed of two-dimensional vectors presenting the screen-space translation of projections of points in field of vision between consecutive frames. While change in three-dimensional space can be estimated using optical flow, the vectors themselves provide two-dimensional information.

The key part of obstacle detection is the ability to react fast enough. If a car is moving at $100\,\mathrm{km/h}$ and uses cameras working at 25 frames per second, the distance it travels between two frames is $1.1(1)\,\mathrm{m}$. Each frame corresponds therefore to an important part of vehicle's braking distance. If the obstacle is in fact another car moving in opposite direction, the need for faster calculations is only greater.

2   Methods

2.1   Hardware

The designed collision avoidance system uses NVIDIA Jetson TX2 embedded computer in a module. Jetson TX2 integrates 256-core NVIDIA Pascal GPU and a hex-core ARMv8 64-bit CPU complex. The CPU complex combines a dual-core NVIDIA Denver 2 and a quad-core ARM Cortex-A57. The module contains built-in 8 GB 128-bit LPDDR4 operating memory which offers 59.7 GB/s bandwidth and 32G non-volatile eMMC 5.1 memory. The typical power consumption of the platform is around 10 W, which makes it ideal for mobile application.

Fig. 1. Two consecutive frames used to obtain optical flow values and to detect obstacles

The system uses two cameras connected within the frame of stereo rig, each operating with the resolution of 1280 by 1024 pixels. While only one camera was used to perform calculations for the purpose of this article, the other was used to provide ground truth in form of reconstructed scene. For the purpose of determining the acceleration and orientation change, We used the MEMS-based, industrial-grade inertial navigation systems (INS) based on the Global Navigation Satellite System (GNSS) receiver and high-performance, temperature calibrated inertial measurement unit (IMU). To synchronize frames with INS data we use a timestamp which is assigned to each image and position frame. The INS system sampling rate was around 58 Hz.

Although intrinsic parameters can often be obtained from technical specification of given camera, the possible margin of error can be reduced using calibration. In our case, we use Zhang's method of camera calibration ([10]) which only requires simple pattern to be recorded. This allows for calibration after manual camera mounting on a car by the user and is therefore consistent with our ease-of-use approach.

2.2   Optical Flow Algorithm Selection

Optical flow calculation is quite an expensive operation. While most algorithms operate offline with high accuracy, they are not applicable to real-time, fast paced nature of obstacle detection in case of vehicular transportation. Some algorithms, however, sacrifice a degree of accuracy in order to gain faster calculations. The selection of such algorithm has to take few things into account:

- Algorithm has to be extremely fast due to the fact that the vehicle is moving with high velocity.
- As consequence of above, algorithm can benefit from parallelisation. One has to therefore select algorithm on the basis of available architecture. In our case, we take advantage of 256 CUDA-cores (Compute Unified Device Architecture) available in NVidia Tegra X1.
- While proposed approach is dedicated to single camera systems, one can benefit from using it in stereo setup, where stereovision is not enough to provide desired accuracy. In such a scenario, algorithm has to be capable of using stereo data.
- The vehicle contains onboard accelerometer and gyroscope. The algorithm can, again, benefit from this data.

A comparison of available optical flow algorithms is published with The KITTI Vision Benchmark Suite's database [6,7]. At the moment of writing of this article, there are four algorithms that are faster than 10 frames per second. Dense Spatial Transform Flow (DSTFlow, [9]) and Prediction-Correction Optical Flow (PCOF) combined with Adaptive Coarse-To-Fine stereo (ACTF, [11]) algorithm (PCOF + ACTF, [8]) are however only two, which are referenced and can therefore be reimplemented. For the purpose of optical flow evaluation within this article, PCOF + ACTF algorithm was selected.

The first step of an algorithm presented by Derome et al. in [8] is to estimate ego-motion using visual odometer eVO proposed in [12]. It is also the only part of the algorithm proposed by Derome et al. which is done on CPU rather than GPU. In our case, however, the vehicle contains accelerometer and gyroscope. Instead of estimating camera motion from images, we can, therefore, use already available information, thus reducing the Prediction-Correction Optical Flow algorithm to the part that can be computed on GPU. We employ algorithms presented in [16–18] to estimate orientation based on IMU readings.

The second important step of the solution presented in [8] is Adaptive Coarse-To-Fine matching. In case of obstacle detection task as presented in this article, this is not really necessary. However, if stereo-based approach would be used, such a step could be useful.

The last part of PCOF is computing the optical flow. The obvious performance advantage of eFOLKI ([13,14] and most importantly [15]) over other algorithms, as presented in [8] convinces us, that it is appropriate for fast-moving vehicles as in our case, for which an example is provided in Fig. 1.

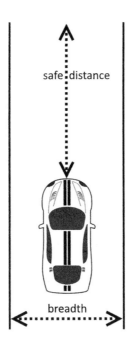

Fig. 2. Safe distance θ and breadth ω illustrated. Safe distance should be set with breaking distance in mind and can actually be a function of velocity. Breadth should be set depending on lane width and desired safety margins.

2.3   Obstacle Detection Conditions

While reconstruction of entire surroundings of autonomous car is possible using stereovision and optical flow, it is unnecessary in terms of avoiding collisions when the motion of the vehicle is the main source of movement. There are three important factors which limit the range of reconstructed scene (see Fig. 2). First is the assumed safe distance. Anything further than that is irrelevant from the collision avoidance point of view and therefore can be easily ignored. Second, is the width of the car including safety margins, typically, breadth of lane. One can assume that the driver will stay within the lane (in case of automated system) or that autonomous vehicle will use vision-based lane recognition (which can be implemented using these same cameras as in case of obstacle detection). Third important factor is the fact that, although close to the camera, the ground is not an obstacle and therefore has to be excluded from obstacle detection, as do objects above the vehicle, such as street lights.

The actual reconstruction process is, however, unnecessary and can be simplified. The important part is recognizing whether given obstacle is within the area defined by safe distance, breadth of the lane and range of altitude rather than where it is exactly. This means that we only require a binary test, which has important implications on the distance calculation.

Let us assume for the sake of simplicity, that the following holds true:

- Camera is positioned along vehicle main axis
- Camera is aligned with main axis of the vehicle
- Lens distortion is either low enough to be irrelevant or is already compensated for
- Vehicle movement is mainly in the direction of its axis.

All of the above are not necessary, however such assumptions simplify the description of idea behind this paper. In first frame, camera is positioned in point C_0. In second frame, camera moves to C_1 along main vehicle axis by distance δ, which leads to change in relative position of obstacles which are captured by the camera and registered in pixel coordinates $(x + x_f, y + y_f)$. (x_f, y_f) is an optical flow vector i.e. the translation vector from coordinates in initial frame f_0 captured from C_0 to coordinates in current frame f_1 captured from C_1. We denote the principal point as (c_x, c_y), while f_x and f_y denote horizontal and vertical focal length in pixel units. When all those variables are known, one can calculate distance d along main vehicle axis between C_1 and object visible in f_1 in coordinates of $(x + x_f, y + y_f)$ using e.g. Eq. 1.

$$d = \delta \left| \sqrt{\frac{(x + x_f - c_x)^2 + (y + y_f - c_y)^2}{(x - c_x)^2 + (y - c_y)^2}} - 1 \right|^{-1} \tag{1}$$

Note that both horizontal and vertical axis have to be used to calculate d. The horizontal offset o_h between main axis of the vehicle and obstacle in given pixel can be calculated using e.g. Eq. 2.

$$o_h = \frac{\delta(x + x_f - c_x)(x - c_x)}{f_x x_f} \tag{2}$$

One can therefore use the defined safe breadth w and safe distance θ to decide whether given pixel contains an obstacle within the safety limits. Similarly, one can find the vertical offset o_v by Eq. 3, which is required to exclude the ground and street lights from obstacle detection.

$$o_v = \frac{\delta(y + y_f - c_y)(y - c_y)}{f_y y_f} \tag{3}$$

Assuming that η is the extent of altitude that has to be obstacle-free, both up and down, one can define conditions for possible obstacle in safety zone using Eq. 4.

$$\begin{aligned}
\theta &> \delta \left| \sqrt{\frac{(x+x_f-c_x)^2+(y+y_f-c_y)^2}{(x-c_x)^2+(y-c_y)^2}} - 1 \right|^{-1} \\
\frac{f_x w}{2} &> \frac{\delta(x+x_f-c_x)(x-c_x)}{x_f} > -\frac{f_x w}{2} \\
f_y \eta &> \frac{\delta(y+y_f-c_y)(y-c_y)}{y_f} > -f_y \eta
\end{aligned} \tag{4}$$

While precalculating the left-side values might provide some performance gains, it is not the extent of usefulness of such information. Since δ is known for entire frame, one can calculate the threshold values in each frame simply by diving by δ, which is relatively fast operation. This also allows to calculate the boundaries of (x_f, y_f) values which result in fulfilling of the collision detection conditions, as in Eqs. 5 and 6

$$
\begin{aligned}
&\frac{f_x \omega}{2|x-c_x|} \geq \delta : \\
&x_f > \max(0, \frac{(x-c_x)\delta|x-c_x|}{-0.5\omega f_x - \delta|x-c_x|}, \frac{(x-c_x)\delta|x-c_x|}{0.5\omega f_x - \delta|x-c_x|}) \\
&\vee x_f < \min(0, \frac{(x-c_x)\delta|x-c_x|}{-0.5\omega f_x - \delta|x-c_x|}, \frac{(x-c_x)\delta|x-c_x|}{0.5\omega f_x - \delta|x-c_x|}) \\
&\frac{f_x \omega}{2|x-c_x|} < \delta : \\
&\frac{(x-c_x)\delta|x-c_x|}{0.5\omega f_x - \delta|x-c_x|} > x_f > \max(0, \frac{(x-c_x)\delta|x-c_x|}{-0.5\omega f_x - \delta|x-c_x|}) \\
&\vee \min(0, \frac{(x-c_x)\delta|x-c_x|}{-0.5\omega f_x - \delta|x-c_x|}) > x_f > \frac{(x-c_x)\delta|x-c_x|}{0.5\omega f_x - \delta|x-c_x|} \\
&\frac{f_y \eta}{|y-c_y|} \geq \delta : \\
&y_f > \max(0, \frac{(y-c_y)\delta|y-c_y|}{-\eta f_y - \delta|y-c_y|}, \frac{(y-c_y)\delta|y-c_y|}{\eta f_y - \delta|y-c_y|}) \\
&\vee y_f < \min(0, \frac{(y-c_y)\delta|y-c_y|}{-\eta f_y - \delta|y-c_y|}, \frac{(y-c_y)\delta|y-c_y|}{\eta f_y - \delta|y-c_y|}) \\
&\frac{f_y \eta}{|y-c_y|} < \delta : \\
&\frac{(y-c_y)\delta|y-c_y|}{\eta f_y - \delta|y-c_y|} > y_f > \max(0, \frac{(y-c_y)\delta|y-c_y|}{-\eta f_y - \delta|y-c_y|}) \\
&\vee \min(0, \frac{(y-c_y)\delta|y-c_y|}{-\eta f_y - \delta|y-c_y|}) > y_f > \frac{(y-c_y)\delta|y-c_y|}{\eta f_y - \delta|y-c_y|}
\end{aligned}
\tag{5}
$$

$$
\begin{aligned}
&(x+x_f-c_x)^2 + (y+y_f-c_y)^2 \geq (x-c_x)^2 + (y-c_y)^2 : \\
&(x+x_f-c_x)^2 + (y+y_f-c_y)^2 > (\tfrac{\delta}{\theta}+1)^2[(x-c_x)^2 + (y-c_y)^2] \\
&(x+x_f-c_x)^2 + (y+y_f-c_y)^2 < (x-c_x)^2 + (y-c_y)^2 : \\
&(1-\tfrac{\delta}{\theta})^2[(x-c_x)^2 + (y-c_y)^2] > (x+x_f-c_x)^2 + (y+y_f-c_y)^2
\end{aligned}
\tag{6}
$$

In typical optical flow algorithm based on Lucas-Kanade approach, one would find correlated pixels within window of given size around the source pixel. This alone results in huge windows that prevent obtaining high performance. In order to avoid this, it is typical to create Gaussian pyramid by image downsampling and then find correlations on subsequent levels of pyramid using upsampled optical flow from previous level as a source point in next one. This allows for smaller windows and results in optical flow as presented in Fig. 3. Still though, calculation of optical flow on downsampled image is quite prone to error, especially in case of boundaries of different objects being represented in the same pixel of downsampled image.

Information about possible range of optical flow values allows for following:

- Less pyramid levels (possibly one)
- Since the range of flow changes smoothly, one can use different amount of pyramid levels for different part of image

Fig. 3. Scaled magnitude and angle of optical flow vectors between frames as shown in Fig. 1

- Smaller search windows due to restricted maximum flow
- Search rings, that is windows that are empty inside, due to the fact that optical flow lower than given value will not be recognized as an obstacle in safe zone.

One might consider whether there is an issue regarding correlations being restricted to potential collision sites. In typical optical flow approach, one would calculate correlation of entirety of search window and then choose the best pixel. This is necessary in case of downsampled image, since such a process can affect pixel values to great extent. Reduction of pyramid levels allows for using of threshold values instead of best match, since the same pixel on pyramid levels with high resolution will be highly correlated, unlike the surroundings. The possible, although not common, errors that might arrive from such behaviour are mostly nullified by smoothing or other filtration steps of common optical flow algorithms, which is aimed at removing incorrect estimations. In presented case, such approach is perfectly reasonable, since in case of obstacle found in the image inside of safety zone, one can expect it to be large enough to be detected in multiple pixels.

Fig. 4. Obstacles detected on the basis of thresholds of optical flow as estimated from images in Fig. 1

3   Results

Presented approach generally does not affect the results of obstacle detection in safety zone in comparison with classification based on 3D scene reconstruction, as can be seen in Fig. 4. However, it allows for reduction of Gaussian pyramid level counts and for using smaller windows to find pixel correspondences. The improvements vary depending on the scene and parameters used (breadth ω and safe distance θ). In case of ten four-minutes long recordings we lowered the number of correlation tests by average of around 31%.

4   Conclusions

In this paper, an approach to detection of possible collision in front of autonomous platform has been presented. The problem has been solved using device capable of highly-parallelized computations on-board car. The authors have presented the necessary steps involved in distinguishing safe and unsafe parts of space in front of the vehicle and the necessary changes to existing algorithms in order to obtain real-time calculations. The results as show that system is capable of detecting obstacles and thus allows for safer journey.

Acknowledgements. This work has been supported by the National Centre for Research and Development, Poland in the frame of project POIR.01.02.00-00-0009/2015 "System of autonomous landing of an UAV in unknown terrain conditions on the basis of visual data".

References

1. European Commission, Directorate General for Transport: European Commission, Annual Accident Report, June 2016
2. Thrun, S.: Toward robotic cars. Commun. ACM. **53**(4), 99–106 (2010)

3. Bonin-Font, F., Ortiz, A., Oliver, G.: Visual navigation for mobile robots: a survey. J. Intell. Robot. Syst. **53**(3), 263 (2008)
4. DeSouza, G.N., Kak, A.C.: Vision for mobile robot navigation: a survey. IEEE Trans. Pattern Anal. Mach. Intell. **24**(2), 237–267 (2002)
5. Aguirre, E., González, A.: Fuzzy behaviors for mobile robot navigation: design, coordination and fusion. Int. J. Approximate Reasoning **25**(3), 255–289 (2000)
6. Menze, M., Geiger, A.: Object scene flow for autonomous vehicles. In: Conference on Computer Vision and Pattern Recognition (CVPR) (2015)
7. Menze, M., Heipke, C., Geiger, A.: Joint 3D estimation of vehicles and scene flow. In: ISPRS Workshop on Image Sequence Analysis (ISA) (2015)
8. Derome, M., Plyer, A., Sanfourche, M., Le Besnerais, G.: A prediction-correction approach for real-time optical flow computation using stereo. In: Rosenhahn, B., Andres, B. (eds.) GCPR 2016. LNCS, vol. 9796, pp. 365–376. Springer, Cham (2016). https://doi.org/10.1007/978-3-319-45886-1_30
9. Ren, Z., Yan, J., Ni, B., Liu, B., Yang, X., Zha, H.: Unsupervised deep learning for optical flow estimation. In: Proceedings of the Thirty-First AAAI Conference on Artificial Intelligence (AAAI 2017) (2017)
10. Zhang, Z.: A flexible new technique for camera calibration. IEEE Trans. Pattern Anal. Mach. Intell. **22**(11), 1330–1334 (2000)
11. Sizintsev, M., Wildes, R.P.: Coarse-to-fine stereo vision with accurate 3D boundaries. Image Vis. Comput. **28**(3), 352–366 (2010)
12. Sanfourche, M., Vittori, V., Le Besnerais, G.: eVO: a realtime embedded stereo odometry for MAV applications. In: IROS (2013)
13. Le Besnerais, G., Champagnat, F.: Dense optical flow by iterative local window registration. In: ICIP (2005)
14. Plyer, A., Le Besnerais, G., Champagnat, F.: Folki-GPU: a powerful and versatile cuda code for real-time optical flow computation. In: GPU Technology Conference (2009)
15. Plyer, A., Le Besnerais, G., Champagnat, F.: Massively parallel Lucas Kanade optical flow for real-time video processing applications. J. Real-Time Image Proc. **11**(4), 713–730 (2016)
16. Szczęsna, A., Skurowski, P., Lach, E., Pruszowski, P., Pęszor, D., Paszkuta, M., Słupik, J., Lebek, K., Janiak, M., Polański, A., Wojciechowski, K.: Inertial motion capture costume design study. Sensors **17**(3), 612 (2017)
17. Szczęsna, A., Pruszowski, P.: Model-based extended quaternion Kalman filter to inertial orientation tracking of arbitrary kinematic chains. SpringerPlus **5**(1), 1965 (2016)
18. Słupik, J., Szczęsna, A., Polański, A.: Novel Lightweight quaternion filter for determining orientation based on indications of gyroscope, magnetometer and accelerometer. In: Chmielewski, L.J., Kozera, R., Shin, B.-S., Wojciechowski, K. (eds.) ICCVG 2014. LNCS, vol. 8671, pp. 586–593. Springer, Cham (2014). https://doi.org/10.1007/978-3-319-11331-9_70

Intelligent Video Monitoring System with the Functionality of Online Recognition of People's Behavior and Interactions Between People

Marek Kulbacki$^{(\boxtimes)}$, Jakub Segen, Sławomir Wojciechowski,
Kamil Wereszczyński, Jerzy Paweł Nowacki, Aldona Drabik,
and Konrad Wojciechowski

Polish-Japanese Academy of Information Technology,
Koszykowa 86, 02-008 Warsaw, Poland
mk@pja.edu.pl

Abstract. The intelligent video monitoring system SAVA has been implemented as a prototype at the 9th Technology Readiness Level. The source of data are video cameras located in the public space that provide HD video streaming. The aim of the study is to present an overview of the SAVA system enabling identification and classification in the real time of such behaviors as: walking, running, sitting down, jumping, lying, getting up, bending, squatting, waving, and kicking. It also can identify interactions between persons, such as: greeting, passing, hugging, pushing, and fighting. The system has module-based architecture and is combined of the following modules: acquisition, compression, path detection, path analysis, motion description, action recognition. The effect of the modules operation is a recognized behavior or interaction. The system achieves a classification correctness level of 80% when there are more than ten classes.

Keywords: IVA · Human action recognition
Human interaction recognition · Motion recognition · Machine learning

1 Introduction

The field of human activity recognition in video [1,10,13,22,23,26,40] has broad practical applications including intelligent video analytics (IVA) [20,37,42], intelligent monitoring systems for ambient-assisted living [5,21,25,27] or man-machine interactions [16,29,30]. Most methods for intelligent motion analysis and behavior recognition of people or objects from monocular camera can be assigned into one from two categories: (a) methods that use the global characteristics of a moving object, such as functions of speed vector or trajectory or spatio-temporal features, (b) a skeletal or structural representations that use models in the form of a skeleton or set of related parts. The first approach

© Springer International Publishing AG, part of Springer Nature 2018
N. T. Nguyen et al. (Eds.): ACIIDS 2018, LNAI 10752, pp. 492–501, 2018.
https://doi.org/10.1007/978-3-319-75420-8_47

works well under real conditions for rigid objects, but has a limited power of discrimination in data containing articulated objects and characters, while the second one gives good results for the characters in the lab environment, but it can completely fail in real conditions with low contrast between the objects and the background. The SAVA system has been developed incrementally as a combination of both categories, what is described in Sect. 2.

The project has been focused on resolving the problems related to two industrial concepts: (1) IVA for video content analysis on the higher level of details for people and moving objects; (2) *streaming analytics* intended for automatically detecting events on-the-fly in monitoring systems.

The law enforcement and personal safety are very important application areas and developers of IVA systems try to address these uses of video surveillance solutions. There were proposed various approaches for selecting and securing sensitive information in surveillance systems: content access control [4], privacy preserving filters [3,28], by tracking sensitive objects [43], in the cloud environment [41]. The first comprehensive review about visual privacy protection mechanisms [24] was provided in 2015. In the SAVA project, methods for data anonymization have been developed that conform with the Polish law and adhere to the regulations and standards of Act on the Protection of Personal Data. The personal privacy problem have been addressed by introducing and implementing an optical flow based module for face anonymization in video sequences [38].

The rest of the paper is organized as follows. Section 2 discusses the design and implementation of SAVA modules and related elements of IVA environment. Section 4 presents an effective solution for a simple action recognition and discusses the optimization elements. Section 5 describes the compound action recognition component of the SAVA system, required for recognizing interactions.

2   System SAVA

The goal of the SAVA project was the design and verification of real-time IVA system prototype able to recognize and classify behavior and actions of individuals and groups and identify situations requiring the generation of on-line and off-line alerts. The results of the SAVA project include improvements and extensions of behavior recognition technology and related technologies such as enhanced motion tracking, background separation, motion representation, and optimized learning methods as well as technology transfer from 6th to 9th Technology Readiness Level.

System SAVA has a modular structure. In the base version each module acts as a separate process, executing a specific task. It is possible to extend the system with new functionality by adding additional modules. The base version of SAVA pipeline in the form of high level system architecture is shown in Fig. 1. The acquisition module retrieves video stream in the MJPEG and H264 format from Milestone environment or directly from PTZ cameras in MJPEG format by a proprietary driver. Conformity with ONVIF standard enables the system to collect video streams from surveillance cameras or from Milestone video server.

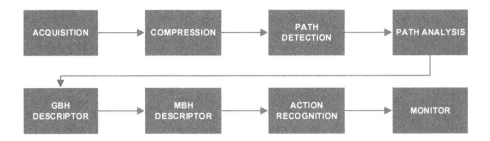

Fig. 1. High level SAVA system architecture in the base version

The system has been tested and shown to work correctly with WISENET Samsung and Axis cameras. The compression module makes use of the algorithms for detection of background and movable areas, implemented on the GPU, and basic image processing operations. The path detection module searches for paths along which selected key points move. SURF/SIFT detectors and descriptors are used for detecting and comparing the key points, and the Hungarian algorithm is applied for the pairing of these points. The detected paths that exceed the thresholds on the minimum number of frames and the trajectory length are transferred to the subsequent module. The path analysis module clusters the detected paths into trajectories of objects and determines a bounding box for each object in every frame. The result of the modules operation is a 2D trace of the positions in time of each analyzed, a unique identifier, and a reference to the source data. On the basis of the paths, the GBH and MBH descriptor modules determine descriptors of the movement performed by the detected object, constituting the grounds for the behavior classification. The Action Recognition module classifies the object's behavior on the basis of the determined descriptors, using a trained SVM classifier. The effect of the modules operation is a recognized action. A visualization module allows operator to calibrate, configure and view the camera image mixed with the automatically selected moving objects. This module shows the detected objects for specific actions, and reports alerts set for selected action classes. It also allows replays of stored video data with notification and configured alerts. The modular design allows the operator to use the scalable configuration of the system in a distributed environment. The individual processes can be run on different machines, but it is also possible to run the system on a single workstation. Some modules allow operator to run multiple instances, and speed up the processing of the video stream.

The SAVA system can work in the cloud and it is more stable than the previous MIS prototype and much faster, enabling the recognition function to be faster than real-time video, which enables multiple camera streams to be processed simultaneously. Using the improved detection methods for descriptors, recognition and learning the accuracy of action recognition has been increased and the frequency of false alerts has been reduced. One of the most important improvements is the parameter optimization process, which uses the technique of cross validation to set the parameters of the classifier, descriptor modules and the feature selector to optimize the recognition performance, based on a part of the training data. Other improvements include complete and clean software integration, data visualization and editing, and remote control of the system with a client application.

3   Tracking by Clusterization of Motion Paths

Tracking is the method used to effectively increase the level of information about each moving object in a video sequence. An effective method for tracking moving objects in the video stream in real-time by finding paths of moving local features and then clustering the paths has been developed, with an objective to identify and separate moving objects in a video sequence. Grouping paths by observing the similarity of key point descriptors, neighborhood in the image space and coexistence over time leads to determination the trajectories of the objects (they are different from trajectories of the centers of moving areas). The object trajectories are used to identify objects and set of paths assigned to a single object. A distinctive approach to tracking and behavior recognition has been developed and implemented in the previous project named More Informative Surveillance (MIS). It created a system for recognition and classification of actions, which introduced a representation of moving objects called Structured Non-Skeletal Representation (SNSR), that was used for tracking multiple people in a video sequence, together with a method of clustering motion paths of local features in video [11].

The SNSR motion representation has a structural form, which avoids some of the rigid requirements of a skeletal model, and that in turn reduces or eliminates some problems of working in real, live environments and realistic conditions, such as low contrast video or object occlusion. In the tests conducted on video streams recorded in real environments and close-to-realistic conditions, the prototype system provides good results recognizing a selected set of human motions relevant to practically important alerts. The basis of the motion paths clustering approach is the algorithm introduced by Segen and Pingali [31].

To facilitate and accelerate the preparation of training data, a dedicated tool - video editor for annotating human actions and object trajectories has been developed and presented in [19]. For training and testing algorithms in a real environment a massive dataset has been constructed, with video data collected from the surveillance cameras network VMASS [18], where the amount of collected video data exceeds 4000 h. In addition interfaces have been developed to publicly accessible video datasets prepared for human action and activity recognition, to check and compare our algorithms against other methods. A survey and comparison of publicly accessible datasets for video-based human activity recognition is presented by Chaquet et al. [7]. The research prototype developed by the MIS project was the starting point for the SAVA project. For more effective work using video streams from PTZ cameras a method of camera calibration and navigation based on continuous tracking [12], a result of the MIS project, has also been utilized (Fig. 2).

Fig. 2. Visualization of motions tracks generated by SNRS

4   Simple Actions Recognition Solution

Simple action recognition concerns classification of a temporally and spatially trimmed video. To solve this problem an advantage is taken of the Gradient Boundary Histogram method [33]. It is a very efficient approach and the authors obtained good results on challenging datasets, namely 62% on HMDB51 [17] and 86.6% on UCF101 [34]. Instead of time consuming feature detection the method uses *random sampling* combined with *local part model* [32]. There is no need for additional interest region searching. Features are computed within a *root part* cuboid placed around a randomly chosen point. The local part model spans other cuboids in theirs surroundings. As a result, each randomly chosen point has a representation of features computed within couple of neighboring volumes. Two descriptors are computed for each cuboid separately - first, *gradient boundary histogram (GBH)*. It computes derivatives on horizontal and vertical directions, and then between consecutive frames, thus capturing both appearance and motion features. Next, *motion boundary histogram (MBH)* is computed, as described by Wang et al. [36]. The descriptor requires a computation of optical flow field, which is time consuming. This is a major bottleneck of the method as reported in [36] and alleviating this problem by using the motion field from a compression have been described in [15]. The *principal component analysis PCA* [14] has been used to reduce the dimensionality. For feature encoding *Gaussian mixture model (GMM)* along with *Fisher vector (FV)* have been employed. These representations were further *normalized, concatenated* and fed into *support vector machine (SVM)* [8] classifier which outputs a class probability vector. Figure 3 shows the architecture flowchart for the single action recognition method.

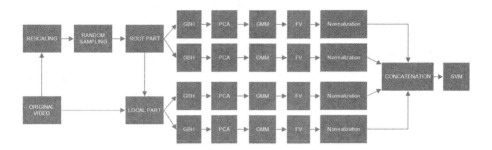

Fig. 3. Simple action recognition process

4.1 Method Optimization

The optimization method has two steps. First, the code is parallelized as much as possible using the OpenMP [6,9] library for this purpose, and a OpenCV-GPU based implementation is used for the optical flow calculation. Second, a parameter optimization toolbox has been developed, and used for the fourteen most important parameters selected from all algorithms, running it multiple times to find the best trade-off between speed and accuracy.

4.2 Localization in Time

In order to localize the action in time the video stream has been divided into segments of varying length which can overlap in time. Each segment was assigned a rank value, based on the output class probability vectors from the simple action classifiers centered at this and the neighboring segments. The segments that obtained a locally highest rank were selected as the centers of a recognized simple action.

5 Compound Actions

Compound action is a complex behavior that can include several objects, a set of simple actions, and a group of time-space relations among the objects. The class of the compound action is the set of all compound actions that were identified with same name - the label of the compound action. The schema describing the class of the compound action is called the scenario of compound action. The example of such scenario is shown in Fig. 4. A compound action representation has been constructed in the form of a time series consisting of vectors

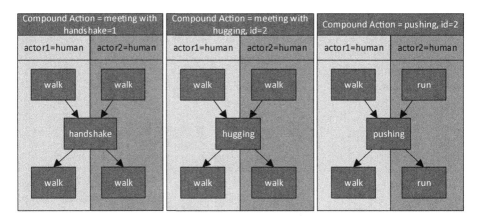

Fig. 4. The example scenarios of three classes of compound actions: greeting, fight and assault. These scenarios were used in the set of compound actions included in the tests

of discrete elements that describe the recognized simple actions and time-space relations identified in short video segments. Two methods for classifying compound actions, using this representation have been developed and tested. The first method is based on HMM (Hidden Markov Model), the second on DTW (Dynamic Time Warping). Both methods provide approximately equal performance. In experiments with four classes of compound actions: "meeting including handshake", "pushing", "close passing", and "meeting including hugging", the classification accuracy ranged from 64 to 78 percent averaged over all classes, measured in tests using 184 training examples and 122 test examples. On selected classes the accuracy reaches 80% and higher.

6   Conclusions

This paper contains an overview of SAVA - IVA system with the functionality of real time recognition of one person actions and interactions between people. The system description is focused on general architecture with insights into the original elements that make SAVA an intelligent, autonomous system and a cost saving tool for operators of surveillance systems. The implemented methods for a single person action recognition are the result of an extensive review and verification of the fastest, most efficient latest methods based on image processing from monocular camera from the area of: character tracking [35], 2D position estimation [2], space-time based methods for action recognition [39], followed by improvements, extension and optimization of the selected methods. The methods for recognition of people interactions represent an entirely novel work on classification of compound actions.

The SAVA environment is scalable, it can work with multiple PTZ cameras in real time and provides the possibility of an easy replacement of pipeline elements, classification algorithms, camera drivers and other modules, according to the user requirements. The system can work as an independent unit deployed on user premises, where all the physical servers and applications are stored on-site. It can also be used as a customized and scalable remote solution where SAVA is deployed in the cloud infrastructure, such as the PJAIT data center located in the R&D Center, PJAIT Bytom, Poland.

Acknowledgments. The project SAVA has been supported by the National Centre for Research and Development (project UOD-DEM-1-183/001 "Intelligent video analysis system for behavior and event recognition in surveillance networks"). We would like to express our gratitude to all the project participants.

References

1. Aggarwal, J., Ryoo, M.: Human activity analysis: a review. ACM Comput. Surv. **43**(3), 16:1–16:43 (2011)
2. Bąk, A., Kulbacki, M., Segen, J., Świątkowski, D., Wereszczyński, K.: Recent developments on 2D pose estimation from monocular images. In: Nguyen, N.T., Trawiński, B., Fujita, H., Hong, T.-P. (eds.) ACIIDS 2016. LNCS (LNAI), vol. 9622, pp. 437–446. Springer, Heidelberg (2016). https://doi.org/10.1007/978-3-662-49390-8_43
3. Barhm, M.S., Qwasmi, N., Qureshi, F.Z., el-Khatib, K.: Negotiating privacy preferences in video surveillance systems. In: Mehrotra, K.G., Mohan, C.K., Oh, J.C., Varshney, P.K., Ali, M. (eds.) IEA/AIE 2011. LNCS (LNAI), vol. 6704, pp. 511–521. Springer, Heidelberg (2011). https://doi.org/10.1007/978-3-642-21827-9_52
4. Castiglione, A., Cepparulo, M., De Santis, A., Palmieri, F.: Towards a lawfully secure and privacy preserving video surveillance system. In: Buccafurri, F., Semeraro, G. (eds.) EC-Web 2010. LNBIP, vol. 61, pp. 73–84. Springer, Heidelberg (2010). https://doi.org/10.1007/978-3-642-15208-5_7
5. Chaaraoui, A.A., Climent-Pérez, P., Flórez-Revuelta, F.: A review on vision techniques applied to human behaviour analysis for ambient-assisted living. Expert Syst. Appl. **39**(12), 10873–10888 (2012)
6. Chapman, B., Jost, G., Van Der Pas, R.: Using OpenMP: Portable Shared Memory Parallel Programming, vol. 10. MIT press, Cambridge (2008)
7. Chaquet, J.M., Carmona, E.J., Fernández-Caballero, A.: A survey of video datasets for human action and activity recognition. Comput. Vis. Image Underst. **117**(6), 633–659 (2013)
8. Cortes, C., Vapnik, V.: Support vector machine. Mach. Learn. **20**(3), 273–297 (1995)
9. Dagum, L., Menon, R.: Openmp: an industry standard api for shared-memory programming. IEEE Comput. Sci. Eng. **5**(1), 46–55 (1998)
10. Gavrila, D.: The visual analysis of human movement: a survey. Comput. Vis. Image Underst. **73**(1), 82–98 (1999)
11. Gudyś, A., Rosner, J., Segen, J., Wojciechowski, K., Kulbacki, M.: Tracking people in video sequences by clustering feature motion paths. In: Chmielewski, L.J., Kozera, R., Shin, B.-S., Wojciechowski, K. (eds.) ICCVG 2014. LNCS, vol. 8671, pp. 236–245. Springer, Cham (2014). https://doi.org/10.1007/978-3-319-11331-9_29
12. Gudyś, A., Wereszczyński, K., Segen, J., Kulbacki, M., Drabik, A.: Camera calibration and navigation in networks of rotating cameras. In: Nguyen, N.T., Trawiński, B., Kosala, R. (eds.) ACIIDS 2015. LNCS (LNAI), vol. 9012, pp. 237–247. Springer, Cham (2015). https://doi.org/10.1007/978-3-319-15705-4_23

13. Herath, S., Harandi, M., Porikli, F.: Going deeper into action recognition: a survey. Image Vis. Comput. **60**, 4–21 (2017)
14. Jolliffe, I.T.: Principal component analysis and factor analysis. Principal Component Analysis. Springer Series in Statistics, pp. 115–128. Springer, New York (1986). https://doi.org/10.1007/978-1-4757-1904-8_7
15. Kantorov, V., Laptev, I.: Efficient feature extraction, encoding and classification for action recognition. In: Proceedings of the IEEE Conference on Computer Vision and Pattern Recognition, pp. 2593–2600 (2014)
16. Koppula, H.S., Saxena, A.: Anticipating human activities using object affordances for reactive robotic response. IEEE Trans. Pattern Anal. Mach. Intell. **38**(1), 14–29 (2016)
17. Kuehne, H., Jhuang, H., Stiefelhagen, R., Serre, T.: Hmdb51: a large video database for human motion recognition. In: Nagel, W., Kröner, D., Resch, M. (eds.) High Performance Computing in Science and Engineering, vol. 12, pp. 571–582. Springer, Heidelberg (2013). https://doi.org/10.1007/978-3-642-33374-3_41
18. Kulbacki, M., Segen, J., Wereszczyński, K., Gudyś, A.: VMASS: massive dataset of multi-camera video for learning, classification and recognition of human actions. In: Nguyen, N.T., Attachoo, B., Trawiński, B., Somboonviwat, K. (eds.) ACIIDS 2014. LNCS (LNAI), vol. 8398, pp. 565–574. Springer, Cham (2014). https://doi.org/10.1007/978-3-319-05458-2_58
19. Kulbacki, M., Wereszczyński, K., Segen, J., Sachajko, M., Bąk, A.: Video editor for annotating human actions and object trajectories. In: Nguyen, N.T., Trawiński, B., Fujita, H., Hong, T.-P. (eds.) ACIIDS 2016. LNCS (LNAI), vol. 9622, pp. 447–457. Springer, Heidelberg (2016). https://doi.org/10.1007/978-3-662-49390-8_44
20. Mathur, G., Bundele, M.: Research on intelligent video surveillance techniques for suspicious activity detection critical review. In: 2016 International Conference on Recent Advances and Innovations in Engineering (ICRAIE), pp. 1–8. IEEE (2016)
21. Meinel, L., Findeisen, M., Hes, M., Apitzsch, A., Hirtz, G.: Automated real-time surveillance for ambient assisted living using an omnidirectional camera. In: 2014 IEEE International Conference on Consumer Electronics (ICCE), pp. 396–399. IEEE (2014)
22. Negin, F., Bremond, F.: Human action recognition in videos: a survey. Technical report, INRIA Technical Report, Sophia Antipolis, France (2016)
23. Onofri, L., Soda, P., Pechenizkiy, M., Iannello, G.: A survey on using domain and contextual knowledge for human activity recognition in video streams. Expert Syst. Appl. **63**, 97–111 (2016)
24. Padilla-López, J.R., Chaaraoui, A.A., Flórez-Revuelta, F.: Visual privacy protection methods: a survey. Expert Syst. Appl. **42**(9), 4177–4195 (2015)
25. Pal, S., Abhayaratne, C.: Video-based activity level recognition for assisted living using motion features. In: Proceedings of the 9th International Conference on Distributed Smart Cameras, pp. 62–67. ACM (2015)
26. Poppe, R.: A survey on vision-based human action recognition. Image Vis. Comput. **28**(6), 976–990 (2010)
27. Rafferty, J., Nugent, C.D., Liu, J., Chen, L.: From activity recognition to intention recognition for assisted living within smart homes. IEEE Trans. Hum. Mach. Syst. **47**(3), 368–379 (2017)
28. Rajpoot, Q.M., Jensen, C.D.: Security and privacy in video surveillance: requirements and challenges. In: Cuppens-Boulahia, N., Cuppens, F., Jajodia, S., Abou El Kalam, A., Sans, T. (eds.) SEC 2014. IAICT, vol. 428, pp. 169–184. Springer, Heidelberg (2014). https://doi.org/10.1007/978-3-642-55415-5_14

29. Ramirez-Amaro, K., Beetz, M., Cheng, G.: Transferring skills to humanoid robots by extracting semantic representations from observations of human activities. Artif. Intell. **247**, 95–118 (2017)
30. Rezazadegan, F., Shirazi, S., Upcroft, B., Milford, M.: Action recognition: from static datasets to moving robots. arXiv preprint arXiv:1701.04925 (2017)
31. Segen, J., Pingali, S.G.: A camera-based system for tracking people in real time. In: Proceedings of the 13th International Conference on Pattern Recognition 1996, vol. 3, pp. 63–67. IEEE (1996)
32. Shi, F., Laganière, R., Petriu, E.: Local part model for action recognition. Image Vis. Comput. **46**(C), 18–28 (2016)
33. Shi, F., Laganiere, R., Petriu, E.: Gradient boundary histograms for action recognition. In: 2015 IEEE Winter Conference on Applications of Computer Vision (WACV), pp. 1107–1114. IEEE (2015)
34. Soomro, K., Zamir, A.R., Shah, M.: Ucf101: a dataset of 101 human actions classes from videos in the wild. arXiv preprint arXiv:1212.0402 (2012)
35. Staniszewski, M., Kloszczyk, M., Segen, J., Wereszczyński, K., Drabik, A., Kulbacki, M.: Recent developments in tracking objects in a video sequence. In: Nguyen, N.T., Trawiński, B., Fujita, H., Hong, T.-P. (eds.) ACIIDS 2016. LNCS (LNAI), vol. 9622, pp. 427–436. Springer, Heidelberg (2016). https://doi.org/10.1007/978-3-662-49390-8_42
36. Wang, H., Kläser, A., Schmid, C., Liu, C.L.: Dense trajectories and motion boundary descriptors for action recognition. Int. J. Comput. Vis. **103**(1), 60–79 (2013)
37. Wang, X.: Intelligent multi-camera video surveillance: a review. Pattern Recogn. Lett. **34**(1), 3–19 (2013)
38. Wereszczyński, K., Michalczuk, A., Segen, J., Pawlyta, M., Bąk, A., Nowacki, J.P., Kulbacki, M.: Optical flow based face anonymization in video sequences. In: Nguyen, N.T., Tojo, S., Nguyen, L.M., Trawiński, B. (eds.) ACIIDS 2017. LNCS (LNAI), vol. 10192, pp. 623–631. Springer, Cham (2017). https://doi.org/10.1007/978-3-319-54430-4_60
39. Wojciechowski, S., Kulbacki, M., Segen, J., Wyciślok, R., Bąk, A., Wereszczyński, K., Wojciechowski, K.: Selected space-time based methods for action recognition. In: Nguyen, N.T., Trawiński, B., Fujita, H., Hong, T.-P. (eds.) ACIIDS 2016. LNCS (LNAI), vol. 9622, pp. 417–426. Springer, Heidelberg (2016). https://doi.org/10.1007/978-3-662-49390-8_41
40. Wu, D., Sharma, N., Blumenstein, M.: Recent advances in video-based human action recognition using deep learning: a review. In: 2017 International Joint Conference on Neural Networks (IJCNN), pp. 2865–2872. IEEE (2017)
41. Xu, Y., Gong, J., Xiong, L., Xu, Z., Wang, J., Shi, Y.Q.: A privacy-preserving content-based image retrieval method in cloud environment. J. Vis. Commun. Image Represent. **43**, 164–172 (2017)
42. Yu, C., Zheng, X., Zhao, Y., Liu, G., Li, N.: Review of intelligent video surveillance technology research. In: 2011 International Conference on Electronic and Mechanical Engineering and Information Technology (EMEIT), vol. 1, pp. 230–233. IEEE (2011)
43. Zhang, P., Thomas, T., Emmanuel, S.: Privacy enabled video surveillance using a two state markov tracking algorithm. Multimedia Syst. **18**(2), 175–199 (2012)

Computational Imaging and Vision

Satellite Image Classification Based Spatial-Spectral Fuzzy Clustering Algorithm

Sinh Dinh Mai$^{(\boxtimes)}$ (iD), Long Thanh Ngo, and Hung Le Trinh

Le Quy Don Technical University, Hanoi, Vietnam
maidinhsinh@gmail.com, ngotlong@gmail.com,
trinhlehung125@gmail.com

Abstract. Spectral clustering is a clustering method based on algebraic graph theory. The clustering effect by using spectral method depends heavily on the description of similarity between instances of the datasets. Althought, spectral clustering has gained considerable attentions in the recent past, but the raw spectral clustering is often based on Euclidean distance, but it is impossible to accurately reflect the complexity of the data. Despite having a well-defined mathematical framework, good performance and simplicity, it suffers from several drawbacks, such as it is unable to determine a reasonable cluster number, sensitive to initial condition and not robust to outliers. Owing to the limitations of the feature space in multispectral images and spectral overlap of the clusters, it is required to use some additional information such as the spatial context in image clustering. In this paper, we present a new approach named spatial-spectral fuzzy clustering (SSFC) which combines spectral clustering and fuzzy clustering with local information into a unified framework to solve these problems and also using fuzzy clustering algorithm to converge the global optimization, this method is simple in computation but quite effective when solving segmentation problems on satellite imagery. Making it to find the spatial distribution characteristics of complex data and can further make cluster more stable. Experimental results show that it can improve the clustering accuracy and avoid falling into local optimum.

Keywords: Spectral clustering · Local information · Fuzzy c-means
Satellite image

1 Introduction

Spectral clustering is a technique for finding group structure in data. It is based on viewing the data points as nodes of a connected graph and clusters are found by partitioning this graph, based on its spectral decomposition, into subgraphs that possess some desirable properties. Spectral clustering applies in various fields such as medical image classification [2], satellite image [5, 6, 24, 25], bioinformatics [10]. The widely used similarity measure for spectral clustering is Gaussian kernel function which measures the similarity between data points. However, it is difficult for spectral clustering to choose a suitable scaling parameter in Gaussian kernel similarity measure.

So much research has improved to overcome the drawbacks of the spectral clustering method, the research focus based on distribution characteristics, the structure of

© Springer International Publishing AG, part of Springer Nature 2018
N. T. Nguyen et al. (Eds.): ACIIDS 2018, LNAI 10752, pp. 505–518, 2018.
https://doi.org/10.1007/978-3-319-75420-8_48

the cluster or data density. Some research as, Li et al. [1] constructed a new matrix S based on density as the similarity matrix, and proposed k-Means based on density to converge the global optimization. Making it to find the spatial distribution characteristics of complex data. As we know it, k-means clustering algorithm is hard, with overlapping data, the application of k-Means technique do not often result in high precision. Spectral clustering has shown to be a powerful technique for grouping and/or ranking data as well as a proper alternative for unlabeled problems, so, Peluffo-Ordóñez et al. [3] has an overview of clustering techniques based on spectral clustering problem of dynamic data analysis. Liu and his team [4] improved spectral clustering algorithm (ISC) by applying the hill-climbing techniques to find the cluster centroids and reduce computing costs by merging the pixels have the same gray level. However, merging the pixels have the same gray level will lose the relationship between the spatial pixels, which affect the clustering results.

To increase the accuracy of image clustering, a new Hierarchical Iterative Clustering Algorithm using Spatial and Spectral information (HICLASS) [5] is introduced by Fatemi et al. This algorithm separates pixels into uncertain and certain categories based on decision distances in the feature space. The algorithm labels the certain pixels using the k-means clustering, and the uncertain ones with the help of information in both spatial and spectral domains of the image. The difficulty of this algorithm is knowing exactly the certainty of a pixel belonging to a certain class. Sinh et al. [23] was proposed a interval type-2 fuzzy c-means clustering algorithm with spatial information for land-cover classification. However, the calculation on the interval type-2 fuzzy set is very complex and the algorithm runs quite slowly to get the classification results.

The classic spectral clustering algorithm has a superior performance in the category in any shape of data collection, but the computational complexity of the classic spectral clustering algorithm is very high. In the case of limited computer memory and computing speed, only can process low-dimensional image. To solve these problems, Bo et al. [6] improved Spectral Clustering Algorithm adopts a grid method that the original image was divided into several sub-images first and then processed separately. This is to reduce the computational complexity, but disadvantages of this approach is that on each sub-image, segmentation results easy to fall into local convergence. Zhao et al. [7] developed a fuzzy similarity measure for spectral clustering (FSSC) to replace the traditional measurements of spectral clustering algorithm used is based on Gaussian kernel function. Yang et al. [8] defined the density sensitive similarity measure which can adjust the distance in regions with different density to replace conventional Euclidean distance based and Gaussian kernel function based spectral clustering.

Moreover, Liu et al. [9] proposed a method, novel non-local spatial spectral clustering algorithm for image segmentation. In the presented method, the objective function of weighted kernel k-means algorithm is firstly modified by incorporating the nonlocal spatial constraint term. Then the equivalence between the objective functions of normalized cut and weighted kernel k-means with non-local spatial constraints is given and a novel non-local spatial matrix is constructed to replace the normalized Laplacian matrix. Finally, spectral clustering techniques are applied to this matrix to obtain the final segmentation result. Yan et al. [14] is inspired by density sensitive similarity measure. Making it increase the distance of the pairs of data in the high density areas, which are located in different spaces. And this can reduce the similarity degree among the pairs of data in the same density region, so as to find the spatial distribution characteristics complex data.

Besides, spectral clustering algorithms are sensitive to noise and other imaging artifacts because of not taking into account the spatial information of the pixels in the image [13]. Owing to the limitations of the feature space in multispectral images and spectral overlap of the clusters, it is required to use some additional information such as the spatial information of pixels in image. So the need for improvements in order to overcome the drawbacks mentioned above with spectral clustering method. We present a new approach named spatial spectral fuzzy clustering (SSFC) which combines spectral clustering and fuzzy clustering with local information into a unified framework to solve these problems. This paper constructed a new matrix S based on local information as the similarity matrix.

The paper is organized into five parts: Sect. 1 introduced; Sect. 2 spectral clustering; Sect. 3 a improved spectral clustering algorithm; Sect. 4 experimental and Sect. 5 conclusions.

2 Spectral Clustering

Spectral clustering techniques make use of the spectrum (eigenvalues) of the similarity matrix of the data to perform dimensionality reduction before clustering in fewer dimensions. The similarity matrix is provided as an input and consists of a quantitative assessment of the relative similarity of each pair of points in the dataset.

Let $X = \{x_1, x_2, \ldots, x_n\}$ be the set of n points to be clustered, and S be the $n \times n$ similarity matrix with its elements, s_{ij}, showing pairwise similarities between n points. Let $G = (V, S, X)$ be a weighted, undirected graph with V representing n nodes ($x_i \in X$ to be clustered), and S defining the edges. When constructing similarity graphs the goal is to model the local neighborhood relationships between the data points. There are other several popular constructions to transform the data points with pairwise similarities s_{ij} into a graph. S is usually constructed as a Gaussian function based on (often Euclidean) distances, $d(x_i, x_j)$, between samples x_i, x_j:

$$s_{ij} = \exp\left(-\frac{d^2(x_i, x_j)}{\sigma^2}\right) \qquad (1)$$

with a global parameter σ determining the decay of the similarity. This definition requires either a user-set σ value or a selection among many σ values to find the optimal value.

D is a diagonal matrix and its elements are the degrees of the nodes of G. The degree of each node, d_i, is computed with:

$$d_i = \sum_j s(i, j) \qquad (2)$$

The Laplacian matrix L, is constructed using the similarity matrix S and degree matrix D, depending on the approach for graph-cut optimization [11]. Ng et al. [12] define a normalized Laplacian matrix, L_{norm}, as:

$$L_{norm} = D^{-1/2}SD^{-1/2} \tag{3}$$

Then L_{norm} is used for extraction of k clustering by finding its k eigenvectors with the k highest eigenvalues. The spectral clustering algorithm can be summarized as follows:

1. Calculate a similarity matrix S (Eq. 1), diagonal degree matrix D (Eq. 2), and L_{norm} (Eq. 3).
2. Find the k eigenvectors $\{e_1, e_2, \ldots, e_k\}$ of L_{norm}, associated with the k highest eigenvalues $\{\lambda_1, \lambda_2, \ldots, \lambda_k\}$.
3. Construct the nxk matrix $E = [e_1, e_2, \ldots, e_k]$ and obtain nxk matrix U by normalizing the rows of E to have norm 1, i.e. $u_{ij} = e_{ij}/\sqrt{\sum_k e_{ik}^2}$.
4. Cluster the n rows of U with the clustering algorithm into k clusters.

3   Improved Spectral Clustering Algorithm

3.1   New Similarity Matrix and New Degree Matrix

In image segmentation, the key to determine a pixel belonging to certain area is based on the similarity of these colors, which is calculated through a distance function in the color space $d_{ij} = \|x_i - x_j\|$ e.g. Euclidean distance between the pattern x_i and x_j. However in fact, the shape and structure of the cluster also has a certain influence on the data clustering. Which means that together with information about the color of the pixel, the local information of pixels also need to be considered when clustering data.

We use a mask of size nxn to position on the image, the center pixel of the mask is the considered pixel. The number of neighboring pixels P is determined corresponding to the selected type of mask size i.e. 8 pixels for mask 3×3, 24 pixels for mask 5×5, 48 pixels for mask 7×7, so on.

To determine the degree of influence of the neighboring pixels for the center pixels, a local information measure M_i is defined on the basis of the distance $\|x_i - x_j\|$ and the attraction distance r_{ij}:

$$M_i = \sum_{j=1}^{P} \left(\|x_i - x_j\|r_{ij}\right)^{-1} / \sum_{j=1}^{P} r_{ij}^{-1} \tag{4}$$

in which $\|x_i - x_j\|$ is the distance of the all neighboring element x_j on the mask to the cluster x_i. The distance attraction r_{ij} is the squared Euclidean distance between elements

(x_i, y_i) and (x_j, y_j) about position on the mask. According to the above expression, local information of each pixel comes with a higher value if its color is similar to the color of neighboring pixels and vice versa. We use the inverse distance r_{ij}^{-1} because the closer the neighbors x_j of the center x_i are the more influence they exert on the result and vice versa.

The idea behind the use of this spatial relationship information can be outlined as follows: consider the local $n \times n$ mask, for sliding the mask on the image and calculating the local spatial information of the center pixel x_i based on the location of the center pixel x_i with the pixels x_j in the mask and the distance in color space $\|x_i - x_j\|$. This aims to reduce the effect of noise on the image. From above description, this method of similarity measure fully considers the local information and can avoid the influence of the image noise.

Set $r = max(r_{ij})_{\forall i,j}$ is the radius of the largest circle in which pixels that affect the central pixel.

Next, without loss of generality we standardized similar measurements on the following formula:

$$\overline{M}_i = \frac{M_i - min(M_i)_{\forall i}}{max(M_i)_{\forall i} - min(M_i)_{\forall i}} \tag{5}$$

From above description, a new similarity measure is defined as follows:

$$s_{ij} = \exp\left(-\frac{d^2(x_i, x_j)}{r^2}\right) \tag{6}$$

where s_{ij} showing pairwise similarities between pixels x_i, x_j; $d(x_i, x_j) = \|x_i - x_j\|$ is the Euclidean distance between x_i and x_j; r is the radius of the largest circle in which pixels that affect the central pixel. Similarity matrix S is usually constructed by s_{ij} according to the formula (6).

With degree matrix D, we build by adding local spatial information of each pixel, the degree of each pixel, d_i, is computed with:

$$d_i = \overline{M}_i * \sum_j s(i, j) \tag{7}$$

3.2   Fuzzy Clustering Algorithm

Next, due to the complexity of satellite imagery data, we chose the FCM clustering algorithm to improve the accuracy of the clustering result instead of using the k-Means algorithm. In general, fuzzy memberships in FCM achieved by computing the relative distance among the patterns and cluster centroids. Hence, to define the primary membership for a pattern, we define the membership using value of m. The use of fuzzifier gives a different objective function as follows:

$$J_m(U, v) = \sum_{k=1}^{N} \sum_{i=1}^{C} (u_{ik})^m d_{ik}^2 \tag{8}$$

In which $d_{ik} = \|x_k - v_i\|$ is Euclidean distance between the pattern x_k and the centroid v_i, C is number of clusters and N is number of patterns. Degree of membership u_{ik} is determined as follow:

$$u_{ik} = 1 / \sum_{j=1}^{C} \left(\frac{d_{ik}}{d_{jk}}\right)^{2/(m-1)} \tag{9}$$

Cluster centroids is computed as follows:

$$v_i = \sum_{k=1}^{N} (u_{ik})^m x_k / \sum_{k=1}^{N} (u_{ik})^m \tag{10}$$

In which $i = 1, \ldots, C$; $k = 1, \ldots, N$. Next, defuzzification for FCM is made as if $u_i(x_k) > u_j(x_k)$ for $j = 1, \ldots, C$ and $i \neq j$ then x_k is assigned to cluster i.

3.3   Spatial Spectral Fuzzy Clustering Algorithm (SSFC)

From the above description, the new Laplacian matrix L_{new}, is constructed using the new similarity matrix S and new degree matrix D:

$$L_{new} = D^{-1/2} S D^{-1/2} \tag{11}$$

The main steps of the proposed method are given as follows:

Input: Matrix size used to calculate local spatial information, number of clusters c.

Step 1. Calculate local information measure M_i by (4).
Step 2. Calculate a new similarity matrix S by (6).
Step 3. Calculate a diagonal degree matrix D by (7).
Step 4. Calculate a new matrix L_{new} by (11).
Step 5. Find the c eigenvectors $\{e_1, e_2, \ldots, e_c\}$ of L_{new}, associated with the c highest eigenvalues $\{\lambda_1, \lambda_2, \ldots, \lambda_c\}$ and define the c dimensional space $Y = (y_i)_{i=1,\ldots,n} \in R^c$.
Step 6. Calculates the function value u_{ij} by (9).
Step 7: Update centroids $c_i, i = 1, \ldots, k$ by (10).
Step 8: Calculates the J function value and checks the stop condition $max\{\|J^{(t+1)} - J^{(t)}\|\} \leq \varepsilon$, If satisfied, go to Output, otherwise return to step 6.

Output: Clustering results $C_1, C_2, \ldots, C_c$ with $C_i = \{x_j | u_{ij} \in c_i\}$. Evaluate accuracy, assign color to layers, and display results.

With this approach, the new algorithm will overcome the difficulties in the selection of parameters σ and pepper salt noise reduction in the image. This can increase the accuracy of the image clustering with spectral clustering algorithm.

3.4   Some Indicators Assess the Quality of Image After Clustering

To assess the quality of image segmentation, we use some indexes. $X = \{x_i\} = \{x_1, x_2, \ldots, x_N\}$ and $Y = \{y_i\} = \{y_1, y_2, \ldots, y_N\}$ corresponding to the original image and segment results image.

- Mean Squared Error (MSE) index [15]:

$$MSE(x, y) = \frac{1}{N} \sum_{i=1}^{N} (x_i - y_i) \tag{12}$$

Smaller MSE value is better quality clusters.
- Image Quality Index (IQI) [16]:

$$IQI = \frac{4\sigma_{xy}\bar{x}\bar{y}}{(\sigma_x^2 + \sigma_y^2)(\bar{x}^2 + \bar{y}^2)} \tag{13}$$

With $\bar{x} = \frac{1}{N}\sum_{i=1}^{N} x_i$, $\bar{y} = \frac{1}{N}\sum_{i=1}^{N} y_i$, $\sigma_x^2 = \frac{1}{N-1}\sum_{i=1}^{N} (x_i - \bar{x})$, $\sigma_y^2 = \frac{1}{N-1}\sum_{i=1}^{N} (y_i - \bar{y})$, $\sigma_{xy} = \frac{1}{N-1}\sum_{i=1}^{N} (x_i - \bar{x})(y_i - \bar{y})$. The best value 1 is achieved if and only if $y_i = x_i$, the lowest value of -1 occurs when $y_i = 2\bar{x} - x_i$ with $i = \overline{1, N}$.
- The Dunn's index (DI) [21] is defined as

$$DI = \min_{i \in c}\left\{ \min_{j \in c, j \neq i}\left\{ \frac{\delta(A_i, A_j)}{\max_{k \in c}\{\Delta(A_k)\}} \right\} \right\},$$

$$\text{in which,} \quad \begin{aligned} \delta(A_i, A_j) &= \min\{d(x_i, x_j)|x_i \in A_i; x_j \in A_j\} \\ \Delta(A_k) &= \max\{d(x_i, x_j)|x_i; x_j \in A_k\} \end{aligned} \tag{14}$$

and d is a distance function and A_i is the set whose elements are the data points assigned to the ith cluster. The main drawback with direct implementation of Dunn's index is its computation since calculation becomes computationally very expensive as c and n increases. If a data set contains well-separated clusters, the distances among the clusters are usually large and the diameters of the clusters are expected to be small. Therefore, large values of Dunn's index corresponds to good clustering solution.
- The CS measure is proposed to evaluate clusters with different densities and/or sizes [22]. It is computed as:

$$CSI = \frac{\frac{1}{c}\sum_{i=1}^{c}\left\{\frac{1}{|A_i|}\sum_{x_j \in A_i} max_{x_k \in A_i}\left\{d(x_j, x_k)\right\}\right\}}{\frac{1}{c}\sum_{i=1}^{c}\left\{min_{j \in c, j \neq i}\left\{d(v_i, v_j)\right\}\right\}} \qquad (15)$$

Where $|A_i|$ is the number of elements in cluster A_i and $d(x_j, x_k)$ a distance function. The smallest CS measure indicates a valid optimal clustering.

4   Experiments

4.1   Experiment 1

Synthetic Aperture Radar (SAR) is used to obtain high-resolution images from broad areas of terrain [17]. SAR is capable of operating under inclement weather conditions, day or night. SAR images have wide applications in remote sensing [18] and mapping of the surfaces of both the Earth and other planets [20]. There are many other applications for this technology such as environmental monitoring, earth-resource mapping, reconnaissance and targeting information to military operations, oil spill classification [19], and so on. However, segmentation SAR image is a difficult task in remote sensing applications due to the influence of the speckle noise (see Fig. 1), therefore, using conventional methods will not be inefficient with speckle noise. To testing SSFC algorithm that we proposed, SAR image segmentation with 2 areas with oil spill on the sea.

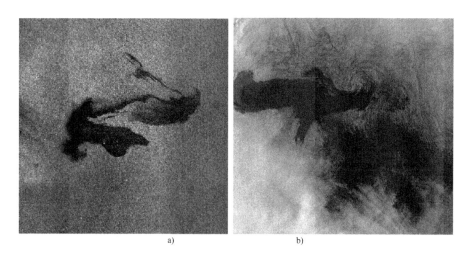

a)   b)

Fig. 1. Spill oil area on Envisat ASAR image in Gulf of Mexico (a) 26/04/2010, (b) 29/04/2010

Test data that Asar envisat images was taken from a spill oil area in Gulf of Mexico on *26/04/2010* (1a) and *29/04/2010* (1b), with coordinates ($0°14'02.75"N$, $0°03'56.39"E$ to $0°04'27.33"N$, $0°22'13.94"E$) with area of *23.32 hectare*. Easily recognized oil stains on Fig. 1a with clearer boundaries, whereas in Fig. 1b oil stains have long existed in the sea, so the contrast with the surrounding waters as well as the boundaries of oil stains is not clarity, many parts mixed with water and oil stains area has spread.

Classification results shown in Fig. 2 in Gulf of Mexico on 26/04/2010, the FCM, SC, ISC and the SSFC algorithm with Fig. 2a, b, c and d, respectively. In Fig. 2, still quite a lot of noise on the Fig. 2a, b and c, especially, in Fig. 2a. Classification results in Fig. 2d shows the noise almost nonexistent on water layer spill area is also clear than other results.

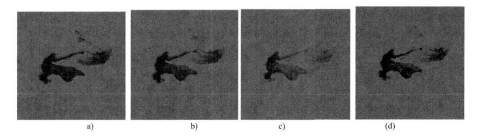

Fig. 2. Result classification oil stains on the Envisat ASAR image in Gulf of Mexico on 26/4/2010

Table 1. Performance of the FCM, SC, ISC and the SSFC algorithms on 26/4/2010

Index	FCM	SC	ISC	SSFC
MSE	0.1871	0.1642	0.1212	**0.0986**
IQI	0.4595	0.5638	0.7851	**0.8968**
DI	0.0186	0.0317	0.0561	**0.0659**
CSI	1.1872	0.9981	0.8725	**0.6521**

Figure 3 shown classification results in Gulf of Mexico on 26/04/2010, the FCM, SC, ISC and the SSFC algorithm with Fig. 3a, b, c and d, respectively. Easy to see, the noise is reduced quite good on all the results in Fig. 3 because SAR image data has quite little pepper salt noise (see Fig. 3b), however a large amount of information about the oil stains on the Fig. 3a, b and c is confused with noise, therefore undetectable. Test result with our algorithm on Fig. 3d shows, not only is the noise reduction possible but that oil stains classification result is also complete and clearer.

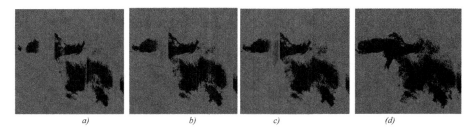

Fig. 3. Result classification oil stains on the Envisat ASAR image in Gulf of Mexico on 29/4/2010

Table 2. Performance of the FCM, SC, ISC and the SSFC algorithms on 29/4/2010

Index	FCM	SC	ISC	SSFC
MSE	0.1761	0.1578	0.1082	**0.0082**
IQI	0.4862	0.5892	0.6823	**0.9447**
DI	0.0372	0.0461	0.0598	**0.0872**
CSI	1.5786	1.2319	0.8873	**0.5619**

Tables 1 and 2 indicate the value of the index assessing the quality of clustering results. Overall, the results classified according algorithm that we propose for better results than the algorithm FCM, SC and ISC. Based on the value of this index, the FCM algorithm for clustering result is the worst, followed by SC and ISC algorithm.

Fig. 4. Lamdong provice (a) Color image; (b) NDVI Image (Color figure online)

4.2   Experiment 2

The second experiments are more visible and could be found from multi spectral remote sensing images. The pixel information in these images is acquired from different temporal sensors. We test the proposed method with a satellite image LANDSAT-7 taken at Lamdong province in the south of Vietnam, see Fig. 4, 107° $28'$ $24.61''$E, 12° $13'04.66''$N to 108° $54'52.39''$N, 11° $37'36.87''$E, with area is 9958.6 km^2 and capacity is 116.89 Mb.

To perform image segmentation into 6 layers: Class1: Rivers, ponds, lakes ▬; Class2: Rocks, bare soil ▬; Class3: Fields, grass ▬; Class4: Planted forests, low woods ▬; Class5: Perennial tree crops ▬; Class6: Jungles ▬.

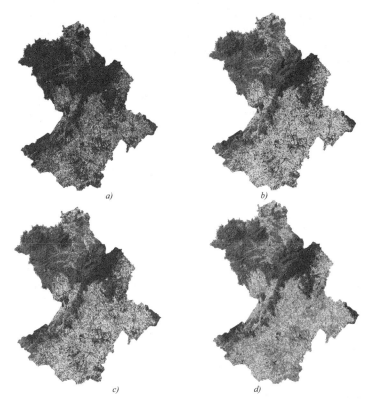

Fig. 5. Result land-cover classification on Landsat 7-TM image in Lamdong province. (Color figure online)

Table 3. Performance of the FCM, SC, ISC and the SSFC algorithms

Index	FCM	SC	ISC	SSFC
MSE	0.1763	0.1452	0.1075	**0.0918**
IQI	0.5623	0.6124	0.6732	**0.8721**
DI	0.0123	0.0109	0.0365	**0.0452**
CSI	1.2512	0.9683	**0.7750**	0.7751

To compare the proposed algorithm with the methods has been studied before, we test empirically on four methods FCM, SC, ISC and SSFC has proposed in this paper. The output images is shown in Fig. 5 above. Easy to see in Fig. 5a, classification result is unclear, there are many water noise appear in the result image. In Fig. 5b and c noise significantly reduced, particularly on the amount of noise in Fig. 5d has fallen strongly and oil stains is clearer.

To assess the performance of the algorithms on the experimental images, we analyzed the results on the basis of several validity indexes. We considered several validity indexes such as the MSE, IQI, DI and CSI. Table 3 shows that IQI index and

DI index are the largest with 0.8721 and 0.0452 on SSFC algorithm, while ISC algorithm is 0.6732 and 0.0365; These indicators decrease with 0.6124 and 0.5623; 0.0109 and 0.0123, respectively on algorithms SC and FCM. With MSE index, the largest is 0.1763 when tested on FCM algorithm, and descending on algorithms SC, ISC and SSFC, while, the lowest value is 0.0918 on SSFC algorithm. The smallest value of CSI index is 0.7750 with ISC algorithm, while, SSFC algorithm is 0.7751 and we easily see that this deviation is insignificant. While CSI index with SC algorithm is 0.9683 and FCM algorithm is 1.2512.

Thus, based on the value of the index clustering quality evaluation, the most of the cases showed SSFC algorithm for clustering results better than the algorithm ISC, SC and FCM.

5   Conclusion

In this study, we have presented a new spatial spectral fuzzy clustering algorithms (SSFC) based on local information. The first one is based on spatial relationships between the pixels and their neighbors in circle to similarity matrix. The second one uses the set of local information values to degree matrix. The advantages of the proposed algorithms are pointed out in reducing noise on image and overcome selection among many σ values to find the optimal value. Test results show that proposed algorithm has high segmentation accuracy and significantly reduces the computational complexity of classical spectral clustering algorithm and through experimental results, according to the visual and validity indexes MSE, IQI, DI and CSI, basically SSFC for sharper image quality, better noise reduction.

There are several further research directions including classification methods based on deep learning in the supervised classification, semi-supervised classification to processing multispectral, hyperspectral satellite images with application in landcover classification, environmental classification and assessment of landcover changes. The issues of speed-ups of the proposed methods based on GPU platforms form another important research direction.

Acknowledgements. This research is funded by Vietnam National Foundation for Science and Technology Development (NAFOSTED) under grant number 102.05-2016.09.

References

1. Li, Y., Liu, X., Yan, X.: A modified spectral clustering algorithm based on density. In: Zu, Q., Hu, B. (eds.) HCC 2016. LNCS, vol. 9567, pp. 901–906. Springer, Cham (2016). https://doi.org/10.1007/978-3-319-31854-7_97
2. Kuo, C.-T., Walker, P.B., Carmichael, O., Davidson, I.: Spectral clustering for medical imaging. In: 2014 IEEE International Conference on Data Mining, pp. 887–892 (2014). https://doi.org/10.1109/icdm.2014.143. 1550-4786/14 $31.00 © 2014 IEEE

3. Peluffo-Ordóñez, D.H., Alvarado-Pérez, J.C., Castro-Ospina, A.E.: On the spectral clustering for dynamic data. In: Ferrández Vicente, J.M., Álvarez-Sánchez, J.R., de la Paz López, F., Toledo-Moreo, Fco.Javier, Adeli, H. (eds.) IWINAC 2015. LNCS, vol. 9108, pp. 148–155. Springer, Cham (2015). https://doi.org/10.1007/978-3-319-18833-1_16
4. Liu, C.-A., Guo, Z., Liu, C., Zhou, H.: An image-segmentation method based on improved spectral clustering algorithm. In: Qi, L. (ed.) ISIA 2010. CCIS, vol. 86, pp. 178–184. Springer, Heidelberg (2011). https://doi.org/10.1007/978-3-642-19853-3_26
5. Fatemi, S.B., Mobasheri, M.R., Abkar, A.A.: Clustering multispectral images using spatial–spectral information. IEEE Geosci. Remote Sens. Lett. **12**(7), 1521–1525 (2015). https://doi.org/10.1109/lgrs.2015.2411558
6. Bo, H., Zhang, J., Wang, X.: Improving spectral clustering algorithm based SAR spill oil image segmentation. In: 2011 IEEE International Conference on Network Computing and Information Security. IEEE. https://doi.org/10.1109/ncis.2011.172
7. Zhao, F., Liu, H., Jiao, L.: Spectral clustering with fuzzy similarity measure. Dig. Signal Process. **21**, 701–709 (2011). https://doi.org/10.1016/j.dsp.2011.07.002
8. Yang, P., Zhu, Q., Huang, B.: Spectral clustering with density sensitive similarity function. Knowl. Based Syst. **24**, 621–628 (2011). https://doi.org/10.1016/j.knosys.2011.01.009
9. Liu, H.Q., Jiao, L.C., Zhao, F.: Non-local spatial spectral clustering for image segmentation. Neurocomputing **74**, 461–471 (2010). https://doi.org/10.1016/j.neucom.2010.08.021
10. Higham, D.J., Kalna, G., Kibble, M.: Spectral clustering and its use in bioinformatics. J. Comput. Appl. Math. **204**, 25–37 (2007). https://doi.org/10.1016/j.cam.2006.04.026
11. Shi, J., Malik, J.: Normalized cuts and image segmentation. IEEE Trans. Pattern Anal. Mach. Intell. **22**(8), 888–905 (2000). https://doi.org/10.1109/34.868688
12. Ng, A., Jordan, M., Weiss, Y.: On spectral clustering: analysis and an algorithm. In: Dietterich, T., Becker, S., Ghahramani, Z. (eds.) Advances in Neural Information Processing Systems, vol. 14. MIT Press (2002)
13. Fowlkes, C., Belongie, S., Chung, F., Malik, J.: Spectral grouping using the Nystrom method. IEEE Trans. Pattern Anal. Mach. Intell. **26**(2), 214–225 (2004). https://doi.org/10.1109/TPAMI.2004.1262185
14. Yan, J., Cheng, D., Zong, M., Deng, Z.: Improved spectral clustering algorithm based on similarity measure. In: Luo, X., Yu, J.X., Li, Z. (eds.) ADMA 2014. LNCS (LNAI), vol. 8933, pp. 641–654. Springer, Cham (2014). https://doi.org/10.1007/978-3-319-14717-8_50
15. Wang, Z., Bovik, A.C.: Mean squared error: love it or leave it? A new look at signal fidelity measures. IEEE Signal Process. Mag. **26**(1), 98–117 (2009). 1053-5888/09/$25.00©2009IEEE
16. Wang, Z., Bovik, A.C.: A universal image quality index. IEEE Signal Process. Lett. **9**(3), 81–84 (2002)
17. Tirandaz, Z., Akbarizadeh, G.: Unsupervised texture-based SAR image segmentation using spectral regression and gabor filter bank. J. Indian Soc. Remote Sens. **44**, 177 (2016). https://doi.org/10.1007/s12524-015-0490-0
18. Ma, M., Liang, J., Guo, M., Fan, Y., Yin, Y.: SAR image segmentation based on Artificial Bee Colony algorithm. Appl. Soft Comput. **11**(8), 5205–5214 (2011). https://doi.org/10.1016/j.asoc.2011.05.039
19. Karantzalos, K., Argialas, D.: Automatic detection and tracking of oil spills in SAR imagery with level set segmentation. Int. J. Remote Sens. **29**(21), 6281–6296 (2008). https://doi.org/10.1080/01431160802175488
20. Boldt, M., Thiele, A., Schulz, K., Hinz, S.: SAR image segmentation using morphological attribute profiles. Int. Arch. Photogram. Remote Sens. Spat. Inf. Sci. **XL-3**, 39–44 (2014). ISPRS Technical Commission III Symposium, Zurich, Switzerland
21. Rendón, E., Abundez, I., Arizmendi, A., Quiroz, E.M.: Internal versus external cluster validation indexes. Int. J. Comput. Commun. **5**(1), 27–34 (2011)

22. Chou, C.H., Su, M.C., Lai, E.: A new cluster validity measure and its application to image compression. Pattern Anal. Appl. **7**, 205–220 (2004). https://doi.org/10.1007/s10044-004-0218-1

23. Mai, S.D., Ngo, L.T.: Interval type-2 Fuzzy C-means clustering with spatial information for land-cover classification. In: Nguyen, N.T., Trawiński, B., Kosala, R. (eds.) ACIIDS 2015. LNCS (LNAI), vol. 9011, pp. 387–397. Springer, Cham (2015). https://doi.org/10.1007/978-3-319-15702-3_38

24. Mai, D.S., Ngo, L.T.: Semi-supervised Fuzzy C-means Clustering for change detection from multispectral satellite image. In: 2015 IEEE International Conference on Fuzzy Systems, pp. 1–8 (2015). https://doi.org/10.1109/fuzz-ieee.2015.7337978

25. Mai, D.-S., Trinh, L.-H., Ngo, L.-T.: Combining fuzzy probability and Fuzzy clustering for multispectral satellite imagery classification. Vietnam J. Sci. Technol. **54**(3), 300–313 (2016). https://doi.org/10.15625/0866-708x/54/3/6463. ISSN 0866-708x

Classification of Food Images through Interactive Image Segmentation

Sanasam Inunganbi, Ayan Seal$^{(\boxtimes)}$, and Pritee Khanna

Computer Science and Engineering,
PDPM Indian Institute of Information Technology,
Design and Manufacturing Jabalpur, Jabalpur, India
`inung.sam@gmail.com`, {`ayan,pkhanna`}`@iiitdmj.ac.in`

Abstract. Food item segmentation in an image is a kind of fine-grained segmentation task, which is comparatively difficult than conventional image segmentation because intra-class variance is high and inter-class variance is low. So, an interactive food item segmentation algorithm using Random Forest is proposed in this work. The first step of the proposed algorithm is interactive food image segmentation, where food parts are extracted based on user inputs. It is observed that some of the segmented food parts may have some holes due to improper distribution of light. So, Boundary Detection & Filling and Gappy Principal Component Analysis methods are applied to restore the missing information in the second step. Local Binary Pattern and Non Redundant Local Binary Pattern are used for extracting features from the restored food parts, which are fed into support vector machine classifier for differentiating one food image from others. All the experiments have been performed on Food 101 database. A comparative study has also been done based on the three existing methods. The obtained results demonstrate that the proposed method outperforms the existing methods.

Keywords: Food images · Interactive image segmentation
Occlusion detection · Image restoration · Classification

1 Introduction

Improper food intake results in unfavorable effects on health by disrupting natural processes. They adversely affect physical as well as emotional strength of individuals. It is thus advisable to understand the calories intake of consuming food items. Self-reporting helps to keep track of food intake. But, it does not accurately reflect eating habits of human in real life. So, self-reporting alone could not be treated as core part of food consumption reporting system. Food images classification might be helpful in keeping track of everyday food consumption and considered as the core part of dietary intake reporting system. In fact, no practical systems for food image recognition and caloric value calculation exist at present. This motivates us to work in this field. In this work, we

© Springer International Publishing AG, part of Springer Nature 2018
N. T. Nguyen et al. (Eds.): ACIIDS 2018, LNAI 10752, pp. 519–528, 2018.
https://doi.org/10.1007/978-3-319-75420-8_49

have proposed a method for extracting food part(s) from whole image and then recognize food items so that caloric value of the food items can be identified easily. However, caloric value calculation will be done in near future.

The organization of the paper is as follows: In Sect. 2, some of the recent and relevant works have been discussed. The details of proposed method has been depicted in Sect. 3. All the experimental results with discussion have been presented in Sect. 4. Finally, the conclusion has been drawn in Sect. 5.

2   Related Work

The availability of a standard database containing large sets of food images with a variety of food items is essential for designing and testing an algorithm. One such database on food images was developed by M. Chen et al. called Pittsburg Fast Food Image (PFI) Dataset [2]. All the images and videos of PFI Database were collected in restaurant and laboratory environment. The authors proposed two methods based on Color histogram and Scale Invariant Feature Transform (SIFT) [3]. In [4], the authors have explored ten GLCM features namely, Angular Second Moment (ASM), Contrast, Variance, Correlation, Inverse Difference Moment (IDM), Entropy, Sum Entropy, Sum Variance, Cluster Shade, and Maximum Probability for classifying food images. A novel method for food classification had been proposed by Bossard et al. [5] using discriminative components and Random Forest (RF). The method was tested on Food 101 dataset and 50.76% average accuracy is obtained. A standard RGB 3D histogram with four quantization levels per channel was exploited in [2]. A $4 \times 4 \times 4$ bin (one for each channel) was used to find 64 features for classifying food images using support vector machine (SVM). For analyzing texture, local binary pattern (LBP) and non redundant local binary pattern (NRLBP) were extracted in [6]. Experiments were conducted on 6 classes namely, sandwiches, meat, salads, donuts, hamburger, and miscellaneous from PFI database. Authors have also performed experiments on their own dataset having 5 categories such as cake, carrot, custard, pizza and pasta. Average accuracy using the two methods on PFI dataset were 68% and 69%, while the same on their own dataset were 53% and 63%. Shape information of bread images was exploited in [7] for classification. Image Binarization was done first to extract its shape and size including area (pixel counts), elongation (ratio of the difference between major and minor axes to their sum), and Minimum Bounding Rectange (MBR). However, shape and size of food items may not be always available specially in case of multiple food recognition system. SIFT was used to form Bag of Feature (BoF), which was fused with color histogram and Gabor features by multiple kernel learning (MKL) method for training purpose of SVM in [8]. The obtained classification rate is 61.34 % for 50 classes. Same authors got 62.52% accuracy for 85 classes using the previous features along with gradient histogram [9]. In [10], authors captured 1453 images of 45 food items and analyzed them [11]. The obtained accuracies were 34 % and 63% for top 1 and top 4 food categories respectively. In [11], scalable color descriptor, dominant color descriptor and texture descriptors such as entropy-based categorization, fractal dimension estimation, gabor-based image decomposition [12] were utilized for classification. In [10], SIFT and Multi-scaled Dense SIFT were integrated with the previously used features in [11] and accuracies improved 22% for top 1 and 10% for top 4 food categories respectively. It is clear from the literature that most of the researchers applied different feature extraction algorithms on the whole image because an image is considered to be a food image only except for [10,11], where multiple food items are there in an image. But, in practical, it is not always true. In most of the cases, food images include other irrelevant object(s) and background. Thus, irrelevant object(s) and background need to be removed before extracting features from food parts since features from the food parts are mainly responsible in distinguishing one food item from other.

3   Proposed Methodology

The proposed system consists of four steps namely, interactive image segmentation, occlusion detection and restoration, extraction of features, and classification. The detailed discussion of these steps are given in the following sub-sections.

3.1   Interactive Image Segmentation

Generally, segmentation is the first step in any attempt to analyze or interpret an image automatically. Segmentation is the process of dividing an image into groups of connected pixels based on some similarity measures. The objective of image segmentation is to represent an image in such a way that it can be simpler to analyze. Numerous segmentation algorithms mentioned in literature which can be categorized into two ways namely, unsupervised/clustering based and supervised/interactive based [13–15].

Clustering based segmentation algorithms require the number of clusters or segments as an input before starting their operation. It is not fixed for images and most of the time, it is difficult to calculate since the objects may be mixed together. Different people interpret a single image differently, and hence what to capture from the image also differs. Therefore, an interactive or supervised based segmentation algorithm is required for the proposed purpose which would segment selective image part based on users' choice on what is important and for what purpose. This task is motivated by the work of [16], which uses user provided scribble to mark foreground and background for automatic segmentation of whole image.

In this work, segmentation has been done at pixel level by reducing it to a two class classification problem, where each pixel of an image is considered to be either foreground (desired object) or background (unwanted part). After marking, unwanted parts need to be discarded since we need to concentrate only on food object for effective classification. Random Forest (RF) ensemble learning method has been adopted to classify each pixel in either food part or background [17]. RF generates many classifiers and takes decision based on their aggregated results. Like other classifiers, RF has two phases namely, training and testing. The training process is conducted based on the user input. It means training samples are being formed based on the portions of an image selected by a user during runtime. One portion is marked from the desired object and the other is selected from the background. Color components from the desired object and the background are considered as features for training and testing. The RGB components of the input image are extracted, and converted to Lab space since Lab color space is believed to be closer to human perception.

After taking user input, image is segmented in such a way that foreground is shown as white and background as black. To do this, all pixels coming under background polygon are labeled as class 1 and the others, as class 2, while the rest are not given class label. For training only those pixels with class labels are used and for testing all pixels of the image are used. After testing, every pixel of the image would be allotted a class label. And based on the class label, a binary image is formed, where pixels value as 1 and 0 for foreground and background region respectively.

Sometimes, more than one object could be there in the obtained binary image. But, the largest object is our region of interest (ROI). So, the other objects except the largest one have to be removed. Morphological Opening is one such operation which is used to remove all unwanted objects from the binary image [18]. After performing opening operation, only the largest object is retained. The image with the largest object is considered as a mask and the RGB components are restored from the original image. So, the final resulting image is colored for the foreground and black for the background. The output of the various steps of the interactive segmentation process has been depicted in Fig. 1. The output of a few images from the food 101 dataset after applying interactive segmentation algorithm have been given in Fig. 2.

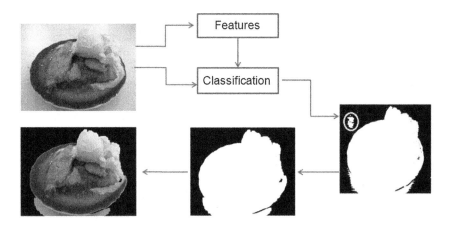

Fig. 1. The output of different steps of segmentation process (Color figure online)

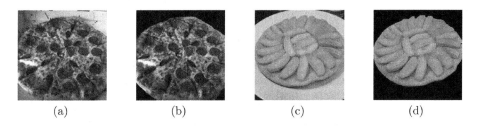

(a)   (b)   (c)   (d)

Fig. 2. Some of the original ((a) and (c)) and segmented images ((b) and (d)) corresponding to them from Food 101 database (Color figure online)

Fig. 3. Segmented image with occluded regions (Color figure online)

(a)   (b)   (c)   (d)

Fig. 4. (a) Boundary of segmented image (b) Filled area (c) Occluded area (d) Filled image (Color figure online)

3.2 Occlusion Detection and Restoration of Image

Segmented images may consist of some holes, which carry no information. It happens due to uneven distribution of illumination. One occluded image is shown in Fig. 3 for a clear visualization of the occluded part. The performance of a classifier deter due to occlusion. As segmentation process depends on color components which depends on illumination. Hence, segmentation process depends on illumination. Two different approaches have been used to restore the occluded parts in an image. First method is proposed to restore the occluded part in an image and second method existing in literature has been applied [19].

The whole process of BDF method is completed in four steps as shown in Fig. 4 and summarized below:

1. The first step of is boundary detection, depicted in Algorithm 1. The findBoundary() algorithm works on binary image. Thus, before applying

findBoundary() algorithm, segmented image needs to be converted to its corresponding binary image. For this purpose, mean of the segmented image is used as a threshold value. If the value of a particular pixel is less than the threshold then the corresponding pixel is treated as background otherwise it is considered as foreground. The boundary of the segmented image of Fig. 3 has been shown in the Fig. 5(a).

Algorithm 1. $findBoundary(SI)$

Input: SI: Segmented image in which boundary has to be found
1: $\forall_{i=1}^{row} \forall_{j=1}^{col} \exists i,j \ s.t \ SI(i,j) == 1 \ then \ //$ row and col are the number of pixels in y and x direction respectively
2:     set $flag_{north}, flsg_{south}, flag_{east}, flag_{west}$ to 0
3:     $\forall_{n=1}^{i-1} \exists n \ s.t \ SI(n,j) == 1 \ then$
           set $flag_{north}$ to 1
4:     $\forall_{s=i+1}^{col} \exists s \ s.t \ SI(s,j) == 1 \ then$
           set $flag_{south}$ to 1
5:     $\forall_{e=j+1}^{row} \exists e \ s.t \ SI(i,e) == 1 \ then$
           set $flag_{east}$ to 1
6:     $\forall_{w=1}^{j-i} \exists w \ s.t \ SI(i,w) == 1 \ then$
           set $flag_{west}$ to 1
7:     $flag = flag_{north} + flag_{south} + flag_{east} + flag_{west}$
8:     $if(flag < 4) \ then \ (i,j) \ is \ considered \ to \ be \ boundary \ pixel$

---

2. A mask has been created based on all the pixels inside the boundary, which has been used further to help in detecting occluded part of the segmented image as shown in Fig. 5(b).
3. The newly created mask is mapped on the segmented image and if the intensity value of a pixel of the mapped region has been below a threshold, then this pixel has been considered to belong to occluded part, otherwise it is considered to be normal pixel. The mean value of the segmented image is considered as threshold value initially. But, this method of finding the threshold value has not been effective. So, first order derivation of the segmented image histogram has been found, and then an index is considered, where the first least positive number has been detected. The index is set as the threshold value for this purpose. The occluded part(s) has been shown in Fig. 5(c).
4. If the pixel has been identified as occluded, then its intensity value is replaced by the intensity value of the original image, otherwise it is kept as it is.

(a)   (b)   (c)   (d)

Fig. 5. Segmented images ((a) and (c)) and restored images ((b) and (d)) using BDF method (Color figure online)

The restored image obtained after applying BDF method has been shown in Fig. 5(d). Some of the restored images using BDF method has been shown in Fig. 5.

Another method, GPCA has been used to detect and restore the occluded part(s) of segmented food image. The detail discussion of GPCA is presented in [19]. A few restored food images obtained using GPCA method have been shown in Fig. 6.

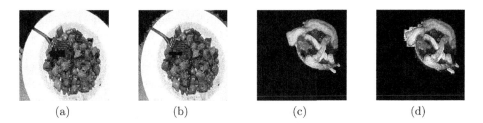

| (a) | (b) | (c) | (d) |

Fig. 6. Segmented images ((a) and (c)) and restored images ((b) and (d)) using GPCA method (Color figure online)

3.3   Feature Extraction and Classification

LBP, introduced by Ojala et al. [20], is widely used to analyse image texture. It may be defined as an ordered set of binary numbers formed by thresholding between central and its neighbor pixels. The generated order set of binary number is converted into decimal by multiplying with the weighted power of 2. The LBP has a number of advantages. First, its computation is simple. Second, it provides highly discriminant features. Moreover, the difference of the central pixel with its neighbors is invariant to the changes of the mean gray value of the image. Input image has been divided into gray-scale and its corresponding gradient. Then, LBP and NRLBP have been applied separately on each of them, giving four sets of feature by concatenating various combination of LBP and NRLBP histogram. NRLBP treats LBP and its compliments as the same. The directional change in the pixel intensities of an image is provided by gradient image and it has been obtained using sobel masks on the restored gray-scale image. Concatenation of information extracted from the gray-scale and corresponding gradient image will increase the discriminative power. The 0th histogram of LBP is neglected as it only represents the background. The combination of the four feature sets are:

Feature Set 1: LBP histogram of the gradient image is concatenated with the LBP histogram of the gray-scale image.

Feature Set 2: NRLBP histogram of the gradient image is concatenated with the LBP histogram of the gray-scale image.

Feature Set 3: NRLBP histogram of the gradient image is concatenated with the NRLBP histogram of the gray-scale image.

Feature Set 4: LBP histogram of the gradient image is concatenated with the NRLBP histogram of the gray-scale image.

These features are fed separately into SVM classifier for classification purpose.

4 Experimental Results and Discussion

Experiments have been carried out on a publicly available Food 101 dataset having 100 images each of 101 classes [21]. In this work, fifty out of hundred and one categories of food, having 20 images each, have been properly segmented and restored for classification. Total twenty experiments have been performed based on features and number of categories of food. Four different feature sets discussed in Sect. 3.3 have been extracted from 10, 20, 30, 40, and 50 categories restored food images separately. Then SVM is used to classify these images. Four fold cross validation has been performed and average accuracy is reported. Table 1 summarizes the classification accuracy for each of these experiments conducted for BDF and GPCA methods.

It is clear from Table 1 that feature set 4 gives best results for the resorted food images using BDF and GPCA. It is also observed that the accuracy decreases with the increase of number of categories.

Comparison has been performed on the combination of proposed interactive image segmentation algorithm, occlusion restoration algorithms, feature set 4, and SVM classifier and some of the existing methods from literature. The results are summarized in Table 2. It can be inferred from the Table 2 that the proposed work is giving better accuracy than the existing methods from literature even if the number of classes increase.

Table 1. Comparison of classification accuracy (in %) for various experiments

Feature set	Method for segmented image restoration	Number of classes				
		#10	#20	#30	#40	#50
#1	BDF	91	62	46.5	38	36
	GPCA	90	81.25	72.17	58.52	50.60
#2	BDF	91.5	58.25	43	38.25	33
	GPCA	88	72	63	53	45
#3	BDF	89	70	61	52	44
	GPCA	88	79	65	53.88	44.4
#4	BDF	89.5	74	67	58.13	51.6
	GPCA	90.5	82	71.83	58	49.9

Table 2. Comparison of the accuracy (in %) of the proposed method with existing works

Name of the feature	Number of classes				
	#10	#20	#30	#40	#50
3D Histogram [1]	17.5	8.75	6	4.88	3.60
GLCM [3]	10	5.25	3.50	2.19	1.70
LBP [5]	18	4.75	3.17	2.13	1.30
NRLBP [5]	11	4.74	3.83	2.87	2.50
BDF+feature set #4	89.5	74	67	58.13	51.6
GPCA+feature set #4	90.5	82	71.83	58	49.9

5 Conclusions

In this work, the effectiveness of an interactive image segmentation algorithm for food item segmentation in an image has been examined, which helps us to recognize food items. The proposed method is validated by four-fold cross validation technique on a publicly available Food 101 dataset. The dataset consists of 101 classes, however, 50 classes have been considered for this work. BDF and GPCA are then used for restoring the missing information due to occlusion in food parts. Then features are extracted using LBP and NRLBP which are fed into SVM classifier for recognizing food images. Overall, the performance of the proposed method is good as compared to existing methods.

Acknowledgment. Authors are thankful to a project entitled Privacy Enhancing Revocable Biometric Identities funded by DAE, BRNS, GOI and PDPM IIITDM Jabalpur for providing necessary infrastructure to conduct experiments relating to this work. Ayan Seal is grateful to Media Lab Asia, MeitY, GOI for providing him Visvesvaraya young faculty research fellowship.

References

1. http://www.who.int/mediacentre/factsheets/fs311/en/
2. Chen, M., Dhingra, K., Wu, W., Yang, L., Sukthankar, R., Yang, J.: PFID: Pittsburgh fast-food image dataset. In: 16th IEEE International Conference on Image Processing, pp. 289–292 (2009)
3. Lowe, D.G.: Distinctive image features from scale-invariant keypoints. Int. J. Comput. Vis. **60**(2), 91–110 (2004)
4. Chen, Q., Agu, E.: Exploring statistical GLCM texture features for classifying food images. In: International Conference on Healthcare Informatics, pp. 453–453 (2015)
5. Bossard, L., Guillaumin, M., Van Gool, L.: Food-101 – mining discriminative components with random forests. In: Fleet, D., Pajdla, T., Schiele, B., Tuytelaars, T. (eds.) ECCV 2014. LNCS, vol. 8694, pp. 446–461. Springer, Cham (2014). https://doi.org/10.1007/978-3-319-10599-4_29
6. Nguyen, D.T., Zong, Z., Ogunbona, P.O., Probst, Y., Li, W.: Food image classification using local appearance and global structural information. Neurocomputing **140**, 242–251 (2014)
7. Pishva, D., Kawai, A., Shiino, T.: Shape based segmentation and color distribution analysis with application to bread recognition. In: IAPR Conference on Machine Vision Applications (2000)
8. Joutou, T., Yanai, K.: A food image recognition system with multiple kernel learning. In: 16th IEEE International Conference on Image Processing, pp. 285–288 (2009)
9. Hoashi, H., Joutou, T., Yanai, K.: Image recognition of 85 food categories by feature fusion. In: IEEE International Symposium on Multimedia, pp. 296–301 (2010)
10. He, Y., Xu, C., Khanna, N., Boushey, C.J., Delp, E.J.: Analysis of food images: features and classification. In: IEEE International Conference on Image Processing, pp. 2744–2748 (2014)

11. He, Y., Xu, C., Khanna, N., Boushey, C.J., Delp, E.J.: Food image analysis: segmentation, identification and weight estimation. In: IEEE International Conference on Multimedia and Expo, pp. 1–6 (2013)
12. Bosch, M., Zhu, F., Khanna, N., Boushey, C.J., Delp, E.J.: Food texture descriptors based on fractal and local gradient information. In: 19th European Conference on Signal Processing, pp. 764–768 (2011)
13. Sridevi, M., Mala, C.: A survey on monochrome image segmentation methods. In: 2nd International Conference on Communication, Computing & Security, vol. 6, pp. 548–555 (2012)
14. Khan, W.: Image segmentation techniques: a survey. J. Image Graph. 1(4), 166–170 (2013)
15. Mageswari, S.U., Sridevi, M., Mala, C.: An experimental study and analysis of different image segmentation techniques. In: International Conference on Design and Manufacturing, vol. 64, pp. 36–45 (2013)
16. Bai, X., Sapiro, G.: Geodesic matting: a framework for fast interactive image and video segmentation and matting. Int. J. Comput. Vis. 82(2), 113–132 (2009)
17. Breiman, L.: Consistency for a simple model of random forests (2004)
18. Gonzalez, R.C., Woods, R.E.: Digital Image Processing. Pearson Education, London (2002)
19. Seal, A., Bhattacharjee, D., Nasipuri, M., Basu, D.K.: UGC-JU face database and its benchmarking using linear regression classifier. Multimed. Tools Appl. 74(9), 2913–2937 (2015)
20. Ojala, T., Pietikäinen, M., Harwood, D.: A comparative study of texture measures with classification based on featured distributions. Pattern Recogn. 29(1), 51–59 (1996)
21. https://www.vision.ee.ethz.ch/datasets_extra/food-101/

A Multiresolution Approach for Content-Based Image Retrieval Using Wavelet Transform of Local Binary Pattern

Manish Khare[1], Prashant Srivastava[2(✉)], Jeonghwan Gwak[3], and Ashish Khare[2]

[1] Dhirubhai Ambani Institute of Information and Communication Technology,
Gandhinagar, Gujarat, India
mkharejk@gmail.com

[2] Department of Electronics and Communication, University of Allahabad,
Allahabad, Uttar Pradesh, India
prashant.jk087@gmail.com, ashishkhare@hotmail.com

[3] Department of Radiology, Biomedical Research Institute,
Seoul National University Hospital, Seoul 03080, Republic of Korea
james.han.gwak@gmail.com

Abstract. The emergence of low cost digital cameras and other image capturing devices has created a huge amount of different types of images. Accessing images easily requires proper arrangement and indexing of images. This has made image retrieval an important problem of Computer Vision. This paper attempts to decompose a Local Binary Pattern (LBP) image at multiple resolution to extract structural arrangement of pixels more efficiently than processing a single scale of the LBP image. LBP descriptors of the 2-D gray scale image are computed followed by computation of Discrete Wavelet Transform (DWT) coefficients of the resulting 2-D LBP image. Finally, construction of feature vector is done through Gray-Level Co-occurrence Matrix. Performance of the proposed method is tested on two benchmark datasets, Corel-1K and Corel-5K, and measured in terms of Precision and Recall. The experimental results demonstrate that the proposed method outperforms some of the other state-of-the-art methods, which proves the effectiveness of the proposed method.

Keywords: Content-Based Image Retrieval · Local Binary Pattern
Discrete Wavelet Transform · Gray-Level Co-occurrence Matrix
Multiresolution LBP

1 Introduction

Nowadays, image capturing is no longer a difficult task due to the availability of low cost smartphones and other image capturing devices. Due to this, huge amount of unorganized images is produced. This has created a new challenge as accessing images from such huge repository of images requires proper indexing. The field of image retrieval provides a solution to this problem. Image retrieval systems can be broadly classified into two categories: (1) text-based image retrieval systems and (2) content-based image

© Springer International Publishing AG, part of Springer Nature 2018
N. T. Nguyen et al. (Eds.): ACIIDS 2018, LNAI 10752, pp. 529–538, 2018.
https://doi.org/10.1007/978-3-319-75420-8_50

retrieval systems. Text-based retrieval systems rely on keywords and text for searching images. Such systems require manual annotation of large number of images and fail to retrieve visually similar images. The second type is Content-Based Image Retrieval (CBIR) systems which require no manual tagging and retrieves visually similar images.

CBIR is the searching and retrieval of images on the basis of features present in the image. CBIR systems take an image or sketch as the query and extract features from the image to construct a feature vector. The feature vector constructed is matched with those of images in the database to retrieve visually similar images [1].

The field of image retrieval has captured a lot of attention of researchers across the world. A number of methods have been proposed based on primary features such as colour [2], texture [3], and shape [4]. These features have been used single or in combination with other features [5]. Most of these methods have been exploited on single resolution of an image. Since natural images have complex structures, single resolution processing of an image tends to fail to gather varying level of details. Hence, the trend shifted to multiresolution processing of an image. Several methods based on multiresolution analysis of an image have been proposed in the past [6, 7]. Local Binary Pattern (LBP) is an efficient texture feature which has been exploited a lot for image retrieval. LBP attempts to extract a texture feature which represents structural arrangement of pixels. Most of the methods have exploited a single scale of an LBP image for constructing a feature vector. However, similar to single resolution of an image, single resolution of an LBP image is not sufficient to efficiently extract structural arrangement of pixels. A number of methods based on multiscale LBP have been proposed where 3×3 window of LBP has been increased to the scales such as 5×5 and 7×7 [8]. However, decomposing a single scale of an LBP image to multiple scales has not been much exploited. This paper attempts to decompose a single resolution LBP image into multiple resolutions. LBP descriptors of a 2-D gray scale image are computed followed by computation of DWT coefficients of the resulting LBP image. Finally, the feature vector is constructed through Gray-Level Co-occurrence Matrix (GLCM). Multilevel decomposition of LBP attempts to extract structural arrangement of pixels at the finer level in order to encode texture features more efficiently as compared to single resolution of an image.

The remaining part of the paper is organized as follows: Sect. 2 discusses related work. Section 3 gives a brief background of DWT and LBP. Section 4 presents the proposed method. Section 5 discusses experiment and results and Sect. 6 includes summary and conclusions.

2   Related Work

Since the term CBIR came into existence, a number of methods have been proposed to improve retrieval accuracy. Several descriptors combining colour and texture features have been proposed such as Multi-Texton Histogram (MTH) [9], Microstructure Descriptor (MSD) [10], Color Difference Histogram CDH [11], Hybrid Information Descriptor (HID) [12], Global Correlation Descriptor (GCD) [13]. These descriptors incorporate multiple features for image retrieval. However, all these features exploit single resolution of an image to construct a feature vector. To overcome disadvantages of single resolution of an image, a number of multiresolution techniques have been developed. Xia et al. [14] proposed Multiscale Local Spatial Binary Pattern for image retrieval. Zhang et al. [15] exploited the concept of curvelet transform for retrieval. Multiresolution analysis techniques tend to decompose images at multiple scales so that features left undetected at one scale get detected at another.

3 Discrete Wavelet Transform and Local Binary Pattern

3.1 Discrete Wavelet Transform

The wavelet series expansion of function $f(x) \in L^2(R)$ relative to the wavelet $\psi(x)$ and scaling function $\varphi(x)$ is defined as [16], which is given as

$$f(x) = \sum_k c_{j_0}(k)\varphi_{j_0,k}(x) + \sum_{j=j_0} d_j(k)\psi_{j,k}(x), \tag{1}$$

where j_0 is an arbitrary starting scale, $c_{j_0}(k)$ is approximation coefficients, $d_j(k)$ is detail coefficients.

The one-dimensional transform can be extended to two-dimensional images. In two-dimensions, a two-dimensional scaling function becomes $\varphi(x, y)$ and along with this there are three two-dimensional wavelets $\psi^H(x, y)$, $\psi^V(x, y)$, and $\psi^D(x, y)$. These three wavelets measure gray-level variations for images along different directions: ψ^H measures variations in horizontal direction, ψ^V measures variations in vertical direction and ψ^D measures variations in diagonal direction.

The DWT of function $f(x, y)$ of size $M \times N$ is given as

$$W_\varphi(j_0, m, n) = \frac{1}{\sqrt{MN}} \sum_{x=0}^{M-1} \sum_{y=0}^{N-1} f(x, y)\varphi_{j_{0,m,n}}(x, y), \tag{2}$$

$$W_\psi^i(j, m, n) = \frac{1}{\sqrt{MN}} \sum_{x=0}^{M-1} \sum_{y=0}^{N-1} f(x, y)\psi_{j,m,n}^i(x, y), \tag{3}$$

where j_0 is an arbitrary starting scale and the $W_\psi^i(j, m, n)$ coefficients define an approximation of $f(x, y)$ at scale j_0. The $W_\psi^i(j_0, m, n)$ coefficients add horizontal, vertical, and diagonal details for scales $j \geq j_0$.

3.2 Local Binary Pattern

LBP was originally proposed by Ojala et al. [4] for texture analysis. The original LBP operator works in a 3×4 pixel block of an image. The pixels in this block are thresholded by the centre pixel of the block. The LBP operator takes 3×3 surrounding of a pixel and

- It generates a binary 1 if the neighbor is greater than or equal to the value of centre pixel.
- It generates a binary 0 if the neighbor is less than the value of centre pixel.
- The values in the thresholded neighbourhood are multiplied by the weights provided to the corresponding pixels.
- Since neighbourhood consists of 8 pixels, $2^8 = 256$ possible texture labels are possible with reference to the value of centre pixels and neighbourhood pixel. LBP has several important properties which are useful for image retrieval.
- It encodes the relationship between gray value of centre pixels and neighbourhood pixels.
- It efficiently extracts local features in an image.

3.3 Gray-Level Co-occurrence Matrix

GLCM was proposed by Haralick et al. [17]. It is a statistical method for texture analysis. GLCM determines how frequently adjacent pixel pairs of specified values and in specified directions occur in an image. This helps in determining spatial distribution of pixel values which other features such as histogram fail to provide it.

3.4 Advantages of the Proposed Method

Texture represents structural arrangement of pixels in an image. Numerous descriptors have been proposed to extract texture feature from an image. These descriptors extract texture feature and are further processed to extract other information from texture such as shape, co-occurrence of pixel values for constructing feature vectors. These texture descriptors are applied on single resolution of an image and produce the texture image which is subjected to further processing in order to construct its feature vector. Since natural images have complex structure, a single scale of a texture image is insufficient for further processing. Hence, there is the need to decompose a texture image into multiple scales so that structural arrangement of pixels can be efficiently analyzed. An important advantage of decomposing a texture feature is that features that are left undetected at one scale get detected at another scale. Also, an image consists of different types of objects with different sizes. Small size objects need high resolution and large size objects can be viewed at a coarse level for further processing. Hence, multiscale decomposition helps in coarse to fine scale analysis of images. This paper extracts the texture feature through LBP and decomposes LBP into five levels of resolution through DWT. The advantages of the proposed technique are as follows:

- It decomposes an LBP image into multiple scales in order to extract structural arrangement of pixels from coarse to fine scales.
- Processing of an LBP image for constructing feature vector may prove to be insufficient in the case of single resolution LBP images as natural images have complex texture information and single resolution processing may not be sufficient to extract complex level of details.
- Features that are left undetected at one scale get detected at another scale.

4   Proposed Method

The proposed method consists of the following steps-

1. Computation of LBP of a 2-D gray scale image.
2. Computation of DWT coefficients of an LBP image.
3. Construction of GLCM of DWT coefficients.
4. Similarity measurement.

4.1   Computation of LBP of the 2-D Gray Scale Image

LBP descriptors of a two dimensional gray scale image is computed and stored in a matrix. This generates a two dimensional LBP image.

4.2   Computation of DWT Coefficients of the LBP Image

DWT coefficients of the resulting LBP image is computed for five levels of resolution. DWT of an image produces coefficients in three different directions: Horizontal, vertical, and diagonal. The directional coefficients are stored in three separate matrices for constructing the feature vector separately.

4.3   Construction of GLCM

GLCM of the three detail coefficient matrices is constructed and stored separately. Each of these three matrices is used for similarity measurement separately. Feature vector constructed through GLCM is used for retrieving visually similar images.

4.4   Similarity Measurement

The purpose of similarity measurement is to retrieve visually similar images. Let $(f_{Q1}, f_{Q2}, \ldots f_{Qn})$ be the set of query images and let $(f_{DB1}, f_{DB2}, \ldots f_{DBn})$ be the set of database images. Then, the similarity measurement between the query image and database image is done using the following formula:

$$\sum_{i=1}^{n} \left| \frac{f_{DBi} - f_Q}{1 + f_{DBi} + f_Q} \right| \text{ where } i = 1, 2, \ldots, n. \tag{4}$$

5   Experiment and Results

For performing experiments using the proposed method, images from Corel-1K [18] and Corel-5K [19] have been used. Corel-1K consists of 1,000 images divided into ten categories, each category consisting of 100 images. The size of each image is either 256×384 or 384×256. Corel-5K dataset consists of 5,000 images divided into fifty categories, each consisting of 100 images. The size of each image is either 128×187 or 187×128. Both these datasets consist of wide varieties of natural images which are widely used for testing the performance of image retrieval methods.

To ease the computation, the size of each image of Corel-1K dataset is rescaled to size 256×256 and that of Corel-5K to size 128×128. Each image of dataset is taken as the query image. If the retrieved images belong to the same category as that of the query image, the retrieval is considered to be successful.

5.1   Performance Evaluation

The performance of the proposed method has been evaluated in terms of Precision and Recall. Precision refers to the ratio of total number of relevant images retrieved to the total number of images retrieved. Mathematically, Precision can be defined as

$$P = \frac{I_R}{T_R}, \tag{5}$$

where I_R denotes total number of relevant images retrieved and T_R denotes total number of images retrieved. Recall refers to the ratio of total number of relevant images retrieved to the total number of relevant images in the database. Mathematically, Recall can be defined as

$$R = \frac{I_R}{C_R}, \tag{6}$$

where I_R denotes total number of relevant images retrieved and C_R denotes total number of relevant images in the database. In this experiment, we set $T_R = 10$ and $C_R = 100$.

5.2   Retrieval Results

LBP descriptors of a 2-D gray scale image are computed and stored in a matrix. DWT coefficients of resulting LBP image are computed. Finally, the feature vector is constructed using each of these matrices through GLCM which is used to retrieve visually similar images.

Application of DWT on the 2-D LBP image produces three detail coefficient matrices: Horizontal detail, vertical detail, and diagonal detail. In this experiment, similarity measurement for each of these matrices is done separately. This produces three sets of similar images. Union of all these sets is done to produce the final set of similar images. Recall is computed by counting total number of relevant images in the final image set. Similarly, for Precision, top n matches for each detail matrix is counted and then union operation is applied on three image sets to produce the final set. Precision is evaluated by considering top n matches in the final set. Mathematically, this can be stated as follows. Let f_H be the set of similar images obtained from the horizontal detail feature vector, f_V be the set of similar images obtained from the vertical detail feature vector, and f_D be the set of similar images obtained from the diagonal detail feature vector. Then, the final set of similar images denoted by f_{RS} is given as

$$f_{RS} = f_H \cup f_V \cup f_D. \tag{7}$$

Similarly, let f_H^n be the set of top n images obtained from the horizontal detail feature vector, f_V^n be the set of top n images obtained from the vertical detail feature vector, and f_D^n be the set of top n images obtained from the diagonal detail feature vector. Then the final set of top n images denoted by f_{PS}^n is given as

$$f_{PS}^n = f_H^n \cup f_V^n \cup f_D^n. \tag{8}$$

The above procedure is repeated for all five levels of resolution. In each level the relevant image set of the previous level is also considered and is combined with the current level to produce a relevant image set for that level. Retrieval is considered to be good if the values of Precision and Recall are high. Table 1 shows the average values of Precision and Recall for all five levels of resolution of the LBP images. Figure 1 shows the plots between average values of Precision and Recall for five levels of resolution. From Table 1 and Fig. 1, it can be observed that the average values of Precision and Recall increase with the level of resolution. It is due to multiresolution analysis that the average values of Precision and Recall increase with the level of resolution.

Table 1. Average recall and precision values for the five levels of resolution

Levels of resolution	Corel-1K		Corel-5K	
	Recall (%)	Precision (%)	Recall (%)	Precision (%)
Level 1	46.85	72.30	17.03	34.79
Level 2	63.14	89.35	24.85	44.80
Level 3	73.27	95.88	29.73	50.30
Level 4	80.60	98.32	33.94	54.77
Level 5	85.33	99.45	38.09	59.85

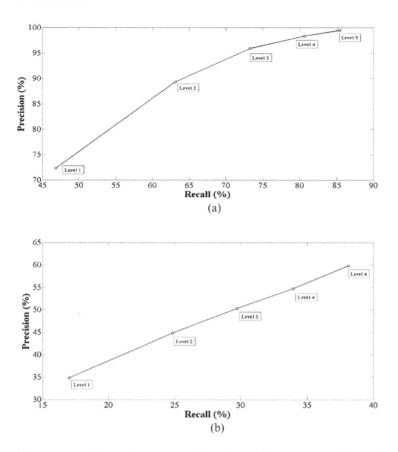

Fig. 1. (a) Average precision vs. Recall on Corel-1K dataset (b) Average precision vs. Recall on Corel-10K dataset

5.3   Performance Comparison

The performance of the proposed method is compared with some of the other state-of-the-art CBIR methods, namely, Srivastava et al. [5], Srivastava et al. [6], MTH [9], MSD [10, 12], and CDH [11, 13], in terms of Precision. Most of the above mentioned methods exploit single resolution of an image and hence fail to gather varying level of details in an image. Srivastava et al. [6] combines DWT with shape feature moments. However, shape being exploited as global feature fails to extract local information and hence produce low retrieval accuracy as compared to the proposed method. The proposed method exploits local texture information at multiple resolutions of an image and hence produces high retrieval accuracy than the above mentioned state-of-the-art CBIR methods, which is supported by the results in Table 2.

Table 2. Performance comparison of the proposed method with other state-of-the-art methods in terms of Precision (%)

Method	Corel-1K	Corel-5K
Srivastava et al. [5]	53.70	32.18
Srivastava et al. [6]	67.16	31.95
MTH [9]	69.32	51.84
MSD [10, 12]	75.67	55.92
CDH [11, 13]	65.75	57.23
Proposed method	**99.45**	**59.85**

6 Conclusion

This paper proposed the concept of multiresolution LBP by combining LBP with wavelet transform. LBP descriptors of a 2-D gray scale image are computed followed by computation of DWT coefficients of the resulting LBP image. Finally, the feature vector was constructed through GLCM which was used to retrieve visually similar images. The advantages of the proposed method are:

- It performs multiresolution analysis of texture features, thereby determining structural arrangement of pixels at fine level of resolution of the LBP image.
- Features left undetected at one level get detected at another level of resolution.

The performance of the proposed method was measured in terms of Precision and Recall. Also, the proposed method was compared with other state-of-the-art image retrieval methods in terms of Precision. The experimental results demonstrate that the proposed method outperformed the other state-of-the-art methods in terms of Precision. The proposed method can be further enhanced by combining other features with LBP and performing multiresolution analysis of the combination to improve retrieval results.

Acknowledgements. This work was supported by the Brain Research Program through the National Research Foundation of Korea (NRF) funded by the Ministry of Science, ICT & Future Planning (NRF-2016M3C7A1905477), and the Basic Science Research Program through the NRF funded by the Ministry of Education (NRF-2017R1D1A1B03036423).

References

1. Dutta, R., Joshi, D., Li, J., Wang, J.Z.: Image retrieval: ideas, influences, and trends of the new age. ACM Comput. Surv. **40**(2), 5:1–5:60 (2008)
2. Smith, J.R., Chang, S.F.: Tools and techniques for color image retrieval. Electron. Imaging Sci. Technol. Int. Soc. Opt. Photonics **2670**, 426–437 (1996)
3. Ojala, T., Pietikainen, M., Harwood, D.: A comparative study of texture measures with classification based on feature distributions. Pattern Recogn. **29**(1), 51–59 (1996)
4. Srivastava, P., Binh, N.T., Khare, A.: Content-based image retrieval using moments. In: Vinh, P.C., Alagar, V., Vassev, E., Khare, A. (eds.) ICCASA 2013. LNICST, vol. 128, pp. 228–237. Springer, Cham (2014). https://doi.org/10.1007/978-3-319-05939-6_23

5. Srivastava, P., Binh, N.T., Khare, A.: Content-based image retrieval using moments of local ternary pattern. Mob. Netw. Appl. **19**, 618–625 (2014)
6. Srivastava, P., Prakash, O., Khare, A.: Content-based image retrieval using moments of wavelet transform. In: International Conference on Control Automation and Information Sciences, Gwangju, South Korea, pp. 159–164 (2014)
7. Youssef, S.M.: ICTEDCT-CBIR: integrating curvelet transform with enhanced dominant colors extraction and texture analysis for efficient content-based image retrieval. Comput. Electr. Eng. **38**, 1358–1376 (2012)
8. Ojala, T., Pietikainen, M., Maenpaa, T.: Multiresolution gray scale rotation invariant texture classification with local binary patterns. IEEE Trans. Pattern Anal. Mach. Intell. **24**(7), 971–987 (2002)
9. Liu, G., Zhang, L., Hou, Y., Yang, J.: Image retrieval based on multi-texton histogram. Pattern Recogn. **43**(7), 2380–2389 (2008)
10. Liu, G., Li, Z., Zhang, L., Xu, Y.: Image retrieval based on microstructure descriptor. Pattern Recogn. **44**(9), 2123–2133 (2011)
11. Liu, G.H., Yang, J.Y.: Content-based image retrieval using color difference histogram. Pattern Recogn. **46**(1), 188–198 (2013)
12. Zhang, M., Zhang, K., Feng, Q., Wang, J., Kong, J., Lu, Y.: A novel image retrieval method based on hybrid information descriptors. J. Vis. Commun. Image Represent. **25**(7), 1574–1587 (2014)
13. Feng, L., Wu, J., Liu, S., Zhang, H.: Global correlation descriptor: a novel image representation for image retrieval. J. Vis. Commun. Image Represent. **33**, 104–114 (2015)
14. Xia, Yu., Wan, S., Jin, P., Yue, L.: Multi-scale local spatial binary patterns for content-based image retrieval. In: Yoshida, T., Kou, G., Skowron, A., Cao, J., Hacid, H., Zhong, N. (eds.) AMT 2013. LNCS, vol. 8210, pp. 423–432. Springer, Cham (2013). https://doi.org/10.1007/978-3-319-02750-0_45
15. Zhang, D., Islam, M.M., Lu, G., Sumana, I.J.: Rotation invariant curvelet features for region based image retrieval. Int. J. Comput. Vis. **98**(2), 187–201 (2012)
16. Gonzalez, R.C., Woods, R.E.: Digital Image Processing, 2nd edn. Prentice Hall Press, Englewood Cliffs (2002)
17. Haralick, R.M., Shanmungam, K., Dinstein, I.: Textural features of image classification. IEEE Trans. Syst. Man Cybern. **3**, 610–621 (1973)
18. http://wang.ist.psu.edu/docs/related/. Accessed Oct 2017
19. http://www.ci.gxnu.edu.cn/cbir/. Accessed Oct 2017

Large-Scale Face Image Retrieval System at Attribute Level Based on Facial Attribute Ontology and Deep Neuron Network

Hung M. Nguyen$^{(\boxtimes)}$ [iD], Ngoc Q. Ly$^{(\boxtimes)}$ [iD], and Trang T. T. Phung [iD]

Faculty of Information Technology,
HCM University of Science - VNUHCM, Ho Chi Minh City, Vietnam
ituni@live.com, lqngoc@fit.hcmus.edu.vn,
trangphung@sgu.edu.vn

Abstract. From the emerging of Deep Convolution Neural Network (DCNN), the Visual Information Retrieval would have good prospects for visual features automatically extracted at high semantic levels. However, the deep features could not be robust to some challenges as one-to-many and many-to-one relationships between face identifiers and facial attributes in querying on the face identifier level. To solve these issues at the large-scale level, we proposed a face retrieval system by using the "divide and conquer" method: query by attributes instead of querying by the identifier. We used Facial Attributes in the Fast-Filter stage, after that, our proposed system would retrieve the Face Identifier from the retrieved candidates. DCNN is very useful in the facial attribute learning because of the same network architecture for multiple-attribute groups. We built the attribute learning model following the bottom-up and top-down process. The bottom-up process uses DCNNs with the corresponding face parts and the top-down process is based on our proposed Facial Attribute Ontology (FAO). FAO supports multi-task learning in DCNN, re-usability for other retrieval tasks, flexibility in intelligent queries. We experimented our proposed method on the LFWA and CelebA dataset; our system achieved the average precision at 85.68%, this result is higher than some state-of-the-art methods. In more details, we also outperformed at 25 on 40 attribute detectors. Moreover, we speeded up the retrieval process based on the multi-attribute space and the indexing method named Hierarchical K-means++. At last, on retrieval experiments, we gathered 0.79 and 0.82 MAP-score average for one attribute query in LFWA and CelebA respectively.

Keywords: Face retrieval · Facial Attribute Ontology
Deep Convolutional Neuron Network · Attribute learning

1 Introduction

The face image datasets of many applications such as surveillance cameras, pedestrian tracking, people counting in the public places, crime retrieval, human resource management in smart cities are rapidly growing in both quantity and diversity. The traditional retrieval methods based on hand-crafted features have had some limitations to meet the requirements of retrieval time and diversity of needs.

© Springer International Publishing AG, part of Springer Nature 2018
N. T. Nguyen et al. (Eds.): ACIIDS 2018, LNAI 10752, pp. 539–549, 2018.
https://doi.org/10.1007/978-3-319-75420-8_51

In the recent years, Deep Learning, especially DCNN plays a significant role to boost the performance of many tasks in Computer Vision such as Detection, Recognition, Classification, Reconstruction, Retrieval. One of the most important reasons is that it could generate deep visual features automatically at the high semantic level.

However, facing the challenges as one-to-many relationships and many-to-one relationships between face identifiers and facial attributes, the deep features could not be the efficient face descriptor, they could decrease the performance of face retrieval system. For example, each person could have some face images on various attributes such as hairstyles, eyeglasses, expressions; otherwise, many face images of different persons could have some same attributes such as skin colors, beard styles, face shapes. So, the deep features are not powerful enough to recognize face identifiers.

The ancient Greek philosopher said that it is much easier to recognize the attributes of the concept than to understand it directly. Nowadays, his descendants who are researching in Face Retrieval employ the strategy "divide and conquer", query by facial attributes instead of face identifiers. Besides, DCNN is very useful in the attribute learning with the same network architecture for multiple attribute groups.

To the best of our knowledge, facial attributes have not organized yet in a Facial Attribute Ontology (FAO). Most of the authors presented them with some alphabetical lists. In this paper, we propose the FAO with two main factors: the community consensus and the realizability of facial attributes. FAO plays a significant role in the attribute learning model based on multi-task learning in DCNN, on re-usability for other tasks, and on flexibility in intelligent queries with its semantics.

In the offline phase, our framework consists of four main stages such as building FAO along with the Attribute Learning Model by a lightweight DCNN, generating the Face Descriptor in the Multi-Attribute space, and the Facial Attributes Indexing. Our online phase consists of three main stages: (i) flexible ways for intelligent queries by using FAO or the Face Descriptor, (ii) efficient and effective search strategy with fast-filtering depend on the indexed results, and (iii) ranking the retrieval results by the proper similarity distance measure, including the relevance feedback technique.

The rest of the paper is organized as follow. In Sect. 2, we review the recent related works. In Sect. 3, we present the proposed face image retrieval system at attribute level based on FAO and DCNN. In Sect. 4, we show experimental results on two standard datasets (LFWA and CelebA) and compare our system with many state-of-the-art systems to prove the high performance of our suggested system. The last section draws the conclusions and some future works.

2   Related Works

2.1   Face Image Retrieval at Identifier Level

Many human face image retrieval systems often retrieve results at the identifier level. Chen et al. [29] used sparse coding with identity constraint, and their system attained the MAP-score is 0.72 in the LFWA dataset. Besides, the method of Wu et al. [13] supports indexing by the inverted index of visual words based on Identity-Based Quantization and Multi-Reference Re-ranking. It also combines local features and global Hamming signatures to improve both the recall and the precision; it could outperform linear face-to-face scan retrieval systems.

However, these authors did not organize the visual words. They treated 175 grid positions equally without considering the weight of each component, and they did not consider the relationships between these visual words.

2.2   Face Image Retrieval Based on Attributes

Chen et al. [14] suggested a cross-age facial search, but this approach only considers the age attribute. In another way, Park and Jain [16] used some soft biometrics such as scars, moles, and freckles with some hard-identifier attributes such as gender and race. The proposed method extracts the face marks in 15 s per face image. It could be able to distinguish some twins with many same global facial appearances. Its disadvantage is just considering a few attributes, and it could not be robust to some challenges such as one-to-many relationships and many-to-one relationships between face identifiers and facial attributes on large-scale face datasets.

Kumar et al. [35] proposed the face retrieval based on attributes with SVM classifiers, and they used the product of attribute's weights with a fitted Gaussian for normalizing the face similarity. Besides, Chen et al. [15] proposed their face retrieval method based on the attribute-enhanced sparse coding and the attribute embedded inverted indexing. Their system attained only 0.19 MAP-score in the LFWA dataset. Moreover, these methods treat attributes related or not related to the identifier in the same way. They also do not consider relationships between attributes.

2.3   Face Attribute Detector: Hand-Crafted Features

Using a hierarchical approach for automatic age estimation, Han et al. [17] analyzed how aging influences individual facial components, they also studied the ability of humans to estimate age using the collected data via crowdsourcing. Their system achieves the cumulative score within 5-year mean absolute error, its age estimation is better than the age estimation of the human. Besides, Moghaddam and Yang [18] employed a Support Vector Machine classifier with 96.6% accuracy in recognizing gender on the FERET dataset. Some other works detect face expressions [1] or aging [4] by hand-crafted features such as Pyramid Local Phase Quantization Descriptor [2] and Joint Shape-Texture cues [3], or even reconstruct face model [5].

However, the main issue is that there is a large variety of facial attributes and appearances in large-scale face datasets, so they need many different solutions. With n facial attributes, in the offline phase, n attribute detectors should be prepared. This approach is inappropriate to implement on compact devices with the low storage and the limited computing power, such as mobile phones, embedded devices, and drones.

2.4   Face Attribute Detector with Deep Neural Network

Many research works have utilized DCNN to detect from 40 [9] to 73 [10] facial attributes. They typically require the substantial computational cost at the training stage for massive datasets although all attributes could share the same architecture of DCNN.

For example, Schroff et al. [19] proposed the FaceNet network model which achieves accuracy rate at 99.63% on the LFW dataset for the face recognition problem. Zhong et al. [21, 23, 24] used several CNN models to extract the CNN off-the-shelf features for saving the training cost. They achieved an encouraging accuracy even though they did not perform the fine-tuning. Moreover, they reached outstanding performance by using a face representation in the deeper level [22]. However, the task of training DCNNs such as FaceNet, LNets+ANet [9], and FaceNet+VGG [10] requires very high computational complexity with the large memory consumption [20].

The other disadvantage of these works is that they do not consider the relationships between facial attributes to identifiers. Some attributes such as accessories or states are named as Non-Identify attributes, they could have existed in many different appearances of the same person. Meanwhile, race, gender, and age are almost invariable or less varied according to each person, so they are named as Hard-Identify attributes. Querying is more complicated with criminal inquiries, where the subjects often use a face mask or some makeup techniques. These situations need a thoughtful consideration of the role of each facial attribute and the relationships between them.

From these observations, we conducted our research to build the facial attribute learning model with the bottom-up and top-down process. The bottom-up process uses DCNN and the top-down process is based on FAO. The high-performance attribute learning model could significantly support the Large-Scale Face Retrieval System.

3 Proposed Systems

3.1 Facial Attribute Ontology (FAO)

Our purpose is to fill the semantic gap in the large-scale face retrieval. We built the FAO from the widely recognized attributes in the community and their ability to be formalized. Firstly, taking into consideration 40 attributes in the CelebA dataset [9] and 73 attributes in the LFWA dataset [10], we classified the attributes into three groups named as Hard-Identify, Soft-Identify, and Non-Identify, which reflect the corresponding dependence level of attributes on the identifiers. Then, we divided attributes regarding the dependence on some face parts such as ear, eye, nose, mouth.

We could express attributes at detailed levels such as color c (hair color), texture t (facial skin), shape s (face shape), and size z (mouth size). From that, we expanded the set of current n attributes (level 1) $A_1 = \{a_i \mid 0 \le i < n\}$ by formula (1):

$$A_2 = A_1 \cup \{ac_{i,j}\} \cup \{at_{i,k}\} \cup \{as_{i,l}\} \cup \{az_{i,y}\}$$
$$0 \le j < nc_i; 0 \le k < nt_i; 0 \le l < ns_i; 0 \le y < nz_i \tag{1}$$

In formula (1), $nc_i/nt_i/ns_i/nz_y$ is the number of different corresponding color/texture/shape/size values for a_i. Thus, $ac_{i,j}/at_{i,k}/as_{i,l}/az_{i,y}$ is the attribute created by combining between a_i and the color/texture/shape/size value numbered as $j/k/l/y$. The proposed FAO also accompanies with some high-level-semantic visual facial attributes A_h such as an attractive woman or an intellectual face. We created them by combining several attributes in A_2, for example, "eyeglasses, female, young" $\rightarrow$ "cultivated lady". Formula (2) shows the final A_3 set and Fig. 1 shows the entire FAO.

$$A_3 = A_2 \cup A_h$$
$$\emptyset \notin A_h, 0 < nh = |A_h| \leq 2^{|A_2|} - 2 = 2^{n + \sum_{i=0}^{n}(nc_i + ni_i + ns_i + nz_i)} - 2 \tag{2}$$

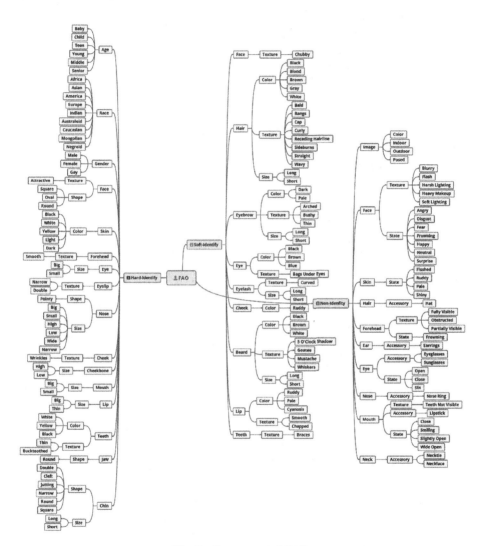

Fig. 1. Our suggested FAO.

3.2   Facial Attribute Detector

We would like to reserve the necessary face parts accordingly to attributes in the stage of facial attribute classifier. Thanks to FAO, our attribute detectors use face parts detected of the corresponding face components by Haar Feature-based Cascade Classifiers [32], instead of the entire face/image, this approach reduces the burden of face parts detected by CNN. It could save the training cost for many attributes that belong to the same facial components. Figure 2 shows the offline stage of our system.

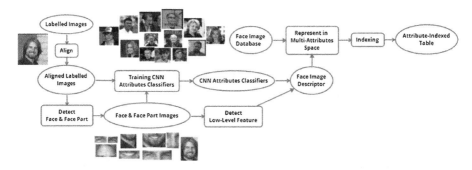

Fig. 2. The offline stage of our retrieval system.

From reviews of some novel network models [20], we used a lightweight CNN model in attribute detectors (see Table 1). It helps to reduce the computational complexity significantly so that we could implement the attribute learning efficiently.

Table 1. The architecture of the lightweight CNN model based on LeNet-5.

Layer	Information / parameters	
Input Image	128x128 pixels RGB	- deep funneled alignment [33]
Convolution1+ReLU1	width = height = 5, filters = 8 stride = 1, pad = 2, bias = 0.1	- LeNet-5-based [6] with large-scale adaptive
MaxPooling1	width = height = 2, stride = 2	Stochastic Gradient Descent [25]
Convolution2+ReLU2	width = height = 5, filters = 16 stride = 1, pad = 2, bias = 0.1	and Cross-Entropy Loss - total weights ≈ 60000
MaxPooling2	width = height = 3, stride = 3	- learning rate = 0.01
FullConnection+SoftMax	classes = 2, binary value output	- epochs = 100

The suggested lightweight CNN model is simple, but it meets requirements of the facial attribute learning, and its performance could be compared to other very deep neuron networks. By using the "divide and conquer" method, we implemented the facial attribute learning based on the face parts. The experimental results show that the accuracy rate in detecting on some attributes related to face parts is higher than the accuracy rate in detecting on some attributes related to the whole face.

We matched some attributes such as hair colors or hairstyles to the low-level features of the corresponding image area instead of training a new CNN. Thus, we further extended each primary attribute i to $nc_i/nt_i/ns_i/nz_i$ attributes (see formula (1)), and with *nh* high-semantic attributes. So we only need train some classifiers for the primary attributes along with using the total of *nc/nt/ns/nz* color/texture/shape/size low-level feature extractors. From formulas (1) and (2), we get formula (3) that compares $|A_1|$ to $|A_3|$, and it shows how much the training time could save.

$$n \ll n + \sum_{i=0}^{n}(nc_i + nt_i + ns_i + nz_i) + 2^{n + \sum_{i=0}^{n}(nc_i + nt_i + ns_i + nz_i)} - 2 \qquad (3)$$

3.3 Indexing and Ranking

Follow the semantic description of Palatucci et al. [26], and with the inspiration from the human action representation of Liu et al. [27], we had created the Face Descriptor f_d after we built the attribute classifiers based on FAO. Formula (4) shows how f_d maps each face image x of the database X to an m-dimension vector v in the multi-attribute semantic metric space M_m. Besides, based on FAO, the query at the high semantic level could be split into some basic attributes, for example, "boy" → "child, male" and "cultivated lady" → "eyeglasses, female, young".

$$f_d: X \rightarrow M_m$$
$$x \in X \mapsto f_d(x) = v = (v_1, v_2, \ldots, v_m) \in M_m, v_i \in [0, 1] \qquad (4)$$

Moreover, in order to not only speed up the query process but also ensure the ability to pre-filter facial attributes in the flexible queries, we used the Extended Reverse Vector (ERV). Firstly, our attribute values are nullable, which enable active/de-active specific attributes. For example, we would like to query all no-beard men, whether they are young or old. In this case, the value of three corresponding elements is 1 (positive for the male attribute), 0 (negative for the beard attribute) and null (the young attribute is neutral). To able to calculate with these nullable vectors, we doubled the dimension of attribute vector as formula (5).

$$f_e: M_m \rightarrow M_{2m}$$
$$v = (v_1, v_2, \ldots, v_m) \in M_m \mapsto f_e(v) = (v_{1,1}, v_{1,2}, \ldots, v_{m,1}, v_{m,2}) \in M_{2m} \qquad (5)$$
$$v_i \neq \emptyset: v_{i,1} = v_i, v_{i,2} = 1 - v_i; v_i = \emptyset: v_{i,1} = v_{i,2} = 0$$

Although we increased the number of data dimension, we could significantly improve the ranking precision of the entire retrieval system by the inversed semantic factors in the second haft of attribute vectors. From Soft Cosine Measure (SCM) of Sidorov and Gelbukh [31], we used similarity measure in formula (6) to express the impact of hard/soft/non-identify attributes with different similarity s_i for attribute i.

$$scm_{FAO}(h, k) = \frac{\sum_{1}^{m} s_i h_i k_i}{\sqrt{\sum_{1}^{m} s_i h_i^2} \sqrt{\sum_{1}^{m} s_i k_i^2}} \text{ with } h, k \in M_m \qquad (6)$$

In order to index attribute vectors in the multi-attribute space, we used the Hierarchical K-means++ (HKM++) method by combining Hierarchical K-means [11] and

K-means++ [12] to speed up the retrieval system, reduce the time of matching and ranking. We also applied the relevance feedback technique [28] to make the system be more and more intelligent. Figure 3 describes our online stage.

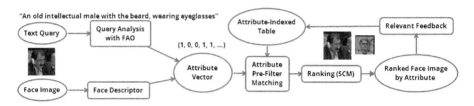

Fig. 3. The online stage of our retrieval system.

4 Experimental Results

We experimented with some high-performance PCs (up to 8500 GFLOPS). For the attribute learning, we split datasets into the training set, validation set, and testing set with the split rate of 60%, 20%, 20% respectively. In formula (6), the corresponding hard/soft/non-identify similarity is 1.0/0.5/0.1. Table 2 shows the experimental results of facial attribute detectors and the corresponding one-attribute retrievals. We achieved 85.68% in the average precision of attribute detections with some best results at 25/40 attributes in the LFWA dataset. The average MAP-score [30] of facial retrievals by one attribute is 0.79/0.82 in the LFWA/CelebA dataset respectively.

Table 2. The precision of attribute detectors and the MAP-score of one-attribute retrievals.

		Attractive	Big Lips	Big Nose	Double Chin	H. Cheekbones	Male	Narrow Eyes	Oval Face	Pointy Nose	Young	5 Shadow	Arch. Eyebrows	Bags Un. Eyes	Bald	Bangs	Black Hair	Blond Hair	Brown Hair	Bushy Eyebrows	Chubby	Goatee
Precision of attribute detector in LFWA	FaceTracer [7]	71	68	73	70	77	84	73	66	74	80	70	67	65	77	72	76	88	62	67	67	69
	PANDA-w [8]	70	64	71	64	75	86	68	64	68	76	64	63	63	82	79	78	87	65	63	65	65
	PANDA-1 [8]	81	73	79	75	86	92	73	72	76	82	84	79	80	84	84	87	94	74	79	69	75
	[29] + ANet [9]	75	70	73	70	79	91	74	66	72	79	78	66	72	86	84	82	90	71	69	68	68
	LNets+ANet(w/o) [9]	80	72	76	74	83	91	77	71	76	82	81	78	79	83	84	86	94	73	79	70	75
	LNets+ANet [9]	83	75	81	78	88	94	81	74	80	86	84	82	83	88	88	90	97	77	82	73	78
	FaceNet+VGG [21]	79	78	83	80	88	94	81	75	83	86	77	83	83	91	91	90	97	76	83	75	83
	Our proposed system	96	90	95	90	88	76	78	75	90	86	90	93	86	89	79	89	77	89	92	84	90
MAP-score of our 1-attribute retrieval	LFWA	0.87	0.99	0.83	0.62	0.92	0.67	0.84	0.68	0.60	0.86	0.83	0.97	0.98	0.89	0.98	0.63	0.99	0.78	0.51	0.77	0.60
	CelebA	0.91	0.99	0.57	0.52	0.51	0.87	0.55	0.76	0.92	0.98	0.93	0.99	0.99	0.84	0.98	0.84	0.99	0.85	0.60	0.72	0.59

		Gray Hair	Mustache	No Beard	Receding Hairline	Rosy Cheeks	Sideburns	Straight Hair	Wavy Hair	Blurry	Eyeglasses	Heavy Makeup	Mouth S. O.	Pale Skin	Smiling	Wear. Earrings	Wear. Hat	Wear. Lipstick	Wear. Necklace	Wear. Necktie	Average
Precision of attribute detector in LFWA	FaceTracer [7]	78	83	69	63	70	71	67	62	73	90	88	77	70	78	88	75	87	81	71	74
	PANDA-w [8]	77	77	63	61	64	68	68	63	70	84	86	74	64	77	85	78	83	79	70	71
	PANDA-1 [8]	81	87	75	84	76	73	89	73	74	89	93	78	84	76	92	82	93	86	79	81
	[29] + ANet [9]	82	79	69	70	71	72	72	65	75	88	89	76	68	82	87	82	93	81	72	76
	LNets+ANet(w/o) [9]	81	87	75	81	72	72	71	73	70	92	91	78	81	88	90	84	92	83	76	79
	LNets+ANet [9]	84	92	79	85	78	77	76	76	74	95	95	82	84	91	94	88	95	88	79	84
	FaceNet+VGG [21]	87	94	86	86	82	82	77	77	88	91	95	81	73	90	94	90	95	90	81	84
	Our proposed system	69	71	80	95	91	97	85	79	83	80	93	86	86	86	90	90	82	71	91	86
MAP-score of our 1-attribute retrieval	LFWA	0.58	0.92	0.99	0.83	0.92	0.68	0.98	0.90	0.82	0.84	0.60	0.58	0.60	0.92	0.65	0.82	0.71	0.98	0.62	0.79
	CelebA	0.86	0.90	0.99	0.90	0.88	0.64	0.96	0.68	0.77	0.76	0.90	0.70	0.69	0.98	0.85	0.96	0.91	0.99	0.64	0.82

Table 3. MAP-score of face identifier retrieval systems and the corresponding query time.

MAP-score and query time	LFWA (13233 images)		CelebA (202599 images)	
Linear Scan [13] *best	< 0.60	≈ 0.50s	< 0.50	≈ 0.65s
Identity-Based [13] *best	< 0.75	≈ 0.05s	< 0.60	≈ 0.10s
Identity Constraint [29] k = 100	0.72	-	-	-
Attribute-Enhanced [15] *best	0.19	≈ 0.03s	-	-
Our work	0.76	0.061s	0.78	0.487s

Table 4. The size of indexed tables and the corresponding query time.

Indexed table size and query time	LFWA (5749 identifiers)		CelebA (10177 identifiers)	
HKM++	105KB	0.005s	1584KB	0.009s
K-means++	52KB	0.004s	792KB	0.037s
Default (non-index)	2443KB	0.062s	26095KB	1.003s
K-d Tree [34]	9358KB	0.179s	96823KB	6.459s

Table 3 shows the MAP-score and the query time of some state-of-the-art retrieval systems (not including face alignments, face/face-part detections, feature extractions time), with 220 queries in one single CPU limited at the same 2.4 GHz speed. Table 4 gives the size of some different indexed tables with the corresponding query time.

5   Conclusion

In this article, we introduced FAO and the corresponding face image retrieval system. FAO describes the properties of facial attributes and the relationships between them. It helps to save the training time significantly with extensions of facial attributes.

Our facial attribute detection system is based on a lightweight DCNN and FAO. It achieved 85.68% average precision on detecting facial attributes with many better results at 25 on 40 attribute detectors in comparison to some state-of-art methods.

Our facial attribute retrieval system is based on the facial attribute detection mentioned above with the HKM++ indexing method. It achieved 0.79 and 0.82 MAP-score average for query by one attribute on the LFWA and CelebA dataset respectively. The face identifier retrieval system outperformed some state-of-the-art methods on these datasets. Moreover, both the size of the indexed tables and query time are adaptive to large-scale datasets. These results prove that the efficiency and the effectiveness of FAO on both Face Attribute Retrieval and Face Identifier Retrieval.

Notably, we could extensively enhance the entire system through developing FAO, and the attribute detectors also would be improved on the accuracy through using some nearly novel techniques of facial component regions detection or using more powerful CNN models such as Generative Adversarial Networks on massive datasets.

References

1. Truong, Q.T., Ly, N.Q.: Building the facial expressions recognition system based on RGB-D images in high performance. In: Nguyen, N.T., Trawiński, B., Fujita, H., Hong, T.-P. (eds.) ACIIDS 2016. LNCS (LNAI), vol. 9622, pp. 377–387. Springer, Heidelberg (2016). https://doi.org/10.1007/978-3-662-49390-8_37

2. Vo, A., Ly, N.Q.: Facial expression recognition using pyramid local phase quantization descriptor. In: Nguyen, V.-H., Le, A.-C., Huynh, V.-N. (eds.) Knowledge and Systems Engineering. AISC, vol. 326, pp. 105–115. Springer, Cham (2015). https://doi.org/10.1007/978-3-319-11680-8_9

3. Le, D.D.M., Nguyen, D.T., Ly, N.Q.: Facial expressions recognition based on joint shape-texture cues in video. J. Sci. **53**, 111–125 (2014)

4. Buu, A.L., Ly, N.Q., et al.: An individualized system for face ageing and facial expressions based on 3D practical faces data. In: ICCAIS, pp. 398–403. IEEE (2012)

5. Le, D.C., Ly, N.Q.: Building a 3D face reconstruction system based on 2D frontal facial image and 3D morphable model in high performance. In: IEEE-RIVF (2015)

6. Le Cun, Y., et al.: Handwritten digit recognition: applications of neural network chips and automatic learning. IEEE Commun. Mag. **27**(11), 41–46 (1989)

7. Kumar, N., Belhumeur, P., Nayar, S.: FaceTracer: a search engine for large collections of images with faces. In: Forsyth, D., Torr, P., Zisserman, A. (eds.) ECCV 2008. LNCS, vol. 5305, pp. 340–353. Springer, Heidelberg (2008). https://doi.org/10.1007/978-3-540-88693-8_25

8. Zhang, N., et al.: PANDA: pose aligned networks for deep attribute modeling. In: CVPR (2014)

9. Liu, Z., et al.: Deep learning face attributes in the wild. In: ICCV (2015)

10. Huang, G.B., et al.: Labeled faces in the wild: a database for studying face recognition in unconstrained environments. Technical report 07-49, Amherst, October 2007

11. Arai, K., Barakbah, A.R.: Hierarchical K-means: an algorithm for centroids initialization for K-means. Rep. Fac. Sci. Eng. **36**(1), 25–31 (2007)

12. Arthur, D., Vassilvitskii, S.: K-means++: the advantages of careful seeding. In: Proceedings of the 18th Annual ACM-SIAM Symposium on Discrete Algorithms (2007)

13. Wu, Z., et al.: Scalable face image retrieval with identity-based quantization and multireference reranking. IEEE Trans. PAMI **33**(10), 1991–2001 (2011)

14. Chen, B.-C., Chen, C.-S., Hsu, W.H.: Cross-age reference coding for age-invariant face recognition and retrieval. In: Fleet, D., Pajdla, T., Schiele, B., Tuytelaars, T. (eds.) ECCV 2014. LNCS, vol. 8694, pp. 768–783. Springer, Cham (2014). https://doi.org/10.1007/978-3-319-10599-4_49

15. Chen, B.-C., et al.: Scalable face image retrieval using attribute-enhanced sparse codewords. IEEE Trans. Multimed. **15**(5), 1163–1173 (2013)

16. Park, U., Jain, A.K.: Face matching and retrieval using soft biometrics. IEEE Trans. Inf. Forensics Secur. **5**(3), 406–415 (2010)

17. Han, H., Otto, C., Jain, A.K.: Age estimation from face images: human vs. machine performance. In: International Conference on Biometrics (ICB). IEEE (2013)

18. Moghaddam, B., Yang, M.-H.: Learning gender with support faces. IEEE Trans. Pattern Anal. Mach. Intell. **24**(5), 707–711 (2002)

19. Schroff, F., Kalenichenko, D., Philbin, J.: FaceNet: a unified embedding for face recognition and clustering. In: 2015 CVPR (2015). https://doi.org/10.1109/cvpr.2015.7298682

20. Sze, V., et al.: Efficient processing of deep neural networks: a tutorial and survey. arXiv preprint arXiv:1703.09039 (2017). https://doi.org/10.1109/jproc.2017.2761740

21. Zhong, Y., Sullivan, J., Li, H.: Face attribute prediction using off-the-shelf CNN features. In: 2016 International Conference on Biometrics, June 2016
22. Zhong, Y., Sullivan, J., Li, H.: Leveraging mid-level deep representations for predicting face attributes in the wild. In: ICIP (2016). https://doi.org/10.1109/icip.2016.7532958
23. Zhong, Y., et al.: Transferring from face recognition to face attribute prediction through adaptive selection of off-the-shelf CNN representations. In: 23th ICPR (2016)
24. Zhong, Y.: Human face identification and face attribute prediction: from Gabor filtering to deep learning. Doctoral thesis, Stockholm, Swedish (2016)
25. Bottou, L.: Large-scale machine learning with stochastic gradient descent. In: Lechevallier, Y., Saporta, G. (eds.) Proceedings of COMPSTAT'2010, pp. 177–186. Physica-Verlag HD, Heidelberg (2010). https://doi.org/10.1007/978-3-7908-2604-3_16
26. Palatucci, M., Pomerleau, D. Hinton, G.E., Mitchell, T.M.: Zero-shot learning with semantic output codes. In: Advances in Neural Information Processing Systems (2009)
27. Liu, J., Kuipers, B., Savarese, S.: Recognizing human actions by attributes. In: IEEE Conference on Computer Vision and Pattern Recognition (2011)
28. Zhou, X.S., Huang, T.S.: Relevance feedback in image retrieval: a comprehensive review. Multimed. Syst. **8**(6), 536–544 (2003)
29. Chen, B.C., et al.: Semi-supervised face image retrieval using sparse coding with identity constraint. In: Proceedings of the 19th ACM International Conference on Multimedia (2011)
30. Manning, C.D., et al.: Introduction to Information Retrieval, vol. 151. Cambridge University Press, Cambridge (2008)
31. Sidorov, G., Gelbukh, A., et al.: Soft similarity and soft cosine measure: similarity of features in vector space model. Computación y Sistemas **18**(3), 491–504 (2014)
32. Lienhart, R., Maydt, J.: An extended set of Haar-like features for rapid object detection. In: Proceedings of the International Conference on Image Processing. IEEE (2002)
33. Huang, G.B., Mattar, M., et al.: Learning to align from scratch. In: NIPS (2012)
34. Bentley, J.L.: Multidimensional binary search trees used for associative searching. Commun. ACM **18**(9), 509–517 (1975). https://doi.org/10.1145/361002.361007
35. Kumar, N., et al.: Describable visual attributes for face verification and image search. IEEE Trans. Pattern Anal. Mach. Intell. **33**(10), 1962–1977 (2011)

Computer Vision and Robotics

Speed-Up 3D Human Pose Estimation Task Using Sub-spacing Approach

Van-Thanh Hoang[(✉)] ⓘ and Kang-Hyun Jo

School of Electrical Engineering, University of Ulsan, Ulsan, Korea
thanhhv@islab.ulsan.ac.kr, acejo@ulsan.ac.kr

Abstract. This paper tackles the problem of reconstructing 3D human poses from given 2D landmarks, which is still an ill-posed problem. The existing works have successfully applied Active Shape Model approach to estimate 3D human poses, but the execution time is quite high. In this paper, we propose a speed-up method by separating data into subspaces to reduce the execution time of existing methods in two steps: (i) Predicting the subspace that the need-estimated 3D shape could belong to. (ii) Estimating 3D shape from given 2D landmarks and predefined basis shapes of this subspace. Compare to existing works; our approach shows a significant reduction in computational time.

1 Introduction

One of a fundamental problem in computer vision field is human action recognition, which has many applications such as human-computer interaction or video surveillance. Usually, a human action can be described by a sequence of 2D and/or 3D human poses. The 3D geometric can not only provide much more information about the scene for next high-level tasks but also improve the performance of action recognition function.

Usually, a 3D human pose is represented as a model that consists of joints locations called skeleton (e.g. Fig. 1). These locations are represented in an internal coordination. It means they do not change for any viewpoint.

Nowadays, estimating 3D pose from a single view is still an inherently ill-posed problem. There are many factors make this task more difficult, e.g., the accuracy of 2D joints estimation, unknown camera parameters, and many 3D shapes may corresponding to the same 2D shape after projection. But this is a possible task for the human, thanks to the ability to memorize object shapes visually.

There are many recent works [6,7,10] have incentivized by this fact to apply the "active shape model" (ASM) approach. According to ASM, the estimated shape can be represented as a linear combination of predefined basis shapes. Furthermore, this must match with the 2D landmarks annotated or detected from the view.

By this way, the estimating-3D-shape problem turns into a 3D-to-2D fitting problem, where the weights (shape parameters) and viewpoints (camera parameters) of predefined basis shape must be simultaneously estimated. This issue

© Springer International Publishing AG, part of Springer Nature 2018
N. T. Nguyen et al. (Eds.): ACIIDS 2018, LNAI 10752, pp. 553–562, 2018.
https://doi.org/10.1007/978-3-319-75420-8_52

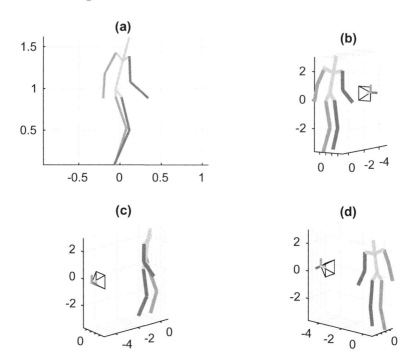

Fig. 1. Examples of an input 2D landmarks (a) and estimated 3D shape of our method in different views (b, c, d). From the input 2D landmarks, the unknown 3D shape and camera position can be estimated by applying sub-spacing approach. This paper uses the $p = 15$ joints skeleton model for the human pose.

can be solved by minimizing an ℓ_1-norm penalty between the projection of 3D shape and the 2D detections [6,7,11]. But this penalty is nonconvex, means we can get a locally-optimal result. Additionally, it is sensitive to initial data.

In [10], Zhou et. al. converted this ℓ_1-norm problem to a convex spectral norm problem by applied a "convex relaxation approach". The result of this new problem is globally-optimal.

The Zhou et al. method has an attribute: if the number of predefined basis shapes increases, with the same regularization parameters, the accuracy and execution time will increase. But when the number of basis shapes quite large, the accuracy will not improve much, while the execution time dramatically increases (as shown in [3]). So one possible solution to speed up this is reducing the number of basis shapes.

In this paper, we proposed the sub-spacing approach to speed up the estimating-3D-shape job of Zhou et al. in two steps:

1. Predict the subspace that the needed-estimated 3D shape could belong to.
2. Estimate 3D shape from given 2D landmarks and predefined basis shapes of this subspace.

2   Related Work

The 3D-to-2D fitting approach has been applied by many works [2,6,7,10] to reconstruct the 3D human pose. They tried to fit a shape-space model to annotated 2D landmarks. Following are a few recent examples.

Ramakrishna et al. [6] proposed a method that bases on sparse representation in an overcomplete dictionary. Using Projected Matching Pursuit (PMP) algorithm, they solved the reprojection error optimization problem to reconstruction 3D human pose.

From the result of [6], Fan et al. [2] proposed a method to make it get better performance. They constructed a hierarchical pose tree when built the pose dictionary; then they clustered a given large set of 3D shape training data into groups such that each group represents a subspace by using the affinity matrix. The shape dictionary consists all of the basis poses of all groups.

Wang et al. [7] method try to reconstruct 3D human pose from a single image. Firstly, they used a 2D human pose detector [8] to automatically locate the 2D landmarks of the input image. Then they initialized a 3D mean pose. And next, they used a robust estimator to improve the accuracy of joint locations of 3D shape by repeating the estimating camera parameters step and re-estimating 3D pose step in turn until convergence.

Zhou and De La Torre [9] formulates the problem of human pose estimation as a spatio-temporal matching (STM) between a 3D motion capture model and trajectories in videos problem. Then solving this problem by using linear programming.

The problem that these above works try to solve is non-convex. So the result may be delicate to the initialization data and can get the locally-optimal. To turn this non-convex problem to a convex problem, Zhou et al. [10] proposed a convex relaxation approach. Then, they solved this convex program by applying an algorithm based on the alternating direction method of multipliers (ADMM) and the proximal operator of the spectral norm.

Hoang et al. [3] proposed a method that uses cascade of neural network to improve the result of [10]. Firstly, they use the output of [10] as the initial estimated 3D shape, and then make this shape more accuracy by using the cascade of neural networks.

Proposed method in this paper decrease the execution time of estimation job of Zhou et al. [10] work by applying sub-spacing approach. At training step, we divide the dataset into subspaces based on geometric features and then learn the basis shapes for each subspace. At evaluating step, firstly, we predict the subspace that the need-estimated 3D shape could belong to. After that, we do estimate 3D shape from given 2D landmarks and predefined basis shapes of this subspace.

3   Our Approach

3.1   Problem Statement

Usually, a skeleton model with p joint locations is used to describe 2D and 3D shapes. The 3D-to-2D fitting problem is described below:

$$X = \Pi \hat{S} \tag{1}$$

where $\hat{S} \in \mathbb{R}^{3 \times p}$ denotes the unknown 3D shape, $X \in \mathbb{R}^{2 \times p}$ denotes the input 2D landmarks. $\Pi \in \mathbb{R}^{2 \times 3}$ is the camera calibration matrix. We can consider Π as a weak perspective camera model with scalar coefficient s thank to the big difference between the depth of object and the distance from camera,

According to ASM, an unknown shape can be represented as a linear combination of predefined basis shapes.

$$\hat{S} = \sum_{i=1}^{K} w_i B_i \tag{2}$$

where $B_i \in \mathbb{R}^{3 \times p}$ for $i \in [1, K]$ denotes a predefined basis shape learned from 3D shape training database, w_i denotes the weight of basis shape, and K is the number of basis shapes.

The relative rotation and translation between the camera and the 3D shape are considered when reprojecting a 3D shape to 2D shape. Since the translation can be eliminated by centralizing the data, our problem becomes:

$$X = \sum_{i=1}^{K} w_i R_i B_i \tag{3}$$

where the scalar coefficient s has been absorbed into the weight of basis shapes, and $R_i \in \mathbb{R}^{2 \times 3}$ denotes the first two rows of the rotation matrix of the basis shape B_i.

Based on the sparse representation of shapes, to estimate an unknown shape, we must solve the ℓ_1-norm optimization problem as below:

$$\min_{w_1,\ldots,w_K,R_1,\ldots,R_K} \frac{1}{2} \left\| X - \sum_{i=1}^{K} w_i R_i B_i \right\|_F^2 + \alpha \|w\|_1$$

$$\text{s.t.} \quad RR^T = I_2 \tag{4}$$

where $w = |w_1, \ldots, w_K|^T$, $\|w\|_1$ represents the ℓ_1-norm of w, and $\|\cdot\|_F$ denotes the Frobenius norm of matrix. In the loss function (4), the former part is the reprojection error and the latter is the sparsity of representation.

The optimization program in (4) is a non-convex problem with a orthogonality constraint. This program is converted into a convex optimization program based on the convex relaxation approach proposed by Zhou et al. [10]:

$$\min_{M_1,\ldots,M_K} \frac{1}{2} \left\| X - \sum_{i=1}^{K} M_i B_i \right\|_F^2 + \alpha \sum_{i=1}^{K} \|M_i\|_2 \tag{5}$$

Algorithm 1. ADMM to solve (5) (proposed by [10])

Require: $X, \alpha, \tilde{B} = [B_1 \quad \cdots \quad B_K]^T$

Ensure: $M_1, M_2, \ldots, M_K$

1: Initialize $Z^0 = D^0 = 0, \mu > 0$

2: **while** not converge **do**

3:      **for** $i = 1$ to K **do**

4:          $Q_i^t =$ the ith column-triplet of $Z^t - \dfrac{1}{\mu}D^t$

5:          $M_i^{t+1} = \mathcal{D}_{\frac{\alpha}{\mu}}(Q_i^t)$

6:      **end for**

7:      $Z^{(t+1)} = \left(X\tilde{B}^T + \mu \tilde{M}^{(t+1)} + D^{(t)} \right) \left(\tilde{B}\tilde{B}^T + \mu I \right)^{-1}$

8:      $D^{(t+1)} = D^{(t)} + \mu \left(\tilde{M}^{(t+1)} - Z^{(t+1)} \right)$

9: **end while**

---

where $\| \cdot \|_2$ denotes the spectral norm of a matrix and $M_i \in \mathbb{R}^{2 \times 3}$ is the product of w_i and the first two rows of the rotation matrix R such that $M_i M_i^T = w_i^2 I_2$.

This convex problem can be globally solved by the Algorithm 1 proposed in [10] which is based on ADMM and the proximal operator of the spectral norm.

In Algorithm 1, Z is an auxialiary variable, D is the dual varible, μ is a parameter controlling the step size in optimization and

$$
\mathcal{D}_{\frac{\alpha}{\mu}}(Q_i^t) = U_{Q_i^t} \ diag \left[\sigma_{Q_i^t} - \frac{\alpha}{\mu} \mathcal{P}_{\ell_1} \left(\frac{\sigma_{Q_i^t}}{\frac{\alpha}{\mu}} \right) \right] V_{Q_i^t}^T \tag{6}
$$

where Q_i^t is the is the i-th column-triplet of $Z^t - \dfrac{1}{\mu}D^t$; $U_{Q_i^t}$, $V_{Q_i^t}$ and $\sigma_{Q_i^t}$ denote the left singular vector, right singular vector, and the sigular values of Q_i^t; and $\mathcal{P}_{\ell_1}$ is the projection of a vector to the unit ℓ_1-norm ball.

After solving (5), the unknown 3D shape is reconstructed based on (3), where w_i and R_i are recovered from M_i.

Zhou et al. method cannot use in the real-time system since it takes around 1 s to reconstruct 3D shape. In this paper, we propose a method uses Algorithm 1 as the baseline to estimated 3D shape, then tries to speed up it by the sub-spacing approach.

3.2 Proposed Method

A possible solution to speed up the 3D shape estimation task based on active shape model is reducing the number of basis shapes. Based on this idea, we propose a new method to address the issues as mentioned above of Zhou et al. work by categorizing the 3D shape training data into subspaces.

An overview of the proposed method has been seen in Figs. 2 and 3. At training step, as in Fig. 2, we separated the massive training database into N subspaces. Each subspace contains related shapes that have similar characteristics. Some characteristics that we can use are the joints location of 3D shape

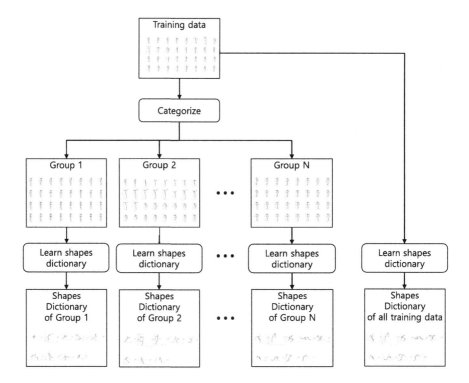

Fig. 2. Training step of the proposed method. At this step, we separated 3D shapes in training database into the N groups and then learned dictionary of basis shapes for each group $B_i(i = 1 \ldots N)$ and all training data B_a.

and/or the geometric features introduced in [5]. We can use geometric features in binary value or absolute value.

After having N subspaces, we learned the basis shapes dictionary of not only N subspace but also all training data (Learning shapes dictionary is detailed in Sect. 3.3). At last, we have $N + 1$ basis shapes dictionary corresponding to N subspaces and all training data.

At evaluating step, as in Fig. 3, from input 2D landmarks and basis shape dictionary of all training data, we reconstruct a temporary-estimated 3D shape, then we find the group that this 3D shape can belong to. After that, we reconstruct the 3D shape from the input 2D landmarks and basis shapes dictionary of that group as the final result.

3.3   Learning Basis Shapes

There are thousands of 3D shapes in training data, so it is impossible to use all of them as basis shapes. One possible solution is to get a basis-shapes-dictionary of this data by solving the below problem:

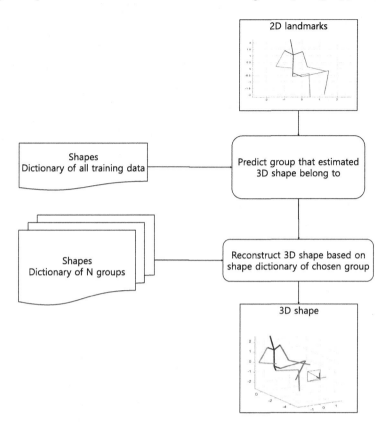

Fig. 3. Evaluating step of the proposed method. At this step, firstly, based on B_a we reconstruct the temporary-estimated 3D shapes, then find the group that this shape could belong to. After that, we use basis shapes of chosen group B_i to generate the 3D shape as the final result.

$$\min_{B,W} \sum_{i=1}^{M} \frac{1}{2} \left\| S_i - \sum_{j=1}^{K} w_{ij} B_i \right\|_F^2 + \alpha \sum_{i,j} w_{ij}$$

$$\text{s.t.} \quad w_{ij} \geq 0, \|B_i\|_F \leq 1,$$
$$\forall i \in [1, K], j \in [1, M] \tag{7}$$

where K is the number of basis shapes to be learned, M is the number of shapes in training database, S_j denotes the j-th training database, B_i denotes the i-th basis shape and w_{ij} denotes the i-th coefficient of training shape S_j.

To solve this problem, we applied projected gradient descent, a strategy commonly used in dictionary learning literature [4]. We alternately updated w_{ij} and B_i until it converges to an optimum.

4   Experiments

We use MoCap database [1] to carry out evaluation. We use all sequences of Subject 86 as training data. And test on the sequences from Acrobatics, Basketball, Jump, Run actions that are carried out by various subject (Table 1). We use the $p = 15$ joints model to represent human pose (Fig. 1). All experiments have been implemented in MATLAB and tested on a desktop with an Intel i5-4770 3.40 GHz CPU and 8 GB RAM.

Table 1. List of actions from MoCap Dataset [1] used in experiments.

Name	Description	No. of shapes
Acrobatics	Acrobatics actions carried out by various subject	2422
Basketball	Actions when playing basketball carried out by various subject	13060
Jump	Jump action carried out by various subject	10303
Run	Run action carried out by various subject	9983
Subject 86	Sports and various activities form subject 86	116137

At training step, we learn basis shapes B_a from all training data. Then we separate this dataset into $N = 4$ subspaces based on the distance from the left hand and right hand to the body plane which contains left hip, right hip, and the central point of left and right clavicle. After that, we learn basis shapes dictionary of all N subspaces. The number of shapes in dictionaries is $K = 8$.

We use a metric called 3D reconstruction error as the evaluation criterion. It is calculated as below:

$$e = \frac{1}{p} \left\| \hat{S} - S \right\|_F$$

With $\hat{S}$ is the result of estimating 3D shape job, S is the ground truth 3D shape.

Table 2 shows the average 3D reconstruction error and execution time of Zhou et al. method [10] and proposed method. For all testing dataset, the proposed

Table 2. Average 3D shape reconstruction error (in cm) and execution time (in milliseconds) of Zhou et al. [10] (with default parameters: $K = 64$ and $\alpha = 0.1$) and proposed method on Mocap Data [1]

		Acrobatics	Basketball	Jump	Run
Zhou et al.	Error	15.58	8.25	5.85	6.85
	Time	277.02	504.45	545.69	448.25
Proposed	Error	15.61	8.69	6.49	7.25
	Time	20.62	22.38	21.58	20.94

method is much faster than Zhou et al. (15–25 times faster) while can achieve similar reconstruction error. It takes around 20 ms to reconstruct a 3D shape, that means it can be used in realtime system.

5   Conclusion

In summary, this paper proposes a method for reconstructing 3D shape from input 2D landmarks and predefined basis shapes. This method improves the Zhou et al. [10] work by doing categorize training data into subspaces at training step. At evaluating step, method predicts the group which estimated 3D shape could belong to and next, use basis shapes of that subspaces to reconstruct the 3D shape corresponding to the input 2D landmarks. The proposed method can achieve a similar result to Zhou et al. work with 15–25 times speed faster. It can be used in realtime system since just takes around 20 ms to reconstruct a 3D shape.

Acknowledgment. This research was supported by the MSIT (Ministry of Science and ICT), Korea, under the Grand Information Technology Research Center support program (IITP-2017-2016-0-00318) supervised by the IITP (Institute for Information & communications Technology Promotion).

References

1. Carnegie mellon university human motion capture database. http://mocap.cs.cmu.edu/

2. Fan, X., Zheng, K., Zhou, Y., Wang, S.: Pose locality constrained representation for 3D human pose reconstruction. In: Fleet, D., Pajdla, T., Schiele, B., Tuytelaars, T. (eds.) ECCV 2014. LNCS, vol. 8689, pp. 174–188. Springer, Cham (2014). https://doi.org/10.1007/978-3-319-10590-1_12

3. Hoang, V., Hoang, V., Jo, K.: An improved method for 3D shape estimation using cascade of neural networks. In: IEEE International Conference on Industrial Informatics, pp. 285–289 (2017)

4. Mairal, J., Bach, F., Ponce, J., Sapiro, G.: Online learning for matrix factorization and sparse coding. J. Mach. Learn. Res. **11**(Jan), 19–60 (2010)

5. Müller, M., Röder, T., Clausen, M.: Efficient content-based retrieval of motion capture data. ACM Trans. Graph. (TOG) **24**, 677–685 (2005)

6. Ramakrishna, V., Kanade, T., Sheikh, Y.: Reconstructing 3D human pose from 2D image landmarks. In: Fitzgibbon, A., Lazebnik, S., Perona, P., Sato, Y., Schmid, C. (eds.) ECCV 2012. LNCS, vol. 7575, pp. 573–586. Springer, Heidelberg (2012). https://doi.org/10.1007/978-3-642-33765-9_41

7. Wang, C., Wang, Y., Lin, Z., Yuille, A.L., Gao, W.: Robust estimation of 3D human poses from a single image. In: Proceedings of the IEEE Conference on Computer Vision and Pattern Recognition, pp. 2361–2368 (2014)

8. Yang, Y., Ramanan, D.: Articulated pose estimation with flexible mixtures-of-parts. In: 2011 IEEE Conference on Computer Vision and Pattern Recognition (CVPR), pp. 1385–1392. IEEE (2011)

9. Zhou, F., De la Torre, F.: Spatio-temporal matching for human detection in video. In: Fleet, D., Pajdla, T., Schiele, B., Tuytelaars, T. (eds.) ECCV 2014. LNCS, vol. 8694, pp. 62–77. Springer, Cham (2014). https://doi.org/10.1007/978-3-319-10599-4_5

10. Zhou, X., Leonardos, S., Hu, X., Daniilidis, K.: 3D shape estimation from 2D landmarks: a convex relaxation approach. In: Proceedings of the IEEE Conference on Computer Vision and Pattern Recognition, pp. 4447–4455 (2015)

11. Zia, M.Z., Stark, M., Schiele, B., Schindler, K.: Detailed 3D representations for object recognition and modeling. IEEE Trans. Pattern Anal. Mach. Intell. **35**(11), 2608–2623 (2013)

Pedestrian Action Prediction Based on Deep Features Extraction of Human Posture and Traffic Scene

Diem-Phuc Tran[1], Nguyen Gia Nhu[1], and Van-Dung Hoang[2(✉)]

[1] Duy Tan University, Da Nang, Viet Nam
phuctd@gmail.com, nguyengianhu@duytan.edu.vn
[2] Quang Binh University, Dong Hoi, Quang Binh, Viet Nam
zunghv@gmail.com

Abstract. The paper proposes a solution for pedestrian action prediction from single images. Pedestrian action prediction is based on the analysis of human postures in the context of traffic in traffic systems. Normally, other solutions use sequential frames (video) motion properties. Technically, these solutions may produce high results but slow performance since the need to analyze the relationship between the frames. This paper takes into account analyzing the relationship between the pedestrian postures and traffic scenes from an image with the expectation that ensures accuracy without analyzing the relationship of motion between frames. This work consists of two phases, which are human detection and pedestrian action prediction. First, human detection is solved by applying aggregate channel features (ACF) method and then predict pedestrian action by extracting features of this image and use the classifier model which is trained by features extracted of pedestrian image dataset in convolution neural network (CNN) model. The minimum accuracy rate is 82%, the maximum is 97%, with the average response rate of 0.6 s per pedestrian case has that been identified.

Keywords: Deep learning · Pedestrian action prediction
Deep-feature extraction · People detection · Linear classifier

1 Introduction

Recently, object recognition technology has been developed based on various frameworks and different applications. Some major applications using object recognition technology are autonomous systems (e.g., robotics, self-driving and so on) [14]. Regarding the autonomous vehicles (AV), there have been many objects that are processed, recognizing and classified such as: (i) objects for on-road vehicles; (ii) obstacles on roads; pedestrians and so on. To different objects, we have found different recognized solutions. Of all the objects that appear during the travel of AV, pedestrian identification is considered to be the most difficult because of the complexity of the identity, the range of travel and the trajectory of pedestrians. Therefore, the ability to accurately predict pedestrian actions and the speed of alerts are placed on the forefront in order to solve problems with high accuracy, ensuring the safety of pedestrians and vehicles. The study shows that there are many different types of pedestrians, three of

© Springer International Publishing AG, part of Springer Nature 2018
N. T. Nguyen et al. (Eds.): ACIIDS 2018, LNAI 10752, pp. 563–572, 2018.
https://doi.org/10.1007/978-3-319-75420-8_53

which are the most common: crossing pedestrians, walking pedestrians, waiting pedestrians. These three types cover extensively all types of pedestrians interacting with AV. When pedestrians move (or stand still) on the road, the features are clearly displayed between the pedestrian posture, the pedestrian's position and the context of each frame (road, edge of the road, …) (Fig. 4). Thus, the features extraction from pedestrian images can be used to train, predict and identify pedestrian action.

2   Related Work

Nowadays, there are many contributions that using "tracking" technologies to detect and recognize the objects [13, 16]. Using tracking technologies can bring high accuracy; however, this approach which takes a long time to process becomes a challenge of AVs, especially in case of emergency.

Recently, there have been some proposed approaches for pedestrian recognition technologies. Histograms of Oriented Gradients (HOG) [6, 15], for instance, is a feature descriptor used in computer vision and image processing. HOG recognizes objects using information about direction and color/grayscale that change in each local area of the image and standardizes the contrast between blocks to improve accuracy. Latent SVM [9] is the algorithm classify objects by looking at their parts and its geometric location constraints. The detector requires a trained model that uses the image dataset including some desired images and some opposite images, or Kanade–Lucas–Tomasi (KLT) algorithm [2, 3].

Among mentioned approaches, CNN is the promising solution to extracting features. There are some CNN models which are used such as AlexNet [1], GoogleNet, Microsoft ResNet, Region Based CNNs (R-CNN, Fast R-CNN, Faster R-CNN). Each model has different features in terms of processing speed and the rate of accuracy. In this paper, for optimizing the process of feature extraction, we propose the models which have already been trained of the algorithm (pre-trained). Subsequently, extracted features from features extracted from CNN models are used for the classifier model. Depending on the model and the actual requirements, different classification algorithms for training such as k-Nearest Neighbor (kNN), SVM, Random Forest and Fully Connection… are applied.

There are a few algorithms for pedestrian action recognition which are proposed in previous works [4, 5, 11, 12]. However, they focus on recognizing pedestrians without basing on specific scenarios when attending traffic system; AVs is not able to detect the levels of alert with different level of dangers. In this way, in this paper, we propose a new approach to predicting the action of pedestrians by analyzing the scenarios when pedestrian appear on the road. The proposed give a high accuracy and an effective speed of process which are suitable for AVs.

3   Generalization

The proposed approach includes two phases which are: (i) training a classifier model, which is used to predict pedestrian actions, with features extracted from CNN models (Fig. 1); (ii) with the frame image from real-time video of AV on the road, the order of process are: detecting pedestrians, extracting region of interest (ROI), extracting features of ROI and predict pedestrian action in this ROI (Fig. 2). To extract features, CNN model of AlexNet is proposed [1]. To detect pedestrian, ACF algorithm is proposed [7, 8, 10] and to train and predict pedestrian action, SMV model is proposed.

Fig. 1. The process of extracted features with CNN model from image dataset

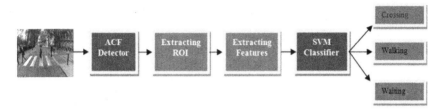

Fig. 2. The process of pedestrian action prediction.

4   Proposed Method

4.1   Extracting Features and Training Classifier Model

In machine learning, a convolution neural network (CNN) is a class of deep which is usually applied in analyzing visual imagery. In the CNN, many models are created and proposed. Each model has its own characteristics of architecture, size and number of layers… Common models such as AlexNet, GoogleNet, Microsoft ResNet, Region Based CNNs (R-CNN, Fast R-CNN and Faster CNN) are low error-rate models. In this paper, CNN model of AlexNet, which reduces the time of process, is proposed.

The AlexNet model extracts and keeps the basic features of input images as stimulated in Fig. 3. 3000 input images are used, including 1000 images of crossing pedestrians, 1000 images of walking pedestrians and 1000 images of waiting pedestrian. With each image, the CNN will extract the rich features which are pedestrian postures, roadways, roadsides and positions of pedestrians on road, Fig. 4. The rich features extracted will be used for training SVM classifier model.

In CNN model, many feature layers can be extracted such as convolution layer or full connected layer but the more advantageous layer is layer 19 (fc7 – 4096 fully connected layer) – the one right before the classification layer.

a) Input image.

b) Rich features simulation.

Fig. 3. Input image and simulate rich features of image.

Literally, in cases of object recognition such as animals, things and vehicles, the rate of recognizing object is higher (90% to 100%). In case of predicting the action of pedestrians, the features of input images focus on not only a specific object but also others such as vehicles, buildings, trees, and things around roadsides as shown in Fig. 4.

Fig. 4. Other objects on the road influence predict action pedestrian.

In this regard, in term of accuracy, ACF algorithm is used to detect pedestrians before extracting ROI, classifying and predicting the action of pedestrians.

4.2   Pedestrian Action Prediction

Pedestrian detection by ACF. ACF classification model, specified as *'inria-100x41'* or *'caltech-50x21'*, is a person detection. The *'inria-100x41'* model was trained using the INRIA Person data set. The *'caltech-50x21'* model was trained using the Caltech Pedestrian dataset. The *'inria-100x41'* model (default) is proposed in ACF. In ACF algorithm, detection scores value - confidence value - return an M-by-1 vector of classification scores in the range of [0..1]. When a pedestrian is detected, a bounding box will appear. The scores on top of bounding box is confidence value (by percentage). Larger score values indicate higher accuracy in the detection. In some complex images, the ACF algorithm sometimes recognizes errors. During the real-time experimental process, the score value 0.25 is proposed to avoid error-recognizing cases. For example, if the score value is 0.1, the result will not be accurate in some cases (Fig. 6(a)) and if the score value is 0.25, the result will be of higher accuracy (Fig. 6(b)).

Fig. 5. Example input image for recognition.

a)   b)

Fig. 6. Pedestrian detection with scores = 0.1 (a) and scores = 0.25 (b).

In particular, when the AV moves on the roads, there are some cases in which so many pedestrians appear in one frame of the video. Therefore, to ensure the accuracy, it is considered that a frame be extracted into many separate frames to be easily recognized in each case. The region extracted is called ROI (Fig. 7). Also, in real-time, the image received from AV is in big size and contains a lot of irrelevant data. Hence, extracting ROI of image at a certain scale, which removes irrelevant objects around, is necessary for each pedestrian detected. Extracting ROI of image helps the CNN model extract the exact features and reduce the error rate in the process of action recognition and classification of the SVM. The size of ROI is proposed as follows:

Supposing that H and W are height and width of the rectangle covering pedestrian object; x and y are the coordinates of the top left of rectangle and Width and Height are the size of input image, the values x1, y1, W1, H1 describe the size of ROI which are defined as follow:

$$
\begin{cases}
x1 = x - H \times 1.5 \\
y1 = y - W \\
W1 = W + H \times 3 \\
H1 = H + W \times 2 \\
if(W1 > Width) \, then \, W1 = Width \\
if(H1 > Height) \, then \, H1 = Height
\end{cases}
\tag{1}
$$

In special cases, when x1, y1, W1, H1 are smaller than the edge value of the frame or bigger than the size of the input image, the values equal the edge values of the image.

$$
\begin{cases}
if(x1 < 0) \, then \, x1 = 0 \\
if(y1 < 0) \, then \, y1 = 0 \\
if(x1 + W1 > Width) \, then \, x1 = Width - W1 \\
if(y1 + H1 > Height) \, then \, y1 = Height - H1
\end{cases}
\tag{2}
$$

On the other hand, when ROI is out of image input size, the offset value of ROI on the opposite side is proposed in Fig. 7.

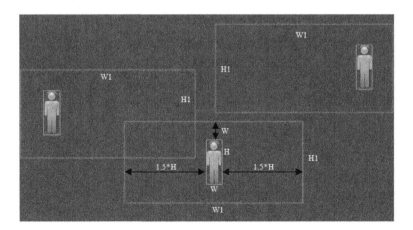

Fig. 7. ROI extraction from pedestrian image.

Pedestrian action prediction. After ROI is extracted into a single image, the features are extracted (by CNN model) to be classified (by SVM classifier model). The outputs are labeled according to values of prediction of pedestrian case (i.e., Pedestrian_crossing, Pedestrian_waiting, Pedestrian_walking).

(i) **Pedestrian_crossing:** When a pedestrian is crossing or walking in the road of other vehicles.
(ii) **Pedestrian_waiting:** When a pedestrian is standing on the roadside and waiting to cross.
(iii) **Pedestrian_walking:** When a pedestrian is walking on the edges of the road.

(1)	(2)	(3)
Pedestrian_crossing	Pedestrian_waiting	Pedestrian_walking

Fig. 8. The order of classifications of pedestrians when there are many pedestrians on the road in an input image.

5 Experimental Results

5.1 Extracting Features and Training Classifier Model

The experiment is carried out with about 3,000 images being extracted by CNN model. There features are used for training of SVM classifier model. Table 1 shows the image and label datasets of extracted and trained features.

Table 1. Image and label datasets of extracted and trained features.

Class	Number	Label
Pedestrian crossing	1,000	Pedestrian_crossing
Pedestrian waiting	1,000	Pedestrian_waiting
Pedestrian walking	1,000	Pedestrian_walking

90% of images from each set is used for the training data and the rest 10% is used for the data validation.

5.2 Pedestrian Detection and Action Prediction

With the input images (i.e., Fig. 5), after using pedestrian detection ACF algorithm, the output is executed as in Fig. 9. In case of the input images with many pedestrians in a frame, we extract ROI into a single image for action prediction by SVM classifier as shown in Fig. 9. Each image in Fig. 9 will be extracted features; finally, the system will rely on the SVM classification model to conduct action prediction of pedestrian and issue appropriate alerts for AV accordingly in Fig. 8.

Fig. 9. Pedestrians detected and ROI extracted.

The maximum results of rate-recognition after training and comparing with dataset in Table 2 are as follow:

Table 2. Maximum confusion matrix for pedestrian action prediction.

	Pedestrian crossing	Pedestrian waiting	Pedestrian walking
Pedestrian crossing	**0.9796**	0.0204	0
Pedestrian waiting	0.0612	**0.9286**	0.0102
Pedestrian walking	0.0102	0.0408	**0.9490**

The result of experiment in real-time video on the road gives minimum accuracy rate of 82%, maximum of 97% and the speed for processing reaching 0.6 s per pedestrian detected. They are promising results for potential self-driving.

6 Conclusion

In this paper, we propose a new approach for pedestrian action prediction which is one of the major problems of autonomous vehicles. As shown in the experiment, although the rate of recognition and classification is not 100% accurate, the contributions of our approach include:

(1) Improving the speed of recognition process using a single frame in video without analyze the relationship between the frames and tracking the object action.

(2) Relatively high accuracy thanks to the combination of the pedestrian detection algorithm and extracting region of interest (ROI) object before the conducting of pedestrian action predictions.

For future works, we take into account some problems of improving the accuracy and speed of action prediction such as increasing the number of input images for extracting features (around 5,000) and filtering the input images to reduce the high interference before extracting features and training. Moreover, when an AV moves on road, the images will be retrieved from videos; consequently, next similar frames can be skipped when a pedestrian is detected.

In general, recognizing pedestrians and predicting their actions are a hard procedure, especially for AVs. Thus, although our approach is not able to achieve the maximum accuracy rate, this proposal promises better potential for self-driving, one of the hottest issues of smart transportation nowadays.

Acknowledgment. This research is funded by Vietnam National Foundation for Science and Technology Development (NAFOSTED) under grant number 102.05-2015.09.

References

1. Krizhevsky, A., Sutskever, I., Hinton, G.E.: ImageNet classification with deep convolutional neural networks. In: Advances in Neural Information Processing Systems (NIPS 2012), vol. 25, pp. 1106–1114 (2012)
2. Lucas, B.D., Kanade, T.: An iterative image registration technique with an application to stereo vision (IJCAI). In: Proceedings of the 7th International Joint conf on Artificial intelligence, vol. 81, pp. 674–679 (1981)
3. Tomasi, C., Kanade, T.: Detection and tracking of point features. Int. J. Comput. Vis. **9**, 137–154 (1991)
4. Hoang, V.-D.: Multiple classifier-based spatiotemporal features for living activity prediction. J. Inf. Telecommun. **1**, 100–112 (2017)
5. Hariyono, J., Jo, K.-H.: Detection of pedestrian crossing road a study on pedestrian pose recognition. Neurocomputing **234**, 144–153 (2016)
6. Dalal, N., Triggs, B.: Histograms of oriented gradients for human detection. In: Proceedings of the 2005 IEEE Computer Society Conference on Computer Vision and Pattern Recognition (CVPR 2005), vol. 1, pp. 886–893 (2005)
7. Dollar, P., Wojek, C., Schiele, B., Perona, P.: Pedestrian detection: a benchmark. In: IEEE Conference on Computer Vision and Pattern Recognition, pp. 304–311 (2009)
8. Dollar, P., Wojek, C., Schiele, B., Perona, P.: Pedestrian detection: an evaluation of the state of the art. IEEE Trans. Pattern Anal. Mach. Intell. **34**, 743–761 (2012)
9. Felzenszwalb, P.F., Girshick, R., McAllester, D., Ramanan, D.: Object detection with discriminatively trained part-based models. IEEE Trans. Pattern Anal. Mach. Intell. **32**, 1627–1645 (2010)
10. Dollár, P., Appel, R., Belongie, S., Perona, P.: Fast feature pyramids for object detection. IEEE Trans. Pattern Anal. Mach. Intell. **36**, 1532–1545 (2014)
11. Stewart, R., Andriluka, M., Ng, A.Y.: End-to-end people detection in crowded scenes. In: 2016 IEEE Conference on Computer Vision and Pattern Recognition, pp. 2325–2333 (2015)

12. Piérard, S., Lejeune, A., Van Droogenbroeck, M.: A probabilistic pixel-based approach to detect humans in video streams. In: IEEE International Conference on Acoustics, Speech and Signal Processing (ICASSP), pp. 921–924 (2011)
13. Mittal, S., Xue, F., Saurabh, S., Prasad, T., Shin, H.: Pedestrian detection and tracking using deformable part models and Kalman filtering. J. Comput.-Mediated Commun. **10**, 960–966 (2013)
14. Hoang, V.-D., Le, M.-H., Jo, K.-H.: Motion estimation based on two corresponding points and angular deviation optimization. IEEE Trans. Ind. Electron. **64**, 8598–8606 (2017)
15. Hoang, V.-D., Le, M.-H., Jo, K.-H.: Robust human detection using multiple scale of cell based histogram of oriented gradients and AdaBoost learning. In: Nguyen, N.-T., Hoang, K., Jędrzejowicz, P. (eds.) ICCCI 2012. LNCS (LNAI), vol. 7653, pp. 61–71. Springer, Heidelberg (2012). https://doi.org/10.1007/978-3-642-34630-9_7
16. Xiang, Y., Alahi, A., Savarese, S.: Learning to track: online multi-object tracking by decision making. In: 2015 IEEE International Conference on Computer Vision (ICCV), pp. 4705–4713 (2015)

Deep CNN and Data Augmentation
for Skin Lesion Classification

Tri-Cong Pham[1,2], Chi-Mai Luong[2,3], Muriel Visani[4],
and Van-Dung Hoang[5(✉)]

[1] ThuyLoi University, Dong Da, Hanoi, Vietnam
phtcong@gmail.com
[2] ICTLab, Vietnam Academy of Science and Technology,
University of Science and Technology of Hanoi,
18 Hoang Quoc Viet, Cau Giay, Hanoi, Vietnam
[3] Institute of Information Technology, VAST, Hanoi, Vietnam
[4] Laboratory L3i, University of La Rochelle, La Rochelle, France
[5] Quang Binh University, Dong Hoi, Quang Binh, Vietnam
zunghv@gmail.com

Abstract. Deep CNN techniques have dramatically become the state of the art in image classification. However, applying high-capacity Deep CNN in medical image analysis has been impeded because of scarcity of labeled data. This study has two primary contributions: first, we propose a classification model to improve performance of classification of skin lesion using Deep CNN and Data Augmentation. Second, we demonstrate the use of image data augmentation for overcoming the problem of data limitation and examine the influence of different number of augmented samples on the performance of different classifiers. The proposed classification system is evaluated using the largest public skin lesion testing dataset, containing 600 testing images, and 6,162 training images. New state-of-the-art performance result is archived with AUC (89.2% vs. 87.4%), AP (73.9% vs. 71.5%), and ACC (89.0% vs. 87.2%). In additional, we explore the influence of each image augmentation on the three classifiers and observe that performance of each classifier is influenced differently by each augmentation and has better results comparing with traditional methods. Thus, it is suggested that the performance of skin cancer classification and medial image classification could be improved further by applying data augmentation.

Keywords: Medical image · Skin cancer · Deep learning · Data augmentation
Melanoma classification

1 Introduction

More than half of cancer diagnoses world-wide are skin cancer [1, 9, 10]. Basically, there are two types of skin cancer called melanoma and non-melanoma. In recent decades, melanoma's incidence and mortality rate has increased dramatically, thus becomes a major problem in public health. However, early detected patients have

© Springer International Publishing AG, part of Springer Nature 2018
N. T. Nguyen et al. (Eds.): ACIIDS 2018, LNAI 10752, pp. 573–582, 2018.
https://doi.org/10.1007/978-3-319-75420-8_54

higher chance of curing, especially if the cancer is detected in its early stages and removed, the cure rate can be over 90% [1, 7]. Frequently, skin cancer diagnosis is conducted using visual examination of skin lesion images, and then clinical analysis is conducted if there is a suspicion signal. Automated classification of skin lesions using images inspires the development of artificial intelligence-computer vision [13]. Especially, deep convolutional neural networks (Deep CNN) [15] has been achieved a high level of accuracy in image classification with large datasets [14] and very well suited to the problem of melanoma classification. But applying Deep CNN in melanoma recognition is still a challenge due to insufficiency of labeled data.

In this research, we apply Deep CNN [14, 15, 19, 20] which was pre-trained on approximately 1.28 million images from the 2014 ILSVR Challenge [4] for the problem of melanoma classification [5]. Deep CNN is the most successfully applied and the most accurate deep learning architecture in image classification tasks. It comprised of two main parts which are feature extractor and classifier. Images' features are analyzed and summarized by different layers in CNN, features in each layer represent the level of abstraction of the object, the lower-level features are extracted by the previous layer, and the higher-level features are extracted by the following layer of CNN. In the present work, CNN is used to extract features of input image, and then those features are trained and classified by three methods including Support Vector Machine (SVM), Random Forest (RF) and Neural Network (NN).

However, the challenge is the limited train skin lesion data, only 6,162 train images, of those only 1,114 (18.08%) images of melanoma when compared to the data collection of more than 1.28 million images of 2014 ILSVR Challenge. With the relatively small data collection, the accuracy rate is affected by overfitting problem [14]. An effective solution to the issue is Data Augmentation and Dropout [14]. Image augmentation artificially enlarges the dataset through different steps of processing or combination of multiple processing, such as random rotation, translation, mirror, scale, crop, shifts, shear and flips [14]. A key concept of image augmentation is that the generated images do not change the sematic meaning of the original image [14]. This technical has been applied in some melanoma classification researches [8–10, 12, 16–18] and achieved state-of-the-art performance at the time of the researches but still need more deeply study to find out pros-cons of this solution. Furthermore, because of imbalanced data (18.08% melanoma vs. 81.92% non-melanoma) its accuracy lost its meaning of accuracy rate like in normal image classification.

Thus, in the article, we propose a solution of combining Data Augmentation, CNN for feature extraction and NN for classification to overcome the lack of labeled data problem and to improve performance of melanoma classification. Furthermore, we explore different types of classifiers such as SVM, RF, NN and their performance influenced by the image deformations. We validate the effectiveness of the algorithms by the area under the curve (AUC), Average Precision (AP), Sensitivity (SEN), Specificity (SPC) and Positive Predictive Value (PPV) instead of Accuracy as a normal classification system.

2 Proposed Classification System

In this study, we propose a melanoma classification system which includes three main components: augmenting data module, extracting features module and classifier as shown in Fig. 1.

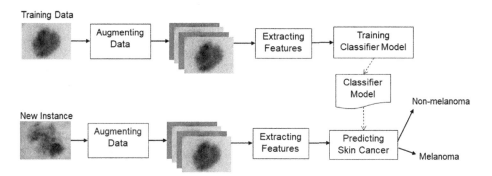

Fig. 1. Proposed classification system.

In the classification system, the data augmentation is applied using traditional transformation of Google's TensorFlow, the feature extractor is implemented by Deep CNN, the classifier is built by traditional algorithms such as NN, SVM, RF. The order of training process is: augmenting skin lesion image, extracting features of generated images, training classifier model, and saving classifier model to model file. In predicting process, these steps are: augmenting skin lesion image, extracting average feature vector of augmented images, predicting label of the image by the average feature vector and classifier's model file.

3 Data Augmentation and Deep CNN

3.1 Data Augmentation

Recently, Data Augmentation has been widely being used by not only natural image classification but also melanoma classification such as Matsunaga et al. [16], González-Díaz [12], Menegola et al. [17, 18], Esteva et al. [10], Codella et al. [9]. This is the easiest and most common method to mitigate overfitting problem of scarcity of labeled data in melanoma classification. The most importance concept of image augmentation is that the deformations applied to the annotated data do not change the semantic meaning of the labels. In melanoma classification, we apply three types of data augmentation.

Geometric augmentation: The skin lesion scale and position within the image still maintains the semantic meaning of the lesion, thus does not alter its final classification. Therefore, input images were transformed to generate new samples with the same label of original one by random combination of cropping and horizontal, vertical flips.

Color augmentation: The skin lesion images are collected from different resources and are created by different types of device. Therefore, it is important to normalize the colors of the images when we use them for training and testing any system to improve performance of classification system [6, 11].

Data warping based on specialists knowledge: The fact that the melanoma specialist diagnosis is performed over the observation of the patterns around the lesion. In machine learning, affine transformations such as shearing, distorting and scaling randomly warp stroke data for image classification [21]. Thus warping is very well suited to augment data for improving performance and mitigating overfitting of melanoma classification.

In current research, the data augmentation module combines these three types of augmentation in two steps. Firstly, we normalize input image by adding multiples a converting all pixels into [−1.0, 1.0] range to create normalized data. Secondly, we combine cropping, scaling, distorting and horizontal, vertical flips processes in one step to augment the normalized data. In this step, we apply random parameters of each function to generate samples from original one.

3.2   Feature Extraction

In the classification system shown in Fig. 1, we use Deep CNN as feature extractor. There are many models of CNN such as AlexNet, GoogleNet, ResNet and so on. Our research uses Inception V4 [19] to implement the feature extraction process.

Inception V4 is a well-known architecture developed base on the GoogLeNet platform, this is an upgrade of Inception V3 [20], the input of this network is an image (299 × 299 pixels), the output depends on how many classes targeted to predict. In the pre-trained model used in this research, the output is the 1000-categories. However, in this study, we use the latest version GoogLeNet (Inception V4); remove full-connected layer and use the Average Pooling layer as the final layer. The output of Average Pooling layer is a 1-Dimension of 1,536 floating numbers. Beside the 1-Dimension features at the Average Pooling layer, three logit features are added to final features from the CNN network, therefore overall of 1-Dimension output is 1,539 floating numbers for each image. The architecture of Inception V4 [19] is demonstrated in Fig. 2.

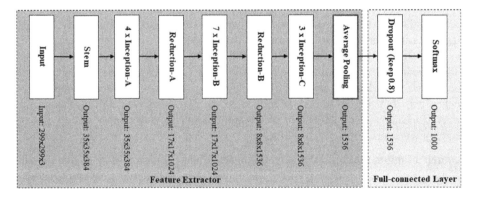

Fig. 2. The overall schema of the Inception-V4 network.

As shown Fig. 2, Inception V4 has two parts, feature extractor and full-connected layer. In detail, the feature extractor has many convolutional blocks include one Stem block, four Inception-A blocks, seven Inception-B blocks, three Inception-C blocks and one Average Pooling layer. The full-connected layer is combined by one dropout block and one softmax layer. Technically, the stem module uses Conv and MaxPool block to convert $299 \times 299 \times 3$ image shape into $35 \times 35 \times 384$ shape which is the input of Inception-A block. In the other hand, Inception-A, Inception-B, Inception-C blocks use only Conv and Avg Pooling to convolute higher abstract features of images. While Inceptions with same type have same shape size and connect directly in sequence, Inceptions with different type need a reduction grid-module to connect together. For instance, Reduction-A grid-reduction module which converts a 35×35 shape to a 17×17 shape is used to connect Inception-A block and Inception-B block. Moreover, Reduction-B grid-reduction module which converts a 17×17 shape to an 8×8 shape is used to connect Inception-B block and Inception-C block. The output $(8 \times 8 \times 1536$ shape) of Inception-C block is converted into 1-Dimension of 1536 features by average pooling layer. In Inception-V4 schema, although the features of average pooling layer is changed by dropout layer (keep 0.8) then trained or classified by softmax as full-connected layer, we save this features to a file as output of feature extractor component in this research as shown in Fig. 1.

3.3   Classifier

In this study, we empirically investigate the influence of data augmentation on three machine learning classifiers, and attempt to answer the question of should we apply data augmentation to improve performance of melanoma classification. The classifiers are trained by the same features which are extracted from augmented images by Deep CNN. In testing process, all of these classifiers are also tested by same test dataset of 2017 ISBI Challenge with three types of data augmentation (without data augmentation - NO DAUG, augments 50 samples each image - DAUG 50, augments 100 samples each image - DAUG 100).

Support Vector Machine (SVM) is a well-known classifier in many image classification, and also showcases outstanding performance [7] in melanoma classification. Therefore, we use LinearSVC library of sklearn-python as SVM model to compare with other classifiers. We do not create new SVM algorithm, we modify the parameters of TOP #3 [18] instead. We find out the best estimators of LinearSVC by hyperparameter tuning with RandomizedSearchCV library of sklearn-python. The best estimators are the same for NO-DAUG, DAUG-50 and DAUG-100 with C = 0.0003668060861257186, Dual = False, Multi Class = "ovr" and Random State = 0.

Random Forest (RF) is a popular classifier in machine learning. Thus, we explore the influence of its performance on image augmentation in this study. While the best estimators of DAUG-50 and DAUG-100 are (n_estimators = 200, max_features = "sqrt", max_depth = 10, min_samples_split = 5, criterion = "entropy"), they are (n_estimators = 50, max_features = "sqrt", max_depth = 10, min_samples_split = 10, criterion = "entropy") for NO-DAUG.

Neural Network (NN) is a traditional algorithm. In this research we apply a simple NN comprised of two hidden layers to be the classifier with activation using sigmoid in Keras. The loss function is the binary_crossentropy. The parameters of NN are shown in Table 1.

Table 1. NN parameters.

Layer	Parameter	Value	Layer	Parameter	Value
1st Activation	Type	ReLU	Dropout	Threshold	0.5
1st Activation	input_shape	(1539)	Optimizer	Type	rmsprop
1st Activation	Output	256	Optimizer	Metrics	sensitivity, specificity
2nd Activation	Type	sigmoid	Optimizer	Loss	binary_crossentropy
2nd Activation	Output	1			

4   Experimental Results

In order to focus exclusively on the improvements obtained by data augmentation, the research implements the classification system in Python, then empirically investigate the influence of data augmentation on three machine learning classifiers, and attempt to improve performance of melanoma classification task with the 2017 ISBI Challenge test dataset.

4.1   Datasets

In this research, we use a data collection of 6,162 train images and 600 testing images which are the same with the test images of ISBI Challenge [5]. Besides, the train dataset includes 2000 images from the ISBI Challenge [5], and the remainder images are collected from the sources such as ISIC Archive [2], PH2 Dataset [3] in Table 2.

Table 2. Data resources.

Resources	Melanoma	Non-melanoma	Total
ISIC 2017 challenge train	374	1,626	2000
ISIC 2017 challenge test	117	483	600
ISIC archive	700	3,262	3,962
PH2 dataset	40	160	200
Total	1,231	5,531	6,762

Table 3 below describes in details the number of images in two types: melanoma and non-melanoma in new train and test datasets. Although we still keep test data of ISBI Challenge and add 4,162 images to the train dataset, the percentage of melanoma images in train dataset is almost same (18.1% in new train dataset vs. 18.7% in train dataset of ISBI Challenge 2017) compare with this percentage in test dataset is 19.5%.

Table 3. Train and test datasets.

Skin lesion	Training	Training %	Testing	Testing %
Melanoma	1,114	18.1%	117	19.5%
Non-melanoma	5,048	81.9%	483	80.5%
Total	6,162	100.0%	600	100.0%

4.2   Evaluation

In this paper, we have chosen to use the ISBI Challenge 2017 [5] dataset for evaluation, which is used by dozens of prior algorithms, in order to archive many result for comparison. The ISBI Challenge provides 2,000 labeled images (374 of melanoma vs. 1,624 of non-melanoma) as train dataset and 600 labeled images (117 of melanoma vs. 483 of non-melanoma) as test dataset. This challenge is the largest standardized and comparative study in this field to date because it has not only the biggest number of training and testing datasets but also 23 finalized test submissions with different algorithms in melanoma classification task. Although the challenge's winner is announced, the train and test datasets are still available for further research and development. In the following, the results of the research are compared with the top ranked participant submissions by the same effectiveness measures of the challenge.

Effectiveness measures: To evaluate the effectiveness of a classification model, we have many effectiveness measures, depending on our data. Our study is using five measures to evaluate classification system such as Area Under the Curve (AUC), Average Precision (AP), Sensitivity (SEN), Specificity (SPC) and Positive Predictive Value (PPV). Mathematically, SEN, SPC and PPV can be expressed base on true positive (TP), true negative (TN), false positive (FP), false negative (FN) as below:

$$\text{SEN} = \frac{\text{TP}}{\text{TP}+\text{FN}} \qquad \text{SPC} = \frac{\text{TN}}{\text{TN}+\text{FP}} \qquad \text{PPV} = \frac{\text{TP}}{\text{TP}+\text{FP}}$$

In melanoma classification of ISBI Challenge [5], the results are ranked and awarded in AUC. The top three submissions are shown in Table 4 as TOP #1 [16], TOP #2 [12] and TOP #3 [18]. The submissions are state-of-the-art performances in melanoma classification at the time of submission (the highest AUC is 87.4%). This research and the three winners use the same test dataset of the challenge. Besides, while this study uses 4,162 external train images, compares with 1,444 images of TOP #1 and 7,640 images of TOP #3, TOP #2 does not use external training images. Technically, all the winners use Deep CNN, while TOP #1 and TOP #2 use 50-layer ResNet [19] with full-connected layer, TOP #2 uses Deep CNN as feature extractor and SVM as classifier. Our research also uses Deep CNN as feature extractor; and explores performances of three classifiers (NN, SVM, RF) with three types of data augmentation (NO DAUG, DAUG 50 and DAUG 100). The final results are shown in Table 4.

Table 4. Performance of classifiers with same test dataset with 600 images.

Classifier		AUC	AP	SEN	SPC	ACC	PPV
ISBI TOP 3	TOP #1	0.868	0.710	**0.735**	0.851	0.828	–
	TOP #2	0.856	0.654	0.103	**0.998**	0.823	–
	TOP #3	**0.874**	**0.715**	0.547	0.950	**0.872**	–
DAUG 100	NN	**0.892**	**0.739**	0.556	**0.971**	**0.890**	**0.823**
	SVM	0.773	0.547	0.581	0.965	**0.890**	0.800
	RF	0.751	0.530	0.530	0.973	0.887	0.827
DAUG 50	NN	0.882	0.736	**0.598**	0.950	0.882	0.745
	SVM	0.775	0.727	0.590	0.961	0.888	0.784
	RF	0.757	0.526	0.547	0.967	0.885	0.800
NO DAUG	NN	0.862	0.696	0.581	0.942	0.872	0.708
	SVM	0.771	0.522	0.590	0.952	0.882	0.750
	RF	0.746	0.517	0.521	0.971	0.883	0.813

For the first experiments, the combined Data Augmentation and NN solution archived new state-of-the-art performance in melanoma classification task with AUC (89.2% vs. 87.4%), AP (73.9% vs. 71.5%), and ACC (89.0% vs. 87.2%). To get there results, the NN is trained on extracted features of DAUG 100 and run on 500 epochs, 15% validation split (923 validation images vs. 5,239 train images), the binary_crossentropy as loss function, and min loss checkpoint. Although the training is run on 500 epochs, the min loss value is reached (0.08236) in the first epoch. The classification scores is normalized between 0.0 to 1.0 and any confidence above 0.5 is considered positive. In additional, at the same time of writing this paper, the new-state-of-the-art solution is published by Codella et al. [8] performed same AUC (89.2%) with the proposed method. Besides, the results are outperformance when compare with prior algorithms of Gutman et al. [13] (AUC: 89.2% vs. 80.4%, SEN: 55.6% vs. 50.7%, SPC: 97.1% vs. 94.1%, AP: 73.9% vs. 63.7%) and Codella et al. [9] (AUC: 89.2% vs. 84.3%, SPC: 97.1% vs. 83.6%, AP: 73.9% vs. 64.9%).

The next experiments show the influence of each image augmentation on three classifiers. We examine the difference between the effectiveness measures by three proposed classifiers with and without augmentation. According to the result of Table 4, we see that overall the classification performances are improved when applied data augmentation for all classifiers. With NN classifier, the more samples are augmented, the higher effectiveness of all measures is performed; and NN is the best classifier when we compare it with SVM and RF. However, although SVM and RF classifiers performed better accuracy with DAUG 50, we observe that image augmentation can also have a detrimental effect when we increase the number of samples (DAUG 100). For instance, SVM and RF algorithms perform the highest AUC, AP and SEN when applied DAUG 50, but they are affected negatively when data augmentation samples are increased to 100 samples each image.

5   Conclusions

In the study, we propose the use of Data Augmentation and Deep CNN to improve performance of melanoma classification and explore the influence of image augmentation on three classifiers performances. As shown in the experimental results, the two primary contributions of the approach include: (1) new cutting-edge performance is archived with the AUC (89.2%), AP (73.9%), and PPV (82.3%) in melanoma classification task on test dataset of ISIC 2017 Challenge [5]. (2) we examined the influence of skin lesion image deformations on performances of three classifiers (NN, SVM, RF). We observed that the performances of all three classifiers are influenced and improved differently by data augmentation. While NN classifier archived the best performance, SVM and RF classifier's performance had detrimental effect when we changed data augmentation from DAUG 50 to DAUG 100.

However, in this research, although AUC achieved the state-of-the-art performance but SEN is still average comparing to TOP #1 [16] and needs more improvement. Besides, the combination of another architectures or lower layer of Deep CNN and data augmentation is still a challenge for researchers. Furthermore, researchers could still carry out the method of fine-tuning at the last layer to reuse the weight of the network trained by 1.2 million images, because skin lesion images have many similarities with trained natural images.

References

1. American Cancer Society: Cancer facts and figures 2016. https://www.cancer.org/content/dam/cancer-org/research/cancer-facts-and-statistics/annual-cancer-facts-and-figures/2016/cancer-facts-and-figures-2016.pdf. Accessed 15 Oct 2017
2. The International Skin Imaging Collaboration (ISIC). https://isic-archive.com/. Accessed 15 Oct 2017
3. PH2 Dataset. https://www.fc.up.pt/addi/ph2%20database.html. Accessed 15 Oct 2017
4. Large Scale Visual Recognition Challenge 2014 (ILSVRC 2014). http://image-net.org/challenges/LSVRC/2014/. Accessed 15 Oct 2017
5. ISIC 2017: Skin Lesion Analysis Towards Melanoma Detection. http://challenge2017.isic-archive.com. Accessed 15 Oct 2017
6. Barata, C., Celebi, M.E., Marques, J.S.: Improving dermoscopy image classification using color constancy. IEEE J. Biomed. Health Inform. **19**, 1146–1152 (2014)
7. Codella, N.C.F., Cai, J., Abedini, M., Garnavi, R., Halpern, A., Smith, J.R.: Deep learning, sparse coding, and SVM for melanoma recognition in dermoscopy images. In: Zhou, L., Wang, L., Wang, Q., Shi, Y. (eds.) MLMI 2015. LNCS, vol. 9352, pp. 118–126. Springer, Cham (2015). https://doi.org/10.1007/978-3-319-24888-2_15
8. Codella, N.C.F., Gutman, D., Celebi, M.E., Helba, B., Marchetti, M.A., Dusza, S.W., Kalloo, A., Liopyris, K., Mishra, N., Kittler, H., Halpern, A.: Skin lesion analysis toward melanoma detection. In: A Challenge at the 2017 International Symposium on Biomedical Imaging (ISBI), Hosted by the International Skin Imaging Collaboration (ISIC). ArXiv e-prints arXiv:1710.05006 [cs.CV] (2017)

9. Codella, N.C.F., Nguyen, Q.B., Pankanti, S., Gutman, D., Helba, B., Halpern, A., Smith, J. R.: Deep learning ensembles for melanoma recognition in dermoscopy images. IBM J. Res. Dev. **61**(4), 5 (2017)

10. Esteva, A., Kuprel, B., Novoa, R.A., Ko, J., Swetter, S.M., Blau, H.M., Thrun, S.: Dermatologist-level classification of skin cancer with deep neural networks. Nature **542**, 115–118 (2017)

11. Ercal, F., Chawla, A., Stoecker, W.V., Lee, H.C., Moss, R.H.: Neural network diagnosis of malignant melanoma from color images. IEEE Trans. Biomed. Eng. **41**, 837–845 (1994)

12. González-Díaz, I.: Incorporating the knowledge of dermatologists to convolutional neural networks for the diagnosis of skin lesions. ArXiv e-prints: arXiv:1703.01976 [cs.CV] (2017)

13. Gutman, D., Codella, N.C.F., Celebi, E., Helba, B., Marchetti, M., Mishra, N., Halpern, A.: Skin lesion analysis toward melanoma detection. In: A Challenge at the International Symposium on Biomedical Imaging (ISBI) 2016, Hosted by the International Skin Imaging Collaboration (ISIC) (2016). ArXiv e-prints: arXiv:1605.01397 [cs.CV]

14. Krizhevsky, A., Sutskever, I., Hinton, G.E.: ImageNet classification with deep convolutional neural networks. In: Advances in Neural Information Processing Systems (NIPS), vol. 25, pp. 1097–1105 (2012)

15. LeCun, Y., Bengio, Y., Hinton, G.: Deep learning. Nature **521**, 436–444 (2015)

16. Matsunaga, K., Hamada, A., Minagawa, A., Koga, H.: Image classification of melanoma, nevus and seborrheic keratosis by deep neural network ensemble. ArXiv e-prints arXiv: 1703.03108 [cs.CV] (2017)

17. Menegola, A., Fornaciali, M., Pires, R., Bittencourt, F.V., Avila, S., Valle., E.: Knowledge transfer for melanoma screening with deep learning. ArXiv e-prints arXiv:1703.07479 [cs. CV] (2017)

18. Menegola, A., Tavares, J., Fornaciali, M., Li, L.T., Avila, S., Valle, E.: RECOD titans at ISIC challenge 2017. ArXiv e-prints arXiv:1703.04819 [cs.CV] (2017)

19. Szegedy, C., Ioffe, S., Vanhoucke, V., Alemi, A.A.: Inception-v4, Inception-ResNet and the impact of residual connections on learning. In: Artificial Intelligence, pp. 4278–4284 (2017)

20. Szegedy, C., Vanhoucke, V., Ioffe, S., Shlens, J., Wojna, Z.: Rethinking the inception architecture for computer vision. In: Computer Vision and Pattern Recognition (CVPR), vol. 2016, pp. 2818–2826 (2016)

21. Wong, S.C., Gatt, A., Stamatescu, V., McDonnell, M.D.: Understanding data augmentation for classification: when to warp? In: 2016 International Conference on Digital Image Computing: Techniques and Applications (DICTA), pp. 1–6 (2016)

Stationary Object Detection for Vision-Based Smart Monitoring System

Wahyono[1(✉)], Reza Pulungan[1], and Kang-Hyun Jo[2]

[1] Department of Computer Science and Electronics, Universitas Gadjah Mada,
Yogyakarta, Indonesia
{wahyo,pulungan}@ugm.ac.id
[2] School of Electrical Engineering, University of Ulsan, Ulsan, Korea
acejo@ulsan.ac.kr

Abstract. This work proposes a method for detecting stationary objects in a vision-based smart monitoring system. A stationary object is defined as an object that previously moves, but currently remains stable in a certain location. The case samples for such objects include abandoned objects and illegally parked vehicles, whose surveillance is a crucial task for ensuring public safety and security. The proposed method is based on dual background modeling for separating foreground from background. To extract the candidates of stationary objects, a Gaussian mixture model-based cumulative dual foreground difference is implemented. An SVM-based object classifier is then integrated to verify the region candidates whether they are vehicle, human, or other objects. If the object is classified as either vehicle or baggage, the duration of the object being stable is counted using detection-based tracking. If the duration exceeds a certain value, an alarm will be triggered. In experiment, the method is evaluated using public datasets iLIDS and ISLab.

Keywords: Stationary object · Illegally parked vehicle detection
Abandoned object detection · Background segmentation
Smart monitoring system · CCTV

1 Introduction

In recent years, the utilization of Close Circuit Television (CCTV) is growing significantly for monitoring purpose in order to ensure the safety and security of public areas, such as stations, airports, schools, shops, etc. Although public spaces have been covered with many CCTVs, unfortunately, they have not been equipped with automated analysis of abnormal situation in the monitored area. Therefore, it is necessary to build a system that can analyze CCTV directly and can detect any suspicious event in the monitored area. This system is referred as a smart monitoring system.

Detecting stationary objects in a monitored area is one of the most important tasks for smart monitoring system. A stationary object is defined as an object

© Springer International Publishing AG, part of Springer Nature 2018
N. T. Nguyen et al. (Eds.): ACIIDS 2018, LNAI 10752, pp. 583–593, 2018.
https://doi.org/10.1007/978-3-319-75420-8_55

that previously moves and enters the monitored area, but it currently remains stable in that area [1]. This object can be either an abandoned baggage/luggage or an illegally parked vehicle. Abandoned objects are necessary to be detected for bombing prevention, while illegally parked vehicles may cause traffic problems, such as traffic jams and traffic accidents. Thus, smart monitoring system should be capable of detecting stationary objects.

A recent comprehensive and up to date survey of stationary object detection has been published by Cuevas [2]. Typically, monitoring systems use static cameras, so most approaches employs a background modeling to extract candidate regions of stationary objects by implementing background subtraction (BS) and foreground analysis. The candidate regions are further analyzed using object detector. Generally, the method of stationary object detection is divided into two major steps: stationary object extraction and tracking. Hassan et al. [3] proposed combination of three-step pixel-based classification method and adaptive edge-based tracking to detect stationary objects on single model. However, using only single background model to detect stationary regions is unstable due to imperfect BS, which is affected by illumination condition of outdoor scenes. Porikli [4] overcame this problem by maintaining two background models with different learning rates on the same sequences, named dual foreground segmentation. However, it could not deal with moving objects at a very slow speed. A similar approach, proposed in our previous work [5], with less computational cost has been implemented for the task of illegally parked vehicle detection. In practice, this can also be adopted for general stationary object detection [1], as both tasks exploit the existence of static objects on the scene. Hence, a robust solution to detect general stationary objects in a monitored area is proposed. Object classifier based on the support vector machine (SVM) and the improved histogram of oriented gradient are also integrated. It classifies the candidate regions into four possible classes, namely human, vehicle, baggage, and other objects. In addition, in this work, we also implement an improved version of histogram of oriented gradient as feature extraction for training and testing the object classifier.

2   Proposed Method

2.1   Overview

Figure 1 shows the process diagram of our proposed method for detecting stationary objects. The method is divided into three main stages: candidate region extraction, region verification with object classifier, and decision based on tracking. In the first stage, candidate regions of stationary objects are extracted using dual background modeling combining spatial and temporal information [5]. These candidates are then verified by classifying them into four possible classes using object classifier. The object classifier is trained by a cascade support vector machine [6] using scalable histogram of oriented gradient feature [7]. If the

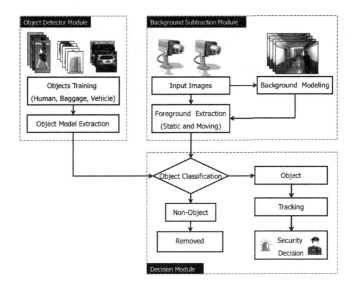

Fig. 1. Process diagram of the stationary object detection system

object is classified as baggage and vehicle, further analysis is required by performing detection-based methods. The final stage is to decide whether the object is stationary or not.

2.2 Object Candidate Extraction

In many computer vision applications, background modeling is an important stage for region extraction. Typically, it is used for separating the object from background scene in pixel level. Here, it classifies the image pixels into either background or foreground. In literature, there are many background modeling methods, such as statistical-based [8], single Gaussian, or mixture of Gaussian. Due to its effectiveness, we build background modeling using Gaussian Mixture Model (GMM) [9]. Note that background modeling is not our main contribution. Thus, although we use GMM, in practice, other background modeling methods are applicable as well.

In background modeling, the updating process is one of the crucial steps in which the system updates the background model B for each incoming frame I_t. Mathematically, updating process in pixel location (x, y) is modeled by:

$$B(x, y) = \alpha I_t(x, y) + (1 - \alpha)B(x, y) \tag{1}$$

where α is the updating rate. In Eq. (1), if the value of α is bigger, the background model is updated faster. In this sense, if a moving object stops in a certain area, within short period of time the object will be identified as background, as shown in the center image of Fig. 2. On the other hand, using a small value of α, the background model will be updated very slowly. In this case, if a moving object

Fig. 2. Sample results of dual background modeling with different updating rates

stops in the monitored area, the object will still be detected as foreground for longer period, as shown in the right image of Fig. 2. The first model with the bigger value of α is defined as short-term model B_S, while the second model is defined as long-term model B_L. By applying background segmentation, the short-term and long-term foreground models, F_S and F_L, are formulated by:

$$F_S = \begin{cases} 1, & \text{if } |B_S - I| < T_B, \\ 0, & \text{if } |B_S - I| \geq T_B, \end{cases} \quad \text{and} \quad F_L = \begin{cases} 1, & \text{if } |B_L - I| < T_B, \\ 0, & \text{if } |B_L - I| \geq T_B. \end{cases} \tag{2}$$

By analyzing the result of short-term and long-term foreground models, we can extract the object candidate by computing the difference between these two models. This is called dual foreground difference, which is defined by:

$$P(x, y) = F_L(x, y) - F_S(x, y). \tag{3}$$

According to Eq. (3), there will be three possible values of $P(x, y)$. $P(x, y) = 0$ represents that the pixel in location (x, y) is either background or moving object. $P(x, y) = 1$ indicates that the pixel is part of a stationary object. $P(x, y) = -1$ indicates an uncovered background. As we only focus on stationary objects, only $P(x, y) = 1$ is used for extracting stationary objects.

By assuming the value of $P(x, y) = 1$ exists in a long period, the cumulative value in time domain should be high enough and more than a certain threshold. Thus, instead of using a single value of $P(x, y)$, we calculate the cumulative value of $P(x, y)$ within a period of time in order to analyze the pixel type. We call this method cumulative dual foreground difference [5], which is formulated by:

$$H_t(x, y) = \sum_{i=t-K}^{t} P_i(x, y), \tag{4}$$

where $P_i(x, y)$ is the foreground difference at time i, and K is the number of frames. The value of K is set to be $f \times s$, where f is the video frame rate, and s is the analysis time period, e.g., 30 s. Next, the pixel $P(x, y)$ is classified into either stationary or moving by using thresholding as follows:

$$O(x, y) = \begin{cases} 1, & \text{if } H_t(x, y) \geq T_h K, \\ 0, & \text{otherwise,} \end{cases} \tag{5}$$

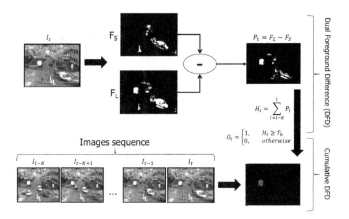

Fig. 3. Cumulative dual foreground difference for extracting stationary objects

where 1 and 0 represent stationary objects and moving objects, respectively. We then apply connected component labelling to form candidate regions of stationary objects. As post processing, we also apply geometric filtering to reject false positives due to noise and imperfect background modelling. The geometric filtering is conducted based on geometric properties of regions, such as aspect ratio, occupation ratio, and region area. Figure 3 shows an illustration of object candidate extraction process based on cumulative dual foreground difference.

2.3   Object Verification

Object classifier is one of important stages in detecting stationary objects. It is used to determine whether the object candidates are target objects in the system. There are four classes used in our classification module, namely human, vehicle, baggage, and other objects. Note that although we have four classes in object classifiers, each object is assigned as target for a certain monitoring task. For instance, in abandoned object detection, the target object is baggage, while in illegally parked vehicle detection, the target object is set to be vehicle. Here, we use object vehicle as illustration in this subsection.

The region candidates, which were previously detected, need to be verified using object classifier. In this work, we build a cascade support vector machine [6] based on an improved version of histogram of oriented gradient, called scalable histogram of oriented gradient [7]. First, Sobel operator [10] is applied in a candidate region to compute the gradient magnitude and the orientation map. Second, the region is divided into $n \times n$ blocks, where n is the block size; for instance, 2×2 blocks, 4×4 blocks, or 16×16 blocks. In each block, the gradient-based local features are extracted using a similar way as the extraction of histogram of oriented gradient [11]. By applying quantization of gradient orientation into 9 bins, we will have a single histogram with 9 values for each block. If the block size is set to 2×2 blocks, we will then have 36 feature values. We repeat this

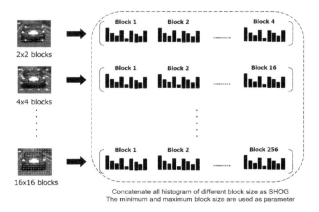

Fig. 4. An illustration of feature extraction process

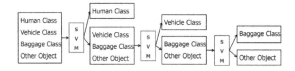

Fig. 5. Cascade support vector machine

process using different block sizes. Our final feature will be a concatenation of several normalized locale features [12] in different block sizes. Figure 4 shows the illustration of how feature extraction is carried out. Once the object's feature has been extracted, the object is classified into four possible classes, namely vehicle, human, baggage, and other objects. Here, we use a cascade support vector machine as shown in Fig. 5.

2.4   Tracking and Decision System

In the previous stage, a stationary object has been detected and verified. The tracking method is applied for determining how long the object is being stable in the monitored area. Based on Home Office United Kingdom regulation, if a baggage has been abandoned for more than 30 s, an alarm should be triggered. On the other hand, if a vehicle has been parked illegally for more than 60 s, the system should notify a security office.

There are several available tracking methods that can be used for duration counting, such as Kalman filter, particle filter, etc. However, since the tracked object is not moving, these methods are computationally expensive to track such an object. Thus, we implement a simple tracking method based on detection. When we have detected a stationary object in the current frame, the location of the object is stored by the system. The method then checks whether the same object exists in the next frame using the sum of squared difference-based template matching.

3   Experiment

3.1   Setting and Dataset Specification

The proposed method was implemented using C++ programming language with OpenCV library for basic image processing with computer specification Core i7-4770 CPU 3.40 GHz 8 GB RAM. For evaluation purposes, each video input is resized to 640 × 480 pixels. The method was evaluated using standard evaluation protocol, such as precision, recall, and F-measure. Precision is the ratio between the number of detected stationary objects and the number of real stationary objects, while recall is the ratio between the number of stationary objects in the experiment, namely true positives (TP), with the number of stationary objects in the ground truth (GT). F-measure is defined as the harmonic mean between precision and recall. We evaluated our method using datasets collected from Imagery Library for Intelligent Detection System (i-LIDS) [13] and ISLab dataset [5]. In total, there are 40 video data, which are recorded in different locations and times.

3.2   Results and Discussion

Tables 1 and 2 show the evaluation results of stationary object detection for both cases of abandoned object detection (AOD) and illegally parked vehicle detection (PVD), respectively. As shown in both tables, scenarios for AOD vary: indoor and outdoor environments, day and night times, true-color and infrared camera. On the other hand, PVD only focuses on outdoor environment in both day and night times.

As shown in Table 1, the proposed method successfully detects almost all scenarios (20 of 21) of abandoned objects. However, unfortunately, it produces one incorrect stationary object, namely false positive (FP), in video #8 because of the object's reflection in the water spill. In addition, it also misses one object in video #11 because of dark object. It is quite difficult to extract object with dark color in infrared images. Overall, our method achieves 95%, 95%, and 95% for precision, recall, and F-measure, respectively. These results proved that our method is effective and acceptable to be implemented for abandoned object detection. Figure 6 shows several visual results of abandoned object detection.

Table 2 shows the result of illegally parked vehicle detection. It can be observed that almost parked vehicles in all scenarios have been detected successfully, except for videos #15 and #20. In video #15, the method misses to detect illegally parked vehicle due to the small size of the object region in the image, so that it is rejected during the geometric filtering. In video #20, the illegally parked vehicle could not be detected because of background modeling problem. Overall, the proposed method obtains 90%, 100%, and 94.7% in precision, recall, and F-measure, respectively. Although the proposed method cannot detect all scenarios, these results are still acceptable for smart monitoring system. Figure 7 shows several visual results of illegally parked vehicle detection.

Table 1. Evaluation results for abandoned object detection

No	Video	Duration	Scenario	GT	TP	FP
1	D01	00:02:09	Day time, Outdoor	1	1	0
2	D02	00:03:03	Day time, Outdoor	1	1	0
3	D03	00:02:53	Day time, Outdoor, Far	1	1	0
4	D04	00:02:14	Day time, Indoor	1	1	0
5	D05	00:03:11	Day time, Indoor	1	1	0
6	D06	00:01:53	Day time, Outdoor	1	1	0
7	D07	00:01:04	Day time, Indoor	1	1	0
8	D08	00:01:56	Day time, Indoor	1	1	1
9	D09	00:01:19	Day time, Indoor	1	1	0
10	D10	00:02:29	Day time, Indoor	1	1	0
11	N01	00:03:53	Night time, Infrared, Outdoor	2	1	0
12	N02	00:01:21	Night time, Infrared, Outdoor	1	1	0
13	N03	00:02:04	Night time, Infrared, Outdoor	1	1	0
14	N04	00:03:30	Night time, Indoor	1	1	0
15	N05	00:04:34	Night time, Indoor	1	1	0
16	N06	00:03:26	Night time, Infrared, Indoor	1	1	0
17	N07	00:03:26	Night time, Infrared, Indoor	1	1	0
18	N08	00:03:07	Night time, Infrared, Indoor	1	1	0
19	N09	00:01:22	Night time, Infrared, Indoor	1	1	0
20	N10	00:01:15	Night time, Infrared, Indoor	1	1	0
			Total detection	21	20	1

Fig. 6. Selected sample results of abandoned object detection

Table 2. Evaluation results of illegally parked vehicle detection

No	Video	Duration	Scenario	GT	TP	FP
1	D01	00:01:49	Day time, Outdoor	1	1	0
2	D02	00:02:10	Day time, Outdoor	1	1	0
3	D03	00:01:57	Day time, Outdoor	1	1	0
4	D04	00:02:17	Day time, Outdoor	1	1	0
5	D05	00:02:32	Day time, Outdoor	1	1	0
6	D06	00:03:29	Day time, Outdoor	1	1	0
7	D07	00:02:19	Day time, Outdoor	1	1	0
8	D08	00:01:49	Day time, Outdoor	1	1	0
9	D09	00:01:25	Day time, Outdoor	1	1	0
10	D10	00:03:28	Day time, Outdoor	1	1	0
11	D11	00:03:11	Day time, Outdoor, Road	1	1	0
12	D12	00:02:09	Day time, Outdoor, Road	1	1	0
13	D13	00:02:02	Day time, Outdoor, Road	1	1	0
14	D14	00:01:55	Day time, Outdoor, Road	1	1	0
15	D15	00:04:54	Day time, Outdoor	1	0	0
16	N01	00:02:20	Night time, Outdoor, Infrared	1	1	0
17	N02	00:02:10	Night time, Outdoor, Infrared	1	1	0
18	N03	00:03:04	Night time, Outdoor, Infrared	1	1	0
19	N04	00:03:42	Night time, Outdoor, Infrared	1	1	0
20	N05	00:01:45	Night time, Outdoor	1	0	0
			Total detection	20	18	0

Fig. 7. Selected sample results of illegally parked vehicle detection

To evaluate the effectiveness of the method for real-time processing, the processing speed is calculated. Based on experiment, our method spends around 65.6 ms in processing time. Overall, our method can perform at around 15 frame per seconds (fps). This result is still acceptable for automated smart monitoring systems that require at least 10 fps, as the results can still be improved by implementing parallel processing.

Nevertheless, our method has limitation when handling full illumination changes and weather condition; for instances, the changing from evening to night time, dawn to morning time, cloudy to sunny, sunny to rainy, etc. This is because the background modeling could not adapt perfectly to this situation. Thus, integrating more powerful background modeling will be considered in future works.

4 Conclusion

In this work, a method for detecting stationary objects has been successfully implemented. It uses cumulative dual foreground difference, which also integrates a support vector machine-based object classifier. The proposed method was evaluated using public datasets in two cases: abandoned object detection (AOD) and illegally parked vehicle detection (PVD). In evaluation result, our method achieved good performance: around 95% and 94.7% for AOD and PVD cases, respectively. However, the proposed method depends on the background modeling results. If the background modeling gives a good performance, the results of our method will also achieve a good result; otherwise the method may produce many false positives.

Acknowledgment. This work was supported by the 2017 Research Fund of University of Ulsan and the 2017 Research Fund of Faculty of Mathematics and Natural Sciences, Universitas Gadjah Mada (No. 94/J01.1.28/PL.06.01/2017).

References

1. Branch, H.O.S.D.: Imagery library for intelligent detection systems (i-LIDS). In: 2006 IET Conference on Crime and Security, pp. 445–448, June 2006
2. Cuevas, C., Martnez, R., Garca, N.: Detection of stationary foreground objects: a survey. Comput. Vis. Image Underst. **152**, 41–57 (2016)
3. Hassan, W., Birch, P., Young, R., Chatwin, C.: Real-time occlusion tolerant detection of illegally parked vehicles. IJCAS **10**(5), 972–981 (2012)
4. Porikli, F.: Detection of temporarily static regions by processing video at different frame rates. In: 2007 IEEE Conference on Advanced Video and Signal Based Surveillance, pp. 236–241, September 2007
5. Wahyono, Jo, K.H.: Cumulative dual foreground differences for illegally parked vehicles detection. IEEE Trans. Ind. Inf. **13**(5), 2464–2473 (2017)
6. Cortes, C., Vapnik, V.: Support-vector networks. Mach. Learn. **20**(3), 273–297 (1995)
7. Wahyono, Hoang, V.D., Kurnianggoro, L., Jo, K.H.: Scalable histogram of oriented gradients for multi-size car detection. In: MECATRONICS2014-Tokyo, pp. 228–231, November 2014
8. Garca, J., Gardel, A., Bravo, I., Lizaro, J.L., Martnez, M., Rodrguez, D.: Directional people counter based on head tracking. IEEE Trans. Ind. Electron. **60**(9), 3991–4000 (2013)
9. Stauffer, C., Grimson, W.E.L.: Adaptive background mixture models for real-time tracking. In: CVPR, pp. 2246–2520. IEEE Computer Society (1999)

10. Sobel, I., Feldman, G.: A 3 × 3 isotropic gradient operator for image processing. Presented at a talk at the Stanford Artificial Intelligence (SAIL) Project (1968)
11. Dalal, N., Triggs, B.: Histograms of oriented gradients for human detection. In: Proceedings of IEEE CVPR, Washington, DC, USA, pp. 886–893. IEEE Computer Society (2005)
12. Horn, R.A., Johnson, C.R.: Matrix Analysis. Cambridge University Press, Cambridge (1990)
13. i-LIDS dataset for AVSS 2007 (2007). ftp://motinas.elec.qmul.ac.uk/pub/iLids. Accessed 30 July 2017

CNN-Based Character Recognition
for License Plate Recognition System

Van Huy Pham$^{(\boxtimes)}$ (iD), Phong Quang Dinh, and Van Huan Nguyen

Faculty of Information Technology, Ton Duc Thang University, Ho Chi Minh City, Vietnam
phamvanhuy@tdt.edu.vn

Abstract. License Plate Recognition is a practical use of computer vision based application. With the increase in demand of automation transportation systems, this application plays a very big role in the system development. Also, the use of vehicles has been increasing because of population growth and human needs in recent years makes the application is more challenging. Moreover, license plates are available in diverse colors and style and that the presence of noise, blurring in the image, uneven illumination, and occlusion makes the task even more difficult for conventional recognition methods. We propose an approach of using a Convolutional Neural Networks (CNN) classifier for the recognition. Preprocessing techniques are firstly applied on input images, such as filtering, thresholding, and then segmentation. Then, we train a CNN classifier for character recognition. Although the performance of a CNN is very impressive, it costs much time to complete the character recognition step. In this study, a modified CNN is proposed to help the system run in real-time. Experimental results have done and analyzed with other methods.

Keywords: Convolution Neural Network · Character recognition
License Plate Recognition System

1 Introduction

License Plate Recognition (LPR) systems are very popular and studied all over the world. A LPR system is a combination of several modules and involves object detection, image processing, and pattern recognition. Beside image acquisition and preprocessing, the process of reading a license plate goes through 3 main phases. The first phase is plate localization or plate extraction. The second one is the character segmentation which each character is detected and separated from the others. The last one is the character recognition which the extracted characters are recognized. Because of quality of image and the variance of plate's shape, size, color, orientation, it is difficult to finding the plate location. The character segmentation may be in trouble due to quality of image, partially connected characters, noise, and rotation of plate. The last phase also has to deal with variation in fonts, similarity among characters and non-characters.

    In the past two decades, many different methods have been proposed and tested. The purpose of this study is to provide a comparison for some of existing method and then propose an improvement in performance of our system which is based on different

© Springer International Publishing AG, part of Springer Nature 2018
N. T. Nguyen et al. (Eds.): ACIIDS 2018, LNAI 10752, pp. 594–603, 2018.
https://doi.org/10.1007/978-3-319-75420-8_56

methods of character recognition: Artificial Neural Network (ANN) and Convolutional Neural Networks (CNN).

The fact is that ANN-based OCR can work in real-time, but the performance of the system is not as good as CNN-based systems. Although the use of CNN yields better performance, it is very slow in both training and recognition phase. In our experiment using CNN Lenet-5, the license plate system recognized one character in about 11 ms, but the overall system needs to process in 30 ms to avoid false detection. Therefore, in this study, we proposed a CNN structure which processes 20× faster than ordinary CNN Lenet-5 that can help the system run in real-time.

2 Related Works

There are many difficulties that a LPR system must overcome. Due to the camera zoom factor, the extracted characters do not have the same size and thickness. Resizing the characters into one size before recognizing helps to overcome this problem. The character's font is not the same all the time since different states, different countries use different fonts. Extracted characters may have some noise or may be broken. In the following, we categorize some existing character recognition methods.

Ahmed et al. [1] have presented a template matching. Template matching is a simple method. The similarity between a character and the template is measure. The template that is the most similar to the character is recognized as the target. Most template matching methods use binary images because the gray-scale is changed due to any change in the lighting. Template matching is performed after resizing the extracted character into the same size. This method is useful for recognizing single font, non-rotated, non-broken. If a character is different from the template due to any font change, rotation, noise, the template matching produces incorrect recognition.

LeCun et al. [2] used HOG-feature for character recognition. At the training phase, training data was generated from a high-solution image of each letter and distribution of each letter in the HOG-feature space was then obtained. At the recognition phase, each character is cut out from the image, calculated the HOG-feature vector, and recognized characters with the distribution in HOG-feature space obtained above. Siddharth et al. [3] used Support Vector Machine (SVM) classifier. SVM classifier is trained by a given set of training data and a model is prepared to classify test data based upon this model. For multiclass classification problem, we decompose multiclass problem into multiple binary class problems, and we design suitable combined multiple binary SVM classifiers. According to how all the samples can be classified in different classes with appropriate margin, different types of kernel in SVM classifier are used. Commonly used kernels are: Linear kernel, Polynomial kernel, Gaussian Radial Basis Function (RBF) and Sigmoid (hyperbolic tangent).

Sharma and Singh [4] has applied Artificial Neural Network for character recognition. This method simulates the way human neural system works to create intelligent behavior. The idea is to take a large number of characters, known as training set, and then develop a system which can learn from those training. In other words, the neural network uses the training to automatically infer rules for recognize character.

These methods can achieve impressive result on good dataset but still produce incorrect recognition in hard dataset. However, we can do much better if we introduce some concept from CNN theory. Bounchain [5] has applied Lenet-5 in character recognition task. This network was tested with a database containing more than 50,000 hand-written digits, all normalized in the input image. An error rate of about 0.95% was achieved. This is an impressive result on OCR task.

In this study, we will apply our CNN for OCR for the system running in real time by reducing some unnecessary convolutional layers.

3   The Proposed Method

3.1   Plate Detection and Segmentation

Plate detection. In order to localize license plates in a given image, different methods have been used. The simple and fast method is primarily based on identifying the vertical edges of each license plate in the input image [6]. After extracting vertical edges from the image, morphological filtering is applied to obtain candidate regions. Then spatial features such as area, aspect ratio and edge density are considered to discard wrong candidate regions.

In our study, we use the methods in [1, 7–9] for plate detection and segmentation. A combination of Hough Transform and a counter algorithm in [10] is used to detect license plate region. In the first step, the counter algorithm is applied to detect close boundaries of the objects. These counter lines are transformed to Hough coordinate to find interacted parallel lines that are considered as license plate candidates. To filter out the candidate plates, the aspect ratio and the horizontal cross cuts are used. In [7–9], Connected Component Labelling (CLL) is used license plate detection. CCL scans the image the labels the pixels according to the pixel connectivity. In [8], a feature extraction algorithm is used to count the similar labels to distinguish it as a region. The region corresponding with the maximum area is considered as a possible license plate region. Likewise, in [7], two detection methods are performed to detect white frame and to detect black characters. To determine the candidate frames, aspect ratio of license plate, height and width of characters have to be known. In [9], after the successful CCL on the binary image, measurements such as orientation, aspect ratio for every binary object in the image are calculated. Criteria such as orientation $< 35°$, $2 <$ aspect ratio < 6 are considered as candidate plate regions.

Character segmentation. In [11], adaptive thresholding is applied to binary the input image. Then connected component analysis is applied for character segmentation. The components obtained from the process may be character or non-character. Character's aspect ratio is considered to suppress non-character. The ratio is set based upon the observation from different images.

The method of row-column scan is chosen to segment characters in [10]. Firstly, the line scan method is used to scan the binary image and lower-upper bounds are located. Secondly, the column scan method is chosen to scan binary image and the left-right bounds are located. Based on these, each character can be accurately segmented.

Experimental results show that this method can even handle license plate images with fuzzy, adhering, or fractured characters with high efficiency.

In [1], the strategy is based on pixel count (i.e., vertically projecting and counting the number of pixel in each column). First, the image is converted into binary form and vertical projections are made. Then, the number of black pixels in each column is counted, and a histogram is plotted. The characters are segmented depending on the transition from a crest to its corresponding trough. To avoid unnecessary segmentation, some thresholds are taken properly.

3.2 Convolutional Neural Networks for Character Recognition

Convolutional neural networks have been successfully applied in the field of computer vision. Unlike artificial neural network, the layers of a CNN have neurons arranged in 3 dimensions: width, height, depth (Fig. 1). The neurons in a layer will only be connected to a small region of the layer before it, instead of all of the neurons in a fully-connected layer.

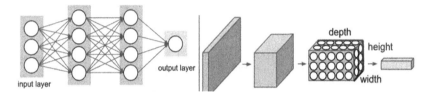

Fig. 1. Left: A regular 3-layer neural network. Right: A Convolutional Neural Networks arranges its neurons in 3 dimensions (width, height, depth) as visualized in one of the layers. Every layer of a CNN transforms the 3D input volume to a 3D output volume of neuron activations. In this example, the red input layer holds the image, so its width and height would be the dimensions of the image, and the depth would be 3 (red, green, blue). (Color figure online)

A general CNN is a sequence of layers, and every layer of a CNN transforms one volume of activations to another through a differentiable function. There are 3 main types to build CNN architecture: convolutional layer, pooling layer, and fully-connected layer (exactly as seen in artificial neural network). These layers will be stacked to form a full CNN architecture. In more detail, a simple CNN classification could have the architecture [INPUT - CONV - RELU - POOL - FC]. INPUT holds the raw pixel values of the image with 3 color channels. CONV layer computes the output of neurons that are connected to local regions in the input, each computing a dot product between their weights and a small region they are connected to in the input volume. RELU layer applies an elementwise activation function, such as the $max(0, x)$ threshold at zero. POOL layer performs a down-sampling operation along the spatial dimensions (width, height). FC layer computes the class scores. As with artificial neural networks, each neuron in this layer will be connected to all the numbers in the previous volume. In this way, CNN transforms the original image from original pixel values to the final class score.

In this study, we follow the work of using Lenet-5 in [2]. Lenet-5 comprises 7 layers, not counting the input, all of which contain trainable parameters (Fig. 2). In the following, convolutional layers are labeled Cx, subsampling layers are labeled Sx, and fully-connected layers are labeled Fx, where x is the layer index. Layer C1 is a convolutional layer with 6 features maps. Each unit in each feature is connected to a 5 × 5 neighborhood in the input. The size of the feature maps is 28 × 28. Layer S2 is a subsampling layer with 6 feature maps of size 14 × 14. Each unit in each feature map is connected to a 2 × 2 neighborhood in the corresponding feature map in C1. The 2 × 2 receptive fields are non-overlapping, therefore feature maps in S2 have half the number of rows and columns as feature maps C1. Layer C3 is a convolutional layer with 16 feature maps. Each unit in each feature map is connected to several 5 × 5 neighborhoods at identical locations in a subset of S2's feature maps. Layer S4 is a subsampling layer with 16 feature maps of size 5 × 5. Each unit in each feature map is connected to a 2 × 2 neighborhood in the corresponding feature map in C3, in a similar way as C1 and S2. Layer C5 is a convolutional layer with 120 feature maps. Each unit is connected to a 5 × 5 neighborhood on all 16 of S4's feature maps. Layer F6 contains 84 units and is fully connected to C5. As in artificial neural networks, units in layers up to F6 compute a dot product between their input vector and their weight vector, to which a bias is added. This weighted sum is then passed through a sigmoid function. Finally, the output layer is composed of Euclidean Radial Basis Function units (RBF), one for each class, with 84 input each. The output of each RBF unit y_i is computed as follow.

$$y_i = \sum_j \left(x_j - w_{ij} \right)^2$$

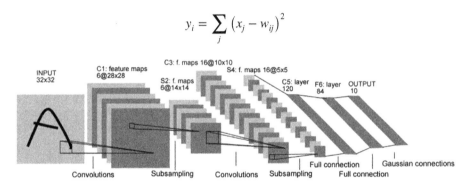

Fig. 2. Architecture of Lenet-5 [2]

In other words, each output RBF unit computes the Euclidean distance between its input vector and its parameter vector. The further away is the input from the parameter vector, the larger is the RBF output.

Lenet-5 is designed to extract local geometric feature from the input field in a way that preserves the approximate relative locations of these features. This is done by creating feature maps that are formed by convolving the image with local feature-extraction kernels. Lenet-5 has several advantages that make it attractive for recognizing characters when high variability is expected. First, Lenet-5 has state-of-the-art accuracy by its impressive performance. Second, Lenet-5 runs at high speeds on specialized hardware. Lenet-5 can also be readily trained to recognize new character styles and fonts. It works well for both handwritten and machine printed characters.

3.3 Our Modification of Lenet-5

The modified Lenet-5 comprises 6 layers, not counting the input. Layer C1 and C3 has 4 features maps. Each unit in each feature C1 is connected to a 4 5 × 5 neighborhood in the input. Each unit in each feature map C3 is connected to 4 3 × 3 neighborhoods at identical locations in a subset of S2's feature maps. Layer S2 and S4 is a subsampling layer with 4 feature maps. Layer F5 is a fully-connected layer with 1204 units. Each unit is connected to all units in S4's feature maps. Output layer contains 35 units and is fully connected to F5. As can be seen, we remove the last convolutional layer and we just use 4 feature maps in C1, C3 convolutional layer. The reason is there are only about 3 or 4 gray levels on each real character image. Therefore, we don't need many features as the original Lenet-5. The time-consuming tasks occur mostly in the convolutional layers, so reducing in the number of features makes our CNN faster. Moreover, each unit in each feature map C3 is only connected to 3 × 3, instead of 5 × 5 neighborhoods, as the thickness of the characters is about 3 pixels. This can help to reduce the processing time (Fig. 3).

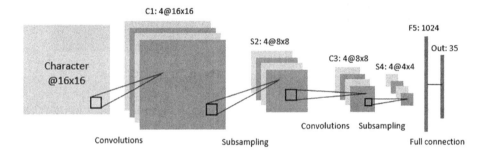

Fig. 3. The proposed Lenet-5 with less feature maps in convolutional layers

4 Experimental Results

For experiments, we used two datasets for evaluation: MINIST used in [2] and our collection. The MINIST database (Modified National Institute of Standards and Technology database) is a large database of handwritten digits commonly used for training various image processing system. Figure 4 shows some example characters in the database. The MNIST database contains 60,000 training images and 10,000 testing images.

For comparison, we built our own dataset from real-life traffic car images. On each of collected images, we extracted the regions which have a single character. The segmented characters are sorted into 35 categories: 0–9, A–Z. All the characters are resized to 16 × 16 before training. Figure 5 shows some examples of license plate characters in our dataset. The dataset was collected from US traffic images using a simple LPR commercial product in our company.

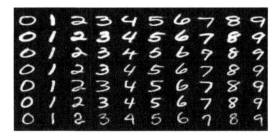

Fig. 4. Examples from MNIST dataset

Fig. 5. Car license plates from some traffic images

We use a dataset of 112,000 samples to train, and test on various states in US (Cali test set). The models used for comparsion are ANN Lenet-5 and our proposed CNN, and we evaluate on the accuracy and processing time for each of the models.

Firstly, these models was trained with MNIST dataset and tested with our test set. Table 1 shows the accuracy and time process of ANN, CNN Lenet-5 and our proposed CNN in the MNIST dataset. The CNN-based models have best performance, and our proposed method is at a bit lower accuracy at 98.77% compared with 99.10% of Lenet-5 but the processing time is much lower at 0.38 ms compared with 10 ms of Lenet-5.

Table 1. The accuracy and time process of ANN, CNN Lenet-5 and our proposed CNN in the MNIST dataset.

Data test	ANN		CNN Lenet-5		Proposed CNN	
	Accuracy (%)	Time (ms)	Accuracy (%)	Time (ms)	Accuracy (%)	Time (ms)
MNIST	96.820	0.050	99.190	10.158	98.770	0.386

For better and more practical evaluation, we divide the test set into 3 categories: normal and good dataset containing characters in common and good conditions; and hard dataset containing characters challenging the detection. Some examples are shown in Fig. 6.

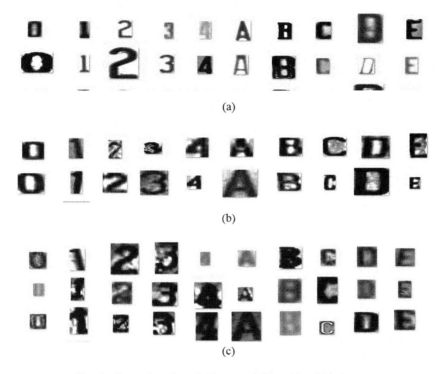

Fig. 6. Examples of good (a), normal (b) and hard (c) dataset

Tables 2, 3 and 4 show the accuracy and processing time of the 3 testing models on 3 levels of the testing datasets.

Table 2. The accuracy and processing time on ANN, CNN Lenet-5 and our proposed CNN with good dataset.

Data test	ANN		CNN Lenet-5		Proposed CNN		Number of sample
	Accuracy (%)	Time (ms)	Accuracy (%)	Time (ms)	Accuracy (%)	Time (ms)	
Arizona	96.160	0.055	99.745	11.239	98.162	0.507	1500
Idaho	94.900	0.054	99.565	11.502	97.819	0.503	1500
Missouri	93.378	0.055	99.454	11.309	97.640	0.547	1500
Texas	94.277	0.055	99.135	11.439	97.801	0.573	1500
Average	94.679	0.055	99.475	11.372	97.856	0.533	

Table 3. The accuracy and processing time on ANN, CNN Lenet-5 and our proposed CNN with normal dataset.

Data test	ANN		CNN Lenet-5		Proposed CNN		Number of sample
	Accuracy (%)	Time (ms)	Accuracy (%)	Time (ms)	Accuracy (%)	Time (ms)	
British Colombia	89.382	0.078	94.547	11.532	92.010	0.479	1500
WashingtonDC	92.117	0.054	95.343	11.396	93.885	0.509	1500
Georgia	90.429	0.054	95.149	11.372	93.804	0.595	1500
Vermont	88.147	0.058	94.987	11.276	91.653	0.667	1500
Average	90.019	0.061	95.007	11.394	92.838	0.562	

Table 4. The accuracy and processing time on ANN, CNN Lenet-5 and our proposed CNN with hard dataset.

Data test	ANN		CNN Lenet-5		Proposed CNN		Number of sample
	Accuracy (%)	Time (ms)	Accuracy (%)	Time (ms)	Accuracy (%)	Time (ms)	
Delaware	84.942	0.054	89.894	11.892	88.153	0.499	1500
Maryland	86.983	0.055	90.510	11.388	88.468	0.526	1500
Average	85.963	0.055	90.202	11.640	88.310	0.512	

Evaluation of the overall system is based on averaging the accuracy and the processing time of each models on the datasets. Table 5 shows the evaluation on our collected dataset.

Table 5. The average in accuracy and processing time of ANN, CNN Lenet-5 and our proposed CNN in our collected dataset

Data test	ANN		CNN Lenet-5		Proposed CNN	
	Accuracy (%)	Time (ms)	Accuracy (%)	Time (ms)	Accuracy (%)	Time (ms)
Average	90.220	0.057	94.894	11.469	93.001	0.536

The accuracy of our proposed CNN is higher about 3% than ANN and less 2% than CNN Lenet-5. Although the time processing is still slower 10× than ANN, it is 20× faster than CNN Lenet-5. Therefore, our proposed CNN can still run in real time at a practically acceptable accuracy.

5   Conclusion

Starting from a basic classifier, various concepts for character recognition were introduced. In this study, we proposed a CNN model for character recognition task. The CNN lenet-5 is modified by reducing the number of convolution layer and the size of neighborhood to become our model. The model was tested with 36,500 images of character.

Although the accuracy is less 2% than Lenet-5, our proposed CNN gave satisfactory of performance and real-time processing.

References

1. Ahmed, M.J., Sarfraz, M., Zidouri, A., Al-Khatib, W.G.: License plate recognition system. In: Proceedings of the 2003 10th IEEE International Conference on Electronics, Circuits and Systems, 2003, ICECS 2003, pp. 898–901. IEEE (2003)
2. LeCun, Y., Bottou, L., Bengio, Y., Haffner, P.: Gradient-based learning applied to document recognition. Proc. IEEE **86**, 2278–2324 (1998)
3. Siddharth, K.S., Jangid, M., Dhir, R., Rani, R.: Handwritten Gurmukhi character recognition using statistical and background directional distribution. Int. J. Comput. Sci. Eng. (IJCSE) **3**, 2332–2345 (2011)
4. Sharma, S., Singh, N.: Optical character recognition using artificial neural networks approach. Int. J. Emerg. Technol. Adv. Eng. **4**, 339–344 (2014)
5. Bouchain, D.: Character recognition using convolutional neural networks. Inst. Neural Inf. Process. 2007 (2006)
6. Zheng, D., Zhao, Y., Wang, J.: An efficient method of license plate location. Pattern Recogn. Lett. **26**, 2431–2438 (2005)
7. Wen, Y., Lu, Y., Yan, J., Zhou, Z., von Deneen, K.M., Shi, P.: An algorithm for license plate recognition applied to intelligent transportation system. IEEE Trans. Intell. Transp. Syst. **12**, 830–845 (2011)
8. Caner, H., Gecim, H.S., Alkar, A.Z.: Efficient embedded neural-network-based license plate recognition system. IEEE Trans. Veh. Technol. **57**, 2675–2683 (2008)
9. Anagnostopoulos, C.N.E., Anagnostopoulos, I.E., Loumos, V., Kayafas, E.: A license plate-recognition algorithm for intelligent transportation system applications. IEEE Trans. Intell. Transp. Syst. **7**, 377–392 (2006)
10. Jin, L., Xian, H., Bie, J., Sun, Y., Hou, H., Niu, Q.: License plate recognition algorithm for passenger cars in Chinese residential areas. Sensors **12**, 8355–8370 (2012)
11. Kumari, S., Gupta, D., Singh, R.M.: A robust method for vehicle license plate recognition based on Harries corner algorithm and artificial neural network. Int. J. Comput. Appl. **148**, 16–19 (2016)

Improving Traffic Signs Recognition Based Region Proposal and Deep Neural Networks

Van-Dung Hoang[1(✉)], My-Ha Le[2], Truc Thanh Tran[3,4],
and Van-Huy Pham[5(✉)]

[1] Quang Binh University, Dong Hoi, Quang Binh, Vietnam
zunghv@gmail.com
[2] Ho Chi Minh City University of Technology and Education,
Ho Chi Minh City, Vietnam
halm@hcmute.edu
[3] Institute of Research and Development,
Duy Tan University, Da Nang, Vietnam
tranthanhtrucl982@gmail.com
[4] Department of Information and Communication, Danang City, Vietnam
[5] Ton Duc Thang University, Ho Chi Minh City, Vietnam
phamvanhuy@tdt.edu.vn

Abstract. Nowadays, traffic sign recognition has played an important task in autonomous vehicle, intelligent transportation systems. However, it is still a challenging task due to the problems of a variety of color, shape, environmental conditions. In this paper, we propose a new approach for improving accuracy of traffic sign recognition. The contribution of this work is three-fold: First, region proposal based on segmentation technique is applied to cluster traffic signs into several sub regions depending upon the supplemental signs and the main sign color. Second, image augmentation of training dataset generates a larger data for deep neural network learning. This proposed task is aimed to address the small data problem. It is utilized for enhancing capabilities of deep learning. Finally, we design appropriately a deep neural network to image dataset, which combines the original images and proposal images. The proposed approach was evaluated on a benchmark dataset. Experimental evaluation on public benchmark dataset shows that the proposed approach enhances performance to 99.99% accuracy. Comparison results illustrated that our proposed method reaches higher performance than almost state-of-the-art methods.

Keywords: Traffic sign recognition · Region proposal · Data augmentation
Deep neural networks

1 Introduction

Nowadays, automatic systems have been developed and applied into many fields on robotics, autonomous vehicles, intelligent transportation systems, and other industrial applications. Autonomous vehicle navigation becomes an important research area within various applications of motion path planning, localization, scene understanding

© Springer International Publishing AG, part of Springer Nature 2018
N. T. Nguyen et al. (Eds.): ACIIDS 2018, LNAI 10752, pp. 604–613, 2018.
https://doi.org/10.1007/978-3-319-75420-8_57

and mapping. There have been many research groups focusing on studying autonomous vehicle/robot, especially intelligent transportation in outdoor environments [1–4]. Scene understanding is typically relevant to pedestrian detection, action prediction, traffic sign recognition and mapping [5–7]. The scene understanding based on vision sensors is fundamental in various recognition applications. However, pattern recognition in outdoor environments encounters many challenges such as various articulate poses, appearances, illumination conditions and complex backgrounds of outdoor scenes, as well as partially occlusion. In the field of traffic sign recognition, there are numerous traffic signs and they are completed by supplementary signals. Usually, the supplementary signal is located at below or inside of main sign plate.

The paper [8] presented an approach for high-performance traffic sign recognition framework to fast recognize multiclass traffic signs in high-resolution images. This method consists of three steps, such as a region of interest extraction based on high-contrast region extraction, the cascade tree detector using the split flow of traffic signs, and robust method of traffic sign classification by an extended sparse representation classification. This system supports the quick and accurate detection and recognition for the multiclass traffic signs. The study in [9] presented a traffic sign recognition method based on weakly supervised metric learning. In that method, the recognition of a traffic sign is based on the use of LIDAR sensors instead of vision sensors as typical. Light detection and combined inputs including 3-D deep point images and 2-D multi-view sign images are used for the traffic sign recognition. Here, sign images plays as a role of supplement sources that improve the detection accuracy. Other works paid their attention to other aspects; e.g., the study in [6] concentrates on the finding for a method that gains the computational efficiency in traffic sign recognition. It was shown that computational savings could be achieved as a result of applying two following techniques. Firstly, histogram of oriented gradient (HOG) is used for feature description. Secondly, classification is based on extreme learning machine (ELM) when performing the relevant trainings and tests. It was proved that HOG could address the unbalance data problem and was robust in representing distinctive shapes of various objects. ELM and HOG in a combination achieve the better trade-off between accuracy and computational time merit. In classification, the extreme learning machine (ELM) method is used for train and test. Through experiment, authors showed that the proposed approach supports to keep a balance between the accuracy and computational time. Authors in [10] presented a method for traffic sign recognition based on kernel extreme learning machines with deep perceptual features. The method used color spaces for representation learning of convolutional neural network by using a kernel-based extreme learning machine. Experimental results show that the proposed method reaches higher precision than most of the state of the art approaches with 99.54% accuracy. In this study, we propose an approach for traffic sign recognition based on data augmentation, region proposal of a traffic sign plate to enhance supplemented signal information, and the use of deep neural network. The experimental results point that our approach is better than almost state of the art approaches.

2 Overview Architecture

In practice, traffic signs are diverse in shapes, images, texts so that recognizing all signs become a very challenging task. Here, supplemental and major signs should be distinguished to reduce the complexity of recognition. The difficulty of traffic signal recognition is potentially sub signal becoming infinite number of traffic classes by as additional information can be written on nearby major information. In this study, we propose a traffic sign recognition system based on the region proposal using segmentation and deep neural network (DNN). The overall of system architecture is illustrated in Fig. 1 for training and Fig. 2 for recognition task.

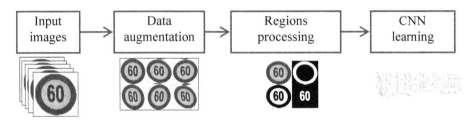

Fig. 1. Architecture of training

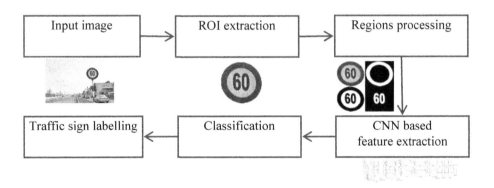

Fig. 2. Overview of traffic sign recognition architecture

There is diverse traffic signs of the same types, for example prohibitory of speed limit sign plate usually consists of variable numbers. The traffic plate displays main signal of restriction plate combining with supplement speed signals, e.g. 40, 50, 60, 80 and 100. In traditional approach, the full traffic signs are used to recognize a meaning of traffic signal. This paper presents an approach for interpretation of main signal with supplemental signals for improving accuracy of recognition system. The method consists of several phases as follows: traffic candidate detection and region of interest

(ROI) extraction based on color filter, segmentation of ROI into several regions. The segmented regions combining original sign plate are feed to the deep neural network for recognition the meaning of traffic signal, as depicted in Fig. 2. This approach is different to the Region convolution neural network (R-CNN) approach [11], which consists of some tasks as follows: Generating a set of region proposals for bounding boxes, implementing the images within bounding boxes to the pretrained AlexNet model [12], the final classification processing based on the Support vector machine (SVM). In our approach, the classification task was evaluated on two machines of SVM and Full-connection neural network. An image augmentation task on training data is also applied for efficiency improvements. This task is important to make powerful recognition models since using small data, especially in the experimental data, some traffic sign classes just consist of several image samples. In augmentation task, training images are reproduced to make a larger data, increasing the effective size of the training data set.

3   Region Proposal

There are several stages for image segmentation and refine results. First, an input sample is converted into Hue Saturation Value (HSV) image one. Then the gamma correction is applied on HSV image. There are many approaches for image segmentation, amount them the Simple Linear Iterative Clustering (SLIC) super pixels approach [13] is a state-of-the-art method. In the special situation of traffic sign recognition, the traffic sign plates are usually homogeneous with several colors in each plate. In order to improve processing speed, k-means approach is appropriate for segmentation instead of clustering of superpixels based on Density-based spatial clustering of applications with noise (DBSCAN) to generate the final segmentation [14]. In this case, the proposed approach is simple and relatively fast than related works.

The channels of hue, saturation information are used for clustering and they are converted to coordinates of data point for segmentation. The objective of this task is that classify regions, which have similar color. The details of this approach are summarized as following: The data points are randomly assigned into k clusters; for each data point, we calculate the distance from the data point to each cluster. Data points are set to their closest cluster; it repeats this step until no change data point to another cluster, that mean data of each cluster are stable. After clustering data, the post-processing is applied to merger small-connected regions to nearby bigger region. An example result is shown in Fig. 3.

Fig. 3. Region proposal results

4   Design Deep Neural Networks for Recognition Traffic Sign

In this study, the DNN learner consists of 25 layers, which are an input layer, convolution layer + rectified linear unit layer, cross normalization, max-pooling layers, fully connection layer. They transform the input image into a serial hierarchical feature descriptor. The input data are a set of pixel intensities of an image, which is fed into deep network. The input data consist of $64 \times 64 \times 3$ original image and $64 \times 64 \times 3$ region proposal. In this model, the filters at the first layer are corresponding to three color channels and second parts of three marks of region proposal. Filters are independent connecting to each other, which related to three channels of the input image. The final layer processes on the extracted feature vector to be classified by three layer of fully connected. A convolutional layer performs a convolution of input maps with a filter size $n_x \times n_y$. The details of the deep neural network architecture are shown in Fig. 4.

1	Image Input	64x128x3 images with 'zerocenter' normalization
2	Convolution	32 7x7x3 convolutions with stride [1 1] and padding [2 2 2 2]
3	ReLU	ReLU
4	CC Normalization	Cross channel normalization with 5 channels per element
5	Max Pooling	3x3 max pooling with stride [2 2] and padding [0 0 0 0]
6	Convolution	64 7x7x32 convolutions with stride [1 1] and padding [2 2 2 2]
7	ReLU	ReLU
8	CC Normalization	Cross channel normalization with 5 channels per element
9	Max Pooling	3x3 max pooling with stride [2 2] and padding [0 0 0 0]
10	Convolution	64 7x7x64 convolutions with stride [1 1] and padding [2 2 2 2]
11	ReLU	ReLU
12	Convolution	64 7x7x64 convolutions with stride [1 1] and padding [2 2 2 2]
13	ReLU	ReLU
14	Convolution	64 7x7x64 convolutions with stride [1 1] and padding [2 2 2 2]
15	ReLU	ReLU
16	Max Pooling	3x3 max pooling with stride [2 2] and padding [0 0 0 0]
17	FConnected	4096 fully connected layer
18	ReLU	ReLU
19	Dropout	50% dropout
20	FConnected	4096 fully connected layer
21	Dropout	50% dropout
22	FConnected	43 fully connected layer
23	Dropout	50% dropout
24	Softmax	Softmax
25	Classification	Crossentropyex with '0', '1', and 41 other classes

Fig. 4. Deep neural network architecture

On the task of recognition, we evaluated using different approaches of original DNN, and another approach using DNN and SVM. The output of a full connection network layer is used as vector features to SVM. The SVM method is significant perform in image classification task. The details of basic SVM could be referred to [15], it becomes a state of the art technique and successfully implemented. The main advantage of the SVM technique is its ability to extract highly discriminative profits.

5   Experiment

The German traffic sign detection benchmark dataset in [16] is used for experiment and evaluation the proposed approach. This dataset was collected in different scenarios of real traffic scenes. Training dataset consists of 43 traffic sign classes, with 39,209 images. The image data contains a test set within about 12,630 images. Region annotations are given in text file with rectangular box (x1, y1, x2, y2) with border of 10% around the sign and their labels. Traffic signs vary size from 15×15 to 250×250 pixels. The configuration of hardware system is CPU Core i7, and graphics cards GPU GTX 950. The details of comparison results on configuration hardware and required computation cost of training time are evaluated and presented in Table 1.

Table 1. Hardware configuration for experiment

	Hardware platform	Training time
CNNs [17]	Core I7-950 (3.33 GHz), 4 GTX 850 GPU	37 h
M2-tMTL [18]	Core I5-4590 (3.3 GHz), no GPU	0.5 h
Our system	Core I7-5500 (2.4 GHz), 1 GTX 950 GPU	21 h

The training set and testing set are resized all images to 64×64 pixels. Consequently, the different scaling of traffic signs with rectangular bounding boxes in original dataset are used to cropped. Resized sample images are forced to archived square images. Preprocessing tasks are applied for contrast normalization by histogram equalization balance using value layer of HSV images. Images from the training set are augmented to make a larger dataset for training. An original image is augmented into maximum of 10 output images. There are many techniques for resampling such as rotation, stretching, shearing: random rotation with angle between $-5°$ and $5°$, random shearing with angle between $-5°$ and $5°$; random shearing stretching with stretch factor between $5°$ and $10°$; flipping is not applied in this situation because it make different meaning of traffic sign plate. Some augmented results of each sample image are illustrated in Fig. 5. In proposal region segmentation task, we applied the method as describer in Sect. 3.

Fig. 5. Example of augmentation images: the original input to 10 outputs.

In the training part, there are many approaches for construction a deep network for traffic sign recognition. Some pretrain models can be applied in the special situation of traffic sign by retrained the model. Amount them, AlexNet model is a nice option. However, in our approach, it is not suitable due to flowing reasons. The size of image input layer is too big in this situation of traffic sign recognition that spends more time for processing.

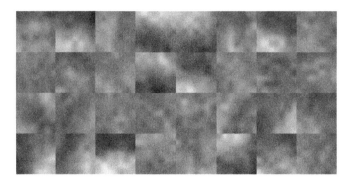

Fig. 6. The weights of kernel filters of the first convolutional layer. This layer consists of 32 kernel (7 × 7) connected to the three channels of the input image

Our method processes on both original image and region proposal that results 1 × 2 input layer size ratio. 32 filters of the first convolution layer are illustrated in Fig. 6. The outputs of some layers are shown in Fig. 7, which responded to input image sample. The results show that the parts of region proposal are significant distinguish then original RGB images.

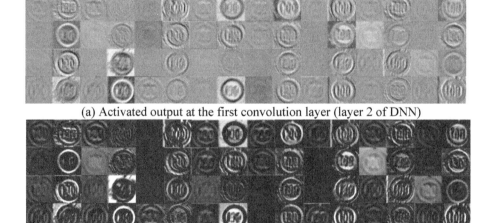

(a) Activated output at the first convolution layer (layer 2 of DNN)

(b) Activated output at the first ReLU layer (layer 3 of DNN)

Fig. 7. Results of the first activations of some DNN layer

We used benchmark dataset for training and testing as following German Traffic Sign Benchmarks (GTRSB) document [16] for evaluation, one training dataset and another testing dataset. The GTRSB is challenging due to variety conditions, reality situations, and the low quality of image. Some images from GTRSB are displayed in Fig. 8.

Fig. 8. Some images of traffic regions from GTRSB dataset.

Our approach is working perfect on GTRSB image dataset. The system archives 99.99% accuracy on overall test set. To show the performance on individual class more clearly, we also output of individual kind of traffic sign in Table 2. Prediction results are separated into subset of traffic sign from class 0 to class 42. There is only one sample, which is miss classification of class 38 and false detection to class 12.

Table 2. Our results on individual class using GTRSB dataset.

Class label	0	1	2	3	4	5	6	7	8	9	10
#Testing samples	60	720	750	450	660	600	120	480	450	480	660
#False recognition	0	0	0	0	0	0	0	0	0	0	0
#Miss recognition	0	0	0	0	0	0	0	0	0	0	0
Class label	11	12	13	14	15	16	17	18	19	20	21
#Testing samples	420	690	720	240	210	120	360	390	60	120	90
#False recognition	0	1	0	0	0	0	0	0	0	0	0
#Miss recognition	0	0	0	0	0	0	0	0	0	0	0
Class label	22	23	24	25	26	27	28	29	30	31	32
#Testing samples	120	150	90	480	180	60	180	90	150	240	60
#False recognition	0	0	0	0	0	0	0	0	0	0	0
#Miss recognition	0	0	0	0	0	0	0	0	0	0	0
Class label	33	34	35	36	37	38	39	40	41	42	
#Testing samples	209	120	390	120	60	690	90	120	60	60	
#False recognition	0	0	0	0	0	0	0	0	0	0	
#Miss recognition	0	0	0	0	0	1	0	0	0	0	

Table 3 shows comparison results of our method to some state of the art methods. The results illustrated that methods based on our DNN is higher performance than other methods, evenly human inspection. In our approach, performance is improved due to data augmentation, region proposal based on segmentation and designed deep neural network.

Our proposed approach reaches the best accuracy with 99.99%. In the case of combination of DNN and SVM, it is archives perfect result with 100% correct rate of classification. For input data of SVM, the output vector extracted from the third full connection network layer (layer 22th) is fed to SVM for training and testing.

Table 3. Comparison results on GTRSB dataset.

	Speed limits	Other prohibition	Derestriction	Mandatory	Danger	Unique	Overall
CNNs [17]	99.47	99.93	99.72	99.89	99.07	99.22	99.46
Human [19]	98.32	99.87	98.89	100	99.21	100	99.22
M2-tMTL [18]	98.33	98.97	98.71	97.26	97.23	99.61	98.27
CNN RF [19]	95.95	99.13	87.50	99.27	92.08	98.73	X
LDA [19]	95.37	96.80	85.83	97.18	93.73	98.63	95.68
DP-KELM [10]	X	X	X	X	X	X	99.54
Our method[a]	100	100	100	99.86	100	100	99.99

[a]The DNN model is available at http://islab.quangbinhuni.edu.vn/data/MyNet_GTSRBModel.mat.

6   Conclusions

This paper introduces a solution for traffic sign recognition. Traffic sign recognition is more challenging due to traffic sign contains more categories and variety of noises.

The designed DNN architecture consists of 25 layers, $64 \times 128 \times 3$ neutrals of input layer, 5 convolution layers with 7×7 kernel size and three full connected layers and other processed layers. The DNN works well on input of both original image and region proposal images with three basic sub-regions. The training data were augmented by rotation, shearing, stretching from -5 to $10°$. We used existing benchmark traffic signs dataset from GTRSB to evaluate the proposed approach.

Experimental evaluation on the benchmark dataset shows that our approach enhances accuracy that is higher performance than almost state of the art methods. The recognition system reaches up to 99.99% and 100% accuracy rate in the case of the use only DNN and the use of both DNN and SVM, respectively.

Acknowledgment. This research is funded by Vietnam National Foundation for Science and Technology Development (NAFOSTED) under grant number 102.05-2015.09.

References

1. Zhang, H., Geiger, A., Urtasun, R.: Understanding high-level semantics by modeling traffic patterns. In: International Conference on Computer Vision (ICCV), pp. 3056–3063 (2013)
2. Hoang, V.D., Le, M.H., Jo, K.H.: Motion estimation based on two corresponding points and angular deviation optimization. IEEE Trans. Ind. Electron. **64**, 8598–8606 (2017)
3. Murillo, A.C., Singh, G., Kosecka, J., Guerrero, J.J.: Localization in urban environments using a panoramic gist descriptor. IEEE Trans. Robot. **29**, 146–160 (2013)
4. Barton, A., Volna, E., Kotyrba, M.: Control of autonomous robot behavior using data filtering through adaptive resonance theory. Vietnam J. Comput. Sci. (2018, to be published). https://doi.org/10.1007/s40595-017-0103-7
5. Hoang, V.-D.: Multiple classifier-based spatiotemporal features for living activity prediction. J. Inf. Telecommun. **1**, 100–112 (2017)

6. Huang, Z., Yu, Y., Gu, J., Liu, H.: An efficient method for traffic sign recognition based on extreme learning machine. IEEE Trans. Cybern. **47**, 920–933 (2017)
7. Hoang, V.-D., Jo, K.-H.: Joint components based pedestrian detection in crowded scenes using extended feature descriptors. Neurocomputing **188**, 139–150 (2016)
8. Liu, C., Chang, F., Chen, Z., Liu, D.: Fast traffic sign recognition via high-contrast region extraction and extended sparse representation. IEEE Trans. Intell. Transp. Syst. **17**, 79–92 (2016)
9. Tan, M., Wang, B., Wu, Z., Wang, J., Pan, G.: Weakly supervised metric learning for traffic sign recognition in a LIDAR-equipped vehicle. IEEE Trans. Intell. Transp. Syst. **17**, 1415–1427 (2016)
10. Zeng, Y., Xu, X., Shen, D., Fang, Y., Xiao, Z.: Traffic sign recognition using kernel extreme learning machines with deep perceptual features. IEEE Trans. Intell. Transp. Syst. **18**, 1647–1653 (2017)
11. Girshick, R., Donahue, J., Darrell, T., Malik, J.: Rich feature hierarchies for accurate object detection and semantic segmentation. In: Proceedings of the IEEE Conference on Computer Vision and Pattern Recognition, pp. 580–587 (2014)
12. Krizhevsky, A., Sutskever, I., Hinton, G.E.: ImageNet classification with deep convolutional neural networks. In: Advances in Neural Information Processing Systems, pp. 1097–1105 (2012)
13. Achanta, R., Shaji, A., Smith, K., Lucchi, A., Fua, P., Süsstrunk, S.: SLIC superpixels compared to state-of-the-art superpixel methods. IEEE Trans. Pattern Anal. Mach. Intell. **34**, 2274–2282 (2012)
14. Shen, J., Hao, X., Liang, Z., Liu, Y., Wang, W., Shao, L.: Real-time superpixel segmentation by DBSCAN clustering algorithm. IEEE Trans. Image Process. **25**, 5933–5942 (2016)
15. Chang, C.C., Lin, C.J.: LIBSVM: a library for support vector machines. ACM Trans. Intell. Syst. Technol. **2**, 1–27 (2011)
16. Houben, S., Stallkamp, J., Salmen, J., Schlipsing, M., Igel, C.: Detection of traffic signs in real-world images: the German traffic sign detection benchmark. In: The 2013 International Joint Conference on Neural Networks (IJCNN), pp. 1–8. IEEE (2013)
17. CireşAn, D., Meier, U., Masci, J., Schmidhuber, J.: Multi-column deep neural network for traffic sign classification. Neural Netw. **32**, 333–338 (2012)
18. Lu, X., Wang, Y., Zhou, X., Zhang, Z., Ling, Z.: Traffic sign recognition via multi-modal tree-structure embedded multi-task learning. IEEE Trans. Intell. Transp. Syst. **18**, 960–972 (2017)
19. Stallkamp, J., Schlipsing, M., Salmen, J., Igel, C.: Man vs. computer: benchmarking machine learning algorithms for traffic sign recognition. Neural Netw. **32**, 323–332 (2012)

Intelligent Computer Vision Systems and Applications

A New Algorithm for Greyscale Objects Representation by Means of the Polar Transform and Vertical and Horizontal Projections

Dariusz Frejlichowski[✉]

Faculty of Computer Science and Information Technology,
West Pomeranian University of Technology, Żołnierska 52, 71-210 Szczecin, Poland
dfrejlichowski@wi.zut.edu.pl

Abstract. Intelligent computer vision systems can be based on various approaches, methods and algorithms. One of them is the usage of object descriptors, devoted to the representation of objects extracted from digital images (or video sequences) mainly by means of low-level features. There are several features that are applied in computer vision theory and practice, e.g. color, context of the information, luminance, movement, shape, and texture. Amongst them shape, color and texture are especially popular. To the contrary, object representation based on greyscale is less popular. The paper proposes and analyses a new simple and fast algorithm for greyscale object representation. The description method is based on the usage of the polar transform of pixels belonging to an object, and projections – vertical and horizontal. Apart from these operations, some additional steps are also applied in order to improve the efficiency of the developed approach, e.g. median and low-pass filtering. The properties of the proposed greyscale object representation algorithm are analyzed experimentally by means of an exemplary application from the computer vision domain. The ear images (taken from The West Pomeranian University of Technology Ear Database) are applied in the experiment. The obtained results constitute the basis for certain conclusions as well as the proposition of future plans and works on the problem.

Keywords: Greyscale objects · Greyscale description · Polar transform
Projections

1 Introduction

1.1 Problem Statement and Motivation

There are numerous approaches, methods and algorithms used in the computer vision theory and applications. One can apply the video sequence or single image for this purpose. Moreover, an image can be analyzed as a whole or it can be divided into smaller sub-parts in order to make the analysis easier and faster on the lower level. However, the selection of the way an image is processed depends on the particular problem or application, e.g. in intelligent computer vision systems. Sometimes, content based image

© Springer International Publishing AG, part of Springer Nature 2018
N. T. Nguyen et al. (Eds.): ACIIDS 2018, LNAI 10752, pp. 617–625, 2018.
https://doi.org/10.1007/978-3-319-75420-8_58

retrieval works better on a higher level – of the whole image. On the other hand, image recognition usually utilizes a lower complexity level, by applying single objects, which are localized and subsequently extracted from an image.

The computer vision approaches based on the use of single objects usually represent them by means of so-called descriptors. As one can easily guess, they are used for appropriate description of the extracted objects. This description should improve the recognition process taking various aspects into account. An effective object descriptor should give high recognition results and be robust to particular modifications and deformations of an object resulting from affine transformations, noise, illumination change, point of view selection, and many others. The object descriptors are based on various features, e.g. color, context of the information, luminance, movement, shape, and texture [1]. These low-level features are very popular in the scientific applications not only thanks to their efficiency, but also because they can be simply and quickly derived for a digital image [2].

Amongst the enumerated low-level features, much effort was made so far in the development and application of the algorithms based on texture, shape and color, while the other ones are less popular nowadays. Meanwhile, in real conditions sometimes the data covered by greyscale is more informative. Moreover, a properly constructed greyscale object descriptor can include additional information about shape or luminance, as a result giving a more expanded and effective object representation. Hence, in this paper, a new representation method for objects extracted from digital images is proposed and analyzed. It is based on the use of the greyscale as a feature, and polar transformation combined with horizontal and vertical projections as the main steps in the developed algorithm.

1.2   Related Works

The number of greyscale descriptors is rather small, especially when compared with algorithms designed for the representation of other features. An example is the application of histogram [3] or moments [4] to greyscale objects. Usually, the representation is applied to the whole image, not a single object placed in it. Moreover, the goal here is rather object localization, not its representation. A popular example of that kind is the SIFT (scale-invariant feature transform) algorithm [5]. Another method used for this purpose was based on the histogram of distances between characteristic points [6]. On the other hand, gender recognition applying local descriptors was described in [7].

Another very popular application of intensity analysis is connected with textures in greyscales. Exemplary algorithms applied for this purpose are: genetic algorithms [8], Local Binary Pattern with Local Phase Quantization [9], Gaussian Markov Random Fields [10], morphological approach [11], Fisher tensors with 'bag-of-words' [12], non-uniform patterns [13], connectivity indexes in local neighborhoods [14], local fractal dimensions [15], Gabor features [16]. However, the main application of the enumerated algorithms is the texture representation, not greyscale object recognition or retrieval.

The rest of the paper is organized as follows. The second section provides a description of the proposed algorithm. The third section presents some experimental results and their analysis. Finally, the last section concludes the paper and provides some future plans.

2   Description of the Proposed Algorithm

The algorithm proposed in this paper was developed in order to represent greyscale objects, previously localized and extracted from digital images. The described objects can be later used in the recognition, retrieval and other applications, using for example particular classification methods as an additional stage. In the experiments described in the next section, for this purpose the Euclidean distance was applied as the dissimilarity measure indicating the template which is the most similar to a test object. The first stages of the proposed description algorithm are similar to the Polar-Fourier Greyscale Descriptor, proposed in [17]. The selected pre-processing steps from the original algorithm are included, since thanks to them the representation can be more robust to various distortions and deformations. The most important part of the developed algorithm is the transformation of the pixels belonging to an object from Cartesian into polar coordinates. It results in a significant change in the object's appearance. In Fig. 1 some objects and their polar-transformed representations are provided as an example. Two completely different applications are presented pictorially. Firstly, the red blood cells are shown, on the left. The recognition of erythrocytes can be applied for example in the automatic diagnosis of some diseases, because some of them (e.g. malaria or anemia) result in significant changes in the red blood cells' appearance. The images of butterflies presented on the right, can be applied in automatic species recognition.

The main difference between the greyscale descriptor developed and investigated here and the above-mentioned Polar-Fourier Greyscale Descriptor, is the second part of the algorithm. Here, instead of the Fourier transform applied at the end, the horizontal

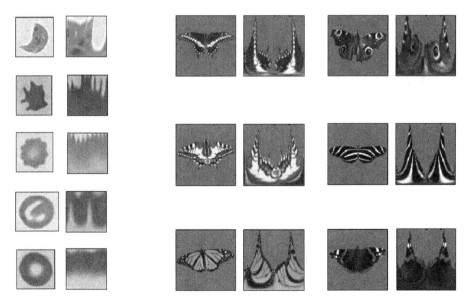

Fig. 1. Examples of polar transformed greyscale objects, belonging to two different classes and applications – erythrocytes in the process of automatic diagnosis and butterflies in the species recognition.

and vertical projections are derived. Thus, the obtained representation is invariant to translation and scaling, but not to the rotation of the represented object in the image plane. As a result, two vectors with 128 elements each, are obtained and together they constitute the final description of a greyscale object.

As it was already mentioned, the proposed algorithm before the crucial stages uses several useful operations for the enhancement of the initial image containing the extracted object. These operations are typical for the pre-processing stage, but it was decided to include them in the developed greyscale descriptor in order to preserve the properties of the algorithm for all applications. Hence, the median and low-pass filters are firstly performed for the enhancement of the image quality. Also, the input image is resized. Thanks to this, the problem of scale change is solved and the constant number of pixels will be used in further steps. The polar transform can be applied now, with the constant size (128×128 pixels was assumed, but this limitation is not strict) of the resultant representation. Later, the horizontal and vertical projections are derived, resulting in two vectors – H and V. Together, they give the final greyscale object descriptor for the representation of the object.

The proposed algorithm is presented below.

Step 1. Median filtering of the input image I, with the kernel size 3.
Step 2. Low-pass filtering, realized through the convolution with mask 3×3 pixels, and normalization parameter equal to 9.
Step 3. Calculation of the centroid by means of the moments [18]. Firstly m_{00}, m_{10}, m_{01} are derived, and later the centroid $O = (x_c, y_c)$:

$$m_{pq} = \sum_x \sum_y x^p y^q I(x, y), \tag{1}$$

$$x_c = \frac{m_{10}}{m_{00}}, \quad y_c = \frac{m_{01}}{m_{00}}. \tag{2}$$

Step 4. Transforming I into polar coordinates (resultant image is denotes as P), by means of the given formulas:

$$\rho_i = \sqrt{\left(x_i - x_c\right)^2 + \left(y_i - y_c\right)^2}, \quad \theta_i = \text{atan}\left(\frac{y_i - y_c}{x_i - x_c}\right). \tag{3}$$

Step 5. Resizing P to the constant rectangular size, $n \times n$, e.g. $n = 128$.
Step 6. Deriving the horizontal and vertical projections of P:

$$H_i = \sum_{j=1}^{n} P_{i,j}, \quad V_j = \sum_{i=1}^{n} P_{i,j}. \tag{4}$$

Step 7. Concatenating the obtained vectors H and V into one, $C = HV$, representing an object.

Taking the problem of affine invariance into account, the above-described algorithm is independent of the influence of scaling and translation, but it is not invariant to rotation. However, there are some applications, in which this limitation is desirable. Moreover, sometimes there are some computer vision problems, where this transformation simply does not occur.

3   Conditions and Results of the Experiment

The goal of the performed experiment, described in this section, was to analyze the behavior of the proposed greyscale object descriptor in real conditions. For this purpose the WPUT (West Pomeranian University of Technology) Ear Database [19] was utilized. The application in biometric identification of persons by means of ear images was selected, because the proposed greyscale descriptor is not invariant to rotation, and in this particular case and in the conditions of the database applied, the problem of rotation change is negligible. Some examples of ear images used in the experiment, as well as polar-transformed representations obtained for them, are provided in Fig. 2.

In the experiment 225 test images were used – 5 images per person, for 45 different persons. The template base included 45 templates, one template per class – person. Obviously, the templates were not included in the test set. Because the greyscale object representation and description are considered here, mainly in the context of testing the developed

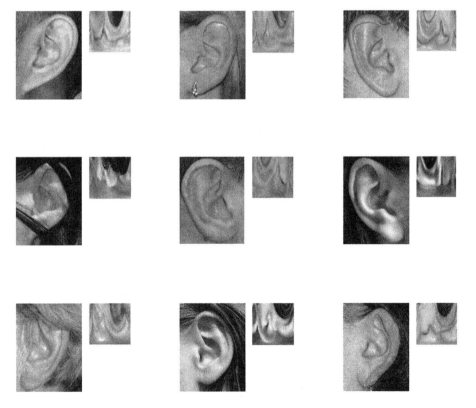

Fig. 2. Examples of ear images applied in the experiment and their polar-transformed representations.

new algorithm, the question of localization and extraction of the objects of interest in the digital image was not investigated. It results from the main goal of this paper, which is the analysis of a proposed new greyscale descriptor (by means of the basic research), and not the particular application in ear biometrics, which is only one selected example of potential usage of the developed algorithm and approach in the future.

Each test image was represented by means of the algorithm described in the previous section and matched using the dissimilarity measure – the Euclidean distance – with the templates, represented in the same way. The smallest value of the dissimilarity measure indicated the recognized template class – the most similar to the test object. This was how the process of classification was performed. It was relevant to the typical simple template matching approach for object recognition.

The experimentally obtained efficiency (recognition results) for investigated classes and the average value derived for all classes are provided in Table 1. The detailed information is given as well – not only the obtained recognition rate, but also the number of properly and wrongly matched instances, in order to give more precise results of the experiment.

Table 1. The experimental results obtained for the application of the proposed approach in the ear recognition problem.

Class no.	Correct results	Wrong results	Efficiency	Class no.	Correct results	Wrong results	Efficiency
1	5	0	100%	24	5	0	100%
2	5	0	100%	25	4	1	80%
3	4	1	80%	26	1	4	20%
4	5	0	100%	27	4	1	80%
5	5	0	100%	28	4	1	80%
6	5	0	100%	29	5	0	100%
7	4	1	80%	30	3	2	60%
8	3	2	60%	31	5	0	100%
9	4	1	80%	32	5	0	100%
10	2	3	40%	33	5	0	100%
11	4	1	80%	34	5	0	100%
12	4	1	80%	35	4	1	80%
13	5	0	100%	36	5	0	100%
14	5	0	100%	37	5	0	100%
15	4	1	80%	38	5	0	100%
16	5	0	100%	39	5	0	100%
17	3	2	60%	40	4	1	80%
18	5	0	100%	41	5	0	100%
19	5	0	100%	42	3	2	60%
20	5	0	100%	43	5	0	100%
21	4	1	80%	44	5	0	100%
22	5	0	100%	45	5	0	100%
23	5	0	100%	Total	**198**	**27**	**88%**

The obtained efficiency of the proposed algorithm was equal to 88%. This result is promising, since one should take into account the problems typical for automatic identification based on ear images. Above all, they can be very similar for different people in some cases, so the analysis has to be more detailed. However, the most important problem is the occlusion caused by some factors. The most obvious among them is the presence of hair covering large parts of an ear under analysis. The images from the West Pomeranian University of Technology Ear Database were prepared in a way emphasizing the most difficult cases. Hence, for many persons particular images were significantly different – the hair covering partly an ear or earrings present in an image gave images that could vary for the same person. Also, some image acquisition problems occurred. There were no strict and pre-established conditions of taking photographs for the database. Hence, the images were taken both indoors and outdoors, in varying weather and – what is especially important – light conditions. Sometimes, the quality of an image was also low. When considering all the above-mentioned problems and limitations, the obtained results of the performed experiment can be regarded as promising.

Most of the analyzed classes were recognized perfectly, i.e. all the five test objects were properly assigned to a template representing particular class. This full, 100% recognition rate was obtained for 27 from 45 persons, which corresponds to 60% of the cases used in the experiment. In 12 cases, 4 images were matched properly and 1 was assigned to a wrong class. This efficiency was obtained for 26.67% test classes. Only in 4 classes (8.88%) the recognition rate was equal to 60%. Finally, there were only two classes, for which the efficiency was considerably worse – one with 40% (2 correct results and 3 wrong ones) and one with 20% (only 1 correct result obtained and 4 wrong indications). For both described cases the one wrong result obtained for 45 test classes corresponds to 2.22%.

The results obtained for the algorithm for greyscale objects representation that are proposed and described in this paper can be additionally compared with the Polar-Fourier Greyscale Descriptor, which was previously applied in the problem of ear recognition as well [20]. The results of the proposed descriptor are slightly better, because the above-mentioned algorithm obtained the efficiency equal to 84% using the same experimental data. It seems surprising at the first sight, since the Polar-Fourier Greyscale Descriptor is more advanced and sophisticated. However, the reason for this can be its property of rotation invariance. As it was already mentioned, the algorithm proposed in this paper does not preserve this property. It is possible that in the investigated problem it is an advantage, not the drawback. Also, it confirms the rule that some algorithms are better suited for particular applications than other ones and one can improve recognition results simply by selecting carefully the method for object description and classification.

The proposed algorithm is invariant to translation and scaling. The experiment also confirmed some other useful properties. The lack of invariance to rotation can be sometimes, in particular application, an advantage. Also, the difficult conditions of the acquisition process influence the results obtained using the proposed approach only to a certain degree. In case of the images from The West Pomeranian University of Technology Ear Database there were no pre-assumed limitations made. Most of them were collected outdoors, in varying weather and light conditions. Despite that the efficiency equal to 88% was achieved, which can be considered as a promising result for future works on the problem of greyscale object description.

4   Conclusions and Future Plans

In the paper a new algorithm for greyscale object representation was proposed and investigated experimentally. It is based on the transformation of pixels belonging to an object from Cartesian to polar coordinates and horizontal and vertical projections of the obtained representation. Some additional steps were also applied in order to make the algorithm more robust to some problems and deformations occurring in real situations. In order to indicate the recognized class, the typical template matching approach was applied. According to this approach, the template and test objects were represented using the proposed description algorithm and later matched by means of the Euclidean distance. The smallest value of the dissimilarity measure indicated the recognized class.

The properties of the proposed algorithm were investigated using an exemplary application – person identification based on ear images, an example originated from the biometric domain. The objects for the tests were taken from the West Pomeranian University of Technology Ear Database [19].

In the future, the proposed approach will be developed in several directions. Above all, the simple and sensitive Euclidean dissimilarity measure will be replaced by some more advanced classifiers. The approach used in the works described in the paper confirmed the usefulness of the proposed greyscale object descriptor, but any modification of the classification stage will surely improve the results, e.g. by means of the approaches analyzed in [21]. The second direction covers the modification of the proposed representation algorithm itself. Other steps can be applied in order to make it more efficient. Finally, the problem of greyscale object description will be investigated in general in order to propose the effective methods for the recognition of objects extracted from digital images by means of greyscale as a feature.

Acknowledgements. The paper was supported by The National Science Centre, Poland under the grant no. 2017/01/X/ST7/00347, entitled "Popularization – by means of the presentation at the international scientific conference – of the greyscale descriptors applied for objects extracted from digital images".

References

1. Frejlichowski, D.: An algorithm for binary contour objects representation and recognition. In: Campilho, A., Kamel, M. (eds.) ICIAR 2008. LNCS, vol. 5112, pp. 537–546. Springer, Heidelberg (2008). https://doi.org/10.1007/978-3-540-69812-8_53
2. Verma, M., Raman, B.: Center symmetric local binary co-occurrence pattern for texture, face and bio-medical image retrieval. J. Vis. Commun. Image Represent. **32**, 224–236 (2015)
3. Gbèhounou, S., Lecellier, F., Fernandez-Maloigne, C.: Evaluation of local and global descriptors for emotional impact recognition. J. Vis. Commun. Image Represent. **38**, 276–283 (2016)

4. Paschalakis, S., Lee, P.: Pattern recognition in grey level images using moment based invariant features. In: The 7th International Conference on Image Processing and its Applications, pp. 245–249 (1999)
5. Lowe, D.G.: Distinctive image features from scale-invariant keypoints. Int. J. Comput. Vis. **60**(2), 91–110 (2004)
6. Chin, T.J., Suter, D., Wang, H.: Boosting histograms of descriptor distances for scalable multiclass specific scene recognition. Image Vis. Comput. **29**(4), 241–250 (2011)
7. Castrillón-Santana, M., de Marsico, M., Nappi, M., Riccio, D.: MEG: texture operators for multi-expert gender classification. Comput. Vis. Image Underst. **156**, 4–18 (2017)
8. Delibasis, K., Undrill, P.E., Cameron, G.G.: Designing texture filters with genetic algorithms: an application to medical images. Signal Process. **57**(1), 19–33 (1997)
9. Nanni, L., Melucci, M.: Combination of projectors, standard texture descriptors and bag of features for classifying images. Neurocomputing **173**, 1602–1614 (2016)
10. Dharmagunawardhana, C., Mahmoodi, S., Bennett, M., Niranjan, M.: Gaussian Markov random field based improved texture descriptor for image segmentation. Image Vis. Comput. **32**(11), 884–895 (2014)
11. Aptoula, E., Lefèvre, S.: Morphological texture description of grey-scale and color images. Adv. Imaging Electron Phys. **169**, 1–74 (2011)
12. Faraki, M., Harandi, M.T., Wiliem, A., Lovell, B.C.: Fisher tensors for classifying human epithelial cells. Pattern Recogn. **47**(7), 2348–2359 (2014)
13. Nanni, L., Brahnam, S., Lumini, A.: A simple method for improving local binary patterns by considering non-uniform patterns. Pattern Recogn. **45**(10), 3844–3852 (2012)
14. Florindo, J.B., Landini, G., Bruno, O.M.: Three-dimensional connectivity index for texture recognition. Pattern Recogn. Lett. **84**, 239–244 (2016)
15. Florindo, J.B., Bruno, O.M.: Local fractal dimension and binary patterns in texture recognition. Pattern Recogn. Lett. **78**, 22–27 (2016)
16. Manjunath, B.S., Ma, W.Y.: Texture features for browsing and retrieval of image data. IEEE Trans. Pattern Anal. Mach. Intell. **18**(8), 837–842 (1996)
17. Frejlichowski, D.: Identification of erythrocyte types in greyscale MGG images for computer-assisted diagnosis. In: Vitrià, J., Sanches, J.M., Hernández, M. (eds.) IbPRIA 2011. LNCS, vol. 6669, pp. 636–643. Springer, Heidelberg (2011). https://doi.org/10.1007/978-3-642-21257-4_79
18. Hupkens, T.M., de Clippeleir, J.: Noise and intensity invariant moments. Pattern Recogn. Lett. **16**(4), 371–376 (1995)
19. Frejlichowski, D., Tyszkiewicz, N.: The west pomeranian university of technology ear database – a tool for testing biometric algorithms. In: Campilho, A., Kamel, M. (eds.) ICIAR 2010. LNCS, vol. 6112, pp. 227–234. Springer, Heidelberg (2010). https://doi.org/10.1007/978-3-642-13775-4_23
20. Frejlichowski, D.: Application of the polar-fourier greyscale descriptor to the problem of identification of persons based on ear images. In: Choraś, R. (ed.) Image Processing and Communications Challenges 3. Advances in Intelligent and Soft Computing, vol. 102, pp. 5–12. Springer, Heidelberg (2011). https://doi.org/10.1007/978-3-642-23154-4_1
21. Klęsk, P., Kapruziak, M., Olech, B.: Boosted classifiers for antitank mine detection in C-scans from ground-penetrating radar. In: Wiliński, A., El Fray, I., Pejaś, J. (eds.) Soft Computing in Computer and Information Science. AISC, vol. 342, pp. 11–25. Springer, Cham (2015). https://doi.org/10.1007/978-3-319-15147-2_2

A New Framework for People Counting from Coarse to Fine Could be Robust to Viewpoint and Illumination

An H. Nguyen$^{(\boxtimes)}$ and Ngoc Q. Ly$^{(\boxtimes)}$

Faculty of Information Technology, HCM University of Science - VNU HCM,
Ho Chi Minh City, Vietnam
an.nguyenhung@gameloft.com, lqngoc@fit.hcmus.edu.vn

Abstract. People counting is one of the important tasks in video surveillance. In spite of the significant improvements, this task still has had many challenges such as heavy occlusion in the crowded environment, viewpoint variation, the variety of illumination, etc. People counting process consisted of people detection stage and people tracking stage. This paper focused on boosting the people counting results based on people detection. Our suggested method combines the Deformable Part Models (DPM) and the Deep Convolutional Neural Network (DCNN) to take their advantages and to overcome the shortcomings of each method in people detection. Firstly, to be robust to viewpoint and occlusion, we fuse the people detection results from parts detected by DPM such as head, head-shoulders, upper body, full body. Secondly, to overcome the inefficiency of DPM due to zoom-in view, we use DCNN in detecting head region because the body is often occluded, leaving only head be in full appearance for counting. Finally, we use the late fusion of the detection results from two listed models. PETS 2012 and TUD datasets are selected to experiment and the performance is evaluated by MAE, MRE. The experimental results show that our method could achieve higher performance than the method of Abiol [1], Conte [5] and Subburaman [22] on PETS dataset and especially it could outperform state-of-the-art method as YOLO9000 [10] with parameters fine tuning accordingly to HollywoodHeads dataset. Moreover, it could achieve the high performance in the sparse, medium-density crowd environment and it could be robust to scale, viewpoint, illumination, occlusion, and deformation.

Keywords: People counting · Deformable Part Models
Deep Convolutional Neural Network · Head detection · People detection

1 Introduction

In recent years, people counting systems have been widely used in security applications and marketing research such as public transport security, security in public places, pedestrian traffic management, visitors flow estimation, customer behavior analysis, etc. These systems support the people in cases that the security guards could not present and process simultaneously in many places. They also give the alert about security

© Springer International Publishing AG, part of Springer Nature 2018
N. T. Nguyen et al. (Eds.): ACIIDS 2018, LNAI 10752, pp. 626–637, 2018.
https://doi.org/10.1007/978-3-319-75420-8_59

quickly and accuracy. So, improving the performance of detecting and counting people is a motivation for researchers.

So that the system could be widely applied, it should be robust to some most important challenges as viewpoint, illumination, scale and occlusion in the crowded environment. The performance of people detection is often unstable in these cases.

To solve these challenges, inspired the event detection approach from coarse to fine in the paper [16], we proposed to use late fusion of DPM [18] and DCNN model for human head detection. The DPM could be robust to viewpoint and deformable components in human detection, but to improve the people detection performance facing occlusion, we have modified this model to detect different components in a flexible manner. We performed people detection process from coarse to fine based on detection head, head-shoulders, upper body and full body with different thresholds. The final detection results could be fused from four previous detection results to achieve better detection results. At the same time, to improve detection performance in the zoom-in view, we use the DCNN model to detect the head parts. Finally, we fuse the results from two listed models with some adjustments to improve head detection performance.

The rest of the paper is organized as follows. In Sect. 2, we review the related works. In Sect. 3, we present the proposed approach based on DPM and DCNN model. In Sect. 4 we evaluate our approach on the benchmark dataset. Finally, conclusions and future works are drawn in Sect. 5.

2   Related Works

So far, many methods have been introduced to solve people counting. They are categorized into two main research directions: counting the total number of people in an image, in a specific frame of the video sequence or counting the number of people moving in and out of a certain area from the surveillance camera.

2.1   Counting the Total Number of People in an Image or a Specific Frame of the Video Sequence

There are two main classes of methods in people counting. The first class is based on the construction of classifier models to directly detect regions containing humans and then counting is performed. This class is suitable for environments where the people density is not too high and it could detect individuals separately [6, 22]. The authors often based on the head detection because head region often present in full appearance in the crowded environment. The Adaboost and SVM models are often used for classifying the people. These systems often have the low rate of precision and recall when detecting people in low-resolution images or the crowded environment with many occlusion or from the rear side.

The second class which is based on pixels to perform estimation, counting people is typically used in moderate or dense crowd environments [5, 8]. In these environments, the people are often occluded and they are only shown with a few pixels. The authors often use the ε-SVR model to estimate the number of people.

More recently, many of the researchers have used the DCNN model for crowd estimation [4, 13], combining all the steps of the people counting model into a single network model. They have achieved good results in terms of performance and computational speed.

2.2 Counting the Number of People Moving In and Out of a Certain Area

There are two main kinds of methods used commonly. The first kind which is called indirect approach could count people without human detection [9, 15]. Bypass the training and tracking phase, it could count people based on the size of the objects moving pass through the virtual lines. It would achieve inaccurate results in some complex environments where the system could not distinguish which objects were human or non-human, the size of objects is too large or too small in comparison to the popular size of people. It is appropriate for the environments where sparse density, less occlusion and the moving between objects are not too close together.

The second kind could directly count people based on human detection and motion tracking when passing through the virtual line [2, 11]. It is often used for the environment with the medium-density crowd. It based on the head or head-shoulders detection because they are not often occluded in the crowded environment. The authors often use the SVM model for classifying the people. These approaches often have many errors in case the contrast between the person and the background is low or the number of people in the counting area increased.

In recent years, DCNN model is widely used in many research areas. Some research works could be found for pedestrian detection [3] or people counting [25]. This model has very good performance, it could achieve top performance among the previously announced models. It is optimized system that combines all the stages as features extraction, object representation and object classification into a single model. However, it also encountered a lot of problems in optimizing the runtime, data for the training in human detection, as well as errors in detecting people who are obscured in the crowded environment.

3 Our Approach

From realistic requirements, it is necessary to have people counting method that could be robust to the camera view, the occlusion in the crowded environments and the influence of illumination conditions.

The people detection method should be improved to be robust to occlusion and viewpoint variation. There are two models that have been developed, one model use Poselets for people detection [14] and one model based on the deformation of body parts as DPM [18]. They have good performance and good prospect due to listed challenges of people counting.

Because of the effectiveness of these two models in addressing the challenges mentioned above, we propose to use the DPM for human detection. DPM is developed based on the HOG feature and it is designed to face to deformable human parts, scale and occlusion. It is easier to build training data sets in DPM than in Poselets model. Moreover, training and detection phase of DPM is faster and simpler than Poselets model. In addition, we also modified the DPM to detect the head, head-shoulders, upper body and full body at different viewpoints and scales to increase the human detection results in the environments having severe occlusion.

The DPM has many advantages in the zoom-out view, to overcome its disadvantages in the zoom-in view, we enhanced a DCNN model based on some filters for detecting head region because the body is often occluded, leaving only head be in full appearance for counting. Based on late fusion strategy, we fuse the results from the two models to increase the performance of the people counting system.

This approach based on head detection and fusion of the results of two models DPM and DCNN, so we named it as the DPM(4 attributes)-DCNN(Filtering) method.

3.1 Deformable Part Models with Multiple Attributes

We implemented DPM with some adjustments on it. These adjustments support to detect different combinations of parts, and they could be robust to occlusion. Based on the evidence of Idrees [8], we can focus on detecting a distinct human component without considering the filter, deformation scores of excluded parts. Adjusting the threshold accordingly to part detection plays a significant role in process.

The formula of the DPM score is described as follows:

$$score = \sum_{i=0}^{n} F_i * \phi(H, p_i) - \sum_{i=1}^{n} d_i * \phi_d(d_{x_i}, d_{y_i}) + b$$
$$\phi_d(d_{x_i}, d_{y_i}) = (d_{x_i}, d_{y_i}, d_{x_i}^2, d_{y_i}^2), \ (d_{x_i}, d_{y_i}) = (x_i, y_i) - (2(x_0, y_0) + v_i) \tag{1}$$

where F_i is the filter for the i-th part, $\phi(H, p_i)$ is feature vector at position p_i in the image, d_i is the deformation cost, ϕ_d is deformation features, b is constant bias term, and v_i is the anchor position with respect to root position. The deformation score of a part with displacement (d_{x_i}, d_{y_i}) is given as $d_i * \phi_d(d_{x_i}, d_{y_i})$. Final score is the sum of scores from the root filter, filter, deformation scores from the parts, plus the bias.

We have achieved four different thresholds to flexible detect the following attributes: head, head-shoulders, upper body and full body. We called our suggested model as DPM(4 attributes). It has many pros in comparison to the normal DPM. A person is occluded at the lower body or even full body will be missed by using normal DPM based on the score of all parts. But for DPM(4 attributes), we only focus on the score of which attribute that we would like to detect. So, if the person could not be detected based on the threshold of full body attribute, this person could be detected based on the threshold of the upper body, head-shoulders or head attribute (see Fig. 1). Moreover, for detecting difference parts, DPM could be modified by using only one training task with full body dataset. In while DCNN models must use many different datasets for training and detecting each part of the body. It is the reason that we chose DPM in our approach.

The detection result from DPM(4 attributes) could be defined as in Eq. 2:

$$box_{result} = box_h \cup box_{hs} \cup box_{ub} \cup box_{fb} \qquad (2)$$

where box_h, box_{hs}, box_{ub}, box_{fb} are the detected bounding boxes of head, head-shoulders, upper body, full body respectively. The detected bounding box of each attribute could be empty set in case of missing detection. We implemented grid search to get the detection threshold of head, head-shoulders, upper body and full body which belong the range of $[-3.7, -3.6]$, $[-3.3, -3.1]$, $[-2.4, -2.2]$ and $[-0.7, -0.5]$ respectively to achieve the best results.

Then, if the number of bounding boxes from box_{result} was greater than one, we would use Non-Maximum Suppression (NMS), a post-processing algorithm to ignore redundant as overlapping bounding boxes. We chose a threshold of 0.3 for this purpose as it could give the optimum detection result among the thresholds ranging from 0.1 to 0.9 step 0.1.

Fig. 1. Head detection results based on the normal DPM (on the first row) and the DPM(4 attributes) (on the second row).

3.2 Head Detection with Deep Convolutional Neural Network

To deal with the issue of occlusion in the crowded scenes, we use the DCNN model for head detection because the head could be detected easier than the other parts in the crowded scenes. We used both pre-trained and manually training model.

Pre-trained network: There are many specialized pre-trained DCNN models for detecting face such as DeepFace [24], FaceNet [7] along with multiple objects detection networks such as SSD300 [23], YOLO9000 [10]. However, they still have not given good results for head detecting at the rear side and have had many cons for detecting the object in both cases as zoom-in and zoom-out view. The pre-trained DCNN model [21] is selected because it has high performance on head detection in real time. We just only use the Local model from this paper. Basically, this Local model is a pre-trained network of Oquab [17] and extended by one fully connected layer. So, it consisted of five convolution layers C1...C5 followed by four fully connected layers FC6...FC9, the architecture of this network is described in Fig. 2. It is used for training on the Holly-woodHeads dataset, which consisted of 369,846 human heads annotated in 224,740 video frames from 21 movies. They optimize parameters by minimizing the sum of independent log-losses using stochastic gradient descent with momentum 0.9, weight decay 0.0005 and learning rate 0.01. This network takes input resolution of 224 × 224. It follows the idea of R-CNN [20], using selective search approach [12] for restricting to set of candidates.

Observing the detection results, we found some bounding boxes that their size could be larger than the size of most the remaining bounding boxes many times. So for decrease the detection errors, we applied a filter for the results. We filtered the head bounding boxes based on the highest score in detection results. Supposed that the bounding box having highest score is denoted as box_{hs}. If the remaining bounding boxes area were greater than 2.5 times box_{hs} area, we will drop these bounding boxes. We named this network as the DCNN(Filtering).

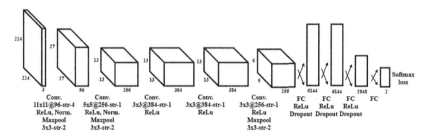

Fig. 2. The architecture of the DCNN model in [20].

Manually train network: Currently, YOLO9000 model [10] is state-of-the-art in detecting multi-objects in real time with high performance and FPS. But, it has still not achieved yet good support for detecting head, face and especially with small objects. This model was trained with ImageNet dataset and we continue on training with HollywoodHeads dataset for detecting head region. We trained with 100,000 iterations with batch size is 16, and keep all configure of YOLO9000. This network takes input resolution of 416 × 416, the architecture of it is described in Fig. 3. We named this network as the YOLO9000-Head.

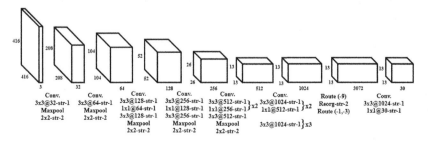

Fig. 3. The architecture of YOLO9000-Head model.

3.3   Fusion of Two Models

The final bounding boxes of the head will be gathered from the head's bounding boxes based on DPM(4 attributes) (Eq. 1) and Deep CNN model (DCNN(Filtering) or YOLO9000-Head). The fusion method is shown in Eq. 3:

$$box_{final} = box_{DPM\_4\_A} \cup box_{Deep\_CNN} \qquad (3)$$

where $box_{DPM\_4\_A}$, box_{Deep_CNN} are head regions detected by DPM(4 attributes) and Deep CNN model. Then we also use the NMS method with a threshold 0.2 to remove the overlap regions from these two models. The selected threshold could give the optimum detection result among the thresholds ranging from 0.1 to 0.9 step 0.1.

This fusion gives a good result that we can see in Fig. 4. By using the DPM(4 attributes), it is possible to detect small and partially obstructed parts better, while the DCNN(Filtering) model effectively detects the objects are seen from the front side and the rear side, fusion the results from the two models give better detection performance.

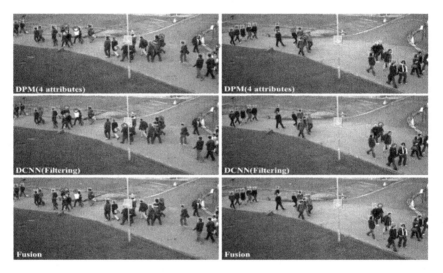

Fig. 4. Head detection based on DPM(4 attributes), DCNN(Filtering) and fusion model. The first row shows head regions (white rect.) detected by DPM(4 attributes); the second row shows head regions (cyan rect.) detected by the DCNN(Filtering); the final row shows head regions (yellow rect.) detected by fusion model. The red circles show head regions are missed. (Color figure online)

3.4   People Tracking

This paper use people tracking for counting people pass through a counting zone. We applied the Extended Kalman method for this purpose.

The head bounding boxes that are gathered from the previous detection stage will be used for tracking. Assume that people go from left side to right side for walking out zone. The number of people walking out would be counted if the centroid of the head bounding boxes from the previous frame is inside the counting zone and at the current frame it is out of the right side of counting zone. The process is performed in a similar manner for counting number of people walking in the zone.

And we also applied a filter for eliminating some false positives by our method and the predicted position from Extended Kalman method. It is only accepted if there were at least six frames people move the same direction on eight previous consecutive frames and in eight previous consecutive frames, this person is not pass through the counting zone.

4 Experiments

4.1 Dataset

We evaluate our approach with PETS[1] 2012 S1 dataset (view 1) consisted of 6 video sequences with the medium, high-density crowd and TUD[2] dataset consisted of 3 video sequences with the low-density crowd for people counting task. These video sequences contain a number of challenges and realistic situations such as occlusion, illumination, shadow, viewpoint and scale. There are four different viewpoints in these video sequences and they are recorded in daylight conditions at different times.

4.2 Experimental Results

Performance measures
The indices used to report the performance are the Mean Absolute Error (MAE) and the Mean Relative Error (MRE) defined as in formula 4 and 5:

$$MAE = \frac{1}{N} \sum_{i=1}^{N} |G(i) - T(i)| \tag{4}$$

$$MRE = \frac{1}{N} \sum_{i=1}^{N} \frac{|G(i) - T(i)|}{T(i)} \tag{5}$$

where N is the number of frames for the test video sequence, $G(i)$ and $T(i)$ are the detected and the true number of persons in the i-th frame respectively. The method had better performance if the value were smaller.

Experimental results on counting people in frames
Table 1 has shown that the YOLO9000-Head model could not give good performance in fusion results in comparison to the DCNN(Filtering) model. Although we retrained

---

[1] http://www.cvg.reading.ac.uk/PETS2012/a.html.

[2] https://motchallenge.net/data/2D_MOT_2015.

YOLO9000 model with the same HollywoodHeads dataset and a low threshold 0.2 is used (best in the range [0, 1], zero is minimum value of this model) for head detecting, the YOLO9000-Head model only effectively detects nearby objects and it could not be robust in the far scenes and small objects. Otherwise, the DCNN(Filtering) model could achieve better results in these cases, the best threshold can be found in the range [−1, 1]. As we can observe the row Average PETS, the MAE score of DPM(4 attributes) is 8.16 and DCNN(Filtering) is 10.5, and after late fusion from two previous results, the final MAE is only 3.81, similarity for case MRE. This evidence has shown that our approach has achieved good performance. So, our suggested method is the combination of DPM(4 attributes) and DCNN(Filtering).

Table 1. The performance of our proposed method. The results based on MAE, MRE.

Video	DPM (4 attributes)		DCNN (Filtering)		YOLO9000-Head		DPM (4 attributes) + DCNN (Filtering)		DPM (4 attributes) + YOLO9000-Head	
	MAE	MRE	MAE	MRE	MAE	MRE	MAE	MRE	MAE	MRE
S1.L1.13-57	4.22	18.86%	7.37	33.25%	20.67	97.30%	**1.82**	**9.58%**	4.14	18.19%
S1.L1.13-59	1.87	13.66%	6.20	43.55%	15.09	98.46%	**1.28**	**10.18%**	1.87	13.77%
S1.L2.14-06	12.37	44.93%	6.21	30.99%	24.91	98.75%	**3.85**	**26.76%**	12.35	44.89%
S1.L2.14-31	6.64	22.97%	12.68	44.00%	28.09	98.94%	**3.27**	**12.48%**	7.03	24.20%
S1.L3.14-17	8.48	31.9%	11.66	49.74%	24.96	100%	**3.98**	**17.52%**	8.68	32.50%
S1.L3.14-33	15.35	40.37%	18.88	53.49%	35.64	99.62%	**8.64**	**22.90%**	15.59	41.03%
TUD-Crossing	1.11	21.44%	3.95	69.99%	1.2	25.13%	0.93	**18.64%**	0.87	18.88%
TUD-Campus	1.56	28.92%	1.69	31.60%	1.31	23.96%	**0.85**	**15.93%**	0.87	16.81%
TUD-Stadtmitte	0.7	12.47%	3.63	64.51%	1.4	24.75%	0.69	12.39%	**0.67**	**12.02%**
Average PETS	8.16	28.78%	10.5	42.5%	24.89	98.85%	**3.81**	**16.57%**	8.28	29.10%
Average TUD	1.12	20.94%	3.09	55.37%	1.3	24.61%	0.82	**15.65%**	**0.80**	15.90%

We also compare our best results with previous best published results in Table 2 on PETS 2012 dataset. Other methods used information from previous frames for subtracting background or some tasks on their methods, but with our method, we could

Table 2. The comparison the performance of our proposed method with the other methods are in [1, 5, 22]. In each cell, the results are evaluated on MAE and MRE (in parenthesis).

Video	Albiol [1]	Conte [5]	Subburaman [22]	Ours
S1.L1.13-57	2.8 (12.6%)	1.92 (8.7%)	5.95 (30%)	**1.82** (9.58%)
S1.L1.13-59	3.8 (24.9%)	2.24 (17.3%)	2.08 (11%)	**1.28 (10.18%)**
S1.L2.14-06	5.14 (26.1%)	4.66 (20.5%)	2.4 (12%)	3.85 (26.76%)
S1.L2.14-31	-	-	7.0 (31%)	**3.27 (12.48%)**
S1.L3.14-17	2.64 (14.0%)	1.75 (9.2%)	2.2 (10%)	3.98 (17.52%)
S1.L3.14-33	-	-	16 (52%)	**8.64 (22.90%)**
Average results	-	-	5.94 (24.33%)	**3.81 (16.75%)**

count people on the frame directly which not depend on the information from other frames. Our method had higher performance in the sparse, medium-density crowd environment in comparison with other methods on MAE, MRE score. The performance will be decreased in high-density crowd environment because in this environment the distance between people-to-people is very small. So, it is very hard or impossible to detect people with a direct approach based only on the people detection.

Experimental results on counting people passing an area

We also use PETS 2012 S1 and TUD datasets for counting people passing an area, except the video sequence S1.L3.14-33 because it is not suitable for this context. In reality, when we set up the camera, based on the angle and height of the camera, we will create the counting zone fit the context and view of the camera. However, we do not have enough information about the camera view in these video sequences. So, we create a counting zone based on the default view of the camera. We consider the frame obtained from the camera as a counting zone. People appear in the video frame will be tracked, a counter will be incremented if this target moves to the left or right side and exit the video frame.

We used the proposed approach DPM(4 attributes)-DCNN(Filtering) for detecting the human head in each frame. Then, we apply the Extended Kalman method for tracking in counting people when passing the counting zone. For evaluating, we list the ground truth in and out for comparing with the counting results. In addition, we also implemented the results on MAE, MRE score.

Table 3. The people counting results (in case of people passing through the counting zone).

Case	Ground truth		Count		MAE		MRE (%)	
	In	Out	In	Out	In	Out	In	Out
L1.13-57	33	1	29	1	4	0	12.12	0
L1.13-59	1	15	1	14	0	1	0	6.67
L2.14-06	0	40	0	41	0	1	0	2.5
L2.14-31	19	0	19	0	0	0	0	0
L3.14-17	0	8	0	7	0	1	0	12.5
Stadtmitte	1	3	1	2	0	1	0	33.33
Crossing	2	11	1	7	1	4	50	36.36
Campus	1	2	1	1	0	1	0	50

Table 3 has shown that the result of counting people passing an area is very well. As we noted earlier, we used the DPM(4 attributes)-DCNN(Filtering) approach for detecting human head and tracking could help to eliminate errors in people counting at the regions without motion. So, it could achieve the encouraging results in people counting. Besides that, the performance would be dropped in a few cases that people crossing from a side-view, moving parallel, overlapping absolutely as in TUD-Crossing and TUD-Campus video sequences.

5 Conclusions

We have presented the DPM(4 attributes)-DCNN(Filtering) method for people counting based on head detection by late fusion of DPM(4 attributes) and DCNN (Filtering). Our experiments have shown the potential of this approach, it has achieved good performance in low, sparse and medium-density crowd environment.

In our future works, we would like to reduce the false positives and false negatives of the results from two methods and improve performance by using other DCNN models trained with head datasets such as Mask R-CNN, GAN, etc. Moreover, we would like to research on people detection in the high-density crowd environment and people counting accordingly to aging, gender, race, apparel.

References

1. Albiol, A., et al.: Video analysis using corner motion statistics. In: Performance Evaluation of Tracking and Surveillance Workshop at CVPR, pp. 31–37 (2009)
2. Zeng, C., et al.: Robust head-shoulder detection by PCA-based multilevel HOG-LBP detector for people counting. In: 20th ICPR, pp. 2069–2072 (2010). https://doi.org/10.1109/icpr.2010.509
3. Tome, D., et al.: Deep convolutional neural networks for pedestrian detection. arXiv:1510.03608 (2016). https://doi.org/10.1016/j.image.2016.05.007
4. Kang, D., et al.: Beyond counting: comparisons of density maps for crowd analysis tasks - counting, detection, and tracking. arXiv:1705.10118 (2017)
5. Conte, D., et al.: A method for counting people in crowded scenes. In: 7th IEEE International Conference on AVSS, pp. 225–232 (2010). https://doi.org/10.1109/avss.2010.78
6. Ling, D., et al.: An automatic people counting method of hotel dining with occlusion. J. Artif. Intell. Pract. **1**(1), 1–7 (2016)
7. Schroff, F., et al.: FaceNet: a unified embedding for face recognition and clustering. arXiv: 1503.03832 (2015). https://doi.org/10.1109/cvpr.2015.7298682
8. Idrees, H.: Visual analysis of extremely dense crowded scenes. Ph.D. dissertation, University of Central Florida, USA (2014)
9. Barandiaran, J., et al.: Real-time people counting using multiple lines. In: WIAMIS, pp. 159–162 (2008). https://doi.org/10.1109/wiamis.2008.27
10. Redmon, J., et al.: YOLO9000: better, faster, stronger. arXiv:1612.08242 (2017)
11. García, J., et al.: Directional people counter based on head tracking. IEEE TIE **60**(9), 3991–4000 (2013). https://doi.org/10.1109/TIE.2012.2206330
12. van de Sande, K.E.A., et al.: Segmentation as selective search for object recognition. In: ICCV (2011). https://doi.org/10.1109/iccv.2011.6126456
13. Boominathan, L., et al.: CrowdNet: a deep convolutional network for dense crowd counting. arXiv:1608.06197, pp. 640–644 (2016)
14. Bourdev, L., Maji, S., Brox, T., Malik, J.: Detecting people using mutually consistent poselet activations. In: Daniilidis, K., Maragos, P., Paragios, N. (eds.) ECCV 2010. LNCS, vol. 6316, pp. 168–181. Springer, Heidelberg (2010). https://doi.org/10.1007/978-3-642-15567-3_13
15. Pizzo, L.D., et al.: Counting people by RGB or depth overhead cameras. Pattern Recogn. Lett. **81**, 41–50 (2016). https://doi.org/10.1016/j.patrec.2016.05.033

16. Ngoc, L.Q., et al.: Event retrieval in soccer video from coarse to fine based on multi-modal approach. In: IEEE RIVF, pp. 308–313 (2010). https://doi.org/10.1109/rivf.2010.5632694

17. Oquab, M., et al.: Learning and transferring mid-level image representations using convolutional neural networks. In: IEEE Conference on CVPR, pp. 1717–1724 (2014). https://doi.org/10.1109/cvpr.2014.222

18. Felzenszwalb, P.F., et al.: Object detection with discriminatively trained part based models. IEEE TPAMI **32**, 1627–1645 (2010). https://doi.org/10.1109/TPAMI.2009.167

19. Felzenszwalb, P.F., et al.: Discriminatively trained deformable part models (2010). Release 4 http://people.cs.uchicago.edu/ ~ pff/latent-release4

20. Girshick, R., et al.: Rich feature hierarchies for accurate object detection and semantic segmentation. arXiv:1311.2524 (2014). https://doi.org/10.1109/cvpr.2014.81

21. Vu, T.-H., et al.: Context-aware CNNs for person head detection. In: IEEE on ICCV, pp. 2893–2901 (2015). https://doi.org/10.1109/ICCV.2015.331

22. Subburaman, V.B., et al.: Counting people in the crowd using a generic head detector. In: 9th IEEE on AVSS, pp. 470–475 (2012). https://doi.org/10.1109/avss.2012.87

23. Liu, W., et al.: SSD: single shot multibox detector. arXiv:1512.02325 (2016). https://doi.org/10.1007/978-3-319-46448-0_2

24. Taigman, Y., et al.: DeepFace: closing the gap to human-level performance in face verification. In: IEEE on CVPR, pp. 1701–1708 (2014). https://doi.org/10.1109/cvpr.2014.220

25. Zhao, Z., Li, H., Zhao, R., Wang, X.: Crossing-line crowd counting with two-phase deep neural networks. In: Leibe, B., Matas, J., Sebe, N., Welling, M. (eds.) ECCV 2016. LNCS, vol. 9912, pp. 712–726. Springer, Cham (2016). https://doi.org/10.1007/978-3-319-46484-8_43

Automatic Measurement of Concrete Crack Width in 2D Multiple-phase Images for Building Safety Evaluation

Hoang Nam Nguyen$^{(\boxtimes)}$ ⓘ, Tan Y Nguyen, and Duc Lam Pham

Faculty of Automotive and Mechanical Engineering,
Nguyen Tat Thanh University, Ho Chi Minh City, Vietnam
nhnam@ntt.edu.vn

Abstract. Most of image processing techniques, that are developed for measuring concrete crack width in building safety evaluation process, have focused on two-phase piecewise constant images. However, the crack images in real applications are usually 2D multiple-phase piecewise constant images, which are defined as having more than 2 piecewise constant intensity phases. In this paper, we proposed a novel technique for automatically measuring crack widths in 2D multiple-phase images, which includes three main steps. Firstly, we used crack enhancement filter to enhance crack and remove other unintended objects in the intensity inhomogeneous image. Secondly, we utilized a B-spline level set model to automatically extract crack locations. Finally, we proposed to use a Savitzky-Golay filter based method to measure crack widths along the whole detected cracks. Several experiments are demonstrated to show the superior of our method in measuring crack width from images of real scene of damaged building.

Keywords: Crack detection · Width measurement · B-spline level set model
Phase congruency · Image segmentation

1 Introduction

Concrete crack could be a warning for the occurrence of forth-coming collapse of concrete structures. In particular, after the attack of an earthquake, civil engineering experts usually need to measure crack width of concrete structures of damaged building for building safety evaluation. Therefore, one of the most important applications of image processing techniques in civil engineering is the measurement of concrete crack width in damaged buildings [1–10].

Currently, the safety of entering damaged buildings is evaluated manually by structural specialists (e.g. structural engineers and/or certified inspectors) [1]. However, this task costs a lot of time and resources. There is an urgent need to support the experts in using computer assisted technique for fast building safety evaluation. Also, the objective inspection nature of computer-assisted technique may help to reduce erroneous judgment.

© Springer International Publishing AG, part of Springer Nature 2018
N. T. Nguyen et al. (Eds.): ACIIDS 2018, LNAI 10752, pp. 638–648, 2018.
https://doi.org/10.1007/978-3-319-75420-8_60

When the building safety evaluation is made after earthquake, cracking is always an important structural damage indicator. For example, the extent and severity of damage to the concrete columns of a reinforced concrete building are quantified primarily by the width and orientation of the cracks that exist in these elements [1]. Although a large number of image processing techniques have been developed for measuring concrete crack width, none of them is really automatic and effective for building safety evaluation.

Barazzetti and Scaioni [3] proposed a method for measuring crack width in the laboratory environment from a color image. Chen and Hutchinson [4] also used level set method and morphological image processing to measure crack width in laboratory environment. Nishikawa et al. [6] proposed a method for measuring crack width under various lighting conditions using genetic programming. However, these methods recognize crack based on the intensity difference between crack and image background, which is piecewise-constant. Therefore, these methods are not applicable for 2D multiple-phase images because the intensity of background has over two piecewise constant phases in this kind of image.

Chen et al. [5] developed a crack measuring system using multi-temporal images by utilizing the first derivative of a Gaussian filter. Quadratic curve-fitting was used to achieve sub-pixel accuracy. Their proposed system can compute the crack width in a controlled environment but requiring users to put nodes along the crack. Therefore, it is not an automatic method.

Adhikari et al. [7] proposed a method for measuring crack width of bridge concrete structure. This method used edge detection algorithms to detect crack. The shortcoming of this approach is edges of dark objects in the images can be misclassified as cracks.

Fujita and Hamamoto [8] utilize line enhancement filter and probabilistic relaxation to segment cracks from noisy real concrete surface images. Lee et al. [9] used crack shape analysis (morphological image processing) technique to detect and measure crack width. However, those two methods employ binarization techniques, which require user to choose an intensity threshold, to detect crack. This limits the application of those methods to 2D two-phase homogenous images.

Yang et al. [10] measure width of thin cracks in reinforced concrete bridge pier test. But this method can only be applied during the testing process.

In this paper, we proposed a novel automatic crack width measurement framework which uses a digital camera and laptop to assist civil engineering expert for automatically measuring crack width of concrete structures of damaged building. Generally, we proposed 3 main steps for measuring crack width from 2D intensity inhomogeneous images. In the first step, we need to enhance crack and remove other unintended objects in the 2D multiple-phase piecewise constant images. In the second step, we adapt a B-spline level set model to segment crack automatically from the filtered crack images. In the third step, we utilised a Saviztky-Golay filter-based method to measure crack width accurately from the original crack images and the crack map.

2 Proposed Method

2.1 Two and Multiple Phases Piece-Wise Constant Crack Images

To understand the differences between two-phase piecewise constant image and multiple-phase piece-wise constant image, we can look at the Fig. 1. The left image, which is called as a two-phase piece-wise constant image, contains only two phases, "bright phase" of the background and "dark phase" of crack. In contrast, the right image is the 2D multiple-phase piece-wise constant image, which has other phases of unintended objects.

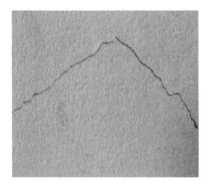

Fig. 1. Examples of two-phase and multiple phase piece-wise constant images

2.2 Overview of the Proposed Method

Overview of proposed method can be shown in following diagram in Fig. 2. Generally, we proposed 3 main steps for crack detection and quantification from 2D image.

Let $I(\mathbf{x})$ be the input image. In the first step, we need to enhance crack and remove other unintended objects in the image $I(\mathbf{x})$ by computing crack-enhancement filtered image $F_C(\mathbf{x})$. In the second step, we use B-spline level-set method to segment crack correctly to obtain the crack map, which is a binary image where "1" pixel represent the crack pixel and "0" pixel belongs to background. In the third step, crack width is measured as the Euclidean distance between two edges in both side of the crack centre-line by using Savitzky-Golay filter-based method.

2.3 Crack-Enhancement Filter

In our previous work [11], we proposed that cracks are dark objects that are more tubular and symmetric (according to local phase analysis) across its center-line than other non-crack objects in the image. In this paper, we utilize the crack-enhancement technique described in [11] to enhance crack and remove unintended objects. Let $I(\mathbf{x})$

be the input image, the crack enhancement filter response $F_C(\mathbf{x})$, which represents a possibility of a pixel $\mathbf{x}(x, y)$ to be inside a crack, is defined as:

$$F_C(\mathbf{x}) = F_L(\mathbf{x}).F_{sym}(\mathbf{x}) \tag{1}$$

where $F_L(\mathbf{x})$ and $F_{sym}(\mathbf{x})$ are line-like filter and symmetry filter response at pixel $\mathbf{x}(x, y)$, respectively. Our basic idea is to divide crack image into regions of high crack enhancement filter response (cracks) and a region of low crack enhancement filter response (background and other non-crack objects).

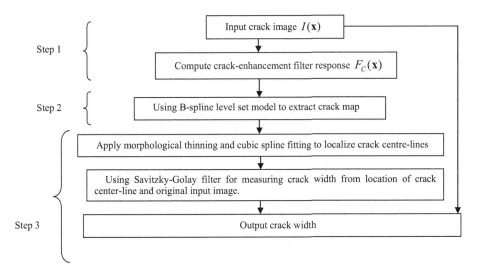

Fig. 2. Overall block diagram of proposed method

2.3.1 Line-Like Filter

We utilize the Frangi's vesselness measure [12] to compute the line-like filter response. The two eigenvectors e_1, e_2 and their corresponding eigenvalues λ_1, λ_2 of the Hessian matrix $\nabla^2 I(\mathbf{x})$ [12] are important shape directors for $I(\mathbf{x})$.

The normalized derivatives proposed by Lindeberg [13] were used to compute the derivatives in the Hessian matrix. The Frangi's vesselness measure $L(\mathbf{x}, \sigma)$ for a point $\mathbf{x}$ at a single scale σ is computed as

$$L(\mathbf{x}, \sigma) = \begin{cases} 0, & \text{if } \lambda_2 \leq 0 \\ e^{\frac{-R_\beta^2}{2\beta^2}}\left(1 - e^{\frac{-S^2}{2c^2}}\right), & \text{otherwise} \end{cases} \tag{2}$$

where $R_\beta = \left|\frac{\lambda_1}{\lambda_2}\right|$; $S = \left(\lambda_1^2 + \lambda_2^2\right)^{1/2}$; β, c are the thresholds that control the sensitivity of $L(\mathbf{x}, \sigma)$ to the measures R_β, S. β is fixed to 0.5 in our work.

For a given set of scales Ω_σ, the maximum response of different scales was chosen to be the line-like filter response as

$$F_L(\mathbf{x}) = \max_{\sigma \in \Omega_\sigma} \{L(\mathbf{x}, \sigma)\} \tag{3}$$

where max{.} is the maximum value of a set of $L(x, \sigma)$ determined for different values of the scale.

2.3.2   Symmetry Filter

It is well known that the symmetry features can be extracted using the local phase information such as transitions and discontinuities of intensity distribution in the image domain [11, 14, 20]. We adapt the method of [11] to build the 2D symmetry filter using the monogenic signal framework [15]. Let $b(\mathbf{x})$ be the isotropic log-Gabor filter given as

$$b(\mathbf{x}) = F^{-1}\left\{ \exp\left(-\frac{(\log(\theta/\theta_0))^2}{2(\log(s/\theta_0))^2} \right) \right\} \tag{4}$$

where $F^{-1}\{.\}$ is the inverse Fourier transform. θ is the spherical co-ordinate of the image in the frequency domain. The center frequency θ_0 is set as $\theta_0 = 1/(2 * 2.1^{h-1})$, h represents the scale of log-Gabor filter (if we used 6 scales then $h = \{1, 2, ..., 6\}$).

The monogenic signal $M(\mathbf{x})$ at $\mathbf{x} = (x, y)$ pixel is computed according to [11, 15] as a three dimensional vector as

$$M(\mathbf{x}) = [M^1(\mathbf{x}), M^2(\mathbf{x}), M^3(\mathbf{x})] \tag{5}$$

where $M^1(\mathbf{x}) = M(\mathbf{x}) * b(\mathbf{x}) * u_1(\mathbf{x})$, $M^2(\mathbf{x}) = M(\mathbf{x}) * b(\mathbf{x}) * u_2(\mathbf{x})$ and $M^3(\mathbf{x}) = M(\mathbf{x}) * b(\mathbf{x})$. $u_1(\mathbf{x})$ and $u_2(\mathbf{x})$ are two monogenic signal filters, computed according to [11, 15]. The two even symmetric and odd symmetric filter responses can be represented, respectively, by the components of the monogenic signal as

$$M_{odd}(\mathbf{x}) = \sqrt{(M^1(\mathbf{x}))^2 + (M^2(\mathbf{x}))^2} \tag{6}$$

$$M_{even}(\mathbf{x}) = M^3(\mathbf{x}) \tag{7}$$

The feature symmetry can be formulated from the monogenic signal as

$$S(\mathbf{x}) = \frac{Max\left\{ \sum_h \left(-M_{even}^h(\mathbf{x}) - \left| M_{odd}^h(\mathbf{x}) \right| \right) - TH, 0 \right\}}{\sum_h \sqrt{\left(M_{even}^h(\mathbf{x}) \right)^2 + \left(M_{odd}^h(\mathbf{x}) \right)^2} + \xi} \tag{8}$$

where ζ is a small constant to prevent division by zero, TH is a noise threshold.

Symmetry filter response is proposed to be calculated from $S(\mathbf{x})$ [11] as

$$F_{sym}(\mathbf{x}) = \max_{o} \left\{ \frac{\sum_{p=1}^{N_p} S(\mathbf{x}_p^o)}{N_p} \right\} \tag{9}$$

where $S(\mathbf{x}_p^o)$ is the 2D local phase-based feature symmetry of the pth pixel on the virtual line segment passing through $\mathbf{x}$ in o^{th} direction. N_p is the total number of pixels on each virtual line segment, N_o is the total number of virtual line segments passing through $\mathbf{x}$.

2.4 B-Spline Level Set Model for Detecting Crack from Crack Enhancement Filter Response

Based on the level-set formalism [16], crack regions are represented by a level set $\psi(\mathbf{x}, t)$ of $\mathbf{x}$ at time t with the closed contours of the cracks embedded in its initial form (time $t = 0$), i.e., zero level set.

$\psi(\mathbf{x}, t)$ must satisfy the following conditions.

$$\begin{cases} \psi(\mathbf{x}, t) < 0, & \text{if } x \in F_C^{in}(t) \\ \psi(\mathbf{x}, t) = 0, & \text{if } x \in \partial F_C^{in}(t) \\ \psi(\mathbf{x}, t) > 0, & \text{if } x \in F_C \backslash F_C^{in}(t) \end{cases} \tag{10}$$

where F_C^{in} is the crack region in the image of F_C. The level set $\psi(\mathbf{x}, t)$ is expressed in terms of a number of B-spline basis functions to achieve a smooth and continuous mathematical model for the cracks as [17]:

$$\psi_B(\mathbf{x}) = \sum_{k_1 \in Z^2} \omega(k_1) C^3\left(\frac{\mathbf{x}}{k_2} - k_1\right) \tag{11}$$

where the knots of the B-splines are located at the nodes of a grid spanning over F_C with a regular spacing of k_2; $C^3(.)$ is the uniform symmetric 2D B-spline of degree 3; $\omega(k_1)$ are the unknown coefficients of the B-spline functions.

Crack is the object where crack-enhancement filter response is high. The cost functional for crack detection is formulated as [17]

$$F_{cost}(f_c, f_b, \psi_B) = \int_{\Omega} (F_C(\mathbf{x}) - f_c)^2 H(\psi_B(\mathbf{x})) d\mathbf{x} + \int_{\Omega} (F_C(\mathbf{x}) - f_b)^2 (1 - H(\psi_B(\mathbf{x}))) d\mathbf{x}$$

$$\tag{12}$$

where $H(.)$ is the Heaviside function; f_c and f_b are averages of the crack-enhancement filter response of crack regions and non-crack regions (background and other unrelated objects), respectively.

B-spline level set ψ_B, that can localize the crack regions, can be found by minimizing the above cost functional using gradient-descent method [16].

2.5   Width and Length Measurement

After minimizing the cost functional, we can achieve the binary crack map $CM(\mathbf{x})$, where pixels in the crack regions have "1" intensity (white pixel) and pixels in the background have "0" intensity (black pixel), by setting $\psi_B(\mathbf{x}) < 0$. The morphological thinning algorithm [18] is then used to remove the exterior pixels from the detected objects to obtain a new binary image of the crack skeleton, which connects the white pixels (value 1) running along the crack centerlines.

We then need to divide the crack skeleton into different crack segments. The center-line of each potential crack segment is fitted by a cubic spline so that accurate first and second derivatives (and hence crack orientation) at any location can be determined.

The cross-sectional pixel intensity profile along a line perpendicular to crack segment, is determined from the original grayscale image as shown in the Fig. 3. The first and second derivatives of 1D cross-sectional pixel intensity profile can be calculated using Savitzky-Golay smoothing and differentiation algorithm [19]. We can quickly compute the second derivative of the cross-sectional pixel intensity profile by implementing the convolution between the intensity profile and a Savitzky-Golay filter [19].

Crack width in a cross-sectional pixel intensity profile containing is determined as following:

- Find the edge points, which are zero-crossings from negative to positive of the second derivative on the left side of the center column or zero-crossings from positive to negative of the second derivative on the right side of the center column.
- Crack width is the Euclidean distance between two edge points.

Note that zero-crossing of the second derivative of the intensity is widely considered as image edge [21]. In Fig. 3, a typical cross-sectional pixel intensity profile is smoothed by the 1D Savitzky-Golay smoothing filter. Two zero-crossings of the second derivative are also illustrated.

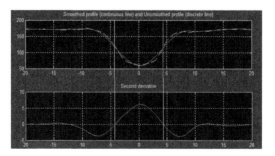

Fig. 3. Measuring crack width from cross-sectional pixel intensity profile of a crack segment

3   Results

The proposed method was implemented in Matlab 7.8 and tested on a computer with Intel Core I3 2.1 GHz CPU and 4 GB RAM. The following settings are applied to crack images:

- The image has an eight-bit dynamics and 2 dimensions.
- $\beta = 0.5, c = 15$, $\Omega_s = \{0.5, 1, 1.5, 2, 2.5, 3\}$, $F_{sym}(\mathbf{x})$ are computed over 6 scales with $s/\theta_0 = 0.65$, $\zeta = 0.0001$, $N_p = 15$, $N_p = 12$, $k_2 = 2$.

In our experiments, camera plane is adjusted to be parallel with the concrete surfaces and the widths of crack are measured in pixel unit. If we know the exact distance from the camera plane to the concrete surface (by using a laser range finder), we can easily convert the width in pixel unit to mm [22]. Please refer to [12] to know how to calculate the pixel/mm conversion ratio.

Figure 4 shows the results of width measurement for the crack in 2D multiple-phase intensity inhomogeneous images. 323 widths along the crack center-line are measured (in pixel unit) from the crack image. The length of the crack contains 323 pixels. To check the accuracy of these measured widths, we manually measure crack widths of 20 points (which are chosen randomly along the crack center-line) and compare with the nearest crack widths which are measured automatically by the program. We have found that all the differences are lower than 1 pixel. This has shown that our method can achieve sub-pixel accuracy. Figure 5 is the diagram plotting the widths measured in pixels along the crack of Fig. 4.

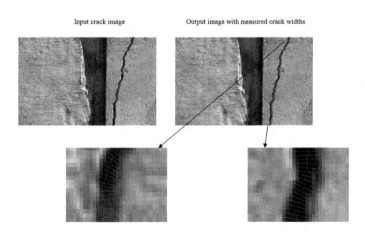

Fig. 4. Measuring crack width

To further demonstrate the capibility of crack width measurement in a real scene of a building, we tested our method on an image of a concrete wall as shown in Fig. 6, which contains not only the surface concrete but also a steel column.

Width (px)

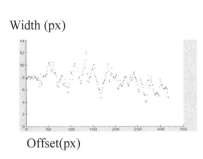

Offset(px)

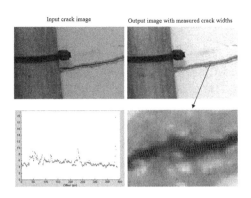

Fig. 5. Plot of crack widths along the crack in Fig. 4

Fig. 6. Measuring crack width in real scene of a building

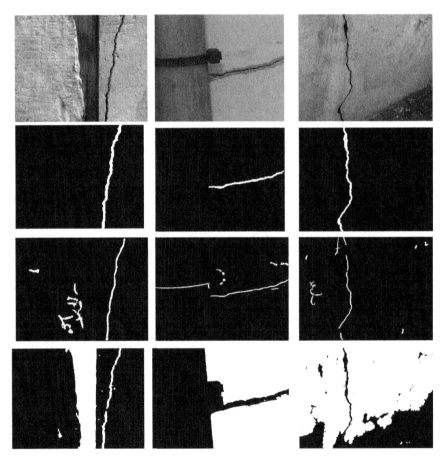

Fig. 7. Comparison between proposed method, Fujita's method [8] and Hutchinson's method [4]. From first to fourth row: input image, binary image result from the proposed method, result of Fujita's method, result of Hutchinson's method.

In the final experiment, we make a comparison between our proposed method and Fujita's method [8] which is well-known for being able to automatically measuring crack widths in noisy images of concrete surface, and Hutchinson's method [4]. We use the robust set of parameters for Fujita's method as described in [8]. Similarly parameters for Hutchinson's method is referenced from [4]. From the results in Fig. 7 we can observe that Fujita's method, Hutchinson's method have generated many wrong detections while the present method has performed the tasks effectively. Please note that Fujiata's method also wrongly detect damages of the wall painting (for ex., the elongated stripe above the crack in the right-hand side of Fig. 7), which is not the real concrete structure's crack.

4   Conclusions

We have proposed an automatic method for measuring crack widths from image of scenes in the building without human intervention. Unlike available methods that measure crack width from images containing only concrete surface and cracks (2D piece-wise constant images), our method measured directly and automatically the widths of cracks from 2D multiple-phases intensity inhomogeneous images of scenes inside building.

Our method can assist the civil engineering experts to measure crack width more effective. This advanced method creates the possibility for a monitoring camera system to automatically detect and measure widths of cracks of the building walls because images from these camera systems usually contain many unintended objects of the building. A number of experiments have been given to demonstrate the applications of the proposed method.

In the near future, we will extend our method to be capable of measuring crack width from 3D images.

References

1. Zhu, Z., German, S., Brilakis, I.: Visual retrieval of concrete crack properties for automated post-earthquake structural safety evaluation. Autom. Constr. **20**, 874–883 (2011)
2. Yamaguchi, T., Hashimoto, S.: Fast method for crack detection surface concrete large-size images using percolation-based image processing. Mach. Vis. Appl. **21**, 797–809 (2010)
3. Barazzetti, L., Scaioni, M.: Crack measurement: development, testing and applications of an automatic image-based algorithm. ISPRS J. Photogram. Remote Sens. **64**, 285–296 (2009)
4. Chen, Z.Q., Hutchinson, T.C.: Image-based framework for concrete surface crack monitoring and quantification. Adv. Civ. Eng. **2010**, 18 (2010)
5. Chen, L., Shao, Y., Jan, H., Huang, C., Tien, Y.: Measuring system for cracks in concrete using multi-temporal images. J. Surv. Eng. **132**, 77–82 (2006)
6. Nishikawa, T., Yoshida, J., Sugiyama, T., Fujino, Y.: Concrete crack detection by multiple sequential image filtering. Comput.-Aided Civ. Infrastruct. Eng. **27**, 29–47 (2012)
7. Adhikari, R.S., Moselhi, O., Bagchi, A.: Image-based retrieval of concrete crack properties for bridge inspection. Autom. Constr. **39**, 180–194 (2014)

8. Fujita, Y., Hamamoto, Y.: A robust automatic crack detection method from noisy concrete surfaces. Mach. Vis. Appl. **22**, 245–254 (2011)
9. Lee, B.Y., Kim, Y.Y., Yi, S.T., Kim, J.K.: Automated image processing technique for detecting and analysing concrete surface cracks. Struct. Infrastruct. Eng. **9**, 567–577 (2013)
10. Yang, Y.S., Yang, C.M., Huang, C.W.: Thin observation crack in a reinforced concrete bridge pier test using image processing and analysis. Adv. Eng. Softw. **83**, 99–108 (2015)
11. Nguyen, H.N., Kam, T.Y., Cheng, P.Y.: Automatic crack detection from 2D images using a crack measure-based B-spline level set model. Multidimension. Syst. Sig. Process. **29**, 1–32 (2016)
12. Frangi, A.F., Niessen, W.J., Vincken, K.L., Viergever, M.A.: Multiscale vessel enhancement filtering. In: Wells, W.M., Colchester, A., Delp, S. (eds.) MICCAI 1998. LNCS, vol. 1496, pp. 130–137. Springer, Heidelberg (1998). https://doi.org/10.1007/BFb0056195
13. Lindeberg, T.: Edge detection and ridge detection with automatic scale selection. Int. J. Comput. Vis. **30**, 117–156 (1998)
14. Kovesi, P.: Symmetry and asymmetry from local phase. In: Tenth Australian Joint Conference on Artificial Intelligence, Australia, pp. 185–190 (1997)
15. Felsberg, M., Sommer, G.: The monogenic signal. IEEE Trans. Sig. Process. **49**, 3136–3144 (2001)
16. Osher, S., Fedkiw, R.P.: Level set methods: an overview and some recent results. J. Comput. Phys. **169**, 463–502 (2001)
17. Bernard, O., Friboulet, D., Thevenaz, V., Unser, M.: Variational B-spline level-set: a linear filtering approach for fast deformable model evolution. IEEE Trans. Image Process. **18**, 1179–1191 (2009)
18. Gonzalez, R., Woods, R.: Digital Image Processing, 2nd edn. Prentice Hall, Upper Saddle River (2003)
19. Gorry, P.A.: General least-squares smoothing and differentiation by the convolution (Savitzky-Golay) method. Anal. Chem. **62**, 570–573 (1990)
20. Kovesi, P.: Image features from phase congruency. Videre: J. Comput. Vis. Res. **1**, 1–26 (1999)
21. Marr, D., Hildreth, E.: Theory of edge detection. Proc. R. Soc. Lond. B: Biol. Sci. **207**, 187–217 (1980)
22. Li, G., He, S., Ju, Y., Du, K.: Long-distance precision inspection method for bridge cracks with image processing. Autom. Constr. **41**, 83–95 (2014)

Intelligent Systems for Optimization of Logistics and Industrial Applications

Community of Practice for Product Innovation Towards the Establishment of Industry 4.0

Mohammad Maqbool Waris[1(✉)], Cesar Sanin[1], and Edward Szczerbicki[2]

[1] The University of Newcastle, Callaghan, NSW, Australia
MohammadMaqbool.Waris@uon.edu.au, cesar.sanin@newcastle.edu.au
[2] Gdansk University of Technology, Gdansk, Poland
edward.szczerbicki@newcastle.edu.au

Abstract. The aim of this paper is to present the necessity of formulating the Community of Practice for Product Innovation process based on Cyber-Physical Production Systems towards the establishment of Industry 4.0. At this developing phase of Industry 4.0, there is a need to define a clear and more realistic approach for implementation process of Cyber-Physical Production Systems in manufacturing industries. Today Knowledge Management is considered as the next arena of global competition. One of the most promising areas where Knowledge Management is studied and applied is product innovation. This paper explains the efficient and systematic methodology for Knowledge Management through Community of Practice for product innovation, thus connecting manufacturing units at global level.

Keywords: Smart Innovation Engineering · Product innovation
Cyber-Physical Production Systems · Set of experience
Community of Practice · Industry 4.0

1 Introduction

The process of Product Innovation is a series of knowledge activities, such as collecting, assimilating, creating, storing, and reusing knowledge at different stages. Knowledge Management (KM) has become the only sustainable competitive advantage for manufacturing organizations in the current turbulent context [1], also product innovation is a continuous learning process rather that a sporadic event. These organizations are also facing continuous market changes and need for short product life cycles [2]. Proper knowledge management therefore plays an important role in product innovation. The fourth industrial revolution, originally initiated in Germany as Industry 4.0, has attracted much attention in recent times. It is closely related with Cyber-Physical Production System (CPPS), Internet of Things, Information and Communication Technology and Enterprise Integration. Knowledge Engineering (KE) and KM are important role players in CPPS.

The concept of Smart Innovation Engineering (SIE) system proposed by Waris et al. [3, 4] is a semi-automatic tool for facilitating product innovation process. The SIE system uses a collective, team-like knowledge developed by past experiences of the

© Springer International Publishing AG, part of Springer Nature 2018
N. T. Nguyen et al. (Eds.): ACIIDS 2018, LNAI 10752, pp. 651–660, 2018.
https://doi.org/10.1007/978-3-319-75420-8_61

innovation-related formal decisional events. This knowledge is stored in the form of a combination of Variables, Functions, Constraints and Rules that comprehensively represents the product's information including its properties, manufacturing, requirements, structure and changes that were made in the past, for more details see [4]. Whenever any innovation-related query is presented to SIE system, it uses this experience-based knowledge to find the top similar experiences. These past experiences help the user to take proper decisions. It is same like seeking advice from the experts before taking the final decision. The goal here is to show how SIE system can be used by the common Community of Practice formed by global group of manufacturing organizations in dealing with the issues of product innovation process and that will indeed be a substantial step towards the establishment of Industry 4.0.

In the context of manufactured products, product innovation can be defined as the process of making required changes to the already established product by introducing something new that adds value to users and also providing expertise knowledge that can be stored in the organization [5]. Strategy for implementation of product innovation includes the use of better components, new materials, advanced technologies, and new product features/functions. Moreover, the establishment of cross-functional, multidisciplinary teams was found to be vital to the success of the innovation project [6] within the organization. The SIE system can initially be used within a manufacturing organization. However, it can certainly be extended for connecting a network of manufacturing units around the globe.

2   Background

To have a clear idea about the integration of different fields viz. KE, CPPS, Industry 4.0 and CoP, a brief discussion about them is presented in this section.

2.1   Knowledge Engineering (KE)

Due to the fact that manufacturing organizations need to perform innovation process frequently and knowledge plays a critical role in it, they need to manage knowledge effectively within and across their organizational borders [1]. Feigenbaum [7] defines KE as an engineering discipline that aims to solve complex problems, normally requiring a high level of human expertise, by integrating knowledge into computer systems. Members of Communities of Practice (CoPs) are interacting on an ongoing basis to deepen their knowledge and expertise in their concerned area. Although CoPs follow the human-oriented KM approach, the use of technology, particularly knowledge management systems (KMS), is however important [8]. It is essentially required in large CoPs where proper and effective interaction is not possible without the support of KMS applications. Moreover, human-oriented CoPs have some drawbacks such as trust, lack of openness, power conflicts among members, emotional effects and so on which are critical aspects that may result in failure.

One of the main external or environmental elements that can be thus considered is the knowledge strategy pursued by the organization aiming for the best use of the knowledge-based resources in the view of the organization's competitive advantage [9]. The knowledge strategy defines aims and tools of KM programs and, thus, of CoPs. Moreover it is associated strictly with the competitive strategy of the organizations [10].

2.2 Cyber-Physical Production Systems (CPPS)

Cyber Physical Systems (CPSs) can be described as the transformative technologies for managing interconnected systems between its physical assets and computational capabilities with the possibility of human machine interaction [11]. CPSs has drawn a great deal of attention from academia, industry, and the government due to its potential benefits to society, economy, and the environment. Some of the practical examples in the present world are autonomous cars, robotic surgery, smart manufacturing, smart electric grid, implanted medical devises, and intelligent buildings. Application of CPS in the manufacturing industry leads to CPPS and hence the ability for continuous viewing of product, production equipment and production system under consideration. The introduction of CPPS in any production system promises social, economic and even ecological benefits.

Due to the competitive nature of today's industry and recent development resulting in higher availability and affordability of computer networks, sensors, and data acquisition more and more industrial organizations are forced to move toward implementation high-tech methodologies. Consequently, the ever growing use of networked machines and sensors has resulted in the continuous generation of high volume data which is known as Big Data [12]. In modern manufacturing organizations, especially high-tech industries, CPPS can be further developed for managing knowledge and experience in the form of Big Data and leveraging the interconnectivity of machines to reach the goal of intelligent factories. Furthermore by integrating CPPS with logistics and services in the current industrial practices would transform today's factories into an Industry 4.0 factory with significant economic potential [13, 14].

2.3 Industry 4.0

Systematic infusion of newest developments in computer science, information and communication technology, and manufacturing science and technology into CPPS may lead to the fourth Industrial Revolution. The term "Industry 4.0" is used for the fourth industrial revolution which is about to take place right now. According to the Federal Ministry of Education and Research, Germany (BMBF): "Industry is on the threshold of the fourth industrial revolution". Driven by the Internet, the real and virtual worlds are growing closer and closer together to form the Internet of Things. Industrial production of the future will be characterized by the strong customization of products under the conditions of highly flexible production as the manufacturing units of Industry 4.0 will have to deal with so called radical innovations frequently. This transformation is based on the extensive integration of customers (by collecting continuous feedback) and stake-holders along with value-added processes, and the linking of production and high-quality services leading to so-called hybrid products.

The concept of Industry 4.0 came into existence in 2011, when an association of representatives from academia, business, and politics promoted the idea as an approach for strengthening the competitiveness of the German manufacturing industry [15]. Industry 4.0 is a collective term for technologies and concepts of value chain organization. Industry 4.0 makes factories more intelligent, flexible, and dynamic by equipping manufacturing with autonomous systems, actors, and sensors [16]. Consequently, industries including smart products, machines and equipment will achieve high levels of automation and self-optimization. In addition, production of complex and customized products with high standards can be manufactured as per expectations [16]. Thus, intelligent factories and smart manufacturing are the major goals of Industry 4.0 [17].

Nowadays, product customization resulting from the preferences and demands of consumers tends to be one more variable in the manufacturing process, and smart factories will have to be able to innovate the manufactured products adapting to the preferences of the users [18]. Thus, Use of experience-based knowledge, is one of the main characteristics of Industry 4.0 to support the integration and virtualization of product design/innovation and production process using KBS to create smart products. This will lead to early launch of innovated products with the objectives to meet the growing needs and expectations of the users. The use of SIE system in manufacturing organizations seems to be promising in this scenario. It will also help in handling the continued flow of new product innovation projects in the industry resulting from shorter lifecycle of products. Therefore implementation of SIE system in Industry 4.0 is a concept that will transform current industries into Smart Factories having a well-defined network of intelligent machines, smart products, systems and processes creating real world virtualization into a huge information system. Potential benefits of Industry 4.0 are flexibility, reduced lead times, reduced costs and customization of products.

2.4   Community of Practice (CoP)

The most popular definition of CoP is that of Wenger et al. [19] who defines CoP as a group of people who share a concern, a set of problems, or a passion about a topic, and who deepen their knowledge and expertise in this area by interacting on an ongoing basis. From the above definition it is clear that CoP focuses on a specific domain, and its members interact around a common pool of knowledge and develop their practice to find the possible solutions to problems. Most of the CoPs are developed among large organizations, in various sectors, such as: *Automotive industry*: Ford, DaimlerCrysler, Caterpillar. *Oil industry*: Shell, ChevronTexaco, ENI. *Computer*: HP, IBM and *Consulting*: Accenture, CAP Gemini.

In all these cases, the CoPs are created for sharing and diffusion of knowledge for improving the problem solving capability and innovative potential of their employees. For example, engineers of DaimlerChrysler working at different plants participate in the "Tech Clubs" CoPs for knowledge sharing about similar problems and solutions. CoPs by Shell, Turbodude, were formed to establish a platform for sharing knowledge of different teams involved in deep water exploration. Virtual CoPs at ANI were created in order to provide guidance and support to exploration managers from more experienced colleagues working elsewhere. The CoPs are considered key components of systematic and deliberate KM strategies [20]. So, the primary function of CoP is to promote knowledge sharing in order to improve the overall performance of the organization [21]. Secondly to provide knowledge database and builds norms, trust and assessment in favor of knowledge sharing. The performance can be improved at the individual, group, and organizational level by improving employees' working experience, reducing the learning curve, accumulating professional talents for the organization, and avoiding overlapping investment on new products and services [22, 23].

3   Smart Innovation Engineering (SIE) System

The importance of the aforementioned aspects creates the necessity for systems that collectively work together for Industry 4.0. One such system is SIE system developed by Waris et al. [4]. This system is a prominent tool to support the innovation processes in a quick and efficient way. The SIE System is based on the Set of Experience Knowledge Structure (SOE) and Decisional DNA (DDNA), which were first presented by Sanin and Szczerbicki [24]. It is a Smart Knowledge Management System (SKMS) capable of storing formal decision events explicitly [24, 25]. The architecture of SIE system consists of three main modules: Systems, Usability and SIE_Experience (see [4] for more details). Sets of experience are created for each module individually that allows the experienced-based knowledge to be stored more systematically for a wide range of similar manufactured products. Set of experience is a unique combination of Variables, Functions, Constraints and Rules. Combination of all the individual Sets of experience are combined under the SIE system that represents complete knowledge and experience necessary for supporting innovation process of manufactured products.

Graphical User Interphase (GUI) for the SIE System is shown in Fig. 1. This GUI will allow the user to interact with SIE System in a user-friendly language. The user can select the set of values from the drop-down menu and is also able to define the required Constraints and preferences in the form of set of variables with selected values.

The query based on innovation objectives is converted into SOE and compared with the similar Sets of Experience that were generated by the SIE System from Comma Separated (CSV) files for each module containing complete information about the product in the form of Variables, Functions, Constraints and Rules. The CSV files contains data in standard format so that the parser collects information as required. The SIE System provides a list of proposed solutions (say five) that is displaced in the GUI. At this time, the user/entrepreneur/innovator has the privilege to select the best possible solution from that list. This selected final solution is stored in the SIE System as SOE that can be used in future for similar query. In this way, the SIE System is a semi-automatic system that facilitates the process of Product Innovation. The SIE System gains experience with each decision taken that increases its expertise and behaves as an expert in its domain.

SIE is an expert system that can facilitate CPPS and it can play a vital role towards the establishment of Industry 4.0 and has the potential to be used by large enterprises or group of SMEs, manufacturing similar products and sharing data among themselves.

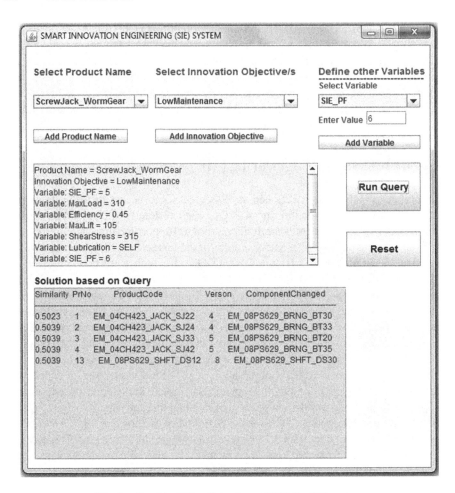

Fig. 1. Graphical User Interphase (GUI) for SIE system

4   Community of Practice for Product Innovation

In this section, we will try to explain how the SIE system itself can be used as a CoP for PI by the means of virtual connection of manufacturing organizations at a common platform (i.e. SIE system). The importance of the aforementioned aspects proves that the cognitive role of a CoP is focused on the knowledge transfer process. Some central elements must be considered here, for instance, what kind of knowledge is to be exchanged, i.e. the nature of the shared knowledge, what procedure and tools are to be used, and more importantly for what purpose. The value of CoPs is more significant in a long-term perspective. "Tech Clubs" communities in DaimlerChrysler help to solve problems on a daily basis, which means benefits in a short-term perspective, but, simultaneously developing the expertise of members, which means benefits in a long-term perspective.

Similarly, SIE system is used to facilitate PI process frequently to meet the demands for customized products and short product life cycles. But, at the same time, it is gathering knowledge by storing all the experiences of the decisional events that can be used for decision making when a similar query is presented in future. Thus presenting long-term benefits. As already proved that SIE system behaves like a group of experts and the experience-based knowledge is stored in different modules (chromosomes-containing experiences of certain category) in the form of particular SOEs. And these chromosomes are grouped together to form what is called as a Decisional DNA (DDNA) of the manufacturing organization. Connecting DDNAs of various organizations through the SIE system will bring the experience-based knowledge of various groups of experts at the same platform. This will help these smart factories in PI process as they can perform innovation process systematically and quickly due to fast computational capabilities of SIE system.

For the successful transformation of current manufacturing organizations into smart factories of Industry 4.0, the knowledge possessed by various actors needs to be sought, elaborated, and mixed to boost innovation and flexibility [8]. This task can be accomplished by using a smart knowledge management system as a CoP where experience-based knowledge can be stored, shared and reused among the groups of manufacturing organizations. For this purpose, SIE system can be implemented in each organization and interconnected with each other to form a group of expert systems, calling it Community of Practice for Product Innovation. The proposed architecture of the CoP for PI is shown in Fig. 2.

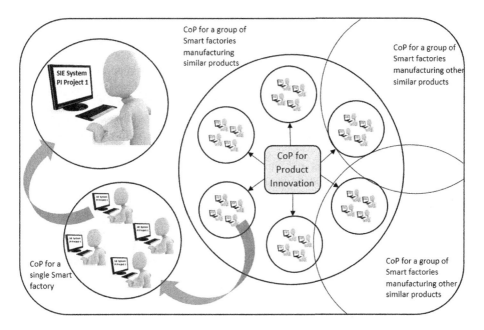

Fig. 2. Architecture of community of practice for product innovation.

According to Ji et al. [26], the innovation sources of enterprises mainly include three aspects under the mode of communities of practice: Internal CoP of the enterprises, CoP outside the enterprise, and the interaction among diverse CoPs. We used the same methodology in defining the proposed CoP for PI (see Fig. 2). At the root level, the CoP for PI starts by implementing the SIE system in different PI projects inside the same organization. This will bring experience-based knowledge of all PI projects at a single platform, thus building a cognitive connection between them. This will allow them to share and reuse the knowledge from the different units thus increasing the overall expertise of the organization. Similarly, different organizations manufacturing the same or similar products join hands together to form a CoP for PI. These organizations now share and reuse experiences among themselves. Thus expanding their expertise at a broader level. One of the main advantage of this system is that the expert/innovator can take quick and systematic decisions that are based on real experiences. There is no need for scheduling the group of experts' meeting or any other formal procedure. Here, it should be made clear that the SIE system and the CoP for PI are basically the same. In fact, CoP for PI is interconnected extension of various Production units inside and outside the organizations.

Further extension of CoP with cross-border integration of industries in different fields will lead to the establishment of highly expert CoP at a global level. The member smart factories will communicate with those of other fields. User preferences in other products, demographic and economic factors will be used to create new ideas leading to innovation.

These CoPs are very beneficial for high-tech start-ups as they can manufacture products confidently with a highly successful probability. At the same time, CoP for PI is also beneficial for already successful organizations as they can regularly update their knowledge system, can create new ideas and modify regular practice. For example, a manufacturing organization who plans to establish another unit in another country can use this system to modify the product according to the user preferences, demographic factors, economic conditions, resources and technologies available in that country, government norms, economic factors and other such conditions. Obviously they cannot manufacture exactly the same product for users of other countries having different taste and economic conditions.

The role of government is also very crucial in forming such CoPs. It should be committed to create and provide favorable environment for proper and legal communication among its members at global level. Making and implementing effective industrial policies and providing incentives for new members will be fruitful both for industries and nation.

The importance or contribution of SIE system can be justified from the potential benefits that it offers towards the establishment of Industry 4.0 that itself is a complex system. Some of them are more flexibility, quick and systematic, customization and Reduction in costs.

5   Conclusion

SIE system is a semi-automatic tool to facilitate the process of product innovation. This paper presented the concept of Community of Practice for product innovation, its methodology for implementation towards the establishment of Industry 4.0, and its advantages for manufacturing organizations and nations as a whole. It was further explained that how the Smart Innovation Engineering (SIE) itself behaves as an expert Community of Practice. Manufacturing organizations around the globe can store, share and reuses the experience-based knowledge among themselves at a common platform. Implementing this system in manufacturing organizations will allow them to take quick and systematic innovation-related decisions. The user get the advice from the SIE system similar to that of group of experts from the click of a mouse making the innovation process quick and systematic. The analysis of basic concepts and implementation method proves that SIE is an expert system that can facilitate Cyber Physical Systems (CPS) and it can play a vital role towards the establishment of Industry 4.0 and has the potential to be used further for lean innovation and sustainable innovation in future. The SIE system has the potential to be used by large enterprises or group of SMEs manufacturing similar products or even by new high-tech start-ups.

References

1. Corso, M., Martini, A., Paolucci, E., Pellegrini, L.: Knowledge management in product innovation: an interpretative review. Int. J. Manag. Rev. **3**(4), 341–352 (2001)
2. Verhagen, W.J.C., Bermell-Garcia, P., van Dijk, R.E.C., Curran, R.: A critical review of knowledge-based engineering: an identification of research challenges. Adv. Eng. Inf. **26**(1), 5–15 (2012)
3. Waris, M.M., Sanin, C., Szczerbicki, E.: Enhancing product innovation through smart innovation engineering system. In: Nguyen, N.T., Tojo, S., Nguyen, L.M., Trawiński, B. (eds.) ACIIDS 2017. LNCS (LNAI), vol. 10191, pp. 325–334. Springer, Cham (2017). https://doi.org/10.1007/978-3-319-54472-4_31
4. Waris, M.M., Sanín, C., Szczerbicki, E., Shafiq, S.I.: A semiautomatic experience-based tool for solving product innovation problem. Cybern. Syst. **48**(3), 231–248 (2017)
5. Waris, M.M., Sanin, C., Szczerbicki, E.: Toward smart innovation engineering: decisional DNA-based conceptual approach. Cybern. Syst. **47**(1–2), 149–159 (2016)
6. Jayaram, J., Okeb, A., Prajogo, D.: The antecedents and consequences of product and process innovation strategy implementation in Australian manufacturing firms. Int. J. Prod. Res. **52**(15), 4424–4439 (2014)
7. Feigenbaum, E.A., McCorduck, P.: The Fifth Generation: Artificial Intelligence and Japan's Computer Challenge to the World. Addison-Wesley, Boston (1983)
8. Scarso, E., Bolisani, E.: Communities of practice as structures for managing knowledge in networked corporations. J. Manuf. Technol. Manag. **19**(3), 374–390 (2008)
9. Zack, M.H.: Developing a knowledge strategy. Calif. Manag. Rev. **41**(3), 125–145 (1999)
10. Akhavan, P., Jafari, M., Fathian, M.: Critical success factors of knowledge management systems: a multi-case analysis. Eur. Bus. Rev. **18**(2), 97–113 (2006)
11. Baheti, R., Gill, H.: Cyber-physical systems, pp. 161–166 (2011)
12. Lee, J., Lapira, E., Bagheri, B., Kao, H.-A.: Recent advances and trends in predictive manufacturing systems in big data environment. Manuf. Lett. **1**(1), 38–41 (2013)

13. Lee, J., Lapira, E., Yang, S., Kao, A.: Predictive manufacturing system - trends of next-generation production systems. In: IFAC Proceedings Volumes, vol. 46, no. 7, pp. 150–156 (2013)
14. Thiede, S., Juraschek, M., Herrmann, C.: Implementing cyber-physical production systems in learning factories. Procedia CIRP **54**(Supplement C), 7–12 (2016)
15. Kagermann, H., Helbig, J., Hellinger, A., Wahlster, W.: Recommendations for Implementing the Strategic Initiative INDUSTRIE 4.0: Securing the Future of German Manufacturing Industry; Final Report of the Industrie 4.0 Working Group. Forschungsunion (2013)
16. Roblek, V., Meško, M., Krapež, A.: A complex view of industry 4.0. SAGE Open **6**(2), 1–11 (2016)
17. Sanders, A., Elangeswaran, C., Wulfsberg, J.: Industry 4.0 implies lean manufacturing: research activities in industry 4.0 function as enablers for lean manufacturing. J. Ind. Eng. Manag. **9**(3), 811–833 (2016)
18. Santos, K., Loures, E., Piechnicki, F., Canciglieri, O.: Opportunities assessment of product development process in industry 4.0. Procedia Manuf. **11**, 1358–1365 (2017)
19. Wenger, E., McDermott, R.A., Snyder, W.: Cultivating Communities of Practice: A Guide to Managing Knowledge. Harvard Business Press, Boston (2002)
20. Smith, H.A., McKeen, J.D.: Creating and facilitating communities of practice. In: Holsapple, C.W. (ed.) Handbook on Knowledge Management 1. INFOSYS, vol. 1, pp. 393–407. Springer, Heidelberg (2004). https://doi.org/10.1007/978-3-540-24746-3_20
21. Pattinson, S., Preece, D.: Communities of practice, knowledge acquisition and innovation: a case study of science-based SMEs. J. Knowl. Manag. **18**(1), 107–120 (2014)
22. Rongo, D.: Managing virtual communities of practice to drive product innovation. Int. J. Web Based Commun. **9**(1), 105–110 (2013)
23. Chu, M.-T., Khosla, R., Nishida, T.: Communities of practice model driven knowledge management in multinational knowledge based enterprises. J. Intell. Manuf. **23**(5), 1707–1720 (2012)
24. Sanin, C., Szczerbicki, E.: Towards the construction of decisional DNA: a set of experience knowledge structure Java class within an ontology system. Cybern. Syst. **38**(8), 859–878 (2007)
25. Sanin, C., Szczerbicki, E.: Genetic algorithms for decisional DNA: solving sets of experience knowledge structure. Cybern. Syst. **38**(5–6), 475–494 (2007)
26. Ji, H., Sui, Y.-T., Suo, L.-L.: Understanding innovation mechanism through the lens of communities of practice (COP). Technol. Forecast. Soc. Change **118**(Supplement C), 205–212 (2017)

Improving KPI Based Performance Analysis in Discrete, Multi-variant Production

Rafał Cupek[1]([✉]), Adam Ziębiński[1], Marek Drewniak[2],
and Marcin Fojcik[3]

[1] Institute of Informatics, Silesian University of Technology, Gliwice, Poland
{Rafal.Cupek,Adam.Ziebinski}@polsl.pl
[2] AIUT Sp. z o.o., Gliwice, Poland
mdrewniak@aiut.com.pl
[3] Western Norway University of Applied Sciences, Førde, Norway
Marcin.Fojcik@hvl.no

Abstract. Discrete production systems face the challenge of moving from a mass to a mass-customised production model. Classic methods for analysing Key Performance Indicators (KPI) that are based on a statistical approach are difficult to apply in the case of short series, multi-variant production. A new approach for KPI analysis that is based on machine learning and data mining methods has to be applied. The authors propose a new approach that is based on K-means clustering that can be useful for performance analysis in the case of short series, multi-variant production. The presented research is focused on discrete production systems with KPI data traceability on the work cell level. The main advantage of the presented solution is its ability to automatically estimate a number of technological variants that affect a given performance indicator.

Keywords: Production tracking · KPI-analysis · Multi-variant production
Data mining · Clustering · K-means

1 Introduction

Production process traceability and Key Performance Indicators (KPI) analysis that are based on the Statistical Process Control (SPC) approach are widely used in many continuous and batch production systems. There are a number of proven methods that are used by industry [1]. Unfortunately, in the case of discrete manufacturing, especially in mass-customised production systems, such solutions can rarely be used. Mass-customised manufacturing [2], which nowadays is forced by the global economy, means that discrete manufacturing systems have to be able to execute customised and multi-variant production with the quality of the product and production costs equal to mass-production [3]. In such a case, the KPI analysis that is based on classical SPC methods is useless.

On the other hand, the production performance has to be traced more and more accurately. According to ANSI/ISA-95 (IEC/ISO 62264) norms a production performance analysis together with data collection and production tracking are key

© Springer International Publishing AG, part of Springer Nature 2018
N. T. Nguyen et al. (Eds.): ACIIDS 2018, LNAI 10752, pp. 661–673, 2018.
https://doi.org/10.1007/978-3-319-75420-8_62

information streams that are exchanged between control and business systems. This include analysis of the information about production unit cycle times, resource utilisation, equipment utilisation, equipment performance, procedure efficiencies and production variability. Relationships between these analyses and others may also be utilised to develop KPI reports [4].

The main contribution of this paper is a new method that uses a data mining approach that is based on the K-means algorithm to automatically estimate a number of different technological variants that are performed by a work cell. The data clustering is focused on preparing the set of production profiles that are relevant to a given KPI and that take into account the different variants of production. The profiles are created during the learning phase of the algorithm and can be repeated and performed simultaneously with standard production activities. The learning phase includes the automatic collection of the selected process variables that are related to the KPI being analysed. Selected data are gathered during every production cycle. Next, the production profiles are created in accordance with the KPI being analysed. The proposed method is based on the following assumptions:

- the production is carried out under unknown number (k) of technological variants;
- k is not explicitly specified and may not reflect the technical specification of the production order. It is possible that two different products will be produced in a very similar manner according to similar technological profiles. It is also possible that the same type of product can have different technological profiles that cannot be indicated directly by its technical specifications.

The research results presented in this paper prove that:

- it is possible to automatically determine the production profiles that affect analysed KPI and therefore determine production data that affects given KPI;
- the learning process is resistant to variations in the production parameters and can be performed during standard production.

The paper is organised as follows: in the second section the authors present the state of the art in production data clustering solutions especially those based on the K-means algorithm. In the section three, the process of the conversion of raw production data into feature vectors convenient for data mining analysis was described. Section four presents the original methodology that was proposed in order to create data clusters that describe separated production variants created against selected KPI. The proposed approach has been proved by simulation. The final conclusions are given in section five.

2 Production Data Clusterisation Based on K-Means Algorithm

The production data analysis in the case of multi-variant discrete production is very complicated and often even impossible due to the complexity of a process. Instead, a repetitive comparison of similar operations that are done by production can be performed. The authors focus on the work cell levels that are standard components of a production line according to the production unit classification proposed by the ISA95 norm. The similarity of operations, which are defined as the production variants, not only permits the general behaviour of a work cell to be observed but can also indicate the best technology to be selected for the production and also permits KPI to be assessed. In this case, modern diagnostics and data mining methods [5] offer assistance. One of the main problems in such an analysis is unknown dependency between different production variants and related KPI. On the one hand, the same product can be produced in different ways, on the other hand the same operation can be used for different products. For this reason, tracking can't be based solely on technical product specifications, but has to include the observation of actual production operations.

Taking into account the reasons mentioned above, the authors focused their research on the problem of work cell level production data tracing and its clustering based on a K-means [6] analysis. K-means clustering is based on finding the centroids for each cluster. A centroid is an artificial object that represents the mean of the cluster. As was shown in [7] in the case of the objects represented in the Euclidean space, the centroid c_i of the i_{th} cluster C_i, can be calculated as (1):

$$c_i = \frac{1}{m_i} \sum_{x \in C_i} x \qquad (1)$$

where all objects that belong to the cluster C_i are instances of feature vector and are denoted as x and m_i is the number of objects in cluster C_i. The dimension of the centroid c_i is the same as for all of the feature vectors x and c_i is placed in the same Euclidean space. However, centroids are contractual points and in reality do not exist among the objects in the analysed data set.

In order to use the K-means algorithm, it is necessary to know k, which is the number of clusters. In the first step of the algorithm, all points are assigned to the nearest centroid from k centroids that are randomly selected from the data set during initialisation, after which the points are assigned to the nearest centroids and the centroids are then updated. In the next step, points are reassigned to the nearest centroid and again the centroids are updated and the points reassigned. When the K-means algorithm terminates because no more changes occur, the centroids have identified the natural grouping of points [6].

The sum of the distances between the c_i centroid and all points x that belong to cluster C_i can be used as a measure of the purity of the cluster. It is denoted as SSE_{Ci} Since, SSE refers to the sum of the squared error that characterises the distance metric between each point in the cluster and its associated centroid. In the case when the Euclidean space is used, the sum of the squared errors SSE_{Ci} for every i_{th} cluster can be calculated as (2):

$$SSE_{C_i} = \sum_{x \in C_i} dist(c_i, x)^2 \qquad (2)$$

where dist is the standard Euclidean distance between two objects in the Euclidean space. For each cluster the mean is calculated based on the accumulated points and the number of points in that cluster.

The sum of the squared errors (SSE) for all of the clusters can be used to assess the overall quality of the clustering. The SSE is calculated as the total sum of all of the squared errors in all of the clusters as (3):

$$\text{SSE} = \sum\nolimits_{i=1}^{k} \text{SSE}_{C_i} = \sum\nolimits_{i=1}^{k} \sum\nolimits_{x \in C_i} \text{dist}(\mathbf{c}_i, \mathbf{x})^2 \qquad (3)$$

where k is the number of clusters.

K-means clustering is one of the data mining techniques that is widely used to extract useful information from datasets. The K-means clustering method is also used to determine the fault and alarm conditions [8] or for energy efficiency analysis [9, 10]. The concepts of adaptive fuzzy-K-means clustering [11] and belongingness are applied to produce a more adaptive clustering [12]. In order to obtain better performance, some solutions are realised as embedded real-time systems [13] that use Field Programmable Gate Arrays [14]. There many different methods for solving K-means cluster [6–8, 10, 11] but they do not guarantee an optimal solution in multi-variant production.

Is not always easy to indicate the best value for k. Clustering algorithms usually require the user to specify the number of clusters. However, in a dynamic environment, determining the correct k is very difficult and therefore it is important to search for solutions to automatically determine k. The X-means algorithm [15] searches many values of k and scores each clustering model with the best Bayesian Information Criterion [16] score on the data. In the Minimum Description Length [17] principle, the description length is a measure of how well the data are fit by the model. The algorithm starts from a large value for k and reduces k when a choice reduces the description length. After every reduction of k, the K-means algorithm is used to optimise the model to fit to the data.

G-means [18] determines an appropriate k by using a statistical test to decide whether to split a K-means centre into two centres. That algorithm uses the K-means with increasing k until the test accepts the hypothesis that the data assigned to each K-means centre is Gaussian. Kernel clustering and spectral clustering methods produce nonlinear separating hypersurfaces among clusters. Kernel clustering allows data to be implicitly mapped into a high-dimensional space. Computing the linear partitioning in this space results in a nonlinear partitioning in the input space. This method has been applied in many solutions such as Kernel-based fuzzy clustering [19], the self-adaptive kernel clustering algorithm [20] or automatic kernel clustering methods [21].

Unfortunately all of analysed data mining approaches and methods have a surprisingly high degree of complexity and also need large training sets in order to obtain reliable results. This is an essential drawback in the case of mass-customised manufacturing. The authors decided to start research on a new approach to production data clustering algorithm that handles both small training sets and the occasional errors that are typical in the case of mass-customised production. The result of the work is as an extended K-means clustering algorithm that is based on the automatic estimation of the k value. The data preparation process is presented in the Sect. 3 and actual algorithm in the Sect. 4.

3   Production Data Selection and Feature Vector Definition

In order to start any effective data mining algorithm the raw process data has to be converted into a form that is convenient for the selected approach. Since the authors use an algorithm that is based on K-means, it is necessary to define the *feature vector*, which gathers the collections of parameters that are related to the production cycle of a given work cell and that can be observed and recorded in successive production cycles. In most discrete production systems, the start of the production cycle can easily be identified by the proper control signal. The authors assume that such information is also available in the examined case and that only one KPI needs to be analysed.

The authors limit the KPI analysis to the operations that are performed by the work cell. In most cases, such an assumption is sufficient because even in the case of external causes such as the quality of raw materials or semi-finished products, the control system must be able to counteract the loss of production reproducibility. Therefore, in the case of authors' method, it was proposed that a detailed analysis of all raw materials that arrive at the input of the production process be replaced with an analysis of the operations that are performed by the control system that operates in a given work cell. The input data set for analysis is composed of the working time of the individual executive circuits that are responsible for particular operations. At this stage it is important to use an engineering knowledge for the relationship between the work of the individual devices that are controlled by the control system and the KPI being analysed. The actual selection of traced executive circuits is made not with regard to a product that occurs in many variants, but to a process that is defined by the constant characteristics of the work cell (drilling, grinding, etc.). The time of device operating per production cycle for n tracked devices will be denoted as $T_1, T_2 \ldots T_n$.

Then, the *feature vector* analysed for each production cycle is formed as (4):

$$V = KPI, T1, T2 \ldots, Tj, \ldots Tn \tag{4}$$

where KPI is the selected Key Performance Identifier and T_j is the operating time for the production cycle for j^{th} of n traced devices. Since, selected KPI can be represented in different domain than time, its value should be normalised by multiple factor b that can be adjusted during learning process (5).

$$\mathbf{V} = [b * KPI_i, T_1, T_2 \ldots T_n] \tag{5}$$

During the learning phase of the algorithm it is required to observe m production cycles and register them as *m* samples created according *feature vector* **V**. The data collected can be represented as matrix **X** (6) that is composed from *m* rows.

$$\mathbf{X} = \begin{matrix} b * KPI_1 & T_{11} & T_{1j} & T_{1n} \\ b * KPI_k & T_{k1} & T_{kj} & T_{kn} \\ b * KPI_m & T_{m1} & T_{mj} & T_{mn} \end{matrix} \tag{6}$$

Our goal is to create data clusters C in X that contain production cycle data described by set of samples collected according to vector V. Each cluster should contain technological cycles performed according to similar technology $T_1, T_2 \ldots T_n$ and similar KPI value. Since K-Means algorithm minimises the sum of the squared errors (SSE) for each cluster [22] the authors have decided to use SSE to assess the overall quality of the clustering for a given k. A lower value of SSE indicates a set that is more concentrated around the centroid, while a higher SSE indicates a more dispersed cluster. In considered use case k equals to number of different production variants that is unknown and has to be find. Figure 1 shows why SSE cannot be used directly to estimate k.

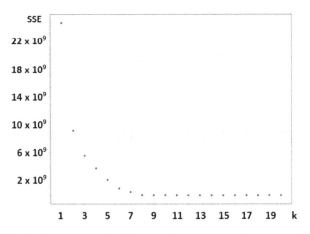

Fig. 1. SSE calculated for a rising k in test data set with eight production variants.

When trying different values of cluster numbers k, a monotonic non-increasing function SSE is received. When k is higher, the SSE will be equal or lower than k and it will eventually reach 0 when k is equal to the number of samples in the data set. The example trend of SSE (k) that was calculated for the data set collected during one of tracing of different production variants can be seen in Fig. 1. In the following section the authors propose the algorithm that solves the problem of monotonic non-increasing relation between k and SSE and can be used for the estimation of number of technological variants related with given KPI.

4 KPI Oriented Process Data Clustering in Multi-variant, Discrete Production

Since the authors are interested in finding of the natural clusters that reflect the repetitive behaviour of a work cell, it is necessary to find a proper measure of clustering quality. The authors propose the approach that examines data density represented by weighted sum of SSE for each cluster for different possible values for k parameter that is used by K-Means algorithm. The verification process starts from k = 1 and ends on k = m, where m is equal to the number of production cycles observed in the learning phase.

For each ith run of K-Mean clustering (the authors use i as the value for k in subsequent runs of clustering algorithm) it is needed to verify the weighted data density represented by SSE. Therefore, for each ith run of the algorithm the authors receive i clusters. The idea of the calculation of the weighted data density can be simplified by representing data clusters using $n + 1$ dimensional hyperballs that contain observed *feature vectors* (n is the number of observed executive circuits of a work cell, $n + 1$ element is KPI of interests after its normalisation by factor b). For a given i^{th} run, it is possible to create one major hyperball and one minor hyperball for each cluster.

The major hyperball contains all of the data covered by i^{th} cluster. The radius of the major hyperball was denoted by R_i and can be calculated in the n-dimensional Euclidean space. R_i is equal to the maximum error value that is calculated for all of the samples that belong to the ith cluster C_i. All sample data that belongs to i^{th} cluster are inside of the major hyperball. To calculate its radius denoted as R_i, the Euclidian distance between all of the samples and the centroid c_i of the ith cluster C_i has to be calculated, than the maximum value has to be selected (7):

$$R_i = \max_{1 \le j \le |C_i|} \left(\text{dist}\left(c_i, x_j\right) \right) \tag{7}$$

Where $|C_i|$ is the number of the samples in the cluster C_i, c_i is a centroid of the cluster and x_j is the j-th sample that belongs to the cluster C_i.

Then, it is possible to create the minor hyperball for the ith cluster. The surface of minor hyperball creates the sphere that can be used to represent average distance between samples and the centroid of ith cluster (C_i). The radius of minor hyperball is denoted r_i and can be calculated as (8):

$$r_i = \frac{1}{|C_i|} * \sum_{j=1}^{|C_i|} \text{dist}\left(c_i, x_j\right) \tag{8}$$

The idea behind the calculation of R_i and r_i for major and minor hyperballs is illustrated graphically in Fig. 2. In the case when i^{th} cluster contains samples from one technological variant, samples should be concentrated around one kernel of the hyperball (as is illustrated on the right side of Fig. 2) and the distance from its centre should correspond only to the measurement errors and the instability of the process parameters. In the case in which cluster contains samples that belongs to more than one technological variant, they will be concentrated around two or more kernels (as is illustrated on the left side of Fig. 2).

Therefore, the smaller value of r_i/R_i ratio indicates that the samples are concentrated around rather one than many kernels determined by the different technology variants. Of course this radio depends on the repeatability of given technology but its minimum can be used for estimation of k for K-means clustering.

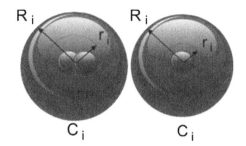

Fig. 2. Clusters with a multi-kernel and a single-kernel.

The r_i/R_i ratio takes its values in the range of $(0;1]$. When the points inside the ball are centred round a single kernel, the quality measure given by the ratio will be smaller than in the case of cluster that contain two or more kernels.

Since the r_i/R_i ratio shows only the purity of ith cluster, the function of the total Quality of K-means clustering for given k, $Q(k)$ has to be calculated as a weighted sum of the r_i/R_i ratios. Each purity measure is multiplied by the $r_i/|C_i|$ ratio that is proportional to the average distance between the sample and the centroid and that is inversely proportional to the size of the ith cluster. Since the aim is to look for the smallest possible number of clusters, the $1/k$ scaling factor is used (9):

$$Q(k) = \frac{1}{k} * \sum_{i=1}^{k} \left(\frac{r_i}{R_i} * \frac{r_i}{|C_i|} \right)$$ (9)

Finally, the desired value of the parameter k is the minimum value of the function Q and reflects the number of clusters in the case of the best possible division of samples (10):

$$K = \mathrm{argmin}_k \left(Q(k) \right)$$ (10)

Calculated K denotes the number of clusters that fits to the best representation of production samples collected during learning phase of the algorithm. Based on the value of K data can be separated by last run of K-mean algorithm. This final run is performed after removal of samples affected by production disturbance (outliers). Finally, proposed algorithm produce the set of reference clusters that reflect the dependency between technology of production and given KPI and can be feather applied in next steps of KPI analysis.

In order to verify the proposed methodology, the authors prepared a number of simulation data sets. Although the production parameters were generated by random function with Gaussian distribution (that reflect instability typical for discreet production) for each data set, each scenario was based on a use case that was taken from an actual work cell. The distribution of the KPI and operating times are shown in Table 1.

Table 1. Mean values and standard deviations of the KPI and nine features

	KPI (l)		Feature 1 (ms)		Feature 2 (ms)		Feature 3 (ms)		Feature 4 (ms)	
	Mean	StDev	Mean	StDev	Mean	StDev	Mean	StDev	Mean	StDev
Variant 1	17.50	0.29	20496.61	1470.67	1518.06	291.83	1499.02	296.53	3523.99	287.40
Variant 2	20.44	1.95	20480.57	1463.45	2952.64	286.95	3503.35	304.88	4516.35	304.13
Variant 3	24.86	0.86	24995.95	588.95	2055.26	275.70	3044.24	287.44	29971.79	3026.40
All	21.08	3.37	22094.85	2408.63	2172.78	675.52	2619.60	906.38	12662.25	12376.35

	Feature 5 (ms)		Feature 6 (ms)		Feature 7 (ms)		Feature 8 (ms)		Feature 9 (ms)	
	Mean	StDev	Mean	StDev	Mean	StDev	Mean	StDev	Mean	StDev
Variant 1	3448.39	849.76	758.38	146.52	6988.85	597.21	0.00	0.00	0.00	0.00
Variant 2	6572.10	893.21	1750.37	150.40	8992.75	581.16	5336.41	451.78	0.00	0.00
Variant 3	5614.09	796.61	1246.46	136.98	3963.66	571.32	5721.96	430.24	1498.42	277.26
All	5184.50	1539.56	1254.49	452.27	6719.57	2161.16	3653.58	2617.88	508.55	737.70

One of tests that was performed is shown in Figs. 3, 4, 5 and 6. In this use case, the KPI reflects the energy consumption of the work cell [in litres of compressed air] and the next nine parameters of the feature vector, which reflect the operating time of nine devices [in milliseconds]. The simulation data were generated for three hundred production cycles that were performed according to three production variants. Since different devices are used in a different manner for each of production variant, the dependencies between a single device and the KPI are not trivial as is presented in Figs. 3, 4 and 5.

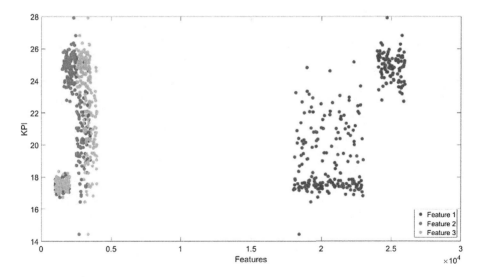

Fig. 3. The dependence between the KPI and features 1–3.

The authors calculated the Q value according to Eq. (9). The dependency between the quality of clustering Q and the number of clusters is presented in Fig. 6. The Q(K) has its minimum for K = 3. Therefore, three is the number of production variants that were found by the authors' method.

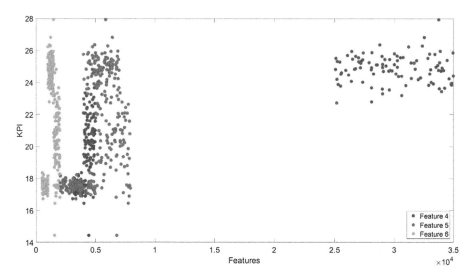

Fig. 4. The dependence between the KPI and features 4–6.

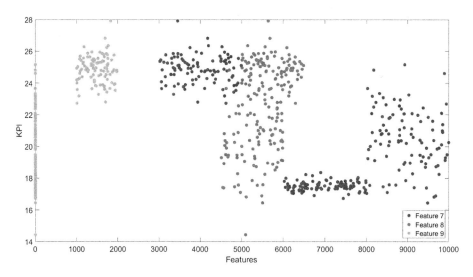

Fig. 5. The dependence between the KPI and feature 7–9.

The K value that was found is consistent with the assumptions on which the simulation data was prepared. The above experiment was repeated in more than a dozen use case scenarios. Each time, the minimum for the Q function reflected the number of production variants as was assumed in the Eq. (10).

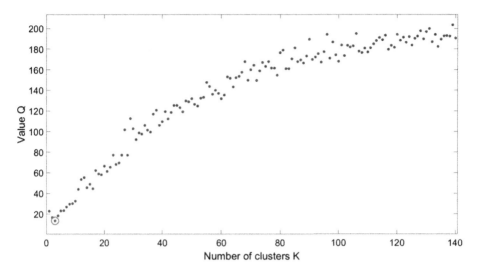

Fig. 6. The Q value calculated according to Eq. (9).

5 Conclusions

The authors have presented an original algorithm that can be used for automatic clustering of production data in the case of multi-variant discrete production when the number of variants of productions is unknown. Such clustering is necessary for advanced KPI analysis in the case of multi-variant production and can be used in the case of mass-customised manufacturing. The proposed method supports the estimation of the number of variants. It can be applied on the work cell level and does not require explicit knowledge about the number of variants that are being produced. The heart of the proposed method is a new algorithm that can support K-means clustering by automatically estimating the value of k parameter. The dependencies that were derived theoretically in the form of patterns (9) and (10) were verified and proven by a number of simulations. The simulation results fully confirm the method for finding the number of production variants proposed by Eqs. (9) and (10). The authors believe that the presented approach can support KPI analysis by the automatisation of the creation of reference data clusters. Finally, it can increase the possibility of the application of KPI analysis in the case of multi-variant discrete production.

Acknowledgements. This work was supported by Polish National Centre of Research and Development from the project ("Knowledge integrating shop floor management system supporting preventive and predictive maintenance services for automotive polymorphic production framework" (grant agreement no: POIR.01.02.00-00-0307/16-00). The project is realised as Operation 1.2: "B + R sector programs" of Intelligent Development operational program in years 2014-2020 and co-financed by European Regional Development Fund.

References

1. Hawkins, D.M., Zamba, K.D.: Statistical process control for shifts in mean or variance using a changepoint formulation. Technometrics **47**, 164–173 (2005)
2. Bornschlegl, M., Bregulla, M., Franke, J.: Methods-energy measurement – an approach for sustainable energy planning of manufacturing technologies. J. Clean. Prod. **135**, 644–656 (2016)
3. Shipp, S.S., Gupta, N., Lal, B., Scott, J.A., Weber, C.L., Finnin, M.S., Blake, M., Newsome, S., Thomas, S.: Emerging global trends in advanced manufacturing. DTIC Document (2012)
4. Cupek, R., Ziebinski, A., Huczala, L., Erdogan, H.: Agent-based manufacturing execution systems for short-series production scheduling. Comput. Ind. **82**, 245–258 (2016)
5. Yeh, W.-C., Jiang, Y., Chen, Y.-F., Chen, Z.: A new soft computing method for K-Harmonic means clustering. PLoS ONE **12**, e0169707 (2017)
6. MacQueen, J.: Some methods for classification and analysis of multivariate observations. In: Presented at the Proceedings of the Fifth Berkeley Symposium on Mathematical Statistics and Probability (1967)
7. Tan, P.-N., Steinbach, M., Kumar, V.: Introduction to Data Mining. Pearson Addison Wesley, Boston (2006)
8. Zhang, W., Ma, X.: Simultaneous fault detection and sensor selection for condition monitoring of wind turbines. Energies **9**, 280 (2016)
9. Cupek, R., Drewniak, M., Zonenberg, D.: Online energy efficiency assessment in serial production - statistical and data mining approaches. Presented at the June 2014
10. Cupek, R., Ziebinski, A., Zonenberg, D., Drewniak, M.: Determination of the machine energy consumption profiles in the mass-customised manufacturing. Int. J. Comput. Integr. Manuf. 1–25 (2017)
11. Sulaiman, S., Isa, N.A.M.: Adaptive fuzzy-K-means clustering algorithm for image segmentation. IEEE Trans. Consum. Electron. **56**, 2661–2668 (2010)
12. Khamassi, I., Sayed-Mouchaweh, M., Hammami, M., Ghédira, K.: Discussion and review on evolving data streams and concept drift adapting. Evolving Syst. (2016)
13. Saegusa, T., Maruyama, T.: An FPGA implementation of real-time K-means clustering for color images. J. Real-Time Image Proc. **2**, 309–318 (2007)
14. Ziębiński, A., Świerc, S.: The VHDL implementation of reconfigurable MIPS processor. In: Cyran, K.A., Kozielski, S., Peters, J.F., Stańczyk, U., Wakulicz-Deja, A. (eds.) Man-Machine Interactions, pp. 663–669. Springer, Berlin (2009). https://doi.org/10.1007/978-3-642-00563-3_69
15. Pelleg, D., Moore, A.W.: X-means: extending k-means with efficient estimation of the number of clusters. Presented at the ICML (2000)
16. Kass, R.E., Wasserman, L.: A reference Bayesian test for nested hypotheses and its relationship to the Schwarz criterion. J. Am. Stat. Assoc. **90**, 928–934 (1995)
17. Bischof, H., Leonardis, A., Selb, A.: MDL principle for robust vector quantisation. Pattern Anal. Appl. **2**, 59–72 (1999)
18. Hamerly, G., Elkan, C.: Learning the k in k-means. Adv. Neural. Inf. Process. Syst. **16**, 281 (2004)
19. Graves, D., Pedrycz, W.: Kernel-based fuzzy clustering and fuzzy clustering: a comparative experimental study. Fuzzy Sets Syst. **161**, 522–543 (2010)
20. Pu, Y.-W., Zhu, M., Jin, W.-D., Hu, L.-Z.: An efficient similarity-based validity index for kernel clustering algorithm. In: Wang, J., Yi, Z., Zurada, J.M., Lu, B.-L., Yin, H. (eds.) ISNN 2006. LNCS, vol. 3971, pp. 1044–1049. Springer, Heidelberg (2006). https://doi.org/10.1007/11759966_153

21. Das, S., Abraham, A., Konar, A.: Automatic kernel clustering with a multi-elitist particle swarm optimization algorithm. Pattern Recogn. Lett. **29**, 688–699 (2008)
22. Steinley, D., Brusco, M.J.: Initializing K-means batch clustering: a critical evaluation of several techniques. J. Classif. **24**, 99–121 (2007)

Estimation of the Number of Energy Consumption Profiles in the Case of Discreet Multi-variant Production

Rafał Cupek[1]([⊠]), Adam Ziębiński[1], Marek Drewniak[2], and Marcin Fojcik[3]

[1] Institute of Informatics, Silesian University of Technology, Gliwice, Poland
{Rafal.Cupek,Adam.Ziebinski}@polsl.pl
[2] AIUT Sp. z o.o., Gliwice, Poland
mdrewniak@aiut.com.pl
[3] Western Norway University of Applied Sciences, Forde, Norway
Marcin.Fojcik@hvl.no

Abstract. An analysis of energy efficiency at the machine level has become an important element of contemporary control and measurement systems. The results of such an analysis can not only be used as information about energy consumption but can also be used for predictive maintenance. The authors present a novel approach that is dedicated to the classification of machine-level energy efficiency that can be applied in the case of multivariate production. The concept was proven by research on the use case of an assembling station that consisted of a number of pneumatic devices. The proposed approach does not require detailed analysis about the production technology that is being used and also does not require additional knowledge about the order of the production variants that are being executed. The algorithm is based on observing the behaviour of the machine and then clustering the machine cycles that are observed.

Keywords: Energy efficiency · Predictive maintenance
Production data clustering · Pneumatic systems

1 Introduction

Energy efficiency of production process is becoming more and more important factor in an economic analysis [1]. Monitoring and the control of energy-related indicators may be obtained in order to monitor and control vital, energy-related performance indicators and the beginning phase of improving energy efficiency [2]. Such approach enable to verify predefined Key Performance Indicators (KPI) and serve as a decision support to enhance the energy management in manufacturing plants. A classical analysis based on Value Stream Mapping (VSM) approach is now extended to Energy Value Stream Mapping (EVSM) methodology [3], which integrates data related to energy consumption with VSM analysis.

Although, the need for monitoring and control the energy consumption [4] is well understood by the industry, however, there is a lack of tools able to precisely determine

© Springer International Publishing AG, part of Springer Nature 2018
N. T. Nguyen et al. (Eds.): ACIIDS 2018, LNAI 10752, pp. 674–684, 2018.
https://doi.org/10.1007/978-3-319-75420-8_63

a contribution of particular production operations executed by subsequent production operations into overall energy efficiency KPIs. Such tools are particularly necessary in the case of multi variant and flexible production where energy consumption depends strongly on production variant executed and production path selected.

The main problem with contemporary discreet production systems is the variability of the production process that is caused by the multitude of product variants that have to be performed by a given production stand. Energy consumption depends not only on the production technology but is also affected by the required technological operations that are defined for different variants of a product. The dependency between the product version and energy consumption is not trivial. Sometimes, different versions of a given product require exactly the same set of technological operations, while in other cases, the same variant of the product can be only be achieved by using different operations. Therefore, the authors propose a new approach for defining energy efficiency in discrete, multi-variant production. A comparative analysis of energy efficiency should not be based on the similarity of products being manufactured but on the similarity of technological operations that are performed during the production. The set of similar technological operations performed by given production stand that is reflected in similar energy consumption is called an energy consumption profile.

The authors propose a new approach for analysing energy consumption profiles that is based on modern diagnostics, data mining methods [5] and tools that are dedicated to the creation of a pneumatic air consumption model that takes into account the individual production technology and the production path being executed. The proposed model is dedicated for production cell level (process step) in discrete manufacturing but received information can be aggregated across production line level (process segment), up to whole production plant (overall energy efficiency KPI). The model preparation is based both on energy consumption meters that is performance indicator and controlled by PLC pneumatic actuators that form aggregate consumption of compressed air by the production cell. The research part contains a case study and experimental verification of the model.

The state of the art energy efficiency analysis is presented in the second section. In section three, the selected use case and experimental test stand was described. Section four presents the experimental results of application of the original methodology that was detail described in [6] and was the subject for verification in this paper. The final conclusions are given in section five.

2 Measurement and Analysis of Energy Efficiency in Discrete Manufacturing

The environmental impact of manufacturing activities is now one of the major issues affecting industrial communities. In the European Concept, the Manufacture Sub-Platform in the description of the Strategic Research Agenda specified the optimization of consumption using energy and material-efficient technologies as one of the main research areas involved in product/service sustainability.

Manufacturing systems have become more and more complex mechanisms [7] and require the predictive detection of maintenance problems. Even minor and invisible technical problems can significantly cause deterioration in production efficiency. Progressive faults [8] can be caused by aging or the deterioration of the operating environment. Often progressive faults are noticed too late. Slow changes can be especially seen in greater energy consumption and therefore the timely monitoring [6] of conditions and modern methods of fault diagnosis are required. Time domain based signal processing methods [9] are popular for diagnosing faults. More sophisticated methods such as intelligent condition monitoring and fault diagnosis methods generally require more complex computational methods [10] such as fuzzy models [11] or data mining methods.

The use of machine learning methods in order to detect a drift in the system parameters can be useful in reducing the delay of a diagnosis that occur in the approaches proposed in [12]. Moreover, it can be interesting to provide the proposed approach with an adaptive capacity in order to learn new fault modes. Therefore, the use of machine learning in dynamic environments can be a solution to provide the classifier with an adaptive capacity [13].

Due to the lack of a thorough analysis of manufacturing processes, it is difficult to obtain data about the emissions of machines [14] or the energy density of individual processes. Most solutions are focused on local measures and do not consider a holistic view of the whole factory system [15]. The measurements that are mapped could be used to generate a consumption profile of the machines on a production line. To support accurate data machine level analysis, it is important to take into account any technological differences between particular machines and different production variants [16].

In the considered use case, the authors focus on a production station with pneumatic air actuators. Compressed Air Systems are widely used on a daily basis in even the most difficult production environments. Pneumatic systems are known for their poor energy efficiency and therefore their analysis and the optimisation of compressed air consumption is one of the main issues in factory environments [17].

The modelling and simulation of the energy and resource consumption of production machines require complicated feedback mechanisms. Some solutions are based on analytical methods and allow a theoretical summary of compressed air consumption to be calculated [18]. The analytical approach can be extended using simulation methods, which allow variabilities in the process parameters to be introduced and also take into account environmental variables [19]. The main disadvantage of the analytical approach is that it requires a very precise description of the analysed object. Using the reverse engineering process, an accurate compressed air consumption model can easily be created [20]. Online information about the summary of air consumption in conjunction with online information about the state of the endpoint devices could be used for a detailed assessment of air consumption [16].

Although an assessment of the energy efficiency of such systems can be done through a comparison of uniquely prepared production stands, such an approach is very complicated. Due to its complexity, using this process is often impossible. A repetitive comparison of the similar operations that are performed by a stand can replace this process. The similarity of operations, which are defined and grouped as production variants, permits the general behaviour of a machine to be observed. Additionally, this solution permits the actuators to be compared, the manual operations to be analysed, the influence of control algorithm on the process to be checked and the efficiency of the technology to be assessed. Moreover, it can also indicate the best technology to be selected for the production. The approach is based on a comparison of the consumption distribution between individual end-point devices, which is why it offers the possibility to prepare a more effective method for detecting anomalies, especially air leakages. Modern diagnostics and data mining methods offer assistance in this case [5]. The main problems in this analysis is the energy (including compressed air) consumption variability for which the many variants of production need to be considered.

3   Methodology and Considered Use Case

One of the problems related to the assessment of energy efficiency in multi-variant, discrete production is the hidden value of the energy efficiency profiles that can be observed on a given production stand. The authors solved this problem by applying a novel methodology for KPI analysis that is based on an extended K-Means approach, which can be applicable for performance analysis in the case of short series and multi-variant production.

According to the methodology described in [6] the authors have defined KPI as pneumatic air consumption collected from pneumatic air meter that was described by eight features that represent operating times of eight pneumatic actuators that were used on our test production cell. We collected this information in learning set as follows (1):

$$\mathbf{X} = \begin{matrix} b*V_1 & T_{11} & T_{1j} & T_{1n} \\ b*V_k & T_{k1} & T_{kj} & T_{kn} \\ b*V_m & T_{m1} & T_{mj} & T_{mn} \end{matrix} \tag{1}$$

where V is the volume of the compressed air that was consumed for a every production cycle in litres and Tx is the operating time for each of eight pneumatic actuators in milliseconds. Since there are different domains for time and air volume, air volume should be normalized by factor b. Presented in research part results were prepared with multiple factor b equal 10^4. Although the choice was arbitrary, the method was verified in order to check whether it provided the same results for the scaling factor by a larger order of magnitude and a smaller order of magnitude. Based on that, the authors did not consider the problem of the optimal normalization of the feature vector and followed an arbitrarily selected scale for normalization.

The authors have prepared a special H-K-Means operator necessary for the preparation of the feature vector were performed using the RapidMiner tool. The operator is based on the standard K-Means algorithm [21] but extends it by procedure of k estimation according based on cluster purity function Q(k) calculated as (2):

$$Q(k) = \frac{1}{k} * \sum_{i=1}^{k} \left(\frac{r_i}{R_i} * \frac{r_i}{|C_i|} \right) \tag{2}$$

where $|C_i|$ is the number of samples in cluster C_i, x_j is the j-th sample that belongs to cluster C_i. R_i (3) and r_i (4) are calculated as follows:

$$R_i = \max_{1 \le j \le |C_i|} \left(\mathrm{dist}(c_i, x_j) \right) \tag{3}$$

$$r_i = \frac{1}{|C_i|} * \sum_{j=1}^{|C_i|} \mathrm{dist}(c_i, x_j) \tag{4}$$

Finally, K is the minimum value of the function Q(k) as (5):

$$K = \mathrm{argmin}_k \left(Q(k) \right) \tag{5}$$

The operational principle of the prepared solution is presented as an algorithm in Fig. 1 and has been presented in detail in [6].

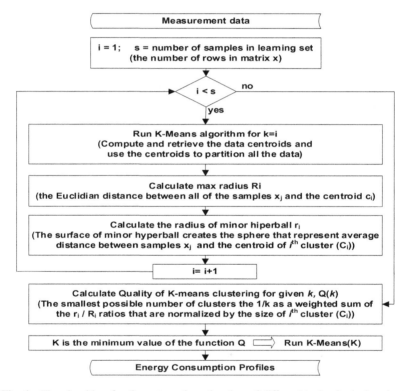

Fig. 1. The algorithm for the automatic estimation of different technological variants

4   Experimental Results

The experimental test stand called DC40 (Fig. 2) was built of actual industrial components including its metal construction, electrical equipment and pneumatic elements such as a complete compressed air preparation set, actuators and the solenoid valve module. The stand is controlled by a PLC controller from the Siemens S7-300 family, I/O modules and an HMI panel, all of which are mounted in an electrical cabinet. Finally, a flow meter is used for the pneumatic system in order to measure the instantaneous consumption of compressed air. The main tasks of the stand are the transport of the elements in a closed loop using gravity transportation, passing the elements from point to point using the pneumatic actuators or suction nozzle and lifting the elements with a pneumatic lift. The sequence of activities, which is defined as a standard work cycle, consists of seven steps, which generally alternate one after another but which can sometimes be performed simultaneously. All of the pneumatic end-point devices are controlled directly by the controller's program. The assignment between pneumatic actuators and PLC digital outputs (Q) is as follows: INDEX UP - Q20.0, INDEX DOWN - Q20.1, SLIDE FORWARD - Q20.2, SLIDE BACKWARD - Q20.3, PRE-STOP DOWN - Q20.4, STOP DOWN - Q20.5, SUCTION - Q20.6, BLOWING - Q20.7, ADDITIONAL VALVE 1 - Q21.0, ADDITIONAL VALVE 2 - Q21.1.

Two different DC40 software setups were prepared for testing purposes. Each allowed four different technological variants to be produced. During the first phase of the experiment, data was collected for 139 production cycles. Different variants of production were interspersed. Figure 2 presents a small portion of the gathered data set. The trend value shows the current compressed air flow [in litres per minute]. It can clearly be seen that it covered eight production cycles. It is far more difficult to determine that they were executed in four different production variants.

In order to convert the raw measurements into the data that is available for the clustering algorithm, they had to be transformed into the format of the feature vector. In the authors' case, there were ten dimensional feature spaces that reflected the production parameters. Nine of them were created based on the operation time (per cycle) of each of the nine pneumatic devices. The values of these nine features are given in milliseconds. The last feature describes the compressed air consumption per cycle. This feature is presented in Fig. 3 for each of the 139 cycles that were performed during this phase of the experiment. The PLC internal time that was counted during the experiment

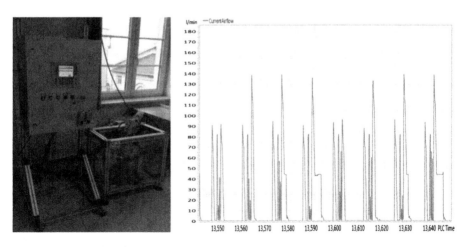

Fig. 2. DC40 experimental stand and example of air flow measurements for 8 production cycles

(in ms) is shown on the horizontal axis and the compressed air consumption for each cycle is shown as dots on the vertical axis (in litres). It is now easy to determine that the DC40 executed eight different production variants during the experiment. The gap between the left and right sections of the trend results from the fact that some of the setup operations for the DC40 were necessary between the first and second phases of production.

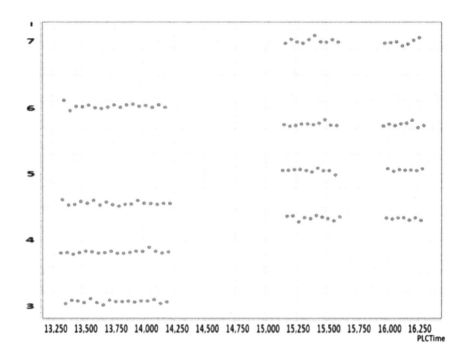

Fig. 3. Compressed air flow for the selected sample data

The average values for the features for every production variant are shown in Table 1. Since there is no natural order of the production variants, they are sorted based on increasing compressed air consumption.

The RapidMiner setup is shown in Fig. 4. A new classifying operator that performs the extension of the K-Means clustering algorithm is visible as the H-K-Means operator (in green). Then, an experiment was performed to determine the number of different variants of production and to build the reference clusters based on these.

The intermediate results of the clustering algorithm were connected to an operator's outputs. The value of the Q1 function and the number of outliers for each test that was performed in a range from 1 to 20 for k are presented in Fig. 5.

The mechanism for creating the outliers can be observed (denoted as Out in Fig. 5). There are eight production variants as expected from Table 1 but Q takes the minimum value for K = 9, but there is one outlier and therefore the final number of production variants that were found is K2 = 9 − 1 = 8. The authors verified the input data. The

Table 1. The average values of the features that were measured during the 8 variants of production (Var_1, ..., Var8)

	Var_8	Var_7	Var_6	Var_1	Var_4	Var_3	Var_2	Var_5
Runs	17	18	17	18	17	18	17	17
Q20_0_tc	1041.9	1082.0	1736.1	1057.2	1733.4	1747.3	1042.4	1750.7
Q20_1_tc	908.5	908.4	9896.4	918.4	9939.4	9726.8	910.6	9806.1
Q20_2_tc	0.0	0.0	1612.1	0.0	1632.1	1622.3	0.0	1628.2
Q20_3_tc	0.0	0.0	6219.3	0.0	6247.6	6061.0	0.0	6151.2
Q20_4_tc	311.1	322.2	1411.1	282.2	1515.9	1396.7	310.2	1502.4
Q20_5_tc	244.2	288.8	3379.7	297.3	3334.8	3289.9	283.4	3301.9
Q20_6_tc	0.0	0.0	315.1	0.0	307.8	307.9	0.0	307.5
Q20_7_tc	99.4	100.2	0.0	96.1	0.0	0.0	89.6	0.0
Q21_0_tc	0.0	821.6	0.0	1795.4	770.0	1780.2	3802.4	3804.1
cycle_air	30772.6	38270.9	43407.1	45667.3	50632.5	57606.3	60352.1	70156.8

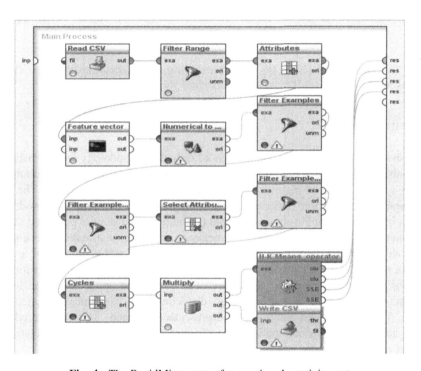

Fig. 4. The RapidMiner setup for creating the training set

longer duration of the execution of a cycle was caused by mechanical problems with one of the pneumatic actuators. The feature of the automatic detection and elimination of outliers from the algorithm was also confirmed in the subsequent experiments. Although the small perturbations in the execution of the process did not affect the reference data, they were automatically rejected from the training set.

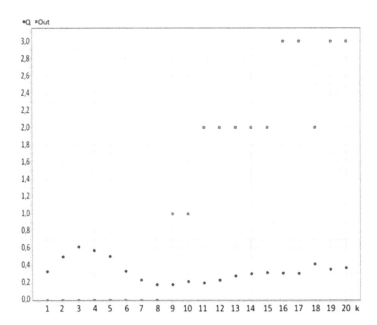

Fig. 5. The value of the quality Q(k) function (in blue) and the number of outliers Out for K-Means with given k (in red) for eight technological variants (Color figure online)

5   Conclusions

The authors have presented an application of original methodology for estimation of the number of energy consumption profiles in the case of discreet multi-variant production systems. A novel algorithm was proposed and proven by research. The main advantage of the proposed algorithm is its ability to find the number of production variants in the case of multi-variant discrete production. The authors presented the use of the new algorithm that can support K-means clustering by automatically estimating the k value. The proposed method can be applied on the production station level. It does not require explicit knowledge about the number of variants that are being produced or about the production technology. The considered application field for the algorithm is automotive assembly station that consists of a number of pneumatic devices. The research part shows that the presented approach can support predictive maintenance solutions by automatically creating reference clusters that characterise the different profiles of production.

The biggest advantage of the proposed method is its robustness in the case of a minor process instability that is registered in the training set. In such a case, the production cycles that have the outstanding parameters are simply removed and do not affect the determination of the variants of production. The research results shown in this paper show that the proposed method can work well with both an unknown technology and an unknown number of production variants with relatively small training sets that can contain sporadic errors.

Acknowledgements. This work was supported by Polish National Centre of Research and Development from the project ("Knowledge integrating shop floor management system supporting preventive and predictive maintenance services for automotive polymorphic production framework" (grant agreement no: POIR.01.02.00-00-0307/16-00). The project is realized as Operation 1.2: "B + R sector programs" of Intelligent Development operational program in years 2014–2020 and co-financed by European Regional Development Fund.

References

1. Garza-Reyes, J.A.: Lean and green – a systematic review of the state of the art literature. J. Cleaner Prod. **102**, 18–29 (2015)
2. Cupek, R., Ziebinski, A., Zonenberg, D., Drewniak, M.: Determination of the machine energy consumption profiles in the mass-customised manufacturing. Int. J. Comput. Integr. Manuf. 1–25 (2017)
3. Bornschlegl, M., Bregulla, M., Franke, J.: Methods-energy measurement – an approach for sustainable energy planning of manufacturing technologies. J. Cleaner Prod. **135**, 644–656 (2016)
4. Brossog, M.P., Bornschlegl, M., Franke, J.: Reducing the energy consumption of industrial robots in manufacturing systems. Int. J. Adv. Manuf. Technol. **78**, 1315–1328 (2015)
5. Yeh, W.-C., Jiang, Y., Chen, Y.-F., Chen, Z.: A new soft computing method for K-harmonic means clustering. PLoS ONE **12**, e0169707 (2017)
6. Cupek, R., Ziebinski, A., Drewniak M., Fojcik M.: Improving KPI based performance analysis in discrete, multi-variant production. In: 10th Asian Conference on Intelligent Information and Database Systems, ACIIDS 2018, Dong Hoi City, Vietnam, 19–21 March 2018
7. Cupek, R., Ziebinski, A., Huczala, L., Erdogan, H.: Agent-based manufacturing execution systems for short-series production scheduling. Comput. Ind. **82**, 245–258 (2016)
8. Shen, C., Wang, D., Kong, F., Tse, P.W.: Fault diagnosis of rotating machinery based on the statistical parameters of wavelet packet paving and a generic support vector regressive classifier. Measurement **46**, 1551–1564 (2013)
9. Wang, D., Tse, P.W., Tsui, K.L.: An enhanced Kurtogram method for fault diagnosis of rolling element bearings. Mech. Syst. Sig. Process. **35**, 176–199 (2013)
10. Dong, S., Chen, L., Tang, B., Xu, X., Gao, Z., Liu, J.: Rotating machine fault diagnosis based on optimal morphological filter and local tangent space alignment. Shock Vib. **2015**, 1–9 (2015)
11. Simani, S., Farsoni, S., Castaldi, P.: Residual generator fuzzy identification for wind TurbineBenchmark fault diagnosis. Machines **2**, 275–298 (2014)
12. Toubakh, H., Sayed-Mouchaweh, M.: Hybrid dynamic classifier for drift-like fault diagnosis in a class of hybrid dynamic systems: application to wind turbine converters. Neurocomputing **171**, 1496–1516 (2016)
13. Khamassi, I., Sayed-Mouchaweh, M., Hammami, M., Ghédira, K.: Discussion and review on evolving data streams and concept drift adapting. Evolving Systems (2016)
14. Steiner, R., Frischknecht, R.: Metals processing and compressed air supply. Ecoinvent report (2007)
15. ISO 55000: Asset management (2014)
16. Cupek, R., Drewniak, M., Zonenberg, D.: Online energy efficiency assessment in serial production - statistical and data mining approaches, June 2014

17. Harris, P., Nolan, S., O'Donnell, G.E., Meskell, C.: Optimising compressed air system energy efficiency - the role of flow metering and exergy analysis. In: Nee, A.Y.C., Song, B., Ong, S.-K. (eds.) Re-engineering Manufacturing for Sustainability, pp. 129–134. Springer Singapore, Singapore (2013)
18. Yang, A., Pu, J., Wong, C.B., Moore, P.: By-pass valve control to improve energy efficiency of pneumatic drive system. Control Eng. Pract. **17**, 623–628 (2009)
19. Harris, P.G., O'Donnell, G.E., Whelan, T.: Modelling and identification of industrial pneumatic drive system. Int. J. Adv. Manuf. Technol. **58**, 1075–1086 (2012)
20. Cupek, R., Folkert, K., Huczala, L., Zonenberg, D., Tomczyk, J.: End-point device compressed air consumption analysis by reverse engineering algorithm, November 2013
21. MacQueen, J.: Some methods for classification and analysis of multivariate observations. In: Presented at the Proceedings of the Fifth Berkeley Symposium on Mathematical Statistics and Probability (1967)

Online Monitoring System of the Enrichment Factory Input Ore Flows Quality on the Base of Temporal Model

Viktor Toporov[1], Valery Axelrod[2], and Ualsher Tukeyev[3(✉)]

[1] Systemotechnika Company, Amangeldy Str. 40/112,
050000 Almaty, Kazakhstan
viktor.toporov@gmail.com
[2] TST-16 Company, Amangeldy Str. 40/112, 050000 Almaty, Kazakhstan
gsns2016com.va@gmail.com
[3] Information Systems Department, Al-Farabi Kazakh National University,
Al-Farabi Av., 71, 050040 Almaty, Kazakhstan
ualsher.tukeyev@gmail.com

Abstract. This article describes the solution of the problem of determining the ore belonging to a particular quarry in the multi-flow technological process of ore crushing at the ore enrichment fabric of a mining and processing plant. The paper proposes a new temporal model for solving this problem and describes its implementation in the form of an online monitoring system of ore quality and it belonging to a concrete quarry.

Keywords: Online · Monitoring · System · Ore · Flow · Enrichment
Factory · Temporal · Model

1 Introduction

The technology for processing of ore raw materials entering to enrichment factory (EF) of the ore mining and processing plant (MPP) starts with its unloading from the railway wagons to the receiving bunkers of the enrichment factory and crushing to a consistency acceptable for the enrichment process. At this stage of ore preparation, one of the most important production and analytical tasks that determine the efficiency of the follow technological parts is an on-line assessment of the volume and quality of the ore delivered by rail transport in real time from each particular quarry.

It should be emphasized that obtaining such estimates on-line in the context of an appropriate diagnostic input control system is not an easy task. This is due, first of all, to the difficulties of conducting reliable and credible measurements in real time of quality indices and volume-weight characteristics of a large-lump ore mass flow (the size of a piece up to 1200 mm) at the time it enters the enrichment factory. The use of laboratory data and surveying measurements made in a quarry is also of little use because of the large intervals between measurements and differences in assessments of the quality of ore from mining and enrichment factories [1].

Thus, the existing arsenals of methodology, software and hardware does not allow reliable measurements of the qualitative and quantitative characteristics of the input ore

© Springer International Publishing AG, part of Springer Nature 2018
N. T. Nguyen et al. (Eds.): ACIIDS 2018, LNAI 10752, pp. 685–695, 2018.
https://doi.org/10.1007/978-3-319-75420-8_64

flows of the multi-flow enrichment factories in real time. In this connection, within the framework of this article, the possibility of determining the qualitative-quantitative characteristics of the input streams from the corresponding measurements of the output flow of ore crushing using the paradigm of temporal logic was investigated. Temporal logics are a powerful means of describing events and have great expressive capabilities in representing real time structures. The main elements of such logics are considered in [2–4].

With regard to the assessment of the volume and quality of ore pit flows, taking into account the temporal features of the process of entering and processing the ore mass for large crushing, it is possible to restore the characteristics of these ore-streams on the basis of the chronology of the unloading process of mobile units, and also on the basis of chronology and measurements of qualitative and quantitative indices of the resulting ore flow after a large crushing. Moreover, the temporal model takes into account the peculiarities of the solution of such a task within the framework of the corresponding monitoring system, namely:

- the need to obtain a solution in conditions of time constraints determined by a real controlled process;
- the need to take into account the time factor (dependencies) in describing the problem situation and in the process of finding a solution;
- impossibility of obtaining all objective information necessary for the decision, and, in this connection, the use of subjective, expert information;
- the need to use a significant amount of data that changes with time (sensor readings, values of control parameters performed by operators of actions, etc.).

The main scientific contribution in this paper is the new temporal description of ore preparation in the multi-flow technological process of ore crushing. That new approach allows to decide the problem of defining ore quality in real-time and increase the management quality in chain "quarry - enrichment fabric - plant".

Below, in Sect. 2 is described new approach for formal description of the ore preparation, in Sect. 3 is presented the implementation of proposed approach as a real-time monitoring system, in Sect. 4 is described results of experimental studies.

2   Temporal Description of Ore Preparation

2.1   Set of Temporary Model Primitives

As it was noted, it is quite problematic to perform direct reliable measurements of the qualitative and quantitative characteristics of the input ore flows of the EF in real time and with the accuracy necessary for practice. In this connection, we consider the possibility of obtaining these characteristics by calculating them by the corresponding measurements of the output flow. It is clear that without the additional information on the moments of initiation of the flows at the input of the concentrating redistribution, the dynamics of their motion along the crushing complex, and the chronology of the measurements of the integral output flow after crushing, this problem does not have a unique solution.

The composition of such temporal information ensuring uniqueness of the solution of the task of the incoming control of the volume and quality of the ore mass entering the concentrator is determined by the technological scheme of crushing. In accordance with this scheme, the ore of gross production with the size of pieces of not more than 1200 mm, arriving by rail (dump trucks with a lifting capacity of 105 tons) from various quarries is fed to the receiving bunkers of the large crushing body.

The ore from each receiving bunker is crushed to the size of 350-400 mm, and then fed to the second stage of crushing (not shown in the diagram), where it is crushed to a fineness of not more than 22 mm. The product of the second stage of crushing is fed to belt conveyors for further processing. Thus, for a time interval ΔT, the discrete flows of wagons with ore from different quarries after large crushing are transformed into an integrated continuous ore stream for which instantaneous weight and quality parameters can be measured with sufficient accuracy and reliability. The time interval ΔT includes various events characterizing the rate of advance of the ore mass to the place of measurement of its qualitative and quantitative characteristics, over which, in the final analysis, a correspondence can be established between the real discrete units of the input streams of the formatting object and the virtual segments of the output integral stream.

2.2   Temporal Model of the Ore Preparation

To construct a temporal model of ore preparation and to determine on its basis the ore segment of the output stream belonging to a particular wagon of one of the input discrete ore streams at the input of the EF, we use the Timed Interval Calculus (TIC) apparatus [2, 3].

The main elements of the TIC, used further in the formalization of the temporal model of ore preparation, define the following concepts:

- The time domain (T) is a non-negative real number and the interval is a sequence of time points, for example, the time interval $[x\ldots y]$ is defined as:

$$\forall x, y: R * [x\ldots y] = \{z: T \mid x \leq z \leq y\}$$

- Constants. For example, the maximum weight (MaxWeight) can be described as a real number (MaxWeight; R, where R is the real number).
- Timestamp (trace) is a function of the time domain of the variable definition. For example, the weight of the ore on the conveyor can be represented by the time dynamics (Weight) of the variable real-world range. So, Weight: $T \rightarrow R$.
- Interval operators. There are three primitives of interval operators: α, ω, σ having type $I \rightarrow T$, where I denotes all intervals and they return the starting point, end point and length of the interval.
- Interval brackets. A pair of interval brackets returns all the intervals that are defined by the predicate inside the parentheses. A predicate is usually a first-order predicate. For example, for the following TIC expression, $[Weight(\alpha) \leq Weight]$, indicating that the value of the variable Weight is not less than the value obtained at the beginning of the interval.
- Rules. Rules define the time properties of intervals and their connections.

All listed elements of the TIC-model, except the last one, are intuitively understandable and can be numerically identified for the technological object of ore preparation at the input of a specific formatting object. The rules of the TIC-model of the ore segment belonging to the output stream segment to the concrete wagon of one of the input discrete flows, and hence the determination of the ore belonging to the career, will be considered in more detail.

Rule 1. Binding events to the timeline.

   if S_i^{dw} = true then T_i^{dw} = t,

where: S_i^{dw} is the sign of the dump of the i-th wagon, T_i^{dw} is the time of the i-wagon's dump, t is the current time.

   This rule means that if an accident occurred as the wagon dump, the time for the wagon dump is the same as the current time.

Rule 2. Determination of the time interval for crushing ore in a bunker.

   if $[\![W_i^{ew} - P^b \neq 0]\!]$ = {x, y: T | $\forall$t: [x ... y] * (W_i (α ([x ... y])) = W_i^{ew} $\wedge$ W_i (ω ([x ... y])) = 0 $\wedge$ W_i = W_i- P^b)}
   then $t^b = \sigma$ ([x ... y]),

where: W_i^{ew} is the estimated weight of the ore of the i-th wagon; P^b - bunker capacity per second; W_i - the current weight of the ore of the i-th wagon, t^b - time interval of crushing of the ore of the bunker.

   This rule means that in each cycle (1 s) of the total weight of the ore in the hopper, the quantity equal to the capacity of the hopper per second is subtracted. This rule allows to calculate the time spent by the bunker on crushing the dumped wagon and fix the expected moment of the end of the wagon ore exit from the bunker.

Rule 3. Determination of the time interval from the moment of appearance of the ore of the wagon on the conveyor until the moment of measurement on the conveyor scales.

$$t_i^c = L^c / V^c,$$

where L^c is the length of the conveyor to the conveyor scales (meters); V^c - the speed of the conveyor (meters/second).

Rule 4. Determination of the expected time of the beginning of the flow of ore in the wagon through the scales.

$$T_i^{etb} = T_i^{dw} + t^b + t_i^c.$$

This rule means that the expected time for the beginning of the flow of the wagon ore through the scale is determined by the sum of the time of the wagon dump with the time interval for crushing the ore of the bunker and the time interval for the appearance of the wagon ore on the conveyor.

Rule 5. Calculation of the length of the time interval of the wagon's flow.

if $[\![S_i^{dw} = true \ \& \ t \geq T_i^{etb}]\!] = \{x, y: T \mid \forall t: [x ... y] * (S_i^{dw} = true \wedge t \geq T_i^{etb})\}$
then $\{W_i^{sum} = 0;$
        while $(W_i^{sum} \leq W_i^{ew})$
        $W_i^{sum} = W_i^{sum} + W_i;$
        $t = t + 1;$
        end
        $\{T_i^{int} = t - T_i^{etb}; \ T_i^{ete} = T_i^{etb} + T_i^{int}\}$
$\},$

where T_i^{int} - the time interval of the wagon's flow, T_i^{etb} – the beginning of the wagon's flow, T_i^{ete} – the ending of the wagon's flow.

This rule means that if the wagon event occurred and the current time is greater than or equal to the time of the expected start of the wagon unloading, then the condition is checked whether the total weight of the wagon is less than the expected weight of the wagon and, as soon as this condition is not fulfilled, wagon and time of the end of the wagon flow.

Rule 6. Correction rule for the beginning of the wagon flow.

If $[\![W(t) - W(\alpha) \geq 0.05 W(\alpha)]\!] = \{x, y: T \mid \forall t: [x ... y] * (W(t) - W(\alpha ([x ... y]))) \geq 0.05$
$W(\alpha ([x ... y])) \wedge \sigma[x ... y] \geq 5\}$
then $t_i^{etb} = \alpha ([x ... y]).$

This rule means that as the start time of the flow, the system captures a significant (more than 5 s) and stable (more than 5%) increase in the signal from the weight sensor.

Rule 7. Correction rule for the end of the wagon's flow.

If $[\![W(t) - W(\alpha) \leq 0.03 W(\alpha)]\!] = \{x, y: T \mid \forall t: [x ... y] * (W(t) - W(\alpha ([x ... y]))) \leq 0.03 W(\alpha ([x ... y])) \wedge \sigma[x ... y] \geq 5\}$
then $t_i^{ete} = w ([x ... y]).$

This rule means that as the end-of-flow time, the system records a stable (more than 5 s), close to zero (not more than 3%) value of the signal from the weight sensor.

3 Implementation of a Monitoring System

Practical use of the above rules of temporal logic was carried out in the development of a system for monitoring ore flows at the input of iron ore of EF. The monitoring system was implemented taking into account the canons of hard real-time systems on a duplicated Siemens controller in the form of functionally complete program blocks united by a common algorithm of operating environment of the controller. The functional description of these blocks is given further in the text.

3.1 Fixing the Time of Entry/Exit of Wagons

The ore enters the crushing site on the railway tracks. To fix the time of entry and exit of unloaded wagons on the tracks, track occupancy sensors are installed. When the signal from the occupancy sensor comes in, the system records the time for entering the train's composition on the crushing site. The fixation takes into account the stability of the signal. If the signal appears for a short time and disappears, the system identifies the event as a false signal and does not fix the input of the train composition. The exit time is fixed when the busy signal is removed.

3.2 Determination of the Number of Dumped Wagons

The task is to identify the moment of dumping the wagon into the bunker and counting such scores for the composition of wagons on the tracks. To identify the moment of a dumping, the system uses the following discrete signals: occupation of the paths, movement of the train along the way, the operator's command to dump the wagon, signal of the wagon dump,

The signal of the wagon's dump should most accurately reflect the moment of the wagon's dump. But because of the highly noisy environment (dust), the sensor does not always provide accurate information. Therefore, the system provides additional algorithmic processing, which allows filtering the appearance of false alarms of the sensor, as well as identify the dump even in the absence of the signal "Dump".

3.3 Accounting of the Ore Weight and Quality

Conveyors feeding ore from bunkers on the queue are equipped with conveyor scales and sensors of magnetic susceptibility of ore. The iron content in the ore is calculated on the basis of the signal from the magnetic susceptibility sensor according to the formula:

$$Fe = A * X + B,$$

where: Fe is the iron content, X is the magnetic susceptibility of the ore; A, B - coefficients of the regression model for the quarry, from which the ore came.

The coefficients A and B are obtained as a result of statistical processing of samples of ore taken from the mine. Obtaining these coefficients is not included in the tasks that the system solves. The system provides interfaces for entering them and uses the entered values in the calculations. Ore comes for redistribution from several mines. The coefficients are given for each of them. In order to calculate the iron content in the ore as precisely as possible, the system identifies the ore that goes through the conveyor to belong to a particular mine and uses the corresponding coefficients.

3.4 Identification of the Ore Passing Through the Conveyor
for Belonging to the Mine

Unloading into the bunker can be conducted in parallel with two paths. Obviously, the unloaded trains could come from different warehouses. Therefore, the solution of this problem is reduced to determining whether ore belonging to the conveyor is belonging to one of the dumped wagons. The task becomes nontrivial if unloading is conducted in a dense mode and the ore is conveyed by a continuous flow. The system determines the time limits of the wagons in the flow relative to the scales. That is, the system determines the ore from which wagon, passed at a certain point in time by scales. Knowing the magnitude of the transport lag between the weights and the magnetic susceptibility sensor, the system determines the time boundaries of the wagons in the flow relative to the magnetic susceptibility sensor. The determination of the time bounders of the wagons in the flow relative to the weights is carried out in four stages:

1. determination of the time of the beginning of passing the wagon through the scales;
2. determination of the end time of passing the wagon through the scales,
3. correction of the wagon boundaries in the flow, taking into account the total weight of the ore in the flow;
4. secondary adjustment of the wagon boundaries in the flow, taking into account the extremes of the signal from the weight sensor.

3.5 Determination of the Start Time of the Passing of the Wagon Through the Scales

At this stage, the system determines the expected time of occurrence of ore from the unloaded wagon on the scales. This time is calculated according to rule 4.

The calculated T_i^{etb} is adopted by the system as the first (rough) approximation of the border of the wagon and in the future is subject to refinement taking into account the actual situation at the facility. If, when T_i^{etb} comes on, there is no ore flow (lumen) on the scales, then the nearest occurrence of ore flow on the scales will be accepted as the beginning of the wagon. Other cases of correction of the moment of the beginning of the car are closely connected with the moment of determining the end of the previous wagon.

3.6 Determination of the End Time of Passing the Wagon Through the Scales

When determining the moment of the end of passage of the wagon through the scales, the system relies on the expected (known a priori) weight of the wagon. The algorithm for determining the end time of the passage of the wagon through the scales is carried out according to rule 5. The algorithm sums the weight of the ore passed on the scales from the moment of the beginning of the wagon and, when the expected weight of the wagon is reached, fixes the moment of the end of the passage of the wagon.

If the expected weight for the wagon is already summarized, but for the next wagon, the inequality is not valid:

$$T_i^{etb} \leq t,$$

where t is the current moment, then the system continues to count the weight for the current wagon until the following condition is satisfied for the next wagon. In this case, T_i^{etb} of the next car will be adopted as the end of the wagon.

If the expected weight of the wagon is not reached, but the end of ore flow on the conveyor (lumen) is observed, the system fixes the end of the wagon, but additional analysis is performed. If the collected weight is much less than expected (less than 70%), then the system analyzes the moment of the end of the exit from the bunker of the previous wagon and the moment of the beginning of the current one. If the difference between these moments makes up a time interval sufficient for the appearance of a lumen on the scales, then the end of the flow is fixed as the moment of the end of the passage through the scales of the previous wagon. If the time interval is too small or absent, the end of the flow is fixed as the end of the current wagon.

3.7   Correction of the Wagon Boundaries in the Flow

Under the flow is understood the time interval from the moment of occurrence of ore flow on the scales until the moment of its termination. Determination of the boundaries of wagons in the flow is carried out according to rule 6. When fixing the end of the flow, the system corrects the boundaries of wagons inside the flow defined at the previous stages. The system determines the total weight of the ore in the flow and the number of wagons in it. Next, for each wagon in the stream, the system summarizes its expected weight. The result of this operation is the expected weight of the stream. Further, the expected flow weight is compared with the actual one and the discrepancy coefficient is determined:

$$k_{disc.} = \frac{W_{exp.} - W_{act.}}{W_{act.}},$$

where: $k_{disc.}$ is the discrepancy coefficient, $W_{exp.}$ - expected flow weight, $W_{act.}$ is the actual weight of the stream.

For each wagon in the flow, the actual weight is calculated by the formula:

$$W_{wag.act.} = W_{wag.exp.} * (1 - k_{disc.}),$$

where:

$W_{wag.act.}$ – the actual weight of the wagon,

$W_{wag.exp.}$ – the expected weight of the wagon.

Then the system arranges the wagon boundaries in the stream in such way that the weight of each wagon corresponds to the $W_{wag.act.}$ received for it.

3.8   Secondary Adjustment of the Wagon Boundaries in the Flow

The appearance of ore flow on the scales is not characterized by an instantaneous abrupt change in the signal from the weight sensor. The signal has some inertia. This is due to the inertia of the crusher, which, when it enters the ore, goes to full capacity with a little delay, and also damping the signal itself in order to suppress noise. A similar inertia is observed at the end of the flow.

With a certain porosity of sites, this leads to situations where the flow of ore for the wagon has already begun to decline and at that time the ore begins to come on the scales from the next wagon. This situation is characterized by the presence of an extremum in the signal from the weight sensor.

The processing of such extremes allows more accurate determination of the wagon boundaries in the flow. The algorithm of secondary correction searches for extremums in the analyzed flow. In finding those, the system analyzes the extremum for proximity to the wagon boundaries obtained at the previous stage. If the extremum found is near one of the boundaries, the system adjusts the boundary, taking the extremum moment as the new boundary. If the extremum is at a considerable distance from the boundaries obtained at the previous stage, it is accepted by the system for changing the amount of ore on the conveyor, possibly related to the interruption of ore feed from the crusher, and is ignored.

3.9 Combining Wagon Compositions

The system provides the operator with an interface for performing the operation of combining the two consecutive trains of wagons into one train. The need for such an operation can occur when a failure occurs in the workload sensor of the path, as a result of which the signal disappears for a while. The system, when the busy signal disappears, fixes the output of the composition, and the next time it appears, it fixes the input of the new composition. If, at the time, there was only one composition on the way, the operator performs the unification operation for the compositions recorded by the system. To combine, the operator specifies the number of the first and the following composition and confirms the operation. As a result, the system assigns the wagons fixed to the second train to the first train and removes the second train. For reconnected wagons, the iron content is recalculated taking into account the parameters of the warehouse of the first composition.

3.10 Separation of the Wagons Composition

Similarly, the composition combining operation allows the system to perform a composition separation operation. The need for such an operation can occur when a failure occurs in the workload sensor of the path, as a result of which the work path sensor does not respond to the output of the train and the busy signal remains unchanged. In this case, the system does not fix the output of the train and the entrance to the unloading of the next one, as a result of which the cars unloaded from the second train are tied to the first train by the system. To correct this erroneous situation, the operator performs a splitting operation. To perform the operation, the operator enters the number of the segregated structure, the number of wagons that should be left in it, and the time to enter the unloading of the next convoy. As a result of the operation, the system fixes the input of the new composition with the entry time entered by the operator and re-ties the wagons indicated by the operator to it. For reconnected wagons, the content of iron is recalculated taking into account the parameters of the warehouse of the second composition.

4 Results of Experimental Studies

The results of experimental studies carried out in real time on the industrial site showed that the proposed scientific and technical solutions make it possible to improve the efficiency of monitoring the qualitative and quantitative characteristics of the ore at the entrance of the large crushing body and to ensure the formation of objective data for effective operational management of the ore preparation and enrichment processes.

The process of approbation of the pilot online monitoring system of the characteristics of the input ore at the mine processing plants was carried out at the industrial plant in Kazakhstan. At the same time, both direct measurements of the characteristics of the ore were carried out, and their analytical evaluation was carried out on the basis of the developed online monitoring system (OMS).

The average values of the characteristics of the ore in the wagons coming from different mines to the crushing department is given in Table 1.

Table 1. The average values of the ore characteristics from different quarries defined by developed online monitoring system and direct laboratory measurement.

Number of quarry	OMS, the average weight of ore in wagon (tons)	The average weight of ore in wagon. Direct measurement.	OMS, weight measurement error (deviation/%)	OMS, average Fe% content in the wagon	Average Fe% content in the wagon. Direct measurement.	OMS, the error in the content of Fe% measurements (deviations)/%
1	94.3	98.8	−4.5/4.5%	32.4	32.0	0.4/1.25%
2	95.0	100.0	−5.0/5.0%	27.9	29.3	−1.4/4.77%
3	91.8	89.1	2.7/3.0%	37.7	40.2	−2.5/6.2%
4	90.5	91.0	−0.5/0.5%	27.9	29.7	−1.8/6.0%
5	102.5	99.6	2.9/2.9%	38.5	39.7	−1.2/3.0%

It can be seen from Table 1 that the relative error in monitoring of the input characteristics of the ore at the input of the enrichment factory does not exceed 6.2%, which will allow the effective use of the results obtained for the operational management of the mine processing plant technological processes.

Comparing proposed approach with other existing techniques. Above proposed approach was compared with traditional laboratory direct measurement. The accuracy of measurement within the permissible range, but traditional laboratory direct measurement is not real-time. Existing technologies for assessing the quality of ores in real-time for shallow-fractions ore flows uses ore-controlling station (OCS), which is commonly used for a single-flow ore crushing process [5]. For a multi-flowed ore crushing process an OCS is required in terms of the number of flows. Since the cost of an OCS is quite high (several hundred thousand dollars), the cost of the ore quality assessment system is proportional to the number of flows. For comparison, the technology proposed in this paper assumes the use of only one OCS for a multi-flowed process. Accordingly, there is a significant benefit in the cost of the proposed technology for real-time assessment of the quality of ores.

5 Conclusion and Future Works

In this paper is proposed to use the temporal model to determine the ore belonging to a quarry in the multi-flow technological process of ore crushing in the mining and processing plant. The proposed temporal model and online monitoring system allows the real-time control of the enrichment fabric input ore flows quality and the effective management of the beneficiation process, providing feedback to the quarries supplying the ore.

As a future works are planed the investigation of proposed approach on other mining plants and on production of building materials.

References

1. Piven, V.A., Romanenko, A.V., Shepel, V.V., et al.: Investigation of the influence of the variability of the qualitative parameters of ore pit flows on the efficiency of iron ore enrichment. J. Metall. Min. Ind. (2), 64–68 (2007) (in Russian)
2. Fidge, C.J., Hayes, I.J., Martin, A.P., Wabenhorst, A.K.: A set-theoretic model for real-time specification and reasoning. In: Jeuring, J. (ed.) MPC 1998. LNCS, vol. 1422, pp. 188–206. Springer, Heidelberg (1998). https://doi.org/10.1007/BFb0054291
3. Chen, C., Dong, J.S., Sun, J.A.: Verification System for Timed Interval Calculus (2007). http://www.comp.nus.edu.sg/~chenchun/TIC2PVS
4. Karpov, Y.G.: Model Checking. Verification of Parallel and Distributed Software Systems, 560 p. BHV-Petersburg, Saint Petersburg (2010). (in Russian)
5. Ore-controlling station RKS-KM. http://www.technoros-kras.ru/products/13/36/

Modeling of Position Control for Hydraulic Cylinder Using Servo Valve

Ngoc Hai Tran$^{(\boxtimes)}$, Quang Bang Tao⬭, Tan Tien Huynh,
Xuan Tuy Tran, and Nhu Thanh Vo

University of Science and Technology,
The University of Danang, Da Nang, Vietnam
{tnhai,tqbang,httien}@dut.udn.vn,
tranxuantuy@yahoo.fr, thanhvous@gmail.com

Abstract. In this conference, the authors present the research on position control of hydraulic cylinders that has piston on one side controlled by servo valves through oil flow. This study includes the dynamic modeling of a hydraulic cylinder with constant load, establishing the mathematical model and controller model, selecting the structural parameters of the system and assembling the experimental model, then finding the control parameters on the experimental model, and writing the program for the fuzzy PID control controller. The authors conduct the simulation of the hydraulic cylinder position using Matlab/Simulink software and also the experiment with piston displacement of 50, 100, 150, 200 (mm) to derive the delay time, settling time, overshot, and position error at steady state. From there, we evaluate the exact displacement of the piston. The results show that the accuracy of displacement of piston is as expected and it can be applied in the current machine cutting tools.

Keywords: Electro-hydraulic force servo systems · Hydraulic accumulator
Force control · Potential energy · Hydraulic actuator position control

1 Introduction

Today machining equipment and CNC machines that are applied the automatic transmission and control of hydraulic power is once among the main research directions the machine manufacturers. For universal machine tools and CNC machines, the dynamic system is a complex elastic system which has the working process due to the characteristics of cutting (continuous cutting, discontinuous), friction, as well as other non-linear processes occur that cause the elastic system to deform and oscillate, thereby affecting the quality of the work piece, the stability of the cutting tools movement, the lifespan of the machine and the cutting tools. All of these issues are closely related to machine dynamics. The loads acting on these elastic systems are very complex, but can be extracted into three main types of cutting characteristics which are linear load, harmonic load, and constant load.

Forental [1] published a study investigating the positioning of hydraulic cylinders by proportional valves, determining the position of the cylinder by position sensors. The mathematical model was set up including the characteristics of the pump, loading

© Springer International Publishing AG, part of Springer Nature 2018
N. T. Nguyen et al. (Eds.): ACIIDS 2018, LNAI 10752, pp. 696–706, 2018.
https://doi.org/10.1007/978-3-319-75420-8_65

inertia, the characteristic of the proportional valve (flow characteristic of the flow through the position control of the valve slide). The results of the cylinder displacement were obtained experimentally and modeled from 0.05 to 5 Hz. This study showed that with a frequency of 0.05 to 3 Hz, the minimum system error is about 2 to 3%.

Adenuga et al. [2] introduced the synchronous control position of three hydraulic transmission cylinders for the bending machine (RBPM) using the Fuzzy - PID control. In this study, they assumed nonlinear systems due to friction, load and the complex relationship between flow and pressure of the valve. The results of their simulation showed that the displacement between the cylinders was negligible (1%) and considered this as a system loss.

Shen et al. [3] announces real-time control of electro-hydraulic servo systems by using feedback control and control adaptation with the assumption of linearized loads. They apply a proportional-integral (PI) controller for their system. They conducted several comparative experiments on a real-time (EHFS) system and proposed a controller that improved performance.

From the above analyses, we find that all the assumptions of the problem are the linearized loads, the system is controlled by applying the classical PID or fuzzy – PID controller. However, in addition to the linear change load, there are constant loads and harmonic loads existing in machining equipment. Therefore, in this study we assume the load is constant and apply the self-tuning fuzzy PID controller for position control of the hydraulic cylinder.

2 Experimental Procedures

2.1 Research Model

In this study, we propose a control model for the position of the hydraulic cylinder with the constant load controlled by a servo valve (see Figs. 1 and 2), the parameters of the system are indicated in Table 1.

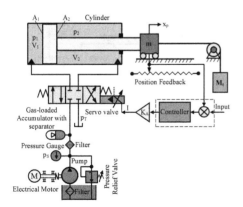

Fig. 1. System configuration

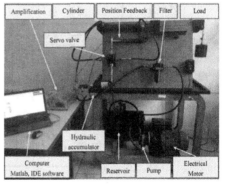

Fig. 2. Picture of experimental system

Table 1. System parameters

Symbol	Description	Unit	Value
L_{max}	Piston displacement	mm	304.8
H_{max}	Position sensor working distance	mm	304.8
m	Mass	Kg	5
M_t	Load	Kg	5
f	Viscous damping coefficient	Ns/mm	250×10^5
λ_1	Internal leakage coefficient	$mm^3/(s/Pa)$	4.6×10^{-10}
λ_2	External leakage coefficient	$mm^3/(s/Pa)$	4.6×10^{-10}
β	Oil effective bulk modulus	Pa	6.9×10^8
$V_{01} + V_1$	Oil volume from valve to working chamber	mm^3	15×10^5
A_1	Area of piston (left side)	mm^2	2826
A_2	Area of piston (right side)	mm^2	2446
K_{SV}	Gain of servo valve	$mm^3/s/A$	40×10^8
ω_{SV}	Angular natural frequency of servo valve	Hz	100
ξ_{SV}	Damping ratio of servo valve		0.7

2.2   Mathematical Model

The mathematical model is described in Fig. 3 with the following assumptions: a hydraulic cylinder with a piston rod on one side, an absolutely rigid cylinder (i.e., does not change cylinder volume under pressure), good lubrication conditions, friction in the system is wet friction and depends on the speed of movement, the viscosity of the oil is fixed (there are control of the working temperature of the oil), the system has a constant load.

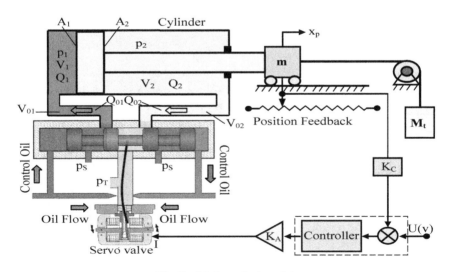

Fig. 3. Mathematical model

From the assumptions and the mathematical model (Fig. 2), we describe the mathematics of the system with the straight stroke of the piston rod as following:

The fluid flow from the servo valve to piston coordinates x_p, [4]:

$$Q_{01} = A_1 \frac{dx_p}{dt} + \lambda_1 p_L + \lambda_2 p_1 + \frac{V_{01} + V_1}{\beta} \frac{dp_1}{dt} \quad \left(where : V_1 = A_1 x_p, \ p_L = p_1 - p_2\right) \quad (1)$$

Servo valve flow [5]:

$$Q_V = K_V I - K_C p_L \tag{2}$$

Equilibrium equation on piston rod:

$$A_1 p_1 - A_2 p_2 = m \frac{d^2 x_p}{dt^2} + f \frac{dx_p}{dt} - M_t \tag{3}$$

Take:

$$R = A_1/A_2 \ then \ Q_1 = R Q_2 \quad with \ Q_1 = K\sqrt{p_s - p_1}, \ Q_2 = K\sqrt{p_2 - p_T} \tag{4}$$

From (4), we have (5):

$$p_2 = \frac{p_s - p_1}{R^2} \tag{5}$$

Replace (5) into (3) and (1), we have (6) and (7):

$$p_1 \left(A_1 + \frac{A_2}{R^2} \right) = m \frac{d^2 x_p}{dt^2} + f \frac{dx_p}{dt} - E; \ \left(E = M_t - \frac{A_2}{R^2} p_s \right) \tag{6}$$

$$p_L = p_1 \left(1 + \frac{1}{R^2} \right) - \frac{1}{R^2} p_s \tag{7}$$

From (1), (6) and (7) we derive Laplace Eqs. (8, 9) with initial condition $p_s = 0$:

$$Q_{01}(s) = A_1 x_p(s) s + \left[\lambda_1 \left(1 + \frac{1}{R^2} \right) + \lambda_2 + \frac{V_{01} + V_1}{\beta} s \right] p_1(s) \tag{8}$$

$$x_p(s) \left(m s^2 + f s \right) = \left(A_1 + \frac{A_2}{R^2} \right) p_1(s) + E(s) \tag{9}$$

Take:

$$A = A_1 s; B = \lambda_1 \left(1 + \frac{1}{R^2} \right) + \lambda_2 + \frac{V_{01} + V_1}{\beta} s; C = m s^2 + f s; D = A_1 + \frac{A_2}{R^2} \tag{10}$$

Equations (8, 9) can be rewritten:

$$\begin{cases} Q_{01}(s) = Ax_p(s) + Bp_1(s) \\ Cx_p(s) = Dp_1(s) + E(s) \end{cases} \tag{11}$$

Present Eq. (11) with block diagram (see Fig. 4):

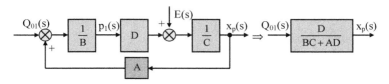

Fig. 4. Block diagram

We has transfer function (12):

$$W_{XL}(s) = \frac{x_p(s)}{Q_{01}(s)} = \frac{D}{BC + AD} \tag{12}$$

from (10) and (12) we derive (13):

$$W_{XL}(s) = \frac{x_p(s)}{Q_{01}(s)} = \frac{K_{XL}}{s\left(T_1^2 s^2 + T_2^2 s + 1\right)}$$

$$\text{or } W_{XL}(s) = \frac{x_p(s)}{Q_{01}(s)} = \frac{K_{XL}}{s\left[\left(\frac{s}{\omega_0}\right)^2 + 2\xi\left(\frac{s}{\omega_0}\right) + 1\right]} \tag{13}$$

Where:

$$K_{XL} = \frac{A_1 + \frac{A_2}{R^2}}{\lambda_1 f\left(1 + \frac{1}{R^2} + \lambda_2\right) + A_1^2 + \frac{A_1 A_2}{R^2}} \tag{14}$$

$$T_1 = \sqrt{\frac{V_{01} + V_1}{\beta\left[\lambda_1 f\left(1 + \frac{1}{R^2} + \lambda_2\right) + A_1^2 + \frac{A_1 A_2}{R^2}\right]}}; T_2 = \frac{m\left[\lambda_1\left(1 + \frac{1}{R^2}\right) + \lambda_2 f \frac{V_{01} + V_1}{\beta}\right]}{\lambda_1 f\left(1 + \frac{1}{R^2} + \lambda_2\right) + A_1^2 + \frac{A_1 A_2}{R^2}} \tag{15}$$

$$\omega_0 = \frac{1}{T_1} = \sqrt{\frac{\beta\left[\lambda_1 f\left(1 + \frac{1}{R^2} + \lambda_2\right) + A_1^2 + \frac{A_1 A_2}{R^2}\right]}{V_{01} + V_1}} = \sqrt{C_H} \tag{16}$$

$$\xi_{XL} = \frac{T_2}{2T_1} = \frac{m\left[\lambda_1\left(1+\frac{1}{R^2}\right) + \lambda_2 f\frac{V_{01}+V_1}{\beta}\right]}{\sqrt{\frac{(V_{01}+V_1)\left[\lambda_1 f\left(1+\frac{1}{R^2}+\lambda_2\right) + A_1^2 + \frac{A_1 A_2}{R^2}\right]}{\beta}}} \tag{17}$$

With:

w_0 -   is the frequency of oscillation;

C_H -   is the hydraulic hardness;

ξ_{XL} -   is the damping coefficient.

Thus the open transfer function from the supply of the servo valve (Q_{01}) to the coordinates of the piston (x_p) carrying the mass (m) and constant load (M_t) is the product of the two modules which are integration and oscillation.

2.3 Application of the Controller and Construction of a Closed Circuit Position Control Circuit

Servo valves can be considered as an amplification or inertia or oscillation module for the calculating of the dynamics. According to [5], it is considered as an oscillating module with the transfer function as following:

$$W_{SV}(s) = \frac{Q_{01}(s)}{I(s)} = \frac{K_{SV}}{\left(\frac{s}{\omega_{SV}}\right)^2 + 2\xi_{SV}\left(\frac{s}{\omega_{SV}}\right) + 1} \tag{18}$$

In this study, we applied a self-tuning PID controller [6].

From (2), (13), (18) and the self-tuning PID controller [6, 7], we set up the block diagram of the control system as shown in (see Fig. 5).

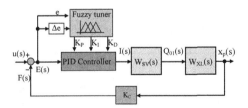

Fig. 5. Block diagram of a fuzzy self-tuning PID controller

The fuzzy controller [7] has two inputs, e and Δe and has three outputs corresponding to each control parameter K_p', K_I' và K_D'. We apply the Mamdani fuzzy rule with the processor to get optimal values K_P, K_I and K_D. The range of PID controller parameters are (K_{Pmin}, K_{Pmax}), (K_{Imin}, K_{Imax}), (K_{Dmin}, K_{Dmax}) and are determined experimentally. The range of parameters are as followed: $K_P \in (13, 33)$, $K_I \in (0, 0.0001)$ and $K_D \in (0, 0.04)$. Therefore, they can be normalized in terms of $(0, 1)$ as follows:

$$K_P = 20K_P' + 13; \quad K_I = 0.0001K_I'; \quad K_D = 0.04K_D' \tag{19}$$

Fuzzy rules for K_P, K_I and K_D are shown in Tables 2, 3 and 4; the self-tuning PID controller is shown (see Fig. 6). The fuzzy PID controller adjusted PID parameters to ensure the system has no overshoot, very small error with systematic biases ($\leq 5\%$).

Table 2. Fuzzy rules of K_P gain

e	Δe				
	NB	N	Z	P	PB
NB	S	S	S	MS	M
N	S	MS	MS	MS	M
Z	S	MS	M	MB	B
P	M	MB	MB	MB	B
PB	M	MB	B	B	B

Table 3. Fuzzy rules of K_I gain

e	Δe				
	NB	N	Z	P	PB
e	B	B	B	MS	S
N	B	MB	MB	M	S
Z	MB	MB	M	MS	S
P	M	MB	MS	MS	S
PB	M	MS	S	S	S

Table 4. Fuzzy rules of K_D gain

e	Δe				
	NB	N	Z	P	PB
Nrt	S	S	S	MB	B
N	S	MS	MS	M	B
Z	MS	MS	M	MB	B
P	M	MS	MB	MB	B
PB	M	MB	B	B	B

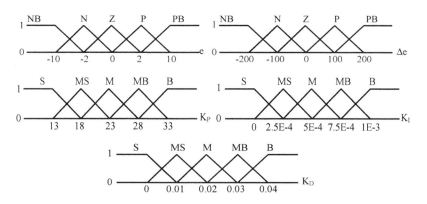

Fig. 6. Fuzzy sets of a fuzzy self-tuning PID controller

Arduino Mega2560 [14] differs from all previous processors because it does not use FTDI control chips to transfer signals from USB to the processor. Instead, it uses ATmega16U2 programmable as a signal converter from USB. In addition, the Arduino Mega2560 has a built-in library for MatLab [16, 17], which makes it easier to interface and process data. Comes with the Arduino Mega2560 is an integrated IDE [15] development environment that runs on common PCs and allows users to write and import system control programs on the IDE programming language software such as C or C++. The Arduino Mega2560 connects to the computer via a virtual COM port with a physical USB connection. Also, Matlab R2015a has built-in functions and procedures that make it easy to transfer data to Arduino Mega2560 via COM ports.

From the mathematical model (see Fig. 3) mathematical Eqs. (13) to (18) and self-tuning fuzzy PID controller (19) and control system block diagrams (Fig. 5). we conduct simulation for the position of the pistons 50, 100, 150 and 200 (mm) on the Matlab/Simulink software as shown in (Figs. 7 and 8).

Based on the numerical values of the self-tuning fuzzy PID controller (18), we write the control programs on the IDE software and interface with Matlab/Guide. The experiments are repeated 10 times at different times with the positions 50, 100, 150 and 200 (mm) of the pistons as shown in the two graphs of (see Figs. 9 and 10). The simulated and experimental results are presented in (Table 5).

3   Results and Discussion

Figures 7 and 8 show a theoretical simulation of position control of hydraulic pistons at four positions of 50, 100, 150 and 200 mm.

When simulating the position of the piston rod at 50 mm we get the following result: the actual position of 49.8 mm, set time 2.04 s, and the error of 0.4%.

When simulating the position of the piston rod at 100 mm we get the following result: the actual position of 99.8 mm, set time of 2.89 s, and the error of 0.2%.

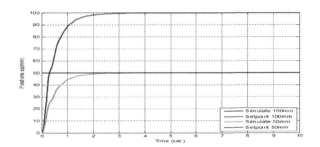

Fig. 7. Simulation control of piston at 50 mm and 100 mm

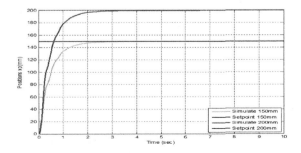

Fig. 8. Simulation control of piston at 150 mm and 200 mm

When simulating the position of the piston rod at 150 mm we get the following result: actual position of 149.7 mm, set time of 3.05 s, and the error of 0.2%.

When simulating the position of the piston rod at 200 mm we get the following result: actual position of 199.7 mm position, the setting time of 4.24 s, and the error of 0.15%.

Figures 9 and 10 show the experimental result of the four different position control of the hydraulic pistons at 50, 100, 150 and 200 mm.

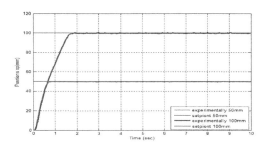

Fig. 9. Actual experiment control of piston at 50 mm and 100 mm

When controlling the position of the piston rod to get 50 mm: the actual position is 49.9 mm, the setting time is 0.8 s and the error is 0.2%.

When controlling the position of the piston rod to get 100 mm: the actual position is 99.9 mm, the setting time is 1.85 s and the steady state error is 0.1%.

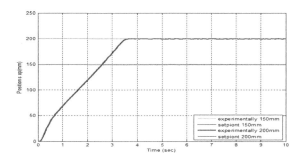

Fig. 10. Actual experiment control of piston at 150 mm and 200 mm

When positioning the piston rod at 150 mm in the actual experiment we get the following results: the actual position of 149.9 mm, the settling time of 2.72 s, and the error of 0.06%.

When positioning the piston rod at 200 mm in the actual experiment we get the following results: the actual position of 199.8 mm, the setting time of 3.68 s and the steady state error 0.1%.

The simulation and experimental results show that the system doesn't have over-shoot, indicating the control parameters work very well in automatically adjusting the controlled parameter to achieve the control positions.

As shown in Table 5, we find that the simulation response time is slower than the actual response time in experiment. However, the other parameters at steady state (response times, overshoot, and settling error) are almost the same. Thus, the experimental results are quite consistent with the simulation results.

Table 5. Simulation and actual experiment parameters

Set point x_p (mm)	Methodology of research	Real value (mm)	Delay time (sec)	Overshoot (%)	Error (%)
50	Simulate	49.8	2.04	0	0.4
	Experiment	49.9	0.8	0	0.2
100	Simulate	99.8	2.89	0	0.2
	Experiment	99.9	1.85	0	0.1
150	Simulate	149.7	3.05	0	0.2
	Experiment	149.9	2.72	0	0.06
200	Simulate	199.7	4.24	0	0.15
	Experiment	199.8	3.68	0	0.1

4   Conclusion

The results of the research on position control of hydraulic cylinders are summarized as follows:

(1) Set up a modeling method for analyzing and solving the dynamic problem of hydraulic cylinder servo position control by assuming the constant load. Select the structural parameters to assemble the empirical model and find the parameters of the self-adjusting PID controller.

(2) Write a program that controls the position of the piston (x_p) 50, 100, 150 and 200 mm. Through simulation and experiment, we found that the fuzzy PID controller adjusted itself for the desired results, namely: No overshoot, very small set error (see Table 5) with systematic biases ($\leq 5\%$).

(3) The results can be applied to drive and control in coordinated motions on universal machine tools and CNC machine tools which requires a low steady state error in trade off with a relatively large setting time.

In the next article we will publish the results of the study when applying linear and harmonic loads; thereby, experimenting to find the optimal set of control parameters to propose the scope of application on the cutting tools motion of machining equipment.

References

1. Forental, V.I., Forental, M.V., Nazarov, F.M.: Investigation of dynamic characteristics of the hydraulic drive with proportional control. In: International Conference on Industrial Engineering, vol. 129, pp. 695 –701 (2015). Procedia Engineering
2. Adenuga, O.T., Mpofu, K., Adeyeri, M.K.: Analysis and control design of Tri-electrohydraulic cylinders synchronization for RBPM. Int. J. Control Autom. **10**, 253–270 (2017)
3. Shen, G., Zhu, Z., Zhao, J., Tang, Y., Li, X.: Real-time tracking control of electro-hydraulic force servo systems using offline feedback control and adaptive control. ISA Trans. **67**, 356–370 (2017)
4. Liu, Y., Gong, G.F., Yang, H.Y., Han, D., Wang, H.: Regulating characteristics of new tamping device exciter controlled by rotary valve. IEEE/ASME Trans. Mechatron. **21**(1), 497–505 (2016)
5. Shen, G., Zhu, Z.C., Li, X., et al.: Real-time electro-hydraulic hybrid system for structural testing subjected to vibration and force loading. Mechatronics **33**, 49–70 (2016)
6. Ding, W.-H., Deng, H., Xia, Y.-M., Duan, X.-G.: Tracking control of electro-hydraulic servo multi-closed-chain mechanisms with the use of an approximate nonlinear internal model. Control Eng. Pract. **58**, 225–241 (2017)
7. Sinthipsomboon, K., Hunsacharoonroj, I., Khedari, J., Po-ngaen, W., Pratumsuwan, P.: A hybrid of fuzzy and fuzzy self-tuning PID controller for servo electro-hydraulic system. In: Iqbal, S., Boumella, N., Garcia, J.C.F., (eds.) Fuzzy Controllers - Recent Advances in Theory and Applications, pp. 299–314. InTech (2012)
8. Tran, N.H., Le, C., Ngo, A.D.: Experimental investigation of speed control of hydraulic motor using proportional valve. In: IEEE International Conference of System Science and Engineering, pp. 350–355 (2017)
9. Truong, D.Q., Ahn, K.K.: Force control for hydraulic load simulator using self-tuning grey predictor – fuzzy PID. Mechatronics **19**(2), 233–246 (2009)
10. Cerman, O., Hušek, P.: Adaptive fuzzy sliding mode control for electro-hydraulic servo mechanism. Exp. Syst. Appl. **39**(11), 10269–10277 (2012)
11. Aly, A.A.: Self tuning fuzzy logic control of an electrohydraulic servo motor. ACSE J. **5**(4) (2005)
12. Peter, C.: Principles of Hydraulic Systems Design. Momentum Press, New York (2015)
13. Merritt, H.E.: Hydraulic Control Systems. Wiley, New York (1967)
14. https://store.arduino.cc/usa/arduino-mega-2560-rev3
15. https://www.arduino.cc/en/Main/Software
16. https://uk.mathworks.com/products/MATLAB.html
17. http://playground.arduino.cc/Interfacing/MATLAB

A Large Neighborhood Search Heuristic for the Cumulative Scheduling Problem with Time-Dependent Resource Availability

Nhan-Quy Nguyen[(⊠)], Farouk Yalaoui, Lionel Amodeo, and Hicham Chehade

ICD, LOSI, University of Technology of Troyes, UMR 6281, CNRS,
12 rue Marie Curie, CS 42060, 10004 Troyes Cedex, France
nhan_quy.nguyen@utt.fr

Abstract. This paper addresses the cumulative scheduling problem with time-dependent total resource availability. In this problem, tasks are completed by accumulating enough amounts of resources. Our model is time-indexed due to the variation of the total resource availability. A Large neighborhood search (LNS) approach is developed. As an extension of our previous work, this paper contributes a new request removal method and an integration of a simulated annealing procedure to the LNS. Through the computational results, those modifications reduce the derived gap between the worst solutions and also the average derived gap.

1 Introduction

The cumulative problem (CuSP) is a class of the machine scheduling problem with additional resources where the number of tasks is equal to the number of machines. Also, the processing times of tasks are controllable by the rate of resource consumption of those tasks. The more resource a task consumes at a time, the faster it can complete. Baptiste *et al.* [1] considered the CuSP as a scheduling problem with a constant total amount of resource available. Moreover, each task has a release date and a deadline (time-windows constraints), requires a fixed quantity of resources. Nattaf *et al.* [5] studied a more general problem, named Continuous Energy-Constrained Scheduling Problem (CECSP). In the CECSP problem, the resource consumption of each task is bounded between a minimum and maximum value. This problem tackled a broader domain of practical issues notably the energy reasoning problem where the power rate of electrical equipment is usually bounded. To cope with more real-life issues, we define a more general problem which considers the variation of the total available resource. The direct application of our study is the charging scheduling of electric vehicles. The power available for the charging procedure is time-dependent. Each charging tasks is limited by the parking time of the client. The electrical demand is proportional to the daily traveled distance. The charger accepts only a limited charging rate and the bigger the rate, the faster the battery can be fully charged. The objective is to minimize the total weighted completion times.

© Springer International Publishing AG, part of Springer Nature 2018
N. T. Nguyen et al. (Eds.): ACIIDS 2018, LNAI 10752, pp. 707–715, 2018.
https://doi.org/10.1007/978-3-319-75420-8_66

The Large Neighborhood Search (LNS) approach has yielded many successes in coping with scheduling problems [3,8]. A detailed description of the method could be found on [9]. The method can be succinctly described as follows. The requirement of an LNS heuristic is an initial solution. Partial destruction and reconstruction then improve this solution. Those operators increase the size of the neighbor exploited. Hence the neighborhood can be called *large*. The work presented in this paper is an extension of the work done on our previous paper [7], tenting to improve the performance of the developed heuristic by modifying the destruction operator and the solution acceptance.

We present in Sect. 2 the mixed-integer-linear-programming formulation of the problem. In Sect. 3 we precise the mechanism of the LNS to solve the considered problem, followed by the removal operators and the reallocation operators. We conduct numerical tests to investigate the algorithms in Sect. 4 and we also present how to calibrate the parameters. In Sect. 5, we conclude the effects of the improvements made and precise our future work as the perspective of the article.

2   Problem Formulation

Given a set of n jobs in $\mathcal{J} = \{1, 2, \ldots, n\}$ to be scheduled in H decision intervals, each interval lasts 1 unit of time. Let k be the interval time index, $k \in \mathcal{K} = \{0, \ldots, H\}$. A job i starts to process as soon as it consumes an amount of resource $u_{i,k}$ at time k. Jobs are non preemptive and when they are processing, their resource consumptions are bounded between a lower bound $\underline{u}_i$ and an upper bound $\bar{u}_i$. Job i has a release date r_i, a deadline d_i and a demand $\tilde{x}_i$. The total amount of available resource at time k is denoted by U_k. We use the Mixed-Linear-Integer-Programming formulation introduced by Nguyen *et al.* [6] to formulate the problem. Variable $\alpha_{i,k}$ marks the starting time of job i, hence $\alpha_{i,k} = 1$ if job i starts at time k, otherwise it is 0. In the same way, variable $\beta_{i,k}$ marks the completion times of job i. $\beta_{i,k} = 1$ if job i completes at time k, otherwise it is 0.

MILP formulation

$$\text{Minimize} \quad z = \sum_{i \in \mathcal{J}} w_i C_i = \sum_{i \in \mathcal{J}} \sum_{k \in \mathcal{K}} (w_i k \beta_{i,k}) \tag{1}$$

subject to

$$\sum_{k=e_i}^{d_i} \beta_{i,k} = 1, \ i \in \mathcal{J} \tag{2}$$

$$\sum_{k=r_i}^{l_i} \alpha_{i,k} = 1, \ i \in \mathcal{J} \tag{3}$$

$$u_{i,k} \leq \bar{u}_i \sum_{\tau=r_i}^{k} (\alpha_{i,\tau} - \beta_{i,\tau}), \ i \in \mathcal{J}, \ k \in \{r_i, \ldots, d_i\} \tag{4}$$

$$u_{i,k} \geq \underline{u}_i \sum_{\tau=r_i}^{k} (\alpha_{i,\tau} - \beta_{i,\tau}), \ i \in \mathcal{J}, \ k \in \{r_i, \ldots, d_i\} \tag{5}$$

$$\sum_{k=r_i}^{d_i} u_{i,k} \geq \tilde{x}_i, \ i \in \mathcal{J} \tag{6}$$

$$\sum_{i \in \mathcal{J}} u_{i,k} \leq U_k, \ k \in \mathcal{K} \tag{7}$$

$$\alpha_{i,k}, \beta_{i,k} \in \{0,1\} \ \forall i, k \tag{8}$$

A job is processing at time k if $\sum_{\tau=r_i}^{k} (\alpha_{i,\tau} - \beta_{i,\tau}) = 1$. Constraints (2) and (3) restraint the processing of a job to be in the time-window. Constraints (3) and (5) bounded the resource consumption of a job when it is processing. Constraint (6) assures all the demands are satisfied. Constraint (7) keeps the resource consumption of all tasks at any moment be lower than or equal to the resource capacity (Fig. 1).

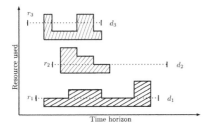 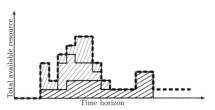

(a) Illustration of jobs' resource alloca- (b) Illustration of total time-varying re-
tion   source

Fig. 1. The illustration of the CuSP problem with time-dependent total available resource

3   LNS Heuristic to Solve CuSP

Figure 2 describes the LNS overall structure. It starts by finding a feasible solution. This initial solution construction is limited by a timeout. After having a feasible solution, one begins iterations of destruction and reconstruction to improve the quality of the solution. Those three components of the LNS are detailed as follows.

Initial solution. The initial solution is an iterated randomized heuristic named Cut and Layering taken from [7]. At first, one generates a randomized order of jobs. In every iteration, one allocates resources to a job in a way that minimizes the resources peak throughout repeated guillotine cuts.

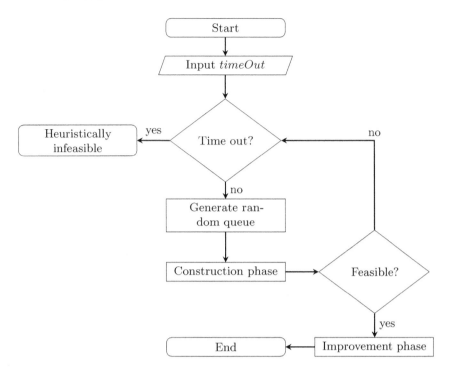

Fig. 2. The LNS structure

3.1 Resource Allocations Removal

Random removal. The heuristic developed in [7] is based on the random removal heuristic. At each iteration, q jobs' resource allocations are removed from the strip. We suggest that this random removal strategy may cause the difference between the worst and the best gap found for small instances.

Shaw removal. To make the heuristic more parameterized and less randomized, we adapt the Shaw removal heuristic for the resource allocation [10]. This removal heuristic choses a random job i. Then for the next job to chose, the choice is based on a relatedness function. The priority is on the job which is more related to the reference job i. In the following iterations, a job will be taken randomly among the chosen ones as a reference. Another job will enter the chosen jobs set based on its relatedness to the reference job. It is somewhat easy to shuffle similar jobs to get a better solution.

We define the relatedness function between two jobs i and j based on the difference in terms of time-windows, resource demands and resource consumptions. Let J' denote the set of jobs that have not been chosen yet.

$$R(i,j) = \lambda_1 * norm(r_i, r_j) + \lambda_2 * norm(d_i, d_j) + \lambda_3 * norm(\tilde{x}_i, \tilde{x}_j)$$
$$+ \lambda_4 * norm(\bar{u}_i, \bar{u}_j) + \lambda_5 * norm(\underline{u}_i, \underline{u}_j) \qquad (9)$$

with $norm(x_i, x_j) = \frac{|x_i - x_j|}{max_{l \in J'}|x_i - x_l|}$ and $\sum_{k=1}^{5} \lambda_k = 1$ to normalize the value of the relatedness in $[0, 1]$. The removal heuristic is detailed in Algorithm 1.

Randomly chose a job i and add it to the removed jobs set;
Remove i from J';
q = q-1;
while $q > 0$ **do**
    Randomly chose a already removed job i;
    Rank all the job in J' by $R(i, j)$;
    $r = random(0, 1)$;
    $k = \lceil r^p * |J| \rceil$;
    Add the kth related job in set J' to the removed jobs set ;
    $q = q - 1$;
end

Algorithm 1. The Shaw removal heuristic

Besides the λ parameters, we also have the number of removed jobs q and a determinism parameter p. If $p = 1$ then the choice of the job to be removed is entirely random. If $p = +\infty$ then one always chose the most related job to remove.

3.2 Resource Reallocation

To re-insert the removed jobs into the resources strip, we apply the same greedy resources reallocation introduced in [7]. There are two operators: greedy free-form repackage and greedy rectangular form repackage. In the former operator, the re-entering tasks consume the most resource possible left, to finish the soonest possible. In the latter operator, the re-entering tasks should also complete the soonest possible, but in rectangular form. In other words, their resource allocation is constant throughout their processing times.

In our previous work [7] the acceptance of new solution after each reallocation is only done when it improves the incumbent. In this paper, we integrate a classic simulated annealing acceptance for the resource allocation operator. The Algorithm 2 precises the detail procedure:

Data: Actual objective function z, new solution's objective function z'
        and incumbent objective function z^*
if $z' < z$ **then**
    Accept new solution;
    if $z' < z^*$ **then**
        | Save new incumbent;
    end
else
    Accept the new solution with the probability $e^{\frac{z'-z}{T}}$;
    Cooling down $T := \alpha T$;
end

Algorithm 2. New solution acceptance

The simulated annealing accept criteria helps us to avoid local optimum by accepting worst solutions during the search procedure. If the new solution is worst than the actual one, the acceptance probability is controlled by $e^{\frac{z'-z}{T}}$ where $T > 0$ is the actual temperature, and $z'-z$ is the difference of the objective values of the two solutions. In the beginning, the value of T is big enough to let the search procedure can have *big skips*. After each acceptance, this temperature is cooling down by the rate α so that the search can converge.

4   Computational Experimentation

The test protocol used is identical to one in [7]. The numbers of tasks tested are $n \in \{10, 15, 20, 30\}$ and the time-horizons tested are $H \in \{20, 40, 60, 80, 100, 150, 200\}$. The exact result is found by solving the MILP model in solver IBM ILOG CPLEX 12.6. All the algorithms are implemented in C++. The testing computer uses CPU Intel Core i5 3.20 GHz with 8 GB of RAM. We test the two order of removal-reallocation: RGA and ARG [7]. In the RGA order, the rectangular form shaking (shuffle) will be made first, then it is followed by the free form shaking (shuffle) and the greedy occupation. In the ARG order, we perform the greedy occupation first, then the rectangular form shaking and the free form shaking after.

4.1   Parameters Tuning

The LNS heuristic is parameterized by the Shaw removal heuristic and the solution acceptance (based on simulated annealing principles).

Tuning Shaw removal heuristic. The parameter for the Shaw removal are $\lambda = \{\lambda_1, \lambda_2, \lambda_3 \lambda_4, \lambda_5\}$, the number of removed tasks q and the determinism parameter p. We use a middle size instance $n = 15$ and $H = 80$ to tune the parameters. The parameters used in the computation experimentation are: $\lambda_1 = 0.35, \lambda_2 = 0.05, \lambda_3 = 0.35, \lambda_4 = 0.125, \lambda_5 = 0.125, p = 3$ and $q = 4$.

Temperature control of Simulated Annealing. We calibrate the initial temperature T and the cooling down rate α of the simulated annealing by procedure introduced by Ben-Ameur [2] and Munakata and Nakamura [4]. The initial acceptance probability P_1 is expected to be 0.5 and the stationary probability P_2 is expected to be 0.05.

4.2   Computational Results

Each instance $\{n, H\}$ is executed 30 times. We calculate the average, the best and the worst derived gap of each removal operator for the two LNS order RGA and ARG. The timeout of the solver CPLEX is set to 1800 s then it cannot solve the four instances (in grey) to optimal. For that reason, the heuristic finds better solution than CPLEX (negative gap). Since the Shaw removal mechanism and the simulated annealing are more complex than the random removal, then, for the same number of shuffle, the old heuristic takes fewer times to solve the instances. To have a fair comparison, we set the timeout for the heuristic with random removal to be the same to the Shaw removal, hence increasing the number of shuffles.

Table 1 shows the results of the worst, the best and the average derived gap found from both removal strategies in two LNS orders. The smaller value of the gap between the two removal operators is set in bold. One can notice that the Shaw removal reduced the worst derived gap in the majority of the tested instances. Concerning the best gap, there is no significant difference between the

Table 1. The worst, average and best derived gap (in percentage) from different removal operators

Orders		RGA						ARG					
Removal		Random			Shaw			Random			Shaw		
n	H	Wrs	Avg	Bst	Wrs	Avg	Bst	Wrs	Avg	Bst	Wrs	Avg	Bst
10	20	7,0	4,4	**2,8**	**4,5**	**2,9**	**1,4**	5,9	4,2	**2,8**	**4,9**	**3,6**	**2,4**
	40	**5,2**	**3,6**	**1,8**	5,3	3,6	2,2	6,7	**3,8**	**1,3**	**6,2**	4,1	2,5
	60	7,0	4,3	1,9	**5,8**	**3,3**	**1,1**	5,3	4,1	**2,0**	**5,9**	**3,7**	2,2
	80	6,1	3,4	**1,8**	**4,1**	**3,2**	2,5	12,7	5,1	3,3	**4,7**	**3,8**	**3,0**
	100	8,9	3,8	1,8	**3,1**	**2,3**	**1,8**	8,1	4,2	**2,0**	**4,5**	**2,9**	**2,0**
	150	12,6	5,8	2,8	7,5	4,4	2,2	**4,8**	**3,3**	**2,0**	8,6	5,0	2,6
	200	3,3	1,9	**0,9**	**3,1**	2,0	1,2	2,6	**1,5**	**0,8**	2,9	1,6	1,0
15	20	5,3	4,0	**2,7**	**4,2**	**3,1**	1,8	**2,9**	**2,3**	1,8	4,0	2,8	**1,8**
	40	4,5	2,8	**1,8**	**3,4**	**2,5**	**1,5**	4,1	2,9	**2,0**	**3,5**	**2,4**	1,9
	60	5,4	3,6	2,2	**3,6**	**2,5**	**1,6**	4,5	3,7	2,8	**3,7**	**3,0**	**2,1**
	80	7,8	3,9	2,1	**3,7**	**2,5**	**1,2**	2,9	2,1	**1,3**	**2,7**	**1,9**	**1,0**
	100	3,4	2,3	**1,1**	**3,2**	**2,3**	1,3	4,1	3,1	**1,8**	**3,8**	**2,8**	2,1
	150	**5,9**	**4,3**	**2,6**	6,1	4,7	2,8	6,6	4,4	2,8	6,7	4,7	3,3
	200	6,3	2,1	**-1,4**	**3,1**	**0,2**	-1,7	6,2	2,2	**-1,2**	**4,1**	**0,2**	**-1,7**
20	20	4,3	2,9	**1,6**	**3,4**	**2,6**	1,6	4,8	3,7	2,4	**4,5**	**3,1**	**1,4**
	40	4,0	2,8	**1,9**	**3,7**	**2,3**	1,8	3,9	2,7	**1,5**	**3,8**	**2,3**	1,6
	60	7,0	4,3	2,6	**5,1**	**3,2**	2,4	6,4	4,4	**2,7**	**4,7**	**3,8**	2,8
	80	7,1	4,0	**1,8**	**3,7**	**2,7**	2,0	3,9	2,5	1,8	**3,3**	**2,4**	**1,8**
	100	3,3	2,1	1,5	**2,4**	**1,7**	**1,3**	2,4	1,4	0,4	**2,4**	**1,0**	**0,3**
	150	4,4	1,4	**0,0**	**1,8**	**0,4**	-0,5	1,4	**0,0**	-1,3	**1,5**	**-0,2**	**-1,4**
	200	4,9	3,9	2,5	**4,8**	**3,2**	**2,1**	3,5	**1,7**	**0,8**	**3,4**	2,2	1,5
30	20	3,8	2,5	**1,7**	**2,5**	**2,1**	1,8	3,3	2,5	2,0	**3,0**	**2,5**	**1,9**
	40	2,8	1,9	**1,4**	**2,6**	**1,9**	1,6	3,4	2,1	**1,3**	**3,5**	**2,2**	1,6
	60	3,3	2,0	**1,3**	**3,1**	**1,8**	1,3	1,3	0,5	**0,0**	**2,0**	**0,6**	**0,0**
	80	4,0	2,3	**1,3**	**3,6**	**2,1**	1,5	3,7	2,4	**1,6**	**2,8**	**2,3**	1,6
	100	3,2	2,3	1,8	**2,4**	**1,9**	**1,3**	2,5	1,4	**0,6**	**1,8**	**1,2**	0,7
	150	2,6	1,1	**0,1**	**0,8**	**0,2**	-0,4	2,0	0,7	-0,6	**1,3**	**0,2**	**-0,5**
	200	3,0	2,0	**1,4**	**2,6**	**1,9**	1,2	2,3	1,6	**1,2**	**2,2**	**1,7**	**0,9**
AVG		5,2	3,1	**1,6**	**3,7**	**2,4**	1,4	4,4	2,7	**1,4**	**3,8**	**2,4**	1,4

Table 2. The worst, average and best CPU times of the two LNS orders compared to the CPU times of solve CPLEX

Order		RGA			ARG			CPLEX
n	H	Wrs	Avg	Bst	Wrs	Avg	Bst	
	20	0,4	0,3	0,3	0,3	0,3	0,3	0,08
	40	0,5	0,5	0,5	0,4	0,4	0,4	1,56
	60	0,7	0,7	0,6	0,6	0,6	0,6	8,06
10	80	1,0	0,9	0,9	1,0	0,9	0,8	8,28
	100	1,3	1,3	1,2	1,2	1,2	1,1	22,77
	150	2,1	1,9	1,8	1,8	1,8	1,7	15,59
	200	3,1	2,8	2,6	2,8	2,6	2,4	34,92
	20	0,8	0,8	0,8	0,6	0,6	0,6	0,09
	40	1,3	1,2	1,1	0,9	0,9	0,9	0,63
	60	1,2	1,2	1,2	1,2	1,0	1,0	6,09
15	80	1,9	1,7	1,6	1,4	1,4	1,4	21,86
	100	2,3	2,1	2,0	2,3	2,2	2,1	24,08
	150	3,8	3,6	3,3	3,5	3,4	3,2	284,71
	200	8,2	7,8	7,3	8,4	8,1	7,7	1800,00
	20	1,4	1,4	1,3	1,3	1,1	1,1	0,36
	40	2,1	1,9	1,8	1,5	1,4	1,4	4,56
	60	2,6	2,4	2,3	2,2	2,0	1,9	5,72
20	80	3,3	3,1	2,9	2,8	2,6	2,5	21,69
	100	4,1	3,8	3,5	3,8	3,6	3,5	138,55
	150	7,7	6,4	5,9	7,3	6,9	6,4	1800,00
	200	9,3	8,6	7,6	9,5	8,9	8,5	1583,80
	20	3,2	3,1	3,0	2,7	2,5	2,4	0,74
	40	4,4	4,0	3,9	3,6	3,3	3,2	16,35
	60	5,5	5,1	4,8	4,6	4,1	3,8	11,80
30	80	6,7	6,5	6,2	5,2	5,0	4,8	11,85
	100	8,0	7,6	7,1	7,1	6,5	5,9	171,76
	150	15,4	13,9	12,8	13,1	12,4	11,8	1800,00
	200	18,3	16,2	14,6	17,0	16,1	15,2	1800,00
Sum		120,7	110,9	103,0	108,2	101,8	96,6	9595,9

two removal operators. However, the average gap is also reduced sharply. By reducing the worst gap and the average gap, the Shaw removal helps to stabilize the algorithm. In the RGA order, the worst gap is cut to half in average, and the average gap is reduced by 25%. In the ARG order, the impact of the Shaw removal is less significant. However, it helps to cut the worst gap from 4.4% to 3.8% and the average gap from 2.7% to 2.4%. With the Shaw removal operator, the result from two LNS orders become less or more identical.

Table 2 shows the CPU times of the heuristics. One can notice that even adding a more complicated operator; the execution time is still reasonable. In the biggest instance $n = 30$ and $H = 200$ the execution times is less than 19 s.

The *ARG* order has a slightly better performances in terms of execution times, but in overall, there is no big difference between the two orders.

5   Conclusion

In this paper, we consider a Large Neighbor Search approach to solve the cumulative scheduling problem with the time-dependent total available resource. We aim to reduce the randomness of our previous heuristic [7] by adding the Shaw removal operator and integrating a simulated annealing to the solution acceptance. Numerical tests show that those modifications help to stabilize the heuristic by reducing the worst and the average derived gap while maintaining a rapid execution time. Our future works concentrate on the developing of a further adaptive large neighbor search to cope with the problem, hence the choice of the operator would be more flexible and adaptive at each iteration.

Acknowledgments. This research has been supported by ANRT (Association Nationale de la Recherche et de la Technologie, France).

References

1. Baptiste, P., Le Pape, C., Nuijten, W.: Satisfiability tests and time-bound adjustments for cumulative scheduling problems. Ann. Oper. Res. **92**, 305–333 (1999)
2. Ben-Ameur, W.: Computing the initial temperature of simulated annealing. Comput. Optim. Appl. **29**(3), 369–385 (2004)
3. Godard, D., Laborie, P., Nuijten, W.: Randomized large neighborhood search for cumulative scheduling. In: ICAPS, vol. 5, pp. 81–89 (2005)
4. Munakata, T., Nakamura, Y.: Temperature control for simulated annealing. Phys. Rev. E **64**(4), 046127 (2001)
5. Nattaf, M., Artigues, C., Lopez, P., Rivreau, D.: Energetic reasoning and mixed-integer linear programming for scheduling with a continuous resource and linear efficiency functions. OR Spectr. **38**(2), 459–492 (2016)
6. Nguyen, N.-Q., Yalaoui, F., Amodeo, L., Chehade, H., Toggenburger, P.: Solving a malleable jobs scheduling problem to minimize total weighted completion times by mixed integer linear programming models. In: Nguyen, N.T., Trawiński, B., Fujita, H., Hong, T.-P. (eds.) ACIIDS 2016. LNCS (LNAI), vol. 9622, pp. 286–295. Springer, Heidelberg (2016). https://doi.org/10.1007/978-3-662-49390-8_28
7. Nguyen, N.Q., Yalaoui, F., Amodeo, L., Chehade, H., Toggenburger, P.: Total completion time minimization for machine scheduling problem under time windows constraints with jobs linear processing rate function. Comput. Oper. Res. **90**, 110–124 (2017)
8. Pacino, D., Van Hentenryck, P.: Large neighborhood search and adaptive randomized decompositions for flexible jobshop scheduling. In: International Joint Conference on Artificial Intelligence (2011)
9. Pisinger, D., Ropke, S.: Large neighborhood search. In: Gendreau, M., Potvin, J.Y. (eds.) Handbook of Metaheuristics, vol. 146, pp. 399–419. Springer, Boston (2010). https://doi.org/10.1007/978-1-4419-1665-5_13
10. Shaw, P.: Using constraint programming and local search methods to solve vehicle routing problems. In: Maher, M., Puget, J.-F. (eds.) CP 1998. LNCS, vol. 1520, pp. 417–431. Springer, Heidelberg (1998). https://doi.org/10.1007/3-540-49481-2_30

Solving the Unrelated Parallel Machine Scheduling Problem with Additional Resources Using Constraint Programming

Taha Arbaoui[(✉)] and Farouk Yalaoui

Laboratory of Industrial Systems Optimization,
Charles Delaunay Institute ICD-LOSI, UMR CNRS 6281,
University of Technology of Troyes,
12 rue Marie Curie, CS 42060, 10004 Troyes, France
{taha.arbaoui,farouk.yalaoui}@utt.fr

Abstract. This work studies the Unrelated Parallel Machine scheduling problem subject to additional Resources (UPMR). A set of jobs are to be processed by a set of unrelated parallel machines. The processing time and the number of needed resources for each job depend on the machine processing it. Resources are renewable and available in a limited amount. The objective to minimize is the maximum completion time. We formulate the problem using a constraint programming model and solve it using the state-of-the-art solver. We compare the results of this model against the existing approaches of the literature on two sets of small and medium instances. On the set of small instances, we show that the proposed model outperforms existing approaches and optimality is attained for all instances of the set. We further investigate its performance on the medium instances and show that it is able to reach more optimal solutions than any performing approach.

Keywords: Parallel Machine Scheduling · Constraint programming
Resources

1 Introduction

Parallel Machine Scheduling problem (PMS) is one of the most studied scheduling problems in the literature due to its importance, application and complexity. Three types of PMS problems exist: unrelated, uniform and identical. In the identical type, the processing time of a job is identical on all machines. Essentially, this creates symmetry between the machines and an ordering is possible. As in the identical type, the processing time of a job in the uniform type is identical for all machines. However, machines have different speeds and the processing of a job depends on its processing time and the machine's speed. The unrelated type is the most general one. Jobs' processing times completely depend on the machine.

© Springer International Publishing AG, part of Springer Nature 2018
N. T. Nguyen et al. (Eds.): ACIIDS 2018, LNAI 10752, pp. 716–725, 2018.
https://doi.org/10.1007/978-3-319-75420-8_67

Different objective functions are considered in the literature. The maximum completion time (makespan) is the most studied objective function. Several approaches have been proposed to tackle PMS problems with different constraints and objectives. Vallada and Ruiz [14] studied the unrelated PMS problem with sequence-dependent setup times and proposed a genetic algorithm to solve it. Tran et al. [13] studied the same problem and proposed decomposition approaches to attain optimal solutions. Yalaoui and Chu [15] studied the identical PMS with setup times and the job splitting property and proposed an efficient heuristic to solve the problem. The heuristic was later improved by Tahar et al. [12] using a matheuristic. Arbaoui and Yalaoui [3] proposed a bender's decomposition approach for the same problem. For a detailed review on scheduling problems in general and PMS in particular, we refer the reader to [2,10].

The PMS problems have been thoroughly studied with different constraints, especially with setup times, no-wait and unavailability constraints. Fewer works, however, tackled the PMS problems with resource constraints. There exist two types of PMS problems with resource constraints: static and dynamic. In the static type, the resource allocation is fixed throughout the schedule, whereas in the dynamic type resources can be assigned and reassigned depending on jobs' assignments.

Among the earliest works that studied PMS problems subject to resource constraints is the one of Blazewicz et al. [5]. The authors proposed a classification for the scheduling problems according to the type and number of resources considered. Ruiz-Torres et al. [11] studied two variants of the problem. The first considers a pre-assignment of jobs to machines, a configuration that the second ignores. To solve the first variant, an integer programming model was introduced while the second variant was tackled using multiple heuristics. Edis and Oguz [6] studied the unrelated PMS with resource constraints and developed a Lagrangian-based constraint programming method. Edis and Ozkarahan [8] introduced a hybrid integer and constraint programming approach to tackle the resource-constrained identical PMS problem with machine eligibility restrictions. They showed that the combination of integer and constraint programming yields better results and allows to attain more optimal solutions on a benchmark of 200 instances.

The PMS problem studied in this paper was introduced by Fanjul-Peyro et al. [9] and denoted UPMR. The problem involves assigning jobs to unrelated parallel machines while respecting needed resources for each job. Resources are renewable, i.e. when a job finishes being processed, resources can be used again by another job. Such resources represent operators, tools or dies that are necessary in certain manufacturing fields. This problem is a dynamic UPMR since resource allocation is not fixed before solving the problem. The authors proposed two MIP models, one adapted from an existing work and another based on the strip packing problem. Using these two MIP models, they also introduced three matheuristics and analyzed their performance against each other on two sets of small and medium instances. We introduce a Constraint Programming model and compare it to their approaches. We show that the proposed CP model outperforms all their approaches by attaining optimal solutions for all small instances and more optimal solutions than all matheuristics for medium instances.

The remainder of the paper is organized as follows. Section 2 describes the studied problem. In Sect. 3, we recall the mixed integer formulation presented in [9] and used to obtain lower bounds for the problem. Section 4 details the Constraint Programming model that was designed for the problem and Sect. 5 discusses its results and a comparison between existing approaches and the CP model. Finally, the conclusion is to be found in Sect. 6.

2 Problem Description

The Unrelated Parallel Machine Scheduling Problem with additional Resources (UPMR) is described follows. A set of jobs are to be scheduled on a number of unrelated parallel machines while consuming renewable resources. Each job requires a number of scarce resources that depends on the machine and the job. At each moment, the number of consumed resources must not exceed a fixed amount of available renewable resources. Once a machine finishes processing a job, the resources consumed by the finished job are again available for use. The objective considered in work is the minimization of the maximum completion time.

One or more types of resources can be considered for UPMR (operators, tools, materials, etc.). In this work, we focus on the UPMR with one additional resource type. Hence, all jobs share the same type of resources and the same total amount. Moreover, it is assumed that the amount of the resources needed for each job on each machine does not affect the processing time.

The set of machines is made of M machines and the set of jobs of N jobs. For the sake of simplicity, let i denote a machine ($i \in \{1, \ldots, M\}$) and j denote a job ($j \in \{1, \ldots, N\}$). The unrelated PMS problem implies that the processing time of job j depends on machine i, thus referred to as p_{ij}. Each job requires a number of resources r_{ij} that depends on both machine i and job j. The total number of available resources is expressed by R_{max}.

Scheduling problems become more complex when resource constraints are considered. For the studied problem, the difference is noticeable. To illustrate it, we provide a numerical example. Consider a set of six jobs to be placed in

Table 1. A numerical example of six jobs, three machines and four units of resources.

Jobs	Processing times						Jobs	Needed resources					
	1	2	3	4	5	6		1	2	3	4	5	6
M1	3	7	4	8	6	3	M1	4	2	2	1	2	2
M2	4	5	3	2	7	2	M2	1	1	3	3	3	2
M3	2	1	4	5	1	10	M3	2	3	2	1	3	1

three machines and the total number of the renewable resources is four units. The processing times and needed resources for each job are given in Table 1.

Solving the example with and without considering resources gives two completely different solutions. When it is solved without considering the resources, i.e. solving the classical unrelated PMS (UPM), one may find an optimal solution with a $C_{max} = 4$. When resources are considered, the optimal solution has $C_{max} = 8$.

3   Mixed Integer Linear Model

In [9], the authors introduced two mixed integer linear models, the first is adapted from the UPMR presented in [7] denoted UPMR-S, and the second is based on the strip packing problem denoted UPMR-P. Recall that when resource constraints are not considered, UPMR becomes an Unrelated Parallel Machine scheduling problem (UPM). In [9], authors showed that the UPM MIP model can be used to obtain a lower bound and was used to evaluate the solutions of the proposed matheuristics.

Model UPMR-S uses the following decision variables: $x_{ijk} = 1$ if job j is assigned to machine i and finishes being processed at time k. Since job j can never be finished before time p_{ij}, $k \geq p_{ij}$. Model UPMR-S is described as follows:

$$Minimize \quad C_{max} \tag{1}$$

$$\sum_{i=1}^{M} \sum_{k \geq p_{ij}} k \, x_{ijk} \leq C_{max} \qquad j = 1, \ldots, N \tag{2}$$

$$\sum_{i=1}^{M} \sum_{k \geq p_{ij}} x_{ijk} = 1 \qquad j = 1, \ldots, N \tag{3}$$

$$\sum_{j=1}^{N} \sum_{s \in \{max(k,p_{ij}),\ldots,k+p_{ij}-1\}} x_{ijs} \leq 1$$
$$i = 1, \ldots, M, \quad k = 1, \ldots, K_{max} \tag{4}$$

$$\sum_{i=1}^{M} \sum_{j=1}^{N} \sum_{s \in \{max(k,p_{ij}),\ldots,k+p_{ij}-1\}} r_{ij} \, x_{ijs} \leq R_{max}$$
$$k = 1, \ldots, K_{max} \tag{5}$$

$$x_{ijk} \in \{0, 1\} \qquad i = 1, \ldots, M,$$
$$j = 1, \ldots, N, \quad k = 1, \ldots, K_{max} \tag{6}$$

Equation (1) is the objective function. Constraints (2) are used to obtain the maximum completion time. Constraints (3) state that a job can be assigned to exactly one machine and constraints (4) force it to finish at one specific time.

Finally, constraints (5) are used to avoid exceeding the total number of available resources.

K_{max} is an important parameter of the formulation. It is inherently related to the maximum completion time. Setting it too small or too big might have consequences on the performance of the formulation. If K_{max} is set too small, solving the formulation becomes impossible since some jobs might not be assigned to a certain time. Setting it too big makes solving the formulation difficult since this leads to a large number of variables. Contrary to M and N, the maximum numbers of machines and jobs, K_{max} can be calculated in different manners. Fanjul-Peyro et al. [9] practically observed that setting $K_{max} = \min_{i \in \{1,...,M\}} \sum_{j=1}^{N} p_{ij}$ provides the best results. They tested other tighter values but the solver faced difficulties in finding feasible solutions.

4   Constraint Programming Model

Constraint Programming (CP) is considered as an alternative modeling tool for linear and Integer Programming (IP). It has seen an increasing interest due to the facility and the flexibility one finds when using it to model combinatorial problems. While CP and IP have their strengths and weaknesses, CP approaches are mostly used to search for feasible solutions while IP approaches have an emphasis on optimality and improving lower bounds. When solving a constraint programming model, the solver focuses on reducing variables' domains so as to obtain multiple feasible solutions. Solving an integer programming model requires both improving the lower bounds (among others LP relaxations) and upper bounds (feasible solutions) to reduce the gap between both bounds.

CP has seen an increasing interest in the scheduling community due to its successful application on different problems [8]. Solvers are improved along the years and some solvers offer a special framework for scheduling problems that allows efficient modeling of complex and specific scheduling constraints like the no overlap constraint, which is typically imposed between the jobs of the same machine. For a detailed review on the use of CP in scheduling, we refer the reader to [4].

Contrary to linear and integer programming, the constraint programming model largely depends on the modeling framework used to express the variables, the constraints and the objective function. We choose to utilize the IBM CP Optimizer in this study. We only provide the description of the structures used in the CP model as the details of the modeling framework can be found in [1].

On top of the usual continuous and integer variables, CP Optimizer offers a special variable called the *interval* variable. This variable is characterized by two integers representing its start time and its end time. Moreover, an interval variable can be optional i.e. it can be ignored (or deleted) in the solution. When the interval variable is considered in the solution, it is said to be *present*. Otherwise, it is considered *absent*. The length of an interval variable is defined as the difference between its end time and its start time.

In our model, v_{ij} is an interval variable referring to the processing of job j on machine i. Thus, we have $M \times N$ optional interval variables v_{ij}. The length of variable v_{ij} is equal to p_{ij}.

The optionality of interval variables allows us to select only one interval variable for each job so that a job is assigned to a single machine. This is achieved using the *alternative* constraint. The alternative constraint is applied to a set (at least two) of interval variables and states that if one of the interval variables is present, all the other interval variables should be absent. This constraint is applied to all variables v_{ij} of job j to enforce that only one of them should be present and the others have to be absent, i.e. the job j can be assigned to only one machine.

Jobs scheduled on the same machine should not overlap. To avoid this situation, we use the *noOverlap* global constraint applied to a set of interval variables to avoid overlapping. The *noOverlap* constraint is applied to all v_{ij} variables of the same machine i.

Machine completion times are expressed using interval variables O_i. Naturally, the start time of variable O_i is equal to the start time of the first job scheduled on this machine and its end time is equal to the end time of the last job scheduled on it. To accomplish this synchronization, the *Span* global constraint is utilized. It states that one interval variable (O_i in our case) should be spanned over a number of variables ($v_{ij}, \forall j$), i.e. it starts with the first interval variable v_{ij} and ends with the last interval variable v_{ij}.

To be able to model the resource constraint, the cumulative constraint is used. By applying this constraint, denoted *Pulse* constraint in the used framework, each time an interval variable v_{ij} is present, r_{ij} (the amount of consumed resources by job j on machine i) is taken into account. At any time, the total consumed resources should be no more than R_{max}.

The CP model is comprised of two types of interval variables v_{ij} (optional) and O_i and can be written as follows:

$$Minimize \quad max \quad O_i \tag{7}$$

$$Alternative(v_{1,j}, \cdots, v_{M,j}) \quad j = 1, \cdots, N \tag{8}$$

$$noOverlap(v_{i,1}, \cdots, v_{i,N}) \quad i = 1, \cdots, M \tag{9}$$

$$Span(O_i, v_{i,1}, \cdots, v_{i,N}) \quad i = 1, \cdots, M \tag{10}$$

$$\sum_{i=1}^{M} \sum_{j=1}^{N} Pulse(v_{ij}) \leq R_{max} \tag{11}$$

Equation (7) express the makespan that corresponds to the maximum of machines' completion times. Equation (8) enforce the presence of one v_{ij} variable, i.e. job j is assigned to at least one of the machines. Equation (9) ensure that jobs assigned to the same machine do not overlap. Equation (10) establish the link between variables v_{ij} and O_i. Finally, the resource constraint is respected using Eq. (11).

5   Experimental Results

This section details the different experimentations conducted on the CP model and a comparison with existing results in the literature on an established benchmark. It particularly illustrates the strengths and difficulties faced when dealing with the CP model and analyzes its performance with regards to the lower bounds.

5.1   Environment and Benchmarking

The MIP and the CP models were implemented in C++ using the Concert library of the IBM Optimization Studio (CPLEX and CP Optimizer respectively) and compiled using GCC 6.3.1. The experimentations were run on an Intel(R) Xeon(R) CPU E5-2650 v2@2.60 GHz with 8 GB of memory. The number of threads in each run was set to one. The time limit set for each run is one hour.

Fanjul-Peyro et al. [9] introduced a benchmark of 900 instances divided into two sets: small and medium instances. The small set contains instances with 8, 12 and 16 jobs. The medium set contains instances with 20, 25 and 30 jobs. The numbers of machines considered in both sets are 2, 4 and 6. For a detailed description of the instances, please refer to [9].

5.2   Lower Bounds

As previously stated in Sect. 2, the solution of UPM represents a lower bound for UPMR. To solve UPM, model UPMR-S can be used without constraints (5). We practically observed that solving this formulation without constraints (5) and end times of jobs (variables x_{ij} are used instead) requires a few seconds and is therefore used to obtain the lower bound denoted LB. It should be noted that this lower bound was the best among three tested in [9], especially for medium instances.

5.3   Results and Comparison with Existing Approaches

In this section, we compare the results obtained by the proposed CP model against the existing approaches in the literature. Since the problem was introduced by Fanjul-Peyro et al. [9], the only existing approaches proposed to solve the problem are the ones described in their work. As presented previously, they proposed two MIP models and derived six matheuristics from them. We first compare the CP model to both models, UPMR-S and UPMR-P, since the former allows to reach optimality more often whereas the latter obtains more feasible solutions. We later compare the CP model to matheuristics MAF-S and MAF-P, the best matheuristics in terms of obtaining optimal solutions.

To check the optimality for a given solution of the UPMR, the authors used lower bound LB obtained by solution the Unrelated Parallel Machine (UPM) problem as described in Sect. 5.2.

Table 2 presents a comparison between the CP model and the existing MIP models (UPMR-S and UPMR-P). The first column represent the number of jobs considered. For each model, a set of four columns are provided. Column 0%*Gap* gives the number of instances for which optimality was proved within the time limit (one hour). Column Opt provides the number of instances for which the optimal solution was reached but optimality was not proved by solving the models. Instead, a solution is qualified as optimal if its fitness is equal to LB. Column Feas presents the number of instances for which a feasible non-optimal solution is obtained. Finally, column NoSol gives the number of instances for which no solution was found.

Table 2. Comparison of the CP model against existing formulations in the literature.

	UPMR-S [9]				UPMR-P [9]				CP model			
N	0%Gap	Opt	Feas	NoSol	0%Gap	Opt	Feas	NoSol	%Gap	Opt	Feas	NoSol
8	137	147	13	0	146	150	4	0	150	150	0	0
12	84	87	65	1	47	85	103	0	150	150	0	0
16	53	65	78	19	1	26	149	0	150	150	0	0
20	21	45	108	21	0	7	150	0	114	20	36	0
25	6	18	114	30	0	1	150	0	56	36	94	0
30	1	5	106	43	0	0	150	0	49	40	101	0

For the small instances, one can remark that the CP model obtains the optimal solution for all the 150 instances. It is clear that, on the set of small instances, the MIP models UPMR-S and UPMR-P provide lower performances compared to the CP model, particularly model UPMR-S which struggles to find feasible solutions for instances with 16 jobs. It is also worth noting that both MIP models were not able to reach the 0%Gap optimality for all the instances, even for smallest instances (with 8 jobs).

For medium instances, the CP model faces difficulties as the number of jobs increases. For example, for 20-job instances, it is still able to reaches optimality for 136 instances whereas it reaches optimality for 89 instances in the case of 30-job instances. It is worth noting that the CP model is also able to obtain feasible solutions for all 900 instances. When comparing UPMR-S to UPMR-P, one can see that the former remore optimal solutions than the latter. However, it faces difficulties in finding feasible solutions as the size of instances increases, which is not the case of UPMR-P. The CP model overcomes both difficulties. It is able to find feasible solutions for all the instances, regardless of their size. Moreover, it finds more optimal solutions than both models combined.

Table 3 presents a comparison between matheuristics MAF-S and MAF-P (the best performing heuristics) and the proposed CP model. For each size of instances N in the first column, three columns Opt, Feas and NoSol are shown for each approach. Column Opt presents the number of optimal solutions found by the approach while columns Feas and NoSol show respectively the number of feasible non-optimal solutions and the number of instances for which no solution was found.

One can clearly remark that, on the set of small instances, the CP model outperforms both matheuristics by finding optimal solutions for all the 450 instances while both matheuristics MAF-S and MAF-P find at maximum optimal solutions for half instances (75 for 8-job instances).

Table 3. Comparison of the CP model against existing matheuristics in the literature.

N	MAF-S [9]			MAF-P [9]			CP model		
	Opt	Feas	NoSol	Opt	Feas	NoSol	Opt	Feas	NoSol
8	75	75	0	75	75	0	150	0	0
12	72	78	0	72	78	0	150	0	0
16	64	85	1	71	79	0	150	0	0
20	64	85	1	67	83	0	134	16	0
25	48	99	3	60	90	0	92	58	0
30	50	98	2	76	74	0	89	61	0

For the medium instances, the difference is noticeable. In the case of the 20-job instances, the proposed CP model reaches optimality for 134 instances whereas matheuristics MAF-S and MAF-P respectively find optimality for 64 and 67 instances.

From the presented results, one can clearly see that the proposed CP model outperforms existing exact and heuristic approaches in the literature. Most importantly, it is able to prove optimality for the set of small instances and reaches optimality for more medium instances than all existing approaches. However, one can clearly see that, despite finding all feasible solutions, the proposed CP model finds less optimal solutions as the number of jobs increases.

6   Conclusion

We studied the unrelated PMS problem with additional resources with an objective of minimizing the maximum completion time. The problem, though very practical and realistic, was recently introduced and studied by Fanjul-Peyro et al. [9]. Resources are renewable and consumed by jobs as they are processed on the machines. Like the processing times, the number of needed resources by the job depends on the machine on which it is processed.

We proposed a constraint programming model that was implemented using the scheduling framework of the IBM CP Optimizer. Using this approach, optimality was reached for all the small instances. The approach was also able to outperform existing approaches as it reached more optimal solutions than all approaches.

The proposed approach confirms the important and interesting application of constraint programming in scheduling. Despite the promising results, it is expected that the approach faces increasing difficulty as the size of the instances grows. Therefore, further research on larger instances should be directed towards using metaheuristics, hyper-heuristics and decomposition approaches.

References

1. IBM Software: IBM ILOG CPLEX Optimization Studio 12.6 (2016)
2. Allahverdi, A.: The third comprehensive survey on scheduling problems with setup times/costs. Eur. J. Oper. Res. **246**(2), 345–378 (2015)
3. Arbaoui, T., Yalaoui, F.: An exact approach for the identical parallel machine scheduling problem with sequence-dependent setup times and the job splitting property. In: 2016 IEEE International Conference on Industrial Engineering and Engineering Management (IEEM), pp. 721–725. IEEE (2016)
4. Baptiste, P., Le Pape, C., Nuijten, W.: Constraint-Based Scheduling: Applying Constraint Programming to Scheduling Problems, vol. 39. Springer Science & Business Media, New York (2012). https://doi.org/10.1007/978-1-4615-1479-4
5. Blazewicz, J., Lenstra, J.K., Kan, A.H.G.R.: Scheduling subject to resource constraints: classification and complexity. Discret. Appl. Math. **5**(1), 11–24 (1983)
6. Edis, E.B., Oguz, C.: Parallel machine scheduling with additional resources: a Lagrangian-based constraint programming approach. In: Achterberg, T., Beck, J.C. (eds.) CPAIOR 2011. LNCS, vol. 6697, pp. 92–98. Springer, Heidelberg (2011). https://doi.org/10.1007/978-3-642-21311-3_10
7. Edis, E.B., Oguz, C.: Parallel machine scheduling with flexible resources. Comput. Ind. Eng. **63**(2), 433–447 (2012)
8. Edis, E.B., Ozkarahan, I.: A combined integer/constraint programming approach to a resource-constrained parallel machine scheduling problem with machine eligibility restrictions. Eng. Optim. **43**(2), 135–157 (2011)
9. Fanjul-Peyro, L., Perea, F., Ruiz, R.: Models and matheuristics for the unrelated parallel machine scheduling problem with additional resources. Eur. J. Oper. Res. **260**(2), 482–493 (2017)
10. Pinedo, M.: Scheduling: Theory, Algorithms and Systems, 5th edn. Springer, New York (2012). https://doi.org/10.1007/978-1-4614-2361-4
11. Ruiz-Torres, A.J., López, F.J., Ho, J.C.: Scheduling uniform parallel machines subject to a secondary resource to minimize the number of tardy jobs. Eur. J. Oper. Res. **179**(2), 302–315 (2007)
12. Tahar, D.N., Yalaoui, F., Chu, C., Amodeo, L.: A linear programming approach for identical parallel machine scheduling with job splitting and sequence-dependent setup times. Int. J. Prod. Econ. **99**(1), 63–73 (2006)
13. Tran, T.T., Araujo, A., Beck, J.C.: Decomposition methods for the parallel machine scheduling problem with setups. INFORMS J. Comput. **28**(1), 83–95 (2016)
14. Vallada, E., Ruiz, R.: A genetic algorithm for the unrelated parallel machine scheduling problem with sequence dependent setup times. Eur. J. Oper. Res. **211**(3), 612–622 (2011)
15. Yalaoui, F., Chu, C.: An efficient heuristic approach for parallel machine scheduling with job splitting and sequence-dependent setup times. IIE Trans. **35**(2), 183–190 (2003)

Author Index

Printed in the United States
By Bookmasters